Progress and Prospects in Evolutionary Biology

The Drosophila Model

JEFFREY R. POWELL

New York Oxford
Oxford University Press
1997

Oxford University Press

Oxford New York
Athens Auckland Bangkok Bogota Bombay Buenos Aires
Calcutta Cape Town Dar es Salaam Delhi Florence Hong Kong
Istanbul Karachi Kuala Lumpur Madras Madrid Melbourne
Mexico City Nairobi Paris Singapore Taipei Tokyo Toronto

and associated companies in
Berlin Ibadan

Published by Oxford University Press, Inc.
198 Madison Avenue, New York, New York 10016

Library of Congress Cataloging-in-Publication Data
Powell, Jeffrey R.
Progress and prospects in evolutionary biology : the Drosophila model / by Jeffrey R. Powell.
p. cm.
ISBN 0-19-507691-5
1. Drosophila—Evolution. 2. Evolution (Biology) 3. Biological models. I. Title.
QL537.D76P68 1996 96-12478
575—dc20

9 8 7 6 5 4 3 2 1

Printed in the United States of America
on acid-free paper

For Gisella

L'amor che move il sole e l'atre stelle.

Nel suo profondo vidi che s'interna
legato con amore in un volume,
ciò che per l'universo si squaderna:
sustanze e accidenti e lor costume,
quasi conflati insieme, per tal modo
che ciò ch' i' dico è un semplice lume.

Dante, *Paradiso*

Preface

A primary impetus for writing this book is a conviction that if a major advance in understanding evolution is to be achieved (what some have called the "newer synthesis"), it is most likely to come from integration of levels of knowledge centered on a single closely related set of species. It is unlikely that knowledge of development of nematodes, ecological principles learned from small mammals, population genetics of *Drosophila*, and molecular mechanisms of bacteria are soon, if ever, going to be integrated. Such integration will be greatly facilitated by having all these aspects of biology available for one organism. *Drosophila* has the greatest chance of being this organism. Further, I submit that empirical studies are, and will continue to be, the driving force in directing development of the theoretical, and generalizable, principles that will lead to a more comprehensive understanding of evolution. Thus this book is about empirical studies of *Drosophila*.

When I first began to study *Drosophila*, in 1969, the use of these flies was at a nadir and the use of bacteria and viruses at its zenith. In the last ten years this has changed, with an explosion in the number of laboratories employing "fruit flies" as their research subject. Not only have the numbers of *Drosophila* workers increased, but the diversity of studies has expanded far beyond what was done before 1970. In response to this explosion of knowledge, efforts have been made to collate and organize the knowledge into an accessible format. The volumes edited and written by Michael Ashburner, the computerized database called FlyBase, and the *Encyclopedia of Drosophila* (EofD) are manifestations of these efforts. While finding these sources extremely useful, as an evolutionary geneticist I felt there was a void in regard to our subject. The last published single book on *Drosophila* evolution was Patterson and Stone's 1952 *Evolution in the Genus Drosophila* (Macmillan). Volume 3 (five books) of *Genetics and Biology of Drosophila* (Academic Press) is a multiauthored compilation of many aspects covered in the present book. There still seemed to be a need for a single volume presenting in a logical and coherent manner the progress in *Drosophila* evolutionary biology.

I have written this book with particular audiences in mind. One is that first-year graduate student who walks into your office and says "I think I want to do a project on *Drosophila* evolution, but I don't know anything about *Drosophila*." Having heard this on many occasions, I knew there was no single starting place to send the student. The literature, especially the most recent material, was scattered in many journals and edited volumes. I hope this book is now that single starting place. Because beginning graduate students were in my mind, often throughout this volume I point to areas I feel are especially ripe for further work; hence the word "prospects" in the title. I have assumed the reader knows almost nothing about *Drosophila*, although I do assume a basic understanding of evolution and genetics, including population genetics.

But there are three other constituencies I have sought to serve here, as well. Many of my colleagues studying development, molecular biology, and neurobiology of *Drosophila* have come to me asking an evolutionary question relevant to their research. This growing appreciation for the insights an evolutionary perspective can provide to all biological research has been most heartening. I hope *Drosophila* workers in areas other than evolution will find answers to many of their questions in this volume.

Evolutionary biologists not working on *Drosophila* should also find the book useful. What is all the excitement with this little fly? Why so much effort on it? Has all this work on a single model yielded insights that can be useful to me working on some other organism? Many of the principles first fully documented in *Drosophila* have proven to be general to the biological world. By chance, the book has just about the right number of chapters to cover one a week in a one-semester seminar. This book is not just for *Drosophila* researchers.

Finally, I hope my colleagues working directly in *Drosophila* evolution will find use for this book as a reference, if nothing else. While I have been working on evolution of these flies for more than 25 years, in writing this book I learned a lot. Every time I started a chapter I thought, this is going to be fast and easy. After a week or so, I was inevitably humbled to realize how much vaster is the literature than I, a "professional," ever realized. By the end of each chapter, I had to (arbitrarily, sometimes) just call a halt; each one could be a book unto itself. While the bibliography of this book is large, with more than 1,750 references, as noted in the first chapter well over 60,000 publications on *Drosophila* exist. Assuming a quarter of these are relevant to the subject of this book, this means the bibliography contains about 12% of all relevant citations. This is an oblique way of apologizing to my colleagues for not citing all their work. This is not an encyclopedia. However, I have tried to include sufficient references to allow entrance into the literature on each subject; often I have used tables as vehicles for citing literature without interrupting the flow of the text.

Acknowledgment of assistance in writing this book must go first to Michael Ashburner. He not only provided the inspiration by his own heroic efforts to bring together knowledge on *Drosophila*, but he was an ideal host during a sabbatical stay in Cambridge when this book was initiated. He sponsored me as an Overseas Fellow of Churchill College, which provided me with the peace and stimulation to write about half the book in four months in residence at Churchill. It has taken twenty months to write the second half.

Colleagues who read, critiqued, and corrected various chapters include: David Hale, Mercedes Ebbert, Jennifer Gleason, Hampton Carson, Rob DeSalle, Junhyong Kim, Steven Schaeffer, Chung-I Wu, and W. Joe Dickinson. Needless to say, the errors and idiosyncrasies remaining are all my own as I have not always followed their good advice, although I was always grateful to have it. Etsuko Moriyama cheerfully provided data and analyses whenever I asked. Eli Daniel, Samantha Bertini, and Carol Hwang assisted in many mechanical ways such as preparing figures, finding references, obtaining permissions, and so forth. Without them, the book would not yet be completed.

It is safe to state that this book would not have been written without two other individuals. George B. Craig, Jr. was my inspiration to enter biological research and his enthusiasm and encouragement came at a crucial time. Theodosius Dobzhansky introduced me to *Drosophila* as well as to many other aspects of biology. His devotion to understanding evolution and his energetic discipline in producing scientific work has sustained me even though I last saw him twenty years ago this month.

Finally and foremost, Gisella Caccone contributed to this book in more ways than even she knows. The dedication to her is a small symbol of my deep appreciation and love.

November, 1995 JRP
New Haven, Connecticut

Contents

Progress and Prospects in Evolutionary Biology

1

An Overview

All animals are equal but some animals are more equal than others.
George Orwell, *Animal Farm*

The Problem

Of the millions of species that inhabit the earth, biological researchers tend to concentrate on relatively few organisms that subsequently become "model systems." The reason is obvious: Research builds on past research. To advance the forefront of knowledge, the system one studies must be known up to that forefront. Obviously not all the millions of species can be so well known, so a few have been chosen for intense "vertical" study. The faith is that what holds for one species holds for all, a mindset captured in the famous (or infamous) dictum, "What is true of *E. coli* is true of elephants." While today this naive faith is rarely embraced in a literal sense—given the abundant evidence that the dictum is just plain wrong—the vertical, in-depth approach to understanding biological systems still rests on confidence that what is found for one species applies in some degree to all (or at least many) living systems. Indeed, the great advances in biology of the last thirty years attest to the fact that the vertical approach is extremely powerful in advancing biological knowledge at a certain level of organization.

Often the actual choice of the model organism is an historical accident, although seldom totally accidental. Many organisms have been studied, and by a process not unlike natural selection, certain organisms come to the fore as popular models. Examples include the house mouse, yeast, *Escherichia coli*, corn, and *Drosophila.* In recent times, organisms such as *Caenorhabditis elegans, Arabidopsis*, and the zebrafish have been chosen more deliberately based on criteria that make them especially suitable for the problems at hand.

Of all these models, it is arguable that none has received as much attention as has *Drosophila.* At the beginning of this century, genetics was the most exciting new area of biological investigation, and *Drosophila* played the major role in its early development. In the middle of the century, the synthesis of genetics and evolutionary theory was at the forefront of biological work and again *Drosophila* was a primary model for empirical studies. In recent years, *Drosophila* has attained perhaps its widest popularity ever as a model system in the study of development and molecular biology. Thus there is information on the ecology, population biology, systematics, behavior, genetics, development, and molecular biology of this fly. No other model has been so thoroughly studied at all these levels.

To most biologists, "*Drosophila*" means *Drosophila melanogaster*. This species has been more extensively and intensively studied than any other. Consequently, to a large degree the model system that has received so much attention for vertical study is a single species, which presents a problem in discussing its evolutionary biology: *Horizontal investigation is crucial to evolutionary biology.* It is not possible to fully understand the evolution of a single species by studying just that species in isolation. All species are members of groups of more or less closely related species, and each species can only be understood in the historical context of belonging to groups united in the past into single lineages. Furthermore, the cornerstone of evolutionary investigation is *variation*, not constancy. So the vertical approach that seeks constancy of biological processes is at odds with the variation among individuals and species that is the cornerstone of evolutionary studies. (See Lewontin [1981] and Mayr [1982] for further discussion of these points.)

This is the goal I attempt to achieve in this book: to place the in-depth knowledge of *Drosophila* into an evolutionary framework. In a sense, I will try to "horizontalize" essentially vertical knowledge. One hope is that we (the reader and author) will gain insights into evolutionary processes in general. This hope, ironically, is somewhat akin to the faith of the vertical approach to biology, namely that by understanding in some detail the evolutionary biology of one particularly well-studied group (*Drosophila*) we will learn principles applicable to other organisms. The second major hope for this book is that workers on all aspects of the biology of *Drosophila* will gain insights into their own particular fields by considering the evolutionary context of their research. Dobzhansky's often-quoted dictum captures the spirit: Nothing in biology makes sense except in the light of evolution.

History

Drosophila in the Laboratory

The earliest reference to *Drosophila* in the scientific literature carries the date 1684, and 358 citations are found prior to 1900 (Drosophila Information Service, 1994). However, *Drosophila melanogaster* was first used as a research organism in the laboratory of W. E. Castle in 1901, just one year after the beginning of the field of genetics. He conducted inbreeding experiments and developed the banana-based medium that served for so long as the medium of choice. When Fernandus Payne came to Thomas Hunt Morgan's laboratory at Columbia University in 1907, he proposed an experiment to use *Drosophila* to test the Lamarckian notion that the size of organs could be altered over generations by use or disuse. Payne had been a student at Indiana University with William J. Moenkhaus, who had been a student of Castle; thus Payne's proposal to use the fast-breeding fly was not surprising. Payne carried out his experiment by keeping *D. melanogaster* (then called *D. ampelophila*) in the dark for 49 generations to see if the eyes would decrease in size—they didn't. He published his work in 1910, the same year the first mutant (*white eye*) was discovered in the Morgan lab. *So the first evolutionary studies on Drosophila predate the discovery of the first mutant!* (Lutz [1911] had also been carrying out evolutionary studies of sexual selection on *D. ampelophila*, although they were published a year after *white* was discovered.) Of course it was the development of genetics by Morgan and his students that really put *Drosophila* on the map as a widely used laboratory animal. Histories of this period of early *Drosophila* genetics can be found in Sturtevant (1965), Oliver (1976), Allen (1975, 1978), and Kohler (1994).

The growth in numbers of individuals and laboratories engaged in *Drosophila* research since these early beginnings is remarkable. One need only attend one of the several annual *Drosophila* research conferences to appreciate how many people are working on such a wide spectrum of problems. The exponential growth of knowledge is also evident from analysis of the numbers of publications on *Drosophila*; Fig. 1-1 shows the trends. The exponential growth of publications throughout the 20th century shows no indication of leveling. Now there are well over 60,000 works, of which I estimate about a quarter concern directly or indirectly the subject matter of this book.

This growth in popularity of *Drosophila* has been due partly to the suitability of the organism for a variety of studies. However, another factor very important in the

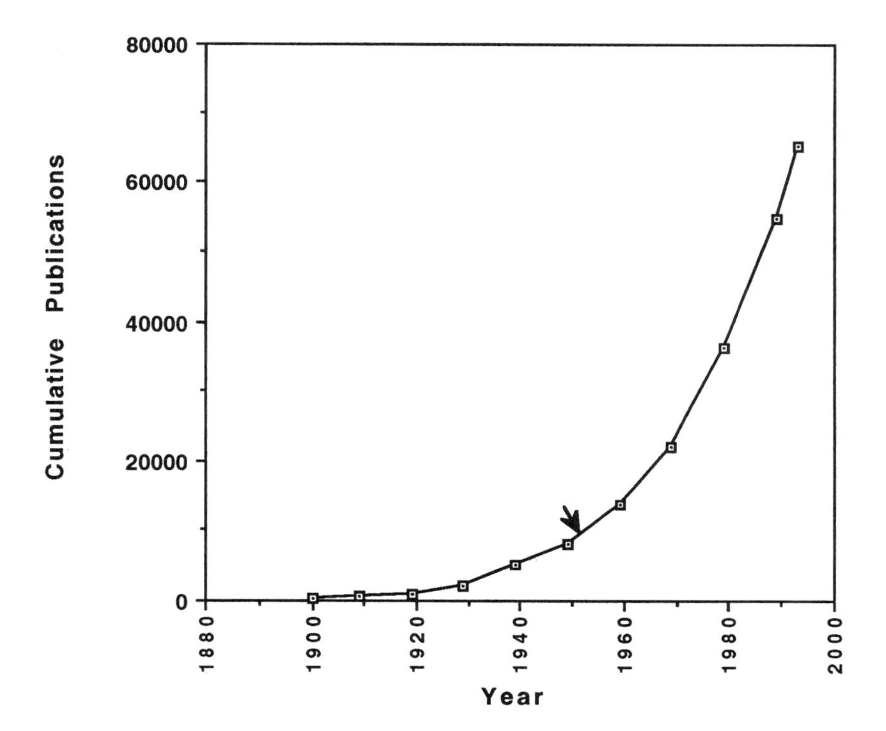

FIGURE 1-1 Cumulative numbers of publications about *Drosophila*. The arrow indicates the publication of Patterson and Stone (1952), the most recent general book on *Drosophila* evolution. Data compiled from Drosophila Information Service (1994).

spread of *Drosophila* research is a tradition that goes back to the earliest days and is best described by the first paragraph in the series titled *Drosophila Information Service*, presumably written by the editors, C. B. Bridges and M. Demerec, in 1934:

> An appreciable share of credit for the fine accomplishments in Drosophila genetics is due to the broadmindedness of the original Drosophila workers who established the policy of a free exchange of material and information among all actively interested in Drosophila research. This policy has proved to be a great stimulus for the use of Drosophila material in genetic research and is directly responsible for many important contributions. In over twenty years of its use no conspicuous abuse has been experienced.

Drosophila in Evolutionary Studies

The use of *Drosophila* in explicitly evolutionary studies of natural populations began in Russia. Three individuals can be singled out: N. W. Timofeeff-Ressovsky (e.g., Timofeeff-Ressovsky and Timofeeff-Ressovsky, 1927), N. P. Dubinin (1931), and perhaps most importantly, S. S. Chetverikov (1926; English translation, 1959; this name is also sometimes transliterated Tshetverikov). These initial attempts to begin a program in experimental population genetics of *Drosophila* were extinguished by the tragic rise of T. D. Lysenko. Lysenko, with the cooperation of Stalin, virtually eliminated modern genetics in the Soviet Union through a combination of draconian measures (Medvedev, 1969). Undoubtedly the history of experimental evolutionary genetics would have been very different had Lysenko not come to power. Nevertheless, the efforts of these early workers had a lasting influence on the field primarily through the work of the Russian emigrant, Theodosius Dobzhansky.

Dobzhansky was born in Ukraine in 1900 and grew up during a tumultuous time of war and revolution. Though experiencing a less than ideal formal education, he acquired an early love of biology. Initially, he studied ladybird beetles (Coccinelidae) and wrote papers on their natural history and systematics in the traditional framework of early 20th century entomology. He loved collecting and working with organisms in nature and developed a strong interest in evolution. He read Darwin's *Origin of Species* at the tender age of 13 and later in life claimed to have understood at least its essentials at that first reading. In his early 20s he recognized that genetics (patterns of inheritance) was crucial to understanding evolution and so became a student of the nascent field of Mendelian genetics. He tried applying the principles learned from *Drosophila* to his beloved beetles, but the polygenic basis of variation in colors and patterns made Mendelian genetics nearly impossible. By default he turned to *Drosophila*, the one organism for which the background information existed that would allow him to delve more deeply into genetics. (A biography with fuller details of Dobzhansky's life has been published [Land, 1973] and considerable information exists in the Oral History Collection at Columbia University. Other sources of biographical information are Provine [1981] and Glass [1980].)

Recognizing that Morgan's group at Columbia University in New York City was the center of *Drosophila* genetics, Dobzhansky obtained a two year fellowship to study with Morgan beginning in 1927; he was never to return to his native land, much to his regret. Quite naturally, Dobzhansky's mentor in the Morgan group became Alfred H. Sturtevant because, of all the early members of the group, Sturtevant had the broadest training in classical systematics and natural history, so there was some common ground. However, Dobzhansky wanted to learn genetics and concentrated on basic genetics for about seven years, putting his interest in evolution on the back burner. It wasn't until Morgan moved his group to the California Institute of Technology, bringing Dobzhansky and Sturtevant with him, that Dobzhansky moved evolution from the back burner to the front burner.

The Western U.S. and California in particular presented Dobzhansky with a fascinating landscape in which to free his instincts to travel and study nature. In the laboratory, all of the research still focused on *D. melanogaster*, a species not well-suited for many evolutionary studies because it is a human commensal. Fortunately, Dobzhansky learned of a species of *Drosophila* (*D. pseudoobscura*) that had natural populations in just the localities that most fascinated him. Here then was the ideal: natural, free-living populations of a species that was amenable to detailed genetic study. Provine (1981) details this period including the role played by Sturtevant in the initiation of studies on natural populations of *D. pseudoobscura.*

In 1937 Dobzhansky published the first edition of *Genetics and the Origin of Species*, arguably the most important book on evolution in the 20th century. This book was the stimulus for the development of experimental population and evolutionary genetics. In it, Dobzhansky demonstrated the essential compatibility of Mendelian genetic concepts and Darwinian evolution. The theoreticians, particularly R. A. Fisher, S. Wright, and J. B. S. Haldane, had been demonstrating the same from a solely mathematical approach. What Dobzhansky did was to show how well actual data fit the mathematical theories, and reciprocally, how the mathematical theory lends insight into the meaning of the data. One could collect data and interpret it critically in the context of a growing body of quantitative theory. Because *Drosophila* population genetics was in its infancy, data from *Drosophila* played a relatively minor role in the initial edition; however, subsequent editions (1941, 1951) came to rely increasingly on data from *Drosophila* as the prime examples of various evolutionary genetic principles.

Why was Dobzhansky's 1937 book so influential? Others, Morgan included, had also written on the implications for evolution of the new findings in genetics, yet their contributions were not nearly so lasting. The answer probably lies in Dobzhansky's roots in Russia. Because he had come from a tradition of natural history—working with natural populations—and taxonomy, Dobzhansky, of all the early geneticists (with the possible exception of Sturtevant), was the one who could most fully appreciate the implications of genetics for understanding populations in nature, speciation, and systematics. He was an exceptionally clear writer and was willing to put speculation on paper, something Sturtevant was loath to do. His book established a research agenda that continues to the present. (See Gould [1982] and Powell [1987] for more discussion of the impact of the first edition of *Genetics and the Origin of Species*; Levine [1995] provides evidence of the continuing influence of Dobzhansky; and Mayr and Provine [1980] provide an overview of the history of the evolutionary synthesis in the mid-20th century.)

Evolutionary studies of *Drosophila* grew rapidly from these beginnings. Dobzhansky himself established a very strong school after he moved back to Columbia University in 1940. He continued to influence the field until his death in 1975, and his many students and their students continue to represent a large segment of *Drosophila* evolutionary genetics. His influence was not confined to the United States; he greatly influenced the development of evolutionary genetics of *Drosophila*, especially in continental Europe (discussed by Krimbas, 1995) and South America.

Other groups developed independently. H. J. Muller established *Drosophila* research groups at both Indiana University and the University of Texas, the latter group being particularly active in evolutionary and especially systematic studies. H. L. Carson and H. D. Stalker established a group at Washington University (St. Louis) and later at the University of Hawaii. A British group including J. B. S. Haldane began to work on the European counterpart to *D. pseudoobscura, D. subobscura.* This British school was particularly influential in establishing an active and productive group of researchers in Australia.

While much of the evolutionary work on *Drosophila* can be traced to individuals with academic ties to one of the groups mentioned, others have come to the field quite independently, as the following chapters will attest. The above is in no way meant to be a complete history of the field; nor is it without the bias inherent in the writer. A fuller history remains to be written.

The Origin of the Flies

In 1952, Patterson and Stone listed 705 described species of *Drosophila.* Presently there are more than 1,000 described species excluding Hawaiian endemics. While still mildly controversial, the bulk of the evidence indicates that the Hawaiian endemics belong in the genus *Drosophila* (see chapter 8 for more detailed discussions). It is estimated that 800 to 1200 endemic species inhabit the Hawaiian archipelago, although only about 500 have been formally described (Hardy and Kaneshiro, 1981). Today it is safe to say there are about 2,000 known species in the genus; undoubtedly, many more occur in poorly studied areas such as sub-Saharan Africa, Southeast Asia, and the South Pacific Islands. Bächli (1994) has compiled a list of all described species for the family Drosophilidae, which contains some 3,341 valid Latin binomials. Figure 1-2 places the most studied model species into the context of this large group of flies.

Of these many species, relatively few have received much attention from biologists. Most of the work has been on two subgenera within the genus *Drosophila, Sophophora* and subgenus *Drosophila.* The latter name leads to some confusion. In this book when the word *Drosophila* appears it will refer to the genus; when referring to the subgenus the modifying appellation will precede *Drosophila.* We (Powell and DeSalle, 1995) suggest placing the Hawaiian endemic fauna into a separate subgenus (*Idiomyia*) within the genus *Drosophila.* Table 1-1 is an overall summary of the species most discussed in this book. Without presenting the evidence at this point, fig. 1-3 shows the phylogenetic relationship as well as approximate time estimates of origin of major groups. The systematics of these flies and the evidence for the information presented in fig. 1-3 will be detailed in chapter 8. The information is presented at this point simply to familiarize the reader with an overview of the group.

In general, Drosophilids are flies specialized to breed in rotting plant and fungal material. Much of the nutrition of most species derives from yeasts and bacteria decomposing this material. Exceptions to this generalization do occur (see chapter 5). *Drosophila* exist virtually throughout the world: in temperate regions, deserts, and the tropics, where they reach their greatest diversity. However, there are no truly Arctic

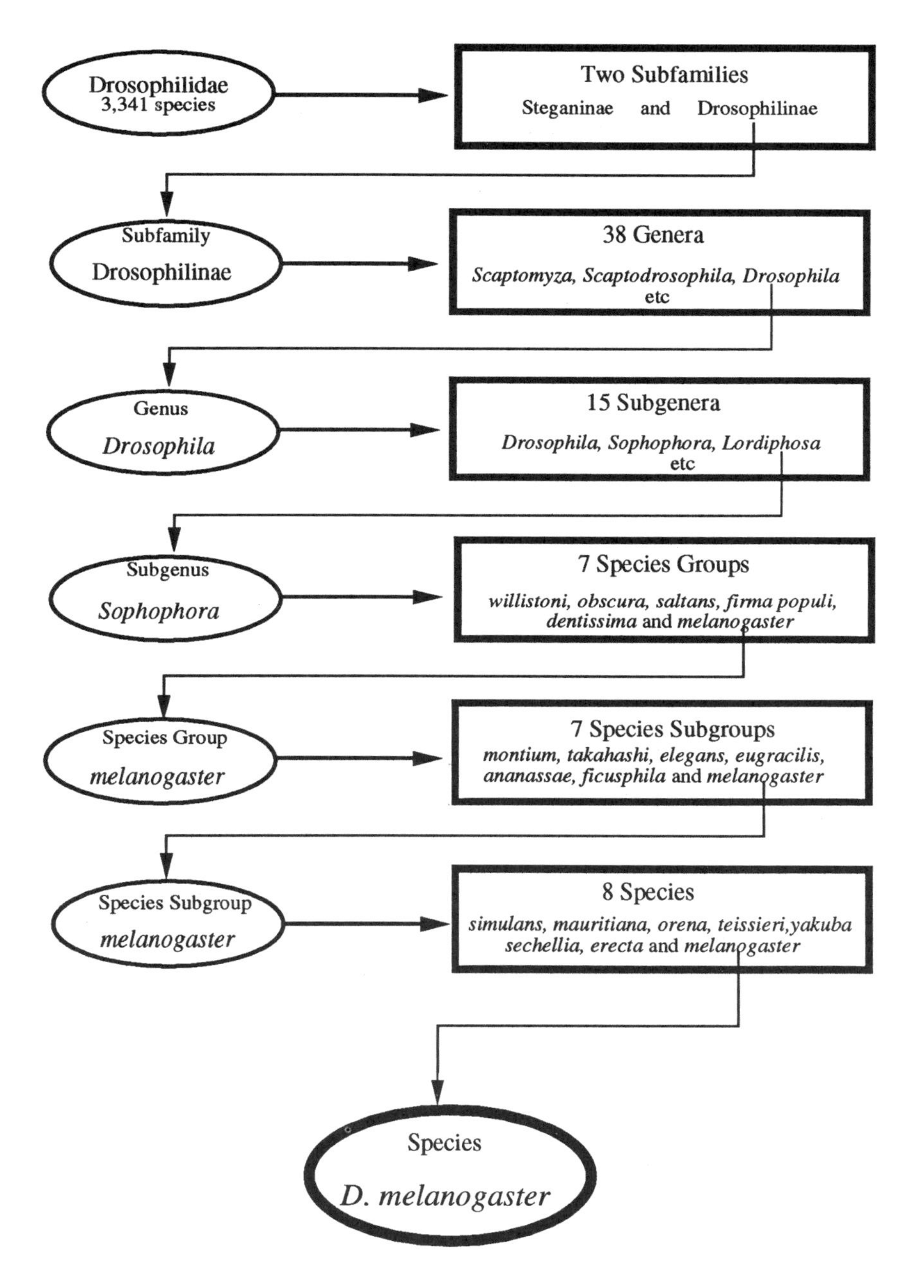

FIGURE 1-2 Placing *D. melanogaster* into the context of the family Drosophilidae. Not all authors agree on the precise number of taxa in each box, but the scheme here generally follows Wheeler (1981) and Ashburner (1989).

TABLE 1-1 Overview of species groups that contain most of the species discussed in this book.

	Total No. Species	Major Groups
Genus *Drosophila*		
Subgenus *Sophophora*	298	*melanogaster*
		obscura
		willistoni
		saltans
Subgenus *Drosophila*	681	*virilis*
		repleta
		robusta
		funebris
		immigrans
Subgenus *Idiomyia*	800–1200	Hawaiian endemics
		Picture wing
		Modified mouthparts
		spoon tarsi
		White-tip scutellum
		"Scaptomyza"

species. Only a few, such as *D. melanogaster*, have become human commensals; Dobzhansky (1965) lists nine other "domestic" species: *simulans, ananassae, montium, funebris, immigrans, virilis, repleta, hydei*, and *busckii.*

Throckmorton (1975, 1982a) presents an overview of evolution and biogeography, and a likely scenario of the origin of the major groups of Drosophilidae. The following is a brief summary of his views, the only major treatises on the subject.

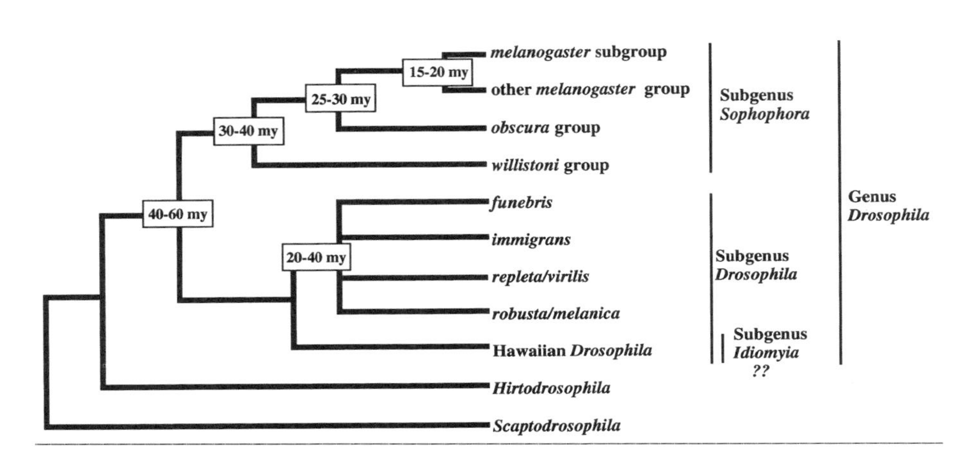

FIGURE 1-3 Summary of phylogenetic relationships among major groups in the genus *Drosophila*. The justification for this is given in detail in chapter 8. The dates are all based on a combination of biogeographic and/or fossil data; no molecular clock has been assumed. From Powell and DeSalle (1995).

Some 80–120 million years ago, two lineages of acalypterate Diptera appeared, which exploited rotting plant material. One was the ephydrids, which became more specialized on wetter substrates and aquatic habitats. The other was drosophilids. Originally drosophilid larvae probably fed on fermenting leaf litter and rotting vegetation in general. A shifting dependence on the microorganisms responsible for decomposition rather than on the plant material directly allowed drosophilids to expand to fruits, sap, and fungi, which support the growth of appropriate yeasts and bacteria. The initial colonization of fungi led to the first major split in Drosophilidae, between the subfamilies Steganinae (fungus breeders) and Drosophilinae. The drosophilinid lineage gave rise to several groups, some of which have also become fungus breeders; adaptation to fungi evolved independently several times. The lineage giving rise to the genus *Drosophila* probably remained a plant breeder, at least initially. The early radiation of the genus *Drosophila* occurred in the tropics, Southeast Asia being the likely geographic origin of the genus.

The first major split in the genus occurred in the Old World tropics giving rise to the two subgenera *Sophophora* and *Drosophila* approximately 50 million years ago. The pattern of diversification of each subgenus is similar. First, there is separation of Old World and New World tropical lineages. The origin of the North American flies must be no later than 30 million years ago, because there is a fossil *Drosophila* in amber from Chiapas, Mexico, that is dated to be about 30 million years old (Wheeler, 1963). In the case of *Sophophora*, the tropical split gave rise to the present day Old World *melanogaster* group and the New World *willistoni* and *saltans* groups. Presumably this split must have occurred by invasion of Asian forms through the Bering connection rather than when South America and Africa split. This is for the reason that the lineages split about 30–40 million years ago and South America and Africa were separated 85 million years ago. The Bering region had a considerable amount of truly tropical plants at least until 40 million years ago (L. Hickey, personal communication), so a migration of a tropical fly is not out of the question.

The *obscura* group evolved from the African tropical group to exploit more temperate regions. An amber fossil is known from Europe dated to about 40 million years ago (Hennig, 1965), which coincides with the origin of temperate forests. The *obscura* group subsequently colonized north temperate regions of both the Old and New Worlds. The split between the Old and New World *obscura* group species was originally thought to have occurred long ago, by about 20 million years ago, which Throckmorton concluded was the last time any temperate forest might have connected North America and northern Europe. More recent molecular data (e.g., Goddard et al., 1990) indicate the split may be much younger and it is conceivable that the Bering Strait provided the origin of the North American group as it has for many other insects; this would allow for a more recent split of the *obscura* group into Old and New World lineages. This pattern of an early tropical split of groups followed by a temperate split was repeated at least five times within Drosophilidae, two more of which are most relevant to studies discussed in this book.

Subgenus *Drosophila* also originated in the Old World tropics. The initial tropical radiation of the ancestral *virilis/repleta* branch is, in Throckmorton's words, "a nondescript cluster of species, species groups, genera and subgenera (1975, p. 458)." However, the New World tropical lineage, the *repleta* group, was one of the most conspicuous and successful drosophilid radiations. This was due to its adaptation to cactus breeding, a wide-open niche. The *repleta* group is now found throughout the New World, in both hemispheres, in the tropics, semi-tropics, and temperate regions. The temperate counterpart in this subgenus was the *virilis/melanica/robusta* lineage. Like the *obscura* group, both Old and New World temperate species of these groups are known today. Throckmorton is less certain as to whether the temperate split occurred when eastern North America and northern Europe ceased to be connected, or whether Beringia continued to allow migration more recently. Nevertheless, it does seem clear that the temperate split, as in *Sophophora*, occurred well after the tropical split.

The third major group showing a very similar pattern is the *immigrans* group, also within subgenus *Drosophila*. Present day Old World tropical *immigrans* group species represent the early tropical origin, while the *tripunctata* group spread in the New World tropics. The *quinaria* group is the temperate lineage which, more recently, became disjunct between the Old and New Worlds.

How the origin of the final major lineage of interest, the Hawaiian *Idiomyia*, fits into these scenarios is not clear. Somehow, early in the subgenus *Drosophila* lineage before the initial Old-New World split, presumably a tropical member colonized an island in the Pacific, which eventually gave rise to the monophyletic huge assemblage of today's Hawaiian *Drosophila*. As shown in fig. 1-3, this founding in Hawaii may predate the origin of all extant groups in subgenus *Drosophila*, so it was much older than the present oldest Hawaiian island (Kauai, 5–6 million years old). Thus it is necessary to hypothesize that the founding event occurred on an island of the chain that eroded and is now covered by ocean; this is further discussed in chapter 8.

Some Basic Biology

Life History

Like all holometabolous insects, *Drosophila* occupy two very different habitats during a life cycle. Figure 1-4 is an outline of a life cycle. Females lay eggs, often in clutches, on a soft substrate that is suitable for larval development. This usually means an actively fermenting patch of plant or fungal material. After embryogenesis, the larva hatches from the egg into a first instar. The larva spends its time burrowing through the soft substrate, almost constantly feeding. Three larval instars occur, followed by pupation. The length of this embryonic and larval period varies from species to species and with environmental factors such as temperature. Under ideal laboratory conditions, *D. melanogaster* completes the embryonic stage in about 20 hours, and the first larval molt occurs 25 hours after hatching. The second larval molt takes place 24 hours later, and the puparium is formed 48 hours after that. Histolysis of larval organs and tissues and formation of adult structures in the pupal stage require about 100 hours. After eclosion, another two to three days are necessary for females to develop mature eggs. Thus a minimum of about 11 days is required to proceed from egg to egg. This is the shortest developmental period known for any drosophilid species, and undoubtedly most species in nature require considerably longer. Because lower temperatures slow down the developmental process appreciably, temperate species generally have longer life cycles than tropical species. Even some tropical species, such as many of the Picture Wing Hawaiian species, have very long life cycles, extending to months in some cases. But as far as is known, all species in the genus execute

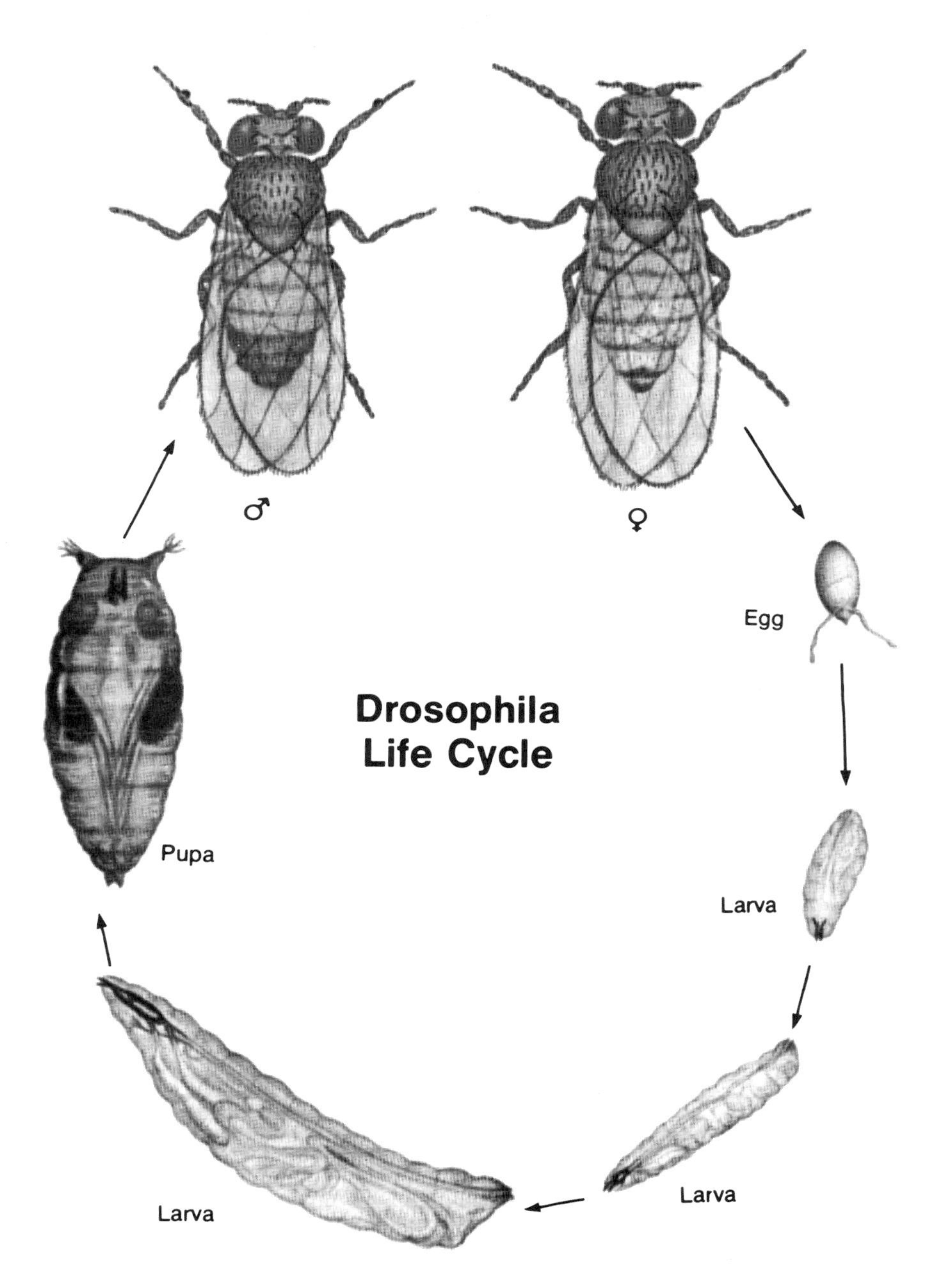

FIGURE 1-4 Outline of *Drosophila* life cycle. Note that *D. melanogaster* and other species have minor differences, such as different numbers of filaments on the egg. All species have three larval stages, however. From Carolina Biological Supplies, Burlington, North Carolina.

the same developmental program: They undergo embryogenesis in the egg capsule and have three larval instars and at least a four-day pupal stage. Details of the molecular events occurring during embryogenesis also seem very conserved in the genus, at least to the level of detail attained by present day technologies (see chapter 11).

Once adults eclose, obviously they inhabit a very different environment than did the larvae. They are reasonably strong fliers who spend their adulthood looking for food and mates. Females in particular need nourishment to develop eggs. Males need less energy, just enough to find mates and carry out courtship. Generally the way mates are brought into close proximity is through attraction to a feeding site, as there are no known long-range pheromones in *Drosophila.* Adults often feed on the same substrate in which larvae develop. Yeasts and bacteria are favored sources of nutrition for adults as well as larvae. Carson (1971a) pointed out that often adults feed on a much broader range of substrates than the oviposition substrate (discussed further in chapter 5). Females store sperm in special organs called *spermathecae* and a single insemination can supply sperm for several days and several hundreds of eggs.

Culturing

The ease with which species can be cultured in the laboratory varies considerably. There is a distinct correlation between ease of culturing and breadth of niche, as well as with adaptation to be human commensals. Those domestic species listed previously, including the most-studied model, *D. melanogaster*, have a broad range of habitat preference and do not have particularly narrow nutritional requirements, which is precisely why they could relatively easily become human commensals adapted to human refuse. Such species are perfectly happy on a variety of laboratory media: bananas, cornmeal/molasses, wheat cereals, and others. This ease of culture and availability near human habitats are the reasons these species have become favored laboratory animals. But these attributes are precisely why commensal species are not appropriate for many kinds of evolutionary studies. There is no way of knowing whether natural populations or, more accurately, free-living populations, of these domestic species have very odd histories (from being founded recently by movement of human garbage) or whether they are well-established populations. The whole problem of population genetics equilibrium, a necessary assumption for many types of analysis, is greatly exacerbated in domestic species. (See chapter 10 for further discussion of *D. melanogaster* biogeographic history.)

For studies of truly natural populations, species still living in their native habitats are preferred. This is why many of the population and evolutionary genetics studies have involved "wild" species, *D. pseudoobscura, D. subobscura, D. robusta*, and *D. willistoni* being good examples. These species are somewhat unusual in that they exist naturally in their native habitats yet can be easily cultured in the laboratory.

Many other species can be cultured, but with great difficulty, as exemplified by most of the Hawaiian species and most fungus breeders. Often the culture medium needs to be supplemented with specific plant extracts or the natural substrate itself to get the flies to breed. Even with great care and bother, strains of such species often die out. It must be kept in mind that those species that have been singled out for extensive study, those used in most of the work to be discussed in this book, represent a selected subset of all *Drosophila.*

What's in a Name

"*Drosophila*" is Greek for dew (or liquid) lover, presumably referring to these flies' attraction to moist fruits and vegetation. "*Melanogaster*" means black belly, due to the black tip of the male abdomen in this species. The older name for *D. melanogaster* was *D. ampelophila*, which means a lover of grapevines. Thus the original full name would be translated "dew lover, grapevine lover," a rather awkward appellation; perhaps it was wise to drop this name (although the change was not for aesthetic reasons, but rather to conform to standard rules of nomenclature, *ampelophila* having been misapplied to a species already named *melanogaster.*) The "black-bellied dew lover" is much more euphonious.

The most widely used common name for *Drosophila* is "fruit fly"; however, strictly speaking, this is incorrect. *Drosophila* are not true fruit flies; true fruit flies are Tephritidae, a related Dipteran family. True fruit flies breed in fruit that is still on trees and thus cause economic damage. *Drosophila* almost never attack fresh fruit, but are polite enough to wait until it has started rotting. About the only economically important role *Drosophila* may occasionally play is as a pollinator of some crops. (However, it has been reported that visitors to tasting rooms at some of the California wineries have been bothered by *Drosophila* competing for the liquid in the glasses, thereby decreasing the sale of wine.)

Other common names have been used for *Drosophila*: vinegar fly, pomace fly, pickled-fruit fly, and wine fly. All of these names clearly refer to *Drosophila's* attraction to fermentation products, an accurate description of its major habit.

Sources of Information

For many years, three books, still extremely valuable today, served as the primary sources of information. Demerec's 1950 book, *Biology of Drosophila*, was the source for basic information on general biology and development; it has recently been reprinted. Information on mutants of *D. melanogaster* has been compiled periodically beginning with Bridges and Brehme (1944) followed by Lindsley and Grell (1967) and Lindsley and Zimm (1992). The only book devoted to evolutionary subjects was Patterson and Stone's 1952 book, *Evolution in the Genus Drosophila*, a valuable source for systematic and especially biogeographic information.

Fortunately, recent years have seen a number of excellent compilations of information. The series of volumes titled *Genetics and Biology of Drosophila* edited by

TABLE 1-2 Internet addresses to access *Drosophila* databases.

FlyBase	
Web sites	http://flybase.bio.indiana.edu:82/
	http://cbbridges.harvard.edu:7081/
Gopher sites	flybase.bio.indiana/edu (Port 72)
	cbbridges.harvard.edu (Port 7071)
FTP (anonymous)	flybase.bio.indiana/edu
EofD Web site	http://shoofly.berkeley.edu

Michael Ashburner and others is a broad treatment of a large number of aspects of *Drosophila*, volumes 3a to 3e being particularly relevant to the topics addressed in this book. Ashburner's (1989) encyclopedic *Drosophila: A Laboratory Handbook* is a marvelous compilation of experimental techniques as well as much other information. The development of *Drosophila* is thoroughly covered in the two volumes of *The Development of Drosophila melanogaster*, edited by M. Bate and A. Martinez Arias. For those connected to computer networks, there is a series of highly useful electronic documents under the name FlyBase (described by Ashburner and Drysdale, 1994) and *Encyclopedia of Drosophila*, EofD. These can be accessed on the Internet at the addresses indicated in table 1-2.

2

Population Genetics—Genes

Population genetics has come to occupy a rather special place in biology. It represents the interest in the processes of evolution and in the improvement of domesticated plants and animals which strongly motivated the early students of genetics. It made possible the great renaissance of evolutionary biology which began about 1930. Population genetics then tended to reunite fields of biology such as genetics, ecology, paleontology and systematics, which had tended to take separate paths. It has thus been referred to, and with cause, as the core subject of general biology.

L. D. Dunn, 1965, p. 192

From the earliest days after the rediscovery and confirmation of Mendel's rules of inheritance in 1900, there has been a fundamental split in biologists' view of how Mendelian principles operate in populations and through evolutionary time. T. H. Morgan wrote three books with evolution as the primary subject (1903, 1925, 1932). The last book was written after many of the fundamentals of population genetics had been elucidated by the triumvirate Fisher, Wright, and Haldane, yet Morgan's view incorporated little if any populational thinking. From a solely genetic point of view, Morgan thought evolution was a very simple process: Individuals homozygous for the "normal" most-fit allele, that is, the wild type in classical genetics, were in the vast majority in natural populations. Occasionally a new mutation arose that was fitter than the predominant wild type; it quickly swept through the population, and a new wild type was established. I have purposely used the term "population" above, rather than "species," as Morgan did not believe in the existence of species, considering them merely scholastic constructs needed by simple-minded taxonomists.

Almost simultaneously the Russian school was developing a very different view, with Chetverikov the intellectual leader. His view of populations and species, of *Drosophila* in particular, was one of variation in natural populations being the normal state. While individuals and populations of a species superficially resemble one another (which is why they are considered the same species, after all), there is "hidden" in populations a vast store of genetic variability. Populations are able to "soak up" variation somewhat sponge-like—and somewhat mystically in Chetverikov's original papers! Because of these ample stores of variation, considerable natural selection and adaptation could be effected without any further input of new mutations (Chetverikov, 1926).

These fundamentally divergent views of the genetic basis of evolutionary change, while always lurking consciously or subconsciously in geneticists' minds, came into sharp focus in the middle of the century. H. J. Muller's 1950 presidential address to the American Society of Human Genetics was concerned with the widespread and almost frivolous use of X-rays. He had been a major figure in demonstrating the mutagenic effects of X-rays, and given the venue of a human genetics society meeting, he discussed the potential damage to human populations. He introduced the term "genetic load," which was meant to describe this damage, somewhat dramatically. Muller's argument was that since it was well known that the vast majority of new mutations were deleterious with respect to the wild type, every new mutation that entered a population due to an increased mutation rate (e.g., as caused by the indiscriminate use of X-rays) put a load on the population that it would have to remove by eliminating the deleterious mutation by natural selection. Selection means either death or sterility. Unless society allowed the elimination by selection to occur, the deleterious mutations would accumulate and the species as a whole would degenerate.

How seriously one took such arguments depended on one's view of genetic variation in populations, such as the Morgan versus Chetverikov view. If the vast majority of individuals of a species were homozygous for all the best fit alleles, each new mutation would represent a lowering of fitness of the species, and would indeed represent a load. On the other hand, if genetic variation were the norm and populations were replete with variation already, one might not be so concerned with an increased mutation rate—at least it would not be quite so unconditionally bad for a species if what are thought of as mutations are normally occurring in natural populations and may even increase the population's fitness.

A few years later, Dobzhansky (1955) actually named these two points of view, and as often happens, by naming them, their reality was assured. Perhaps because it was associated with the founder of modern genetics, Dobzhansky named Morgan's perspective the *classical view*. He named Chetverikov's perspective the *balance view*, because it implied that populations were in a balance of many genotypes segregating in populations, hypothesized to be maintained by some form of balancing selection such as heterosis. Dobzhansky clearly subscribed to the latter view. The Morgan/Chetverikov views had evolved into the Muller/Dobzhansky dichotomy, now called the classical versus balance views. (It is beyond the scope of this book to delve into all the ramifications of these opposing perspectives of the living world. Suffice it to say they have fundamental implications not only for the esoteric mechanisms of evolution, but also for several issues of direct interest to human societies: medical genetics, the meaning of racial differentiation, the inheritance of mental capabilities, conservation biology, and more. Some of these issues are discussed in Dobzhansky [1973a].)

The Epistemological Paradox

To resolve this dispute over the nature of variation in natural populations, it would seem to be a simple solution to simply go out into nature and see what really exists! However, this simple solution is fraught with difficulties. In 1974, R. C. Lewontin wrote a highly insightful and influential book titled *The Genetic Basis of Evolutionary Change*. In it he details just why a definitive answer to the question of amounts of genetic variation naturally occurring in populations has been so elusive. While I will address many of the same issues and data in this chapter, the much more detailed treatment in Lewontin's book is highly recommended.

Near the beginning of his book, Lewontin presents the major crux of the problem in what he called an epistemological paradox. Natural selection, in the traditional Darwinian view, acts at the level of survival and reproduction, which is determined by morphology, physiology, and behavior. These traits are often amply variable in natural populations, but are generally distributed in a continuous manner with subtle variation among individuals. Population genetics, on the other hand, deals with discrete variation, as all of population genetics theory is based on gene and genotype frequencies. One must be able to enumerate genotypes in populations to acquire the observational data to put into the theory. Discrete variation is seldom found for those traits (morphology, physiology, behavior) that make up the stuff of adaptive evolution. Almost invariably such traits are highly susceptible to environmentally induced variation (e.g., nutrition of developing stages) and when there is genetic influence, it is almost certain to be multigenic, with each locus having a small effect. As will be documented in the remainder of this chapter, discrete variation in populations can be measured and gene and genotype frequencies calculated for such traits as known morphological mutations and recessive lethals. However, these are either very rare (in the case of morphological mutants) or so deleterious (recessive lethals) that they are unlikely to constitute the general basis for the continuous variation of the adaptively important variation. So the paradox consists of the fact that *what is measurable is uninteresting, and what is interesting is unmeasurable*.

Lewontin (and many others at the time) believed this epistemological paradox could be overcome by the application of molecular technologies. Twenty years after Lewontin's book, this has not yet been fully achieved. I will, however, argue in subsequent chapters that for the first time in nearly a hundred years, evolutionary genetics is beginning to move beyond the preoccupation with the fundamental dichotomy epitomized by the classical and balance views. It is not that the issue has been unambiguously resolved by evidence convincing to all practitioners. Rather, in a sense, we have simply gotten over it.

Why bother, then, to discuss these issues at all? Why not just get on to where the field is today and discuss the modern problems? Unfortunately, while such an approach may be possible with some kinds of molecular biology subjects (reductionist sciences), it is impossible to understand the field of evolutionary genetics without at least a passing knowledge of its history and the kinds of data on which our present views rest. In particular, the subject of this book is *Drosophila*, and since a major part of the interplay and partial resolution of the problem has come from empirical data from this organism, it is only fitting to discuss these "old data." I am also convinced that one avenue of future work will be a reconsideration of these older (though no less important) observations, as they require integration with the newer molecular work, a subject barely touched on by any worker. One cannot simply ignore them, as they are part of the empirical data base any population genetics paradigm, old or new, needs to incorporate.

Morphological Mutants

In Natural Populations

By 1944, Bridges and Brehme could list 402 morphological mutants known in *Drosophila melanogaster*. A natural question arose as to the frequencies of these mutants in natural populations. Because so many single-gene mutations had been characterized for this species, it meant that a knowledgeable *Drosophila* geneticist would know what kind of morphological variants to look for in natural populations. The answer to this question is important in three respects. First, there were serious doubts early in the history of *Drosophila* genetics whether laboratory work was relevant to real populations. Were all the mutants studied by Morgan and colleagues simply artifacts of laboratory culturing with little or no relevance to the natural world? Second, if the frequencies of the known mutants were high in natural populations, this might begin to shed light on the classical/balance problem. Third, if one could detect recognizable allelic variation in sufficiently high frequency in natural populations, this would provide the material to conduct population genetics studies and test theoretical expectations.

Table 2-1 is a listing of some of the studies on morphological mutants in natural populations of *Drosophila*. Generally such studies were conducted with the recognition that most mutations are recessive, so that some form of inbreeding was needed to reveal the presence of mutant alleles. Females inseminated in nature were placed singly into culture bottles and allowed to establish *isofemale* (isolated female) strains. These were inbred and examined for the presence of mutants among F_2 progeny. Table 2-2 summarizes one of the more extensive such studies, done by Spencer (1957) on *D. mulleri*; the results are more or less typical for such studies. About a third of the strains revealed recognizable visible mutants in the F_2 generation. The observation that most (85%) had a single type of visible mutant, 13% had two, and 3% had three, does not deviate from the expectations of a Poisson distribution, that is, there is no

TABLE 2-1 Species examined for morphological mutants in strains derived from natural populations.

Species	Locality	Reference
D. melanogaster	Russia	Chetverikov, 1926
	Berlin	H. A. and N. W. Timofeeff-Ressovsky, 1927
	Slough, England	Gordon, 1936
	Russia	Dubinin et al., 1934, 1936, 1937
	Russia	Berg, 1941, 1942
D. subobscura	England	Gordon, 1936
		Gordon, Spurway and Street, 1939
D. affinis	Eastern U.S.	Sturtevant, 1940
D. repleta	New York City	Sturtevant, 1915
D. immigrans	Several U.S. locales	Spencer, 1946
D. robusta	Ohio	Spencer, 1947
D. hydei	Ohio	Spencer, 1932
	California	Spencer, 1944
	Tennessee	Spencer, 1994
	Several U.S. locales	Alexander, 1949
D. macrospina limpiensis	Western U.S.	Alexander, 1949
D. mulleri	Texas	Spencer, 1957
D. virilis group complex		Alexander, 1952
americana		
texana		
novamexicana		

TABLE 2-2 Summary of results of visible mutation screen of a natural population of *Drosophila mulleri* (Spencer, 1957).

	No. Strains Examined	No. with Mutant in F_2	Total No. Mutations
	736	223	263
	No. of Mutations per Strain		
	one	two	three
No. of strains	189	28	6
Poisson expectation	184	33	4

indication of nonrandom clustering or nonclustering of the mutations. The types of mutants found (affecting eye color, bristles, wing shape, wing veins, etc.) spanned the range known from laboratory studies of *D. melanogaster.* So the first question was answered by these studies. The mutants studied by laboratory geneticists are not artifacts; they exist in natural populations, albeit in low frequency. This is perhaps predictable since all the mutants of *Drosophila* that had been so carefully cataloged and studied up to about 1940 were so-called spontaneous mutations. That is, before the use of irradiation or chemical mutagens, mutants simply popped up in strains. Unless banana medium was a mutagen, there really should be no reason not to expect them to exist in real populations.

The answer to the second question is much more ambiguous. The very low frequency of the mutations would seem to argue in favor of the classical view. Each of the 736 strains studied by Spencer represented four haploid genomes from nature, two from the female and two from her mate. So the average number of mutations observed per haploid genome studied was 0.09 (263 mutants on 2,944 genomes). Spencer acknowledges that his methods might have underestimated the true number of mutants by a factor of 6, so the actual number might be around 0.54 per haploid genome. This number is concordant with other studies listed in table 2-1. We can estimate that about 2,000 loci in *Drosophila* can mutate to recognizable morphological variants, so the average frequency of a mutation is around 0.5/2000 or 0.00025. This low level of morphological genetic variation is not at all surprising and is consistent with a classical view of population genetics. After all, it was the preponderance of wild type phenotypes in most cultures of *Drosophila* that led to this view. If it were not the case, we would not have the concept of wild type at all!

Implications

Why didn't such observations put the nail in the coffin of the balance view of populations? First, almost certainly the proportion of genes that can mutate to viable morphologically variant phenotypes is small. Lewontin (1974) estimated this to be on the order of 10%. Another way to estimate this is to compare the numbers of loci identified in the 1968 compilation of *Drosophila* loci with that done in 1992 (Lindsley and Grell, 1968; Lindsley and Zimm, 1992). In the 1968 compilation virtually all the loci were identified as visible mutants. Few new visible mutants have been identified since 1968, so the major additions to the 1992 compilation have come from biochemical, molecular, and genetic fine structure studies. In 1968 some 362 X-linked loci are listed, while in 1992, 1,358 are recognized. If the 1,358 X-linked genes now known represent close to saturation, then we could estimate that about a quarter of X chromosome genes can give rise to visible mutants. However estimated, it is clear that the study of frequencies of viable visible mutants does not examine the majority of genes.

More importantly, such studies only identify the most discrete kind of changes, those recognized by the geneticist peering through a microscope. Obviously mutations leading to lethality are missed. Subtle variation leading to no easily recognized phenotype would go undetected. The whole concept of isoalleles—different alleles yielding identical phenotype—places serious limitations on such studies. This also applies to wild-type alleles: Simply because two alleles give identical wild-type phenotypes does not mean they are identical. This was clearly known even before the advent of molecular technologies. One classic example is that of Timofeeff-Ressovsky (1932), concisely summarized in Wallace (1968). Timofeeff-Ressovsky's experiment used crosses to transfer a wild-type allele of the *white* (w^+) locus isolated from a strain collected in America into a Russian strain, and, conversely, to transfer the Russian w^+ into the American strain. Thus there were four strains, the American allele in the American and Russian backgrounds, and the Russian allele in the Russian and American backgrounds. Timofeeff-Ressovsky then measured the rate at which the wild-type alleles mutated (spontaneously) to either true *white* or to one of the many intermediate alleles. The results are in table 2-3. Clearly the American w^+ allele mutated at a higher frequency than the Russian allele independently of the genetic background; the proportion of mutants that were true *w* was also higher. The two wild-type alleles were not the same genetic entities. Green (1959, 1963) followed with detailed analyses of this

TABLE 2-3 Results of experiments of Timofeeff-Ressovsky (1932) indicating that white-eye wild type alleles from different localities are genetically different in *D. melanogaster*.

Allele	Background	No. of Tests	No. of Mutations	Frequency
American	American	31,000	27	0.00087
American	Russian	28,200	28	0.00100
	Total:	59,200	55	Mean 0.00093
	41 of the 55 were to true *white*			
Russian	Russian	49,200	26	0.00053
Russian	American	26,100	14	0.00054
	Total:	75,300	40	Mean 0.00053
	19 of the 40 were to true *white*			

Difference in frequencies of total mutations: 0.00040 ± 0.00013.

**Note:* "American" and "Russian" indicate the origin of the allele or the genetic background. Numbers of spontaneous mutations at the locus are indicated, some of which are true white and other less extreme alleles.

After Wallace (1968).

locus and confirmed the presence of more than one wild-type "gene" in what turns out to be a gene complex.

Although these studies of visible mutants clearly did not solve the classical/balance problem, they did provide a definitive answer to the third question. The frequency of visible mutants in natural populations is too low in general to be of any use to the experimental population geneticist trying to characterize gene and genotype frequencies to determine the nature of forces affecting them. When this became clear from the studies listed in table 2-1, little or no further work was done along these lines.

There is at least one interesting exception to the fact that morphological variants are too infrequent to be practical for study. This is the *abnormal abdomen* syndrome in *Drosophila mercatorum*, studied by A. R. Templeton and colleagues (see Templeton et al., 1990b, 1993, and references therein). This is an analog mutant of the bobbed syndrome in *D. melanogaster* and similarly involves changes in the rDNA genes. The abnormal genes have been found in extremely high frequencies, up to over 50% in some Hawaiian populations. While the visible expression is much rarer due to suppresser loci, it nonetheless still shows phenotypic effects in nature. This will be discussed in more detail in chapter 6.

Recent studies have provided one major insight into the basis for morphological variation in *Drosophila* populations. Many of the loci that can mutate to the classical visible mutations have been molecularly characterized. The most interesting generality is that *in the vast majority of cases, the molecular basis of the mutant phenotype is an insertion or deletion, most often with some involvement of a transposable element* (Finnegan and Fawcett, 1986; Green, 1987). Presumably, the molecular basis of the visible mutants in nature is the same as that of the spontaneous mutants that appear in culture. Thus we must conclude that visible mutants in nature are likely the consequence of insertions or deletions of the sort caused by transposable elements. To the extent that these kinds of mutants play a role in evolution (discussed in chapter 9), transposable elements are important factors in the evolution of these flies.

It would be of great interest to confirm that most visible mutations isolated directly from natural populations are due to insertions or deletions caused by transposable elements. This is particularly intriguing considering that species may vary greatly in the numbers of transposable elements they harbor (see chapter 9). A testable prediction is that the frequency of such mutants should reflect the density of transposable elements in the genome. Does this mean species vary in their evolutionary potential?

Fitness Variation

Obviously, for Darwinian natural selection to occur, there must be genetic variation that affects fitness in the population. This is the basis of R. A. Fisher's modestly termed "fundamental theorem" of evolution: The rate of change in fitness of a population is equal to the variance in fitness. Thus the crucial data required to assess the role of natural selection in controlling evolutionary change is the level of genetic variation found in populations that directly affects fitness. There is considerable literature on *Drosophila* addressing this issue.

Homozygous Chromosomes

The basis for much of the work to be discussed is illustrated in figure 2-1. H. J. Muller first devised the crossing scheme illustrated here, which allows one to make whole chromosomes homozygous. Assuming that most genetic variation tends to be recessive, this is a method for revealing concealed variability that would not otherwise be measurable. The crucial points to understanding this scheme are that crossing over does not occur in the meiosis of male *Drosophila* (of most species) and that inversion heterozygotes do not produce viable recombinant gametes for the inverted chromosome. By having the marker strain with both a recessive and a dominant marker on the inverted chromosome, all four possible chromosome combinations can be distinguished in the F_2 generation and only the proper combination chosen as parents, that is, individuals showing only the dominant phenotype. Generally, the dominant marker used is lethal in the homozygous state, so in the F_3 one expects 1/3 of the offspring to be homozygous for the chromosome carried by the male in the P generation, a male usually taken from a natural population; 2/3 would be heterozygous for the same chromosome and the laboratory-derived marker chromosome. Because the F_3 off-

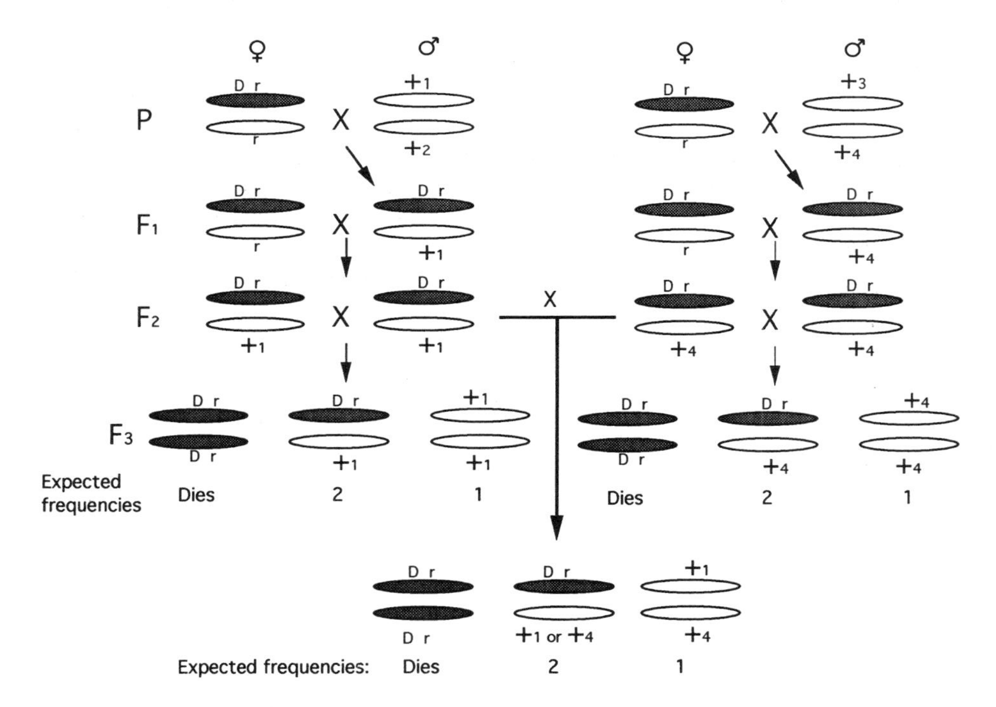

FIGURE 2-1 Scheme for testing fitness effects of homozygous chromosomes. D and r denote dominant and recessive mutations. The shaded chromosome signifies an inversion relative to the open chromosomes. $+_1$, $+_2$, etc. are chromosomes taken from a natural population. The left and right crosses test the relative fitness of the $+_1/+_1$ and $+_4/+_4$ homozygotes relative to the heterozygote with the marker chromosome. The middle cross tests the relative fitness of the $+_1/+_4$ heterozygote against the same standard.

spring are developing in the same culture bottle, they will be competing with one another. Deviation from the expected 1/3 wild type in the F_3 generation will be a measure of the fitness of individuals homozygous for the chromosome relative to the same chromosome being heterozygous with the laboratory strain chromosome. One might wish to know the relative fitness of homozygous chromosomes from nature relative to heterozygotes from nature rather than heterozygotes with laboratory marker strains. This is done indirectly by taking F_2 parents from two different sets of crosses and carrying the crosses through to the F_3 generation. This allows one to compare the fitness of the heterozygote for two chromosomes from nature to the same standard as used for the same chromosomes when homozygous. Other variations on this basic scheme are possible, including producing individuals homozygous for more than one chromosome simultaneously. In *D. melanogaster*, for which the proper set of marker strains is available, the whole genome can even be made homozygous.

From the late 1930s into the 1960s, the use of this technique to reveal concealed variation dominated *Drosophila* population genetics and a considerable literature was generated, far too vast to deal with in detail here. Good summaries and further references can be found in chapter 3 of Wallace (1968), chapter 3 of Dobzhansky (1970), chapter 2 of Lewontin (1974), and chapter 5 of Wright (1978). The following is a brief summary of the essential conclusions from this work.

One of the more remarkable results is the consistency from chromosome to chromosome and from species to species in the outcome of these experiments. The three examples illustrated in figure 2-2 are typical results. The fitness of homozygotes is inevitably bimodal with a peak at zero, that is, lethals, and a second peak just below the 1/3 expected for viability equal to the heterozygotes, and a smattering of chromosomes in between. Dobzhansky called these, going left to right, lethal, semilethal, subvital, and quasi-normal chromosomes. Tables 2-4 and 2-5 indicate the extent to which these results have been generalized in a large number of studies.

Clearly there is considerable variation in viability due to recessive genes in these populations. From 10% to 50% of chromosomes tested were lethal or semilethal (viability less than 50% generally being the cutoff for the semilethal category) with a mean of 28% for all studies listed in these two tables. Even those chromosomes not in this category, on average, carry deleterious recessive genes so that the second mode of the distribution is always below that for heterozygotes (the mean averages about 86% that of heterozygotes). Furthermore, recall that these experiments only test one component of fitness, egg to adult viability. A few tests have been designed to determine other fitness components. Table 2-6 shows a test of the percentage of each sex that are sterile when different chromosomes are made homozygous for two species. Of the carriers that appear quasi-normal, 10% to 66% are genetically lethal in that they are sterile. Marinkovic (1967) studied other fitness components with similar results.

Sved and Ayala (1970) and Sved (1971) devised a clever scheme to test the fitness effects of homozygous chromosomes more thoroughly. Simply by mixing the F_2 individuals in figure 2-1 and beginning population cages, all components of fitness could be measured in aggregate. The flies were allowed to breed freely for several generations, and the phenotype frequencies were monitored. Recall that the homozygous marker chromosome is lethal. If, as expected, the heterozygote with the marker chromosome has a higher fitness than the homozygote, a stable balanced polymorphism due to heterosis should emerge. Knowing the fitness of one homozygote (zero)

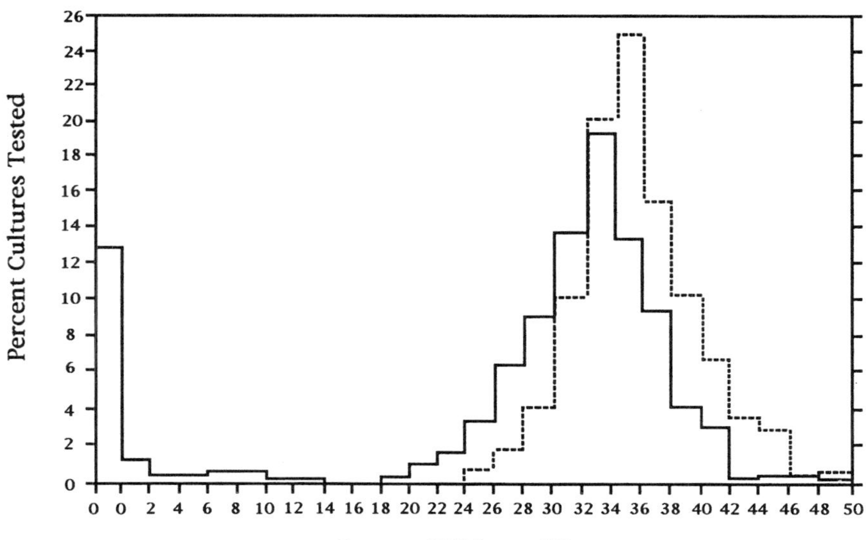

Percent Cultures Tested
Percent Wildtype Flies
0 0 2 4 6 8 10 12 14 16 18 20 22 24 26 28 30 32 34 36 38 40 42 44 46 48 50
0 2 4 6 8 10 12 14 16 18 20 22 24 26

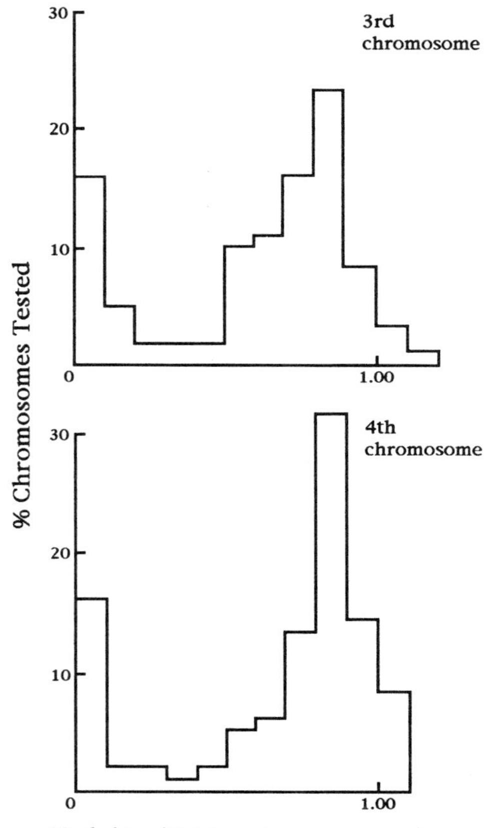

3rd
chromosome
4th
chromosome
% Chromosomes Tested
Viability (% Mean Heterozygote)
0
1.00
10
20
30

TABLE 2-4 Summary of studies testing for viability effects of homozygous chromosomes of several species and populations of *Drosophila.*

Species	Chromosome	Locality	No. Chromosomes Tested	Percentage Lethals and Semilethals	Average No. Lethals per Genome
D. prosaltans	First	Brazil	304	32.6	0.99
	Second	Brazil	284	9.5	0.50
D. willistoni	First	Brazil	2004	41.2	1.33
	Third	Brazil	1166	32.1	1.94
	Second	Florida	109	31.1	0.93
	Third	Florida	122	32.8	1.99
	Second	Cuba	25	36.0	1.12
	Third	Cuba	39	25.6	1.48
D. melanogaster	Second	N. Caucasus	795	12.3	0.33
		S. Caucasus	2738	18.7	0.52
		Ukraine	2700	24.3	0.61
		Pennsylvania	117	28.2	0.83
		Amherst	3549	36.3	1.13
		Ohio	343	43.1	1.41
		Florida	468	61.3	2.37
D. pseudoobscura	Third	Yosemite, CA	109	33.0	2.00
		San Jacinto	326	21.3	1.20
D. persimilis	Third	Yosemite, CA	106	25.5	1.47
			Mean	30.3	1.23

Taken from Dobzhansky and Spassky (1954), which lists complete primary citations.

allows one to calculate the fitness of the other. The most important results from these studies is that *the decrease in fitness due to reduced viability is only about a third of the total decrease in fitness.* Evidently fertility (or development time) plays a larger role than simple egg-to-adult viability in determining fitness, a result consistent with other studies (see chapters 3, 4, and 6). So when you contemplate the fitness distributions illustrated in graphs like figure 2-2, keep in mind that the total fitness reduction due to homozygosity is on the order of three times greater than such graphs indicate.

Synthetic Lethals

Undoubtedly, then, there exists considerable genetic variance in fitness within *Drosophila* populations. What is not clear is how many loci are involved in such effects. A homozygous chromosome could be lethal due to a single locus or to the epistatic interaction of many loci. Evidence for epistatic fitness interactions comes from studies of "synthetic" lethals and the effects of recombination. The experiment is to take two

FIGURE 2-2 (FACING PAGE) Typical result of making *Drosophila* chromosomes homozygous. Upper graph shows result for the third chromosome of *D. pseudoobscura*; the dashed line shows the relative viability of the heterozygotes and the solid is for homozygotes. The two lower graphs show the same for the third and fourth chromosomes of *D. persimilis*; the 1.00 on the x-axis is the average viability of heterozygotes. From Dobzhansky and Spassky (1953).

TABLE 2-5 Summary of several studies testing viability effects of homozygous chromosomes.

Species	Chromosome	Locality	Percentage Lethals and Semilethals	Mean Viability of Quasi-normals
D. pseudoobscura	2nd	Bogotá	18.3	89.9
		Yosemite, CA	33.0	75.0
		San Jacinto, CA	21.3	99.2
	3rd	Yosemite, CA	25.0	77.1
		Death Valley, CA	15.0	91.7
		Central America	30.0	89.2
	4th	Yosemite, CA	25.9	85.6
		San Jacinto, CA	25.5	87.6
D. persimilis	2nd	Yosemite, CA	25.5	87.9
	3rd	Yosemite, CA	22.7	83.4
	4th	Yosemite, CA	28.1	77.9
D. prosaltans	2nd	Brazil	32.6	91.7
	3rd	Brazil	9.5	96.6
D. willistoni	2nd	Brazil	41.2	86.7
	3rd	Brazil	32.1	88.8
D. melanogaster	2nd	Wisconsin	25.1	79.0
	2nd	Long Island, NY	20.5	83.1
		Mean	26.3	86.5

Taken from Lewontin (1974) with complete primary references in that book.

TABLE 2-6 Percentage sterility found among carriers of quasi-normal homozygous chromosomes with respect to viability.

Species	Chromosome	Percentage Females Sterile	Percentage Males Sterile
D. pseudoobscura[a]	2nd	10.6	8.3
	3rd	10.6	10.5
	4th	4.3	11.8
D. persimilis[a]	2nd	18.3	13.2
	3rd	14.3	15.7
	4th	18.3	8.4
D. prosaltans[b]	2nd	9.2	11.0
	3rd	6.6	4.2
D. willistoni[c]	2nd	40.5	64.8
	3rd	40.5	66.7

[a]From Dobzhansky and Spassky (1953); Sankaranarayanan (1965)
[b]From Dobzhansky and Spassky (1954)
[c]From Malogolowkin-Cohen et al. (1964)

TABLE 2-7 Frequency of production of synthetic lethals due to recombination of quasi-normal chromosomes.

Species	No. Recombinant Chromosomes Tested	No. Lethal	Reference
D. pseudoobscura	100	3	Dobzhansky (1946)
	450	19	Dobzhansky (1955)
D. melanogaster	4830	95	Wallace et al. (1953)

independently derived quasi-normal chromosomes, recombine them, and observe the viability of recombinant chromosomes in homozygous chromosome tests, that is, take them through the crossing scheme in figure 2-1. Tables 2-7 and 2-8 summarize the results. First (in table 2-7), lethals are generated from nonlethal chromosomes; this can only be explained by epistasis between two or more loci. Second (in table 2-8), a considerable amount of the total variance in viability can be regenerated by the recombination of just two quasi-normal chromosomes. We can only conclude that epistasis among multiple loci produces much of the fitness effects in such studies.

Therefore, we know that several loci are involved in generation of this fitness variance. Is there any way to calculate the actual number of loci involved, and then address the related question of the frequency of the recessive alleles at each locus? Lewontin (1974) estimated that approximately 900 to 1300 loci in *Drosophila* are capable of mutating to lethals. As noted in table 2-4, each genome is estimated to contain about one lethal. On average, therefore, the frequency of any lethal allele is about 0.001. This is consistent with the allelism tests performed by Dobzhansky and Wright (1941), where the vast majority of lethals isolated were not allelic with any other lethals found in the same or more distant populations. Of course this is assuming that most of the lethal and semilethal effects of homozygous chromosomes are generally attributable to single genes being made homozygous rather than to epistatic interactions of many loci, each with relatively small effect. It is impossible to estimate

TABLE 2-8 Regeneration of variance in viability by recombination.

Species	Percent Recovered Relative to Whole Distribution	Percent Recovered Relative to Quasi-normals	Reference
D. pseudoobscura	43	75	Spassky et al. (1958)
D. persimilis	24	27	Spiess (1959)
D. prosaltans	25	28	Dobzhansky et al. (1959)
D. willistoni	35	100	Krimbas (1961)

Note: Shown is the percentage of the variance found in the natural population sample of more than 100 chromosomes that is regenerated by recombination of just two quasi-normal chromosomes. Both the percentage recovered with respect to all the chromosomes (the whole distribution in figure 2-2) or just with regard to quasi-normals (the right hump in figure 2-2).

either the number or frequencies of loci and alleles with fitness effects less drastic than lethals or at least semilethals.

The Load Problem

Before concluding this section, it is worth mentioning for historical reasons, if no other, the controversy generated by these studies of lethals. Simply stated, the question at the time (in the 1950s) was whether all the recessive deleterious alleles indicated by these studies are "good" or "bad" for the populations. That is, was this genetic variation simply the measurable part of the ubiquitous variation envisioned by adherents of the balance school? (Th. Dobzhansky and B. Wallace were the principal proponents of this view.) Or was this part of the genetic load on the population and therefore must be harmful? (H. J. Muller and J. F. Crow represented the principal advocates of this view.) The effect of these variants in the heterozygous state became a crucial question. The relevancy of this issue is due to their rare frequencies: By far the majority of lethal alleles exist in heterozygotes. For example, given the average frequency of lethal recessives as 0.001 and assuming Hardy-Weinberg equilibrium, then there are 1,998 heterozygotes for every homozygote carrier of the recessive. A 1% heterozygous fitness deficit would result in removal of 20 alleles every generation for every 2 removed by lethality of the homozygote. If such lethals had partial dominant deleterious effects, they would be expected to be removed from the population rapidly and would merely be part of the genetic load due to mutation-selection balance. Alternatively, if they had no deleterious heterozygous effect or were even heterotic, then they would be considered much less harmful or even good. Wallace extended these ideas to include all new mutations, both spontaneous and those induced by irradiation. If, on average, new mutations had no heterozygous effect (or were even occasionally heterotic), then populations would be absorbing variation, as envisioned by Chetverikov. Furthermore, if most of the lethal and especially less deleterious effects were not exclusively due to single genes, but to epistatic interaction among many loci, then perhaps the calculations assuming single loci greatly underestimated frequencies in natural populations. Many genes, with small deleterious effects even when several are made simultaneously homozygous (quasi-normals), could reach appreciable frequencies in populations and in fact be beneficial, on average, when present in outcrossing, highly heterozygous populations. One can imagine the technical difficulties with assessing the small heterozygous effect of a gene that is only mildly deleterious when homozygous. Suffice it to say that the heterozygous effects are slight and hard to measure; it seems to depend largely on the laboratory, species used, and so forth. Reviews of this literature include Simmons and Crow (1977) and Wallace (1970a; 1991).

The time and energy expended on this problem and the emotional pitch of the arguments is difficult to understand today. Perhaps the vigor of the arguments was because these were the only kinds of data that were relevant to this major problem in population genetics, and despite (or, more likely, because of) their insufficiency, they necessarily engendered controversy.

Selection Experiments

In natural populations, many traits in *Drosophila* (and almost any other organism) vary quantitatively, rather than discretely. Measurements of traits such as body size, development time, bristle numbers, and so forth, usually take on the classic bell-shaped normal distribution. If, as is often asserted, traits of these sorts are most likely to be the objects of adaptive Darwinian evolution, it is of considerable interest to determine the ubiquity of such quantitative variation and how much of it can be attributed to genetic—as opposed to environmental—factors. If the variation in the trait is evolutionarily significant, it must have some heritable component, referred to as *heritability* in the quantitative genetics literature. The underlying genetic model usually assumed (though seldom unambiguously proved) is termed *polygenic*: It is proposed that many loci with small individual effects are segregating in the population. Generally there is also a large environmentally induced component of the variation (Falconer [1981] can be consulted for details).

Laboratory

How often can one take a trait, even an arbitrarily chosen trait that may have no bearing on the history or ecology of *Drosophila*, and successfully select for heritable change under laboratory conditions? The surprising answer to this question is: almost always! Table 2-9 is a partial listing of successful selection experiments with *Dro-*

TABLE 2-9 Examples of studies demonstrating the selectability of various traits.

Trait	Reference
Body size	Robertson & Reeve, 1952
	Robertson, 1959
Abdominal bristle numbers	Mather & Harrison, 1949
Sternopleuro bristle numbers	Thoday & Boam, 1961
Developmental rate	Clarke et al., 1961
	Prout, 1962
Fecundity and egg size	Bell et al., 1955
Behavior	
Photo- and geo-taxis	Dobzhansky & Spassky, 1967
	Korol & Iliadi, 1994
Mating preference	Knight et al., 1956
Pupation site	de Souza et al., 1968
Pattern	
Ocelli	Maynard Smith & Sondhi, 1960
Posterior crossvein	Milkman, 1964
Invariant patterns by genetic assimilation	Waddington, 1953
	Rendel, 1959
Change in variance rather than mean of a character	Rendel, 1967
	Scharloo, 1964
Genetic modifiers of mutants	Numerous studies
Developmental sensitivity to environment	Waddington, 1960
Resistance to thermal stress	Huey et al., 1991
Insecticide resistance	Bennett, 1960
Novel environmental challenge	Tabachnick & Powell, 1977
Genetic system	
Parthenogenesis	Carson, 1967
Recombination rate	Chinnici, 1971a, b

sophila. It is unlikely that this is a biased sample of the outcome of such experiments, which might happen if workers tended not to publish negative results. In this case, negative results would be more startling than positive results, so a researcher with negative results would be even more eager to publish. The important thing to note in table 2-9 is the broad range of different kinds of selectable traits: morphology, behavior, physiology, and even the genetic system itself (e.g., recombination). These traits span such a range it can hardly be argued they are a biased sample of traits that might have been guessed or known a priori to have a genetic component of variation. (Chapter 8 in Wright [1977] and Mather [1983] can be consulted for details and insightful discussion of some of these studies.)

The single exception to response to selection was Maynard Smith and Sondhi's (1960) attempt to select for flies asymmetrical with respect to occelli, so-called left-handed flies. While the frequency of the left-handed flies increased in the selected lines, this was attributable to an increase in overall asymmetry. That is, what was selected was developmental asymmetry in general rather than specific handedness. Considering what has been learned about development in *Drosophila*, this is not surprising; there are no laterally asymmetrical developmental processes known on which selection could have acted.

Among all these studies, the work of Korol and Iliadi (1994) is noteworthy in addressing the question of the effect of selection on recombination. If the response to selection is enhanced by recombination, then recombination might increase in selected lines as a byproduct of selection for some other trait. They selected for positive and negative geotaxis in *D. melanogaster* and demonstrated increased rates of recombination after selection. Their paper lists a few other similar cases. This interaction between selection for a quantitative trait and its effects on recombination deserves more attention, as it may well be relevant to one of the more enigmatic problems in evolution, the evolution of sex.

Polygenic Variation in Nature

How relevant are these studies to real populations in nature? Is there variation in these kinds of traits in natural populations that can be attributed to selection? Two examples are body size and alcohol tolerance. Several studies have demonstrated that *Drosophila* can be selected for larger and smaller size (e.g., Druger, 1962) and this is a typical polygenic quantitative trait. In natural populations there is often a correlation of body size with regional temperature. Flies from cooler regions of a species range tend to be larger than flies from warmer climates, and the difference has a strong genetic component. This has been shown in *D. robusta* (Stalker and Carson, 1947, 1948), *D. subobscura* (Misra and Reeve, 1964; Prevosti, 1955), and *D. melanogaster* (Tantawy and Mullah, 1961; David et al., 1983), but not in *D. pseudoobscura* (Sokoloff, 1965, 1966). Interestingly, when replicate populations are held in the laboratory for long periods at different temperatures, the same phenomenon occurs. Laboratory populations kept at higher temperatures become genetically smaller than replicate populations kept at lower temperatures. This is true of a temperate species, *D. pseudoobscura* (Anderson, 1966, 1973), a tropical species, *D. willistoni* (Powell, 1974), and the cosmopolitan *D. melanogaster* (Kilias and Alahiotis, 1985). Accordingly, this quantitative character would seem to have direct relevance for adaptive evolution.

A second example of a quantitative trait shown to have direct selective response in natural populations is alcohol tolerance in *D. melanogaster*, especially in relation to adaptation to wineries. (While one hesitates to call these truly natural populations, their niche is typical of that occupied by most populations of this species.) Resistance to the detrimental effects of ethanol is a polymorphic trait with a multigenic inheritance, as demonstrated by conventional quantitative genetics experiments (Cohan and Hoffmann, 1986). McKenzie and Parsons (1974) showed that flies inside and outside a winery are genetically differentiated for alcohol tolerance; flies living inside are more tolerant (see also McKechnie and Geer, 1993). Here is an example of a quantitative trait involved in adaptation to a man-made environment that also affects the distribution of genotypes.

Many other examples exist. Lemeunier et al. (1986) list a variety of quantitative, polygenic characteristics that vary latitudinally, altitudinally, and seasonally in populations of *D. melanogaster.* These include such traits as body weight, wing length, ovariole number, egg production, alcohol tolerance, temperature tolerance, desiccation resistance, oviposition rhythm, light sensitivity, and sexual behavior. David et al. (1983) can be consulted for details and guides to the literature.

The Molecular Basis of Polygenic Variation

The actual molecular basis of quantitative variation in *Drosophila* is not well understood. Mukai and Cockerham (1977) concluded that mutations in structural loci (loci coding for proteins) were unlikely to be the basis for polygenic viability effects. Their reasoning is as follows. In a study of five enzyme-coding genes they estimated the mutation rate per locus to be between 3 and 18×10^{-6} per locus per generation. Estimating the number of such loci on the second chromosome to be about 2,200 (the number of polytene bands), they concluded that mutations in structural genes on the whole chromosome occur at a rate of 0.008 to 0.040 per chromosome per generation. Extensive studies with this same chromosome had indicated that mutations affecting viability occur at a rate of 0.12–0.17 per chromosome per generation (Mukai et al., 1972). While each clearly has some error, these estimates appear much too different to be measuring the same thing. However, it is now thought that many more genes than polytene bands exist in *Drosophila* genomes, perhaps by a factor of three or four. If so, then one would estimate a three- to four-fold greater per-chromosome mutation rate for the structural genes, thus overlapping the estimated rate for viability polygenes.

A more direct approach has been to study the possible role of transposable elements in generating quantitative variation. Mackay (1988) reviews her series of elegant studies using P-elements to generate polygenic variation for bristle number in *D. melanogaster.* She concludes that mobilization of P in dysgenic crosses contributes substantially to the selectable variation for these traits; in fact, the use of dysgenic crosses is more efficient than EMS or radiation in increasing selectable variance. The usual way of expressing such results is V_m/V_e, where V_m is the added variance in the trait due to mutation and V_e is the environmental variance (Lynch, 1988; this paper also contains a table summarizing V_m/V_e for a large number of quantitative traits in *Drosophila*). Table 2-10 summarizes Mackay's conclusions. There is an increase of

TABLE 2-10 Comparisons of V_m/V_e (a measurement of the mutational input contributing to the genetic variance of a trait) for various cases.

Sources of Mutations	Bristle Traits	
	Abdominal	Sternopleural
Spontaneous (M strain)	1.0×10^{-3}	0.7×10^{-3}
Spontaneous (P strain)	6.3×10^{-3}	5.3×10^{-3}
X-ray–induced (per 1000 rads)	3.4×10^{-3}	—
Dysgenesis-induced		
1. PM line	1.5×10^{-1}	3.1×10^{-1}
2. MP line	0.7×10^{-1}	1.9×10^{-1}

From Mackay (1988).

approximately three orders of magnitude in this measure over either spontaneous input or X-ray induced variance. While these studies have been criticized as even the controls (the reciprocal nondysgenic cross) exhibit increased mobilization of P (Moran and Torkamanzehi, 1990), further experiments eliminating this partial flaw in the controls still led to the conclusion that P-element mobilization can contribute greatly to generating polygenic variation (see also Torkamanzehi et al., 1992). Mackay and Langley (1990) demonstrated a correlation between variation in bristle numbers and insertions of DNA sequences in the *achaete-scute* region, a major region controlling bristle number in *Drosophila.* Other studies of physical mapping of positions of P-insertions affecting quantitative traits are Mackay et al. (1992) and Lai and Mackay (1993). It should also be noted there have been negative results from experiments testing whether P-elements induce genetic variance that can be selected (Randall Phillis, unpublished results).

One can only conclude there is considerable genetic variance for a whole host of continuously distributed characters in *Drosophila* populations. Since this variation can be demonstrated by both laboratory experiments and studies of free-living populations, the evidence is this variation is not simply a laboratory curiosity, but that these kinds of traits may often be adaptively variable in nature. While there is evidence that transposable elements may account for some of this variation, it is likely not the whole story. It is impossible to place a quantitative estimate on the number of loci that potentially account for such variation, although the observed diversity of type and ubiquity of genetic variation indicate the involvement of many loci. Unlike the case for visibles and lethals, these loci must be polymorphic for alleles at reasonably high frequencies.

Allozymes

As with virtually all fields of biology, molecular biology has had a great impact on evolutionary biology. In the mid-1960s there was considerable excitement and rejuvenation of evolutionary genetics by the introduction of molecular approaches to studying population genetics. Lewontin and Hubby published their ground-breaking papers in 1966 and the field, quite literally, has not been the same since. Elsewhere (Powell, 1994) I have recounted the history of this period and that paper can be consulted for historical as well as anecdotal discussions.

The value and power of using molecular techniques to study genetic variation in populations can easily be understood, as the conceptual underpinnings are simple and elegant. All the kinds of genes studied prior to the molecular era, virtually all the mutants listed in Lindsley and Grell (1967) and all the other kinds of genetic variation presented in the previous sections were recognized as genetic loci *precisely because they were variable.* Were there not a discrete, relatively frequent, white-eyed mutant we would never define a white locus. Estimates of the number of genes in the *Drosophila* genome vary, but certainly more than 5,000 exist, yet only about 1,000 visible mutants are known. Is this set of variable visible loci a biased sample of the genome and therefore not a fair measure of the overall genetic diversity in populations? Are the frequencies of the visible mutants typical of genic diversity in general? If so, population geneticists are in trouble as it is difficult to perform population genetics on gene frequencies less than 0.1%.

The most important advance in molecular biology in the 1960s was to understand what a gene is and how it functions. Simply stated, genes are stretches of DNA that code for a protein or polypeptide. Of course, there are exceptions, such as ribosomal RNA genes, and other complexities such as introns, but these need not distract us yet. Could one look directly at variation at a locus by looking at variation in its molecular structure without the necessity of a recognizable morphological phenotype? A second line of research was also developing at this time, namely, technical advances in separating biological macromolecules. Electrophoresis in particular developed in the late 1950s and early 1960s. This technology allowed efficient separation of proteins, the primary gene products, based on charge and other factors affecting rates of migration through a medium (gel). Markert and Moller (1959) had demonstrated that the proteins coded by different loci but producing the same enzyme activity could be separated on such gels; they coined the term "isozyme" to describe such enzymes. If protein products (enzymes) from different loci could be separated by electrophoresis, why not the products of different alleles at a single locus?

Finally, here was a method that might solve Lewontin's epistemological paradox. One could simply study a population and assay a large number of individuals for allelic variation at a number of loci for which histochemical stains had been developed that allowed visualization on the gel. There was no a priori reason to believe the loci studied would or would not be variable in any given species. The only bias in sampling the genome was that only protein-coding genes were studied and someone had developed a means of staining gels for the enzymatic activity. There was no reason to believe the allelic variation had large, small, or any phenotypic effects. Thus, there should be no bias with respect to having discrete phenotypes associated with the allelic variation (which is the case with visibles or lethals). And equally important, individuals (even of such small creatures as *Drosophila*) could be genotyped so that one could calculate gene and genotype frequencies to put into the theoretical machinery of population genetics. It was a heady time.

Level of Variation

What were the results of the application of protein electrophoresis to the study of populations, and *Drosophila* populations in particular? Lewontin and Hubby (1966) and Hubby and Lewontin (1966) published the first such results for any species (although in the same year Harris made similar calculations for humans). Similar to the story with homozygous chromosome studies, the initial report yielded a result that has been repeatedly observed in literally hundreds of successive experiments. Somewhere between 1/3 and 2/3 of all loci studied are polymorphic for naturally occurring alleles detectable by electrophoresis. Knowing that the technique of electrophoresis is imperfect at separating all protein differences, that is, detecting the effects of all amino acid substitutions, the results would indicate that probably well over half of all loci were polymorphic in a species. Prakash, Lewontin, and Hubby (1969) introduced the term *allozyme* to describe the allelic isozymes detected in such studies to distinguish them from *isozymes*, which are multiple forms of enzymes that arise for any reason (e.g., encoded by different loci).

Table 2-11 is a summary of allozyme studies performed on *Drosophila.* In addition to the proportion of polymorphic loci, another favorite statistic calculated from such studies is average expected heterozygosity. That is, one can calculate the heterozygosity at each locus and simply take the mean over the loci studied. If this was an unbiased sample of the genome, it provides a good estimate of the total genomic heterozygosity—with the important limitation of sensitivity of the technique to detect amino acid substitutions. There is some difference in the level of variation detected from study to study, but the initial heterozygosity estimated by Lewontin and Hubby (1966) for *D. pseudoobscura* (12.8%) is very near the mean for all species. It is unclear how much of the variation in heterozygosity estimates is due to variation in choice of loci, laboratory procedures, and so forth, but the tenfold variation, from a low of 0.025 to 0.25, would seem to indicate some real differences. This is not surprising considering the very different ecologies and population histories experienced by different *Drosophila* species.

On the face of it, it would seem that the classical/balance conflict had been resolved in favor of the balance view. As predicted most boldly by Wallace (1959), it would appear that no individual is homozygous, that is, there is no true wild type. Knowing the limitations of protein electrophoresis, it is likely that nearly all loci are polymorphic—a prediction confirmed by more recent DNA studies (see next section and chapter 10). But, given the positions strongly held by the opposing schools, it is not surprising that these data did not resolve the conflict; rather, it had shifted ground. While the proponents of the classical view could not deny that heterozygosity abounded in natural populations, they could claim was that it was meaningless. They contended that the large number of segregating alleles detected by protein electrophoresis meant nothing to the physiology or fitness of the flies. That is, all the variation was neutral with respect to selective importance. As Lewontin (1974) characterized it, we had obtained the right answer to the wrong question. The question never really was simply, "How much variation is there in natural populations?" Rather, the question is, "How much potentially adaptive variation is there in populations?" Thus, the next stage in the evolution of the debate had emerged, now called the "selectionist-neutralist" controversy. This controversy in one way or another dominated evolution-

TABLE 2-11 Studies of allozyme variation in various *Drosophila* species, listed alphabetically.

Species	Populations Studied	No. Individuals	No. Loci	*H*	*P*	Reference
affinis	15	1343	16	0.193	0.652	Kojima et al., 1970 Richmond et al., 1977
aldrichi	1	162	16	0.121	0.481	Richmond et al., 1977 Zouros, 1974
algonquin	7	390	16	0.210	0.930	Richmond et al., 1977
ananassae	11	1705	13	0.135	—	Gillespie & Kojima, 1968 F. Johnson, 1971 Cariou & Da Lage, 1993
arizonensis	2	227	17	0.126	0.383	Zouros, 1974
athabasca	17	890	16	0.140	0.451	Kojima et al., 1970
bifasciata	23	600	21	0.242	0.620	Saura, 1974
bipectinata	19	196	23	0.241	0.493	Yang et al., 1972
busckii	18	90	30	0.044	—	Prakash, 1973
buzzattii	52	10190	29	0.065	0.241	Barker & Mulley, 1976
engyochracea	2	1026	20	0.012	0.300	Steiner, 1975 Steiner et al., 1976
equinoxialis	5	207	31	0.165	0.624	Ayala & Powell, 1972 Ayala & Tracey, 1974 Ayala et al., 1974a, b
guanche	3	340	61	0.054	0.329	Gonzalez et al.. 1982
heteroneura	3	541	25	0.090	0.678	Sene & Carson, 1977
immigrans	1	51	17	0.115	0.710	Steiner, 1975 Steiner et al., 1976
maderiensis	2	225	61	0.086	0.516	Gonzalez et al., 1982
malerkotliana	10	316	23	—	0.503	Yang et al., 1972
mauritiana	2	40	55	0.045	0.182	Gonzalez et al., 1982
melanogaster	29	1080	34	0.135	0.562	Cabrera et al., 1982 Gonzalez et al., 1982 Kojima et al., 1970 Langley et al., 1974 Singh & Coulthart, 1982 Singh et al., 1982
mercatorum	2	156	20	0.132	0.446	Clark et al., 1981
mimica	2	1225	20	0.194	0.480	Steiner, 1975 Steiner et al., 1976
mojavensis	8	765	17	0.076	0.266	Zouros, 1974
mulleri	1	157	16	0.113	0.481	Zouros, 1974
nebulosa	6	151	25	0.170	0.589	Ayala & Tracey, 1974 Ayala et al., 1974b
obscura	57	787	33	0.109	0.580	Lakovaara & Saura, 1971b
ochrobasis	2	116	14	—	0.357	Carson et al., 1975
orthofascia	1	63	13	0.025	0.310	Steiner, 1975 Steiner et al., 1976
parabipectinata	12	220	23	0.134	0.343	Yang et al., 1972
paulistorum	32	1277	18	0.177	0.595	Ayala et al., 1974b Richmond, 1972
pavani	14	1000	24	0.192	0.552	Nair & Brncic, 1971 Nair et al., 1971
persimilis	6	225	28	0.104	0.336	Prakash, 1969, 1977a, b
pseudoananassae	4	16	19	0.203	—	Yang et al., 192

(*continued*)

TABLE 2-11 *Continued.*

Species	Populations Studied	No. Individuals	No. Loci	*H*	*P*	Reference
pseudo-obscura	18	1792	27	0.119	0.430	Lewontin & Hubby, 1966 Prakash, 1977b, c Prakash et al., 1969 Singh & Coulthart, 1982
robusta	8	527	40	0.110	0.390	Prakash, 1973
setosimentum	6	785	14	—	0.857	Carson et al., 1975
silvestris	3	834	25	0.084	0.342	Sene & Carson, 1977
simulans	9	327	51	0.055	0.342	Cabrera et al., 1982 Gonzalez et al., 1982 Kojima et al., 1970 Steiner et al., 1976
sporoati	1	125	15	0.083	0.470	Steiner, 1975 Steiner et al., 1976
subobscura	39	2584	23	0.145	0.610	Cabrera et al., 1980 Lakovaara & Saura, 1971a Marinkovic et al., 1978 Pinsker et al., 1978 Saura et al., 1973 Zouros et al., 1974
tropicalis	6	185	26	0.168	0.532	Ayala & Powell, 1972 Ayala et al., 1974b
willistoni	56	4000+	31	0.177	0.557	Ayala & Powell, 1972 Ayala et al., 1971, 1972, 1974a, b

Note: *H* is average heterozygosity, and *P* is proportion of loci found polymorphic (1% criterion).

From Nevo et al. (1984) with additions from Powell (1975).

ary genetics for 20 years. Books have been written on the subject and it is not the purpose of this book to rehash the arguments yet again. The reader can consult such references as Nei (1975 and 1987) and Kimura (1983) for neutralists views and Gillespie (1991) for a selectionist view. Inevitably, however, we will return to the issue in chapter 10 on molecular evolution as well as in chapter 12.

Leaving aside the question of the selectivity of the allozyme variation, what have allozyme studies taught us about *Drosophila* populations? First, it is extremely important to realize that the introduction of molecular techniques allowed real population genetic analysis of natural populations. There are naturally occurring alleles at appreciable frequencies that can be determined. Table 2-12 presents an example from *D. willistoni*, an example chosen because this writer is most familiar with it, although it could be repeated in numerous species. The level of variation at these three loci is very high, with some exceeding 50% expected heterozygosity. These are numbers and frequencies that population geneticists can use in the equations of the large body of theoretical population genetics. This result held not only for *Drosophila*, but when other organisms were studied for electrophoretic variation, they too were found nearly (though in general not quite) as variable as *Drosophila*. Nevo et al. (1984) list some 1,111 species that had been studied at that time for at least 10 loci (mean of 23)

TABLE 2-12 Typical allozyme results from *Drosophila willistoni*.

Gene	Allele	Mirassol, Southern Brazil	Guyana	P. Lopez, Colombia	Grenada	Martinique
Lap-5	0.98	0.07	0.10	0.09	0.04	0.06
	1.00	0.25	0.32	0.30	0.49	0.53
	1.03	0.57	0.48	0.57	0.42	0.40
	1.05	0.09	0.09	0.04	0.04	0.01
Sample size		1806	606	402	266	264
Est-7	0.96	0.01	0.02	0.02	0.02	0.05
	0.98	0.13	0.19	0.15	0.13	0.14
	1.00	0.57	0.50	0.58	0.65	0.63
	1.02	0.26	0.22	0.20	0.17	0.16
	1.05	0.01	0.07	0.05	0.03	0.01
Sample size		720	442	294	233	153
Pgm-1	0.96	0.05	0.06	0.03	0.02	0
	1.00	0.88	0.79	0.88	0.97	0.77
	1.04	0.08	0.15	0.08	0.02	0.23
Sample size		40	190	188	198	286

Note: Populations range from southern Brazil to the Caribbean. Allele designations in left column indicate relative electrophoretic mobility; allele frequencies are in main part of table.

From Ayala et al. (1971) and Ayala et al. (1972).

demonstrating that experimental population genetics was no longer the sole purview of workers using *Drosophila*, mice, and a few other organisms. Virtually any and all species could be, and were, studied for population genetic variation.

Geographic Homogeneity

A second point from table 2-12 is the geographic pattern of genetic differentiation among populations of a species. I have chosen to illustrate the point with five populations that span nearly the entire range of *D. willistoni*, the most distant populations being some 6,000 kilometers apart. The most remarkable result from this and almost all other studies of *Drosophila* is the homogeneity of allele frequencies from population to population despite great distances between them. *The same alleles are found in all populations at about the same frequency.* This is true even of island populations, such as Grenada and Martinique in table 2-12. This was a most remarkable and completely unexpected finding considering what was known about the only other well-studied polymorphism previous to 1966, chromosomal inversions (chapter 3). Inversions vary considerably from population to population not only in frequency, but in their presence and absence.

This homogeneity in allozyme frequencies is even more starkly illustrated by the application of improvements in electrophoretic technique. Several workers developed refinements in allozyme analysis that detected more of the amino acid substitutions than did the standard technique. One approach was to vary the conditions of electrophoretic separation by varying gel concentration and/or pH of the buffer (e.g., Coyne et al., 1978). Alleles resolved into even more classes, so the assumption that standard electrophoretic studies seriously underestimate the total allelic variation was empirically demonstrated. For the present discussion of geographic homogeneity, the results of Keith (1983) are most dramatic. Table 2-13 shows the distribution of alleles at the *Est-5* locus found in two populations of *D. pseudoobscura* about 650 km apart. The virtual identity of distributions of allele frequencies in the different populations is truly remarkable. Two alleles, *T* and *BB*, are in high and virtually identical frequency in both populations; the moderately frequent alleles (e.g., *S, AA, FF*, and *LL*) were detected in both populations as were some very rare ones. Given sample sizes of a little over 100 per population, one can only conclude these populations are genetically identical at this locus. That this is not a peculiarity of this locus was demonstrated by virtually identical observations for a second locus, *Xdh* (Keith et al., 1985). We will return to the issue of causes of such identity, but surely migration must be playing a large role.

The homogeneity of allele frequencies across populations has one major exception: Alleles associated with inversions are often found to be geographically differentiated in parallel with their inversions. Linkage disequilibrium between specific alleles and inversions will be discussed in the following chapter.

Metabolic Function

Before leaving the subject of allozymes, two more points merit discussion. First, after several studies had been reported, it became evident that not all enzymes showed the same level of variation. From species to species, the same enzymes tended to exhibit high or low levels. For example, in virtually every *Drosophila* species studied, esterases are highly variable, whereas α-glycerophosphate dehydrogenase is virtually monomorphic in every species. In examining these data, Gillespie and Kojima (1968) noted a pattern with respect to the function of enzymes and their level of naturally occurring variation. Enzymes that were involved in glucose metabolism were less variable than enzymes with nonspecific or multiple substrates. The hypothesis was that if an enzyme acted on a single specific substrate, one form could be selected to perform the catalytic function most efficiently. Enzymes with multiple substrates (such as esterases, phosphatases, peptidases, etc.) may be selected to be more versatile in their action. Thus, selection keeps them polymorphic. Johnson (1974) extended this thinking by suggesting that what is most important in generating metabolic diversity is variation in genes controlling metabolic pathways, as opposed to genes in pathways that exert no controls. Johnson defined a series of these regulatory enzymes and demonstrated that the available data supported his notion that genes whose products exerted control of metabolic flux in a pathway, even when acting on a single substrate, are about as variable as nonspecific enzymes. Data from *Drosophila* corroborating Gillespie and Kojima's (1968) and later Johnson's (1974) hypotheses are discussed in Kojima et al. (1970), Ayala and Powell (1972), and Powell (1975).

Null Alleles

Finally, while it is well-established that many enzymes have electrophoretically detectable variation in natural populations, how frequent are alleles with no enzymatic

TABLE 2-13 Frequencies of alleles at *Est-5* of *D. pseudoobscura* from two populations in California approximately 650 kilometers apart.

Allele	James Reserve	Gundlach-Bundschu	Total
A	1	0	1
B	1	2	3
C	1	0	1
D	2	0	2
E	2	6	8
F	1	0	1
G	0	1	1
H	0	1	1
I	1	0	1
J	0	1	1
K	0	1	1
L	1	2	3
M	1	0	1
N	2	0	2
O	2	0	2
P	0	1	1
Q	1	2	3
R	0	1	1
S	5	5	10
T	40	44	84
U	1	6	7
V	1	3	4
W	1	0	1
X	1	0	1
Y	2	0	2
Z	1	1	2
AA	4	3	7
BB	24	26	50
CC	0	2	2
DD	2	0	2
EE	1	0	1
FF	3	4	7
GG	1	0	1
HH	1	1	2
II	0	1	1
JJ	1	0	1
KK	1	0	1
LL	5	7	12
MM	1	0	1
NN	1	0	1
Totals	33 alleles (19 unique)	22 alleles (8 unique)	41 alleles
Sample size	116 genes	121 genes	237 genes

Note: Five sequential conditions of electrophoresis were used. Twelve alleles were detected by standard single-condition electrophoresis.

From Keith (1983).

activity, that is, null alleles? If null alleles are frequent, then apparently selection at each enzyme locus is not as strong as one might expect. A high frequency of nulls would indicate metabolic redundancy, that is, there must be more than one way to perform a particular chemical transformation or the transformation is not crucial. To address this possibility, Langley, Voelker, and colleagues (Voelker et al., 1980; Langley et al., 1981) conducted extensive screens for nulls in two populations of *D. melanogaster*, one from North Carolina and one from London, with virtually identical results. Of five X-linked loci, with samples of tens of thousands of chromosomes, null alleles were not detected in either population. Of the 20 autosomal genes examined, 13 had naturally occurring null alleles with average frequencies of 0.0025 and 0.0023 in the two populations. Because hemizygous males are expected to exhibit strong selection against recessive deleterious X-linked genes, these results are perfectly consistent with selection-mutation balance maintaining an extremely low frequency of nulls in populations. There is no evidence of redundancy of function.

DNA Variation

The obvious next step, once the technology had developed, was to assess variation in genes themselves, rather than in the gene product. The problem of hidden protein variation would be overcome simply because it is impossible to delve deeper into genetic variation than nucleotide sequence variation. Kreitman (1983) published the first detailed view of genetic variation in *Drosophila* populations. It is difficult to overstate the importance of this seminal work, as it provided the first detailed picture of genetic variation in a population. Kreitman studied the alcohol dehydrogenase locus (*Adh*), the structure of which is schematically shown in figure 2-3. We will have reason to return to this gene often, including cases where this structure varies. Kreitman completely sequenced 2,721 bases in and around this gene for 11 alleles from 5 different populations. Table 2-14 is a summary of his findings. A total of 43 polymorphisms were detected in this sample of only 11 alleles. *No two alleles were identical!* While one would have called all these strains "wild type" as they were perfectly

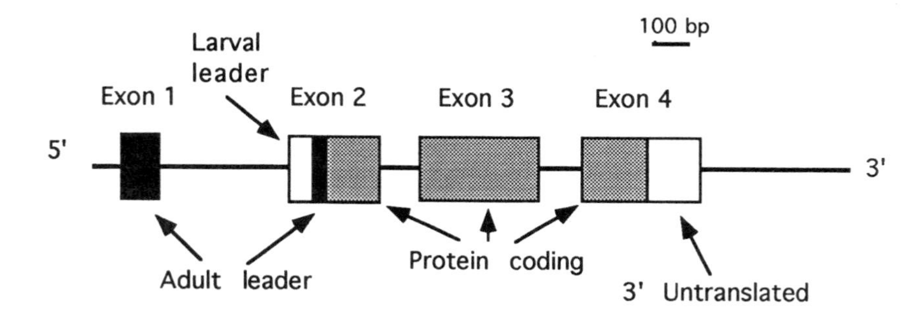

FIGURE 2-3 Structure of the alcohol dehydrogenase gene (*Adh*) in *D. melanogaster*. Two different transcripts are produced in the larvae and adults, although the protein-coding regions are identical. In chapter 10, variation in this structure among species is discussed.

TABLE 2-14 Summary of results from Kreitman's (1983) nucleotide sequence data for 11 alleles of *D. melanogaster Adh.*

	Introns	Translated	Transcribed Nontranslated	3′ Nontranscribed
Number of nucleotides	789	765	335	767
Number of polymorphisms	18	14	3	5
Percent polymorphisms	2.8%	1.8%	0.9%	0.7%

Note: It is now known that another structural gene lies just 3′ to *Adh* in this species, so the category called 3′ nontranscribed is probably misnamed and the low variation in this region is attributable to the presence of the other gene. In addition to the nucleotide variation represented in the table, six insertions or deletions ranging from 1 to 37 base pairs were found outside the coding region.

normal in all respects, clearly the wild type concept is nonapplicable at the level of DNA sequence. *Every individual is a heterozygote.* The average difference between any two alleles was about 0.5%, which can be thought of as the nucleotide heterozygosity. We will see in chapter 10 that this level of nucleotide variation, or even higher, has been repeatedly observed for a number of genes in a number of different species. Given that an average gene is probably at least 1,000 nucleotides in length, there should be at least five nucleotide differences between any two alleles on average. Except for inbred populations with high frequencies of alleles identical by descent, one would never expect to see a homozygote. Among the nucleotide substitutions, only one changed the amino acid coded for, an A/C polymorphism at a site that causes a threonine/lysine protein polymorphism already detected as the "Fast" and "Slow" alleles in allozyme studies (Kreitman, 1983).

Aquadro et al. (1986) provided a broader look at nucleotide variation in this same region. Their study included a 13 kilobase (kb) region with the 2.5 kb *Adh* locus occupying a position near the middle. They also used a less direct method of detecting DNA sequence variation, restriction enzymes. Nevertheless, their results provided further important insights into variation in *Drosophila* genomes. Figure 2-4, taken from their paper, summarizes the result. They analyzed 48 alleles, all derived from the eastern United States. The remarkable feature of their results is the high level of insertions and deletions in the vicinity of the *Adh* gene.

The conclusion from the combined picture of Kreitman's nucleotide sequence data and figure 2-4 is that *Drosophila melanogaster* genomes are replete with naturally occurring variation of all sorts. Further work has, for the most part, confirmed these initial findings, although with important differences and exceptions, all of which will be discussed at length in chapter 10.

Mutation

Before ending this chapter on population genetics at the genic level, it is instructive to consider the origin of all the variation just discussed, that is, mutation in *Drosophila.* From an evolutionary standpoint, we are more interested in mutation rates than in the molecular details of the mutation process. The considerable data on mutation rates in *Drosophila* are thoroughly reviewed by Woodruff et al. (1983). The following is a brief summary of that paper.

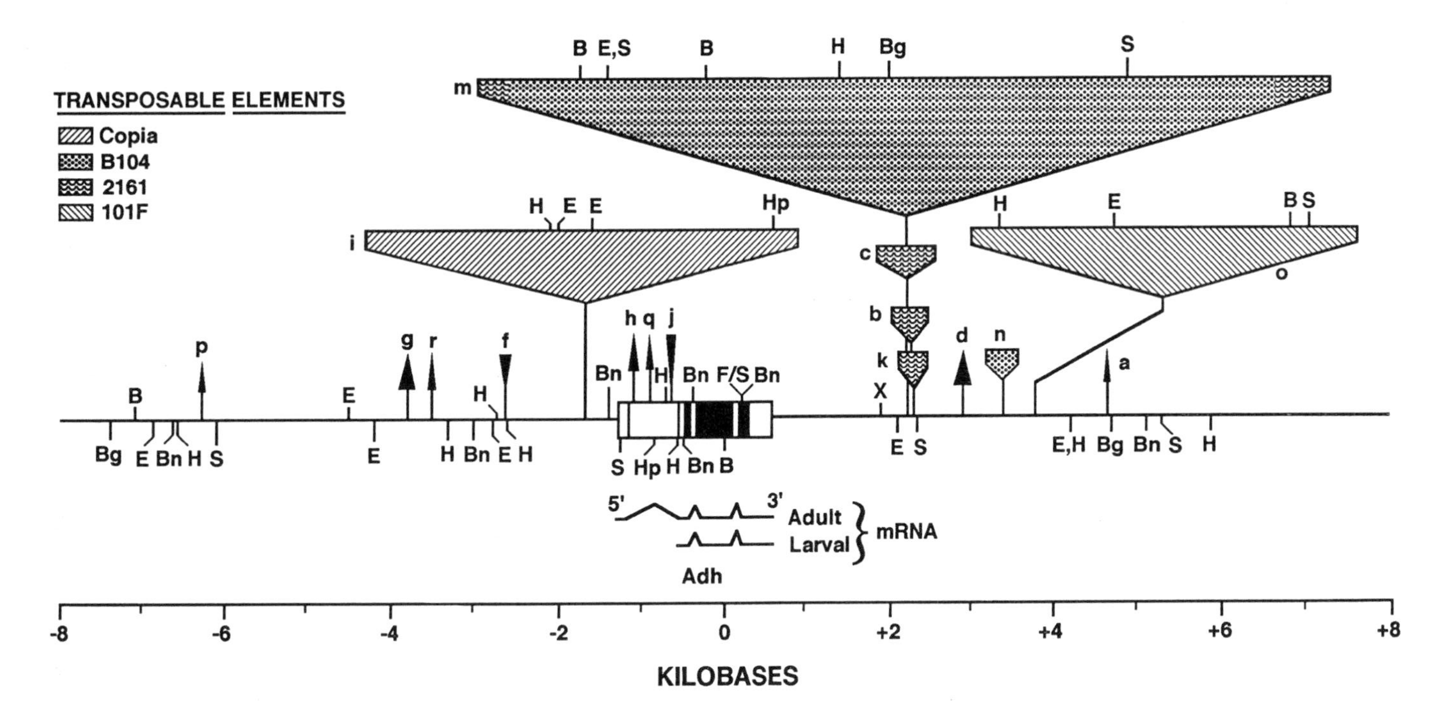

FIGURE 2-4 Summary of restriction site variation around the *Adh* gene of *D. melanogaster.* The *Adh* gene is in the center with the structure as in Figure 2-3. Restriction enzyme site recognition variation is indicated by the capital letters with B = *Bam* HI, Bg = *Bgl* III, Bn = *Ban* II, E = *Eco* RI, H = *Hind* III, Hp = *Hpa*I, S = *Sal*I, and X = *Xho*I. Invariant sites are indicated below the line and variable sites above. Triangles indicate insertions and deletions, with insertions (relative to the most common sequence) indicated by downward pointing triangles, and deletions by upward pointing triangles. In some cases known transposable elements are involved, whereas other insertions or deletions are of unknown function and range from 27 bp (h) to 700 bp (c). From Aquadro et al. (1986).

Rates

Table 2-15 summarizes what is known about mutation rates in *Drosophila*. I have excluded from this table a few exceptional studies indicating very high rates of mutation or reversion, probably due to some unusual molecular events (although given our knowledge today, it isn't exactly clear what is usual and what is unusual). The studies summarized here reflect spontaneous mutation rates in the absence of such phenomena as dysgenesis; an exception to this is the allozyme data where some studies, unknown to the authors at the time, may have involved dysgenic crosses. Also note that in the upper half of the table, the rate is given *per chromosome*, while in the lower half it is *per gene*.

The rates of mutation to lethals seem to be fairly evenly distributed across the genome. The second and third chromosomes of *D. melanogaster* are bi-armed meta-

TABLE 2-15 Mutation rates in *Drosophila* for various types of variants discussed in this chapter.

Type of mutation	Number Mutants/Sample Size	Frequency
Recessive lethals		
Sex-linked		
Old lab stocks:		
Males	2,906/1,828,144 chromosomes	0.16% per chrom.
Females	532/328,614	0.16%
Newly collected:		
Males	1,096/433,920	0.25%
Females	42/13,489	0.31%
Autosomal		
2nd Chromosome		
Males	657/107,675	0.61%
Females	139/25,929	0.54%
3rd Chromosome		
Males	6/2,797	0.57%
Females	22/3,067	0.72%
Visibles		
Forward		
X-linked (20 loci)	120/31,270,348 genes	0.38×10^{-5} per gene
Autosomes (13 loci[a])	10/2,751,115	0.36×10^{-5}
Reverse		
19 loci[b]	9/15,333,180	0.06×10^{-5}
Electrophoretic		
Mobility shifts	3/669,904[c]	0.44×10^{-5}
	4/3,111,598[d]	0.13×10^{-5}
To nulls	12/3,111,598[d]	0.39×10^{-5}

[a]*dumpy* is excluded from this sample as it had exceptionally high rates.
[b]*forked* and w^{j} are excluded as they had exceptionally high rates.
[c]Tobari and Kojima, 1972.
[d]Mukai and Cockerham, 1977; Voelker et al., 1980.

Most of the data come from Woodruff et al. (1983), which can be consulted for literally hundreds of references to the original literature. For the electrophoretic rates, separate studies are designated, as Woodruff et al. suspect the Mukai and Cockerham and Voelker et al. studies may have involved dysgenic crosses.

centrics about twice the size of the acrocentric X-chromosome; the mutation rate per chromosome for these autosomes is about twice that for the X-chromosome. For unknown reasons, there is apparently a higher rate of mutation to lethals in newly collected flies than in old laboratory stocks. Earlier in this chapter it was estimated that on the order of a few hundred to a thousand loci per chromosome arm can mutate to lethals. If the number is 1,000, then the lethal mutation rate per locus would be around 0.3×10^{-5}, comparable to the rate for visibles.

Visible mutation rates vary greatly from locus to locus, but little from chromosome to chromosome. The highest rate reported in Woodruff et al. (1983) is for *dumpy* at 17.3×10^{-5}, and the lowest rates are cases where no mutations were observed out of sample sizes of nearly two million genes. Whether this large variation in mutation rates among genes is due to differences in gene size, their sensitivity to mutating, or the number of changes that lead to the mutant phenotype is not known. However, it is clear that for all visibles, reversion is much less frequent, on the order of 1/7 the frequency of forward mutations. This is not surprising, as almost all the mutants studied were recessive, many known to be loss-of-function mutations. Recovery of a function is a less probable event than its loss.

Allozymes seem to be neither more nor less mutable than other types of genes. Considering that null mutations can arise by either change in amino acid composition of the protein, nonsense codons, or change in the DNA-regulating expression of the gene, whereas change in band mobility on gels can come about only by amino acid changes, it is not surprising that null mutations occur more frequently than mobility-shift mutations. Because standard methods of electrophoresis used in these mutation studies detect only a fraction of all amino acid substitutions, these rates represent minimal estimates with actual rates of amino acid substitutions being perhaps two to four times greater.

There are no direct studies of mutation rate at the nucleotide level in *Drosophila*. However, indirect estimates have been made by extrapolating from data such as that in table 2-15 and making various assumptions about gene sizes, number of germline cell divisions per generation, and so forth. Both Drake (1969) and Propping (1972) have argued that *Drosophila* are similar to all other organisms, from bacteria to humans, in having a per-base per-replication mutation rate of about 7×10^{-11}. While such rare events are difficult to confirm by experiments, there is no reason to think this estimate is very greatly inaccurate.

Causes of Mutations

What causes mutations? Obviously errors in DNA replication and/or repair can cause base changes, so-called point mutations. On the other hand, many of the classical visible mutations, when studied at the molecular level, have been shown to be the consequence of insertions or deletions of pieces of DNA, including transposable elements a few kb in size. So-called mutator strains have been found multiple times in natural populations of *Drosophila* (documented in Woodruff et al., 1983; see also Green, 1976), which undoubtedly are strains with very active mobile elements. Crosses between ostensibly normal strains can lead to mobilization of such elements, as exemplified by *P* element–induced hybrid dysgenesis. It is not clear at this time whether the relatively low mutation rates reported in table 2–15, which presumably reflect the normal situation, are really the crucial rates when considering the whole population or species over evolutionary time. Some authors have argued that what is important in evolution is not so much this low-level background mutation rate, but rather the increased mutation rates brought about by stress, which often mobilizes transposable elements (discussed further in chapter 9).

Heat and viral infection are two extrinsic factors known to increase mutation rates in *Drosophila.* Generally, increased temperature raises the mutation rate. As Woodruff et al. (1983) correctly point out, some of these studies are confounded by higher rates of cell division at higher temperatures, but overall the pattern is almost certainly valid. The role of viruses in inducing mutation in *Drosophila* has been studied (reviewed in L'Heritier, 1970; Brun and Plus, 1980). However, the data are too few to really assess their potential role in evolution, a subject deserving more study.

There is little evidence in *Drosophila* that the age of either sex has an effect on mutation rate. Nor is there any compelling evidence that mutation rates differ in males and females. There is some evidence, however, that length of storage of sperm in spermathecae may increase mutation rates. This last point indicates that not all mutations occur at replication.

Epilog

In this chapter we have proceeded through 70 years of analysis of genetic variation in *Drosophila* populations. The picture now available from direct DNA studies is that populations are remarkably variable at the nucleotide level, more variable than all but the most ardent believers in the balance school would have thought possible. Genetic variation in natural populations is the rule, not the exception.

It is interesting to imagine what might have been the case had the molecular studies not revealed this wealth of variation. It is conceivable that the initial allozyme and DNA studies would have revealed monomorphism predominating in natural populations. Then the problem posed by the homozygous chromosome studies and selection experiments might have been solved. After all, the variation revealed by those studies must have a basis in protein and DNA variation. If molecular variation were rare, then we could have concluded that very few loci must account for the fitness variation and polygenic quantitative variation. Or, if variation only occurred outside coding regions, then we might have concluded with Mukai and Cockerham (1977) that the genetic basis of polygenic traits lies outside coding genes. As it is, molecular variation is so ubiquitous, it provides few clues to what aspects of this variation account for the fitness variation revealed by homozygous chromosome studies and additive genetic variation revealed in selection studies.

In addition to being ubiquitous, DNA variation patterns are often far from random, as we will see in chapter 10. At almost every level on which there is sufficient data, there is evidence for the action of selection—often cited as a constraint on the type of variation present. It remains one of the major open questions as to how to match or map the variation from one level of analysis to the next. Yet for any complete picture of genetic variation in populations, ultimately we need to understand how all the levels of variation reviewed in this chapter relate to one another.

3

Population Genetics—Inversions

Faith is a fine invention
For gentlemen who see;
But microscopes are prudent
In an emergency
Emily Dickinson

[I]t is clear that study of salivary gland material is going to play an important part of the future genetic studies on Drosophila.
Theophilus S. Painter, 1935

General Features

Crossover Suppression

Naturally occurring chromosomal inversions are widespread in *Drosophila* and were indirectly detected very early in the history of genetics. Sturtevant (1917) found strains of *D. melanogaster* that exhibited reduced recombination in certain chromosomes and chromosome regions. He designated these strains *crossover suppressors* and deduced they might be due to an inversion in the order of genes on a chromosome, a proposal he subsequently verified (Sturtevant, 1926) with genetic mapping data. The physical reality of chromosomal inversions was confirmed with the discovery of giant polytene chromosomes (Painter, 1933) from which one could read the gene order by the morphology of the banding patterns (Patau, 1935; Tan, 1935; Koller, 1936). The reasoning behind Sturtevant's deduction was that if synapsis of homologous chromosomes during meiosis related to the gene order, then in an inversion heterozygote, maximization of homologous pairing required some contortions of the chromosomes to give the now familiar reverse loop (figure 3-1). If one traces out the consequences of a crossover event within the region included by the two breakpoints, it becomes apparent that the chromatids that were involved in the crossover event are unbalanced; that is, they do not contain a single copy of all genes, nor do they carry a single centromere. The gametes ultimately formed from such chromatids would have an abnormal genetic constitution, which would lead to nonfunction and/or zygotic inviability. Inversions, then, do not actually suppress crossovers; rather, the gametes produced by single crossover events between the breakpoints would never be recovered in progeny. The importance of this is that *inversions are inherited intact as single simple Mendelian units.* In a population genetics sense then, we can treat the various gene arrangements differing by inverted segments as alleles at a single locus.

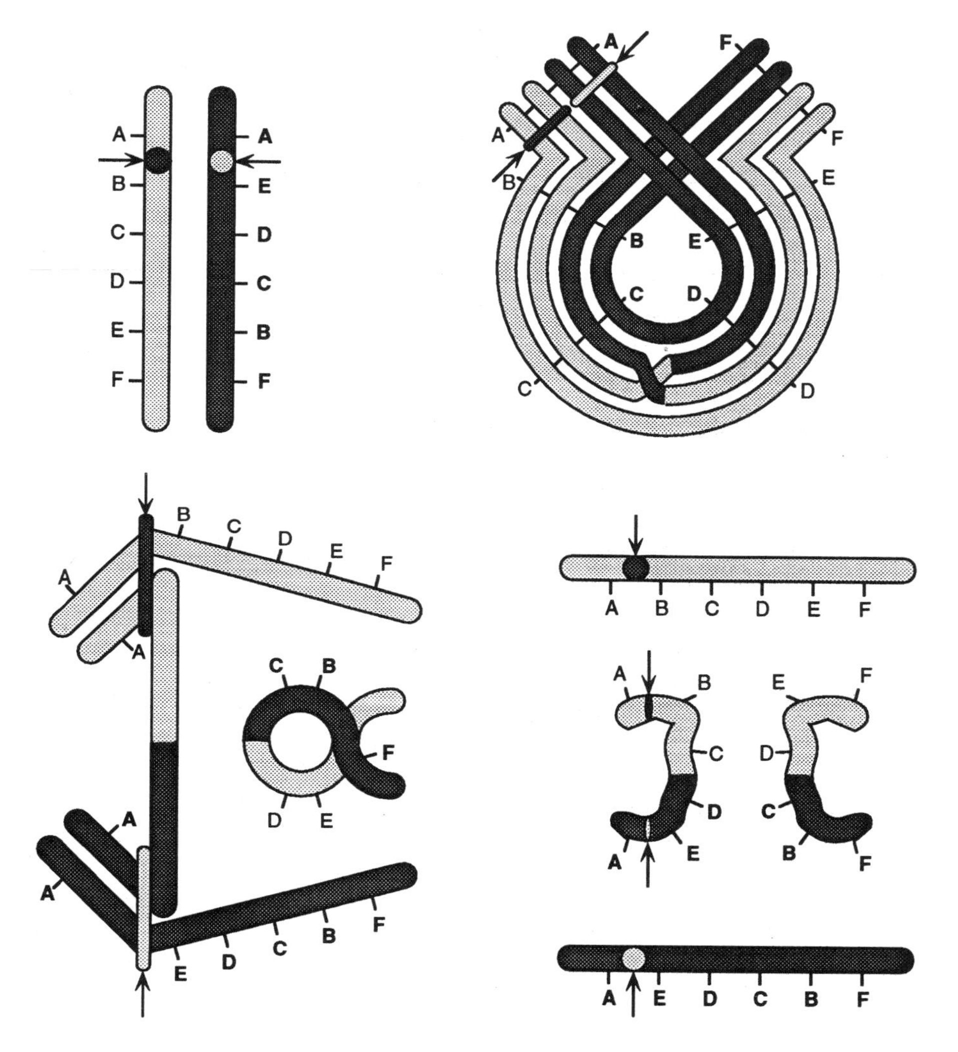

FIGURE 3-1 Illustration of the reason why crossover products in inversion heterozygotes do not produce functional gametes. The centromeres are indicated by arrows. Letters indicate gene order.

Ubiquity of Inversion Polymorphism

More than half of *Drosophila* species examined in any detail are naturally polymorphic for inversions in one or more chromosome arms. This is unusually high for any type of chromosomal variation, for the simple reason just noted: Heterozygotes for such chromosomal rearrangements have abnormal meiosis leading to at least some reduction in fertility due to the production of nonfunctional gametes. Since all new mutations (including inversions) in diploids arise first in the heterozygous state, one would expect such negatively heterotic mutations to be quickly eliminated. Obviously this is not the case in *Drosophila*, the reasons for which are known. First of all, in the case of paracentric inversions (inversions not involving the centromere, by far the most common in *Drosophila*) it is only recombinant gametes that suffer. Recombination does not occur during spermatogenesis in most (though not all) species, so males do not suffer any infertility from inversion heterozygosity. In females, only one of the four meiotic products becomes the functional egg, while the other three become polar bodies. Probably due to the fact that the dicentric bridge slows the separation of the recombinant chromatids, the faster-migrating noncrossover chromatids are incorporated into the pronucleus and become the functional egg, while the recombinant products form the polar body nurse cells (Sturtevant and Beadle, 1936; Carson, 1946). Thus, at worst, one would expect that in this particular insect inversions would be more or less neutral with respect to selection, at least in regard to reduced production of functional gametes.

The extent and pattern of inversion polymorphisms in *Drosophila* is well documented by Sperlich and Pfriem (1986); the interested reader is referred to that paper and its extensive bibliography for more details. Sperlich and Pfriem list 182 species in the subgenera *Drosophila* and *Sophophora* for which at least 10 independently derived strains from natural populations have been examined for inversions. They do not include the Hawaiian *Drosophila*, which we would place in subgenus *Idiomyia*; however, Carson (1992 and personal communication) has made this compilation, summarized in table 3-1. While overall, about 60% of all species are naturally polymorphic for inversions, subgenus *Sophophora* seems to contain a higher proportion of polymorphic species than do subgenera *Drosophila* and *Idiomyia*. Even within each subgenus the distribution of inversions is not random. For example, one might expect the total inversions to be spread across the genus or a subgenus in a Poisson distribution. Levitan (1958) was the first to demonstrate that this is not the case, and the more recent compilation by Sperlich and Pfriem is consistent with Levitan's conclusions. The deviation from randomness is due to an excess of species with many inversions and too few species with a small number of inversions. A similar nonrandom pattern is evident with regard to distribution on chromosome arms: All arms of all chromosomes are polymorphic in some species (well-studied examples are *D. willistoni* and

TABLE 3-1 Numbers of species polymorphic for naturally occurring inversions.

Subgenus	Number Polymorphic	Number Monomorphic
Drosophila	41	44
Sophophora	38	2
Idiomyia (Hawaiians)	27	31
Total	106	77

Only species for which at least 10 independently derived strains have been examined are included.

From Sperlich and Pfriem (1986) and Carson (1992, and personal communication).

D. subobscura), some have only one arm of one chromosome that is polymorphic (largely the case in *D. pseudoobscura* and *D. persimilis*), and some species have no inversions in any chromosomes (essentially the case for *D. simulans*). Autosomes and X chromosomes may be equally or unequally polymorphic. In *D. cardini*, for example, 22 out of the 29 known inversions are in the X-chromosome. Several species exhibit the opposite pattern. For example, *D. guaramunu, D. mediostriata*, and *D. rubida* have, respectively, a total of 31, 21, and 19 inversions with none in the X. Thus, it is hard to make any generalizations about the distributions of inversions with respect to species or chromosome.

Krimbas and Powell (1992b) present a more detailed discussion of the general properties of inversions in *Drosophila* with specific regard to evolutionary considerations, including effects on recombination, constancy of banding patterns, and so forth.

Origin of Inversions

Considering the just-documented nonrandom distribution of naturally occurring inversions, it is natural to ask about the origin of inversions. Is the nonrandom distribution due to nonrandom production or is the nonrandomness due to a nonrandom retention? At the end of this chapter, we will see that theoretical work indicates a minority of all newly occurring inversions are retained.

Although it is not presently clear what causes inversions in natural populations of *Drosophila*, there is some evidence implicating transposable elements. It is well established that transposable elements can cause chromosomal breaks and lead to inversions, as in dysgenic crosses involving P-elements (Engels and Preston, 1984). In *D. robusta*, an inversion-inducing mutator stock has been found that is due to unusually frequent movement of transposable elements (Levitan, 1992). Lyttle and Haymer (1992) have found a transposable element (hobo) near the breakpoints of inversions of *D. melanogaster* endemic to Hawaii. However, in a study of *D. pseudoobscura* middle-repetitive sequences, we made extensive efforts to find elements that were concentrated on the inversion-containing third chromosome and/or at known breakpoints (J. R. Powell, unpublished data). No such elements were detected in a screen of approximately 20 of the most common middle-repetitive sequences. The only direct molecular studies of breakpoints provide no evidence of transposable element involvement (Wesley and Eanes, 1994; Cirera et al., 1995). There is presently no direct evidence of involvement of movable elements in the generation of naturally occurring inversions in populations of *Drosophila*; the only direct studies (Wesley and Eanes, 1994; Cirera et al., 1995) must be taken as evidence against the involvement of mobile elements. Nevertheless, given the Levitan and Lyttle and Haymer findings, it may be that in some cases transposable elements are involved in generating inversions.

The rate of spontaneous generation of inversions has been ascertained by experiments of near-heroic dimensions by Yamaguchi and Mukai (1974) and Yamaguchi et al. (1976). In the first study, they used five different strains, three of which generated no inversions during some 22,000 chromosome generations. The two strains that did generate inversions did so at a frequency of 0.00198 and 0.00041 per chromosome generation in samples of 45,000 and 46,000, respectively. In the second study, three lines varied in their frequency of generation of inversions from 0 to 0.0098. The distribution of breakpoints was not random and the variation among strains is too great to explain by random fluctuation. Such data are consistent with a role for transposable elements if one hypothesizes that the strains differ in the types and/or numbers of elements they carry. But whatever the cause, we can conclude that the rate of generation of new inversions is low, on the order of the mutation rate of genes, 10^{-5}.

Uniqueness of Inversions

From the earliest discussion of inversions in natural populations, arguments were presented in favor of a monophyletic origin for these chromosomal rearrangements; that is, each inversion had a unique origin, having arisen in a single fly, and has since been maintained in the species as a polymorphism or become fixed. The arguments were essentially probabilistic (Sturtevant and Dobzhansky, 1936; Dobzhansky, 1937). First, as we have just seen, the generation of an inversion is a rare event, on the order of 10^{-5} or less. Second, the probability that two inversions will contain exactly the same two breakpoints is also very small; assuming randomness of breaks along the length of a chromosome, Sperlich and Pfriem (1986) calculate this probability to be on the order of 10^{-6}. Third, after being generated the newly inverted chromosome has a low probability of being retained in the population; while there is no way to calculate what this probability is, the theoretical work would indicate it is low. So, retention of the identical inversion in the population more than once (polyphyly) requires the concatenation of three rare events, with the overall probability being the product of three small probabilities. In all likelihood, then, inversions are monophyletic.

The monophyly hypothesis has been strengthened by molecular studies. First, there are numerous cases of allozymes being nonrandomly linked with inversions. This is as expected because a unique event would capture one allele of each locus and would be expected to retain those alleles, barring any mutation, double crossovers, or gene conversions. It would not matter from where in the distribution of the species the inversion was taken. All carriers of the same inversion should be genetically more similar to one another for genes within those inversions than they are to carriers of different gene arrangements, even when from their own population. This is what has been found, at least in some cases (see later in this chapter). However, there is evidence of gene conversion between inverted forms (see chapter 10), so inversions do not completely repress gene exchange.

Even more direct evidence for uniqueness of inversions comes from the groundbreaking work of Wesley and Eanes (1994). They studied the cosmopolitan *Payne* inversion in the 3L of *D. melanogaster*. The technique they used to isolate the breakpoints molecularly involved a library containing the standard arrangement and pools of PCR-amplified sequences taken by microdissection of the inversion breakpoints. The prediction was that any clone from the standard library that would cross-hybridize to both pools from the breakpoints of the inverted stock must contain the breakpoint. This strategy worked. Subsequent sequencing of the breakpoints from six standard chromosomes and seven inverted chromosomes from throughout the range of the species indicated the breakpoints were identical. Furthermore, the inverted and standard arrangements did not share any polymorphisms indicating complete suppression of gene exchange, at least in the 2.4-kb region right at the breakpoints. The breakpoint sequences showed no similarity to known transposable elements, but surprisingly seemed to interrupt coding sequences. Northern blot analysis indicated the distal break interrupts three transcripts of unknown function. The only other molecular study of the breakpoints of naturally occurring *Drosophila* inversions involved characterization of a fixed inversion between *D. melanogaster* and *D. subobscura* (Cirera et al., 1995). As in Wesley and Eanes' study, there was no evidence of either complete or defective transposable elements in the region, nor did the break disrupt a transcriptional unit in this instance.

At this time at least, the overwhelming evidence favors the traditional view that inversions are monophyletic. Each gene arrangement was generated in a single gamete and has been retained as a polymorphism or become fixed in a species. This conclusion is not only important for the work on population genetics to be discussed here, but also for the phylogenetic studies presented in chapter 8. However, given the hints of possible involvement of transposable elements and the sometimes distinct nonrandom distribution of breakpoints, it would seem prudent to remain open to the possibility that inversions can be multiply generated and retained in evolutionary time. Obviously, more studies comparable to Wesley and Eanes (1994) are called for; similar analyses of other inversions and species should determine the generality of their findings.

Geographic Distributions

Geographic Heterogeneity

In contrast to allozymes, discussed in the previous chapter, inversions display interesting and revealing patterns of geographical differentiation. The first species to be studied in this regard remain among the best studied, *Drosophila pseudoobscura* and its very closely related sibling species, *D. persimilis*. Because of the importance of these two species in understanding inversion polymorphisms, a short digression introducing their more general biology is in order. More detailed reviews of these species are Dobzhansky and Powell (1975) and Powell (1992).

D. pseudoobscura is fairly continuously distributed in temperate forests and chaparral throughout the western third of North America (figure 3-2). A geographically isolated population in the vicinity of Bogota, Colombia is partially reproductively isolated from North American populations (more on this in chapter 7). Its sibling species, *D. persimilis*, was first recognized due to production of sterile male offspring in crosses with *D. pseudoobscura*; initially the two taxa were called Races A and B until Dobzhansky and Epling (1944) raised Race B to species level by naming it *D. persimilis*. *D. persimilis*'s distribution is within the distribution of *D. pseudoobscura* (figure 3-2).

The inversion polymorphism in these species is primarily in the acrocentric third chromosome with some 52 gene arrangements, all paracentric inversions. Sturtevant and Dobzhansky (1936) were the first to demonstrate, using these species as examples, that one can deduce the phylogenetic relationship of inversions simply by assuming monophyly and parsimony. This is especially clear with overlapping inversions, as in *D. pseudoobscura* and *D. persimilis*. These principles will be discussed in more detail in chapter 8 on phylogenies, and the uninitiated reader is referred to that discussion. At this point, we simply present the inversion phylogeny to aid in the discussions in the present chapter (figure 3-3). Perhaps the most remarkable aspect of this phylogeny

FIGURE 3-2 Approximate distributions of *D. pseudoobscura* (solid lines) and *D. persimilis* (dashed line). Note the isolated population in the vicinity of Bogota, Colombia. Within these ranges, the species are quasi-continuously distributed in appropriate ecological settings, primarily temperate forests and chaparral.

is its completeness. With a single exception (Hypothetical), every inversion in the 53 allele system has been observed in nature; each can be related to its nearest relative by a single overlapping inversion. Note also that the tree has a distinct trunk with terminal branches. Dobzhansky used the term *phylad* to describe the major components of the trunk along with its branches. He recognized the Tree Line Phylad, Santa Cruz Phylad, and Standard Phylad as those named inversions along with all their immediate (and secondary) descendants. We will have reason to refer to these phylads later. Arrows are drawn to indicate temporal directionality, although strictly speaking,

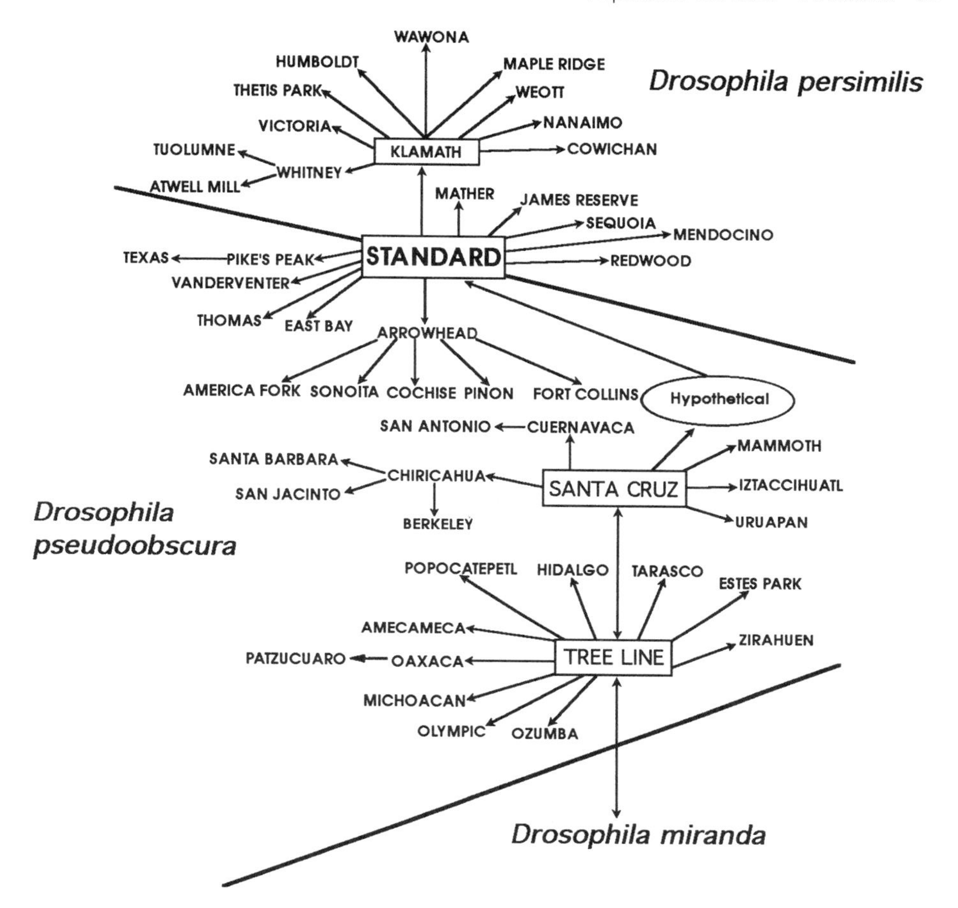

FIGURE 3-3 The deduced phylogeny of third chromosome inversions of *D. pseudoobscura* and *D. persimilis*. Tentative directionality of evolution is indicated by arrows. Because the evidence indicates that either Santa Cruz or Tree Line is the ancestral gene arrangement, these are connected by a double arrow.

one cannot assign directionality in inversion phylogenies without some external information. The evidence now seems in favor of Tree Line (TL) or Santa Cruz (SC) as the ancestral third chromosomes in these species, so we have placed double arrows connecting these two arrangements, and single directional arrows on the rest of the phylogeny. Some of the evidence for TL as the ancestor includes:

1. It is the most widespread gene arrangement, being the only one to occur throughout the species range.
2. More gene arrangements are directly derived from it than from any other.
3. Some evidence indicates its banding pattern is most similar to *D. miranda*, the outgroup for this phylogeny (see Olvera et al. [1979] and Powell [1992] for more details).

The evidence in favor of SC is based on molecular analysis and will be discussed in chapter 10; those data eliminate Standard (ST) and Hypothetical (HY) as ancestral and are consistent with either TL or SC, with the latter being weakly favored.

For the issue of geographic distribution of inversions in these species, the most important single reference is Dobzhansky and Epling (1944). The only major additions to these data include more information from Mexico (Powell et al., 1972; Guzman et al., 1975; Arnold, 1982) and from the isolated Bogota population discovered in the 1960s (Dobzhansky et al., 1963). Table 3-2 is a representative sample of variation in frequencies seen in this species. In general, the geographic differentiation corresponds fairly closely to the phylogeny of the inversions, that is, more closely related inversions tend to be geographically clustered. Mexico, Guatemala, and Bogota populations are predominantly Santa Cruz and Tree Line phylads, while northern populations are primarily Standard and Santa Cruz phylads. Tree Line persists throughout the distribution. More specifically, in order of frequency, ST, AR, CH, and TL characterize all populations in the northwest part of the distribution. AR comes to dominate in the drier areas of Arizona, Utah, and Colorado. PP dominates in Texas. CH dominates in northern Mexico. The most southerly populations are predominantly TL and CU. Clearly, this is in stark contrast to what is observed with allozymes (tables 2-12 and 2-13).

Inversions of *D. persimilis* are not geographically distributed so neatly with respect to phylogeny. Table 3-3 is an example of *D. persimilis* geographic data. KL predominates in the far northwest area, accounting for 94% of the chromosomes in British Columbia, Oregon, and Washington. California can be conveniently divided into the Coast Range on the west and the Sierra Nevada in the east. In the former ME and KL predominate and in the latter ST and WT predominate. Yet phylogenetically the combinations ME/KL and ST/WT are not closely related (figure 3-3). Therefore

TABLE 3-2 Inversion frequencies (in percent) from across the range of *D. pseudoobscura*.

Locality	ST[a]	AR	CH	PP	TL	SC	OL	EP	CU
1. Methow, WA	70.4	27.3	0.03	—	2.0	—	—	—	—
2. Mather, CA	34.5	35.5	11.3	5.7	10.7	0.9	0.5	0.1	—
3. San Jacinto, CA	41.5	25.6	29.2	—	3.4	0.3	—	—	—
4. Fort Collins, CO	4.3	39.9	0.2	32.9	13.2	—	2.1	7.2	—
5. Mesa Verde, CO	0.8	97.6	—	0.5	—	—	—	0.2	—
6. Chiricahua, AZ	0.7	87.6	7.8	3.1	0.6	—	—	—	—
7. Central Texas	0.1	19.3	—	70.7	7.7	—	2.4	—	—
8. Chihuahua, Mexico	—	4.6	68.5	20.4	1.0	3.1	0.7	—	—
9. Durango, Mexico	—	—	74.0	9.4	3.1	13.5	—	—	—
10. Hidalgo, Mexico	—	—	—	0.9	31.4	1.7	13.5	1.7	48.3
11. Tehuacan, Mexico	—	—	—	—	20.2	1.1	—	3.2	74.5
12. Oaxaca, Mexico	—	—	10.3	—	7.9	—	0.9	1.6	71.4

Note: Localities are listed from north to south.

[a]Abbreviations for inversions: ST = Standard; AR = Arrowhead; CH = Chiricahua; PP = Pikes Peak; TL = Tree Line; SC = Santa Cruz; OL = Olympic; EP = Estes Park; CU = Cuernavaca.

From Powell et al. (1972).

TABLE 3-3 Frequencies in percent of *D. persimilis* third chromosome inversion types taken by region.

Region	No. of Localities Sampled	ST[a]	WT	ME	KL	Others	No. Chromosomes
British Columbia, Washington, Orgeon	7	1.0	0.7	3.1	93.7	1.4	290
Coast Range, California	11	11.0	1.6	40.8	45.8	1.7	426
Northeast California, Sierra Nevada	8	32.4	60.1	0.1	5.2	1.4	444

[a]ST = Standard; WT = Whitney; ME = Mendocino; KL = Klamath.

Modified from Dobzhansky and Epling (1944).

one cannot invoke historical origin to explain the geographic distribution of inversions in *D. persimilis*, although this is a possibility with *D. pseudoobscura*.

In recent years, *D. persimilis* has been undergoing a range expansion (Dobzhansky, 1973b; Moore et al., 1979). It has been moving easterly and especially southerly. Little is known of the inversion frequencies in the recently colonized area with the exception of Mt. San Jacinto. Moore et al. (1979) found *D. persimilis* to be about as frequent as *D. pseudoobscura* during 1974 and 1975, while all records indicate it was absent before 1966. The frequency of ST in *D. persimilis* was found to be 84% and WT was 15%. These are the two arrangements typical of the Sierra Nevada, but the frequency of ST has never been observed to be this high in any locality previously. Three other gene arrangements (MD, KL, and SE) were also found in very low frequencies, indicating that founder effects in this expansion are not particularly severe if present at all.

Clines are another feature of these species inversions. Figure 3-4 is an example for *D. pseudoobscura*. Curiously, where the two species coexist, they often show contrasting patterns in this regard (figure 3-5). Clearly, if the clines are due to selection, the two species are responding differently to environmental changes. While initially accepted as evidence that selection operates on these inversions, it is now recognized that clines can arise for other reasons, including historical stochastic processes (e.g., Endler, 1977). Nevertheless, given all the other data indicating strong selection on these inversions, it is not unreasonable to think selection may also play a large role in establishing clines.

D. subobscura is in many ways the Old World counterpart to *D. pseudoobscura*. This species occurs in temperate woodlands, is easily cultured, and is replete with naturally occurring inversions. One major difference is that it is polymorphic for inversions in every arm of every chromosome. This species has been extremely well studied by European scientists and reviews of this work are in Krimbas and Loukas (1980) and Krimbas (1992; 1993). In this species the five acrocentric chromosomes are designated by the vowels, A, E, I, O, and U; A is the X chromosome and J is often substituted for I.

Figures 3-6A and 3-6B show contrasting frequency clines for different inversions in the same chromosome, A. In the case of A_1 the cline is east-west and for A_{st} the

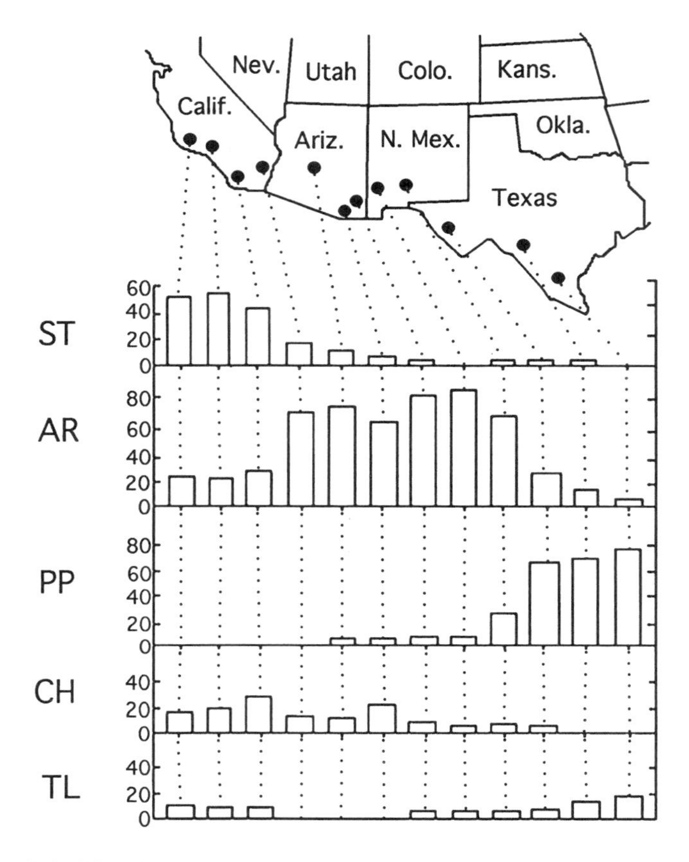

FIGURE 3-4 Example of an east-west cline in inversion frequencies of *D. pseudoobscura*. Frequencies are in percent on y-axis. From Dobzhansky and Epling (1944).

cline is north-south. The identical pattern holds for another chromosome, J (figures 3-6C and 3-6D). Finally, a third clinal pattern is one radiating from the center of the distribution, as exemplified by inversions in chromosome U (figure 3-6E). These three types of clines are repeated for each chromosome. It is difficult to imagine a scenario that could account for independent chromosomes showing such similar, but contrasting, patterns of clines. That is, not all clines are in the same direction, but all three contrasting patterns (east-west, north-south, and emanating from the center) are repeated in independent chromosomes. A recent study of clines in this species is that of Larruga et al. (1993), in which inversion, allozyme loci, and mtDNA polymorphisms were studied simultaneously. The inversions displayed the expected north-south cline, as did some of the loci associated with the inversions; the mtDNA did not vary clinally.

Until 1980, the evidence that inversion polymorphism in *D. subobscura* was subject to strong selection was equivocal (e.g., Krimbas and Loukas, 1980). In February 1978, *D. subobscura* was found in Puerto Montt, Chile at a south latitude of 41° 30′

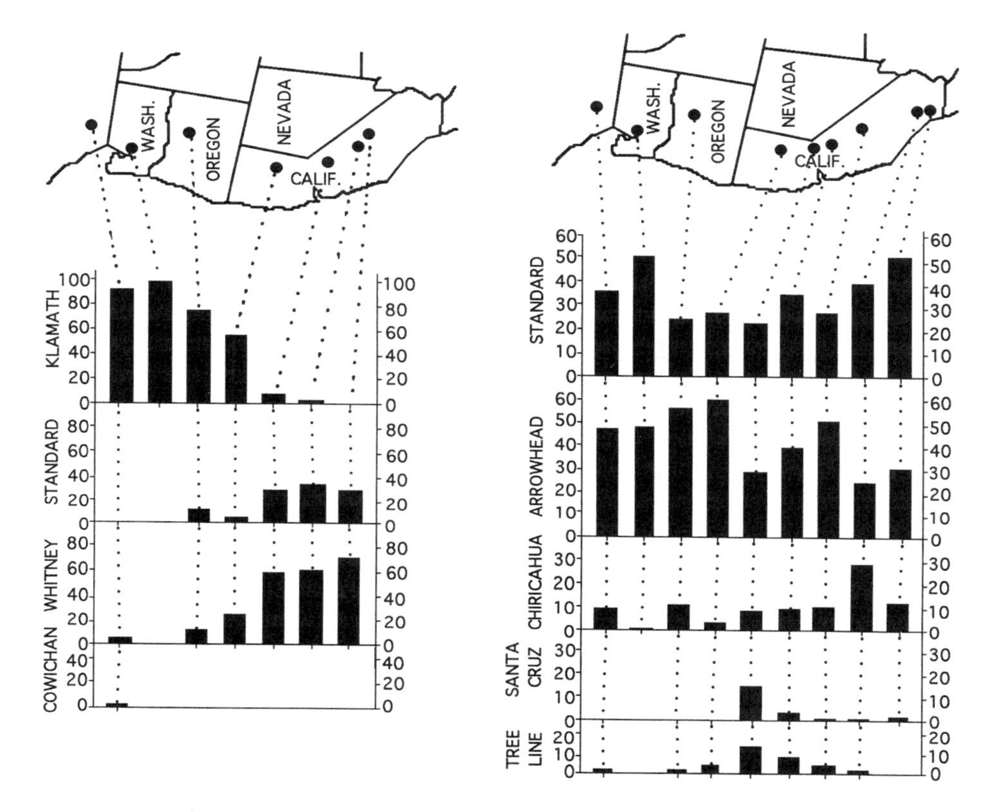

FIGURE 3-5 Contrasting examples of *D. persimilis* on the left, which shows a north-south cline in inversion frequencies, while the closely related species, *D. pseudoobscura*, from the same localities displays no clinal patterns (on the right). From Dobzhansky and Epling (1944).

(Budnik and Brncic, 1982; Prevosti et al., 1985). Since 1925, this area had been surveyed frequently by experienced *Drosophila* workers and no sign of the species was ever noted. Thus it is very likely that the introduction occurred in the mid to late 1970s. In just a few years *D. subobscura* spread throughout Chile and as far east as the coast of Argentina by 1984. In 1982 it was found in the northern coast of Washington State in the United States (Beckenbach and Prevosti, 1986) and has since spread through the Central Valley of California. All indications are that the founders of the New World populations were 4 to 6 European flies (Ayala et al., 1989; Rozas and Aguadé, 1991b; Krimbas, 1992). The data also argue strongly that the source of the North American invasion was South America, rather than a second introduction from the Old World. This invasion of the New World by an Old World species has been called "a grand experiment in evolution" (Ayala et al., 1989, p. 246).

The amazing finding, and the one most relevant to our present focus, was that the north-south latitudinal gradients so apparent in the Old World (e.g., figures 3-6A and 3-6D) have been established in South America in exactly the reverse order (Pre-

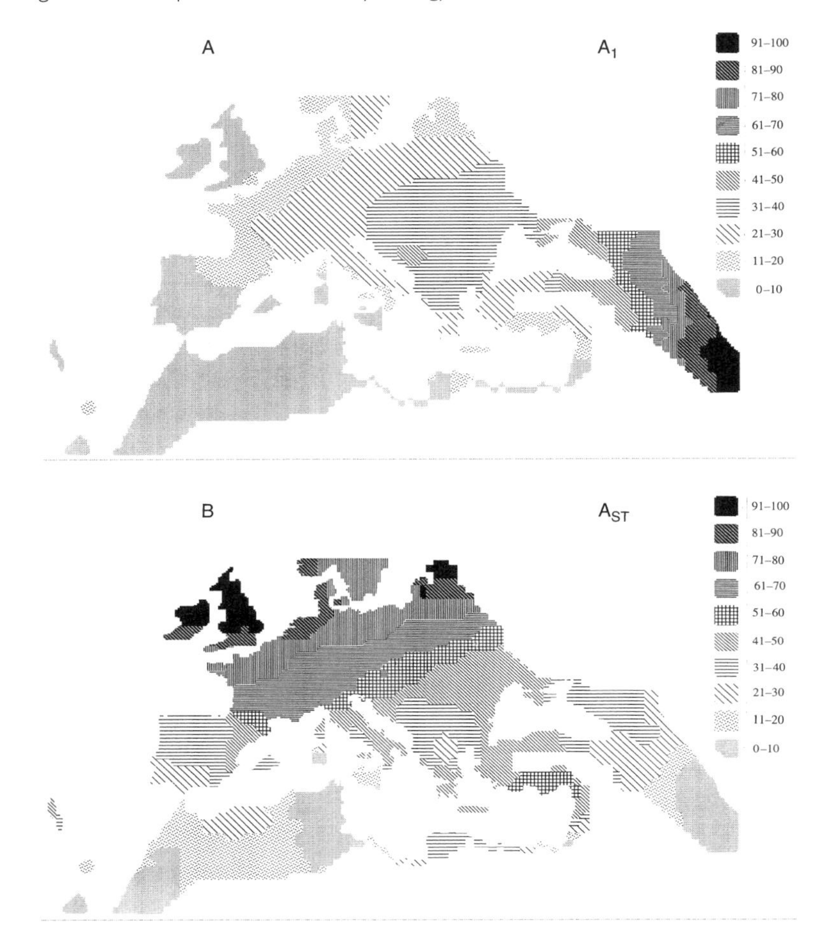

FIGURE 3-6 (A–E) Example of clines in *D. subobscura* inversions. The darker to lighter shading indicates higher to lower frequencies. A and B are examples of the A chromosome (X chromosome) having inversions with both east-west and north-south clines. C and D are the same for an autosome, the J chromosome. E is an example of a center outward cline for the U autosome. From Krimbas (1992).

vosti et al., 1985; 1988); that is, the cline tracks the latitude regardless of hemisphere! While the effect is not quite so striking, the same seems to be occurring in North America; the same inversions are forming clines in relation to latitude in the same direction as in the Old World. Table 3-4 is a compilation of the data. It is virtually impossible to explain these patterns except to invoke selection related to latitude, probably temperature or some factor (e.g., vegetation) related to temperature. This "grand experiment" ended the doubt over selection for the inversions in this species.

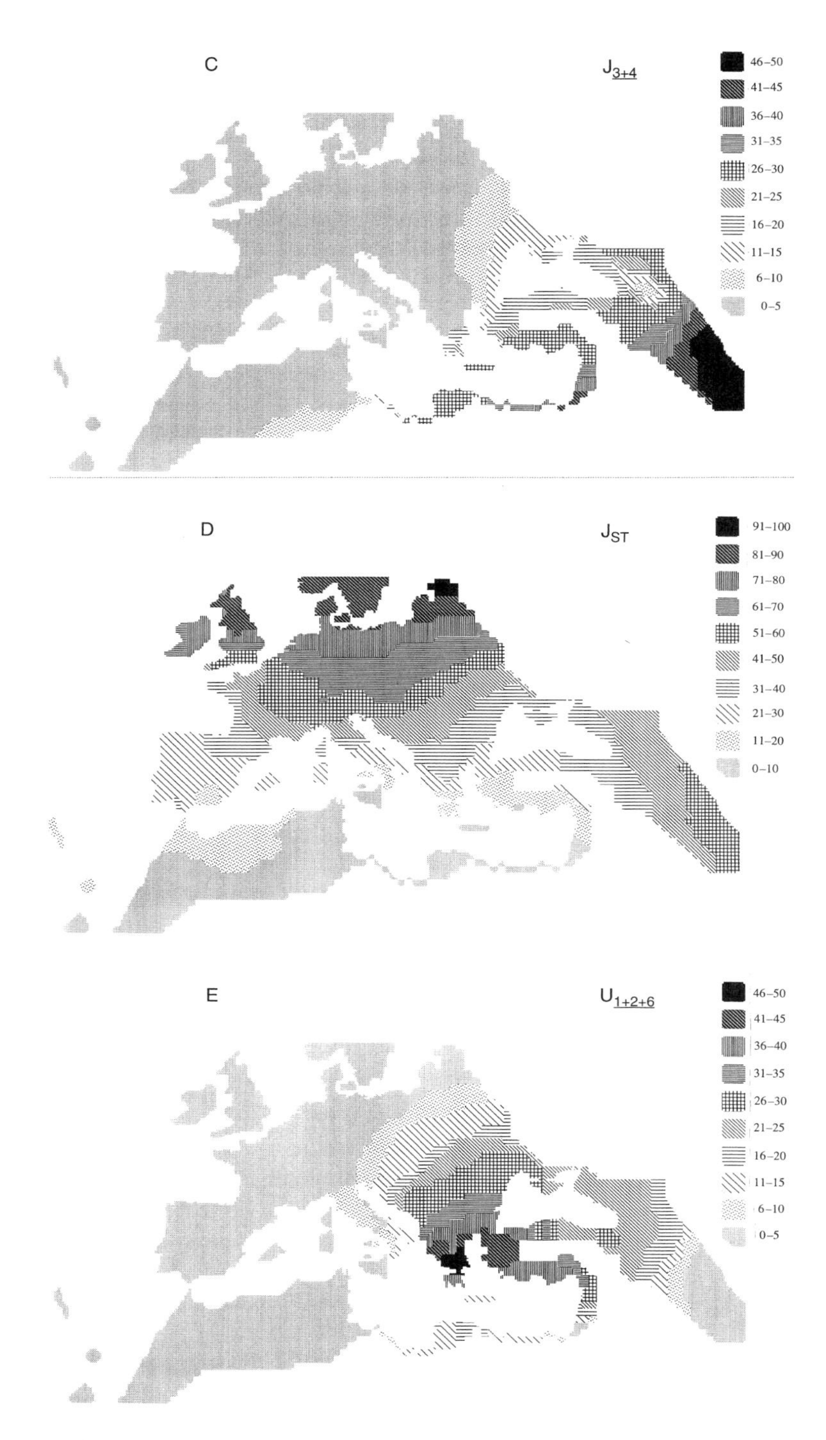

FIGURE 3-6 Continued.

TABLE 3-4 Correlation coefficients between latitude and inversion frequencies in *D. subobscura* populations in northern and southern hemispheres.

	Correlation Coefficient, *r*		
Chromosomal Arrangement	Old World	South America	North America
A_{st}	0.880***	0.492	0.347
J_{st}	0.972***	0.457	0.433
U_{st}	0.974***	0.727**	0.850*
E_{st}	0.973***	0.838***	0.868**
O_{st}	0.870***	0.219	0.793*

Note: Only the standard arrangement is shown for each chromosome, all of which have positive correlations with latitude. The Old World correlations are significant more often, possibly due to larger sample sizes.

*$p < 0.05$; **$p < 0.01$; ***$p < 0.001$

From Prevosti et al. (1988).

A similar pattern of latitudinal clines in both hemispheres exists for *D. melanogaster* (Lemeunier and Aulard, 1992). Only four inversions in this species are common cosmopolitan inversions, that is, they are found in reasonably high frequencies in many populations throughout the world. Each arm of the two major metacentric autosomes has one of these inversions (figure 3-18). For populations in America, Europe, and Australasia, Knibb (1982) showed that the inverted chromosomes for all four arms were negatively correlated with latitude.

D. robusta has also been extensively studied for geographic patterns of inversion frequencies (Levitan, 1992). This species is found in temperate forests of the eastern half of the United States; the larval breeding site is slime fluxes of the American Elm and other deciduous trees. This species also has inversions in all arms of all chromosomes, although in this case the chromosomes are metacentric rather than acrocentric as in *D. subobscura.* Levitan (1992) reviews the literature and documents north-south and east-west clines. Both smooth clines and step clines can be detected. One step cline is extremely impressive: the XR-1 inversion goes from about 4% to 60% over 1° longitude! Levitan (1992) also conducted an experiment on a wild population, artificially shifting inversion frequencies by the mass release of laboratory-reared flies. These flies entered the gene pool (i.e., successfully reproduced), but within a year, the population had returned to its previous inversion constitution.

Because inversions exist in both arms of metacentric chromosomes, and because Levitan used test crosses to a laboratory stock of known karyotype, he was able to determine gametic frequencies for both arms of each chromosome. He detected considerable linkage disequilibrium between inversions (Levitan, 1958; 1992). Interestingly, the disequilibria sometimes changed from population to population. That is, in one population the repulsion gametes are in excess and in another the coupling gametes for the same inversions are in excess. This clearly rules out an historical chance explanation. Disequilibrium could have been due to chance association at the origin of the inversions and subsequent low recombination. However, the reversal of the disequilibrium is not consistent with this explanation unless one discounts monophyly of inversions.

Altitude is another variable that has been studied in regard to spatial clines in inversion frequency. Figure 3-7 is an example for *D. pseudoobscura*. Data from altitudinal transects exist for several other species and references are in table 3-5. The case of *D. robusta* is especially noteworthy, as this species displays similar altitudinal clines on mountains over a wide geographic range (Etges, 1984), very good evidence that selection is responsible for altitudinal clines in that species. *D. silvestris* on the youngest Hawaiian island is another noteworthy example, as this is undoubtedly a

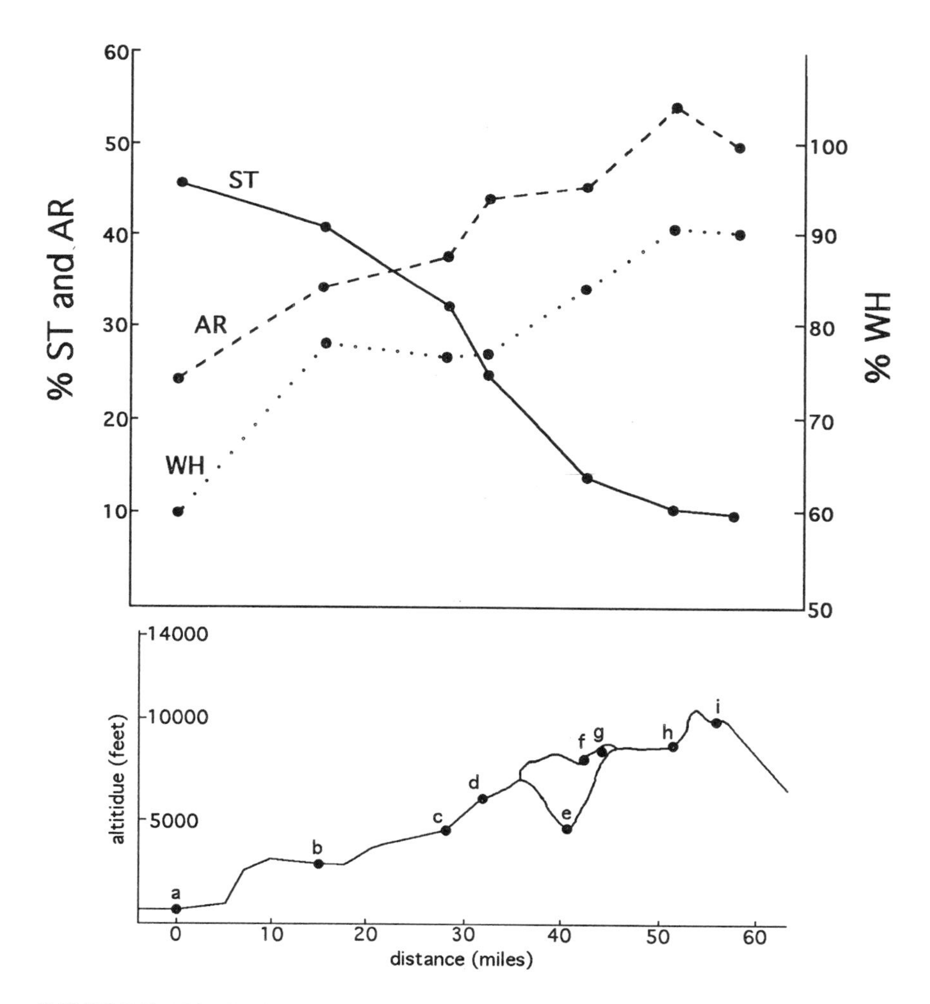

FIGURE 3-7 Altitudinal clines for the inversions ST and AR, in *D. pseudoobscura* and WH in *D. persimilis*. The lower figure is a schematic progressing from the Pacific Ocean on the left to the Sierra Nevada Mountains on the right. Localities are a = Jacksonville, b = Lost Claim, c = Mather, d = Aspen, e = Pate, f = Porcupine, g = Benson, h = Tuolumne, i = Timberline. Redrawn from Dobzhansky (1948a).

TABLE 3-5 Examples of studies of altitudinal changes in *Drosophila* inversion frequencies (in addition to *D. pseudoobscura*, illustrated in figure 3-7).

Species	Reference
D. persimilis	Mohn and Spiess, 1963
	Dobzhansky, 1971
D. robusta	Stalker and Carson, 1948
	Etges, 1984
	Levitan, 1992
D. flavopilosa	Brncic, 1972
D. nasuta	Ranganath and Krishnamurthy, 1978
D. annanassae	Reddy and Krishnamurthy, 1974
D. silvestris	Craddock and Carson, 1989
	Carson et al., 1990

relatively young species, yet its geographic differentiation with regard to inversions, including altitudinal clines, is remarkable (Craddock and Carson, 1989). Even populations that have been made extinct by volcanic activity in the past 2100 years and become recolonized have restored the clinal pattern (Carson et al., 1990).

Geographic Homogeneity

The above picture is the general one for *Drosophila* inversions; geographic differentiation often correlates with environmental factors. However, rare cases of geographic homogeneity are also known. The most remarkable instance is that of *D. pavani* in Chile (Brncic, 1973). This well-studied species exhibits considerable ecological tolerance, is geographically widespread, and has a high level of inversion polymorphism. Brncic studied 53 samples collected over hundreds of kilometers, from sea level to 1800 meters, in subtropical green valleys, semiarid temperate zones, and humid cold rain forests. Table 3-6 presents an example of his data. In this species, inversion polymorphism is clearly responding very differently from all the examples given above, and indeed compared to almost all other studies of geographic distribution of inversion polymorphism in *Drosophila.*

This contrast among species in how the inversion polymorphism system responds to environmental variability led Dobzhansky (1962) to describe these differences as "rigid" and "flexible" polymorphisms. Rigid systems vary neither geographically nor temporally, whereas flexible systems respond to environmental changes by shifts in inversion frequencies spatially and/or temporally. While sometimes convenient for discussion, this dichotomy is not always easily applied. An example will be given later of inversion polymorphism, which sometimes acts as a rigid polymorphism and sometimes as a flexible one.

Marginal/Central Patterns

Of all the geographic patterns of inversions observed in *Drosophila*, none has raised more interest than the comparison of central populations with those from the margins of species' distributions. The general pattern is that populations sampled from the center of a species distribution are high in inversion heterozygosity, and as one approaches the margins, it decreases nearly to chromosome monomorphism in many cases. The general assumption is that the center of a species distribution represents the optimal ecology for the species and that approaching the periphery, conditions for the species become progressively less favorable until it is no longer found. Several explanations for this pattern have been proffered, but before presenting those, four examples will be given to illustrate the pattern. These four are chosen as they are the best documented, and because they have been studied for allozymes, so we can contrast the pattern of the chromosome structural variation with single-gene variation.

Carson (1956; 1958a; 1958b) was among the first to note the distinct decline in inversion heterozygosity near the margins of the distribution of *D. robusta.* Figure 3-8 is an example of his data; subsequent collections have confirmed these initial findings (Levitan, 1992). This species exists in the eastern half of the United States. Northwest Nebraska is marginal and the Missouri Ozarks are near the center of distribution; a similar pattern holds going south to nearly monomorphic Florida populations. Carson introduced what he called the *index of free recombination* or IFR, which is the percentage of euchromatin in an individual that is free to recombine, that is, structurally homozygous. The IFR decreases from 99.7% in marginal populations (virtual monomorphism) to 67% in geographically central populations. Prakash (1973) studied this same species for allozyme frequencies and found no such pattern for single-gene polymorphisms; heterozygosity was as high in marginal populations as in central populations.

Da Cunha, Dobzhansky, and colleagues have reported similar observations on the neotropical species, *D. willistoni* (da Cunha and Dobzhansky, 1954; Dobzhansky,

TABLE 3-6 Geographic homogeneity for *D. pavani* inversions in South America.

	Chromosome 4R		Chromosome 4L	
Locality	Standard	Inverted	Standard	Inverted
Copiapó	40.0	60.0	31.2	68.8
Vallenar	34.4	65.6	33.6	66.4
La Serana	43.5	56.5	21.7	78.3
Vicuna	42.4	57.6	38.1	61.9
El Tabo	31.0	69.0	25.0	75.0
Leyda	44.8	55.2	36.2	63.8
Melipilla	39.4	60.6	34.5	65.5
Bellavista	39.0	61.0	34.5	65.5
S. J. Maipo	37.5	62.5	27.5	72.5
Volcán	33.9	67.0	38.1	61.9
Rancagua	33.0	67.0	38.2	61.8
Santa Cruz	43.4	56.6	38.2	61.8
San Fernando	37.8	62.2	37.8	62.2

Note: Localities extend over several hundreds of kilometers and represent divergent ecological conditions.

From Brncic (1973).

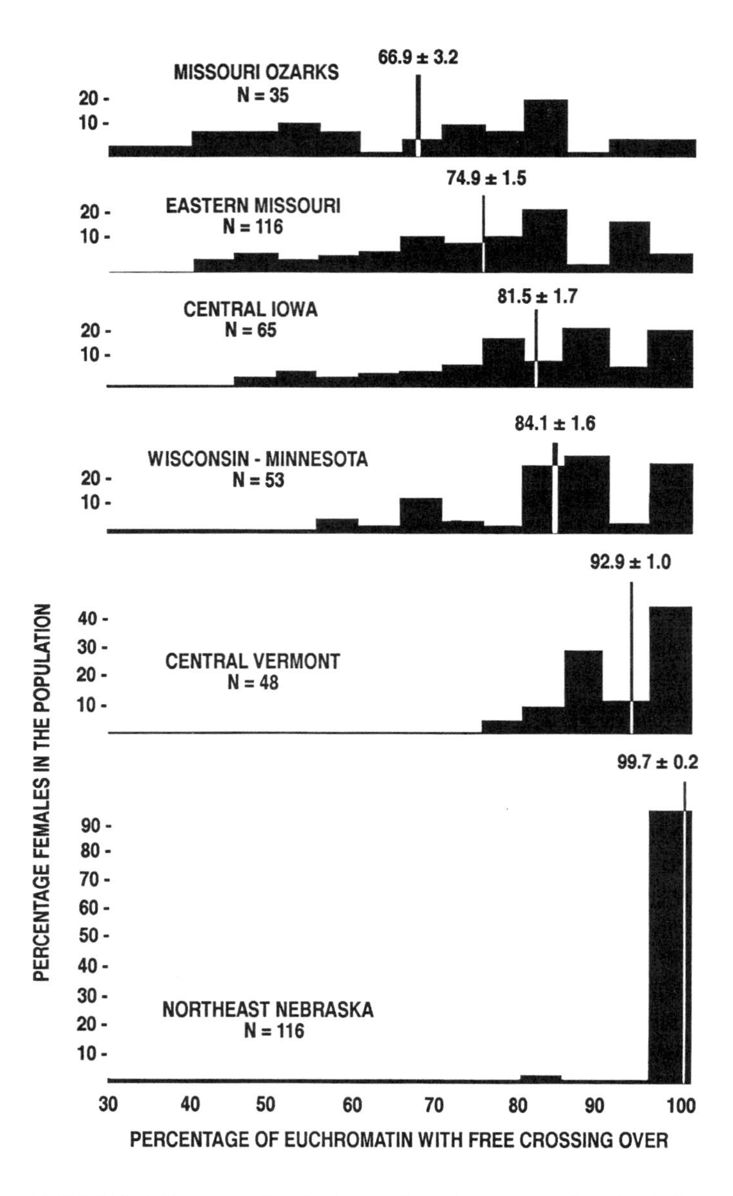

FIGURE 3-8 Pattern of level of recombination suppression due to inversions in *D. robusta*. The distribution of individuals with the indicated percentage of euchromatin free of inversion heterozygosity is indicated from populations ranging from the center of the species distribution in the Ozarks to the margins of the distribution in northwest Nebraska. From Carson, 1958b.

1957; da Cunha et al., 1959). Unfortunately, the polytene chromosome preparations for this species are rather poor; consequently, it is virtually impossible to distinguish homozygous gene arrangements from one another. Therefore, the data consist of heterozygote and homozygote frequencies for some 50 different gene arrangements, polymorphic in all arms of all chromosomes. Figure 3-9 illustrates the geographic pattern of inversion heterozygosity, which is very similar to that in *D. robusta.* Central (Brazilian) populations have the highest inversion heterozygosity and moving either south into Argentina or north into Mexico and the Caribbean, it declines. In the previous chapter, allozyme studies on this species were presented (table 2-12), indicating the constancy of these single-gene frequencies across the species range including Caribbean islands. Table 3-7 contrasts the inversion heterozygosity with single-gene (allozyme) heterozygosity in this species. Again, while inversion heterozygosity decreases toward the margins, allozyme heterozygosity remains virtually identical with the same alleles in about the same frequencies in all populations.

The temperate Old World species *D. subobscura* has also been analyzed with respect to degree of inversion polymorphism in marginal and central populations (Krimbas and Loukas, 1980; Krimbas, 1992) and the conclusion is similar to the two previous cases. Figure 3-10 is based on calculation of IFR for 87 populations. The highest level of inversion heterozygosity (lowest IFR) is centered in Asia Minor and the Caucasus, which does not actually correspond to the center of the known distribution of the species; however, this species has not been at all well-studied in the former Soviet Union, and consequently the eastern boundary of the species has not been firmly established. Saura et al. (1973) and Lakovaara and Saura (1971a) studied the allozyme polymorphism in *D. subobscura* nearly throughout its range. Heterozygosity in central populations ranged from 0.18 to 0.23; in intermediate populations, 0.20; and in peripheral populations, 0.22. Again, the same alleles were found in about the same frequency throughout the species range.

The picture for *D. pseudoobscura* is not nearly as clear. As indicated from the data in tables 3-2 and 3-10, this species is fairly polymorphic for inversions throughout its range, with the exceptions of populations in Arizona, New Mexico, southern Utah, and western Colorado. Where *D. pseudoobscura* is found in this dry region (usually in higher, moister mountains that support suitable vegetation), it is nearly monomorphic for AR. This area is in the center of the species distribution. Figure 3-11 illustrates the situation using an index of inversion diversity. How can one square these observations with those for the previous species? Arguably, this is simply a case of geographic centrality or marginality not coinciding with levels of ecological stress. It is difficult to really define ecological marginality in the west as the species simply stops at the ocean. The semidesert central region may well present an ecologically marginal habitat, much more so than elsewhere. The southern part of the species distribution presents a more consistent picture: populations from the Mexican highlands are the most polymorphic for inversions (up to 14 in a single population), while populations from Guatemala and Bogota harbor only 2 and 3 inversions, respectively. Geographically homogeneous allozyme variation is also the rule in this species (Lewontin, 1974).

Soulé (1973) lists a total of 16 species of *Drosophila* for which there is data on inversion polymorphisms, and he concluded that 15 exhibited the central-marginal pattern; the only exception is the already-noted homogeneous *D. pavani.* However,

FIGURE 3-9 Pattern of variation in the level of inversion polymorphism in *D. willistoni* populations. Diameters of circles are proportional to the average number of inversions heterozygous per individual from the different populations. Note the progressive decrease in inversions heterozygosity from the central regions to the margins of distribution. From da Cuhna et al. (1959).

TABLE 3-7 Contrasting heterozygosity of inversions and allozymes in *D. willistoni*.

Locality	Mean No. Inversions Heterozygous/Individual	Allozyme Heterozygosity
Island		
St. Lucia	1.60 ± 0.16	0.152
Grenada	0.84 ± 0.12	0.156
St. Vincent	0.76 ± 0.13	0.184
Bequia	0.68 ± 0.09	0.161
Martinique	0.56 ± 0.10	0.175
Carriacou	0.48 ± 0.14	0.142
Continental		
Brazil	2.75–6.56	0.144–0.214
Colombia	5.09–5.45	0.172–0.210
Panama	4.73 ± 0.51	0.184
Costa Rica	4.38 ± 0.44	0.205–0.214

Note: Not all continental populations were studied for both inversions and allozymes, so typical localities and ranges are given in some instances.

Data from Ayala et al. (1971, 1972).

some of the examples cited by Soulé are somewhat ambiguous or incomplete and, therefore, the four species discussed above are by far the clearest examples, although Soulé's review can be consulted for other possible cases. Brussard (1984) has also reviewed this issue.

How can this central/marginal pattern be explained? Much of the discussion on this point predated allozyme and other molecular data, so some ideas seem a bit outmoded. For example, the observation that inversion-depauperate populations are just as replete with allozyme variation as central populations would seem to argue against genetic drift due to founder effects on the margins. Evidently, migration is sufficient to introduce all the single-gene alleles to these populations and keep them at about the same frequencies as elsewhere. The fact that the diversity in the margins is due to precisely the same alleles as elsewhere argues strongly against regeneration of new variation by mutation, after a founder effect (Soulé, 1973). This was the favored explanation for the *D. willistoni* island data given by Ayala et al. (1971), but I believe this is no longer tenable as a general explanation. The repeated observation of the same alleles in the periphery for a number of species would require that mutation never generated alleles with new electrophoretic properties, a bit hard to believe for so many cases. The data are more consistent with the selective elimination of inversions in the periphery counter-acting migration.

Da Cunha and Dobzhansky (1954) proposed that the level of inversion polymorphism in a population of *D. willistoni* is directly related to the diversity of the habitat occupied by the populations. They devised a habitat diversity index based on biotic (e.g., number of plant species) and abiotic (e.g., temperature) variation and demonstrated a high positive correlation between inversion heterozygosity and this diversity index. The idea is that the greater environmental diversity a population faces, the more inversions it can maintain due to diversifying selection (the Ludwig effect): each

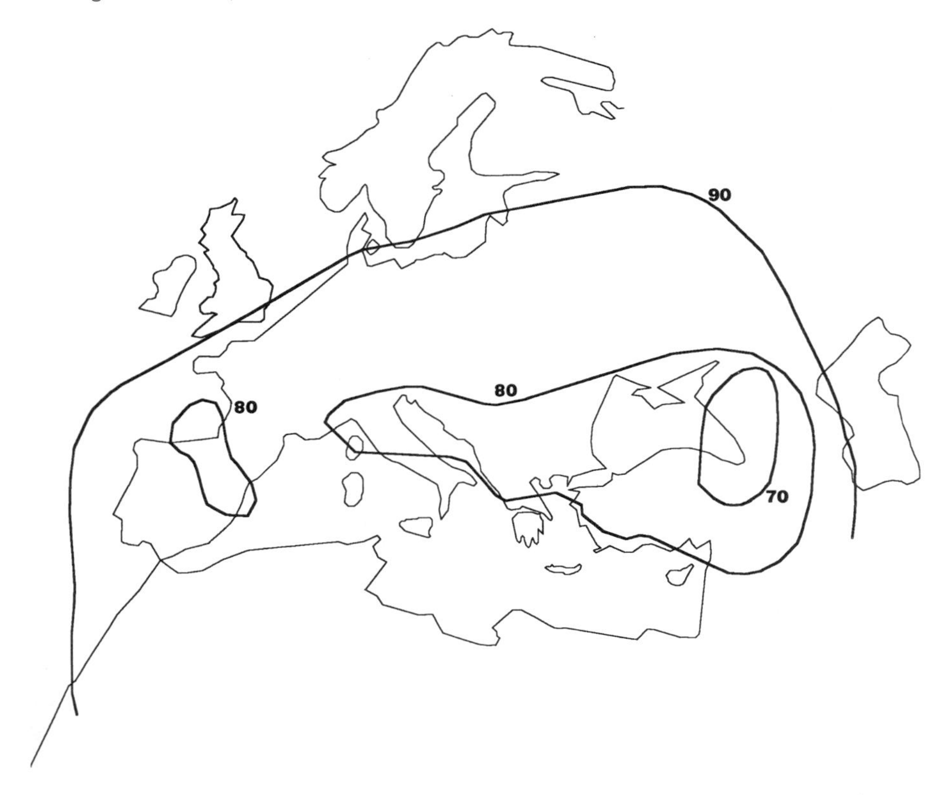

FIGURE 3-10 Isoclines for the index of free recombination (IFR) in *D. subobscura*. The higher index (toward the northern margin of distribution) indicates less inversion heterozygosity. From Krimbas (1992).

karyotype would have a part of the niche in which it was superior in fitness. Toward the margins of a species distribution, the environmental diversity narrows such that fewer and fewer inversions are maintained. Of course, correlations are not causations and many authors have criticized this theory. The greatest difficulty is the lack of evidence for inversions having any kind of direct relationship with any single environmental factor; taken to its extreme, the theory predicts that inversion A is best on papaya, inversion B on breadfruit, and so forth. On the other hand, this theory has not been rigorously tested, especially in the species for which it was proposed, *D. willistoni*.

Carson (1958a; 1958b) presented a different theory emphasizing the role of inversions in reducing rates of recombination. He reasoned that in central populations, conditions for the species are relatively stable and favorable; therefore, stabilizing selection would act to maintain a modal adaptive phenotype. Inversion heterozygotes represent such stable modal phenotypes, which are well buffered from developmental instability and exhibit overall vigor in favorable stable environments. The relative lack of recombination in central populations protects these "tried and true" genotypes from disruption. Approaching the margins of distributions, the environment becomes less

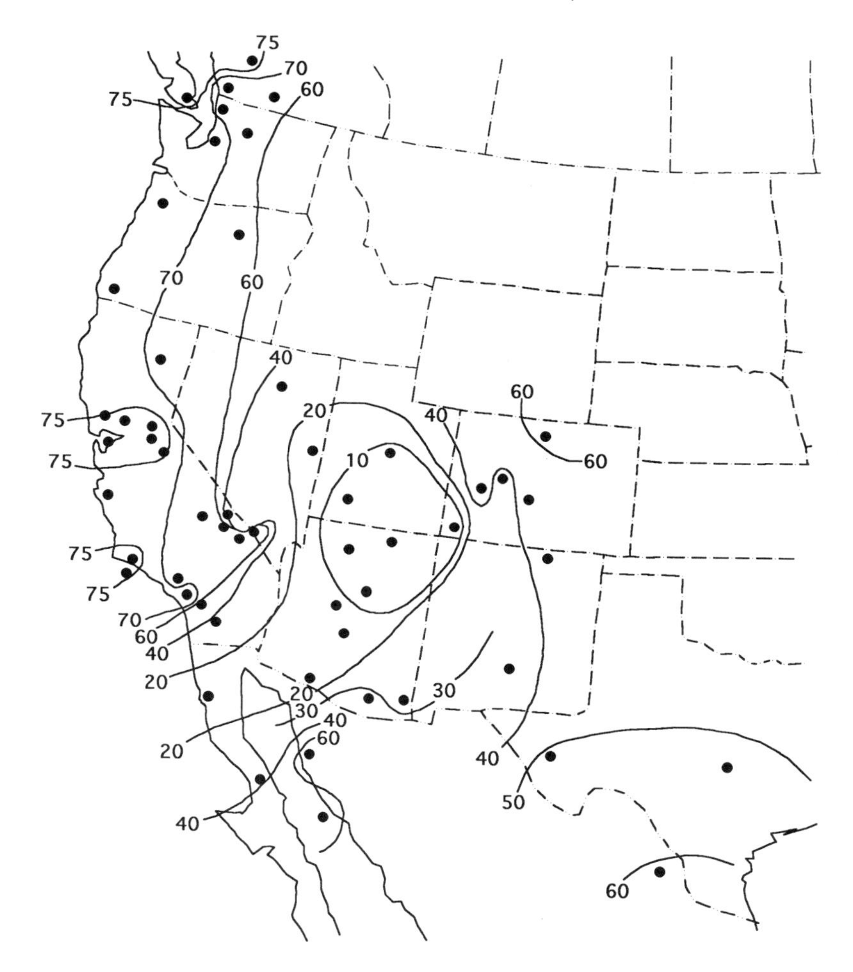

FIGURE 3-11 Isoclines for inversion diversity as measured by $H = 1 - \Sigma p_i^2$ where p_i is the frequency of the i inversion. Note that in contrast to figures 3-12 to 3-14, the central part of the distribution exhibits the lowest inversion diversity; this region may well represent ecological rather than geographic marginality. From Anderson et al. (1991).

favorable and predictable, and extreme phenotypes produced by inversion homozygotes may have an advantage. Furthermore, such homozygotes would experience greater recombination of their genomes, increasing the probability of generating genotypes capable of occupying extreme environments. Carson called this phenomenon *homoselection*, whereby chromosome structural homozygosity is actively favored in the margin.

Both the da Cunha/Dobzhansky and Carson theories are hard to rectify with the allozyme data. Both would predict a decrease in genic and inversion diversity in marginal populations. In the da Cunha/Dobzhansky scenario, the one or few inversions whose niche remains at the margin should contain less genic diversity than would a mixture of many gene arrangements (assuming, of course, it is the genes in the inversions that are controlling the fitness for the niche). Likewise, Carson's concept of more extreme phenotypes produced by homozygotes must also apply to single genes as well. Lewontin (1957, 1974) rescues a modified version of this theory by re-emphasizing the recombination factor rather than genic homozygosity for extremes. The temporal instability of marginal habitats places a premium on adaptive flexibility that would be enhanced by structurally homozygous, but genically heterozygous, chromosomes such as are found at the distribution margins. Experimental laboratory work tends to agree with this modified version (see chapter 4).

Soulé (1973) presents yet another alternative, hypothesizing essential functional differences between inversion polymorphism and single-gene (allozyme) polymorphism, a theory he calls an *epistatic cycle.* He emphasizes the role of inversions in maintaining linkage of epistatically interacting coadapted alleles; this epistatic fitness interaction is postulated to occur among genetic units not involved in coding for protein structure. Genes coding for proteins, those detected by allozyme studies, are independent of the epistatic system, but are themselves individually heterotic in a general sense. When an extreme phenotype is favored, as in an extreme (marginal) habitat, directional selection occurs, as opposed to the stabilizing selection for the mode in the center. Such directional selection leads to fixation of one or a few of the epistatic interacting blocks (the inversions). However, individuals from the extreme of the phenotype distribution can still benefit from general heterosis, so protein-coding gene heterozygosity would be favored in the margins as well as in the center. There is something appealing about Soulé's theory in that it does emphasize very different roles for the two types of polymorphisms, something that surely must be true. On the other hand, it is not at all clear just how to separate protein-coding gene loci from epistatic interactions given our poor knowledge of gene regulation and interactions. Furthermore, there is little evidence in support of individual heterosis at most allozyme loci. Carson's (1975) proposal that there are "closed" and "open" genetic systems may have some relationship to Soulé's proposal, at least in emphasizing that not all polymorphisms behave in the same manner. Carson argues there is an open system that allows for minor adaptive adjustments (e.g., allozymes). Other more epistatically interactive blocks of genes (e.g., inversions) are difficult to change, but when changed, have more drastic consequences for the organism, including speciation.

Finally, Wallace (1984) weighs in with yet one more theory. He postulates that the essential difference between central and marginal populations is the presence or absence of strong intraspecific competition. In the center of a species distribution, densities are high and intraspecific competition is strong. This produces density-dependent selection, which favors the maintenance of inversion heterozygosity. At the margins, however, populations are small and density is low; the main problem facing a gravid female is not competition from conspecifics, but simply locating any suitable substrate for oviposition. If a female finds a patch of favorable larval substrate at the periphery, she may have a bonanza and produce many offspring in the absence of competition from other families. This density-independent selection operates largely by chance and a chromosome homozygote or heterozygote is equally likely to find the rare patch. The advantage of inversion heterozygosity is lost, and chromosome homozygosity is likely to become fixed.

Whatever the reasons for the contrasting genetic makeup of *Drosophila* populations inhabiting different parts of a species range, this phenomenon may have implications that transcend just understanding this group of insects. Many endangered species live to one degree or another in marginal, stressful environments. If we can understand in any organism what types of genetic attributes contribute to persistence in such habitats, this might facilitate the design of programs to enhance the survival probability of such populations.

Temporal Changes

Short-Term Temporal Changes

The first unambiguous indication that inversions were subject to strong selection came from studies of temporal shifts in frequencies. Until that time, it was taken as self-evident that inversions must be neutral variants; after all, no genes were changed, merely the order of the genes. Dobzhansky was attracted to the system *precisely because he thought it was neutral to selection.* At the time (the 1930s) Sewall Wright was developing his theories of genetic drift, and Dobzhansky thought the inversions of *D. pseudoobscura* were the perfect naturally occurring variants to test predictions of Wright's theories. A study of temporal change in a population on Mount San Jacinto in California fundamentally changed his view. Some thirty years later, Dobzhansky wrote the following:

> In 1938 there was initiated repeated sampling at approximately monthly intervals during the breeding season of populations of *D. pseudoobscura* in three ecologically rather different localities on Mount San Jacinto, in California. The purpose of this work was a study of the frequencies of allelism of recessive lethals in the third chromosomes of these populations, in samples collected simultaneously and collected at different times. The samples were submitted also to cytological examination, to determine the frequencies in them of the different gene arrangements in the third chromosome. *Neither in my memory nor in my notebooks can I find what working hypothesis was the cytological study intended to test* [emphasis added]. Anyway the results obtained seemed startling: not only were the populations of the three localities, about 15 miles apart, clearly different in the chromosome frequencies, but in two of the localities the chromosome frequencies were changing significantly from month to month. That these changes could be caused by natural selection seemed hard to believe; although we do not know even now precisely how many generations of *D. pseudoobscura* occur per year in the natural habitats of this fly, the selective forces that were necessary to assume to account for the changes seemed, at that time, too unlikely to operate in these natural habitats. The possibility that the changes may have resulted from random genetic drift appeared more plausible. A coup de grace to this surmise was administered by the finding that at least some of the changes are regularly cyclic. (Dobzhansky, 1971)

The importance of this shift in attitude toward *Drosophila* inversions had a profound effect on the whole field of evolution, as inversions had become a cornerstone example in experimental work. With the evidence that these a priori obviously neutral characters could be subject to such strong selection, new respect for, and emphasis on, the power of selection in evolution emerged, what Gould (1982) called the "hardening of the synthesis." The 30 "hard years" of evolutionary biology began on Mount San Jacinto.

The three localities Dobzhansky referred to were Andreas Canyon, Piñon Flats, and Keen Camp, varying in altitude from 250 to 1350 meters. The samples from two lower sites, Andreas and Piñon, exhibited seasonal cycles, whereas the Keen samples did not; fewer samples were obtained for Keen due to its high elevation. Figure 3-12

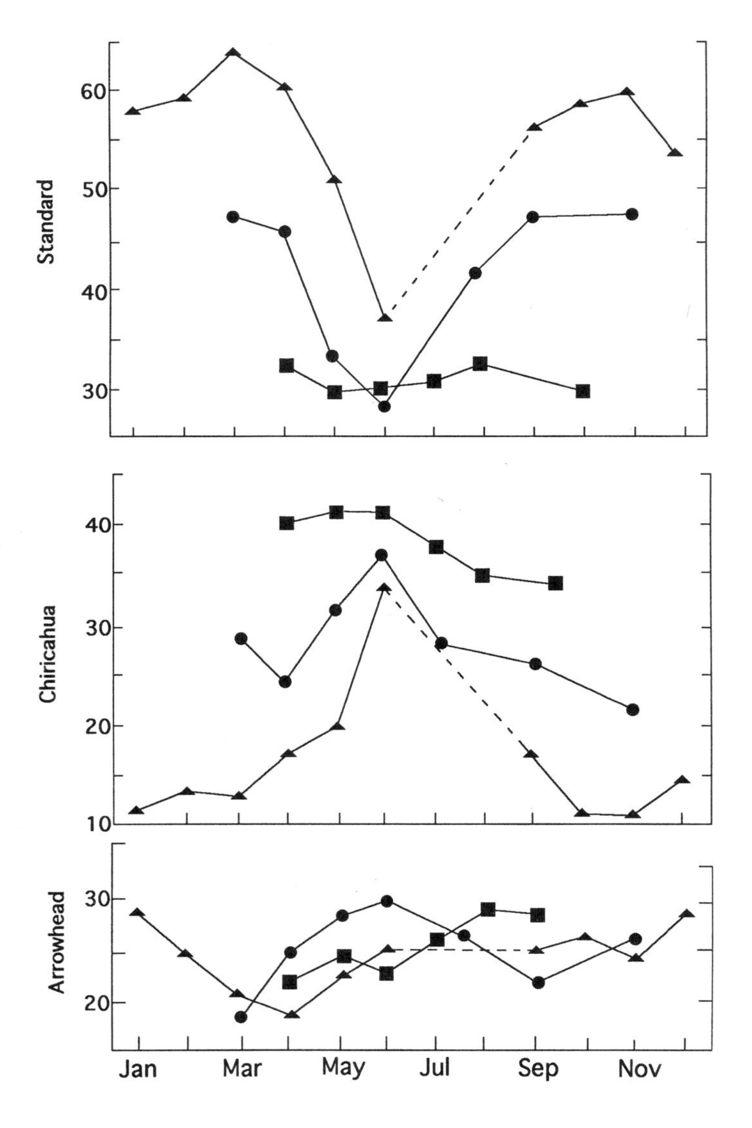

FIGURE 3-12 Seasonal changes in inversion frequencies of *D. pseudoobscura* at three localities on Mount San Jacinto, California. Triangles are for Andreas Canyon, circles for Piñon Flats, and squares for Keen Camp. From Wright and Dobzhansky (1946).

shows the data graphically. The frequency shifts at Andreas and Piñon are very similar. That these shifts are truly cycles, repeated each year, can be seen for the most extensive data, that for Piñon Flats (figure 3-13); the data for this figure are presented in table 3-8. Clearly these are recurring cycles. It is very difficult to explain such cycles by any force except selection. And the strength of selection must be great. These species probably average about one generation a month (more at warm times, less at cold times). ST frequencies are decreasing 25% in the four months from March to June, then turning around and gaining 25% in five months. The magnitude of selection coefficients needed to explain such shifts are on the order of 0.2 or greater, strong selection indeed. As we will see in the next chapter, in laboratory studies selection coefficients of this magnitude can be directly measured for this polymorphism. Here at last was a naturally occurring polymorphic system that was subject to such strong natural selection that it could be detected by studies lasting only months. An experimenter could study it within a lifetime. Experimental population genetics of natural populations was given a great boost by these findings. A system had been found that Darwin had not imagined possible.

Several other studies of seasonal changes in *D. pseudoobscura* inversions have been made. In his publication reporting the altitudinal gradient in the Sierra Nevada

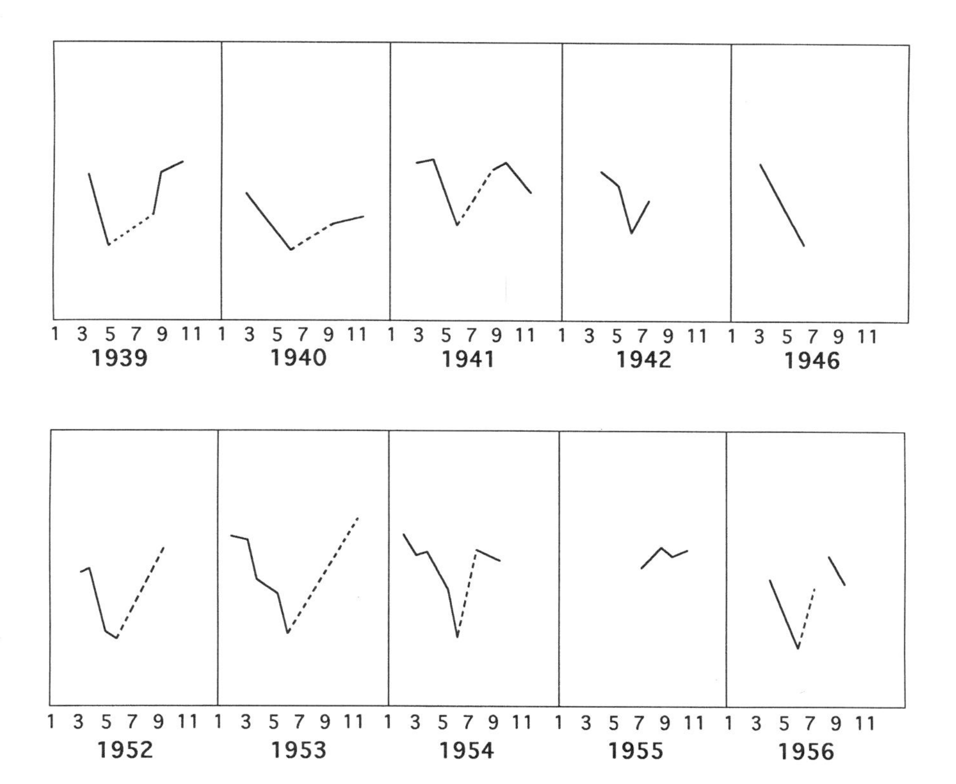

FIGURE 3-13 Frequency of the ST gene arrangement of *D. pseudoobscura* at Piñon Flats over several years. Y-axis is months beginning with January as 1. Data are from Dobzhansky (1971) and figure is after Wright (1978).

TABLE 3-8 Percentage frequencies of the three most common third chromosome gene arrangements of *D. pseudoobscura* at Piñon Flats, Mount San Jacinto.

Month	Year	ST	AR	CH	*n*	Month	Year	ST	AR	CH	*n*
Apr.	1939	51	29	13	61	Mar.	1953	60	19	5	412
May	1939	28	36	30	240	Apr.	1953	48	23	14	588
June	1939	30	35	31	154	May	1953	42	21	15	342
Aug.	1939	36	33	26	156	June	1953	28	37	18	60
Sept.	1939	51	23	23	190	Nov.	1953	70	13	2	180
Oct.	1939	55	25	17	284	Feb.	1954	61	18	5	164
Mar.	1940	45	20	30	386	Mar.	1954	56	19	6	86
Apr.	1940	35	28	34	176	Apr.	1954	56	23	5	304
May	1940	28	27	40	202	May	1954	44	27	11	128
June	1940	24	30	42	170	June	1954	27	41	11	56
Sept.	1940	35	25	38	104	July	1954	58	20	7	168
Nov.	1940	37	33	26	80	Aug.	1954	57	18	10	410
Mar.	1941	56	11	24	110	Sept.	1954	54	9	30	106
Apr.	1941	58	20	17	110	July	1955	51	15	18	192
May	1941	44	28	24	100	Aug.	1955	58	22	12	96
June	1941	33	32	32	192	Sept.	1955	55	16	20	146
Aug.	1941	52	21	26	108	Oct.	1955	57	22	11	250
Sept.	1941	56	19	16	80	Apr.	1956	47	26	12	248
Nov.	1941	45	24	24	100	May	1956	37	30	16	228
Apr.	1942	51	22	20	102	June	1956	23	40	27	52
May	1942	48	17	25	100	July	1956	49	18	20	228
June	1942	30	23	40	114	Aug.	1956	57	17	15	108
July	1942	42	22	31	124	Sept.	1956	45	20	18	56
Mar.	1946	56	18	18	558	Mar.	1963	80	6	3	114
June	1946	26	24	44	500	Apr.	1963	76	11	3	300
Mar.	1952	49	16	19	74	May	1963	63	14	5	190
Apr.	1952	51	20	21	156	Mar.	1970	76	18	2	80
May	1952	30	23	33	40	Apr.	1970	62	22	5	1000
June	1952	26	33	29	140	—	1978	58	18	11	201
Sept.	1952	58	19	14	104	—	1980	70	10	12	146
Feb.	1953	61	19	6	444						

Note: This is probably the most complete record for any locality.

From Dobzhansky (1971) with additions from Anderson et al. (1991).

mountains (figure 3-7), Dobzhansky (1948a; see also Dobzhansky 1971 for data) also reports seasonal inversion frequencies from four of the localities. Interestingly, the results are opposite to those found on Mount San Jacinto in that the lower localities (Lost Claim and Jacksonville) showed no shifts in frequencies through a season, while the higher localities (Mather and Aspen) displayed cycles. Again, in contrast to San Jacinto, ST rose in frequency from spring to fall while AR was falling in frequency. These results gave the first hints of another important aspect of this system, namely that each population's set of inversions could behave differently from other populations. The implication is that ST, AR, CH, and so on, are not identical in all populations. As we will see in the next chapter, the existence of population-specific coadaptation of each gene arrangement became an important finding for this system.

Seasonal changes in inversion frequencies in *D. pseudoobscura* at a site in Napa County, California (McDonald Ranch) have been reported by Dobzhansky and Ayala (1973). The climate at this site allows collecting all year; two years of data were presented, and while some indications of cycles exist, they are not totally convincing.

Strickberger and Wills (1966) studied a population near Berkeley, California for three years. The climate here also allows year-round collecting of *D. pseudoobscura*. Contrary to previous studies, these investigators found ST reaching a maximum in summer (June to August) and falling off in both spring and fall; CH mirrored this pattern. This is the best case of cycles in a population with such stable temperature that there is no overwintering or "oversummering" (estivation).

Crumpacker et al. (1977) studied *D. pseudoobscura* from several sites in Colorado along the border between the Great Plains and the Rocky Mountains. Seasonal collections were begun both north and south of the latitude of Denver. Populations south of this latitude showed seasonal cycles especially in AR, which reached a maximum of 85% in spring and early summer and dropped to 18% by fall. Populations north of Denver displayed no temporal shifts. The authors remark on how differently populations respond despite being only 100 km apart; some are behaving as classical rigid polymorphisms, while others are flexible.

With American and Mexican colleagues, I have studied seasonal changes in *D. pseudoobscura* populations in southern Mexico where the inversion types (table 3-2) are very different from those in the other localities studied for seasonal changes. Unfortunately, these data have never been published in full; partial data are in Levine et al. (1995) and access to the full data set can be made from Arnold (1982). Likewise, J. A. Moore, B. A. Moore, and C. E. Taylor have unpublished seasonal data from other sites on Mount San Jacinto, taken in the late 1970s and early 1980s.

D. persimilis has been less well-studied for seasonal cycles. Most studies of this species show temporally invariant frequencies, and therefore, *D. persimilis* might be classified as a rigid polymorphism. However, the study of Dobzhansky and Ayala (1973) at McDonald Ranch produced as convincing a textbook example of cycles as any previous studies (figure 3-14). This demonstrates that the classification of a polymorphic system as either rigid or flexible is subject to change depending on what populations are studied. Also, in this locality, *D. pseudoobscura* collected at the same time did not show any cycles, again emphasizing differences in these two species' inversion polymorphism as noted for geographic patterns (figure 3-5).

Several other *Drosophila* species have been studied for short-term seasonal changes in inversion frequencies in natural populations; table 3-9 is a partial listing of these studies. Some species, such as *D. robusta*, do display seasonal cycles while others, such as *D. subobscura*, do not. As we have seen, this latter species has considerable geographic flexibility, but apparently little or no temporal flexibility.

Long-Term Temporal Shifts

Longer-term temporal changes in inversion frequencies have been monitored in a number of species. Data for *D. pseudoobscura* go back nearly 50 years. Even though the record is relatively long for this species, the major finding has been a general

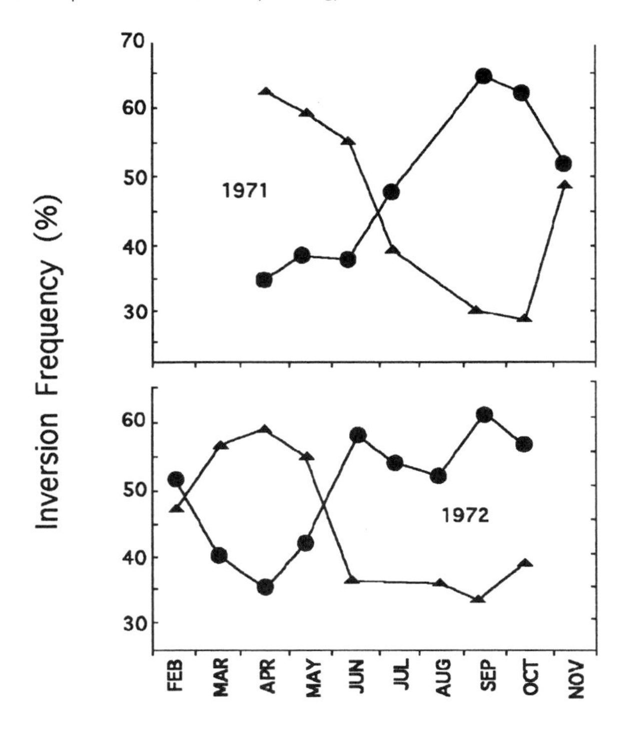

FIGURE 3-14 Seasonal cycles in inversion frequencies of *D. persimilis* at McDonald Ranch, Napa County, California. Circles are for the KL gene arrangement and triangles are for MD. From Dobzhansky and Ayala (1973).

stability in frequencies in the 22 populations best-studied (the latest report is Anderson et al., 1991). Table 3-10 presents examples of the data. Some small statistically significant shifts have occurred in individual populations, but perhaps the more interesting pattern has been a parallel shift in frequencies among a number of populations in the same region. For example, the frequency of PP rose in several populations from the 1940s to 1960s and has since retreated; this phenomenon is apparent in populations from British Columbia to central California. TL has also been climbing in frequency in this same region, although with little sign yet of abating. Other than these two cases, the picture of stability holds.

Two, not mutually exclusive, explanations can account for the PP and TL patterns. First, the environment might be changing and the *Drosophila* populations responding to the changes. One possible widespread environmental change that was considered for the coordinate shifts in the 1950s and 1960s was the insecticide use that began then in this region. However, laboratory experimental data are negative or ambiguous with regard to the sensitivity of *D. pseudoobscura* inversion polymorphism to insecticides (Anderson et al., 1968; Cory et al., 1971). Attempts to correlate other climatic factors such as temperature and rainfall with inversion-frequency shifts have been unsuccessful (e.g., Dobzhansky et al., 1966).

TABLE 3-9 Studies of seasonal changes in inversion frequencies in species other than *D. pseudoobscura* and *D. persimilis*.

Species	References
D. melanogaster	Stalker, 1976
	Inoue et al., 1984
	Masry, 1981
D. robusta	Carson, 1958a
	Levitan, 1992
D. subobscura	Kunze-Muehl et al., 1958
	Pentzos-Duponte, 1964
	Prevosti, 1964
	Burla and Goetz, 1965
	Krimbas and Alevisos, 1973
D. flavopilosa	Brncic, 1972
D. nasuta	Ranganath and Krishnamurthy, 1978
D. melanica	Tonzetich and Ward, 1973
D. willistoni	Hoenigsberg et al., 1977
D. rubida	Mather, 1964
D. funebris	Dubinin and Tiniakov, 1947
	Borisov, 1970a, 1970b

TABLE 3-10 Examples of long-term changes in third chromosome inversions in *D. pseudoobscura*.

Locality	Year	*n*	ST	AR	CH	PP	TL
Okanagan, British Columbia	1940	30	37	47	10	3	3
	1964	80	55	30	1	9	5
	1972	160	54	32	—	1	13
	1981	180	32	37	1	2	28
Mather, CA	1945	308	36	36	17	—	10
	1957	316	45	33	4	10	6
	1963	466	55	22	6	7	10
	1971	390	34	33	12	3	17
	1972	576	22	38	17	6	15
	1975	746	29	33	7	6	19
	1976	222	18	38	14	5	22
Santa Cruz Island, CA	1936	42	55	17	—	—	—
	1940	72	43	18	7	—	—
	1963	400	62	13	10	1	5
	1970	204	41	14	15	1	11
	1980-1981	612	38	6	15	1	7
Chiricahua Mountains, AZ	1940	192	1	89	6	4	1
	1957	400	—	85	12	3	—
	1959	200	—	84	12	4	—
	1964	198	2	89	6	3	1
	1973	262	—	92	5	2	—
	1980	275	—	85	11	3	—

Note: The most common inversions are given in percent. More data on many more localities are in Anderson et al. (1991).

Alternatively, the gene arrangements themselves may be evolving. If sometime during the 1950s a new adaptively superior variant of TL and/or PP arose by mutation, double-crossover, or gene conversion, this event likely occurred in a single population and would have had to spread very rapidly to other distant populations. This scenario was favored by Wright (1978, chapter 4) to explain these shifts as it is consistent with his shifting-balance view of evolution. He points out that when the new TL or PP appeared it would have been rare and primarily in the heterozygous state with alternative gene arrangements. There would be little opportunity for the new complex to recombine with preexisting TL or PP. This view is consistent with the fact that both these arrangements were initially rare which, in this scheme, would enhance the probability that they would survive the initial establishment stage.

D. robusta has also been monitored for long periods in some localities. The most interesting changes recorded have occurred in the Smoky Mountains of Tennessee. Here there has been remarkable stability in autosomal inversion frequencies over a 20-year period, while the X-chromosome arrangements have changed (table 3-11); this is especially clear at the 1,000 meter locality. Recall that the inversion polymorphisms in this species exhibit altitudinal clines. Therefore, the data presented in table 3-11 are grouped for each altitude. One could explain the differences between autosomes and the X as due to the obvious fact that X chromosomes are often subject to selective forces different from autosomes due to their hemizygous state in males; however, there are no relevant data addressing this possibility.

The general stability in regard to inversion frequencies over times on the order of 10s of years is also exhibited by *D. subobscura* (Krimbas, 1992) and *D. persimilis* (Coyne et al., 1992). Populations of *D. melanogaster*, on the other hand, may exhibit more significant shifts as exemplified by some long-term studies in Japan (Inoue et al., 1984), Australia (Knibb, 1986), and Greece (Zacharopoulou and Pelecanos, 1980). This is perhaps not unexpected for a cosmopolitan species that only recently (in evolutionary time) colonized the areas studied; these populations are unlikely to be at equilibrium. All the previously mentioned species exhibiting stability, in general, are wild species studied in their native habitats and thus probably nearer to equilibria.

TABLE 3-11 Long-term changes in *D. robusta* inversion frequencies (in percent) in the Great Smoky Mountains of Tennessee.

Altitude	*n*	XL	XL-1	XR	XR-2	2L	2R	3R
446 m	277/278	43	20	17	83	19	92	25
	51/58	37	24	16	84	26	91	25
666 m	245/301	34	32	26	74	19	89	20
	798/977	28	48	32	68	12	89	25
1000 m	66/102	26	41	29	71	23	85	23
	59/95	14	64	54	46	17	88	31

Note: Upper number at each altitude is from 1947, lower from 1958–1959. The sample size for X chromosomes is the first number under *n* and the second is for autosomes.

From Levitan (1992).

Other Inversions

Pericentric Inversions

Pericentric inversions, those involving the centromere, are much rarer than paracentric inversions, although not unknown in *Drosophila* populations. Ashburner and Lemeunier (1976) found five in their survey of *D. melanogaster* populations and Stalker (1976) found an additional four in the same species. Lemeunier and Aulard (1992) list a total of 10 and 8 pericentric inversions in *D. melanogaster* on chromosomes 3 and 2, respectively.

In *D. ananassae* there are 6 known pericentric inversions in chromosome 2 and 11 on chromosome 3 (Freire-Maia, 1961; Futch, 1966; Tomimura et al., 1993). Pericentric inversions are also known from natural populations of both *D. algonquin* (Miller, 1939) and *D. robusta* (Carson, 1958a; Levitan, 1992).

Little work has been done on the evolutionary dynamics of pericentric inversions, although on the surface one might expect them to be selected against in heterozygotes. Unlike the case of paracentric inversions, there is no known mechanism that segregates unbalanced recombinant chromatids to the nurse cells. In fact, reduced fertility is often the case for females heterozygous for pericentric inversions (e.g., Roberts, 1967). However, the situation, at least in *D. melanogaster*, is more complex. Evidently, depending on the actual breakpoints, pericentric inversion heterozygotes may completely suppress recombination and thus suffer no reduced fertility (Hawley, 1980; Coyne et al., 1993). This appears to be the case for a common, naturally occurring pericentric inversion on two islands in the Indian Ocean (Coyne et al., 1991b).

Sex-Ratio Inversions

Several species of the *obscura* group are polymorphic for unusual X-chromosome inversions that cause segregation distortion of the sex-chromosomes. Males carrying this inverted X produce sperm primarily or exclusively carrying the X; Y-bearing sperm are produced in numbers much below the expected 50%. These inversions are called sex ratio (SR) as the sex-ratio of the offspring is skewed in favor of females. SR chromosomes have been reported from *D. affinis* (Novitski, 1947; Voelker, 1972) and *D. subobscura* (Hauschteck-Jungen and Maurer, 1976). By far the best studied on the population genetics level is that found in *D. pseudoobscura* and *D. persimilis*. This type of sex-ratio condition should not be confused with those in other species that are caused by single genes (e.g., *D. melanogaster*) or infection with a microbial endosymbiont (e.g., *D. willistoni*).

In *D. pseudoobscura* and *D. persimilis* populations, there are four different types of X chromosomes (figure 3-15). In *D. pseudoobscura* the SR X chromosome differs from Standard X (abbreviated ST, but not to be confused with ST for the third chromosome) by three nonoverlapping inversions, the most distal one being a rare telomeric inversion. The SR in *D. persimilis*, remarkably enough, is identical in gene order to the ST of *D. pseudoobscura*, a fact that makes genetic analysis possible. The ST of *D. persimilis* differs from the ST of *D. pseudoobscura* by a single inversion, completely independent of the SR inversions.

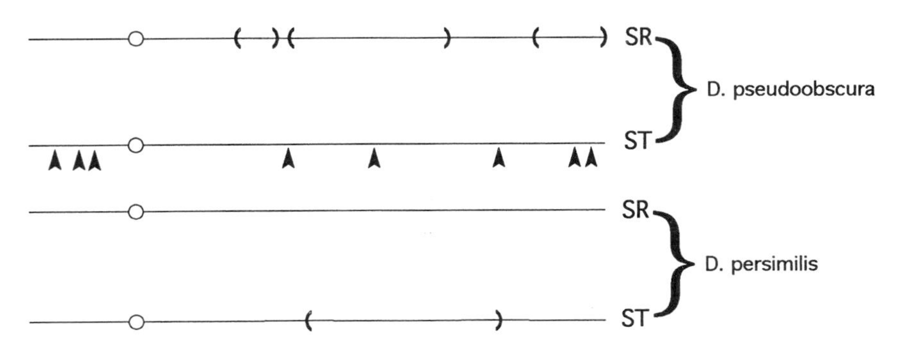

FIGURE 3-15 Schematic diagram of standard (ST) X chromosomes and the sex-ratio (SR) chromosomes of *D. pseudoobscura* and *D. persimilis*. Parentheses indicate regions inverted relative to the ST of *D. pseudoobscura*; triangles indicate positions of genetic markers used by Wu and Beckenbach (1983) in their study of differentiation between STs and SRs (discussed in chapter 4).

The mechanism of SR appears to be degeneration of Y-bearing sperm; SR males produce normal numbers of spermatids but only about half become functional sperm (Policansky and Ellison, 1970). Thus it would appear that SR males should suffer a fertility loss. In the next chapter, we will go into the somewhat complex laboratory studies of SR, which attempt to disentangle the selective forces on the SR system. SR males vary in their production of females, ranging from about 94–99% (Darlington and Dobzhansky, 1942). The few males that are produced are usually due to nullo-X sperm carrying no sex chromosomes, which in *Drosophila* produce sterile phenotypic X/O males. Cobbs (1987) has found an extreme modifier of *D. pseudoobscura* SR, called *msr* (modified sex ratio), which produces about 95% nullo-X sperm and thus many sterile sons.

The geographic distribution of SR in *D. pseudoobscura* is shown in table 3-12. SR is rare or absent in the northwest part of the range. As one goes south, the frequency climbs to nearly 20% at the U.S.-Mexico border, remains high in northern Mexico, and then drops off again going into southern Mexico, Guatemala and Bogota. Curiously, the completely independent SR of *D. persimilis* behaves similarly in being essentially absent in the northern part of the species distribution, and increasing going south (Dobzhansky and Epling, 1944). Such parallel variation of independent systems in two species suggests that selection may be involved in producing the gradients in frequencies, although no plausible hypotheses have been proffered as to just how selection might be doing this.

Dobzhansky (1943) found no evidence for SR seasonal cycles in the San Jacinto experiments with *D. pseudoobscura*. However, in a seminatural population of this species inhabiting an orange grove in southern California, Bryant et al. (1982) did find such cycles. Remarkably, they were able to demonstrate that the frequency of the SR chromosome in this population affected the sex ratio in the population. Figure 3-16 is a summary of their findings. As SR increased and fell in their population, so did the proportion of females collected. This is one of the very rare cases where a

TABLE 3-12 Frequencies in percent of SR X chromosomes in natural populations of *D. pseudoobscura.*

Region	Percent SR
British Columbia	0
Washington State	0
Oregon	0
Idaho, Montana, Wyoming, South Dakota, Nebraska, northern Utah	0
California, Northern Coast Range	9.05 ± 1.88
Northeastern California, Sierra Nevada	6.48 ± 1.26
California, Southern Coast Range	7.00 ± 1.14
Southern California	14.7 ± 0.36
Death Valley region	10.1 ± 0.63
Southern Utah	10.9 ± 3.10
Northern Arizona	16.2 ± 1.97
Southern Arizona	18.6 ± 2.45
Colorado	12.1 ± 2.60
New Mexico	14.4 ± 1.96
Texas	11.5 ± 0.95
Mexico, Nuevo Leon, Sonora	33.3 ± 12.2
Mexico, Chihuahua	6.8 ± 2.9
East-Central Mexico	19.4 ± 3.52
West-Central Mexico	17.2 ± 6.67
Guatemala	1.5 ± 1.5
Bogota, Colombia	0

Note: Localities are listed from north to south.

Modified from Dobzhansky and Epling (1944).

direct populational effect of an inversion has been observed in a natural (or semi-natural) population.

Multiple Inversions

The mechanism by which inversion-heterozygous females avoid aneuploidy and sterility due to recombination (figure 3-1) is that the recombinant products are nonrandomly relegated to nurse cells. However, if more than one arm or especially if more than one independent chromosome is polymorphic, it is possible for females to be simultaneously heterozygous for multiple inversions. Then if recombinants occur in more than one independently segregating chromosome, the problem of relegating all recombinant products to nurse cells becomes much more difficult. Thus one might predict that in species with more than one chromosome polymorphic for inversions some sterility would be associated with multiple heterozygosity. This has been demonstrated for *D. melanogaster* (Cooper et al., 1955), *D. pseudoobscura* (Terzaghi and Knapp, 1960), and *D. paramelanica* (Stalker, 1976). On the other hand, Riles (1965) found no effect of multiple heterozygosity in female *D. robusta*, and she suggested that genetic modifiers of meiosis had evolved in this species to somehow handle the potential difficulties.

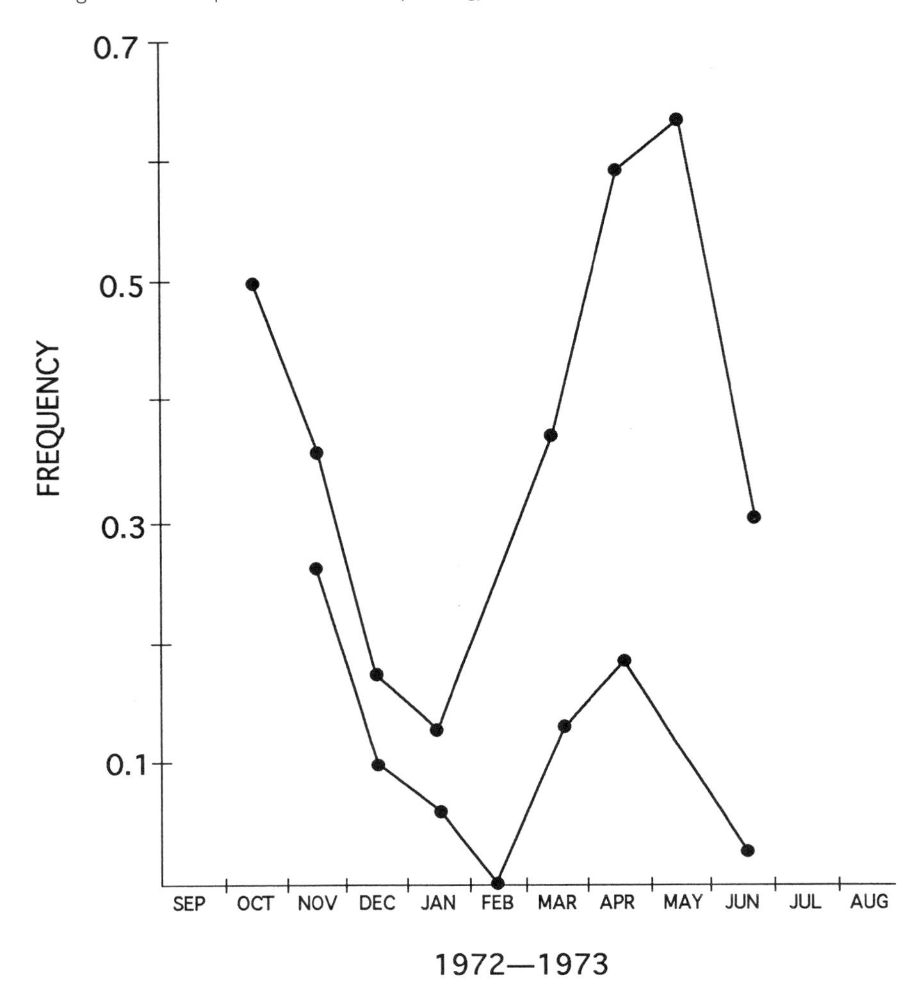

FIGURE 3-16 Frequencies of SR X chromosome (lower line) and sex ratio of collected flies (upper line) in a seminatural population of *D. pseudoobscura* in an orange grove in southern California. From Bryant et al. (1982).

Stalker (1976) studied natural populations of *D. melanogaster* in the southern United States and found a nonrandom distribution of multiple heterozygotes. Females heterozygous for few (zero or one) or many (four or more) were less frequent than expected while females with intermediate numbers of heterozygous chromosomes were more frequent than random expectation. No sterility was observed in females with zero to three heterozygous inversions, while females with all four arms of the major autosomes exhibited varying degrees of sterility, from 0% to 24%. He suggested that there was genetic variation among populations for the ability to avoid aneuploidy during meiosis of multiply heterozygous females.

Allozymes and Inversions

In the numerous studies of allozyme variation in *Drosophila* (table 2-11), authors often attempted to detect linkage disequilibrium between loci, reasoning that if selection was affecting these polymorphisms, then one might expect such disequilibrium at least for some loci. The conclusion from all these studies is that *virtually no linkage disequilibrium exists among allozyme loci* (reviewed in Hedrick et al., 1978; Barker, 1979). This would indicate that any epistatic fitness interactions that might exist are not strong enough to overcome the randomizing effect of recombination. The exceptions to this generalization are allozyme loci located within or very near the breakpoints of inversions. Suppression of recombination extends beyond the breakpoints, presumably because of physical difficulties in pairing due to the contortions associated with loop formation.

Prakash and Lewontin (1968) were the first to note associations between allozymes and the inversion types in which they were located. Table 3-13 is a condensation of their data for two of the four loci located on the third chromosome of *D. pseudoobscura*. There is clearly differentiation among the major phylads of inversions. For both *Pt-10* (coding for a nonspecific protein) and *Amy* (amylase), the ST phylad is characterized by a high frequency of one allele and the SC/TL phylads are characterized by an alternative allele. The associations are not perfect, so some degree of exchange must have occurred between inversions (recall that if monophyletic, each inversion would initially carry a single allele). Despite the evidence for exchange, the nonrandom associations remain intact. Whether this indicates there has not been sufficient time for exchange to randomize associations, or whether in the face of exchange (migration, in a sense), the nonrandom associations are maintained by selective elimi-

TABLE 3-13 Example of nonrandom associations of allozyme alleles with third chromosome gene arrangements in *D. pseudoobscura.*

3rd Chromosome	Allele	MA	SC	MV	AU	CE	BO
Pt-10							
ST Phylad	1.04	1.00	1.00	x	1.00	1.00	—
CH	1.04	0.50	—	—	—	—	—
	1.06	0.50	—	—	—	—	—
SC/TL Phylads	1.04	—	—	—	0.33	—	—
	1.06	1.00	1.00	—	0.66	1.00	1.00
α-Amylase							
ST Phylad	0.84	0.05	0.21	0.21	0.27	0.08	—
	1.00	0.93	0.79	0.79	0.64	0.87	—
CH	0.84	0.36	0.36	—	x	—	—
	1.00	0.64	0.64	—	—	—	—
SC/TL Phylads	0.84	0.90	0.91	—	1.00	1.00	1.00
	1.00	0.10	0.09	—	—	—	—

Note: Average frequencies for phylads are shown; rare alleles are excluded. Localities: MA = Mather, CA; SC = Santa Cruz Island; MV = Mesa Verde, CO; AU = Austin, TX; CE = Cebat, AZ; and BO = Bogota, Colombia. An x means a sample size of one, with that allele.

Modified from Lewontin (1974).

nation of the "wrong" alleles, cannot be determined from these data alone. Interestingly, for both *Pt-10* and *Amy*, CH (a member of the SC phylad [figure 3-3]) seems to hold an intermediate ground, containing reasonably high frequencies of alleles characteristic of all phylads. Arguably, this might indicate CH as ancestral. This contention would be defensible if both SC and ST originated directly from CH. However, this is not the case (figure 3-3); both SC and Hypothetical intervene between CH and ST. Along with two other loci not shown here, *Pt-12* and *Acph-3* (acid phosphatase), all four loci known to map to the third chromosome have been shown to have nonrandom associations. These associations extend beyond the species boundary: *D. persimilis* has either the alleles characteristic of the ST phylad as expected (figure 3-3), or is polymorphic for alleles not found in *D. pseudoobscura*. Lewontin (1974) used the fact that the allozyme-inversion associations predate the split between species to conclude these associations are very old, and thus there has been sufficient time to randomize the alleles between inversions if only chance historical factors were involved.

As expected, given the large differences in phylad frequencies among populations, loci nonrandomly distributed among phylads also show geographic variation in allele frequencies. These are the exceptions to the rule of allozyme geographic homogeneity in *Drosophila*.

Nonrandom allozyme association also occurs with the SR X-chromosome polymorphism (Prakash and Merritt, 1972; Keith, 1983). Initially, using standard electrophoretic techniques, Prakash and Merritt detected strong frequency differences between ST and SR chromosomes for *Est-5*, an X-linked gene; the pattern was similar to that shown for the third chromosome. While there were significant frequency differences, ST and SR also shared some alleles, which suggests some exchange. However, Keith (1983) employed more sensitive allozyme technology, that of sequentially varying electrophoretic conditions. She found that ST and SR did not share any alleles at the *Est-5* locus; the apparent sharing of alleles in the earlier study was due to insensitivity of discriminating electrophoretic alleles. Furthermore, compared to ST, SR was very low in diversity, having only three alleles in a sample of 32 genes (table 3-14). This is in sharp contrast to the case with ST, illustrated in table 2-13. The most likely explanation is that SR is much younger than ST and thus has not had sufficient time

TABLE 3-14 Allele frequencies at *Est-5* on the SR X chromosome of *D. pseudoobscura*.

Allele	No. of Lines
0.97/1.00/1.00/1.00/1.00	7
1.04/1.00/1.00/1.00/1.00	23
1.19/1.00/1.00/1.00/1.00	2
Total	3 alleles 32 genes sampled

Note: The techniques used were the same as for ST X chromosomes shown in table 2-12; none of the three alleles here are the same as any in that table. Allele designations are relative electrophoretic mobility under different conditions.

From Keith (1983).

to accumulate as many alleles, but the few it has accumulated are different from those on ST; alternatively, the smaller population size of SR chromosomes could cause lower allelic diversity. This observation also has relevance to the issue of the patterns of allozyme variation found in island and continental populations of *D. willistoni*, discussed previously. If the *D. willistoni* island populations had been depauperate in allozyme variation due to founder effects, the observations of Keith indicate it is unlikely that mutation would regenerate the identical alleles found in continental populations.

Inversion-allozyme associations have been well-studied in *D. subobscura* with some intriguing findings. Recall that this species has inversions in all arms of all chromosomes, so there is much greater opportunity for single genes to be in disequilibrium with inversions. The patterns of association are complex (reviewed in Krimbas, 1992; Krimbas and Loukas, 1980). Unlike *D. pseudoobscura*, there are instances in *D. subobscura* where some loci within inversions are in linkage disequilibrium, while other loci in the same inversion are not. One explanation might be the *middle gene hypothesis* (Krimbas and Loukas, 1980), which states that loci located nearer the center of an inversion will more often exchange with other gene arrangements due to double crossovers. However, this is not the case, as those loci showing no disequilibrium are in positions no more or less likely to recombine.

A second intriguing observation for *D. subobscura* is that the degree of disequilibrium between allozymes and inversions cycles with the season (Fontdevila et al., 1983). At some times the association is strong, and at other times weaker, but the changes are repeated each year (figure 3-17). The only explanation for this is that the O_{st} arrangement is heterogeneous in this population, the variants being different in

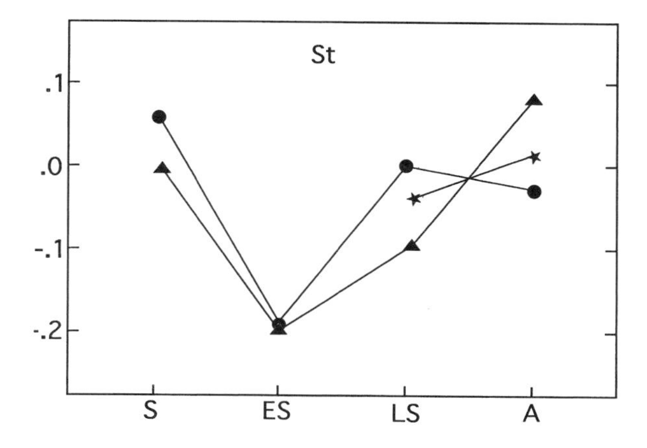

FIGURE 3-17 Seasonal changes in gametic disequilibrium between two allozyme loci (*Lap* and *Pept-1*) in a natural population of *D. subobscura*. The loci are on the O autosome. Seasons on the y axis: S = spring, ES = early summer, LS = late summer, and A = autumn. Triangles, circles, and stars indicate different years. From Fontdevila et al. (1983).

frequencies of alleles at the two loci studied. Evidently the different O_{st} variants vary seasonally in fitness.

D. melanogaster, too, has been studied with regard to linkage disequilibrium between allozymes and inversions; Lemeunier and Aulard (1992) list studies on 62 populations from around the world. By far the best-studied case is that for *αGpdh, Mdh*, and *Adh* on the 2L (figure 3-18). *αGpdh* and *Mdh* lie within the common inversion 2Lt, while *Adh* is located just outside it. The correlation between alleles at these loci and 2Lt varies from population to population, from nearly 0 to 0.6. As already mentioned, clines exist for these inversions in both the northern and southern hemispheres. Not surprisingly, this often causes clines to be found for these loci. Voelker et al. (1978), however, have argued quite convincingly that the cline in *Adh* allozymes cannot be explained solely by their association with inversions. Van Delden and Kamping (1989) have studied this system in a seminatural population in a tropical greenhouse. They too conclude the selection of the allozymes, *Adh* and *αGpdh*, could not be explained solely on the basis of association with the inversion, but are responding to temperature selection consistent with the latitudinal clines.

DNA Studies

The few studies of inversions on the DNA level conducted to date have been highly informative. The detail provided by DNA sequences far exceeds that of allozymes, so we can anticipate greater insights. One such study has provided just that. Aquadro et al. (1991) studied a 26-kilobase section of DNA containing the *Amy* locus. This locus is found within the breakpoints of most naturally occurring inversions of *D. pseudoobscura* and *D. persimilis* and was previously known to have electrophoretically detectable alleles associated with inversions (table 3-13). Aquadro et al. used restriction enzymes to detect differentiation among gene arrangements collected at various locali-

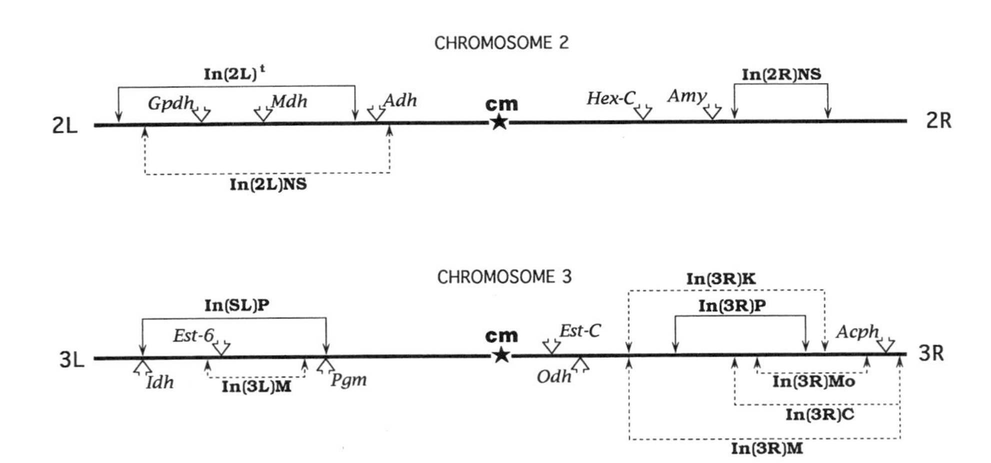

FIGURE 3-18 Schematic maps of *D. melanogaster* second and third chromosomes. Location breakpoints are marked by solid arrows and the position of several allozyme loci indicated by open arrows. From Lemeunier and Aulard (1992).

ties and constructed a phylogenetic tree based on distances (or similarities) that relate the different chromosomes studied. Figure 3-19 summarizes the result. The significant finding was that the same chromosome arrangements (inversions) always clustered with one another regardless of geographic origin. For example, the TL chromosomes occur at the very extremes of the distribution (British Columbia to Bogota, Colombia) and yet they are more similar to one another than to any other gene arrangement, even those from their same populations. Secondly, the topology of the tree coincides remarkably well with that deduced from the inversions alone (figure 3-3). More recent DNA sequence data from this same region has confirmed the restriction site phylogenetic inferences (Popadic and Anderson, 1994; discussed in more detail in chapter 10). *This is the strongest independent confirmation of the monophyly of inversions in these species.* It is difficult to conceive of any other explanation for these data.

Linkage disequilibrium was also detected among variable restriction sites. Out of 153 pair-wise comparisons, 60 (39.2%) were statistically significant at the 0.05 level. Since only 28 chromosomes were studied, the sample size limits the ability to detect

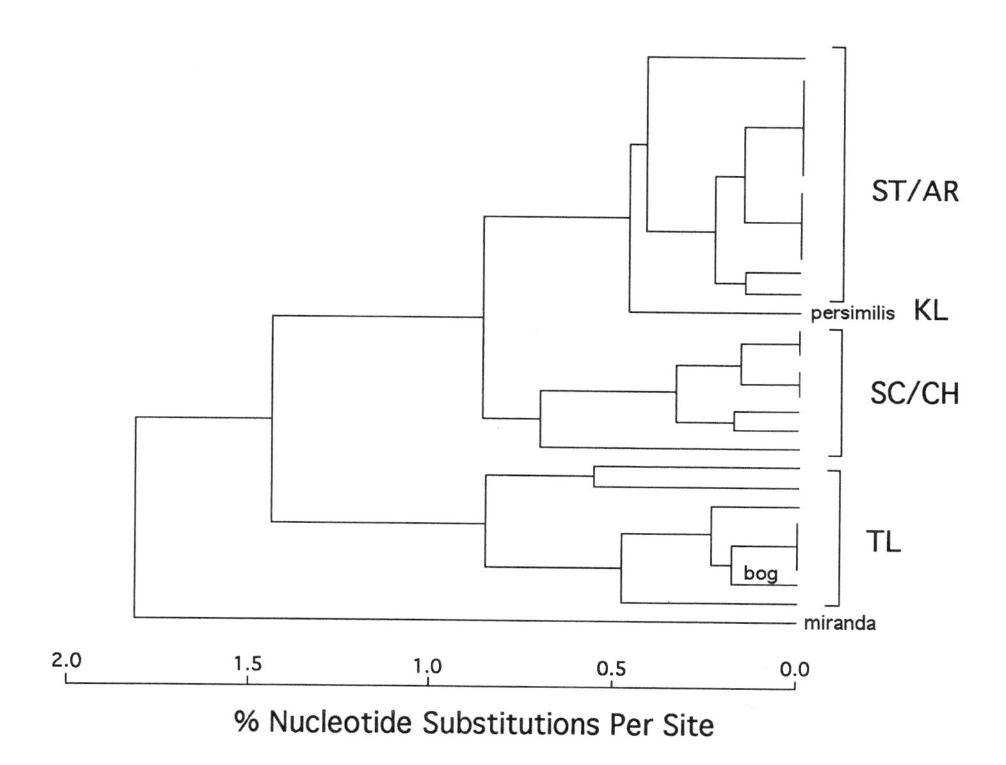

FIGURE 3-19 UPGMA dendrogram of restriction-site variation in and around the amylase locus (*Amy*) of *D. pseudoobscura*. One chromosome each from *D. persimilis* and *D. miranda* was also studied. This locus lies within the breakpoints of inversions of the third chromosome; third chromosome gene arrangements from which the alleles came are indicated on the right. Note that the topology of the relationships is concordant with that deduced from the inversions alone (figure 3-3). Modified from Aquadro et al. (1991).

significant associations, yet associations were found. This contrasts with what Schaeffer et al. (1987) found in a study of a comparably-sized segment of freely recombining DNA (i.e., not associated with inversions) around the *Adh* locus for the same species. This is strong evidence that the recombination suppression effect of inversions has a lasting effect over evolutionary time and is very effective in these species.

These findings with *D. pseudoobscura* contrast with similar studies by Aguadé on *D. melanogaster* and *D. subobscura.* Her first study on *Adh* in *D. melanogaster* yielded no evidence of DNA-level associations with the 2Lt inversion (Aguadé, 1988). While not within the inversion, it is closely linked (figure 3-18) and the allozymes do show disequilibrium. In considering the allozyme difference at *Adh* known as the *F* and *S* alleles, *t-Adh*^{S} chromosomes (the *t* inversion with the slow *Adh* allele) showed greater variation than did *St-Adh*^{F} chromosomes (the standard chromosome with the fast *Adh* allele), leading to the speculation that the inversion polymorphism predated the allozyme polymorphism.

Rozas and Aguadé (1990) studied a ribosomal protein coding locus called *rp49* that lies near the breakpoints of inversions on the O chromosome of *D. subobscura.* They detected several shared haplotypes among the inversions and thus no support for monophyly. Further, in contrast to the *D. pseudoobscura* case, they found no significant disequilibrium over a much smaller distance, 1.6 kb. However, a more detailed DNA sequence study of these inversions in this species did support monophyly (Rozas and Aguadé, 1993, 1994). Evidently the resolution afforded by RFLP analysis was not high enough to accurately deduce the history of the inversions. Also, there is evidence that gene conversion may be occurring, so some degree of exchange is also obscuring matters.

The SR inversion in *D. pseudoobscura* has also been examined for DNA variation in a region within the breakpoints of one of the inversions (Babcock and Anderson, 1996). The results are consistent with the allozyme studies in that the ST chromosome has much more variation than does the SR. The divergence between ST and SR within *D. pseudoobscura* is greater than that between X chromosomes of *D. pseudoobscura* and *D. persimilis*, indicating the origin of SR predates the speciation event. The close clustering of all copies of SR is consistent with their monophyletic origin.

More discussion of DNA studies on inversions will be presented in the context of phylogenies in chapter 8 and in the context of molecular evolution in chapter 10.

Theory of the Origin of Inversions

Now that the reader has a feel for the nature of inversions in *Drosophila*, some discussion of theoretical studies of the origin of inversions is in order. Unlike many chromosomal rearrangements, we have already pointed out that in *Drosophila* heterozygotes, paracentric inversions are not at a selective disadvantage so that the establishment of a new inversion is not so difficult as in other organisms. The question of origin of inversions can be stated as a question of the evolution of recombination modifiers. Under what conditions will a gene (inversion) that reduces the recombination between two or more loci be favored in populations? Several studies have been done assuming an additive model in which each locus acts independently (e.g., Ohta and Kojima, 1968; Fraser et al., 1966; Nei et al., 1967). The common picture emerging from these studies is that the conditions for establishment of an inversion are very restrictive and it would be a very rare event if the loci were strictly additive.

Models assuming nonadditive interactions of loci have also been studied (e.g., Nei, 1967; Feldman, 1972; Feldman et al., 1980). An important result of these studies is that an otherwise neutral recombination modifier would only increase in the populations if disequilibrium was already present between the two interacting loci. That is, the modifier could not cause the disequilibrium, but if present, disequilibrium would enhance the probability of establishment of the modifier.

Charlesworth and Charlesworth (1973) studied a model of interacting loci such that heterosis acts cumulatively; that is, heterozygosity at each locus increases fitness more than it would if the other loci were homozygous. They extended the model to include up to five loci. Selection will increase inversions in such a model again only if disequilibrium is already present and the inversion occurs in chromosomes carrying the excess gametic type. Furthermore, the inversion needs to be introduced at a reasonably high frequency (≥0.05) to be maintained; this in itself would cause disequilibrium. Small population sizes would enhance this process. Charlesworth (1978) and Deaken and Teague (1974) have extended these models to examine the stability and nature of the predicted equilibria. Many variations on the theme were explored with the general conclusion that multiple equilibria are often possible, some of which are unstable.

The above studies all deal with the establishment of a polymorphism. Other studies have been concerned with the fixation of a new inversion, something we will see is quite common in *Drosophila* (chapter 8). Hedrick (1981) considered several factors and concluded that drift alone is very unlikely to fix a new chromosome; factors such as inbreeding or meiotic drive are necessary. Bengtsson and Bodmer (1976) examined other factors and found it very difficult to dissect out the relative importance of each, except to conclude that only rarely would one expect fixation. Meiotic drive has been invoked in many models, but its relevancy to *Drosophila* would seem to be remote, except in obvious cases like SR. There is no evidence for meiotic drive in general for *Drosophila* inversions. The most relevant negative observation is that hybrids between species fixed for alternative inversions do not show distorted segregation of the inversions.

While the models are often complex and the nature of the problem (multilocus with recombination modifiers) is very difficult to handle mathematically, there would seem to be one conclusion from the theoretical work. Establishment of an inversion, either in a polymorphic or fixed state, requires unusual circumstances that are rarely met in natural populations. Thus, these studies are consistent with the conclusion that the vast majority of newly arising inversions do not become established. The present day observed inversions are a small subset of those that have occurred by mutation.

One theoretical issue on which there is some data is the question of the length of successful inversions. Obviously a newly arisen inversion will have a much better chance (perhaps its only chance) if it contains interacting alleles enhancing its fitness, especially in the heterozygous state. Until it reaches an appreciable frequency, it will only be in heterozygotes. The larger the inversion, the more likely it is to capture more favorable alleles. On the other hand, long inversions will not suppress recombination as effectively as shorter inversions, due to a higher probability of double crossovers. There should be a tradeoff in terms of length such that inversions of intermedi-

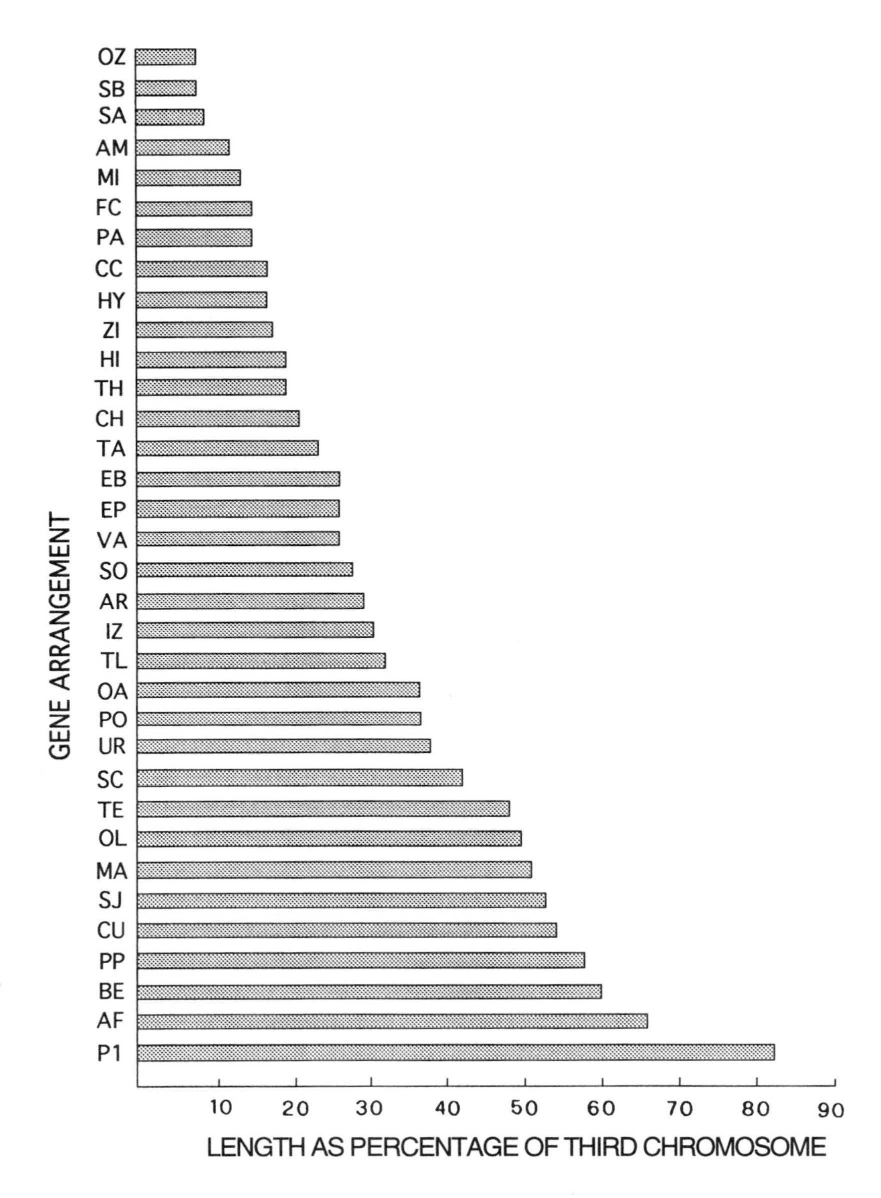

FIGURE 3-20 The lengths of naturally occurring inversions of the third chromosome of *D. pseudoobscura*. Lengths are relative to the gene arrangement from which they arose, as indicated in figure 3-3. Common and widespread arrangements occur between CH and PP, while all outside this region are rare endemics (or nonexistent, in the case of HY). From Olvera et al. (1979).

ate length are favored. In general, that is what is observed. We can use the example of *D. pseudoobscura* illustrated in figure 3-20 (Olvera et al., 1979). The known inversions are placed in increasing order of size relative to the inversion giving rise to it. By and large the most successful inversions are those of intermediate length; we take as a measure of success how widespread the inversions are and their frequency where they are found. The 12 shortest in the figure are all rare endemics, one of which (Hypothetical, HY) has never been recovered from natural populations. Likewise, the three longest inversions are all rare endemics. The successful inversions are those involving 20 to 60% of the chromosome. Most other species conform to this pattern.

Further Information

In the next chapter we take up experiments done in the laboratory that have helped clarify the nature of the selection acting on inversions. Readers interested in a fuller picture of inversion polymorphisms should consult the relevant sections in that chapter. In this chapter we have confined ourselves to describing the patterns of inversion polymorphism found in natural populations. For the fullest single treatment of *Drosophila* inversion polymorphisms, both in natural populations and laboratory studies, the volume *Drosophila Inversion Polymorphism* (Krimbas and Powell, 1992a) should be consulted.

4

Population Genetics—Laboratory Studies

The skeptical ecologist: "But isn't the laboratory environment very different from what the flies experience in nature?"

The cocky geneticist: "Of course it is. So what?"

Experimental populations in a laboratory have several advantages for the empirical population geneticist. One really knows exactly what the population is and can manipulate it more or less at will. L'Heritier and Teissier (1933) were the first to develop a population cage for continuous breeding of *Drosophila* populations. Their cages were boxes 50 cm × 30 cm × 15 cm with 21 holes for food cups that would be replaced periodically, usually one per day. Such cages maintain adult populations of three to six thousand flies with overlapping generations. The larval breeding cups are extremely crowded, with only one or two percent of eggs ever reaching adulthood, making selection quite intense. Many variations on this original design have evolved over the years, but most are similar enough in the essentials that we don't generally need to worry about the details. The most important points are that populations are large enough to minimize the effect of genetic drift and that competition, especially at the larval stage, is intense. Some of the laboratory studies discussed in this chapter were more conveniently conducted in standard rearing bottles (usually about one-half pint in size), but even in these cases selection is very strong; a female can lay several hundred eggs in a lifetime, yet in a stable population, only two on average must survive. Mueller (1985) has discussed the ecology of laboratory populations of *Drosophila* and that reference can be consulted for more details. Here we will emphasize population genetic studies.

Genes

Morphological Mutants

The first studies done with laboratory populations monitored the frequencies of morphological mutations, usually in competition with wild-type alleles. As just one example of a more or less typical outcome, we can consider figure 4-1. *Bar* is a drastic

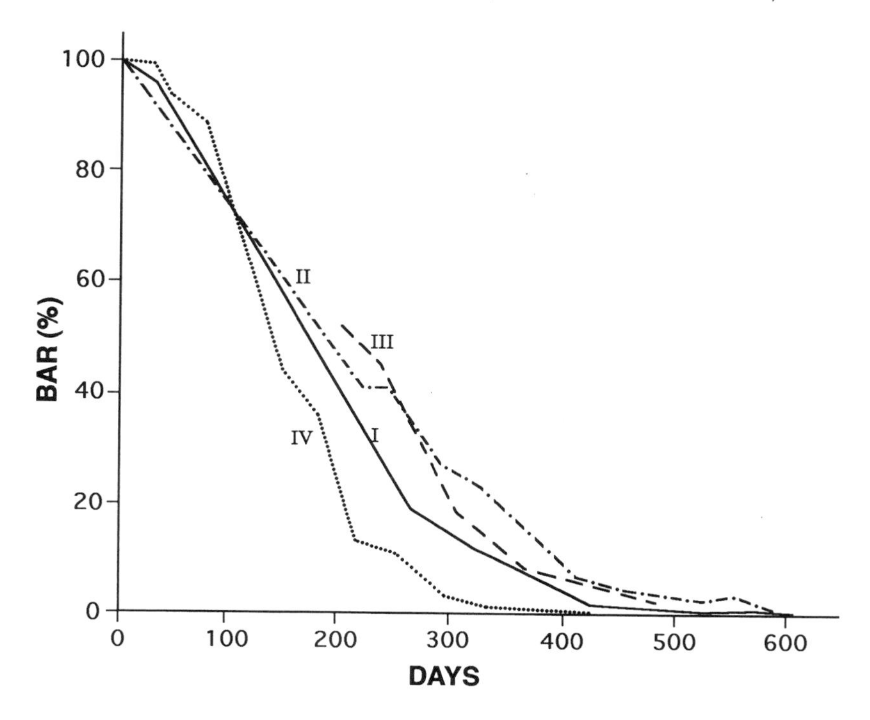

FIGURE 4-1 Course of natural selection in the laboratory on the *Bar eye* mutant of *D. melanogaster*. The graph includes results from four studies: I and II = L'Heritier and Tessier (1937), III = Petit (1951), and IV = Merrel (1965). Redrawn from Wright (1977).

mutation of *Drosophila*, which is due to a rather large, cytologically detectable, deletion. Not surprisingly, in competition with wild type, this mutation is quickly eliminated from the populations. The repeatability of selection against *Bar* in four studies in three different laboratories is quite remarkable. Being a dominant mutation, of course, makes selection much more efficient. Table 4-1 is a listing of several similar studies on a wide range of mutations. The general outcome of these experiments with visible mutations is elimination of the mutant, or at least selection to very low frequency by the time the experiment is terminated. Thus, in keeping with their very low frequencies in natural populations (chapter 2), we can conclude that visible mutations generally confer a lower fitness on their carriers compared to wild type.

However, some exceptions to this rule have been noted. Sometimes visible mutants seem to reach an equilibrium at reasonably high frequency, indicating some kind of balancing selection, such as heterosis. Examples include *ebony* (L'Heritier and Teissier, 1937), *brown* (Nozawa, 1958; Rasmussen, 1958), *Stuble* (Wiedemann, 1936; Helman, 1949; Frydenberg, 1962, 1964), and *spineless* and *rough* (Susman and Carson, 1958). However, in cases where these apparent equilibria have been more carefully studied by either running the populations longer or using other genetic backgrounds, the outcome is almost inevitably elimination (or near elimination) of the

TABLE 4-1 Examples of visible mutants tested for fitness in population cages.

Gene	Inheritance	Reference
Bar (*B*)	X-linked, dominant	L'Heritier & Teissier, 1934
		Wiedemann, 1936
		Petit, 1951
		Merrell, 1965
Beadex (*Bx*)	X-linked, semidominant	Wiedemann, 1936
		Merrell, 1965
white (*w*)	X-linked, recessive	Wiedemann, 1936
		Diederich, 1941
		Reed & Reed, 1950
		Merrell & Underhill, 1956
		Petit, 1958
		Thomson, 1961
yellow (*y*)	X-linked, recessive	Diederich, 1941
		Merrell, 1949
		Barker, 1962
Stubble (*Sb*)	autosomal, dominant	Wiedemann, 1936
		Helman, 1949
		Frydenberg, 1962, 1964
Dichaet (*D*)	autosomal, dominant	Wiedemann, 1936
Curly (*Cy*)	autosomal, dominant	Teissier, 1942
Lobe (*L*)	autosomal, dominant	Merrell, 1965
ebony (*e*)	autosomal, recessive	L'Heritier & Teissier, 1937a
		Rasmussen, 1958
vestigial (*vg*)	autosomal, recessive	L'Heritier et al., 1937
		Wiedemann, 1936
		Merrell & Underhill, 1956
jaunty (*j*)	autosomal, recessive	Wiedemann, 1936
dumpy (*dy*)	autosomal, recessive	Wiedemann, 1936
sepia (*se*)	autosomal, recessive	Wiedemann, 1936
		Rasmussen, 1958
		Susman & Carson, 1958
glass (*gl*)	autosomal, recessive	Merrell & Underhill, 1956
cut (*ct*)	autosomal, recessive	Merrell, 1949
forked (*f*)		Merrell, 1949
raspberry (*ras*)		Merrell, 1949
brown (*bw*)	autosomal, recessive	Wiedemann, 1936
		Nozawa, 1958
		Rasmussen, 1958
		Susman & Carson, 1958
spineless (*ss*)	autosomal, recessive	Susman & Carson, 1958

mutant. For example, Teissier's (1947) follow-up experiments were run longer than the original studies and *ebony* frequencies fell further in several replicate populations. Smathers' (1961) follow-up experiments indicated that changing the genetic background had a very large effect on the original observations of Susman and Carson (1958). These kinds of studies are consistent with the notion that when an apparent stable equilibrium is reached with a visible mutation, it is due to linkage disequilibrium with some other genetic unit that is causing the apparent equilibrium. Either continuing experiments long enough for recombination to destroy the disequilibrium or changing the background leads to elimination of the mutant.

One variation on this theme is worth special mention. L'Heritier et al. (1937) demonstrated how changing the environment can affect the outcome of these experiments. They placed an open population cage outdoors and collected eggs every day. They then determined the frequency of vestigial wing (*vg*) in the egg samples; this recessive autosomal mutation greatly reduces the wings, rendering homozygous flies flightless. The results are in figure 4-2. The first arc on the left shows the percent of *vg/vg* eggs laid over the first few days; evidently the wind had blown away most of the wild-type flies, so those remaining were primarily *vg/vg*. The second arc shows the percentage of adults emerging from the food cups. Finally, the downward arc shows the outcome when the cage was closed and brought back into the laboratory. L'Heritier et al. (1937) associated these results with Darwin's observations that on oceanic islands endemic insects tended to be wingless, presumably due to the danger of being blown off the islands.

Allozymes

Allozymes are another type of single-gene polymorphism. Unlike the generally rare visibles, allozyme alleles are in high frequencies in natural populations. These variants

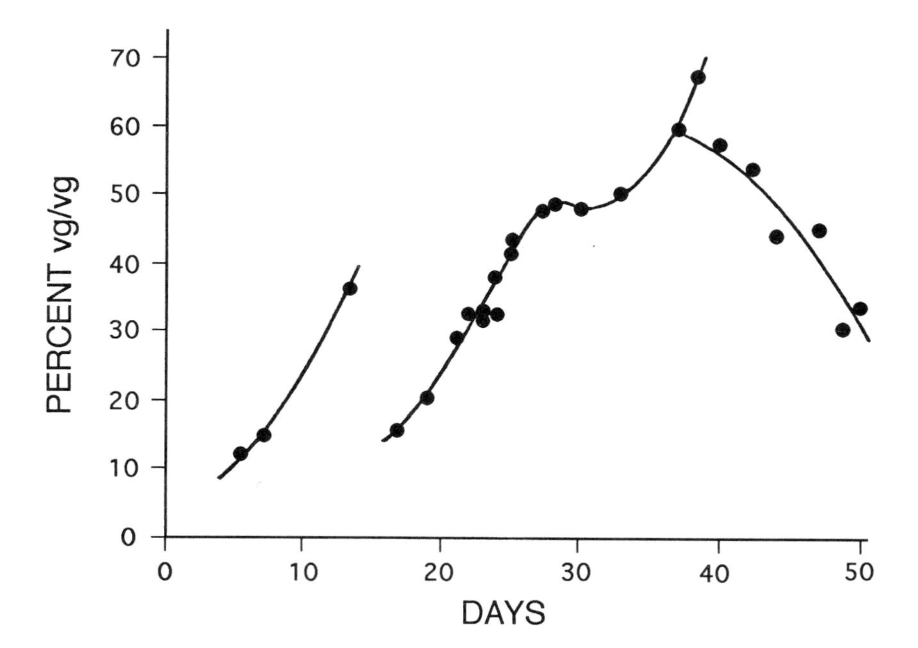

FIGURE 4-2 Selection on *vestigial wing* in an open population left outdoors. The leftmost arc is the number of eggs laid in the first few days. The second curve represents the adults emerging. The rightmost arc shows the outcome when the cage was closed and brought back into the laboratory. From L'Heritier, Neefs, and Teissier (1937), redrawn from Wright (1977).

may be better candidates for being maintained by some sort of balancing selection, rather than mutation–selection balance. Furthermore, considering the controversy over the selective nature of such variants (the selectionist/neutralist controversy) and given the success of demonstrating selection on inversions in laboratory populations (see next section), it was quite reasonable to ascertain whether similar methods might clarify the selective nature of allozymes. The results from several studies have been ambiguous at best (with one exception, alcohol dehydrogenase, *Adh*), and examples will be mentioned briefly. Discussion of *Adh* will be postponed until chapter 10, where this locus will be featured prominently.

Among the first such studies were those of MacIntyre and Wright (1966), Berger (1971), Yamazaki (1971), and Powell (1973) employing four different species, *D. melanogaster, D. simulans, D. pseudoobscura*, and *D. willistoni*. The interpretation of the outcome of any such experiments is greatly complicated by linkage effects, similar to those seen in some visible mutant studies. In allozyme studies, the problem is more serious because selection (if it occurs) is probably much weaker than found for most visibles, and thus linkage or "hitchhiking" is more likely to outweigh selection directly on the locus. Both Berger (1971) and Powell (1973) observed repeatable nonrandom shifts in allozyme frequencies, while Yamazaki (1971) did not. The two former studies had complications of linkage: Berger's study began with relatively few independent chromosomes, and Powell's study involved populations with many inversions segregating on all chromosomes, although inversion frequencies were monitored and shown not to be directly implicated in the allozyme frequency shifts. MacIntyre and Wright observed stable equilibria, but also noted a considerable effect on changing the genetic background.

The importance of linkage was unequivocally illustrated by Powell and Richmond (1974) in their study of an X-linked gene (*To*, later called *Sod*, superoxide dismutase) of *D. willistoni.* In a natural population, this locus displayed an excess of heterozygotes over that expected by Hardy-Weinberg equilibrium, an observation interpreted as evidence of heterosis (Richmond and Powell, 1970). In the laboratory, Powell and Richmond studied two types of populations: those initiated with over 100 independent chromosomes for each of two alleles from a single natural population, and those begun with only four chromosomes for each allele from the same population. Figure 4-3 illustrates the results. When introduced on a more or less randomized background represented by many independent chromosomes, the allele frequencies changed little over nearly three years. On the other hand, populations begun with the alleles on a few chromosomes produced changes that might be interpreted as evidence for selection: despite being begun at different frequencies, the same equilibrium frequency was approached, although erratically. Given the other cages' results, this surely must be due to linkage. Jones and Yamazaki (1974) performed very similar studies with virtually identical results.

Kojima and colleagues took a somewhat different approach in looking for evidence of frequency-dependent selection at allozyme loci (Kojima, 1971, and references therein). The detection of frequency dependence is often complicated by various factors (e.g., Prout, 1965). One study by Kojima and colleagues deserves mention. They studied fitness effects at an esterase locus, *Est-6*, in *D. melanogaster* and designed their experiments to circumvent previous problems associated with detecting frequency dependence. They reared larvae of different genotypes on medium and then

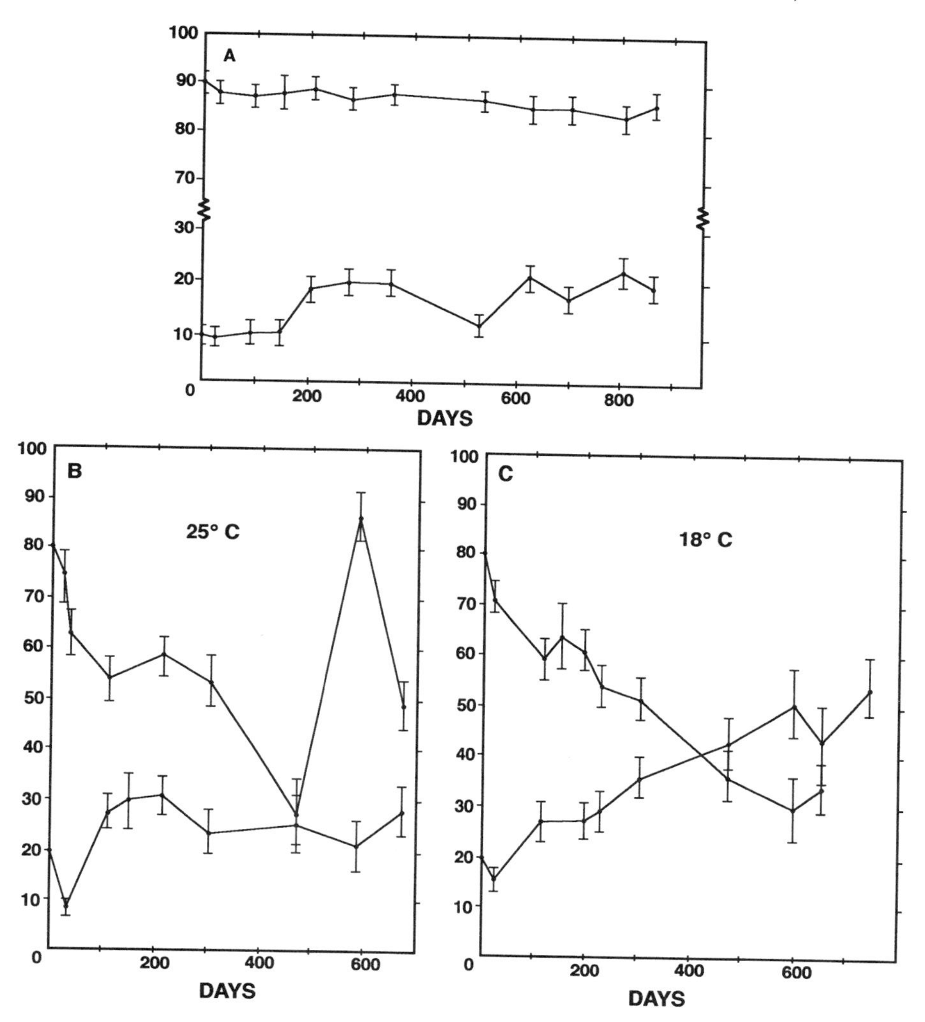

FIGURE 4-3 Dynamics of change at an allozyme locus (*To* or *Sod*) in laboratory populations of *D. willistoni*. The upper graph is for cages initiated with several hundred isofemale lines, while the lower two graphs are for cages begun with a few chromosomes from the same population. Bars indicate ±1 S.E. (standard error). From Powell and Richmond (1974).

killed them by freezing to produce medium preconditioned by certain genotypes. Table 4-2 presents the results. On medium where FF genotype had previously been, the SS genotype had a selective advantage over FF; and *vice versa*. Heterozygotes exhibited intermediate viabilities. Such studies are certainly consistent with frequency dependence, although the chemical details of the experiment have never been clarified, nor have enough similar studies been done to determine the ubiquity of such a phenomenon. Yamazaki (1971) reported negative results with another species.

TABLE 4-2 Evidence for frequency-dependent selection at an esterase locus in *D. melanogaster*.

Tested Genotype	Preconditioning Genotype		
	FF	*FS*	*SS*
FF	0.923 ± 0.053	1.068 ± 0.056	1.130 ± 0.058
FS	1.090 ± 0.057	1.000 ± 0.057	1.087 ± 0.057
SS	1.146 ± 0.059	1.078 ± 0.057	0.928 ± 0.053

Note: The viabilities of genotypes *FF*, *FS*, and *SS* are shown when placed on medium "preconditioned" by activity of larvae of the noted genotype; the standard *FS* on *FS* is set to one.

From Huang, Singh, and Kojima (1971).

A series of laboratory studies of multilocus behavior of both visible mutants and allozymes conducted by M. T. Clegg and colleagues merits special attention. While most laboratory selection experiments attempt to minimize the effects of linkage, Clegg turned the problem on its head and began populations with the *maximum* possible linkage disequilibrium. In all experiments, linked genes on the third chromosome of *D. melanogaster* with known recombination relationships were studied in discrete-generation population cages. One study (Clegg et al., 1976) followed the dynamics of a recessive lethal visible (*Glued*) and two allozyme loci (*Pgm* and *Est-C*; see figure 3-18); another followed the same enzyme loci and the lethal gene *Stubble* (Clegg et al., 1978). Both experiments resulted in the expected loss of the lethal gene with the predicted dynamics. The linked allozyme alleles initially followed the decrease predicted by hitchhiking (although not as rapid as linkage relationships predicted), but then turned around and rose back to intermediate frequencies. Clegg and colleagues concluded that the allozyme markers were likely closely linked to loci being maintained by some form of balancing selection. When recombination allowed these allozyme loci to escape domination by linkage to the deleterious visible mutant, their dynamics became dominated by the linked loci maintained at intermediate frequencies. Of course, one cannot exclude the possibility that the allozyme loci themselves are the target of selection, perhaps in epistatic interaction with linked loci.

Another set of experiments (Clegg et al., 1978) dealt only with the allozyme loci *Pgm, Idh, Est-c*, and *Est-6* (linked as shown in figure 3-18). Figure 4-4 shows an example of the observed and expected decay of the linkage disequilibrium; other pairs of loci showed very similar patterns. In all cases the decay was more rapid than strictly neutral theory would predict given known linkage relationships. Monte Carlo simulations (Clegg, 1978) and other analyses (Asmussen and Clegg, 1982) demonstrated that the results were consistent with the expectation of selection acting on the loci; that is, selection coupled with the known recombination rate could accelerate decay of disequilibrium over that expected from strict neutrality. Furthermore, Clegg and colleagues observed some unexpected behavior of the F statistic measuring the degree of heterozygote excess: $F = 1 - \mathrm{h}/[2p(1-p)]$, where h is the observed heterozygosity and p is the gene frequency. Figure 4-5 shows some results. Early in the experiments (when disequilibrium was maximum), heterozygote excess (negative F)

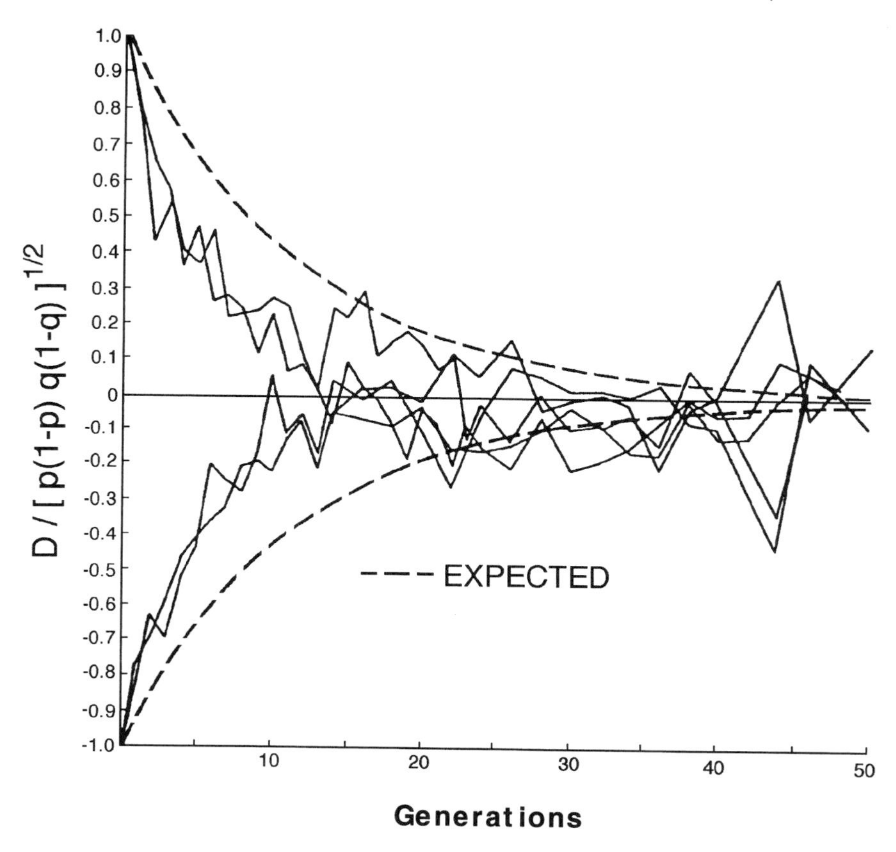

FIGURE 4-4 The decay of linkage disequilibrium between two allozyme loci (*Idh* and *Pgm*) in a laboratory population of *D. melanogaster*. The x axis is time in *t* generations. The expected curve was generated assuming complete neutrality. The statistic used is the correlation in order to remove some of the effect of gene-frequency change. From Clegg (1984).

was common and large. As expected, this decays toward zero as recombination randomizes associations, although much more slowly than expected, and in the case of *Est-c, F* remains negative throughout. Clegg emphasizes the heterogeneity of behavior of the loci, which probably indicates variation along the chromosome in the density of heterotic loci. (Clegg [1984] is a good summary of these studies with further general implications.)

The most important conclusion from this set of studies is the evidence for a *high density of selected loci on the chromosome.* In a sense, the loci studied by Clegg et al. are a random set of points along the chromosome chosen simply because they had polymorphic allozymes. The implication is that one can pick almost any point on any chromosome and find tightly linked selected loci. It is difficult to determine just how many loci are needed for the effect, but it certainly must be many. We will have reason to return to this point later in the book.

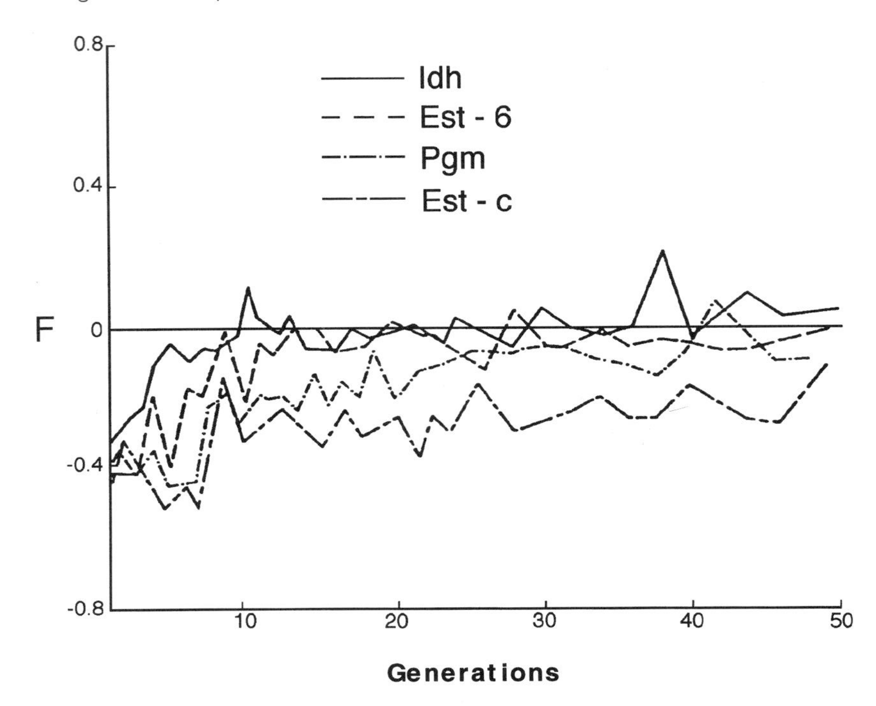

FIGURE 4-5 Plot of the fixation index of Wright (F) for several allozyme loci in laboratory populations of *D. melanogaster*. The lines are averages over replicate experiments. Under neutrality one expects F to be 0. From Clegg (1984).

Inversions

Evidence for Selection

Considering the clear evidence for some kind of strong balancing selection maintaining inversion polymorphisms in natural populations, it is not surprising that inversions have served as model systems for studying the nature of selection in experimental laboratory population cages. By far the most elaborate and complete studies have used *D. pseudoobscura* as the model species and most of the work to be discussed here will be with that species. However, inversions have been studied in laboratory populations of many other species. Table 4-3 is a guide to the literature for these species.

Drosophila pseudoobscura has served as a model for inversion studies in the laboratory for several reasons. First, this species is readily cultured and population cages are easily maintained. Second, it is usually polymorphic in only one chromosome arm, so one can focus on a single Mendelian system without complications of interaction with inversions in other arms. Third, much is known about the system in nature (see chapter 3). Finally, it was the favorite species of Th. Dobzhansky, the most energetic worker in the field.

The first report of inversion selection in laboratory populations of *Drosophila* is that of Wright and Dobzhansky (1946). It is interesting to read this original report as

TABLE 4-3 Species other than *D. pseudoobscura* that have been studied in laboratory populations for changes in inversion frequencies.

Species	Reference
D. persimilis	Spiess, 1966
D. robusta	Carson, 1958a, 1961b
	Krimbas and Loukas, 1980
D. subobscura	Krimbas, 1992
	Moriwaki and Tobari, 1975
D. ananassae	Tobari, 1993b
D. funebris	Filatova, 1973
D. pavanii	Benado and Brncic, 1970
D. melanogaster	Lemeunier and Alaurd, 1992
D. willistoni	Dobzhansky and Pavlovsky, 1953
D. paulistorum	Dobzhansky and Pavlovsky, 1953
D. tropicalis	Dobzhansky and Pavlovsky, 1953

Note: The more recent references are provided for each species, and the citations included within these papers will allow entry into the literature for each species.

these workers studied some 22 different populations with different combinations of inversions maintained under different environments. In fact, no two populations were the same! It is remarkable they were able to make any conclusions. Some results of these studies are in tables 4-4 and 4-5. In table 4-4, the magnitude of changes observed in such a short time is most remarkable. Table 4-5 and figure 4-6 illustrate the repeatability of the changes in seven replicate cages (from a later study). Figure 4-7 demonstrates that the same equilibrium is reached regardless of starting frequency. In almost all cases, a stable equilibrium is established at least under some conditions and combi-

TABLE 4-4 Typical dynamics of inversion frequency changes in laboratory populations of *D. pseudoobscura*.

	Experiment 18			Experiment 19	
Time	ST	AR	CH	ST	CH
Initial	19.9	43.6	36.5	38.3	61.7
Mid Nov. 1944	33.3	27.3	39.3	—	–
Mid Dec. 1944	37.7	28.7	33.7	53.0	47.0
Mid Jan. 1945	39.3	30.0	30.7	63.3	36.7
Late Feb. 1945	44.3	30.0	25.7	65.3	34.7
Late Mar. 1945	42.0	39.0	19.0	65.3	34.7
Late Apr. 1945	46.7	30.3	23.0	65.3	34.7
Early June 1945	56.4	27.3	16.3	70.4	29.6
Late July 1945	50.3	31.7	18.0	72.0	28.0

Note: Flies were derived from 1942 collections at Piñon Flats, California.

From Wright and Dobzhansky (1946).

TABLE 4-5 Starting and ending frequencies of inversions in laboratory populations of *D. pseudoobscura.*

	Initial %				Final %		
Cage	ST	AR	CH	Months	ST	AR	CH
1	27.1	35.0	38.0	6	52.9	27.1	20.1
2	34.5	39.9	25.6	6	53.7	33.0	13.3
5	44.2	37.1	18.7	5	57.3	30.7	12.0
6	29.0	28.3	42.7	3	44.3	21.0	34.7
10	35.7	24.7	39.7	3	48.3	22.3	29.3
12	22.0	38.3	39.7	3	50.7	30.3	19.0
18	33.3	27.3	39.3	9	50.3	31.7	18.0
			Average all cages:		51.1	28.0	20.9
A	25.0	25.0	50.0	12	80.0	16.7	4.3
B	25.0	25.0	50.0	12	81.7	12.7	5.7

Note: The upper part of the table (numbered cages) is for flies collected at Piñon Flats in 1942, and the lower part (lettered cages) from the same locality collected 10 years later.

From Wright and Dobzhansky (1946) and Levene et al. (1954).

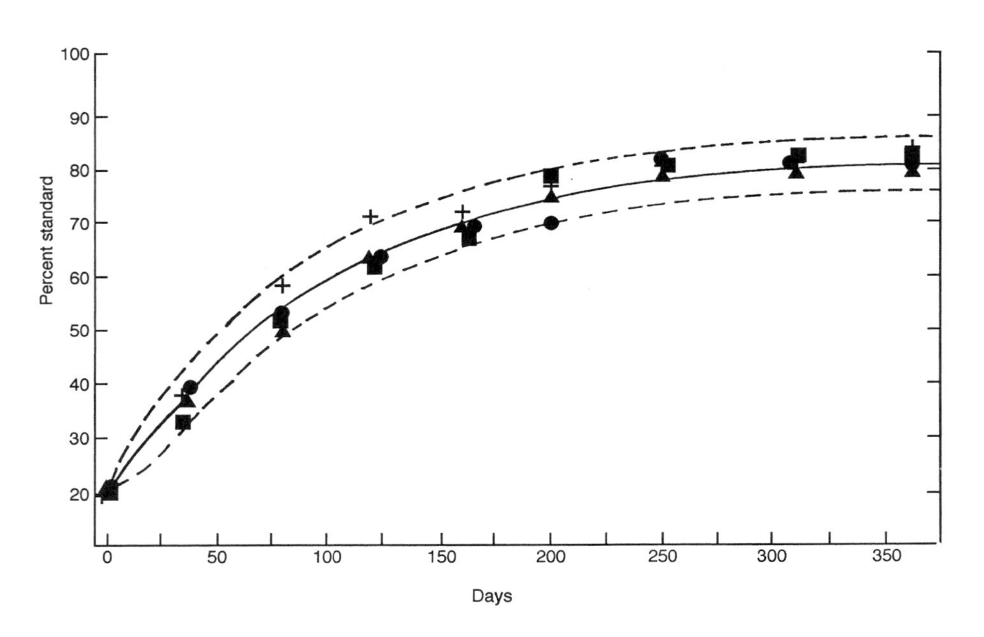

FIGURE 4-6 Typical dynamics of change in inversion frequencies of the third chromosome of *D. pseudoobscura.* Four replicates are shown with the 95% confidence interval indicated by dashed lines. The populations are from Piñon Flats with ST and CH alternative arrangements. From Dobzhansky and Pavlovsky (1953).

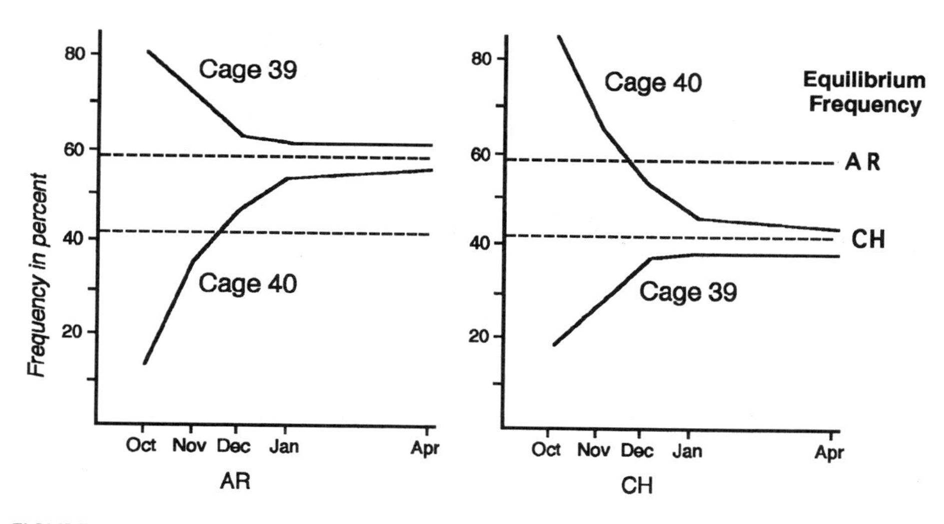

FIGURE 4-7 Dynamics of change in frequencies of third chromosome inversions of *D. pseudoobscura* in cages begun at different frequencies. From Dobzhansky (1948b).

nations of inversions. (An aside about "equilibrium": it is well known that true equilibrium in population genetics is not reached for many generations, on the order of effective population size. In these experiments, researchers follow changes until no further significant changes occur. This could be termed the experimenter's *practical equilibrium.*)

The following other conclusions can be drawn from these studies:

1. Temperature is an important factor in the outcome of these experiments. In *D. pseudoobscura*, virtually no changes are noted at 16°C while the strong selection noted in figures 6 and 7 occurs at 25°C. Curiously, this is the opposite of what Spiess (1966) found with the closely related sibling species, *D. persimilis.*
2. Average changes per month (equal to about a generation under these conditions) were 2 to 12%. This requires selection coefficients on the order of 0.1 or greater (see below for actual estimates).
3. The results conform to expectations for heterotic (overdominance) maintenance. In the initial experiment, the selection against ST/ST and CH/CH homozygotes was estimated to be 0.3 and 0.7, respectively. However, as we will see, the results cannot be distinguished from some kind of frequency-dependent selection.
4. The dynamics of two-chromosome cages cannot always predict the outcome of three-chromosome cages (see below). Wright (chapter 9 in Wright [1977]) interprets this as indicating the inadequacy of heterosis (constant fitness) as the sole explanation for selection in such cages.

Laboratory studies have also shed light on the evolution of inversions in natural populations. Do inversions evolve after they arise or are they static because the set of alleles captured when first arising are so fit as to resist any modification? One approach to answering this question is to take inversions from the same natural population some time after an initial experiment and determine if the same laboratory results are observed. Table 4-6 shows fitness estimates for karyotypes from the same population, Piñon Flats, taken about ten years apart (Levene et al., 1954). Inversions taken from the same natural population at different times behave differently in laboratory populations. This observation has been repeated for flies from Mather; compare Dobzhansky (1948b) and Pavlovsky and Dobzhansky (1966). Recall that some inversions (TL and PP) have risen in frequency in several populations concomitantly (table 3-10). The question was raised as to whether this was due to a change in the environment or a change in the gene content of the inversions (chapter 3). Results such as these demonstrate that inversions do evolve over time. Even though monophyletic (probably), these rearrangements are not static.

Table 4-6 is also instructive in indicating two methods of calculating fitness from these experiments, one based on least squares developed by S. Wright for the original study, and one developed by H. Levene based on minimizing chi-squares. When applied to the same data set (Levene et al., 1954) they give very similar estimates. Dumouchel and Anderson (1968) introduced yet another fitness-estimating procedure based on maximum likelihood, which also allows calculation of confidence intervals for the estimates (which are depressingly large).

Most inversion studies in the laboratory have used two or three alternative gene arrangements artificially set to convenient frequencies. In an attempt to mimic nature a bit better, Anderson et al. (1967) took flies from 11 natural populations and began 11 population cages with the frequencies found in nature. After 18 months at 25°C, all showed significant shifts in inversion frequencies, but all did not attain the same frequencies, although kept in identical conditions. Populations from the Northwest, where ST dominates (table 3-2), developed a high frequency of ST in the laboratory. In three populations where AR dominates, AR remained the most frequent, though much reduced from its frequencies in the natural populations. In two populations from Texas where PP dominates in nature, PP was greatly reduced due to a rise in AR in one case, and AR and ST together in the other. In one population where CH is dominant in nature, CH was eliminated and AR became dominant. These results show clearly that the same gene arrangement in different populations can be very different in gene content; that is, they are population-specific in spite of monophyly.

TABLE 4-6 Estimates of fitnesses of karyotypes of *D. pseudoobscura* from Piñon Flats for experiments performed with flies collected about 10 years apart.

	ST/ST	ST/AR	ST/CH	AR/AR	AR/CH	CH/CH
From Δq least squares (1946)	0.33	1.00	0.77	0.04	0.55	0.16
From Δq least squares (1954)	0.83	1.00	0.67	(0)	0.60	0.43
Minimum chi-square (1954)	0.83	1.00	0.77	0.15	0.62	0.36

Note: All fitnesses are calculated relative to ST/ST, assigned a fitness of 1.00.

From Wright and Dobzhansky (1946) and Levene et al. (1954).

In an attempt to mimic nature even more closely, Anderson et al. (1972) began some cages not only with flies with frequencies of inversions found in the natural population, but also with the other *Drosophila* species collected at the same time, in about the frequencies collected at fermenting banana baits. All species except *D. pseudoobscura* were eliminated fairly quickly, and inversions attained equilibrium frequencies different from those previously observed for flies from the same population, Mather.

The number of inversions in population cages can change estimated fitness values. For flies from Piñon Flats in the 1940s experiments, when only CH and ST were in the cages, the fitness of ST/ST relative to ST/CH was about 0.76. Table 4-6 shows fitness estimates when three inversions are present. For the 1940s experiment, adding AR changed the ratio of fitness of ST/ST : ST/CH to 0.42 (0.33/0.77) or for the 1950s experiment to 1.24.

The question of whether only a single balanced equilibrium exists was addressed by Watanabe et al. (1970). Judging from results pictured in figure 4-7, one might get the impression there is only a single equilibrium point, at least when only two gene arrangements are studied. Especially when more than two alleles exist at a locus, multiple equilibria are often predicted with both constant fitness and frequency-dependence models. Table 4-7 shows that the equilibria reached in four-inversion studies depend on the starting frequencies. Presumably, the populations come to the closest local equilibrium.

The Genetic Basis of Fitness Variation among Inversions

Why do gene arrangements differing by inversions of chromosomal segments have such strong selective differences? What is the genetic basis for the variation in fitness? Two hypotheses have been advanced: position effect and coadaptation. It is known that a gene's expression could be modified depending on where in the genome it is

TABLE 4-7 Results of studies of changes of inversion frequencies in populations of *D. pseudoobscura* begun at different frequencies.

	Initial				Final			
Cage	ST	AR	CH	PP	ST	AR	CH	PP
IA	30.0	5.0	50.0	15.0	83.7	15.7	1.3	0.3
IB	30.0	5.0	50.0	15.0	78.0	20.3	1.0	0.7
IIA	15.0	5.0	30.0	50.0	78.7	18.0	2.7	0.7
IIB	15.0	5.0	30.0	50.0	72.3	25.0	2.7	0
IIIA	5.0	50.0	15.0	30.0	36.7	63.0	0	0.3
IIIB	5.0	50.0	15.0	30.0	34.0	65.0	0.7	0.3
IVA	25.0	25.0	25.0	25.0	63.0	36.3	0	0.7
IVB	25.0	25.0	25.0	25.0	64.3	35.0	0.3	0.3

Note: All cages were monitored for 12 months.

From Watanabe et al. (1970).

located (position effects). F. Mainx and his influential student D. Sperlich (1966) have been the main proponents of the view that, at least initially, the only difference between an inverted and noninverted chromosome is in the positions of genes at the breakpoints (consistent with a classical view of populations). If an initial selective advantage is necessary to keep an inversion in the population when it first arises, this could only be due to some kind of position effect, perhaps one causing heterosis. The alternative is to postulate that when a successful inversion first arose, by chance it captured a set of favorably interacting alleles, which are said to be coadapted (Sperlich has termed this the *pre-adaptation hypothesis*). The term *coadaptation* in this context has two aspects. First, the interacting alleles within an inversion are coadapted with one another to form a particularly fit genotype (epistatic fitness effects). Second, because at least initially the new inversion will be predominantly in the heterozygous state, its nonrecombining block of alleles must be coadapted to those carried on alternative gene arrangements such that heterozygous combinations have high fitness.

Experiments demonstrating that the same gene arrangement taken from different populations has different fitness properties make the position effect hypothesis untenable as the sole factor causing fitness differences. If cytologically identical gene arrangements from different populations have a common origin (i.e., are monophyletic), they should have the same position effects. Dobzhansky (1950) and Dobzhansky and Pavlovsky (1953, 1957) performed a series of experiments on laboratory populations consisting of chromosomes from a single natural population and hybrid populations with a mix of inversions from geographically distant populations. The nice predictable behavior of cage populations, such as those illustrated in figures 4-6 and 4-7, was only obtained if the alternative gene arrangements came from the same population. Hybrid populations behaved erratically, often with no stable equilibrium. Furthermore, the number of inversions used to start the populations could also significantly affect the outcome. Dobzhansky and Pavlovsky (1957) made an F_2 hybrid population between California AR and Texas PP and began a series of cages with either 20 or 4,000 founders each from this hybrid population. Figure 4-8 shows the results. Clearly, the more founders, the more determinate the outcome. It is difficult to find any other reason than differences among the founding gene arrangements and/or their genetic background to explain these results. Clearly, inversions have population-specific properties. This conclusion was confirmed by many follow-up experiments including Dobzhansky and Spassky (1962), Solima-Simmons (1966), and Strickberger (1963, 1965). Strickberger's results are particularly revealing in indicating how fast population-specific coadaptation can develop in inversions. He had maintained replicate populations of *D. pseudoobscura* polymorphic for CH and AR in both a series of bottles and a series of cages. When he mixed flies from independent bottle lines, the inversions behaved "nicely" and formed the classic stable equilibria. When inversions from independent population cage lines were mixed, they too behaved normally. However, mixtures of bottle and cage chromosomes behaved erratically. This indicates that the different laboratory environments over a period of several years had molded different coadapted complexes. (For many years, Dobzhansky kept a dime in a frame on the wall of his office. He had won it from H. J. Muller in a bet over the geographically mixed population experiments. Being a proponent of the classical view of populations, Muller was convinced that the same inversion taken from anywhere it is found would be identical to all others.)

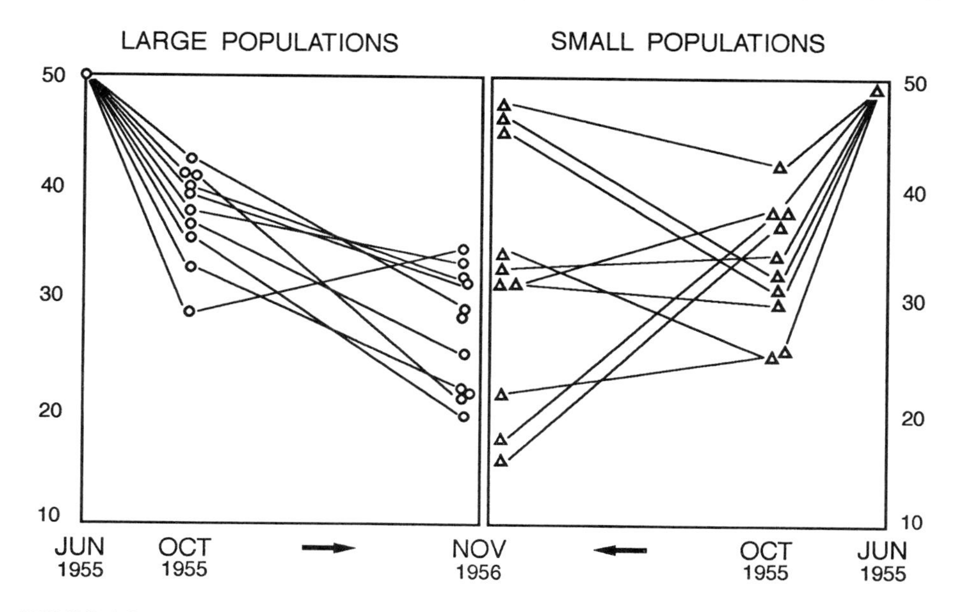

FIGURE 4-8 Changes in PP inversion frequencies in laboratory populations of *D. pseudoobscura* begun with about 4,000 flies (left) and 20 flies (right). Each line is a replicate population. From Dobzhansky and Pavlovsky (1957).

It is clear that coadaptation occurs among loci within the inversions and between different gene arrangements of the same population, at least in *D. pseudoobscura.* The coadaptation hypothesis is well supported. Whether there is any room for a position effect is not clear. In his recent writings, Sperlich (e.g., Sperlich and Pfriem, 1986) has come to accept coadaptation. However, he maintains that coadapted complexes may arise only after the establishment of the inversion. That is, the position effect provides the initial selective advantage to bring a new inversion to high frequency, after which selection may mold epistatically interacting sets of alleles within the breakpoints.

Heterosis or Frequency Dependence?

What kind of selection maintains the stable equilibria so well-documented in these laboratory experiments on *Drosophila* inversions? Two explanations have been offered: heterosis (overdominance) and frequency-dependent selection. While it is well known that the kinds of curves displayed in figures 4-6 and 4-7 are exactly those predicted by heterosis, and the kinds of fitness estimates made from the data usually assume that model (e.g., table 4-6), it is less well known that frequency dependence can predict virtually identical behavior. Figure 4-9 is from an analysis by Wright (1977) in which he used a very simple linear model of frequency dependence and compared the prediction to that of heterosis and to the observed data. Clearly, one cannot distinguish the two types of selection based on the dynamics of change alone. We can summarize the arguments for and against each type of selection:

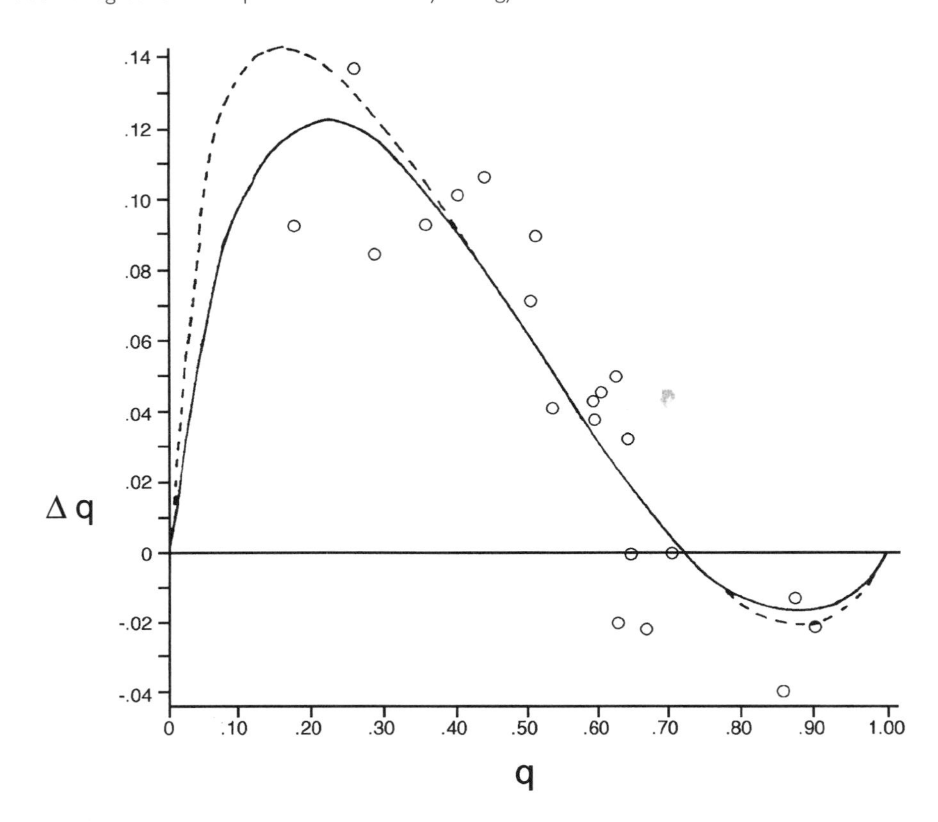

FIGURE 4-9 Theoretical dynamics of change in inversion frequency under a heterosis model (solid line) and a frequency-dependent model (dashed line). Circles are observed data for *D. pseudoobscura* third chromosomes. From Wright (1977).

In favor of heterosis:

1. Inversions often display an excess of heterozygotes over that predicted by Hardy-Weinberg equilibrium. However, this observation is not universal and some of the data are ambiguous at best, and the tests flawed (Lewontin et al., 1981). Interestingly, when Dobzhansky thought the inversions were neutral to selection he went to considerable length to show they conformed to Hardy-Weinberg (in Dobzhansky and Epling, 1944). Only after he found the evidence for selection did he attempt to use the data to argue for heterozygote excess.
2. There are at least three cases of *Drosophila* inversions where heterosis is undoubtedly acting. An essentially balanced lethal system was observed for *D. tropicalis* (Dobzhansky and Pavlovsky, 1955) where homokaryotypes are virtually completely absent. In *D. silvestris*, Carson (1987) found strong evidence of heterotic mating success, especially for males, associated with heterokaryotypes. In *D. ananassae* direct measurement of such fitness components as rate of development, fecundity, and longevity all indicated an undoubted superior-

ity of inversion heterozygotes (Moriwaki et al., 1956; Tobari and Moriwaki, 1993).
3. When initially arising, all new inversions will be found in the heterozygous state with other gene arrangements. To survive this initial precarious period of rarity, some advantage in the heterozygous state would greatly increase the probability of establishment, without which loss is by far the most common result.
4. The adaptive superiority of populations polymorphic for inversions (discussed in chapter 6) would argue for some kind of general heterotic effect associated with these polymorphisms.

Evidence in favor of frequency dependence, or at least nonconstant fitness, includes:

1. Selection experiment dynamics are equally or better predicted by frequency-dependent models than by overdominance models. Wright (1977) argued that the dynamics of the multiple chromosome studies of Pavlovsky and Dobzhansky (1966) especially required model complexity beyond simple overdominance (see also Lewontin et al., 1981). Spiess (1957) argued that his cage experiments with *D. persimilis* were better explained by frequency dependence than by heterosis.
2. The changes in relative fitness of karyotypes dependent on the presence or absence of other chromosomes would argue against constant fitness models. The dependence of equilibrium point on the starting frequencies (Watanabe et al., 1970; table 4-7) is also indicative of frequency-dependent fitness (although overdominance is not necessarily excluded).
3. The theoretical work of Robertson (1962) and Lewontin et al. (1978) must cast doubt on the sole role of heterosis in maintaining stable polymorphisms in general. Robertson showed that under some cases the protection (from fixation or loss) of a polymorphism could in fact be hampered by heterosis, especially if heterosis was bringing the inversion to a stable frequency less than 20%; frequencies this low or lower are often seen in natural populations (e.g., tables 3-2 and 3-10). Lewontin et al. (1978) showed how difficult it is to find any fitnesses that could maintain a stable polymorphism by overdominance with more than 6 or 7 alleles, and then the prediction is the alleles would all come to about equal frequency. Mexican populations of *D. pseudoobscura* with up to 14 third-chromosome inversions in very unequal frequency would be almost impossible to mimic with a constant fitness model of overdominance.
4. Perhaps the strongest argument in favor of frequency dependence is the direct evidence for frequency-dependent male mating success (e.g., Ehrman, 1967; Spiess, 1968) and frequency-dependent viability (Kojima and Tobari, 1969; Anderson, 1989; Tobari, 1993b) for different karyotypes. These studies will be discussed later in this chapter.
5. The association of inversions with microhabitat preferences (discussed in chapter 6) suggests a role for multihabitat selection, which can often be frequency dependent as well as density dependent.

Other Species

On balance, there is evidence for both types of models, with the case for nonconstant fitness models seeming to be the stronger. On the other hand, it must be explicitly pointed out that perhaps it is not necessary to dichotomize the two types of models as being mutually exclusive. There is no reason to exclude a role for each type of selection, and the relative contribution of each may well change over time and among populations. As an example of this latter phenomenon, the work of Kojima and Tobari (1969) on *D. ananassae* inversions is particularly revealing. They observed these inversions attaining a stable equilibrium in laboratory populations and tried to tease apart the effects of heterosis from frequency dependence. Kojima and Tobari looked for viability differences when the karyotypes were in different frequency. The results in table 4-8 are most interesting. When each homokaryotype is rare, it has a viability advantage. However, when equal in frequency, heterosis is displayed (middle entry in table); the equilibrium reached in the populations is about 50%, as would be predicted by the fitnesses for the karyotypes in this middle entry, assuming viability is the dominant factor. It appears the same polymorphism can be both subjected to frequency-dependent selection and be overdominant.

Further discussion of *D. ananassae* inversions is in order as, unlike *D. pseudoobscura* and *D. persimilis*, in this species two chromosomes have high frequencies of inversions (the second and third), and thus one can ask about interactions. Tobari and Kojima (1967) and Tobari (1993b) followed the frequencies in a series of 52 cages in which different frequencies were initiated, with and without segregation of inversions in both chromosomes. Figure 4-10 displays some results. Overall, one can see that the two systems are reasonably independent; that is, the left set of panels shows the second chromosome inversions reaching about the same equilibrium whether the third is fixed for alternative inversions or is polymorphic. The same is more or less true for the third chromosome (right set of panels in figure 4-10), although in the case when

TABLE 4-8 Larval viabilities of *D. ananassae* karyotypes for the 2L when at different frequencies.

	Karyotypes			
	AA	AB	BB	Total Sample
Input frequencies	0.45	0.50	0.05	
Output numbers	254	306	40	600
Relative viability	0.92	1.00	1.11	
Input frequencies	0.25	0.50	0.25	
Output numbers	112	280	105	497
Relative viability	0.80	1.00	0.75	
Input frequencies	0.05	0.50	0.45	
Output numbers	32	285	258	575
Relative viability	1.12	1.00	1.01	

Note: Heterozygote values are arbitrarily set to 1.

From Kojima and Tobari (1969).

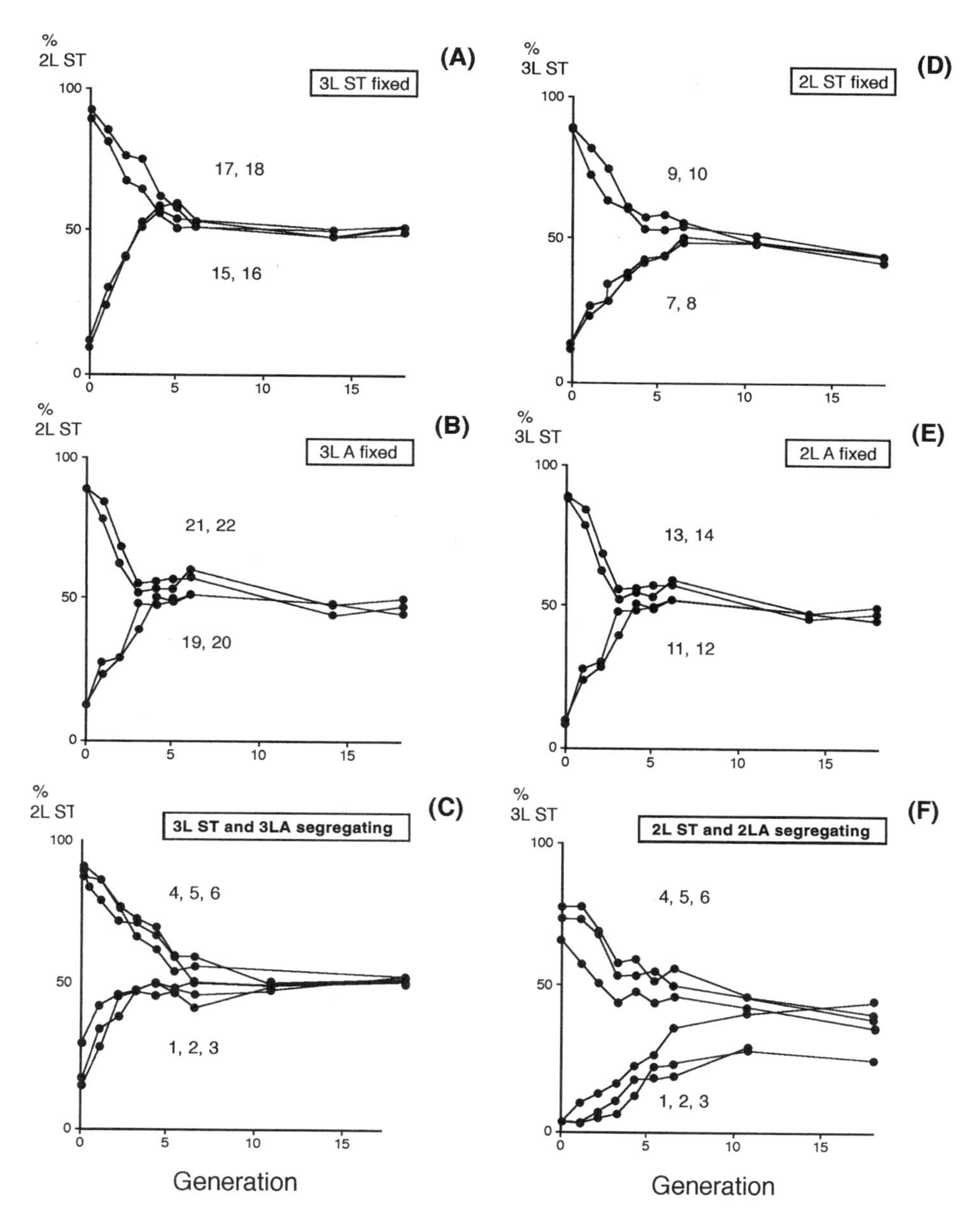

FIGURE 4-10 Graphs showing dynamics of change in *D. ananassae* inversions in two chromosomes. Graphs A, B, and C show the frequency changes of the second chromosome gene arrangements when the third chromosome is fixed for alternative inversions (A and B) or segregating for third chromosome inversions (C). On the right is the complement for the dynamics of change in third chromosome inversions when the second chromosome is fixed for alternative arrangements (D and E) or segregating for these two (F). From Tobari (1993b).

both chromosomes are segregating, some effect on the outcome of the third chromosomes does appear (lower right panel). Tobari (1993b) discusses these interaction effects in more detail.

Here we make a brief mention of why studies with other species have not been so informative, nor have they been as widely used in laboratory cage experiments (table 4-3). The problem with many of these studies is exemplified by two species. In *D. subobscura*, where all arms of five acrocentric chromosomes are highly polymorphic, it is extremely difficult to manipulate the inversion frequencies simultaneously, and/or isolate a sufficient number of copies of a gene arrangement from a single population to avoid founder effects. Studies of *D. subobscura* inversions in laboratory populations have produced erratic or uninterpretable results as a consequence of these problems. When *D. melanogaster* populations are transferred to laboratory environments, the nearly universal observation is elimination of the inversion (relative to standard), which is perhaps not too surprising, as the standard is defined as the arrangement in old laboratory cultures. Thus, *D. pseudoobscura* has been the primary model for these kinds of studies. However, studies on *D. ananassae*, which presents some interesting differences from *D. pseudoobscura* (such as having more than one chromosomes highly polymorphic), have also been highly informative. Unlike most species with multiple inversion polymorphisms, the complexity in this species is not so overwhelming as to preclude careful experimental examination.

Laboratory experiments have been performed that may be relevant to the issue of central versus marginal levels of inversion heterozygosity in nature. One theory, primarily that of Carson, emphasizes the role of recombination in providing adaptive flexibility in marginal environments where the vicissitudes of nature would be particularly marked and unpredictable. Three experiments have shown that populations that are monomorphic for inversions respond more quickly to selective pressures than do polymorphic populations. Carson (1958b) selected *D. robusta* for "motility" and found that inversion monomorphic populations responded better than polymorphic ones. Markow (1975) selected for phototactic behavior in *D. melanogaster* and found that inversion polymorphism in any of the three major chromosomes slowed the rate of selection response. Tabachnick and Powell (1977) challenged populations of *D. willistoni* to survive novel stresses such as high acid or salt in the medium. Again, chromosomally monomorphic populations (from the Caribbean Islands) more often survived the challenges than did polymorphic populations from mainland South America (cf. figure 3-9). Together, these studies provide evidence for the importance of recombination in allowing populations to respond to selection. If, in general, marginal populations in nature face stressful challenges, especially if they are temporally varying, chromosomal polymorphisms that inhibit recombination would be selected against.

Fitness Components

The previous two sections have established that a variety of genetic variants in *Drosophila* are subject to the action of natural selection, that selection occurs in the laboratory under controlled conditions, and that the strength of selection is often great enough that the dynamics of change can be followed over a reasonable amount of time, thus allowing an experimenter to perform replicated studies. These facts make *Drosophila* a convenient model system for studying the nature of selection in more detail. Most such studies deal with the components of fitness; that is, the selection process is dissected to reveal more details of just where and why selection is acting. In *Drosophila* biology in this field, the papers of T. Prout (1971a, b) stand out as seminal contributions.

Prout's approach is conceptually simple, yet analytically complex. Its essence (outlined in table 4-9) is to count the offspring produced when varying the frequencies of genotypes of one sex, holding the other sex constant. As Prout (1971a, p. 130) states, "This experimental system is therefore capable of doing more than simply resolving net fitnesses into components; it is capable of detecting and measuring various modes of selection such as sex dependence, frequency dependence, and mating interactions." As an experimental demonstration of the utility of this approach, Prout used two recessive mutants on the fourth chromosome of *D. melanogaster, eyeless* (the allele ey^2, which we will simply designate *e*) and *shaven naked* (sv^n, which we denote by *s*); the fourth chromosome in this species is very short (the "dot" chromosome) and undergoes no recombination. The mutants were on different chromosomes, so they acted as alleles of one another with all three genotypes being phenotypically distinguishable. Knowing the general selective disadvantage of visible mutants, this system would be expected to form a stable equilibrium due to overdominance. The results of Prout's tests on these mutants are in table 4-10. Selection is resolved into two components, adult component and larval component, separately for each sex. Female fecundity (the adult component) was independent of the genotype of the male mating, so a single estimation suffices for this component of fitness. Males, on the other hand, differed in adult fitness (virility) depending on the genotype of the female, so these parameters were kept separate. Perhaps the most striking feature of these estimates is the importance of male virility; it is the single most important component of fitness.

TABLE 4-9 Summary of strategy to determine fitness components for the fourth chromosome mutations *eyeless* (*e*) and *shaven* (*s*) in *D. melanogaster*.

Fecundity				Virility			
Female Mixture			Male constant	Male Mixture			Female Constant
ee	*e/s*	*ss*		*ee*	*e/s*	*ss*	
30	20	10	*ee*	30	20	10	*ee*
10	20	30	*ee*	10	20	30	*ee*
30	20	10	*e/s*	30	20	10	*e/s*
10	20	30	*e/s*	10	20	30	*e/s*
30	20	10	*ss*	30	20	10	*ss*
10	20	30	*ss*	10	20	30	*ss*
Viability: *e/s* female × *e/s* male, 47 replicates							

Note: Only two frequencies were tested for each sex for a given opposite sex partner, although many more could be tested, especially if one wanted to detect frequency-dependent fitness effects. Fecundity refers to the female adult component and virility to the male adult component.

From Prout (1971a).

TABLE 4-10 Fitness estimates from experiments outlined in table 4-9.

	Viability						
	Females			Males			
	$L_{e/e}$	$L_{e/s}$	$L_{s/s}$	$L_{e/e}$	$L_{e/s}$	$L_{s/s}$	
	0.865 (0.039)	1	0.934 (0.039)	0.839 (0.036)	1	0.777 (0.038)	
	Adult Component						
	Fecundity			Virility			
	$F_{e/e}$	$F_{e/s}$	$F_{s/s}$	$V_{e/e}$	$V_{e/s}$	$V_{s/s}$	Constant Female
	1.037 (0.122)	1	0.458 (0.068)	0.363 (0.074)	1	0.039 (0.033)	*ee*
				0.243 (0.042)	1	0.122 (0.037)	*e/s*
				0.135 (0.036)	1	−0.018 (0.030)	*ss*
	Equilibrium Predictions						
	Genotype Frequencies				Gene Frequencies		
	ee	*es*	*ss*		*e*	*s*	
	0.351	0.506	0.143		0.604	0.396	
H-W exp.	0.365	0.478	0.157				

Note: The larval viabilities are designated *L*; female fecundity, *F*; and male virility, *V*. Standard errors of the estimates are in parentheses. Heterozygote fitnesses are arbitrarily set to 1.

From Prout (1971b).

In the second paper in the series, Prout (1971b) tested whether the fitness components estimated in this manner accurately predict the behavior of these mutants in populations. The results, shown graphically in figure 4-11, indicate that the estimated parameters quite accurately predicted the behavior in populations. However, chi-squared analysis revealed significant deviation from prediction for two replicates. This discrepancy was easily rectified by assuming that the larval viability component of fitness might be greater under the population conditions due to higher density than used in the estimation experiments; a decrease in the viability of the two homozygotes by only 0.05 (less than two S.E.s of the estimates, table 4-10) was sufficient to bring the two outlier replicates into agreement.

Prout's discussion of these studies is illuminating. He points out, for example, that one cannot predict the outcome of selection by measuring a single component of fitness. In the present case, female fecundity alone predicts elimination of *s* and fixation of *e*. Female viability predicts a stable equilibrium with *s* in higher frequency; whereas, taking all components together, the prediction is an equilibrium with *e* in

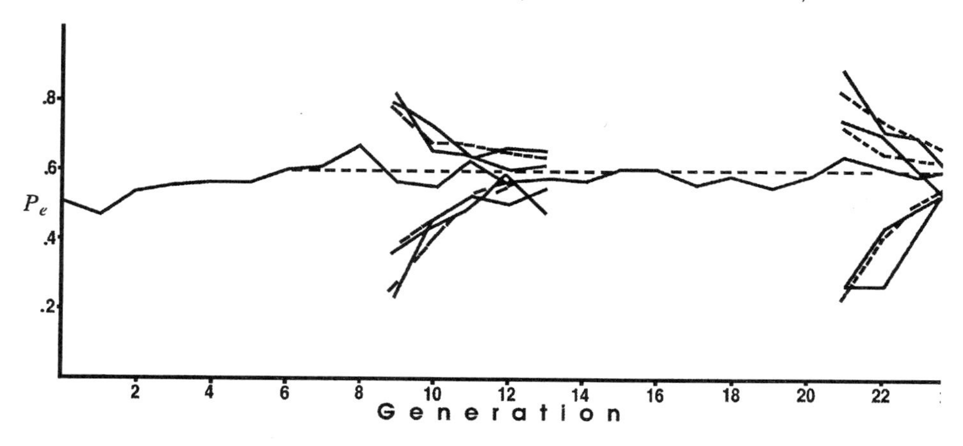

FIGURE 4-11 Summary graphs of results of fitness-components tests. From the fitness components measured by Prout (1971a), the predicted equilibrium is the dashed line. The observed curves are the other lines including several perturbations that quickly collapsed back to prediction. From Prout (1971b).

higher frequency, which was the observed outcome. Another interesting aspect of these data is the conformity of the predicted equilibrium genotype frequencies with Hardy-Weinberg predictions, despite the fact that considerable heterosis exists (bottom of table 4-10). These results highlight the fact that the simplicity inherent in classical population genetics models with single values given for fitness does not reflect nearly the complexity revealed by even relatively simple dissection of components of selection.

Prout (1971b, p. 164) sums up the work by stating that "the reasonably good population fits suggest that one may have a certain degree of confidence that this method of estimating fitness components can provide information concerning a relatively intricate mode of selection occurring in a population whose ecology is somewhat more complicated than conditions of the component experiments."

Bundgaard and Christiansen (1972) present an alternative, though related, method to measure somewhat more directly the components of fitness in *Drosophila* laboratory populations. Their procedures involve controlled mating conditions, egg-laying times, counting eggs from individual females, and monitoring changes in mutant frequencies in each generation. They too used fourth chromosomes of *D. melanogaster* marked by recessive mutants. They divided selection into three components, which they called zygotic selection (viability), sexual selection (mating), and fecundity (numbers of eggs). They recognized there is potentially a fourth component, gametic selection, which might occur in the haploid stage if, for example, multiple matings lead to competition of two sperm pools. Consistent with Prout's results, they concluded that the male mating component was by far the dominant form of selection. The design of the experiments also allowed them to detect that the form of selection on male mating success was frequency-dependent, that is, as a genotype became rare it increased in frequency of mating, a subject to be discussed in more detail in the next section. Bundgaard and Christiansen (cited in Hedrick and Murray, 1983) expanded on these initial studies by utilizing several other mutant combinations. They often found that the male component of sexual selection was an important, though not always the most important, component of fitness, nor could they always demonstrate frequency dependence.

Anderson and colleagues (e.g., Anderson and Watanabe, 1974; Anderson and McGuire, 1978) have shown in similar kinds of experiments with *D. pseudoobscura* that male mating success tends to be the most important component of fitness for the karyotypes of this species. Carson (1987) also found evidence for a strong role of male mating success in causing the heterotic maintenance of an inversion polymorphism in *D. silvestris.* Male mating success, then, emerges as the dominant component in a number of studies on more than one species.

Prout and Bundgaard (1977) performed experiments to explicitly measure gametic selection in *Drosophila.* They asked what was the fate of sperm provided by males that mated in temporal sequence. Females of many species of *Drosophila* are multiply inseminated in natural populations (Milkman and Zietler, 1974; Anderson, 1974; Cobbs, 1977; Levine et al., 1980) so this type of selection is relevant to the real world. They found that when the brown mutant males mated after the wild type, sperm from the brown mutant had a uniform advantage. However, when wild-type males mated after brown, the outcome was highly variable, suggesting competition among sperm with variable outcomes. Prout and Bundgaard also present mathematical theory dealing with this type of selection.

A second type of gametic selection may also occur, although this is rather strange from an evolutionary perspective. Males of several species of *Drosophila* have been shown to produce sperm highly variable in tail length (Joly et al., 1989; 1991a; 1991b), a phenomenon sometimes called polymegaly. All members of the *obscura* group that have been examined exhibit polymegaly. Snook et al. (1994) showed that in *D. pseudoobscura* sperm fall into two fairly discrete size classes, long and short. Virtually all fertilization is accomplished by the long-tailed sperm. The evolutionary meaning of this is not clear, since both types of sperm are produced by the identical genotype.

Because *D. pseudoobscura* inversion polymorphism has played such a prominent role as a model for selection, it is appropriate to select this case for special attention regarding what is known of the components of selection in this system. Birch (1955) attempted to identify a component of selection that might be relevant to the seasonal cycles in this species. He studied the effect of larval density on viability. In uncrowded larval conditions, ST leveled off in frequency at about 30%; under crowded conditions, ST increased to 70%. This was consistent with the seasonal cycles at Piñon Flats, from which Birch's strains came (figure 3-12). In the spring, he reasoned, density would be low in the natural population and ST would be at relative disadvantage; when densities increased in the summer, ST would climb in frequency. Keeping in mind that frequencies in the seasonal studies are for adults in the population which were larvae the previous month, the data in figure 3-12, are in reasonable agreement. The only other study attempting to find a component of selection to explain seasonal cycles is that of Crumpacker et al. (1977) where the cold resistance of different karyotypes from Colorado populations varied in accordance with predictions from the seasonal cycles.

Anderson (1989) summarizes several other studies of components of selection on *D. pseudoobscura* inversions. He and his colleagues demonstrated frequency-dependent larval viability (figure 4-12) as well as density-dependent larval viability. Anderson and colleagues also performed a clever experiment whereby they could obtain inversion frequencies for both larvae and adults by having *Amy* allozyme alleles in complete disequilibrium with the different gene arrangements. The reasoning is that frequency differences between larvae and adults represent viability selection, while differences between adults and the eggs laid represent the fertility component. Figure 4-13 shows some results. Consistently from larvae to adults, ST increased, indicating a viability advantage, whereas from adult to eggs, CH (the alternative arrangement) exhibited a fertility advantage. The two components seem to be about equal in magnitude, although ST climbs in frequency over the course of the experiment. This experiment is one of the few attempts to measure fitness components directly in a freely-breeding population (see also chapter 6).

Hedrick and Murray (1983) discuss more studies of this sort.

The Rare Male Effect

The so-called rare male effect, abbreviated here RME, is one component or form of selection that has received considerable attention, if not notoriety, in *Drosophila* biology. The RME occurs when two (or more) genotypes of males display negative frequency dependence for mating success, that is, mate disproportionately to their frequency when rare. This phenomenon was first noted by C. Petit (1951) while studying why *Bar* was not completely eliminated from population cages. *Bar* males exhibited

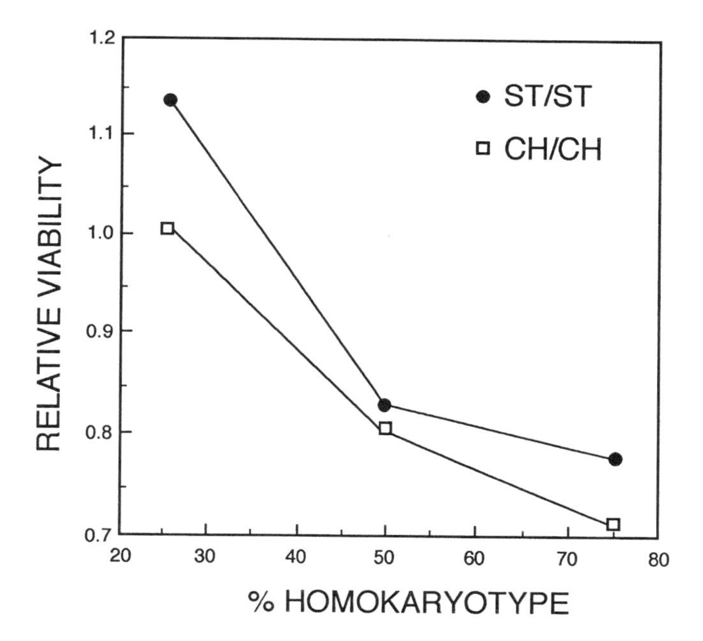

FIGURE 4-12 Evidence for frequency-dependent viability fitness effects for third chromosome inversions from *D. pseudoobscura*. From Anderson (1989).

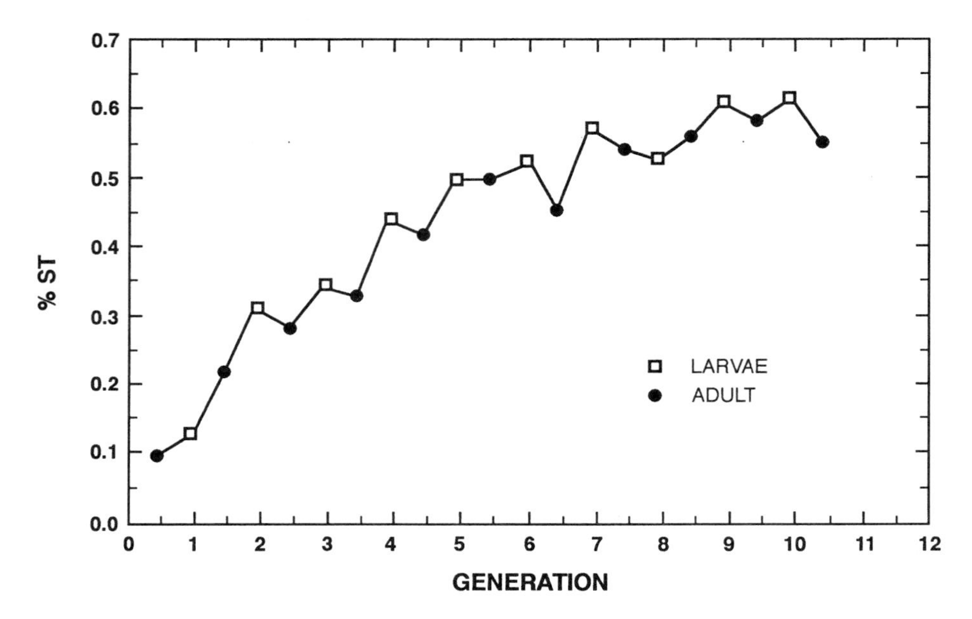

FIGURE 4-13 Frequencies of ST gene arrangement of the third chromosome of *D. pseudoobscura* at two life stages. Open symbol is from egg samples that give rise to the larval population, and the closed symbols represent frequencies in adults in the cage. From Anderson (1989).

a distinct mating advantage when very rare, something she later called the *minority advantage* (Petit, 1958). Independently, in the early 1960s L. Ehrman (e.g., Ehrman et al., 1965) and E. B. Spiess (cited in Ehrman, 1966) rediscovered the phenomenon.

Table 4-11 shows examples of data using different karyotypes of *D. pseudoobscura* and *D. persimilis.* In analyzing data from these experiments, the usual statistic presented is a cross product introduced by Petit, which takes the following form:

$$\frac{A \text{ mating}}{A \text{ present}} \div \frac{B \text{ mating}}{B \text{ present}} = \frac{(A \text{ mating})\ (B \text{ present})}{(B \text{ mating})\ (A \text{ present})}$$

This takes on the value 1 when males are mating in proportion to their frequency; both R and K_m have been used to denote this value and I will use the former. In the case of RME, R should be greater than 1 when rare, and in the case of two-sided RME (where both types of males have a rare advantage), R should become less than 1 when the genotype is frequent. The figures in table 4-11 are this R value. As can be seen for these systems, R varies from about 0.3 for the common male to 2 for the rare male, although different combinations clearly show differences, with Case III being much weaker in this set of examples.

Ayala (1972a) introduced a graphical method to visualize frequency dependence by introducing a log-log plot of a linear regression of the sort:

$$\log\ (A \text{ present}/B \text{ present}) = \log \alpha + \beta \log\ (A \text{ mating}/B \text{ mating})$$

where α and β are constants. Figure 4-14 shows the four examples in table 4-11 graphed in this manner. Obviously, for frequency dependence, $\beta \neq 1$. Sufficient condi-

TABLE 4-11 Examples of the rare-male effect in *Drosophila*.

Frequency of Male	I	II	III	IV
90%	0.13	0.32	0.81	0.45
80%	0.32	0.44	1.11	0.35
70%	0.64	0.59	1.76	0.79
60%	0.47	0.37	1.25	0.50
50%	0.96	0.74	1.32	1.00
40%	1.82	1.36	1.95	1.59
30%	1.83	1.22	1.47	1.42
20%	1.69	2.88	1.88	2.96
10%	1.45	3.71	2.16	2.43

Note: The values shown are the cross product *R*, which is the relative mating success (see text). Case I is for AR and CH gene arrangements in *D. pseudoobscura* from a population selected for positive geotaxis (Ehrman, 1967); case II is the same for a population selected for negative geotaxis; case III is for the same species with gene arrangements PP and AR (Spiess, 1968); and case IV is for two geographic strains of *D. persimilis* that also differed in inversion types (Spiess and Spiess, 1969). In all cases only two types of males were studied, so the frequencies can be given by the single figure in the first column.

tions for a stable polymorphism are $\alpha < 1$ and $\beta < 1$, which are the observed conditions for the data here. The value of this graphical approach is that it allows one to picture the overall pattern of mating as well as providing a statistical test over the whole data set (test for the significance of a linear regression) instead of using a chi-squared or similar test for mating at each frequency. Adams and Duncan (1979) improve on the sensitivity of detecting frequency dependence by introducing a maximum-likelihood test, which successfully detected frequency dependence when Ayala's regression method failed to.

Clearly, the mating success of males carrying different inversions is frequency dependent: when rare, they have an advantage. Because this type of selection can stably maintain a polymorphism (Anderson, 1969; Wright, 1977), it has received considerable attention from evolutionary geneticists. Table 4-12 lists several species of *Drosophila* where the phenomenon has been studied for a variety of genotypic differences among males. Also included are studies where genotypically identical males are morphologically differentiated by such things as rearing at different temperatures. While not all studies yielded positive results (i.e., the observation of an RME), enough did to indicate that the phenomenon is not restricted to a few types of male differences in one or two species. In fact, the RME has been observed in many non-Drosophilid insects and even vertebrates (cited in Knoppien, 1985).

Most studies of the sort discussed here and those listed in table 4-12 were performed in small mating chambers with matings among virgins recorded. Unless a visible mutant is used, to identify the different types of males the end of a wing is clipped. Other details vary among experiments, such as the numbers used, whether mating pairs are removed, the length of observation, whether the genotypes of females

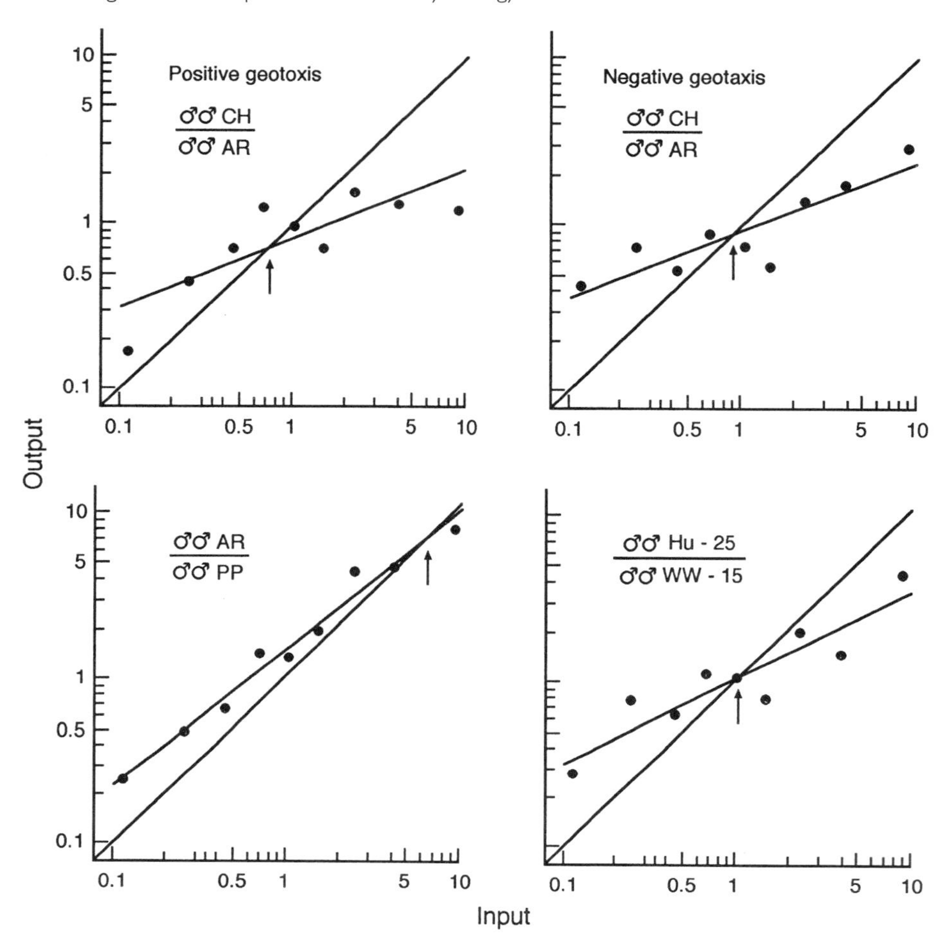

FIGURE 4-14 Illustration of the Ayala method for detecting frequency-dependent mating. The graphs show the log-log plot of males present (x-axis) to those mating (y-axis). The 1 : 1 line, when crossed, would predict an equilibrium at the arrows. Data and references are in table 4-11. Graphs from Ayala (1972a).

vary, and so forth. All these variables may affect the outcomes and they have been discussed extensively (e.g., Markow et al., 1980; Bryant et al., 1980; Knoppien, 1985; Partridge and Hill, 1984).

At least two sets of experiments were performed quite differently than those discussed thus far. Ehrman (1970b) released *D. pseudoobscura* males and females in a room 75 cubic meters in size; the frequency of a visible mutant varied among the males released in different experiments. The mate of the recaptured females was determined by their offspring. She found a significant RME. Anderson and Brown (1984) performed their experiments in large population cages using alternative karyotypes of *D. pseudoobscura* marked by alternative *Amy* alleles. They too found a significant RME for the homozygotes (ST/ST and CH/CH) but not for the heterozygotes, which

TABLE 4-12 Examples of experiments testing for a rare-male effect for various traits in a number of species of *Drosophila*.

Species	Property Studied	Rare-male Effect Detected?	Reference
D. melanogaster	Mutants	No	Alvarez & Fontdevila, 1981
	Genetic background	Yes	Anxolabéhère, 1980
	Mutants	Yes	Bundgaard & Christiansen, 1972
	Environment induced phenotype-scent	Yes	Dal Molin, 1979
	Temperature of rearing—size	No	Knoppien, 1984
	Adh allozymes	No	Kojima, 1971
			Pot et al., 1980
	Mutants	Yes	de Magalhaes & Rodrigues-Pereira, 1976
	Coisogenic mutants	No	Markow, 1978, 1980
	Genetic background	Yes/No	Markow et al., 1980
	Lap allozymes	Yes	Nassar, 1979
	Mutants	No	Partridge & Gardner, 1983
	Geographic origin	Yes	Petit & Nouaud, 1975
	Mutants	No	Rasmuson & Ljung, 1973
	Mutants	Yes	Spiess & Kruckeberg, 1980
	Genetic background	Yes	Spiess & Spiess, 1969
			Tardiff & Murnik, 1975
D. pseudoobscura	Geographic origin	Yes	Ehrman, 1969a
	Temperature of rearing	Yes	Ehrman, 1969a
	Mutants	Yes	Ehrman, 1969, 1970a, 1970b
	Karyotypes	Yes	Ehrman, 1967, 1968, 1969
	Karyotypes/allozymes	Yes/No	Ehrman et al., 1977
	Karyotypes	Yes	Spiess, 1968
	Allozymes	No	Fontdevila & Mendez, 1979
D. persimilis	Karyotypes	Yes	Spiess, in Ehrman, 1966
	Geographic origin	Yes	Spiess & Spiess, 1969
	Temperature of rearing	Yes	Spiess & Spiess, 1969
D. funebris	Karyotypes	??	Borisov, 1970b
D. willistoni	Mutants	No	Ehrman, 1967
	Geographic origin	Yes	Ehrman & Petit, 1968
D. tropicalis	Geographic origin	Yes	Ehrman & Petit, 1968
D. equinoxialis	Geographic origin	Yes	Ehrman & Petit, 1968
D. paulistorum	Geographic origin	Yes/No	Ehrman, 1970a
D. immigrans	Geographic origin	Yes/No	Ehrman, 1972b
			Adams & Duncan, 1979
D. pavani/gaucha	Species and/or mutants	Yes/No	Ehrman et al., 1972

Abbreviated from Knoppien (1985), which also contains references to non-Drosophilid studies.

actually suffered a mating disadvantage at low frequency. Finally we note in passing (and will take it up in chapter 6) that the only type of selection ever detected in a natural population of *D. pseudoobscura* has been one for mating ability of males; the two studies of this sort each indicated the rarer males in natural populations had a mating advantage (Anderson et al., 1979; Salcedà and Anderson, 1988).

The mechanism underlying the RME has been extensively discussed and debated (a good summary is Spiess, 1982). Suffice it to say that little is actually known about the details of the mechanism, except that there is some evidence that olfactory cues may be important. This comes from experiments where a small cheesecloth was placed between two chambers, sometimes with a space and sometimes not (Ehrman, 1970c; Ehrman and Spiess, 1969). Different types of males in the experimental mating chamber were usually in equal numbers, while the adjoining chamber contained an excess of one type. The presence of one type of male in the adjoining chamber could induce the RME for the alternative male in the mating chamber. Ehrman (1972a) further demonstrated that this could also be induced by dead crushed males and/or organic extractions from males (see also Leonard et al., 1974).

There is evidence of one behavioral characteristic of females that may contribute to the RME. This is the observation that females typically do not mate with the first male to court them, a phenomenon that has been observed with three different species (Spiess, 1982, 1987; Spiess and Carson, 1981). Since the more common male is more likely to be the first to court, there may then be a continued discrimination against his type.

It should also be noted that most explanations of the RME focus on females and their ability to discriminate. While good arguments can be made that, by and large, females do make the choice (e.g., Spiess, 1982), this is not universally accepted. H. Spieth, a long-time student of *Drosophila* mating behavior, has maintained that much of the choice of mate is determined by males (Spieth and Ringo, 1983 and references therein).

The reality of the RME has been questioned. The criticisms fall into two categories, experimental artifacts and statistical ambiguities. On the experimental side, the practice of clipping wings is thought to introduce artifacts (Bryant et al., 1980). Markow (1980) has been able to simulate an RME by taking males from different areas of the holding vials. When only a few males of one type are needed for an experiment (the rare type), the tendency is to take those males at the top of the vial, while for the common male one needs to dig deeper into the vial. If more vigorous males tend to be near the tops of vials, an artifactual RME results, as Markow showed empirically. The statistical arguments take on various forms (e.g., O'Donald, 1978, 1980; Knoppien, 1985; Partridge and Hill, 1984). It can be shown that just a difference in mating ability among males can lead to an apparent RME. When females have a set proportion of matings with one type of male, this can also lead to an apparent RME. Many of these critiques have been addressed by proponents of the RME (e.g., Leonard and Ehrman, 1983; Spiess, 1982; Spiess and Dapples, 1981).

Is the RME in *Drosophila* real? For some combinations of species and genotypes (e.g., inversions in *D. pseudoobscura*), the evidence seems compelling to me. The statistical artifacts cannot explain the magnitude of the effect and, furthermore, most of the statistical artifacts lead to a one-sided RME (i.e., only one male genotype showing a minority advantage; Knoppien, 1985), so this cannot explain the two-sided

RME data in table 4-11. The artifact introduced by assuming a fixed proportion of matings by females (e.g., Partridge and Hill, 1984) can lead to only a slight two-sided RME, not nearly to the magnitude observed (table 4-11). The positive results of the cage experiments (Anderson and Brown, 1984) and experiments with release into a large room (Ehrman, 1970b) cannot be explained by any of the criticism pertaining to the small mating chamber experiments; neither can the induction of an RME by placing males or odor extracts in an adjoining chamber.

The critiques of the RME have led to a much fuller appreciation of the complexity of designing and interpreting the experiments. Caution needs to be taken in this regard. Nevertheless, the flaws in some of the experiments—at least recognized as flaws in retrospect—do not invalidate the whole body of data in support of the RME. Of perhaps more general interest is the question of how common is RME, especially in nature, and could it be a major factor affecting the maintenance of genetic variability? While in the laboratory it is often—though hardly universally (table 4-12)—observed for a number of species and genotypes, Lewontin (1974) has pointed out that the ubiquity of the RME in real populations must be questioned. He basis his argument on the (correct) notion that in real populations all males are of a rare genotype, at least at some loci; in fact, as seen in chapter 2 (and again in chapter 10) the evidence is that every individual is genetically unique. So how could RME be universal? On the other hand, recall that in the laboratory experiments male mating success is an extremely important, if not the most important, component of fitness in *Drosophila* and that the only component of fitness ever detected in nature is for male mating success consistent with RME. Thus, it cannot be simply dismissed at this point. It is real (in some cases at least) and it may well be important in natural populations.

Sex Ratio and Segregation Distortion

"Mendelism is a magnificent invention for fairly testing genes in many combinations, like an elegant factorial experimental design. Yet it is vulnerable at many points and is in constant danger of subversion by cheaters that seem particularly adept at finding such points" (Crow, 1988, p. 391). Several cases of this subversion are known in *Drosophila* and have been tabulated by Hurst and Pomiankowski (1991). Clearly, such deviations from Mendelism can have profound effects from an evolutionary perspective, so some discussion of the better studied systems in *Drosophila* is warranted. Sandler and Novitski (1957) introduced the term *meiotic drive* to describe situations where a gene or chromosome is transmitted in higher frequency than predicted by normal 1 : 1 segregation. Reviews of this subject include Hartl and Hiraizumi (1976), Crow (1979, 1991), Sandler and Golic (1985) and Lyttle (1991).

D. pseudoobscura SR

I described the sex ratio (SR) condition of *D. pseudoobscura* (chapter 3), which is caused by an X chromosome differing from the standard X (ST) by three independent inversions. Male carriers of this chromosome produce almost 100% daughters because Y-bearing sperm degenerate. More is known about this system from work in the laboratory.

The first question addressed is the distribution of genes on SR chromosomes that cause the segregation distortion. Initially, it was thought the genes resided in the two proximal inversions (figure 3-15) because a recombinant was found that had only the distal telomeric inversion. It did not cause the SR condition (Dobzhansky and Epling, 1944). Much more detailed studies by Wu and Beckenbach (1983) revealed this interpretation was wrong. They took advantage of the fact that the normal ST X chromosome of *D. persimilis* is identical in gene order to SR of *D. pseudoobscura*, and the fact that female hybrids between the two species are fertile. They used markers on the chromosome (indicated in figure 3-15) to follow the effects of recombinant X chromosomes in the backcross progeny. Their results indicated that all parts of the SR chromosome were required to produce the SR condition; the genes causing the condition are spread throughout the chromosome and missing any of them results in loss of condition. This is consistent with Dobzhansky and Epling's observation, but it is simply that lack of phenotype (SR distortion) is due to *any* part of the SR chromosome being missing.

These results are interesting in light of our discussions about coadaptation in inversions. If one considers the adaptation of SR chromosomes to be the fact they are inherited at a higher rate than ST chromosomes, then Wu and Beckenbach's result clearly supports the coadaptation theory, in that it is epistatic interaction of genes all along the chromosome, held in linkage by the inversions, that causes the high fitness of SR chromosomes.

Considering the segregation advantage, or meiotic drive, of SR chromosomes over ST, one would predict that they should reach high frequency in populations, if not become fixed, at which time the population would go extinct due to lack of fertile males. However, as we saw (table 3-12) SR does not become fixed, but reaches a maximum of only about 20% in some populations, while being completely absent in others. Clearly, something is acting to suppress the predicted rise of this chromosome. This problem has been addressed by a number of laboratory studies, the results of which are somewhat confusing and at times contradictory, except that all show in one way or another that SR chromosomes do have some fitness deficits with respect to ST.

Wallace (1948) was the first to address this problem. When placed in the classical population cages with ST, SR chromosomes are rapidly eliminated, an observation confirmed by Anderson (1968). Given that SR is expected to have about a 50% advantage over ST based on segregation alone, the factor(s) acting to eliminate SR must be strong. Wallace found both male SR and female SR/SR suffered viability and adult component deficits compared to ST. Edwards (1961) studied the theoretical implications of Wallace's fitness estimates and showed they could not lead to a stable polymorphism, in agreement with Wallace's observations.

Policansky and Ellison (1970) produced evidence that the action of SR is to produce nonfunctional Y-bearing sperm, making it logical to test the possibility that SR males suffered in regard to fertility. Policansky (1974) interpreted his results as indicating SR-inseminated females produce about half as many offspring as ST-inseminated females, all of them female. Beckenbach (1978) found, to the contrary, that under most conditions in the laboratory, SR-inseminated females and ST-inseminated females produced about the same number of offspring. The exception was that if males were multiply mated in quick succession, SR males became exhausted of sperm faster than ST males. Policansky (1979) performed a "seminatural" experiment by releasing SR males into a natural population. Females were then collected and the sex ratio and size of their broods recorded. The results agreed with Beckenbach in that SR-inseminated females and ST-inseminated females produced about the same number of offspring. But, Policansky now found that SR males suffered a mating disadvantage, inseminating females at a rate about half that of ST males. Others (e.g., Curtsinger and Feldman, 1980) could detect no significant deviation from random mating of SR and ST males in the laboratory. Wu (1983a), on the other hand, demonstrated that with respect to virgin females (usually used in laboratory mating experiments) SR and ST males were equally successful in mating; however, nonvirgin females preferred ST to SR. Given that multiple insemination is a common occurrence in natural populations of many species of *Drosophila* (including *D. pseudoobscura*), selection based on the mate choice of nonvirgin females is feasible. Beckenbach (1981) found no evidence for differential sperm displacement by SR and ST males, while Wu (1983a) did. Wu emphasizes the importance of this component of selection—what he termed *virility*, which was used to describe both the success of males in mating and the ability of their sperm to displace sperm from previous matings. In a theoretical study, Wu (1983b) showed that this form of selection was sufficient to produce a stable polymorphism.

Contrary to the studies just discussed, Curtsinger and Feldman (1980), in a detailed study of components of fitness, found little evidence for a male adult component of fitness for the SR system in *D. pseudoobscura.* Rather, they detected a heterotic effect for fertility of females: SR/ST heterozygotes produced more eggs. They also found viability differences that theoretically were sufficient to maintain a stable polymorphism. Beckenbach (1983), on the other hand, found no evidence for viability selection. Rather, he found that both SR/SR females and SR males suffered in fertility and, contrary to his previous work, that SR males had a mating disadvantage compared to ST males. Beckenbach argues that viability effects are density-dependent and only seen under limited conditions.

Theoretical treatments of this system have also produced contradictory conclusions. Curtsinger and Feldman (1980) found that their estimates of viability differences led to a predicted stable polymorphism, while the conditions for a stable polymorphism based on adult components of fitness were very restrictive and unlikely to occur. Wu (1983b), on the other hand, found his measurements of virility were sufficient to maintain a stable polymorphism. Thompson and Feldman (1975) treat this subject more generally.

It is clear the literature on the SR system of *D. pseudoobscura* is confusing, and it is difficult to draw any generalizations, except that virtually every study finds some selective disadvantage associated with SR chromosomes. Part of the confusion may stem from variation in experimental design and laboratory environments. Furthermore, little attention has been paid to the origin of the chromosomes studied. If, as in the case with the third-chromosome polymorphism of this species, there are population-specific differences among the identical gene arrangements, this could also account for the apparently contradictory outcome of laboratory studies. Clearly this system merits further study, especially in light of Bryant et al.'s (1982) observation that SR has significant populational impact (figure 3-16) and is not simply a laboratory curiosity.

SR Endosymbionts

A second kind of sex-ratio system caused by an endosymbiont occurs in several species, most notably in *D. willistoni* and its relatives (Williamson and Poulson, 1979; Ebbert, 1993). While this is not a case of meiotic drive as it seems to be caused solely by the bacteria, it is appropriate to discuss it in the context of the sex-ratio distortion. Bacteria are transmitted through the egg cytoplasm; eggs fertilized with X-bearing sperm develop normally into females, while eggs fertilized with Y-bearing sperm fail to develop and die early in embryogenesis. Phenotypically the trait bears a resemblance to the SR condition just discussed in that only female offspring are produced, but differs in its cause as well as in the fact that affected females are the determining factor in producing the condition. This produces an important difference: while with the *D. pseudoobscura* SR conditions, males suffer slight or no decrease in fertility (due probably to excess production of sperm so that even with 50% nonfunctional sperm, males are nearly fully fertile), half of all eggs of females of *D. willistoni* infected with the endosymbiont die. Clearly, there is a selective disadvantage associated with the infection. Ebbert (1988, 1991, 1995) has studied this system in some detail with regard to evolutionary and ecological implications; Ikeda (1970) has done similar experiments for a comparable system in *D. bifasciata.*

The fitness effects of infection with the *D. willistoni* spiroplasma in several combinations were measured by Ebbert (1991). She had one strain of bacteria that induced the sex ratio condition called WSRO ("*willistoni* sex ratio organism") and a spontaneous mutant of this bacterial strain that had lost its effect on the sex ratio but was still maternally transmitted called ARO ("avirulent Recife organisms"). She studied these two bacteria in three different host strains. Obviously, in terms of overall fecundity, WSRO-infected females suffered about a 50% loss compared to uninfected females. Mothers infected with either WSRO and ARO also suffered other fitness deficits such as lower production of daughters, reduced longevity, and so forth. One exception was that with some combinations of host and symbiont, infected females had an initially higher fecundity early in the reproductive period, which in theory (e.g., Lewontin, 1965) could have an important impact on fitness. In this regard Malogolowkin-Cohen and Rodrigues-Pereira (1975) showed that *D. nebulosa* females infected with these bacteria had a significantly earlier tendency for mating than uninfected females. In population cages, females infected with WSRO were quickly eliminated, although the avirulent mutant ARO persisted in laboratory populations (Ebbert, 1988). When placed in an unnatural host, *D. pseudoobscura*, the male-lethal WSRO also persisted in laboratory populations. Ebbert (1995) also ran population cages and found the WSRO-infected females were quickly eliminated, while ARO-infected females persisted in the populations. There were also clear density-dependent effects on fitness, so the system is quite complex.

How such infections are maintained in populations is the subject of considerable speculation. One possibility for maintenance of the spiroplasmas in natural populations is it avoids inbreeding. Because daughters from infected females cannot mate with brothers, outcrossing is assured. Werren (1987) has modeled this situation for cytoplasmically transmitted sex ratio, and his models require knowledge of the inbreeding depression of the host and the maternal transmission rate of the symbiont. These are known for *D. willistoni* and Ebbert (1995) has calculated that the spiroplasmas could reach equilibria between 5 and 80%, although observed infection rates in natural populations are on the lower end of this range, seldom exceeding 10%. Another possibility is that the unhatched male eggs supply nutrients for the daughters (Gregg et al., 1990). Finally, the persistence may be solely a selfish trait of the bacteria; because males cannot transmit the bacteria, it behooves the bacteria to kill males if it increases the survival of females. Other theoretical treatments of infectious sex-ratio conditions can be found in Watson (1960), Uyenoyama and Feldman (1978), Bull (1983), and Hurst and Majerus (1993).

An added dimension of intrigue is that at least two viruses are associated with the bacteria (Oishi et al., 1984; Cohen et al., 1987). These viruses can cause clumping of the bacteria and even temporarily "cure" the sex-ratio condition by killing enough of the spiroplasma organisms. The system, in fact, presents at least a three-layer interaction reminiscent of O'Henry's "flea on the back of the dog." It deserves more attention from evolutionary biologists.

Autosomal Meiotic Drive

Autosomal meiotic drive is best known in *D. melanogaster* and has been called segregation distortion or SD. It was discovered by Y. Hiraizumi in a natural population in Madison, Wisconsin (Hiraizumi et al., 1960; Sandler et al., 1959) and has since been found in many natural populations. The condition is located on the second chromosome; males (but not females) heterozygous for SD transmit the SD-bearing chromosome at a rate greater than 50%. As with SR X chromosomes in *D. pseudoobscura*, the mode of action appears to be inducing sperm carrying the alternative chromosome to become nonfunctional. The genetic basis of the segregation distortion is evidently due to two major genes (or gene complexes), the *Sd* locus located in the euchromatin of 2L and the responder locus, *Rsp*, located in the proximal heterochromatin of the 2R (Temin et al., 1991). The presence of *Sd* will distort the segregation of the chromosome bearing it only if the alternative homologue has a sensitive allele at the *Rsp* locus. *Rsp* has a complex array of alleles producing a continuum of sensitivity to the action of *Sd*.

The molecular structure and evolution of this system have been studied (Lyttle, 1991; Wu and Hammer, 1990). The *Sd* gene encodes what appears to be a topoisomerase-like protein. Intriguingly, *Rsp* appears to be a typical eukaryotic satellite DNA. The sensitivity to the effect of *Sd* depends on the number of 120-bp repeats in this satellite, sensitive alleles having 700 to 2500 copies and insensitive resistant alleles having 100 to 200. This is consistent with the Hoechst staining pattern of second chromosomes: *Rsp* sensitive alleles stain much more darkly in the appropriate region of the heterochromatin than do resistant alleles (Pimpinelli and Dimitri, 1989). This is one of the very few instances of a phenotype being associated with a satellite DNA (see chapter 9).

While this system has been studied extensively from a genetic, and more recently a molecular standpoint, rather less has been done from an evolutionary perspective. One set of experiments performed by Lyttle (1977, 1979, 1981) does deserve mention in this regard. Lyttle translocated the region of the second chromosome carrying the *Sd* gene to a Y chromosome in *D. melanogaster*, producing a pseudo-Y, or pY, which was being driven so that males would predominate in offspring of males carrying pY.

This was not due to elimination of X-bearing sperm, as the segregation target remained the SD^+ autosome; rather, the female zygotes failed to develop due to a lack of the autosome segment translocated to the pY. The strength of the pY-drive varied from 94 to 100% in production of sons. The absolute number of sons produced by these males exceeded that of normal males, so the frequency of pY rose in the populations. Eventually, the pY chromosomes completely replaced normal Y chromosomes in all populations, which meant that in strains where the pY produced 100% sons, the population went extinct, usually within seven to eight generations. In cases where 94% sons were produced, the populations persisted, the low number of females being sufficient to avert extinction. Interestingly, these populations also began to evolve modifiers of the pY-drive so that an increasing number of females was produced. One population evolved an aneuploid sex system with XXY females and XYY males, which came to an equilibrium of about 40% females.

The power of genetic manipulation of *D. melanogaster*, including being able to move the segregation-distorting factors around the genome, along with the molecular characterization of this meiotic drive system, makes it an ideal candidate for evolutionary studies of the evolution of sex ratios, population dynamics of driven chromosomes, and so forth. For example, Lyttle's experiments provide evidence supporting the speculation that meiotic drive may be involved in evolving reproductive isolation and could account for the observation that the heterogametic sex is often the sterile sex in interspecific hybrids, the so-called Haldane's rule (Frank, 1991; Hurst and Pomiankowski, 1991). The argument goes that in cases of meiotic drive, especially for sex chromosomes, there will be strong selection to evolve modifiers to repress the action of the drive, much as occurred in Lyttle's experiments. In an interspecific hybrid the X and Y chromosomes have not coevolved to repress each other's drive systems; sterility is the result. This is discussed further in chapter 7.

Heterogeneous Environments and Habitat Choice

Since the now-classic paper of Levene (1953), there has been considerable interest in the role that heterogeneity of the environment plays in maintaining genetic polymorphisms (see Hedrick 1990 and references therein for development of the theory). Direct laboratory investigation of this issue has taken two forms: monitoring levels of genetic variation in replicate populations held in relatively constant and heterogeneous environments, and studies of habitat choice. This latter issue, namely whether different genotypes choose different parts of the niche space, is especially important as it makes the conditions for stable maintenance of a polymorphism much less stringent.

At least seven experiments manipulating the variation of the environment have been conducted on laboratory populations of *Drosophila* (table 4-13). All of these experiments were monitored by following variability at allozyme loci. The results of one such experiment are shown in table 4-14. In this case there is clearly a positive relationship between the heterogeneity of the environment and the level of polymorphism being maintained in these populations. The variables included spatial variation (food), temporal variation (temperature), and a biotic factor (a competitor). Similar positive associations between environmental heterogeneity and genetic heterozygosity were found by Powell (1971) and McDonald and Ayala (1974).

TABLE 4-13 Experiments to test the effects of heterogeneous environments on maintenance of genetic (allozyme) variability in laboratory populations of *Drosophila*.

Species	Initial Population		Environmental Factors	Reference
	Number of Lines	Time in Culture		
D. willistoni	>500	2 generations	Food Temperature	Powell, 1971
D. pseudoobscura	141	2 months	Food Temperature Light	McDonald & Ayala, 1974
D. pseudoobscura	23	2–4 generations	Food Temperature Competitor	Powell & Wistrand, 1978
D. melanogaster	30	130 generations	Food Temperature	Minawa & Birley, 1978
D. melanogaster	?	?	Alcohol Temperature Yeasts	Oakeshott, 1979
D. melanogaster	400	6 years	Food Temperature Light	Yamazaki et al., 1983
D. melanogaster	100	2 generations	Food	Haley & Birley, 1983

From Hedrick (1990) with additions.

TABLE 4-14 Summary of results from Powell and Wistrand (1978) on the effects of heterogeneous environments on the maintenance of genetic variation.

Treatment	12 Months	18 Months	24 Months	Mean ± S. E.
0 Variables				
PS 16 A	0.245	0.232	0.221	0.223 ± 0.009
PS 16 B	0.280	0.267	0.232	0.260 ± 0.009
PS 25 A	0.253	0.243	0.225	0.240 ± 0.006
PS 25 B	0.236	0.281	0.199	0.213 ± 0.009
1 Variable				
PS 16 A/B	0.318	0.305	0.298	0.306 ± 0.006
PS 16/25 A	0.312	0.304	0.301	0.306 ± 0.005
PS 16/25 B	0.309	0.306	0.305	0.306 ± 0.008
PS 25 A/B	0.266	0.266	0.268	0.267 ± 0.011
PS/PER 25 A	0.280	0.273	0.272	0.275 ± 0.004
2 Variables				
PS A/B 16/25	0.326	0.322	0.318	0.322 ± 0.004
3 Variables				
PS/PER A/B 16/25	0.338	0.333	0.392	0.336 ± 0.005

Note: Numbers shown are the average heterozygosity at nine polymorphic enzyme loci at various times after initiating populations. PS and PER refer to whether only *D. pseudoobscura* was present in the cage (PS) or whether *D. persimilis* was added as a competitor (PS/PER). Under Treatment, numbers refer to temperature and A or B to two types of medium. For each treatment there were 3 to 6 replicates with the average over replicates given in the table.

However, no correspondence between level of environmental heterogeneity and genetic variability was noted in the other four studies listed in table 4-13. In judging the relevance of these studies, it is important to consider the details of the experiments, especially the nature of the starting populations. In all the positive experiments (Powell, 1971; McDonald and Ayala, 1974; Powell and Wistrand, 1978) the flies used to initiate the populations had been in laboratory culture only a few generations. Two of the negative experiments (Minawa and Birley, 1978; Yamazaki et al., 1983) were begun by strains with a long history of laboratory culture, longer in fact than the length of the subsequent experiments. When maintained under standard conditions in a laboratory, the flies would have already experienced a long period in a relatively constant environment and would be poor candidates for beginning such experiments. In fact, as emphasized by Powell and Wistrand (1978), the effect of heterogeneous environments is to slow the loss of variation. Given the length of the experiments (about two years), it isn't clear whether the levels would remain stable. In one case (Haley and Birley, 1983), negative results were observed with freshly collected flies, although in this case only one environmental factor was varied, a spatial one. In the remaining case (Oakeshott, 1979) the published description is not detailed enough to understand the nature of the starting population.

It is a nagging problem that not all experiments are uniformly positive in indicating an association between environmental heterogeneity and genetic variation. However, considering the differences in experimental designs, the nature of starting material, and the different species used, the lack of uniformity is perhaps to be expected. In cases where freshly collected flies of species with truly natural populations (i.e., *D. willistoni* and *D. pseudoobscura*) are utilized, the positive relationship has held. In experiments employing the cosmopolitan human commensal *D. melanogaster*, using strains that had spent considerable time in laboratory culture before the start of experiments, there is no relationship between environmental and genetic variation.

What about habitat choice? Given a choice of environments, do different genotypes nonrandomly assort themselves, and if so, is there a positive correlation with their fitness in the chosen environment? At least five experiments dealing with larval behavior have yielded positive answers to these questions.

De Souza et al. (1970) found population cages of *D. willistoni* had evolved two types of larvae, those that pupated inside the food cups and those that pupated outside the cups. The difference was shown to be genetically determined and positively correlated with survival in the two habitats.

Sokolowski and colleagues (Sokolowski et al., 1986; Rodriguez et al., 1992) made similar observations with natural and laboratory populations of *D. melanogaster*. They found that larvae choose to pupate on fallen fruit in an orchard in one of four places: on the top of the fruit, on the underside of the fruit, on the ground under the fruit, or buried in the soil under the fruit. Two types of larvae were distinguished, "rovers" and "sitters"; the different behaviors have a major genetic control on the second chromosome. The frequencies of these genotypes varied in different pupation sites as well as in laboratory tests. Evidently polymorphism is being maintained by temporal fluctuation in the water content: during dry periods the fruit is the better pupation site and when moisture is high, the soil is better.

Taylor and Condra (1983; see also Powell and Taylor, 1979) performed a series of experiments to test for larval behavior when given a choice of two different types of medium. They used a design summarized in figure 4-15. A petri dish was divided into two parts with different media in each half, with no barrier to movement of larvae across media. They tested 10 different media preparations consisting of different sugars, using 12 different mutant strains of *D. pseudoobscura*; mutants were used for ease of identification after choosing. They first tested the fitness of each strain on each of the different media, then placed eggs in the split dish to see which medium larvae preferred. Table 4-15 summarizes some results. There was a positive correlation between the fitness of a genotype on a medium and its preference for spending time on that medium.

Dodd (1984) studied replicate populations of *D. pseudoobscura* that had been maintained on medium in which the only carbohydrate source was starch or maltose, four cages in each regime labeled I to IV (see Powell and Andjelkovic, 1983, for a description of the populations). Dodd first showed that indeed, the populations had diverged genetically with respect to fitness on the two carbohydrate sources: starch-adapted flies exhibited a higher fitness on starch medium compared to maltose-adapted flies, and vice versa. Using a design similar to Taylor and Condra (figure 4-15), Dodd further demonstrated habitat choice for the appropriate medium on which the flies had higher fitness, summarized in table 4-16. The studies were conducted

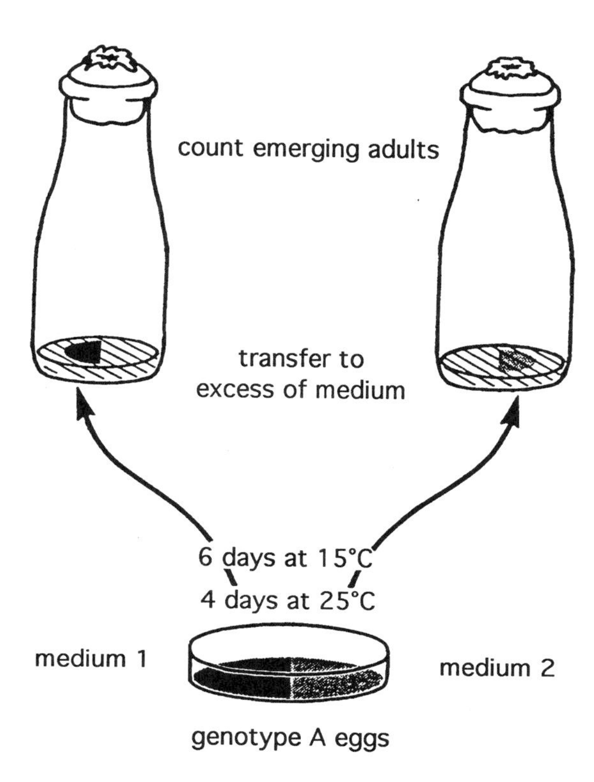

FIGURE 4-15 Design of experiments to detect habitat choice in the laboratory. From Powell and Taylor (1979).

TABLE 4-15 Example of results from Taylor and Condra (1983) on larval choice when placed in a split petri dish with two media (figure 15).

	25°C	15°C
Percent emergence	+0.17	+0.07
Male development time	–0.25	–0.20
Female development time	–0.24	–0.19

Note: The figures given are the mean correlation coefficients between the frequency of choice of medium and three fitness components measured on that medium; correlations above 0.14 are statistically significant at the 0.05 or lower level. The experiments were performed at two different temperatures, 25°C and 15°C. The negative correlations with development time indicate shorter development time, and thus increased fitness.

with larvae one generation removed from the population cages, reared on standard cornmeal-molasses medium, so the response to starch and maltose could not be due to preconditioning or some other environmental factor.

Cavener (1979) studied the response of *D. melanogaster* larvae homozygous for different alleles at the *Adh* locus when given a choice of medium with or without ethanol added. The two *Adh* alleles tested had different levels of enzyme activity using ethanol as a substrate, the F allele having greater activity than the S allele.

TABLE 4-16 Results of habitat choice test for starch-adapted and maltose-adapted *D. pseudoobscura* populations.

		No. Adults Emerging		Percentage
Regime	Population	Starch	Maltose	"Choosing Correctly"
Starch-adapted	I	1747	1663	51.23
	II	1900	1260	59.96
	III	2539	1533	62.35
	IV	1873	1429	56.72
Maltose-adapted	I	2058	2689	56.65
	II	1164	2069	64.00
	III	1527	2348	60.59
	IV	1867	2255	54.71

Note: The first four populations had been maintained for about two years on medium in which starch was the only carbohydrate source. The second four were replicates maintained on medium in which maltose was the only carbohydrate source. The larvae from the populations were given a choice between the two media in a scheme as in figure 4-15. All starch-adapted populations "preferred" the starch medium and all maltose-adapted populations "preferred" the maltose medium. The figures given are the numbers of adults emerging from the larval population after being placed in a bottle with excess medium. The effect of adapted regime is highly significant with a FUNCAT chi-squared value of 789, $p < 0.00001$.

From Dodd (1984).

Cavener's results are in table 4-17. The genotypes with greater *Adh* activity showed an approximate 2 : 1 preference for ethanol supplemented medium, compared to no preference (about a 1 : 1 ratio) for the SS genotypes. Adult oviposition was also studied by Cavener, but no preferences were detected. McDonald (1986) studied the larval behavior of this polymorphism in more detail and found the response of larvae of different genotypes to be dependent on the level of ethanol in the medium. At too high a concentration (10% or more) the higher activity FF genotypes begin to avoid the ethanol. McDonald hypothesizes this is due to more rapid buildup of toxic acetaldehyde, the product of alcohol dehydrogenase.

Finally, there is one experiment that combined a heterogeneous environment with habitat choice. Jones and Probert (1980) studied the white eye mutant (*w*) in populations of *D. simulans.* The premise of the experiment is that wild-type flies are positively phototactic and tend to be essentially blind in low-light conditions. The white mutant, lacking the normal red pigment, is repelled by bright light but tends to have better vision than wild type in low-light conditions. Jones and Probert used double population cages with a vertical partition and a small opening at the top to allow movement between cages. When both cages were illuminated with bright light or both with red light, *w* was eliminated, or nearly so, as expected from previous studies (figure 4-16). However, when one cage had the white light and the other the red light, *w* was not eliminated but was maintained at an appreciable frequency at least through the 30 weeks of the experiment (figure 4-16). Studies of the distribution of the genotypes between cages showed a clear separation of genotypes (figure 4-17). This is perhaps the clearest example of the maintenance of a polymorphism in a heterogeneous environment due to habitat choice.

Migration

Thus far this chapter has dealt almost exclusively with studying selection in the laboratory. We now switch our focus to two other phenomena of considerable interest to population geneticists, migration and genetic drift. These have been less well studied in the laboratory, but a few experiments of interest have been performed.

In a series of experiments, Dobzhansky and Spassky (1967, 1969) selected *D. pseudoobscura* to be positively or negatively phototactic and geotactic. It was found that the realized heritabilities (narrow-sense heritabilities) of these traits was low,

TABLE 4-17 Results of Cavener's (1979) study of preferences of *D. melanogaster* larvae homozygous for different alleles at the *Adh* locus.

Genotype	Ethanol (No. of Larvae)	Control (No. of Larvae)	Approximate Ratio of Ethanol to Control
SS	321	296	1 : 1
FF	376	173	2 : 1

Chi-squared = 32.1, $p < 0.001$

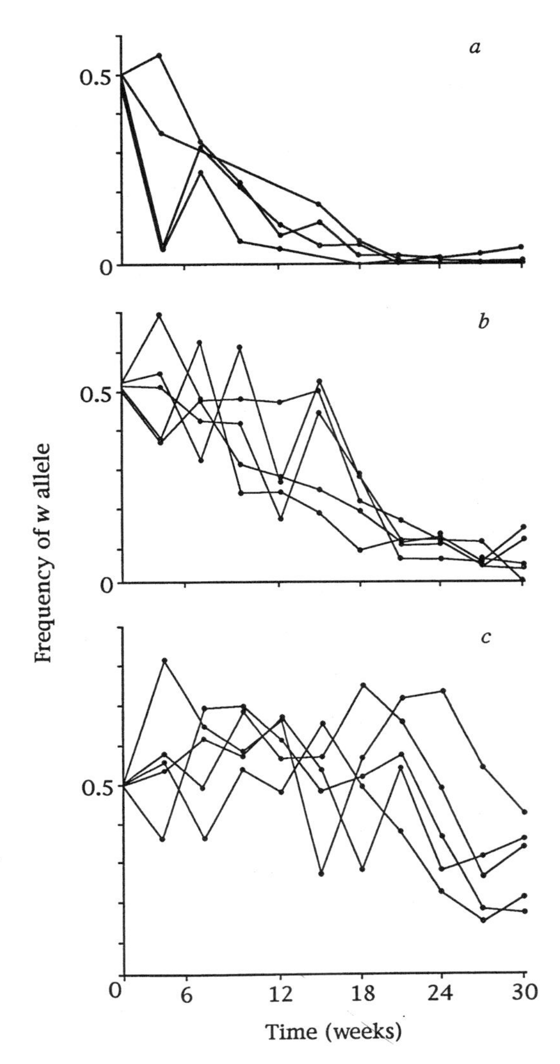

FIGURE 4-16 Changes in frequency of white eye (*w*) in laboratory populations of *D. simulans* kept in normal bright light (a), in red light (b), and in heterogeneous environments with both red and bright light (c). From Jones and Probert (1980).

between 0.02 and 0.10. Dobzhansky, Spassky, and Sved (1969) then studied the effect of migration on the populations. One population, the recipient, was selected for modal behavior; out of 300 individuals of each sex tested, 30 of each closest to the modal score were chosen. The migrants to this recipient population of 60 were 20 flies (10 of each sex) from a selected population that had the worst score for the selected direction. For example, in a donor line being selected for photopositivity, the 20 flies showing the most photonegative behavior were the migrants to the population that was being selected for modal behavior. Table 4-18 shows some results, which were somewhat startling. The recipient population evolved in the same direction as the donor populations, despite the fact that the migrants were phenotypically of the opposite type! The obvious explanation is that the low heritabilities of the traits made the phenotype of any individual a poor predictor of the genes of that individual. (Dobzhansky liked to

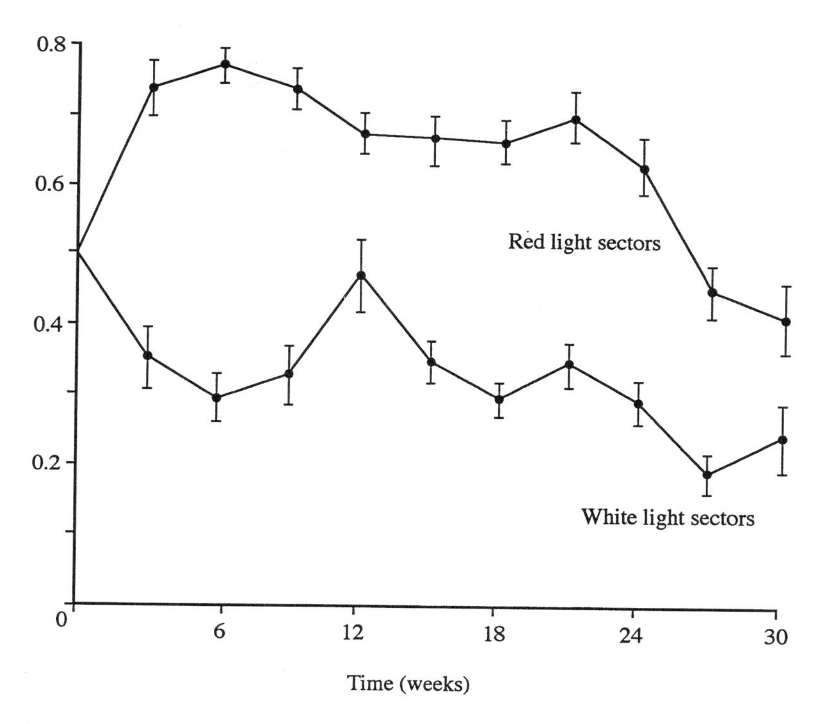

FIGURE 4-17 Distributions of white-eyed adults in the two environments in the Jones and Probert experiments. From Jones and Probert (1980).

TABLE 4-18 The effect of migration on geotaxis and phototaxis in *D. pseudoobscura* populations selected for negative and positive responses.

	Geotactic Selection				Phototactic Selection			
	Plus		Minus		Plus		Minus	
Generation	Donor	Recipient	Donor	Recipient	Donor	Recipient	Donor	Recipient
0	9.3	9.3	9.3	9.3	10.0	10.0	10.0	10.0
10	13.2	12.4	6.4	8.9	13.5	12.0	4.9	7.7
20	13.3	12.9	5.9	7.7	14.3	9.9	2.9	4.1

Note: The recipient populations receive migrants from the selected donor populations whose phenotype (behavior) is opposite that being selected for. For example, the geotactic plus recipient population received geotactically negative flies from the geotactically positively selected population. The numbers given are the scores in the mazes. High numbers mean a high frequency of positive choices toward light or with gravity, and negative numbers indicate a high frequency of choices toward dark or against gravity.

From Dobzhansky et al. (1969).

speculate on the repercussions of these experiments for human behavioral traits that might have similarly low heritabilities. Tongue in cheek, the experimental populations were labeled "plebes" and "aristos" for plebeians and aristocrats.)

Endler (1973) performed a series of migration-selection experiments on *D. melanogaster* populations that at first sight seem quite remarkable. He simulated a selected clinal regime with a series of 15 vials in which the *Bar* eye mutant was selected to be in a frequency of 0.42 in vial 1 to 0.98 in vial 15, with intervening vials differing gradually in increments of 0.04. Natural selection acted to decrease *Bar* in all vials and artificial selection at each generation was needed to maintain the mutation in the populations. In two series of vials, only selection was applied. In two series, selection and migration were imposed. The migration consisted of 20% of the recipient population coming from the vial lower in number and 20% from the vial higher, for a total of 40% migrants. Figure 4-18 shows some results after 35 generations. The surprise was that migration had almost no effect on the cline, despite 40% migration! In fact, Endler went on to model the situation and found the results were not so unpredictable as one might first think. Given that the vials higher and lower were contributing equally to the recipient and that each was being selected to be the same degree different (i.e., ± 0.04), then on average the migrants should resemble the recipient population and thus migration has minimal effect, as observed.

Carson (1961b) performed fascinating experiments designed to detect the effect of migration on population fitness. He used population size as a measure of population fitness (or adaptedness) and maintained a series of populations in small vials, of both wild type and a mutant strain. The populations were equilibrated to a fairly stable size and age distribution with about 100 to 200 adults. He then introduced one haploid set of autosomes; this was done by taking a single F_1 male from a cross between a female of the resident population and a male from the donor (of course, one foreign Y chro-

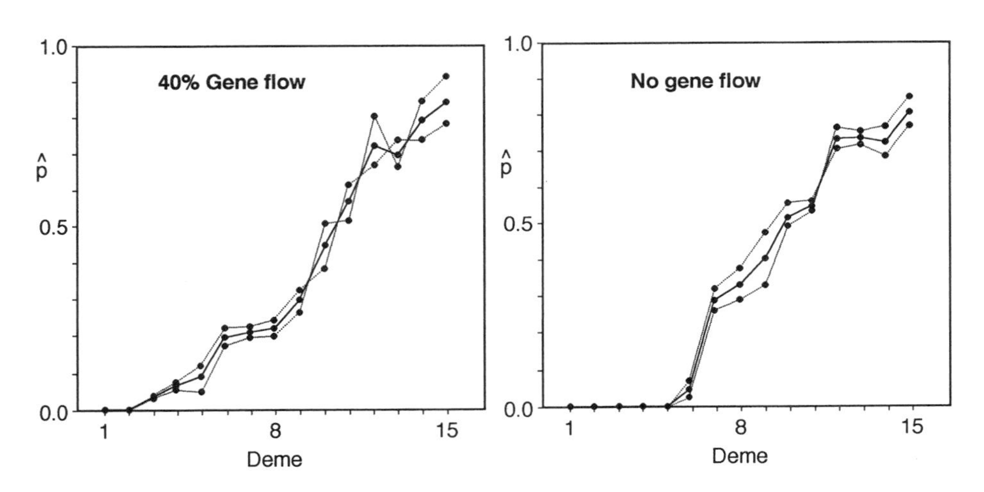

FIGURE 4-18 Example of clines set up in laboratory populations of *D. melanogaster*. A series of demes (vials) were maintained artificially selected for frequencies of mutants along a gradient. In some, no migration was permitted (right) while in others 20% migration occurred from vials on either side of the recipient. From Endler (1973).

mosome was also introduced). Remarkably, whether the single haploid set of autosomes was a mutant set entering a wild-type population, or a wild-type set entering a mutant population, the recipient populations underwent a rapid expansion to about double their previous stable size. This higher fitness was maintained over several generations, after which the populations began to shrink, though never back to their original size. Carson attributed the response to what he termed "luxuriance" rather than to true heterosis; that is, the apparent increase in fitness due to the added heterozygosity introduced by the migrant genome was not due to some kind of overall coadaptation of the migrant genome with the resident; rather, the genomes were simply different. The difference could mask deleterious recessive alleles and the effect would slowly decrease as recombination randomized the migrant genome by interspersing it with the resident chromosomes. This effect of migration of a single haploid genome deserves more attention and further investigation.

Genetic Drift

It is well known that S. Wright was the "father" of genetic drift, a phenomenon sometimes referred to as the Sewall Wright effect. Less well-known is that he participated in critical experiments utilizing *Drosophila* to confirm his theoretical predictions. These experiments consisted of monitoring the frequencies of mutants in populations kept very small, usually four males and four females. The genes studied were *forked* (Kerr and Wright, 1954a), *Bar* (Wright and Kerr, 1954), and *Aristopedia* and *spineless* (Kerr and Wright, 1954b). Wright (chapter 10 in 1977) summarized and reanalyzed this work.

Before presenting an example of their results, we present another virtually identical experiment performed by Buri (1956) using brown eye (bw^{75}) on a genetic background that allowed identification of all three genotypes. This case is greatly simplified from the Kerr and Wright studies, as there was virtually no selection detected for this mutant in these conditions, making the effect of drift particularly clear. Figure 4-19 presents some of Buri's results. Buri used 8 of each sex each generation, so one might think that the effective population size (N_e) for the autosomal gene should be 16. In fact, the dynamics of change and rate of fixation indicate N_e in these experiments was about half the census size, or 7 to 9.

The examples from Kerr and Wright (1954b) and Wright and Kerr (1954) are *aristopedia/spineless* and *Bar*. These give two examples of interaction of selection and drift under two regimes. *Aristopedia* and *spineless* are alleles of a locus giving distinguishable phenotypes for all three genotypes; the two alleles were found to be heterotic in these strains. *Bar* is sex-linked dominant and deleterious and thus should be eliminated. Figure 4-20 shows graphically the predicted distributions of gene frequencies based on genetic drift theory and the observed data points. As with Buri's experiments, the N_e that best fit the data is less than the census size; in the case of the autosomal *aristopedia/spineless* combination the number of parents each generation was 8, but the best fit was for N_e around 5. Given the remarkable congruence of theory and observation evident in figure 4-20, clearly drift theory can predict gene frequency distributions in real populations.

Other studies of *Drosophila* populations in the laboratory have also concluded in virtually all cases that the N_e is always less than the census N. The ratio N_e/N varies

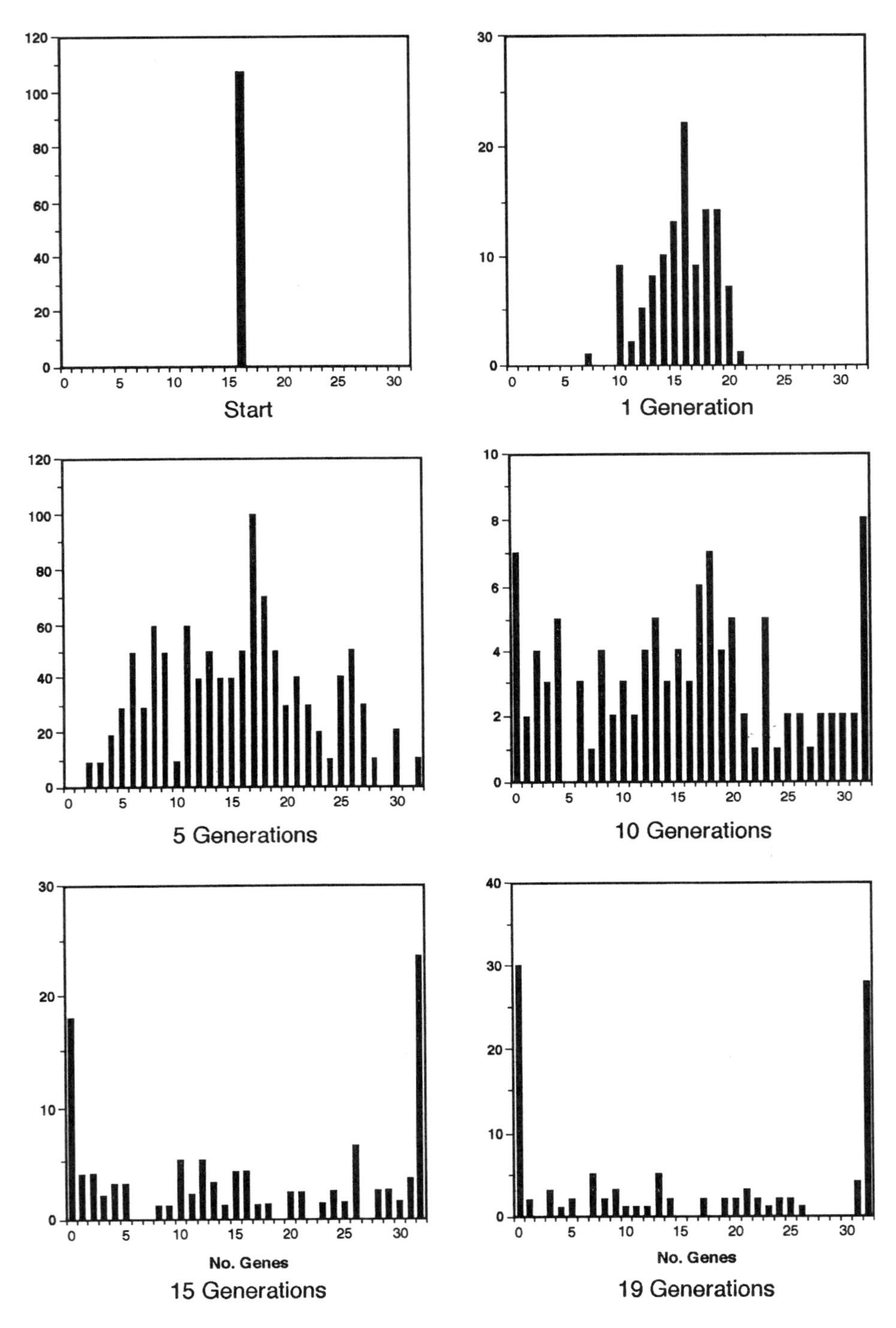

FIGURE 4-19 Example of a genetic-drift experiment in laboratory populations of *D. melanogaster*. 107 populations were begun at a frequency of 50% brown mutant, with a population size of 8 males and 8 females in each generation. Fixation rate was as predicted if N_e were 7 to 9 rather than the census population of 16 (the gene is autosomal). Data from Buri (1956).

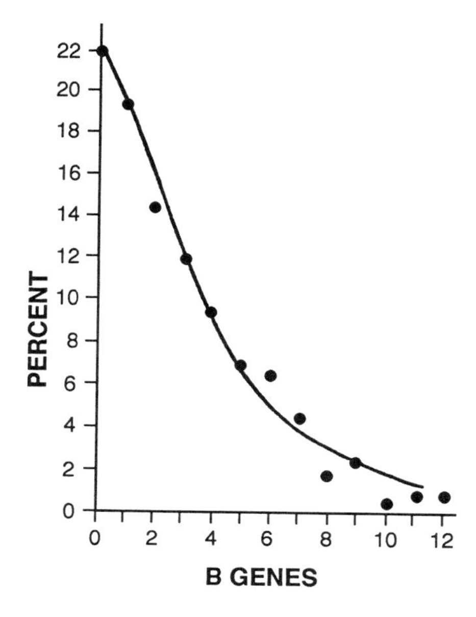

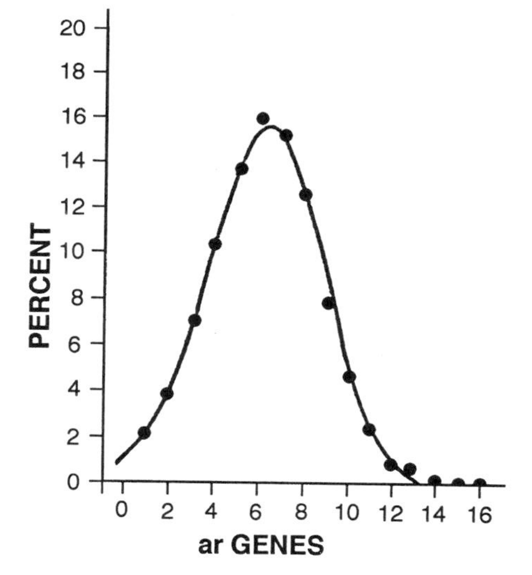

FIGURE 4-20 Examples of results from two experiments of Wright and Kerr. Upper graph is the case of *Bar*, which is expected to be lost due to selection, with drift producing some variance in fixation. The line is that predicted for selection/drift for generations 4–10 and the circles are the observed frequencies of populations with the number of *Bar* genes. The lower graph is for *aristopedia*, which is predicted to form a stable equilibrium with drift causing the variance. Line is that predicted for $2N_e = 10.7$, whereas the census size was 16 each generation (4 males and 4 females). Figures are from Wright (1977).

from about 0.5 as we saw in the Buri experiments to 0.77 in experiments performed by Crow and Morton (1955).

Epilog

Clearly, much has been learned from laboratory studies of *Drosophila* populations. Perhaps most remarkable has been the demonstration and analysis of three major forces in evolution, selection, migration, and genetic drift, and how well theoretical population genetics theory applies to these populations (the fourth factor, mutation, was discussed in chapter 2). The controlled detail possible in these studies is much greater than the essentially descriptive observations of natural populations discussed in the previous two chapters. Yet, as the opening quote hints, there is always a nagging question of just how relevant these laboratory studies are to natural processes. In the case of *Drosophila*, the consistency between observations in the laboratory and in natural populations would indicate the two types of study are not totally disconnected. For example, we saw that morphological variants in natural populations are very rare, and in experimental populations they are at a selective disadvantage. Geographic and temporal patterns of inversions in natural populations give good evidence for natural selection affecting these variants, and laboratory studies have confirmed this. The geographic patterns of allozyme variation in natural populations are consistent with little or no selection on these variants and, by and large, laboratory studies have not been able to detect selection. Male mating success has been the only component of fitness documented for natural populations and this component is most often the major factor in laboratory studies. It has always been assumed that effective population size in nature is less than the census size. Even in the confines of laboratory culture this has been demonstrated to be true. Such consistencies engender optimism that laboratory studies are shedding light on natural processes.

5

Ecology

Perhaps the greatest benefit that can be derived from a consideration of the breeding sites of *Drosophila* is a realistic view of the possibilities of the further investigation of the evolutionary biology of the members of this family. As newer methods of genetic analysis . . . are perfected, it is hoped that those tools will be put to work to answer some of the key questions of micro-evolution. There is indeed a moderately extensive knowledge of the basic ecology of the family on which to build.

H. L. Carson, 1971a

No one can doubt that *Drosophila* are excellent organisms for genetic studies. However, full understanding of evolutionary patterns and mechanisms requires relating the genetics of the organism to the environment in which it is living. That is, the genetic knowledge needs to be placed in an ecological context. Unfortunately, to the majority of *Drosophila* geneticists, the ecology of *Drosophila* can be defined simply as that half-pint bottle in the constant temperature incubator partly filled with a starch–sugar–agar mixture. In this chapter we dispel this notion. In the next chapter we consider studies that connect the kinds of genetic variation discussed in chapters 2 and 3 to the ecology discussed here.

Breeding Sites

As mentioned in chapter 1, most *Drosophila* species breed in rotting plant or fungal material, with microorganisms, especially yeast, as major nutritional sources. It was also pointed out that adults feed at the same site in which larvae develop as well as often having a wider choice of substrates on which to obtain water, sugars, and yeasts. Even if the larvae are monophagic, adults, being opportunists, take appropriate nutrition and liquid wherever it may be found. Another important role of feeding sites is that they often serve as mating arenas. *Drosophila* have no known long-range pheromones and the sexes meet where they come to feed, or as it has been called, "mating at the restaurant" (Lebeyrie, 1978). The exceptions to this rule are Hawaiian *Drosophila*, which often establish separate mating leks. In most species females come to a feeding site to obtain the nutrition needed to develop eggs, and they spend considerable time eating. Males require much less energy and take a bit of moisture and nutrition and then spend their time courting females. Most often females ignore the males and go on feeding and/or ovipositing, although obviously they must occasionally accept a male. The congregation of flies at feeding sites is dependent on temperature, light, humidity, and time of day. Most species of *Drosophila* are inactive below about 12°C and above 30°C, although there are exceptions. For temperate and tropical species, most often active flying and searching for breeding/feeding sites occurs during the first few hours after sunrise and the last few before sunset; however, on particularly cold days, some temperate species may be active only at midday. The rest of the time they are inactive, presumably sequestered in "comfortable" hiding places, although little is known about where adults spend their time when not feeding and breeding. Some desert cactus-breeding species do not leave their feeding/breeding sites, however; there are no other appropriate places for them.

The rest of this section will be concerned with larval breeding sites of selected species. Because the aim is to relate genetics to ecology, the emphasis will be on species that have served as genetic models. However, one cannot resist presenting a few "bizarre" examples.

Tropical Species

D. melanogaster and its close relatives, the *melanogaster* subgroup consisting of *D. simulans, D. yakuba, D. sechellia, D. mauritiana, D. erecta, D. teissieri*, and *D. orena*, are native to sub-Saharan Africa (Tsacas et al., 1981; Lemeunier et al., 1986). Lachaise and Tsacas (1983) review the extensive literature on larval breeding in these species in Africa. Table 5-1 is a summary of that much more extensive review. All the species breed in fruit produced by a number of different plants, although figs (genus *Ficus*) are clearly very important. Some of the species, such as *D. melanogaster* and *D. simulans*, will exploit a very broad range of plants, which is one reason they have evolved to be human commensals and spread around the world. The other six species have not left their native Africa and generally have more larval specificity to native African plants. *D. sechellia* has been found on only a single species of plant, *Morinda citrifolia*, native to the Seychelles Islands; evidently this species has evolved a defense mechanism against an octanoic acid found at toxic levels in this plant (Legal

TABLE 5-1 Larval breeding sites for the *melanogaster* subgroup species in their native habitats in Africa.

Species	Native Plants	Introduced Plants
D. melanogaster	14 species including 6 figs and coffee	6 species
D. simulans	2 fig species	1 species
D. melanogaster	9 species	8 species
D. simulans		
D. sechellia	*Morinda citrifolia* (Rubiacease)	
D. erecta	*Pandanus candelabrum* (Pandanaceae)	
	One fig species	
D. teissieri	7 species including 1 fig	3 species
	1 Euphorbiaceae	
D. yakuba	23 species including 12 figs	5 species

Note: In some studies the sibling species *D. melanogaster* and *D. simulans* were not distinguished.

From Lachaise and Tsacas (1983).

et al., 1994). *D. erecta* is known primarily from one native plant (*Pandamus*) and occasionally a *Ficus*. The ancestral native hosts for *D. mauritiana* are not known as today it "breeds in a great variety of sweet, fermenting resources, most of which are introduced plant species" (Lachaise et al., 1988, p. 176). The breeding site of *D. orena* remains unknown. All these species can be reared in the laboratory and larvae will develop on standard laboratory medium. This shows that the restriction of some species to one or a few plants is not for some absolute nutritional requirement. Much more information on host plants of other African *Drosophila* can be found in Lachaise and Tsacas (1983), including a 32-page table.

One of the more intriguing studies of larval breeding and niche separation of species is the study of Lachaise et al. (1982) on the succession of species on figs in West Africa. Figure 5-1 is a summary of their findings. The fruiting structure of the fig, the syconium, goes through a series of changes that correlate with changes in species of *Drosophila* depositing eggs. The succession begins as soon as the immature flowers appear, and continues through fruit dropping, drying, and decaying, to oblivion. On the order of 25 different species of Drosophilidae utilize this resource, although it must be emphasized that figure 5-1 is a composite for several different species of *Ficus* with no single species harboring all species of *Drosophila.* In fact, there is evidence that some groups of the earliest fig wasp–like Drosophilidae, the *Lissocephala*, may have a one-to-one relationship between species of fig and species of fly (Lachaise and Tsacas, 1983).

Much less is known about the New World tropics, from which the well-studied *willistoni* group emanates. As in Africa, the abundance of fruit in Neotropical forests makes this the prime breeding site. Rotting fruits on forest floors are the breeding sites of the *willistoni* group species (Heed, 1957; da Cunha et al., 1957; Dobzhansky and da Cunha, 1955), which includes *D. willistoni, D. equinoxialis, D. paulistorum, D. tropicalis*, and *D. nebulosa* (among others). Thirty-one species of fruit from 13 families of plants have been documented to be used for larval breeding by the *willistoni* group. It is not only the obvious large fleshy fruits that serve as substrates, but many small dry fruits as well (Heed, 1957; Pipkin, 1965). Little more is known about this group or, indeed, most Neotropical species. As Val et al. (1981, p. 160) conclude in their review of Neotropical *Drosophila*, "there is a surprising lack of information on the environmental situation in which the species are found. Larval breeding sites are known for only a small number of Neotropical species."

Temperate Woodlands

Many of the well-studied *Drosophila* species belong to the *obscura* group, including *D. pseudoobscura, D. persimilis*, and *D. subobscura.* Fortunately, some of these temperate species are well-studied with regard to larval breeding, the review of Shorrocks (1982) being particularly informative. Table 5-2 is a summary of what is known about larval breeding sites for some of the better-studied temperate woodland species. Caution is needed in interpreting this table. At first sight, it would appear that there are large differences in the niche breadth of these species. *D. subobscura* evidently has the broadest niche, breeding on a number of plants, fungi, fruits, fermenting sap (sometimes called slime fluxes), and rotting material. However, the apparent diversity of larval substrates must be at least partly an artifact of the intensity of studies. *D.*

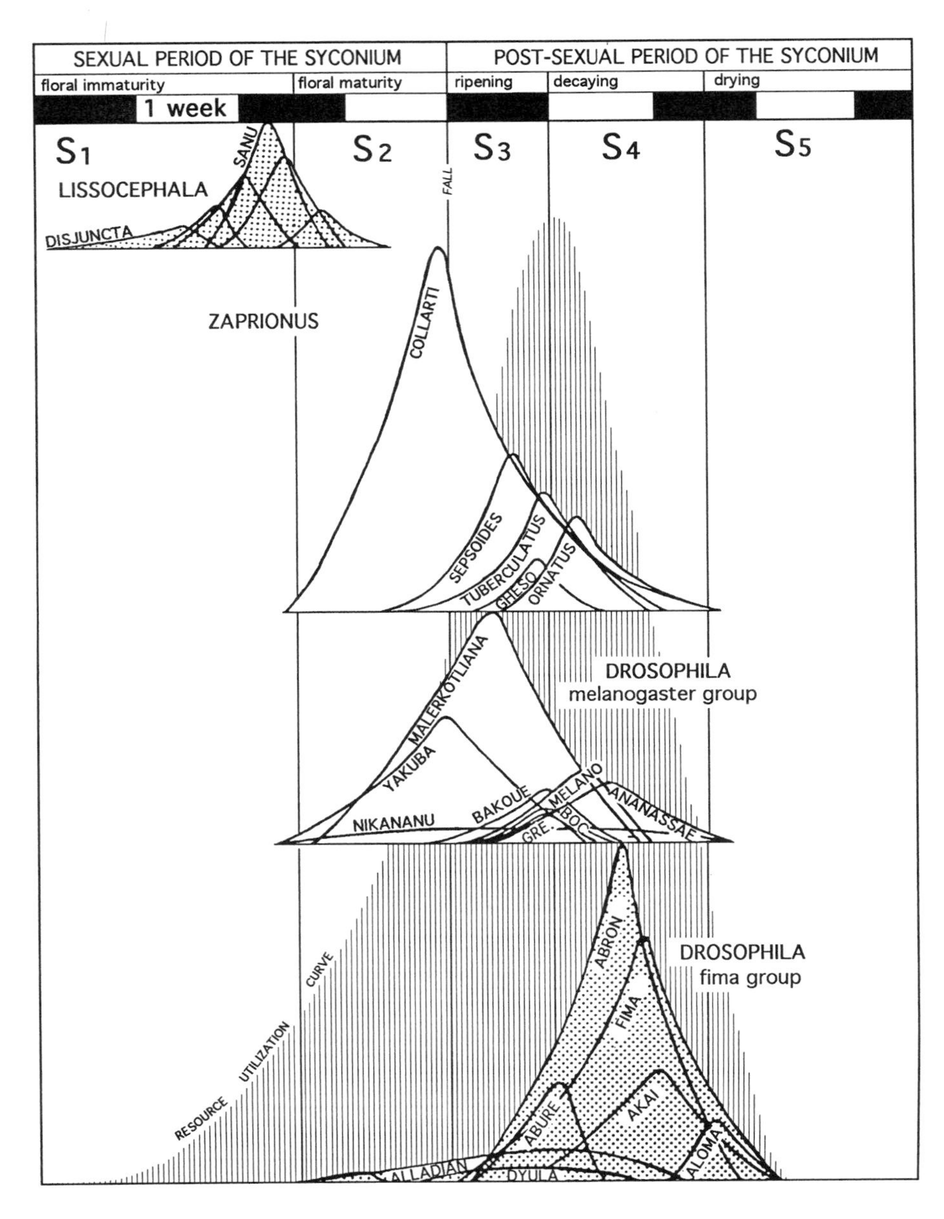

FIGURE 5-1 The temporal distribution of drosophilid eggs on figs in West Africa. This is a composite of different fig species. From Lachaise et al. (1982).

TABLE 5-2 Recorded larval breeding sites for *obscura* group species.

Species	Fruit	Slime Fluxes	Decaying Vegetation	Fungi
D. pseudoobscura		*Quercus kellogii* *Abies concolor* *Vitis californica* *Quercus gambelii* (Heed et al., 1976)	Cacti and Agave (Heed, personal communication)	
D. persimilis		*Quercus kellogii* *Vitis californica*		
D. affinis	*Rubus*, *Podophyllum*	*Vitis californica*		
D. athabasca	*Diosspyros*	*Quercus borealis*		
D. bifasciata		*Betula spp.* *Hydrangea, Zeldova*		*Rhodophyllus*
D. subobscura	*Juglans, Rubus, Sorbus, Malus, Fragaria, Sambucus*, and others	*Ulmus* *Salix* *Acer*	*Allium, Digitalis, Boletis, Lactuca, Magnolia, Oputnia, Taxus, Solanum*, and others	*Amanita, Coprinus, Phallus*, and others
D. obscura	*Rubus, Solanum, Podophyllum, etc.*	*Ulmus* *Salix*	*Magnolia, Taxus*	
D. ambigua	*Sorbus, Solanum*	*Quercus*		

References, where not supplied, are in Shorrocks (1982).

subobscura, a common species in the vicinity of many European laboratories with many favorable attributes for experimental work, has received a considerable amount of attention in this regard. Nevertheless, it is still likely that this species has a broader larval niche than most species studied in the group. Especially noteworthy is that while larval breeding in fallen fruit is generally accepted to be common in the tropics where abundant native fruits exist year-round, fruit breeding can also be very important in temperate species, as documented by Shorrocks (1982) and his colleagues. The small (about 1 cm in diameter), inconspicuous Rowan berry (*Sorbus aucuparia*) is a particularly important breeding site for *D. subobscura*, at least in some localities in England (Begon, 1975).

Just because a larval breeding site has been documented and can be listed in a summary like table 5-2 does not mean the primary breeding site is known. This is especially vexing in the case of *D. pseudoobscura* and *D. persimilis*, two species that have been the subject of a considerable number of genetic studies. Carson's (1951) observation that these species breed in sap fluxes of oaks in the Sierra Nevada Mountains cannot be the whole story for these species. They are found in abundance in areas free of any oaks. Furthermore, extensive efforts to repeat Carson's observations have failed; we (C. E. Taylor, H. L. Carson, K. Kaneshiro, and J. R. Powell, unpublished observations) thoroughly examined oaks and a number of other trees and rotting vegetation in the same localities studied by Carson and, despite a flourishing *D. pseudoobscura* population, were unable to find any larval breeding sites. Furthermore, it is clear this species can be opportunistic, as it has invaded human habitats such as fruit orchards in California where larvae can be found on fallen oranges and other domesticated fruits (e.g., Coyne et al., 1984). Heed (personal communication) has found it breeding in cacti and agave, plants that do not exist in much of the range of the species. This species is clearly polyphagous. It is probable that the primary larval breeding site throughout most of its range has not yet been identified; the same holds for its close relatives, *D. persimilis* and *D. miranda.*

At one time H. Spieth (1988) thought there was evidence that these species might breed in acorns that had been attacked by bruchid beetles, producing a hole for the *Drosophila.* When the beetle pupates and leaves the acorn, considerable frass is left behind in which *Drosophila* could breed. While in the laboratory *D. pseudoobscura* would oviposit in such acorns and the larvae develop, this potential breeding site has never been confirmed to exist in nature, despite efforts by several workers to demonstrate it.

D. robusta likely evolved as a specialist on the sap fluxes of the American elm, *Ulmus americana* (e.g., Carson and Stalker, 1951; Carson, 1971a). The distribution of *D. robusta* coincides very well with the ancestral distribution of this tree, the range of which has shrunk very much due primarily to a fungal disease. *D. robusta* has also been recorded on sap fluxes from a number of other deciduous temperate trees, including other species of elms, *Prunus, Quercus, Morus, Salix, Populus*, and *Robinia* (Carson and Stalker, 1951). Neither it, nor any other species of its lineage (the *robusta-melanica* lineage, see chapter 8) has been found breeding in any substrate other than sap fluxes.

The *D. virilis* group is another well-studied group of 11 species that are almost exclusively slime flux or decaying phloem breeders. Often species are associated with trees living along rivers and ponds such as willows (*Salix*), poplars (*Populus*), and *Alnus* (Carson et al., 1956; Throckmorton, 1982b). Evidently beavers play a role in the ecology of this group by damaging trees, which then become inundated with water and begin to rot (Carson, 1971; Spieth, 1979). Species of the *virilis* group breed not only in fluxes produced by living trees; dead trees with rotting phloem are perhaps even more commonly used. In fact, *D. virilis* has been recorded breeding in various types of wood stored in a lumberyard.

Repleta and Cacti

One of the most successful adaptations and radiations among *Drosophila* occurred for the *repleta* group, which became adapted to breeding in cacti; almost all members of this huge group (see Wasserman, 1982, 1992) breed in one or more species of cactus. The group evolved in the New World tropics and came out of moist rain forests to exploit the drier lowland deserts with many cacti endemic to the Western Hemisphere. The spread of the group out of the Americas has occurred as cacti have been moved to other localities. Cacti in the genus *Opuntia* have been particularly successful in adapting to regions around the world. *Repleta* group species associated with these cacti have likewise spread, prime examples being *D. buzzatii* (Carson and Wasserman, 1965; Fontdevila et al., 1981), *D. aldrichi* (Barker, 1982), and *D. mercatorum* (Templeton and Johnston, 1982). (*D. repleta sensu strictu* has also become a "domestic" species and has spread around the world with human habitats.)

Perhaps the best-characterized ecology of any *Drosophila* group is for *repleta* group species inhabiting the Sonoran Desert of the southwest United States and northern Mexico. This has been due to the efforts of W. B. Heed, his students, and collaborators; much of this work is summarized in Heed and Mangan (1986), Barker and Starmer (1982), selected chapters in Barker et al. (1990), and Etges et al. (1995). J. S. F. Barker, working in Australia, has likewise developed a good knowledge of *repleta* group ecology and genetics of species that have invaded that continent (see Barker's chapters in the previously referenced volumes; and Barker, 1992). Anyone looking for a system to connect ecology with genetics would do well to consider the *repleta* group.

Heed and Mangan (1986) make the case that the ecology of the Sonoran *Drosophila* can best be understood as an adaptive infiltration into a harsh environment. Only four species permanently occupy this ecosystem, so the situation is relatively simple. *D. pachea* and *D. nigrospiracula* are thought to have been adapted to columnar cacti. *D. mojavensis* and *D. mettleri* stem from lineages originally adapted to *Opuntia*-like cacti that later switched to columnar cacti as the thorn forest began to disappear. A brief synopsis of the breeding of *Drosophila* in the Sonoran Desert region is given in table 5-3. As can be seen in this table, each species breeds in rot pockets (caused by injury and microbial invasion) of one or two species of cactus. *Drosophila* species may switch breeding site depending on abundance of particular cacti, an example being *D. mojavensis* shifting from agria cactus in Baja to organ pipe in the mainland (Arizona, Sonora, and western Chihuahua). *D. mettleri* is nearly unique in all *Drosophila* in being a soil breeder; the larvae live in moist soil into which the rot pockets drip. Another case of a soil-breeding *Drosophila* is in Hawaii; see Hawaiian Drosoph-

TABLE 5-3 The major host plants for four endemic *repleta* group species in the Sonoran Desert.

Host Plant	Relative Abundance on Baja Peninsula	*Drosophila* Species Overlap	Relative Abundance on Mainland	*Drosophila* Species Overlap	Resident *Drosophila* Species
Lophocereus schottii (senita cactus)	+++	No	+++	No	*D. pachea*
Machaerocereus gummosus (agria cactus)	+++	No	+[a]	No	*D. mojavensis*
Stenocereus thurberi (organ pipe)	+++[b]	No	+++	No	*D. mojavensis*
Carnegiea gigantea (saguaro)	—[c]	—	+++	Yes	*D. nigrospiracula*
Pachycereus pringlei (cardon)	+++	No	+	Yes	*D. nigrospiracula*
Saguaro soil	—	—	+++	No	*D. mettleri*
Cardon soil	+++	No	+	No	*D. mettleri*

[a]Host plant rare on mainland compared to Baja but regularly used when present.
[b]Host plant common in Baja but rarely utilized.
[c]Host plant absent in Baja.

From Heed (1978).

ila section. The case of *D. pachea* is particularly interesting, as it is the only species that lives in rot pockets of stems of senita cactus. The species has an absolute requirement for some sterols found only in senita; likewise, senita have some toxic alkaloids fatal to species other than *D. pachea* (Kircher, 1982). This is a nice example of specialization and exclusion of competitors explained by the chemistry of the host plant. (*D. pachea* was originally placed in the *repleta* group, but more recently has been placed in the *nannoptera* species group; the mistaken placement was probably due to the convergent morphological similarity of *D. pachea* to *repleta* group species occupying a common environment.) Fitness parameters (viability, development time, and size) have been shown to correlate with the species of cactus on which each *Drosophila* species chooses to oviposit (Heed and Mangan, 1986; Ruiz and Heed, 1988). Genetic studies of these species are featured in the next chapter.

Flowers

Flowers are fairly common breeding sites for *Drosophila*, especially in warmer regions such as Africa and the Neotropics. Brncic (1983) lists some 140 flower-breeding species. In South America the best studied example is *D. flavopilosa* and members of its group (Brncic, 1970; 1983). Generally larvae feed on pollen inside mature flowers of one or a few species; that is, specialization is fairly common. Adults of flower-breeding *Drosophila* have been found to have their guts filled with pollen (Graber, 1957), so pollen serves as a nutritional source for this stage as well. *Hibiscus* have been invaded as a breeding site independently in Africa and Australia (Cook et al., 1977; Lachaise and Tsacas, 1983), although in this case it seems that rotting *Hibiscus* flowers rather than pollen from the living flower are the main breeding site. The possible role of *Drosophila* in pollination of host plants is discussed by Lachaise and Tsacas (1983) and appears to be important in some plants including cacao in Africa. Agnew (1976) reports a case where a plant seems to have specifically evolved to be *Drosophila*-pollinated by evolving an odor similar to fermenting fruit.

Fungus Feeders

More species, primarily of subgenus *Drosophila*, have been reared from fungi than any other single substrate in temperate forests (Shorrocks, 1982). Many of these species are difficult to rear in the laboratory, probably due to the nutritional requirements they obtain from fungi. Often the only way is to place an appropriate mushroom in a breeding container. Okada (1963) and Shorrocks (1982) note a change in the larval mouth hooks from relatively smooth in sap breeders to much greater dentition in fungus feeders, with fruit breeders intermediate (figure 5-2).

One common species of fungus feeder in Europe has been studied from an ecological context in some detail. Figure 5-3 summarizes a study Shorrocks and Charlesworth (1980) performed in England on the substrate specificity and temporal shifts in breeding of *D. phalerata*. Early in the season the bracket fungus *Polyporus squamosus* is used and at this time the major competitor is *D. confusa*. The next generation shifts to the stinkhorn, *Phallus impudicus*, at which time *D. subobscura* becomes a competitor. By August, the third generation of *D. phalerata* exploits a large number of fungi

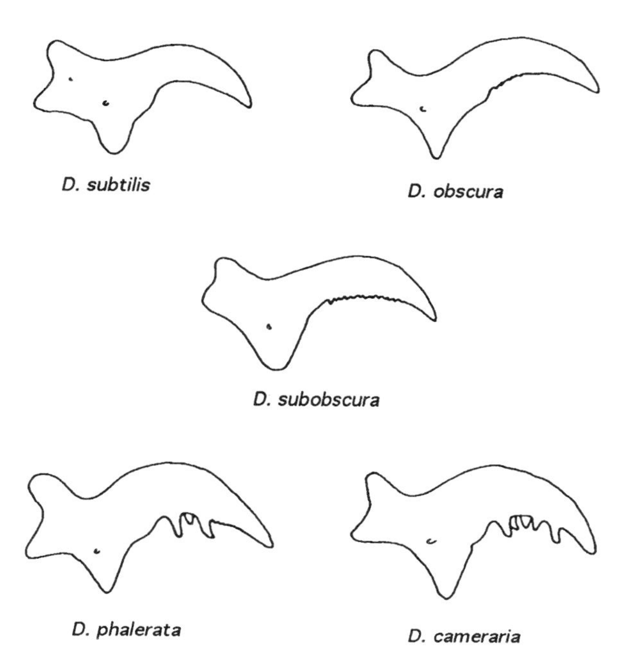

FIGURE 5-2 Larval mouth hooks from five species of temperate woodland *Drosophila*. The top two are sap breeders, the middle a fruit breeder, and the bottom two fungal breeders. Note increasing dentition from top to bottom. From Shorrocks (1982).

when *D. cameraria* becomes its competitor. Finally, the fourth generation goes into diapause to start the cycle again the next spring.

Jaenike and coworkers have studied fungus feeders in North America with some fascinating results. First they (Jaenike et al., 1983) showed that three species of American mushroom breeders, *D. putrida, D. recens*, and *D. tripunctata*, all had much higher tolerance for the toxin α-amanitin found in many mushrooms than did sap and fruit breeders such as *D. melanogaster, D. immigrans*, and *D. pseudoobscura*. Jaenike (1986, 1987, 1989; Jaenike and Grimaldi, 1983) has gone on to study *D. tripunctata* in some detail, as this species occupies at least two distinct larval sites, fungi and fruits. He has shown there is considerable genetic variation for oviposition preference in this species and has even had some success in beginning to identify genes involved in controlling variation in this behavior. One particularly interesting finding has been that the choice of resting site of females in nature and their oviposition preference in the laboratory are negatively correlated (Jaenike, 1986), further evidence on an intraspecific level that feeding/resting sites and oviposition sites may differ. Jaenike (1990) cites other cases where *Drosophila* have been shown to possess intraspecific variation in "host preferences," which can refer to both feeding sites and oviposition sites. He also discusses the interesting case of *D. quinaria*, which coexists in the northeast United States with *D. tripunctata. D. quinaria* can successfully oviposit and

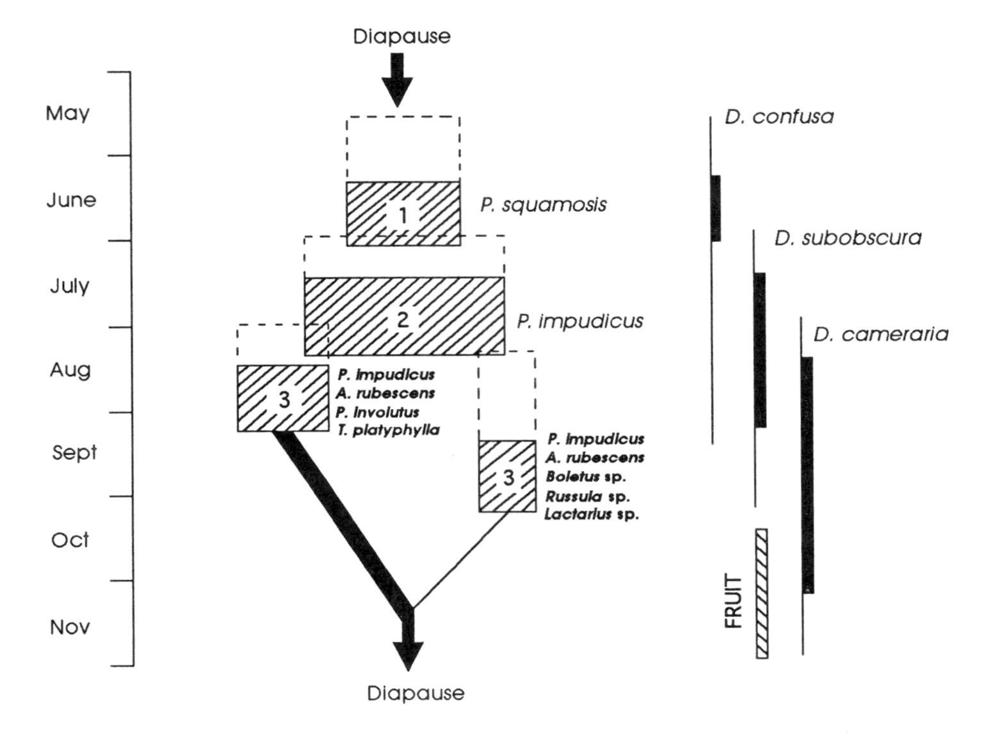

FIGURE 5-3 Temporal changes in the breeding site of *D. phalerata* during a single season. The numbers in the shaded boxes indicate the generation number with the width of each box indicating the numbers of flies emerging. The main periods when other species are present are indicated by the bars on the right. From Shorrocks and Charlesworth (1980).

develop on mushrooms, but females virtually always prefer to lay eggs on the abundant skunk cabbage. This is a case where female behavior, rather than the ability of larvae to grow and mature, has produced a monophagous species. The ecology of these species is made more complicated and more intriguing by the fact that a nematode parasite is also present that infects many, though not all, species of *Drosophila* in the ecosystem.

Hawaiian Species

Hawaii has been the site of an explosive radiation, producing on the order of 800 or more species (Hardy and Kaneshiro, 1981), yet the total land area of the islands is only 15,500 square kilometers. Obviously the breeding sites and niche separation required to sustain such a large number of species on such a small area must be exquisitely balanced. One major factor is the island nature of the area, meaning not only the obvious land masses. Within each island, vegetation is often subdivided into islands called kipukas that are surrounded by lava flows. Heed's (1968) study remains the single most important source of information on breeding sites for Hawaiian *Dro-*

sophila; table 5-4 is a synopsis of this work (see also Heed, 1971, and Montgomery, 1975, for further information and details). The *Drosophila* on the Hawaiian Islands fall fairly distinctly into groups that have been called "picture wing," "modified mouthparts," "spoon tarsi," "light tip scutellum," and so forth, the names being descriptive of each group's unique feature. In addition, the taxonomy of this fauna is complicated by the fact that various subgenera and genera have been proposed to describe the great morphological and behavioral diversity, a diversity greater than all other species from around the world combined. Yet, as far as we know, all the 800-plus species form a monophyletic group which, from molecular studies, is a sister taxon to the subgenus *Drosophila* and is within the genus *Drosophila* lineage (chapters 1 and 8). Despite the different names for genera and subgenera often used in the literature, from a phylogenetic point of view, all these species are members of genus *Drosophila*.

Two genera of trees in Heed's study were found to be the most important breeding sites. *Cheirodendron* (Araliaceae) has large fleshy leaves that ferment when detached and this niche provides larval breeding for 46 species. *Clermontia* (Lobeliaceae = Campanulaceae) has leaves that dry quickly when detached, so they are not favorable for larvae. However, the stems of this tree, when bruised or broken, produce a milky latex under the bark, which is a highly attractive site for larval breeding. About half (22) of the species in Heed's study were found in *Clermontia* stems. While many

TABLE 5-4 Summary of breeding sites for larvae of Hawaiian *Drosophila*.

Group	Leaves	Stems	Flowers	Fruits	Fungus	Flux	Frass	Ferns	Parasite
Picture wings (10)	4	5	4	4	2	3	2	—	—
Modified mouthparts (23)	10	13	3	7	4	—	—	—	—
Ciliated tarsi (10)	10	3	—	4	1	—	—	2	—
Light tip scutellum (9)	—	—	—	—	9	—	—	—	—
Idiomyia (3)	—	3	—	—	—	—	—	—	—
Nudidrosophila (2)	—	2	—	—	—	—	—	—	—
Antopocerus (11)	11	—	—	—	—	—	—	—	—
Bristle tarsi (16)	16	—	—	—	—	—	—	—	—
Fork tarsi (11)	11	—	—	—	—	—	—	—	—
Spoon tarsi (14)	14	—	—	—	—	—	—	—	—
Trogloscaptomyza (19)	6	2	9	8	—	—	1	—	—
Parascaptomyza (1)	1	—	1	—	1	1	—	—	—
Tantalia (3)	3	—	—	—	—	—	—	—	—
Bunostona (1)	—	—	—	—	1	—	—	—	—
Exalloscaptomyza (4)	—	—	4	—	—	—	—	—	—
Titanochaeta (6)	—	—	—	—	—	—	—	—	6
Domestic	1	—	5	2	1	3	1	—	—
Unclassified (8)	6	1	—	1	—	—	—	—	—
Total	93	29	26	26	19	7	4	2	6
Frequency	0.44	0.14	0.12	0.12	0.09	0.03	0.02	0.01	0.03

Note: While different generic and subgeneric names are sometimes applied, the evidence is all species belong to a monophyletic group with genus *Drosophila*. The number of species from each group is in parentheses with the numbers found on each substrate in the main body of the table.

From Heed (1968).

species share larval breeding sites, especially with regard to these two trees, species seem to have no trouble coexisting on them. Heed, for example, cites a case of a single tree of *Cheirodendron* that supported nine species on its leaves. As he emphasizes, the ability of these flies to exploit small isolated substrates and their generally small population sizes help account for their evolution and coexistence. Altogether, Hawaiian *Drosophila* have been reared from 45 genera of plants belonging to 30 families, although over 60% are from Araleaceae and Lobeliacae. The major fungus feeders belong to the "white-tipped scutellum" group, with more than 75 species found on fungus (Hardy and Kaneshiro, 1981).

Kaneshiro et al. (1973) provide an intriguing view of larval breeding by two very closely related (chromosomally homosequential) picture-wing species, *D. heedi* and *D. silvarentis.* These species are the only two found at the high altitude of the saddle between the two major volcanoes on the large island of Hawaii. The environment is very different from what one usually associates with Hawaii; it is very sparsely vegetated with only two species of trees, extremely xeric with powdery fine soil, and cold—frost may occur any month of the year. The larval breeding site for the two species is fluxes from a single species of tree, *Myoporum sandwicense.* Figure 5-4 illustrates how the species are divided on this single resource. *D. silvarentis* is found in fluxes on the tree, while *D. heedi* larvae are only found in the soil soaked by dripping fluxes. This is the only case, other than the already-mentioned Sonoran desert species *D. mettleri*, that breeds in soil. In fact, this particular environment in Hawaii is desert-like. It should be pointed out, however, that while larval soil breeding is very rare, pupation on soil is not, especially in Hawaii. Most of the larger-sized flies come out of their larval vegetation to pupate on soil.

Also listed in table 5-4 is a group of drosophilids called *Titanochaeta*, which are identified as parasites. The larvae of these species parasitize the eggs of the native spider family Thomisidae and after consuming them, pupate in the egg sac. Other examples of entomophagus drosophilids are known (Ashburner, 1981; see below).

Hawaiian *Drosophila* also show many exceptions to the generalization that mating takes place at the feeding/breeding sites. Spieth (1966, 1974, 1984) has amply documented that perhaps a majority of Hawaiian species are relatively quiescent while feeding and males leave the feeding site to establish leks on nearby vegetation. This behavior has led to aggressive male defense of mating territories with accompanying behavioral and morphological modifications. In fact, it may well be this unusual male courtship practice that has driven, by sexual selection, the evolution of much of the morphological diversity seen in the group, a diversity that is often sexually dimorphic.

Finally, we note one other aspect of the Hawaiian flies that is clear from Heed's (1968) study: their exceptionally long generation times. Heed reared larvae from *natural* substrates at temperatures typical of Hawaii. The time to emergence of adults was often very long, for most species on the order of 30 days and up to 76 days. Of course it would depend on the stage of larval development when brought into the laboratory, but still these times are very long compared to most Drosophilids. Furthermore, both sexes often delay sexual maturity up to several weeks after eclosing. For those accustomed to the 11-day life cycle of *D. melanogaster*, these flies would seem to be highly inconvenient laboratory animals. Indeed, this long generation time and the great difficulty, if not impossibility, of finding a suitable artificial rearing medium

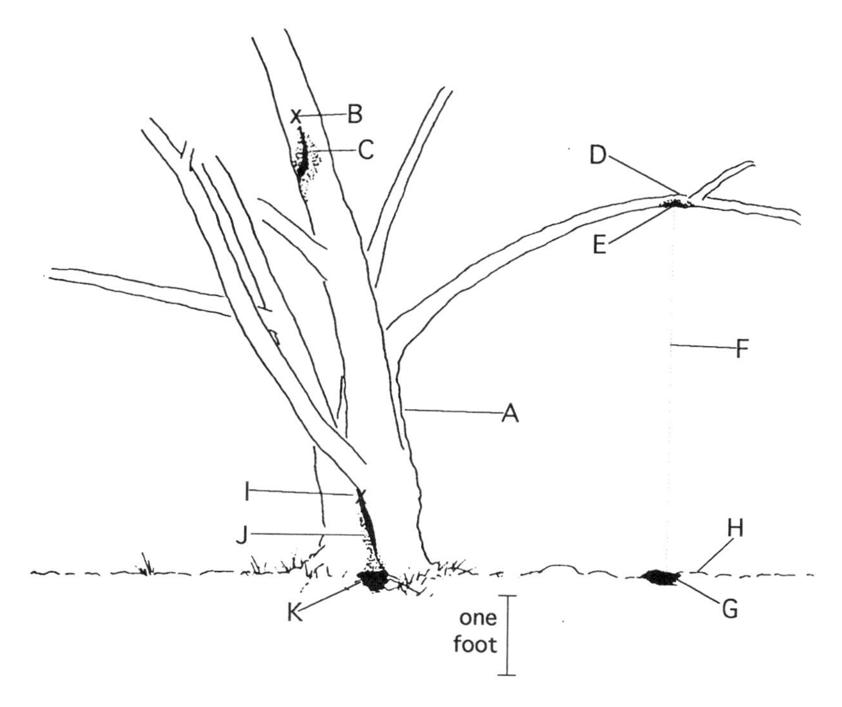

FIGURE 5-4 Breeding sites of *D. heedi* and *D. silvarentis* in Hawaii. The tree is *Myoporum* (A) with a sap flux 7 feet above ground (B), which wets the bark for about one foot (C). The horizontal branch (D) has a flux (E) which is dripping (F), forming a puddle (G) on the soil (H). A flux near the base (I) wets the lower part of the trunk (J) and forms a pool at the base (K). Only *D. silvarentis* larvae are found in C, E, and J, while *D. heedi* larvae are found in G and K. From Kaneshiro et al. (1973).

have severely limited the laboratory studies on the group (which also inhibits, sadly, the attention they have received from workers around the world).

Bizarre Drosophila

We now come to the bizarre category of *Drosophila* breeding (reviewed by Ashburner, 1981). *D. busckii* probably has the broadest niche of any drosophilid; Sturtevant (1921, p. 98) states this species has been found breeding on "bread and milk, moist bran, rotten pigeon egg, stale formalinized chicken, sour milk, spinach leaves, banana, flour paste, decayed onions, rotten fish, rotten potato, tomato and fungi." Furthermore, it has been found breeding in the preserved head of a Hottentot sent from Africa to Germany, and in scum covering formalin-preserved fish, which "are peculiarly nasty habitats even for this fly" (Ashburner, 1981, p. 415). Several cases of entomophagous drosophilids occur outside the genus *Drosophila*, including species that breed in the spittle of the spittle bug (Cercopidae) and in nests of solitary bees.

However, perhaps the strangest example of an unusual adaptation within the genus *Drosophila* is the *D. simulivora* species group (seven species), all of which have aquatic larvae in fresh water streams. The larvae are predators of larvae of blackflies (*Simulium*), midges (Chironomidae), and/or dragon flies (Odonata). Obviously, considerable modification of larvae, especially of their breathing apparatus and mouthparts, have accompanied this shift to being an aquatic predator (discussed in Ashburner, 1981; Lachaise and Tsacas, 1983).

One of the more intriguing cases of parallel evolution has occurred with respect to larval breeding on land crabs (Carson, 1974). Carson cites three cases of independent evolution of flies breeding on the outside of crabs in the excrement produced in the nephritic groove, a "microorganism-laden urinal" (Carson, 1974). One species is of the genus *Lissocephala, L. powelli*, and is found on two crabs in the Christmas Islands in the Indian Ocean. The other two crab-breeding *Drosophila* are found in the Caribbean on a single species of land crab (*Gecarcinus ruricola*), although they never occur sympatrically. One species, *D. carcinophila*, is in the *repleta* group, and *D. endobrachia* is in the *virilis* group. Given the taxonomic spread of the three cases, almost certainly these were three separate evolutionary events.

As a footnote to this crab story, we cite Wallace's (1978) successful attempt to adapt *D. virilis* to a similar habitat in the laboratory. He selected for breeding in cloth soaked with human urine. He was so successful that after a year, the strain could no longer breed on standard laboratory medium!

Yeasts and Other Microbes

Since the pioneering work of Wagner (1944, 1949), it has been known that microbes, particularly yeasts, play a large role in the nutrition and ecology of *Drosophila.* Studies of the microbes associated with several species in various ecological settings have been performed (table 5-5). The usual procedure for these studies is to collect flies in nature, dissect out the crops, and streak the contents onto plates for isolation and identification; sometimes whole flies are streaked after surface sterilization.

One question addressed by these studies concerned the larval and adult breeding sites. It was hoped that these studies could lead to the identification of the elusive breeding site of *D. pseudoobscura* and relatives. However, the studies of Phaff et al. (1956) and Carson et al. (1956) were rather disappointing in this regard. Of 17 species of yeast isolated from the crops of *D. pseudoobscura* and 11 species of yeast isolated from likely breeding sites including sap fluxes, berries, cacti, and soil, only two were in common between the adult fly crops and the potential larval breeding site. Even for species for which the larval breeding site was known (e.g., for *D. occidentalis* in fungi and *D. montana* in rotting logs) there was little correspondence in yeast species isolated from adult crops and larval breeding sites. This lends credence to the previously discussed phenomenon that while adults may feed at the larval breeding site, clearly they are obtaining yeasts from other sources.

Begon and Shorrocks (1978) found somewhat better correspondence between species of yeast isolated from adult crops and those at larval breeding sites for *D. subobscura* and *D. obscura* in England, although the two were different enough that they too had to conclude the two stages' resource utilization was not completely overlapping. Furthermore, they concluded that these two species in a temperate woodland had

TABLE 5-5 Studies performed on yeast and, in some cases, other microbes associated with *Drosophila* and its breeding sites.

Description of Study/Species	Reference
D. pseudoobscura and relatives in California	Phaff et al., 1956
	Carson et al., 1956
D. melanogaster and tomatoes	Camargo & Phaff, 1957
D. willistoni and several other Neotropical species	da Cuhna et al., 1957
D. disticha in Hawaii	Robertson et al., 1968
repleta group in the Sonoran Desert	Heed et al., 1976
	Starmer, 1982
	Starmer et al., 1982
repleta group, *D. aldrichi*, and *D. buzzatii* in Australia	Vacek, 1982
	Barker, 1992
D. subobscura and *D. obscura*, temperate woodland in England	Begon & Shorrocks, 1978
	Begon, 1982
Several native African species	Lachaise et al., 1979
	Pignal & Lachaise, 1979
D. melanogaster, pseudoobscura, and *arizonensis* in an orchard in Arizona	Vacek et al., 1979

Note: This is not a complete listing of studies, but in most cases recent references are given that allow entrance into the literature.

virtually identical yeast floras in their crops, which gave no indication of niche separation with regard to this adult food source.

Correspondence of yeasts in the larval breeding site and adult crops is observed, however, in human-altered environments. Camargo and Phaff (1957) found a good correspondence between yeasts from tomatoes on a northern California farm and the yeasts in crops of the local *D. melanogaster*. Vacek et al. (1979) found the same for an orange orchard in Arizona, and in that study at least one of the species (*D. pseudoobscura*) was the same as one that showed noncorrespondence in more natural settings. Clearly, the behavior of commensal populations can be very different from their behavior in ancestral natural habitats.

Studies in the tropics and the Sonoran Desert also demonstrate good correspondence between yeast in adult crops and larval substrates. As one example, table 5-6 compares the yeast isolates from the *repleta* group flies in the Sonoran Desert and their known specialized larval breeding sites. There is a good correspondence, which would lead one to think adults often feed at the larval site; in fact Heed (1977) has observed *D. nigrospiracula* adults feeding on cardon and saguaro. In the desert environment, there are few alternatives for adult feeding besides the moist microenvironment provided by rotting parts of cacti, a fact also reflected in the sedentary behavior of the adults as long as a rot pocket remains moist and microbe-laden (see next section). Fogelman et al. (1981) have shown that larvae of these Sonoran species eat some species of yeasts preferentially over others. It is not unlikely that adults would as well, so this is another reason for differences between the yeasts in the cactus rots and adult crops, even if adults fed exclusively at the larval site. Starmer et al. (1982) review the yeasts in this ecosystem in considerably more detail.

Another indication of the importance of yeasts as nutritional factors that may be important in separating niches of coexisting *Drosophila* comes from studies of the

TABLE 5-6 Comparison between frequencies of isolation of yeast species from the crops of adult *Drosophila* and the substrate in which their larvae develop.

Drosophila Larval Site	Yeast Species 1	2	3	4	5	6	7
D. mojavensis	31	0	22	19	5	1	0
Organ pipe cactus	21	0	1	16	11	3	0
D. pachea	0	25	0	0	3	12	0
Senita cactus	7	39	1	3	16	9	2
D. nigrospiracula	20	9	0	0	0	1	0
Cardon cactus	16	1	0	5	1	2	2
D. mettleri	16	3	0	1	0	0	0
Saguaro soil	4	3	0	2	2	2	1

Yeast species: 1 = *Pichia cactophila*, 2 = *P. heedii*, 3 = *P. amethionina*, var. *amethionina*, 4 = *Torulopsis sonorensis*, 5 = *Candida ingens*, 6 = *Cryptococcus cereanus*, 7 = *P. amethionina* var. *pachycereana*.

Note: The *Drosophila* species is given and under it the larval substrate.

From Heed et al. (1976) and Starmer (1979).

Drosophila species that are attracted to baits seeded with different kinds of yeasts. This has been done in the tropics (e.g., da Cuhna et al., 1957) as well as in temperate regions (da Cuhna et al., 1951; Klaczko et al., 1983). An example of the last-cited study is in table 5-7. Clearly, in this study yeast F (Fleischmann's commercial *Saccharomyces cerevisiae*), the common yeast used by collectors, was much poorer in attracting *obscura* group flies than was yeast L, *Kloeckera apiculata*, which was a strain originally isolated from the crop of *D. pseudoobscura.* However, F was better in attracting other species, primarily *D. occidentalis* and *D. pinicola.* The study of da Cuhna et al. (1951) was similar in showing distinct species preferences for different

TABLE 5-7 Frequencies of species or species groups attracted to two different yeasts at Mather, California.

	Yeast	Obs	*D. azteca*	Others
Percentages	F	31.5	33.9	34.6
	L	25.5	67.5	7.0
Absolute numbers per	F	1.7	1.8	1.9
day per trap	L	44.6	117.9	12.2

Notes: "Obs" includes both *D. pseudoobscura* and *D. persimilis*, which cannot be distinguished morphologically, although a sample of the total indicated no differences between the two species with regard to yeast choice. "Others" includes several other species but primarily *D. occidentalis* and *D. pinicola*. F yeast is commercial Fleishmann's *Saccharomyces cerevisiae* and L is *Kloeckera apiculata*, the strain used having been isolated from *D. pseudoobscura*.

From Klaczko, Powell, and Taylor (1983).

yeasts, but that study is complex in also demonstrating the apparent preferences may change depending on the competing yeasts used. That is, in one experiment they used two or three different yeasts and then changed combinations of yeasts in consecutive studies.

These studies have other implications. First, using *Drosophila* baits of this sort, namely bananas (or some other fruit) inoculated with a yeast (usually commercially available yeast), to collect flies to determine the species composition of an area may give very biased results. It is clear from data like that in table 5-7 that it is difficult, if not impossible, to have any confidence that species attracted to one kind of bait accurately reflect the relative frequencies in nature. Second, it is clear that the efficiency of collecting varies greatly with yeasts. It may well behoove a collector who is after one species or species group to determine which yeast is the most attractive to that subset of species.

What is it that flies are getting from yeasts? The nutritional needs of *D. melanogaster* have been well-studied (reviewed by Sang, 1978). It is clear that the "minimal medium" developed by Sang and colleagues contains more than just a protein and carbohydrate source. Many vitamins and other minerals are required for development through the larval stage to eclosion, including biotin, folic acid, riboflavin, and so on. It is likely these are supplied by microbes under natural conditions. While adults can survive on a simple sugar source, females cannot mature their eggs without added nutrition, most likely from microbes. Begon (1982) discusses at some length the known compounds produced by yeasts that may be required by *Drosophila.*

Just a brief mention of bacteria and mold follows, brief because they have been given little attention. During the yeast studies, bacteria were often seen in the crops but only rarely isolated and studied. For example, Robertson et al. (1968) actually found more bacteria than yeast in the guts of larvae of *D. disticha.* Adult males of the species always had bacteria in their crops but had yeasts only about half the time. Fogelman (1982) has studied the bacteria *Erwinia carnegieans* in the Sonoran cactus associations. This bacteria may be important as an initial microbial invader to begin the breakdown of tissue in the rot pockets, thereby making it more suitable for yeasts; for example, most of the yeasts associated with these cacti lack cellulase (which the bacteria has) to digest the plant cell walls. Furthermore, *Erwinia* produces a large variety of volatiles that may be important in attracting the flies. Atkinson (1982) studied one of the very few associations between mold and *Drosophila.* He found that the mold *Penicillium italicum* inhibited the development of *D. melanogaster* on citrus fruits in an English fruit market, but that *D. immigrans* actually did better in the presence of this mold. As Begon points out, this is interesting because citrus and *D. immigrans* both originated in Asia, whereas *D. melanogaster* is a relatively recent immigrant from Africa.

It is very important to recognize, as do the workers in this subfield, that the subject is extremely complicated. Most experiments to date have not approached the complexity of the bacteria-yeast-*Drosophila* interactions in nature. Consider the following: it is well known that microbes vary considerably in their metabolism, depending on substrate and other environmental variables such as temperature. Furthermore, microbes exist in communities, so interaction of microbial species is doubtless extremely important. The fact that 24 species of yeast could be isolated from the crop of a single adult *D. pseudoobscura* (Shehata and Mrak, 1952) certainly gives reason to believe microbial communities, not single isolated species grown in laboratory medium, need to be understood. As one example, Starmer and Aberdeen (1990), using just two species of yeast ("bicultures"), demonstrate that development time of *D. mulleri* is greatly enhanced over that on either monoculture.

The importance of yeast in *Drosophila* biology and evolution cannot be in doubt, and from what little work has been done, other microbes also play crucial roles. It is a field that deserves more attention, although the difficulty and complexity of the work requires expertise in both *Drosophila* and microbiology, and therefore is often daunting. Collaboration has been a key to what inroads have been made.

Population Structure

In this section we concern ourselves with facets of population structure that have bearing on evolutionary processes (reviewed in Taylor and Powell, 1983). Probably the most important generality to come from the limited number of studies relevant to this subject is that there are virtually no generalities. Given the great number of species of *Drosophila*—by now the reader should be aware of the diverse ecologies and behaviors they display—it is perhaps not surprising that population processes vary greatly.

Sex Ratio

Sex ratio is just one example of this diversity. Males sometimes outnumber females in natural populations, as in the case of *D. engyochracea* (Fontdevila and Carson, 1978); the sex ratio (females : males) is about 0.5. Begon (1977), on the other hand, found that for *D. subobscura* in a site in England, the sex ratio varied seasonally from 1.5 to 4.63, with females always outnumbering males. *D. pseudoobscura* and *D. persimilis* at Mather, California, have sex ratios very close to even, being 1.09, 0.95, and 1.12 over three seasons of intense collecting (J. R. Powell and C. E. Taylor, unpublished observations). Yet, as we have already seen (figure 3-20), in a *D. pseudoobscura* seminatural population, the presence of the SR chromosome can greatly influence the sex ratio in the population. (Of course, in any such studies we must keep in mind the caveat that sexes of different species may be differentially attracted to the baits used.)

Density

Density estimates have been made for a number of species, generally using one form or another of mark-recapture techniques. Initially, marking of flies was done by releasing laboratory mutants, but in the last 20 years the marking of native flies with UV-fluorescent dusts has become the technique of choice (Crumpacker, 1974). Table 5-8 summarizes several attempts to determine densities by these means. Again, we see considerable variation, not only from species to species, but also within a species, especially for temperate species. Begon's (1976a, b) studies of *D. subobscura* are probably the best documentation of the significant seasonal variation in density.

In considering the implications of the figures in table 5-8 for population genetics one must also take into consideration several other parameters. For example, on the

TABLE 5-8 Estimates of densities of *Drosophila* in native habitats.

Species	Locality	Estimate No./100 m^2	Reference
pseudoobscura	Mt. San Jacinto	3.8–9.8	Dobzhansky & Wright, 1943
	Mather, 1945	0.37	Dobzhansky & Wright, 1947
	Mather, 1974	1.44	Powell et al., 1976
	Colorado	0.38	Crumpacker & Williams, 1973
persimilis	Mather	0.45	Powell et al., 1976
azteca	Mather	0.69	Powell et al., 1976
miranda	Mather	0.07	Powell et al., 1976
subobscura	England	1.3–750	Begon, 1976a, b
obscura	England	3	Begon, 1978
willistoni	Brazil	10–28	Burla et al., 1950
engyochracea	Hawaii	535	Fontdevila & Carson, 1978
nigrospiracula	Arizona	40	Johnston & Heed, 1976
flavipilosa	Chile	37,000	Brncic, 1966

From Taylor and Powell (1983).

face of it, it would seem that *D. engyochracea* must have a much larger population size including effective population size (N_e) than does *D. pseudoobscura*. However, this is almost certainly not the case given the very narrow distribution of the former and the high dispersal rate of the latter (see below). Furthermore, given the temporal fluctuation in population sizes, especially in temperate species, and given the fact N_e is greatly affected by any bottlenecks in population size (being the harmonic mean of N_e for each generation), any single density estimate may give an erroneous impression. Studies like Begon's are necessary to get a good feel for these parameters.

Dispersal

The movement of adult *Drosophila* determines to a large degree the population structure, especially as regards effective population size and genetic differentiation. Many studies have been done on the dispersal behavior of several species. The major differences among studies involves the means by which flies were marked as well as how they were recaptured. Table 5-9 lists dispersal studies done on *Drosophila*. In addition to the already mentioned methods of following movement with laboratory-reared mutants and dusting with UV-fluorescent dusts, Richardson et al. (1969) explored using heavy elements followed by neutron activation of flies that fed on the elements, a technique not widely exploited. Given the great number of species and techniques used, it is difficult to generalize. However, we can draw a few conclusions.

First, it is clear that the technique used to measure dispersal can affect the outcome. The best example is to compare the early work of Dobzhansky and Wright (1947) utilizing laboratory-reared mutants of *D. pseudoobscura* with the later work of Dobzhansky and Powell (1974) and Powell et al. (1976) done at exactly the same locality using the same release site and the same types of baits placed in the same locations. Instead of mutants, the fluorescent dusting technique was used to mark flies from nature. The latter studies gave about an order of magnitude greater estimates of

TABLE 5-9 Dispersal studies of various *Drosophila* species.

Species	Locality	Method of Marking	Reference
D. melanogaster	Berlin	LRM	Timofeef-Ressovsky & Timofeef-Ressovsky, 1940a, b
	New York, Venice	LRM	Wallace, 1970b
	San Jacinto	LRM	Dobzhansky & Wright, 1943
	Texas	NA	Richardson, 1968
	Maryland	LRM	Coyne & Milstead, 1987
D. funebris	Berlin	LRM	Timofeef-Ressovsky & Timofeef-Ressovsky, 1940a, b
	Moscow	LRM	Dubinin & Tiniakov, 1946
D. pseudoobscura	San Jacinto	LRM	Dobzhansky & Wright, 1943
	Mather	LRM	Dobzhansky & Wright, 1947
	Mather	FD	Dobzhansky & Powell, 1974
			Powell et al., 1976
			Dobzhansky et al., 1979
			Taylor and Powell, 1978
			Klaczko et al., 1986
	Death Valley	FD	Jones et al., 1981
			Coyne et al., 1982, 1987
	Colorado	FD	Crumpacker & Williams, 1973
D. persimilis	Mather	FD	Powell et al., 1976
D. azteca	Mather	FD	Powell et al., 1976
D. miranda	Mather	FD	Dobzhansky et al., 1979
			Taylor & Powell, 1978
D. occidentalis	Mather	FD	Taylor & Powell, unpublished data
D. nigrospiracula	Arizona	FD	Johnston & Heed, 1975, 1976
D. mercatorum	Hawaii	FD	Johnston & Templeton, 1982
			Templeton et al., 1990b
D. hydei	Hawaii	FD	Johnston & Templeton, 1982
D. subobscura	England	FD	Begon, 1976b
			Atkinson & Miller, 1980
			Shorrocks & Nigro, 1981
	Yugoslavia	FD	Kekic et al., 1980
D. aldrichi	Texas	NA	Richardson, 1969
D. engyochracea	Hawaii	FD	Fontdevila & Carson, 1978
D. mimica	Hawaii	FD	Richardson & Johnston, 1975
D. willistoni	Brazil	LRM	Burla et al., 1950

Abbreviations: LRM = laboratory-reared mutants, FD = UV-fluorescent dusts, NA = neutron activated elements.

After Taylor and Powell (1983) with additions.

dispersal. One must use caution in extrapolating the dispersal behavior of laboratory-reared mutants to natural populations. That the laboratory-reared mutants used by Dobzhansky and Wright were not totally debilitated was attested by the fact that the mutant gene persisted in the populations for several generations after the release, with a higher concentration nearer the release point. The released flies were healthy enough to inject their genes into the native gene pool, but evidently had impaired mobility.

The magnitude of dispersal observed among species is highly variable. *D. pseudoobscura* is perhaps the most mobile species studied. In favorable habitats (e.g., montane forest) the average movement of flies is 100 to 300 meters per day (Crumpacker and Williams, 1973; Powell et al., 1976). In unfavorable territory like desert, the species greatly increases its movement, presumably in search of favorable habitats, and adults may travel several kilometers in a day (Jones et al., 1981; Coyne et al., 1982, 1987). Even in relatively favorable regions, such as Mather in the Sierra Nevada, the local environment can greatly affect dispersal behavior of *D. pseudoobscura*. Dobzhansky et al. (1979) found that habitats separated by a few hundred meters could induce the flies to disperse an order of magnitude differently. Also, in this species there is good evidence that genotypes may vary in dispersal behavior (Klaczko et al., 1986; chapter 6).

D. melanogaster probably represents the opposite extreme, being a species that tends to be sedentary. Studies by Timofeef-Ressovsky and Timofeef-Ressovsky (1940a, b) and Wallace (1970b) indicate this species moves 10 meters or less per day. Likewise, *D. willistoni* tends to be relatively sedentary (Burla et al., 1950). Perhaps these tropical species that exploit fallen fruits have adapted to the relatively long-lasting resource, the distribution of which is unpredictable.

Coyne and Milstead (1987) present a somewhat different view of *D. melanogaster* population structure. They released about a million flies heterozygous for two recessive mutants. They found the mutants up to 10 km away some two months after release. Equally important, after a winter (the study site was in the northeast United States) no mutants could be found except in a produce stand that operated year-round. The long-distance migration noted by Coyne and Milstead is not necessarily in conflict with the general notion of this species being relatively sedentary. Since it is a human commensal, passive transport by human activity is likely.

D. mercatorum and *D. hydei* present an interesting case of contrasts in dispersal behavior studied in a common locality. Johnston and Templeton (1982) used the fluorescent dust technique to study dispersal in these species cooccurring in a patch of *Opuntia* in Hawaii. They found that *D. mercatorum* had a 0.31 probability of moving from a cactus each day with the average distance moved being 59 meters. *D. hydei* had a probability of movement of 0.68 with an average of 166 meters. They suggest the higher rate in *D. hydei* may be related to its habit of being a colonizing species.

D. nigrospiracula presents a picture of extreme temporal and spatial variability in dispersal behavior (Johnston and Heed, 1976). Generally, this species is relatively sedentary when occupying a cactus rot pocket at the right stage of decomposition for breeding. When the rot dries, the flies become highly mobile, moving in search of a favorable rot pocket. The actual distance moved depends on the density of cacti in the region; the more dense the cacti, the less dispersal before finding a suitable site.

It is often thought that the sexes may disperse differently. Data from *Drosophila* are ambiguous on this point. In our studies of *D. pseudoobscura* and *D. persimilis* at Mather, we could detect no differences between sexes, nor did Johnston and Templeton (1982) in their studies of *D. mercatorum* and *D. hydei.* Begon (1976b) found male *D. subobscura* dispersed more than females, while Fontdevila and Carson (1978) found the opposite for *D. engyochracea.* In these latter two studies the sex differences were not great and could well be within experimental error. Overall, we must conclude there is little evidence for differential dispersal of the sexes.

Endler (1979) made the point that in order to interpret the effect of adult movement on gene flow among populations it is necessary to consider the egg-laying regimen of the females. He suggested that for *Drosophila* this follows approximately a square-root normal distribution. The distance from place of eclosion to where a female deposits eggs will depend on how long she lives and this distribution. He estimated this shortens the gene flow estimate by a factor of 0.6 of dispersal alone. On the other hand, Fontdevila et al. (1977) found evidence that migrant males had a mating advantage over resident males of *D. nigrospiracula* in the Sonoran desert. Wallace (1970b) also speculated this may be occurring for *D. melanogaster* in his studies of dispersion. If these observations are generally true, then dispersion may have a greater effect on gene flow than the numbers of migrants indicate. Clearly the relationship between adult movement and gene flow is a complex one, with no simple one-to-one relationship.

Genetical Structure

So far we have discussed direct ecological and behavioral studies of population structure. One can also use gene frequencies among populations to infer population structuring. The usual statistics for quantitatively comparing genetic structuring are Wright's F statistics (discussed thoroughly in Wright, 1978). The relative population structuring is most often signified by F_{st} or F_{dt}, the subscripts referring to the variation in a subpopulation (s) or deme (d) (the smallest panmictic unit) relative to the total (t) variation in the species. Often the symbol F_{ds} is also used for the relative variation in the deme relative to the sample (s) of populations, as one seldom has the total variation for the species.

In *Drosophila* several types of genetic variants have been used to infer structure. One early attempt was made from consideration of allelism of recessive lethals, ubiquitous in natural populations of *Drosophila* (chapter 2). If a species is highly structured with respect to gene frequencies, the frequency of allelism of recessive lethals should be greater between two lethal chromosomes from the same population than when the chromosomes come from genetically differentiated populations. This prediction is sometimes observed and sometimes not. The most convincing positive results come from studies of *D. melanogaster*, which, we have already noted, is a relatively sedentary species, and thus may be predicted to show more structure. Wallace (1966b), Paik and Sung (1969), and Ives (1970) have all shown a decrease in lethal allelism in this species over distance of a few tens of meters. On the other hand, *D. willistoni* (Pavan and Knapp, 1954) and *D. subobscura* (Loukas et al., 1980) showed no difference in allelism rate among populations. In *D. pseudoobscura* the data conflict somewhat. Initially Dobzhansky and Wright (1941) did find lower allelism between populations than within populations, while Bryant (1976) did not. However, the former study used lethals on the inversion-polymorphic third chromosome and the latter used the inversion-free second. As emphasized in chapter 3, inversions are much more geographically structured than single genes (allozymes). Table 5-10 contains some estimates of F_{dt} calculated from the recessive allelism studies.

Variation in allozyme frequencies among populations has also been used to try to detect population structuring. By and large, allozymes show a relatively even geographical distribution, implying little genetic substructuring for these types of polymorphism. Table 5-10 contains some estimates of F_{dt} for allozymes. The high estimates here for *D. paulistorum* should not be taken as comparable to the others, as this

TABLE 5-10 Estimates of relative genetic differentiation and structure based on three different types of polymorphisms.

	F_{dt}		
Species	Lethals	Allozymes	Inversions
D. pachea		0.007	
D. willistoni, S. America	0.018	0.022	
D. equinoxialis		0.029	
D. pseudoobscura, United States	0.015	0.028	0.270
D. willistoni, Caribbean		0.041	
D. obscura		0.067	
D. subobscura	0.017	—	0.306
D. ananassae		0.116	
D. pavani		0.126	0.012
D. melanogaster	0.010	0.241	
		0.169	0.280
D. paulistorum		0.255	
D. pseudoobscura, including Bogota		0.268	

Note: Average F_{dt} (F_{st}) values are given for all polymorphic allozyme loci, or for all common inversion polmorphisms. The two values for allozymes in *D. melanogaster* come from Wright (1978) and Singh and Rhomberg (1987a); other values from Crow and Temin (1964), Cariou and Da Lage (1993), and Ferrari and Taylor (1981).

study contained several semispecies; likewise, the higher estimate for *D. pseudoobscura* involves the isolated subspecies around Bogota. Excluding these exceptions, *D. melanogaster* again shows the greatest population structuring, with F_{dt} some 4 to 10 times greater than most *Drosophila* species.

Finally, we present some data for inversion polymorphism. Clearly these variants display a much greater degree of population differentiation (chapter 3) and thus much greater F_{dt}'s, discussed and calculated by Wright (1978), Ferrari and Taylor (1981), and Taylor and Powell (1983). We also include these in table 5-10 for comparison. Figure 5-5 is a graphical comparison of F_{st}'s for allozymes and inversions in *D. melanogaster.* Note that the inversions in this figure are those on which the allozymes are located. With the exception of *Adh*, the allozymes show very little structuring while the inversions indicate considerable structure, that is, differentiation (note the scale difference of the two axes). So even in cases where one might predict allozymes should show high geographic structure, they do not, probably because of only a partial linkage disequilibrium with the highly structured inversions.

D. pavani was given as the extreme case of inversion homogeneity over the large range of the species (Brncic, 1973; table 3-6), and indeed the F_{dt} for inversions is the lowest recorded (table 5-10). Yet the allozyme structure is greater than for any species other than *D. melanogaster.* This further emphasizes the independence of the two types of polymorphisms.

Clearly the degree of structure detectable by polymorphisms depends on the particular system (table 5-10). Recessive lethals and allozymes generally lead to the conclusion of little structure, with the exception of *D. melanogaster*. Inversions, on the other hand, usually exhibit highly significant population differentiation, with the ex-

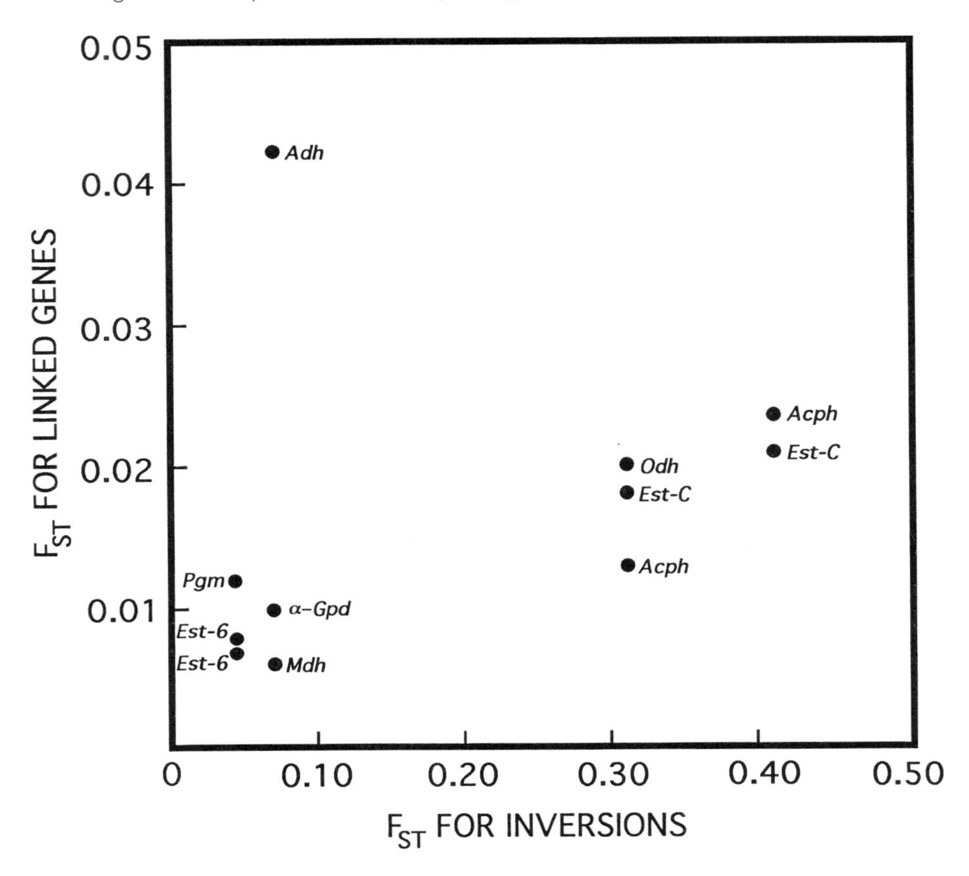

FIGURE 5-5 Correlation between F_{st} for inversions and allozymes linked to inversions in *D. melanogaster*. Generally the two polymorphisms covary but *Adh* is a clear exception. From Singh and Rhomberg (1987b).

ception of *D. pavani*. It is clear that inversions are subject to quite strong selection (chapters 3 and 4) and thus are not particularly good indicators of ecological processes such as migration rates. The more nearly neutral variants like allozymes probably reflect much more accurately such phenomena as population sizes and migration rates.

The relative uniformity of recessive lethals is a bit more difficult to interpret. Clearly they are subject to selection, but since they are measured as whole chromosomes, the nature of the genes being sampled is not clear. However, the study of Bryant (1976) shows recessive allele analysis can aid in understanding population structure. He showed that allelism rates among most populations were about the same as within populations, with the exception of desert oasis populations, where the allelism rate was much higher. However, the rates among flies in the oasis in different years were as low as between most populations. The implication is that the oasis populations are founded anew each winter by a few individuals, and the temporary populations are the descendants of a few individuals. During the summer the populations disappear.

F statistics are just one manner of detecting spatial structure based on gene frequencies. In the next chapter we will discuss spatial autocorrelation (Sokal et al., 1987), which in many ways may be superior to *F* statistics.

Age Structure

Very little is known about the age structure of *Drosophila* populations. Estimates of average daily survival have been inferred from capture-release experiments. For *D. pseudoobscura* this has been estimated to be about 0.91 (Dobzhansky and Wright, 1947). In the Hawaiian study of Templeton and Johnston (1982) it varied from year to year; for *D. hydei* it was 0.92 and 0.88 for 1980 and 1981, respectively, while for the same years the average daily survival of *D. mercatorum* was 0.96 and 0.81. For these species then an estimate of around 0.9 on average would indicate a modal longevity of 6 to 7 days in natural populations. This seems reasonable given that most species reach sexual maturity by 2 days and begin egg-laying by about 3 days. However, obviously for some long-generation species, such as many of the endemic Hawaiian species that do not reach sexual maturity sometimes for weeks after eclosion, survival rates must be much greater.

Boesiger (1968) used an indirect method to attempt to obtain the age structure of *D. melanogaster*. He reasoned that there is uniform longevity under constant optimal laboratory conditions. Flies captured from a natural population were brought into the laboratory, placed in benign conditions, and their time of death recorded. This was subtracted from the maximum longevity expected to estimate the age when brought into the laboratory.

Johnston and Ellison (1982) made the only attempt to directly determine the age of *Drosophila* using the growth layers in the internal thoracic muscle attachments (apodemes). They studied six species in the laboratory and found quite accurate aging especially when flies were kept in a light-dark cycle and a temperature cycle. Presumably the layers formed depend to some extent on these environmental variables cycling daily, as would occur in nature. The application of the technique to natural populations was presented in Johnston and Templeton (1982) for a set of *D. mercatorum* populations in Hawaii. A cline in age structure was noted over an altitudinal transect. In the next chapter we discuss this study in some detail and note the genetic consequences of such differences in age structure.

Overall Patterns of Population Structures

From the above we can begin to discern overall pictures of how various *Drosophila* species are structured. Wallace (1968, 1970b) has advocated what he called the "bed of nails" view (figure 5-6A). Dispersal is very low and local populations are made up of adults who were larvae in the very near vicinity. Locally, densities may be very high with relatively little migration among demes. Wallace arrived at this view from his studies of *D. melanogaster*, both with respect to actual measurements of dispersal and allelism tests. It may well be an accurate view of species such as *D. melanogaster*

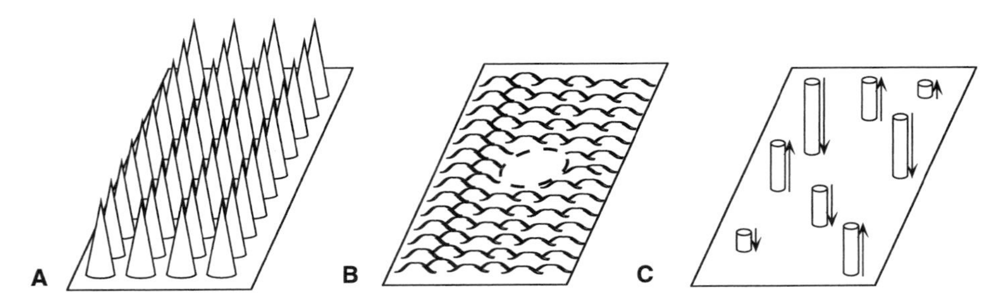

FIGURE 5-6 Three views of population structure. A is the "bed of nails" view proposed by Wallace (1968) for *D. melanogaster*. B is a scheme proposed for long dispersers such as *D. pseudoobscura* where population densities take on a "moderately rough sea" depiction. C is meant to portray a "fluid pegboard," which represents desert-adapted species such as *D. nigrospiracula*; where a cactus is in the appropriate stage of rotting, densities are high, but temporally unstable due to drying of the rot, followed by long distance dispersal to the next appropriate rot. While not all species can be easily categorized into these three population structures, they do represent the diversity of population structures that exist.

that breed on locally abundant, but spatially unpredictable, substrates such as human garbage or fruits in a tropical forest.

An almost diametrically opposed view comes from studies of the American *obscura* group species, especially *D. pseudoobscura.* With this species, dispersal rates are high both within favorable habitats (e.g., Powell et al., 1976) as well as between (Jones et al., 1981). Local differentiation would be low (although see next chapter for evidence it may occur via habitat selection), migration high, and effective population sizes very large. Genetic data such as the striking results of Keith (1983; table 2-13) attest to the effect of migration over long distances in maintaining genetic homogeneity. In the absence of strong selection (such as must be acting on the inversion polymorphism), the species may be nearly a single effective breeding population. This is true despite the rather large geographic range of the species (figure 3-2). Jaenike (1989) has evidence for a similar view for *D. tripunctata*, which also has a considerable geographic range, nearly the eastern half of the United States. Perhaps some kind of modestly rough sea, with uninhabited islands where the species doesn't occur, is an accurate picture of this type of structure (figure 5-6b).

A third view comes from the desert-breeding *Drosophila* studied by Johnston and Heed (1976). Here population densities are sharply discontinuous, being concentrated at cacti that have rot pockets at the proper state of decomposition. Yet migration among populations is high, especially as rots dry up and flies need to find another site. At these times enough migration occurs to make an effectively panmictic population over a great area. Perhaps a "fluid pegboard" picture could describe these species (figure 5-6c).

Obviously these different types of population structures are related to the ecology of the species. Furthermore, these three pictures are meant to be only heuristic and not all species can be easily pigeonholed into one of them. Rather they are extreme views, all represented in the genus *Drosophila*, with doubtless many variations on the themes.

Interspecific Competition

Competition among species of closely related organisms is often thought to be a, or even *the*, major driving force for evolutionary change. Given the 2000 or so *Drosophila* species, many of them coexisting (witness the 800+ species on the small land mass of Hawaii), competition among species must surely play an important role in evolution in the genus. While studies of interspecific competition are difficult to perform in natural populations, considerable attention has been given to laboratory studies of mixed species populations. Barker (1983) is an excellent review of the subject.

Natural Populations

It is notoriously difficult to unambiguously demonstrate interspecific competition in nature. The conventional and convincing procedure of removing one species and observing what happens to other species is impractical with *Drosophila* and to my knowledge has never been attempted. Release of foreign species into nature has been attempted numerous times, but no attempts seem to have been made to monitor the effects on resident species. The introduction of the Old World *obscura* group species into the Americas may present an opportunity to detect the effects of competition, although no studies have yet appeared on this subject. The recent expansion of native American species such as *D. persimilis* and *D. miranda* into areas previously unoccupied by these species seems to have had little impact on the genetics of *D. pseudoobscura* populations (e.g., Moore et al., 1979; Anderson et al., 1991).

Attempts to study interspecific competition in nature, therefore, have generally been more or less indirect. For example, it is often observed that both *D. melanogaster* and *D. simulans* can be bred out of the same fruit and competition is potentially possible. The observation of the absence of *D. simulans* in Japan prior to 1972, followed by a rapid increase at the expense of *D. melanogaster* (Kawanishi and Watanabe, 1977) would seem to verify that the two species compete.

Atkinson and Shorrocks (1977; and Atkinson, 1979) studied the coexistence of seven species in fruit stands in England. They conclude that competition is occurring, largely based on variation in body size that could not be explained by genetic factors. It is well known that crowding and competition leads to smaller body size in *Drosophila*. Both intra- and interspecific competition was inferred. They raised doubts that the species are stably coexisting because such recent and ephemeral habitats as fruit stands may never allow the fly populations to reach equilibrium.

Richardson and Smouse (1975) attempted to quantify competitive interactions among three species of Hawaiian flies, two sibling species (*D. mimica* and *D. kambysellisi*), and *D. imparisetae*. They used the relative frequencies of the species at 151 collecting points over 15 days in a single kipuka. From this they produced a community matrix to infer competition. They conclude that the nonsibling species had acted as an "ecological wedge" to induce the speciation of the siblings while remaining in sympatry.

In a study of the relative abundance of *D. athabasca* and *D. affinis* on the mainland and islands off the coast of Maine, Jaenike (1978) has argued for competition between these two closely related species. However, Barker (1983) points out that such a conclusion may be premature given that it is not even known if the two species share breeding sites. More direct evidence for competition in nature came from Grimaldi and Jaenike's (1984) study on fungus feeders in the northeast United States. Four species coexist: *D. falleni, D. recens, D. putrida*, and *D. testacea.* They showed that the mushrooms used were usually exhausted as a resource. Competition could reasonably be inferred by the fact that supplementing the larval resource with fresh mushroom led to an increase in body size and a decrease in variance. While these studies did indicate competition, the authors point out that it is not possible to distinguish between intra- and interspecific competition from such results.

Sokoloff (1966) concluded it was unlikely that *D. pseudoobscura* and *D. persimilis* were competing in nature. He based this on the relatively small numbers of larvae found in slime fluxes indicating that, at least in the larval stage, the densities were too low to produce competition. Furthermore, the size of these species in nature is very similar to that found when reared in benign uncrowded conditions in the laboratory. Crowding in the laboratory leads to an increased variance in body size, a variance not seen in flies of these species from natural populations.

Mangan (1978; cited in Barker, 1983) carried out a series of direct studies of competition in a seminatural setting using the *repleta* group of the Sonoran desert. He brought saguaro cactus (the natural breeding site of *D. nigrospiracula*) into the laboratory and used flies no more than one generation removed from nature. He studied the effect of density and presence of another species on three fitness components: size, time of development, and preadult survival. He could demonstrate clear interspecific effects on fitness among the three naturally occurring species in the ecosystem, *D. nigrospiracula, D. mettleri*, and *D. mojavensis.* He also noted differences in the effects depending on the state of the substrate, that is, whether the cactus was old or fresh.

Krebs and Barker (1993) have performed similar studies using *Opuntia* cactus and the two *Drosophila* associated with this plant in Australia, *D. buzzatii* and *D. aldrichi.* Overall, *D. buzzatii* was superior in competition with *D. aldrichi* except at a quite high temperature, 31°C. *D. buzzatii* males tended to become sterile at this temperature, while *D. aldrichi* males did not. Interestingly, the geographic distributions of the species accord with this observation; namely, in the cooler south, only *D. buzzatii* is found whereas the two species coexist in the warmer north of Australia.

Laboratory Studies

Many studies have been done competing many different species in laboratory cultures with some fascinating outcomes, only the highlights of which will be presented here; the Barker (1983) review with its references is a more thorough entry to the literature.

Many studies have been aimed at determining whether in fact two *Drosophila* species can coexist in laboratory culture. The answers have been yes and no. One of the first uses of the population cages of L'Heritier and Teissier were to compete two species, *D. melanogaster* and *D. funebris* (L'Heritier and Teissier, 1935). They found coexistence, although with *D. melanogaster* in the majority. Merrell (1951) followed up this observation and found the coexistence was due to changes in the medium over time: when fresh and moist, *D. melanogaster* has a competitive advantage and as the medium dries and hardens, *D. funebris* has the advantage. Mitchell and Arthur (1990) carried out very similar studies with *D. simulans* and *D. funebris* and obtained almost identical results. They found that when food was replaced frequently ("fast-turnover" system) *D. simulans* eliminated *D. funebris*, while they coexisted in a slow-turnover system, that is, when food was left for six weeks before replacing. At the end of this section, we present further evidence that the rate at which resources are made available may affect the outcome of interspecific competition.

Ayala (reviewed in Ayala 1970a, 1972b) performed a series of studies of coexistence in the laboratory, an example of which is shown in figure 5-7. The results of such studies often depend on the exact environmental conditions, in this case temperature. *D. willistoni* eliminates *D. pseudoobscura* at 25°C, whereas the opposite occurs at 19°C. Coexistence occurs at 23°C. Ayala's initial claim for these studies was that they invalidated the principle of competitive exclusion as the laboratory environment was thought to present a single limiting resource. This led to considerable controversy (reviewed in Barker, 1983) over interpretation of both the results and the principle. Alternative models to the classic Lotka-Volterra equations were developed (Ayala et al., 1973; Gilpin and Ayala, 1973) to describe *Drosophila* competition in the laboratory; a four-parameter linear model produced the best realistic fit.

That the outcome of laboratory competition experiments is not always so nicely predictable can be clearly illustrated by Barker's (1963) results with competing *D. melanogaster* and *D. simulans* (figure 5-8). Clearly, there can be considerable indeterminacy in such experiments.

Using a total of eight species, Wallace (1974) determined the effect on productivity for pairs of species. Different combinations of species exhibited competition (lower productivity), facilitation (higher productivity), or neither. In general, intraspecific competition was stronger than interspecific, indicating species are not using precisely the same resources. Thus frequency-dependent selection could account for coexistence, a mechanism also favored by Ayala (1970b). Wallace (1975) carried out further studies meant to mimic and test MacArther and Wilson's (1967) theory of island

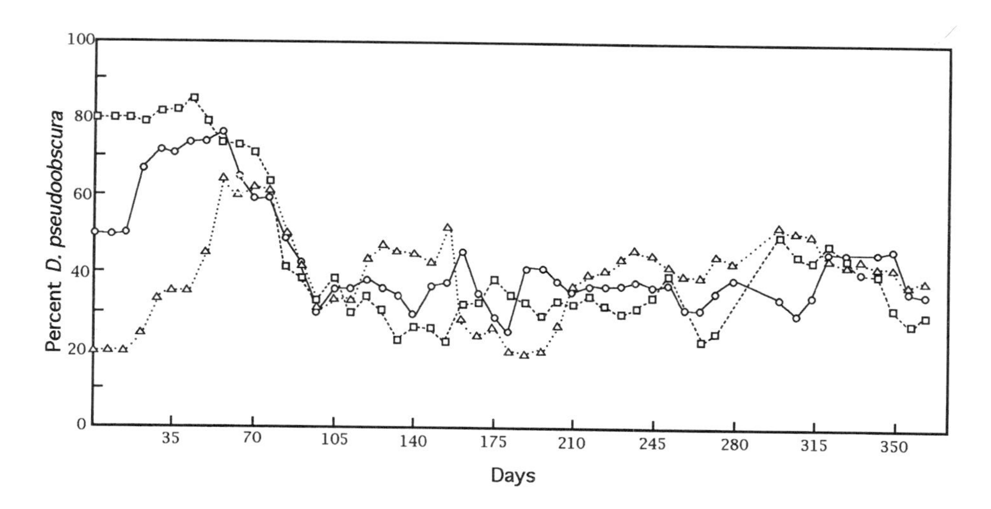

FIGURE 5-7 An illustration of Ayala's competition experiments between *D. pseudoobscura* and *D. willistoni*. Note that independent of the starting frequency, the same apparent equilibrium is reached. From Ayala et al. (1973).

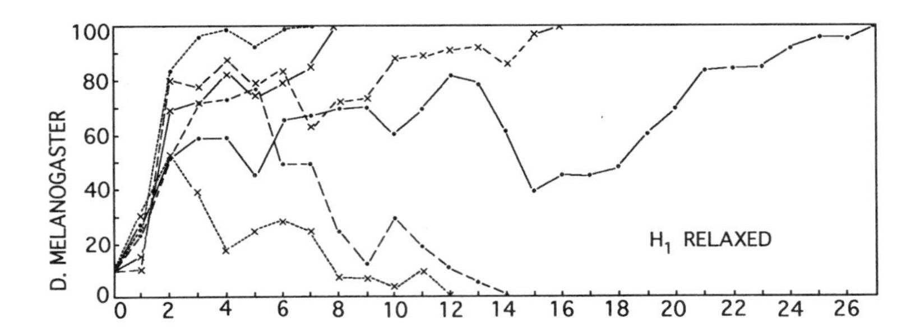

FIGURE 5-8 An example of Barker's competition experiments between *D. melanogaster* and *D. simulans*. In contrast to Ayala's results in figure 5-7, here the outcome is elimination, but which species is eliminated is indeterminant. From Barker (1983).

biogeography in a laboratory situation. He could derive equations describing his findings based on immigration and extinction rates. Perhaps most interesting was the finding that removing *D. melanogaster* from the species pool led to increased numbers of species on his "islands." He relates this to the effect of a weedy species (*melanogaster*) on simplifying communities.

Do populations evolve as a result of competition? Moore (1952) competed *D. melanogaster* and *D. simulans* in 20 replicates. In 19 out of 20, *D. melanogaster* eliminated *D. simulans* quite rapidly, generally by day 100. In the exceptional population, *D. melanogaster* increased in frequency for 73 days, then *D. simulans* increased from day 73 to 218, after which *D. melanogaster* eliminated *D. simulans*. Moore was able to demonstrate that this odd behavior was due to genetic changes in the *D. simulans* strain, and by selecting this species for increased competitiveness, produced strains that could coexist with *D. melanogaster* for up to 500 days. Follow-up studies by Futuyma (1970), Van Delden (1970), and Barker (1973) found ambiguous evidence for any genetic changes brought about by competition. The issue is not so much whether genetic changes can be brought about by selection for competitive ability, but whether the response to such selection is as rapid as implied by Moore's work. This issue requires further investigation.

Several studies have addressed the question of the nature of the competition in laboratory cultures. It is generally thought (i.e., assumed) that most competition occurs at the larval stage, although Ayala (1968a) found that in his serial transfer procedure, adult space may be the limiting resource. Competition may also occur in oviposition. Moth and Barker (1976) showed that in very crowded conditions, larvae may bury or (presumably inadvertently) eat newly laid eggs before they hatch, so choosing a safe oviposition site may be highly important.

In addition to the obvious competition simply for nutrition, other kinds of larval interactions have been demonstrated, in particular for metabolic residues. Weisbrot (1966) preconditioned medium by growing a species on it first, then offering it to a second species. The survival of *D. pseudoobscura* was negatively affected when grown on medium preconditioned by *D. melanogaster*, whereas the reverse occurred in the complementary experiment: medium preconditioned with *D. pseudoobscura* actually increased the survival rate of *D. melanogaster*, an example of facilitation. Budnick and Brncic (1975) carried out more comprehensive studies of this sort, this time adding fresh yeast each day to minimize the shortage of food. They often found, for some pairs of species, a negative effect of previous occupancy by another species, presumably due to some toxicity of accumulated waste products. They point out that this could be an important phenomenon for controlling population size in nature. Such results, however, have not been universally the outcome of this type of experiment (e.g., Dolan and Robertson, 1975).

The effects of different yeasts on competition has also been studied. We already mentioned in a previous section of this chapter that adults of different species are differentially attracted to different yeasts. This has also been shown for larvae of *D. pseudoobscura* and *D. persimilis* (Lindsay, 1958; Cooper, 1960). Fogelman et al. (1981) showed that larvae can preferentially maintain or eliminate species of yeasts. Bos et al. (1977) showed facilitation between *D. melanogaster* and *D. simulans* when grown on medium with a yeast deficient in production of sterols; together the species had greater viability than when either was alone. El-Helw and Ali (1970) and El-Helw et al. (1972) studied the effect of different species of yeast on competition between *D. melanogaster* and *D. simulans* and found considerable effect. The important lesson from these studies is that the microflora needs to be carefully considered in *Drosophila* competition experiments. Unless one uses axenic cultures (which is never done), the number of microbes living in and on laboratory cultures must be very high. Just because there is a single type of medium does not mean there is a single resource.

Goodman (1979) studied another aspect of competition, namely, the transitivity of outcome: if species A eliminates species B, and species B eliminates species C, then if transitivity holds, A should eliminate species C. Goodman studied 19 strains of 15 different species and found a clear hierarchy of competitive ability under laboratory conditions, with virtually no intransitivity. The implications are that all species are competing in the same manner. Goodman speculates this may well be an artifact of using strains that had spent considerable time in laboratory culture (usually many years), meaning they had been selected to exploit laboratory medium. If strains were taken directly from nature, he suspects the outcome would be much more complex and less hierarchical, that is, intransitivity would be observed.

Coexistence in the Tropics

Sevenster and van Alphen (1993a, 1993b) have carried out a particularly elegant combination of theoretical modeling, laboratory studies, and field observation on the coexistence of neotropical *Drosophila* in Panama. Their major premise is that *Drosophila* adopt two different life history strategies: there are fast developers who die young and slow developers who live longer (what Davis and Hardy [1994] called hares and tortoises). In their theoretical modeling they show that when breeding sites are easily found and plentiful, the fast developers out-compete the slow developers. When breeding sites are scarce and hard to find, the slow developers win because they live long enough to find a suitable larval breeding site. Studies in the laboratory confirmed that species with shorter development time had shorter adult lives when faced with starvation, while species that took longer to become adults lived longer under starvation conditions. They studied some 21 different species that coexist in their study site, which represent a large phylogenetic range of species from both subgenera, *Sophophora* and *Drosophila*. Of course, the species did not fall into two nonoverlapping groups; rather, a continuum of development time and adult longevity was found.

Sevenster and van Alphen chose *D. willistoni* as a fast developer (9 days from egg to adult) and *D. sturtevanti* as a slow developer (15 days) as models for laboratory studies. They maintained replicate mixed-species populations, varying the intervals of adding new food from 3 to 21 days. The results were as predicted: when food was provided frequently, *D. willistoni* outperformed and usually eliminated *D. sturtevanti*, while when food was supplied at longer intervals, *D. sturtevanti* did better, sometimes coexisting for the 100 days of the experiment. These results are similar to the *D. melanogaster-simulans* coexistence with *D. funebris* discussed earlier.

Tests in the field relied on the following reasoning: if the slow developers live longer as adults, then adults in nature should be more often the slow-developing species, while larvae breeding in the fruits should more often be the fast breeders. This was observed. Likewise, in areas where fruits were plentiful, the guild of coexisting species skewed toward fast breeders while the opposite was found in areas scarce in larval food sources. They further conclude that the number of fruits used (degree of specialization) was also a factor in these studies. More specialized species tend to be slow developers that lived longer, presumably because it takes longer to find appropriate larval breeding sites. Fast developers tend to be generalists, which makes finding larval resources easier.

This last observation is interesting in regard to the longest developing species known, the Hawaiian flies. Many are specialists that live for very long times as adults. This correlation between longevity and degree of specialization deserves further investigation.

Life Histories, Phenotypic Plasticity, Acclimation, and Aging

One of the more elegant studies of life history parameters in natural populations is that of Kambysellis and Heed (1971). They studied the numbers of ovarioles and mature eggs in a number of Hawaiian species and found the species broke into three more or less discontinuous groups (table 5-11). The first group had but a single mature egg per fly, and in fact in these species the females hold the egg through embryogenesis and "larviposit" rather than oviposit. Group II has 4 to 12 mature eggs on average. Group III has 21 to 101 mature eggs. These three groups differ in oviposition site. Group I larvae breed in flowers, Group II utilizes decaying leaves, and Group III breeds in decaying barks and fruits. The observed oviposition of the flies shows a distinct nonrandom pattern coincident with the ovarian differences. Group I oviposits (larviposits) a single individual at a time. (In another flower-breeder in South America, *D. flavopilosa*, Brncic [1966] also observed a single egg oviposited per flower.) The leaf-breeding Group II oviposit a few eggs per leaf, while Group III bark/fruit breeders oviposit large numbers of eggs in clusters. Montague et al. (1981) have studied these same species with respect to overall reproductive output by relating numbers of eggs produced to size of egg. They found a negative correlation, but not precisely adjusted, so all species have equal reproductive effort.

TABLE 5-11 Relationship of number of ovarioles and mature eggs per female and oviposition substrates in species of Hawaiian *Drosophila*.

Species	Mean No. Ovarioles per Fly	Mean No. Mature Eggs per Ovariole	Mean No. Mature Eggs per Fly	Oviposition Substrate
Group I				
Scaptomyza caliginosa	2.5	0.33	0.8	1
S. mauiensis	4.0	0.25	1.0	1
S. oahuensis	4.1	0.25	1.0	1
D. nasalis	4.7	0.22	1.1	?
D. crassifemer	5.4	0.19	1.0	?
Group II				
D. adunca	11.0	0.63	6.9	2
D. diamphidiopoda	18.4	0.64	11.8	2
D. disticha	11.8	0.33	3.9	3
D. pectinitarsus	12.4	0.75	9.3	3
D. kambysellis	15.0	0.33	5.0	4
Group III				
D. mimica	23.9	0.9	20.8	5
D. setosimentum	35.6	1	35.6	6
D. clavisetae	38.1	1	38.1	?
D. adiastola	45.9	1	45.9	6
D. primaeva	101.3	1	101.3	?

Note: Oviposition substrate codes: 1 = morning-glory flowers, 2 = *Cheirodendron* leaves, 3 = leaves of endemic trees, 4 = *Pisonia* leaves, 5 = fruits and fungus, 6 = leaves, fruits, and stems of endemic plants, ? = unknown.

Simplified from Kambysellis and Heed (1971) in which many more species and other data are reported.

Presumably, the variation in numbers of eggs deposited relates to the substrates. Group I substrates are relatively poor in nutritional value, so fewer larvae per substrate relieves competition; furthermore, the spatial and temporal availability of the substrate is high and predictable. Group III breed in spatially and temporally unpredictable substrates, but when found, they provide very rich and large sources of larval nutrition, so higher densities of larvae can be tolerated. Group II substrates are intermediate. This is a rare case of a life history parameter clearly being associated with ecological factors. Montague (1984) followed up with further observation on one of the flower-breeding species, showing that this resource does not support very many larvae, consistent with the ideas put forward by Kambysellis and Heed.

Drosophila are being increasingly used as a laboratory model for life history evolution. Gebhardt and Stearns (1993a, b) and Stearns et al. (1993) have reported on a series of experiments utilizing *Drosophila* to test some of the theoretical predictions Stearns and others have made regarding evolution of such traits as longevity, fecundity regimes, phenotypic plasticity, and so forth (reviewed in Stearns, 1976, 1989). Stearns et al. (1993) is particularly interesting as an attempt to relate some of these traits to a molecularly-defined locus. Direct r- and K-selection has been applied to *D. pseudoobscura* populations with limited response (Taylor and Condra, 1980). Roff and Mousseau (1987) review much of the *Drosophila* literature and conclude that while generally the heritabilities of life history traits are less than those for morphological traits, they still may exceed 0.20. This high heritability for traits that should be low in heritability may be due in part to negative covariance observed between life history traits.

Resistance to extreme environments is an issue related to phenotypic plasticity as well as to genetic variance. Pioneering work with *Drosophila* in this regard was Levins's (1969) study of Puerto Rican species in regard to dry heat resistance. Much further work has been done on heat and desiccation resistance in *Drosophila*, reviewed in David et al. (1983), Hoffmann and Watson (1993), Hoffmann and Parsons (1991), and Huey and Kingsolver (1993). Considerable genetic variance in both natural and laboratory populations has been demonstrated for heat and desiccation tolerance.

Cold tolerance and overwintering patterns have been studied by Chiang et al. (1962), Crumpacker and Marinkovic (1967), Anxolobéhère and Periquet (1970), Tucic and Krunic (1975), Crumpacker et al. (1977), Tucic (1979), Parsons (1982), Cuesta and Comendador (1982), Izquierdo (1991), Chen and Walker (1993), and Kimura and Beppu (1993). In general, the adult stage is the most cold resistant and is thought to be the stage of overwintering for most temperate species. Many temperate species also undergo diapause (reviewed in Lumme and Lakovaara, 1983). Cases are known in the *virilis* group where variation in the ability to diapause has a simple Mendelian basis (Lumme, 1981; Lumme and Keränen, 1978).

Drosophila as a model in the study of aging has also become a popular theme of several research programs (reviewed in Lints and Hani Soliman, 1988; see also the bibliography of Gartner, 1986). Various factors appear to affect longevity and aging in *Drosophila*. For example, for both males and females mating shortens life span (Partridge and Farquhar, 1981; Partridge et al., 1987). Egg laying by females reduces life expectancy. Daughters produced by females homozygous for a mutant (*grandchildless*) lack ovaries and subsequently have an increased life span equal to that of normal virgins (Maynard Smith, 1958). Starvation as larvae can increase longevity (Zwaan et al., 1991). Selection for delayed senescence has also been successful (Rose 1984; Luckinbill et al., 1984; Partridge and Fowler, 1992), which concomitantly changes various other life history traits. Curtsinger et al. (1992) produced evidence for genetic variation in the form of mortality curves. Much of this work is reviewed in Rose (1991).

Parasites and Symbionts

Drosophila are plagued by a plethora of parasites and symbionts; in fact, by the end of this section the reader may well wonder how some 2,000 plus species have survived to the present. Equally surprising is the relative lack of attention this field has attracted. Given the importance of parasitology for health and economic concerns and the convenience of *Drosophila* as a model laboratory animal, this would seem to be a highly attractive field for further development. Chapter 38 in Ashburner (1989) is a general review.

Viruses

Several viruses are known from *Drosophila*, all of them with RNA genomes, as is the case with most insect-associated viruses. Reviews of *Drosophila* viruses include L'Heritier (1970, 1975) and Brun and Plus (1980).

The best studied virus is the sigma virus, which was first detected by its effect on flies carrying it: they are sensitive to being anesthetized with CO_2 and never recover (L'Heritier and Teissier, 1937). This trait was found to be maternally inherited, and later shown to be a virus (L'Heritier and Hugon de Scoeux, 1947). Brun and Plus (1980) and Fleuriet (1988) describe in some detail the action, transmission, and physiology of the sigma virus.

Natural populations of *D. melanogaster* are found to harbor sigma at a rate of 5 to 20%, and it occurs throughout the distribution of this species. Other species of *Drosophila* also carry sigma and are CO_2 sensitive (Williamson, 1961; Lund, 1959; Felix et al., 1971); these include species from a number of groups of both subgenera *Sophophora* and *Drosophila*. As far as is known, the only manner of transmission of the virus is inheritance through the germline, in most cases only the female's. In some species males may also transmit (Williamson, 1961). If horizontal transmission is impossible (or extremely rare) this implies that the sigma–*Drosophila* association goes back at least 30 million years to the origin of the genus. (There are viruses causing CO_2 sensitivity in mosquitoes [Shroyer and Rosen, 1983; Vazeille-Falcoz et al., 1992], but these are evidently unrelated to sigma.)

Genetic variation in *Drosophila* for the ability to transmit sigma is also known (Gay, 1978). The *Ref(2)P* has been particularly well-studied, as it is polymorphic in natural populations (Fleuriet, 1988). Single allelic substitutions at this locus can cause flies to be sensitive or resistant to transmission. All populations studied were polymorphic with the resistant allele at a frequency of 0.1 to 0.3. In experimental cages, in the presence or absence of the virus, the populations remain polymorphic. As we will see in chapter 10, the *Ref(2)P* locus is unusual in being highly polymorphic for amino acid substitutions and relatively monomorphic for silent substitutions.

The C virus is a picornavirus transmitted by feeding or contact; it is fairly common in *D. melanogaster* populations (Plus et al., 1975). From an evolutionary standpoint, the intriguing aspect of this virus is that it seems to actually be beneficial to the host. Flies infected with it have shorter development time and females are larger with more ovarioles (Thomas-Orillard, 1984). Further studies indicated that the beneficial effects were dependent on the particular population, temperature, and viral dose (Gomariz-Zilber and Thomas-Orillard, 1993).

At least six other RNA viruses have been isolated from natural populations of *D. melanogaster*, most being Picornaviruses and one a Reovirus (Brun and Plus, 1980). Surveys of natural populations around the world and stocks in laboratories attest the widespread nature of these viruses as well. While in general sigma viruses appear to have little effect on fitness of flies, some of these other viruses can be more pathogenic. *D. immigrans* has been shown to harbor yet another Picornavirus called iota, which induces CO_2 sensitivity when injected into *D. melanogaster*, but only in males (Jousset, 1972).

From an evolutionary standpoint, the final paragraph of Brun and Plus's (1980) review sums up the situation very nicely:

> RNA viruses are common in natural, and laboratory, populations of *Drosophila*. Studies of their effects on these populations are badly needed but we note that, in general, they appear to play a role in increasing variability since (a) within a population infected flies differ in a number of ways from uninfected flies (e.g., longevity, fecundity, developmental times), (b) viruses have been shown to be mutagenic in *D. melanogaster* (Gershenson et al., 1975, for review; Paquin and Baumiller, 1978) and (c) natural populations are polymorphic for alleles conferring resistance to particular viruses.

Furthermore, these authors point out the similarity of the biology of *Drosophila* viruses to other arboviruses that cause diseases in humans and economically important animals. *Drosophila* have been underexploited as a model for understanding Dipteran-borne viruses in general.

Bacteria

Several different kinds of bacteria are known to infect *Drosophila* with very different effects. One of the better characterized is the L-form bacteria in *D. paulistorum* (reviewed in Ehrman and Kernaghan, 1972). Apparently, these intracellular cell wall-less bacteria are the causative agent for male F_1 sterility in crosses between the semi-species, a topic to be covered further in chapter 7. We have already had cause to mention the bacteria of the genus *Spiroplasma* (Class: Mollicutes) that infect members of the *D. willistoni* group and cause the sex-ratio condition (reviewed by Williamson and Poulson, 1979; and Ebbert, 1993). These may be transferred among many species of *Drosophila* by injection, invariably causing sex-ratio in the new host. Asymptomatic variants that do not kill male embryos are also known. There are also viruses associated with these bacteria, often specific to the strain of bacteria in which they are found. *Spiroplasmas* from different strains may cause a clumping reaction when mixed in a single host. Evidently this reaction is caused by the viral interactions; only one type survives. With the exception of Ebbert (1988, 1991, 1995, and discussed in chapter 4) this virus-bacteria-*Drosophila* association has received little attention by evolutionists. Extra-chromosomally inherited female-biased sex ratio occurs in a number of other species of *Drosophila* (reviewed in Ebbert, 1993); in many cases the precise nature of the factor involved is not known, though presumably some are similarly caused by bacteria. These include *D. bifasciata* (Ikeda, 1970), *D. borealis* (Carson, 1956), *D. prosaltans* (Cavalcanti et al., 1957), and *D. robusta* (Poulson, 1968).

In some respects, from an evolutionary standpoint, the most intriguing bacteria in *Drosophila* is the Rickettsia-like *Wolbachia.* This bacteria was first shown to be the causative agent for reproductive incompatibility among strains of the common mosquito, *Culex pipiens* (Yen and Barr, 1974), and hence was named *W. pipientis.* A very similar pattern of cytoplasmic incompatibility was known to occur in *D. simulans* (Hoffmann et al., 1986; Hoffmann and Turelli, 1988), for which Louis and Nigro (1989) obtained evidence *Wolbachia* is the causative agent, a finding now well-confirmed (Turelli and Hoffmann, 1991; Turelli et al., 1992). There is some evidence more than one type of *Wolbachia* might exist in different populations of *D. simulans* (Montchamp-Moreau et al., 1991). *D. melanogaster* also harbors *Wolbachia*, although it is not clear if it causes any cytoplasmic incompatibility; Holden et al. (1993) found no evidence for it while Hoffmann et al. (1994) did. O'Neill et al. (1992) confirm the presence of *W. pipientis* in a number of phylogenetically widely separate insects, although the DNA sequences indicate it is probably a single species of bacteria. The possibility that these bacteria may be involved in speciation is discussed in chapter 7.

Bacterial infections in *Drosophila* have also been used as model systems to study insect immune response. Various small antibiotic peptides can be induced in insects, including *Drosophila*, in response to infection. Flyg et al. (1987) demonstrated induced immunity to bacterial infection concomitant with a rise in an antibacterial polypeptide similar in property to some known from the silk moth. Robertson and Postlethwait (1986) have identified three antibiotic peptides that can appear in two hours after infection and persist for 60 days. The cellular response due to encapsulation by blood cells can also occur in response to bacterial infection (Rizki, 1968). Flyg and Boman (1988) provide evidence for genetic variation in the level of virulence of *Serratia marcescens*, as well as genetic variation within the host, *D. melanogaster*, for its susceptibility. Here would seem to be a wonderful model system for experimentally manipulating and studying coevolution between host and parasite.

Molds and Fungi

While it is well known that *Drosophila* utilize yeasts as a food source, other fungi are damaging to flies. Ashburner (1989) cites a large number of fungi known to infect *Drosophila*, some of which are lethal. Even *Penicillium* may be lethal to *Drosophila*. We mentioned earlier that there was evidence that *Penicillium* infection on fruit inhibits growth of *D. melanogaster* but not *D. immigrans*. Several species of parasitic fungi of the Laboulbeniomycetes group infect *Drosophila* in nature (Benjamin, 1973).

Protozoa

Chatton and colleagues (cited in Ashburner, 1989) describe five species of trypanosomes infecting four different species of *Drosophila* (*D. melanogaster, D. busckii, D. confusa, and D. phelerata)* collected around the Institute Pasteur in Paris. Other species of protozoa in *Drosophila* have been identified from flies from the Sudan and India. These flagellates infect the gut and Malphigian tubules (Rowton et al., 1981); presumably they are transmitted through feces.

Microsporidia have often been found in *Drosophila* and are a threat to laboratory culture (e.g., Stalker and Carson, 1963). Infection occurs in all major organs at all stages, including gonads, which causes the most serious problems. It can be transmitted either orally or transovarially. The best studied microsporidian in *Drosophila* is *Nosema kingi*, originally found by Burnett and King (1962) in *D. willistoni*. Armstrong and Bass (1989) include references to previous work, including their own studies of the fitness effects of infection with *Nosema*. Ashburner (1989) cites studies of other sporidian infections in *Drosophila*.

Nematodes

As with virtually all insects, nematodes often infect *Drosophila*, both in the wild and in the laboratory. In fact, the very earliest record of insect parasitism is from a fossil *Drosophila* 26 million years old found with a nematode (Poiner, 1984).

The most thorough study of nematodes infecting *Drosophila* in nature is that of Welch (1959) in southern England. Two species of nematodes (*Parasitylenchus diplogenus* and *Howardula aoronymphium*) were found that infected a mutually exclusive subset of *Drosophila.* The former species was found only in members of subgenus *Sophophora*, namely *D. obscura, D. subobscura*, and *D. subsilvestris*, while the latter was found only in fungus feeders of the subgenus *Drosophila, D. phelerata* and *D. kunztei.* Adult worms enter through the cuticle of larvae and deposit eggs in the hemocoel. The worms go through one or two generations in the host and exit through the anus or genital tract. Considerable damage is done to the flies, both from the nutritional standpoint and by the nematode's exit. *H. aoronymphium* also occurs in North America and infects the system studied by Jaenike (see above), being particularly damaging to *D. phalerata* (Montague and Jaenike, 1985). The α-amanitin mushrooms are nearly devoid of the worms (Jaenike, 1985), so the evolution of the flies to be resistant to this usually toxic compound helps them avoid nematodes. Alternatively it was pressure to avoid nematodes that led to resistance to α-amanitin.

Mites

Every *Drosophila* biologist knows the greatest pest in culturing *Drosophila* is mites. *Drosophila* mites come in two forms, those that actually predate the flies, usually eggs or early larvae, and those that simply breed in the medium, using the flies as the transfer mechanism to a new food source (the phenomenon of phoresy). The nastiness of mites in cultures is caused largely by their very short life cycle, as short as three to five days. In a typical laboratory situation, cultures are routinely transferred only every two to four weeks, so several generations of mites stemming from a single female can very quickly produce an extraordinary number of offspring.

Remarkably little is known, on the other hand, about mites associated with flies in nature. Ashburner (1989) list 18 species of mites from all four major taxa that have been found associated with *Drosophila*, eight of which were found on flies in nature. However, the references to mites in nature are only two in number: Maca (1982) and O'Donnell and Axtell (1965). Obviously field collectors of *Drosophila*, myself included, have paid virtually no attention to the presence of mites on flies from nature. I have regularly but rarely observed mites on field-collected *obscura* group and *willistoni* group flies; I would estimate one in every few hundred flies from nature has a mite clinging to its cuticle, sometimes with a proboscis inserted into the hemocoel. It is interesting to note in Ashburner's compilation that the species of mites known from nature and from the laboratory are mutually exclusive lists; that is, no mite known to infest laboratory cultures has been seen on flies from nature and vice versa. In my experience, I know of no case where flies collected in nature and set up in laboratory culture produced a mite-infested culture independently of an already present mite infestation in the laboratory.

That mites may be playing an unexpected role in *Drosophila* evolution was demonstrated by Houck et al. (1991). They provide evidence that mites may be responsible for horizontal transfer of mobile DNA sequences. Houck et al. found P element DNA sequences in mites (*Proctolaelaps regalis*) that were infesting cultures of flies that had *P* elements, and not in mites from cultures of flies without *P* elements. They purposely chose this particular mite as its feeding habits were such as to increase the probability of picking up and transferring DNA from a host: it very rapidly flits from one egg or small larva to another, inserting its chelicerae for a fraction of a second before moving on.

Hymenoptera

Many small parasitoid wasps of the superfamilies Ichneumonoidea and Chalcidoidea attack *Drosophila.* Carton et al. (1986), the most thorough review of the subject, list some 42 species of wasps that have been found in at least 25 species of *Drosophila.* They exist in tropical, subtropical, and temperate regions. *Drosophila* breeding in fruit, fungi, flowers, and/or rotting vegetation are affected, as are cosmopolitan commensals and narrow endemics. In all cases only immature stages of *Drosophila* are parasitized. This has inhibited their study in nature because most collectors collect adults; one must collect larvae or pupae to find the wasps. Furthermore, because *Drosophila* species are hard or impossible to identify in the larval and pupal stages, it is often not known what species of fly is being parasitized. Most wasps attack either the larvae or the pupae, not both. In all cases, if the parasitism is successful, it often results in the death of the larva or pupa. This is the reason such wasps are called parasitoids rather than parasites; parasites generally do not kill the host, for if they always did so they would be predators. The wasps are somewhere between classical parasites and predators.

Female wasps lay eggs through the cuticle of either young larvae or pupae; the egg hatches and the wasp larva starts feeding, usually killing the host. Development to adult wasp takes about two weeks. Female wasps eclose with eggs already matured and males inseminate them upon emergence. The female finds suitable hosts by being attracted to *Drosophila* breeding sites: wasp parasitoids of fruit-breeding *Drosophila* are attracted to fermenting fruit and ethanol (Carton, 1978; Dicke et al., 1984), those that breed in decaying leaves are attracted to rotting beet leaves (Vet et al., 1983a, b), and wasps specializing on fungus-breeding *Drosophila* are attracted to rotting mushrooms (Vet, 1983).

Another indication of the adaptation of wasp to fly is that in colder temperate regions the wasps diapause as larvae when the temperature is too low to breed *Drosophila.* The wasps come out of diapause and carry through development only when temperatures rise high enough to allow *Drosophila* to begin breeding again (Carton, 1983).

Few intensive field studies have been done on parasitoid wasps of *Drosophila,* although one study in Tunisia is instructive. Three species of *Drosophila* were found breeding in *Opuntia* cactus: *D. melanogaster, D. simulans,* and *D. buzzatii.* Figure 5-9 shows the frequencies of species of fly, species of wasp, and prevalence of parasitism over a year. Both the flies and wasps are temporally fairly well disjunct. The level of parasitism varies greatly, but can reach as high as 50% in *D. simulans.*

Brncic (1972) has similar observations on the flower-breeding *D. flavopilosa* in South America. Figure 5-10 summarizes his observations in one locality in Chile. In this species, the only larval breeding site is the plant *Cestrum,* in which only one larva per flower is ever found. The seasonal variation in frequencies of flowers with larvae and eggs (presumably highly correlated with fly population size) and rate of parasitism (by two species of wasp, *Opius trimaculatus* and *Ganaspis* sp.) and pupal mortality

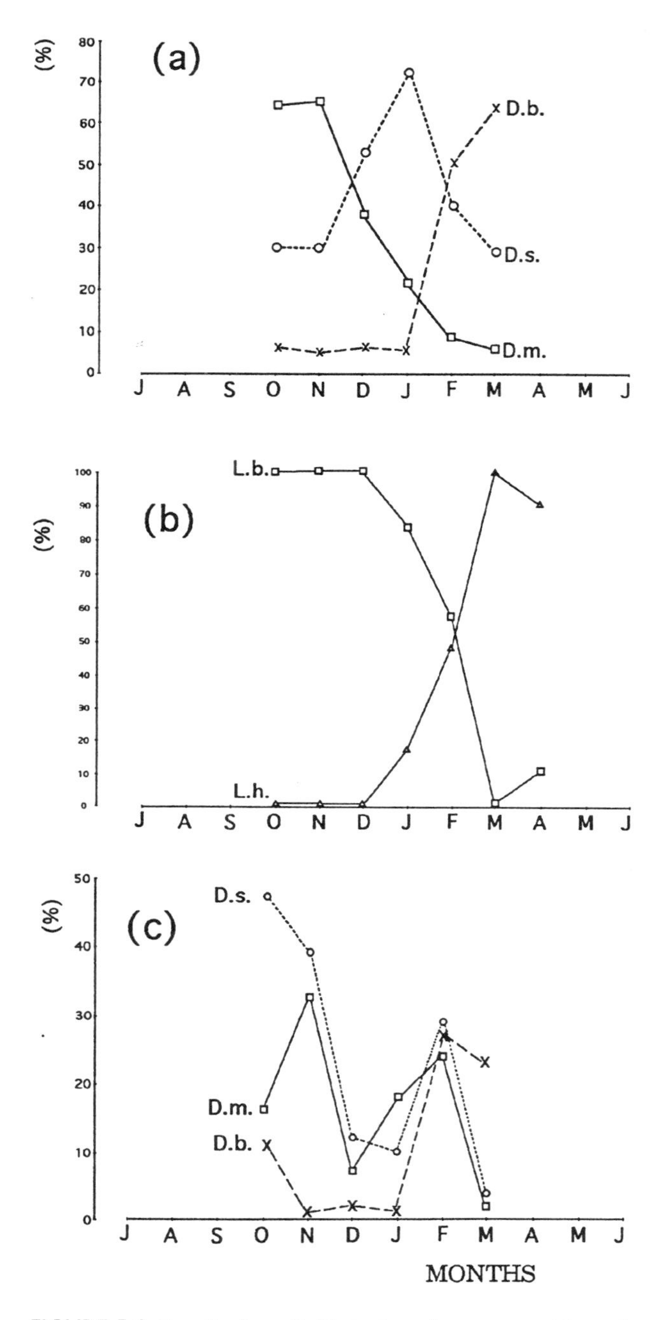

FIGURE 5-9 Results from field studies of wasp parasitism of *Drosophila* breeding in *Opuntia* in Tunisia. The upper graph displays the percentage of *D. buzzatii* (D. b.), *D. simulans* (D. s.), and *D. melanogaster* (D. m.) found on the *Opuntia*. The middle graph shows the percentage of two parasitoids, *Leptopilina boulardi* (L. b.) and *L. heterotoma* (L. h.). The lower graph shows the rate of parasitism for each *Drosophila* species. From Carton et al. (1986).

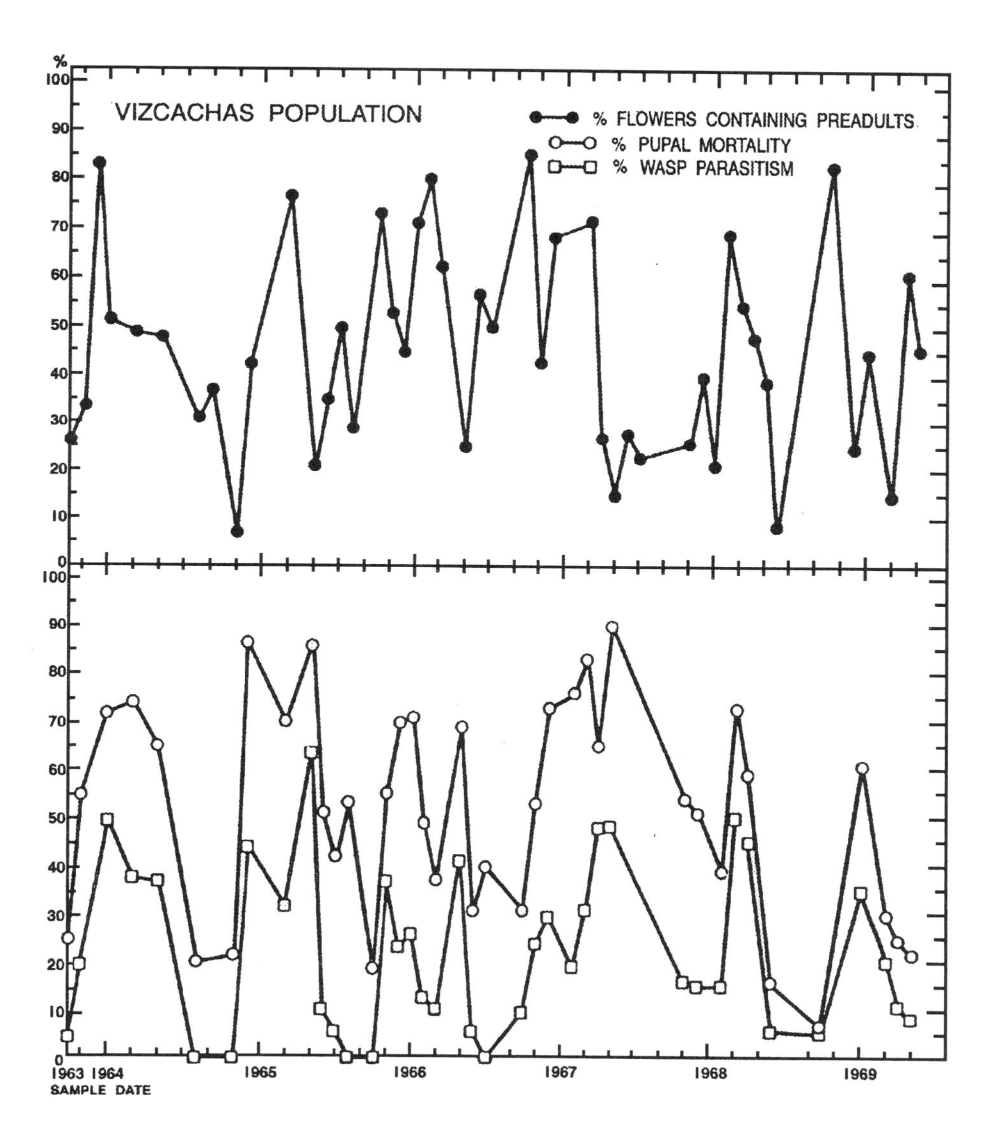

FIGURE 5-10 Seasonal fluctuations in frequencies of flowers with preadult stages of *D. flavopilosa*, the degree of pupal mortality, and the rate of wasp parasitism. From Brncic (1972).

are all closely related. Presumably, there is a causal relationship, that is, wasp parasitism is at least to some extent controlling fly population size.

Other studies have shown both seasonal variation and at times very high levels of parasitism in nature. For example, Baker (1979; cited in Carton et al., 1986) studied larvae found in artificial baits placed outdoors in England; the average parasitism rate was only 6–8%, but reached a maximum of more than 50% in May and June. Prince (cited by Parsons, 1977) found parasitism in Australia to vary seasonally from 0 to 80%. Rouault (1979) found Tunisian *D. melanogaster* and *D. simulans* to vary from 39 to 85% parasitized. Finally, virtually 100% of *Drosophila* larvae breeding in fruit in Switzerland in late September were found parasitized (J. J. M. van Alphen et al., unpublished, cited by Carton et al., 1986). Clearly, wasp parasitism can have a profound effect on natural populations of *Drosophila* and may well be a major factor determining population size and perhaps distributions. This has not been widely appreciated and thus not widely investigated.

Curiously, at least in the laboratory under crowded conditions, larvae parasitized by *Leptopilina boulardi* have a higher viability and eliminate most nonparasitized larvae (Prevost, 1985, cited by Carton et al., 1986). Wasp parasitism also induces a change in *Drosophila* larval behavior; the larvae begin to burrow deeper into the food, presumably to avoid superparasitism (F. Chibani and Y. Carton, unpublished, cited in Carton et al., 1986).

Host *Drosophila* may also mount a defense against the wasps similar to that against bacteria. The presence of the parasitoid in the hemolymph or even its stinging stimulates the development of blood cells (Walker, 1959). The cells aggregate around the parasitoid egg or larvae, eventually encapsulating it and then depositing melanin from the fly's crystal cells (Rizki, 1957). The encapsulated parasitoid dies, either through asphyxiation or inability to move and feed (Salt, 1970). Kraaijeveld and van Alphen (1994a, 1994b, 1995) have studied a particularly interesting case of a braconid parasitoid, *Asobara tabida.* They show that the susceptibility to encapsulation by *D. melanogaster* larvae is geographically variable: parasitoids from the Mediterranean region are more resistant to encapsulation than those from northern Europe. Likewise, the encapsulation ability of the *D. melanogaster* populations varies geographically; this variation is evidently specific for different parasitoid wasps.

Epilog

The best summing up of this chapter is to say that I hope the reader will appreciate the aptness of part of the opening quote, that "there is indeed a moderately extensive knowledge of the basic ecology of the family on which to build." It would also seem there is more than a moderately extensive opportunity to explore in much more depth many aspects of ecology using these flies as models. The possibility of adding a genetic dimension to such studies is particularly attractive and is the subject of the next chapter.

6

Ecological Genetics

> The ecological problem of populations has to do with the numbers of animals and what determines these numbers. The genetical problem of populations has to do with the kind or kinds of animals and what determines kind. These two disciplines meet when the questions are asked, how does the kind of animal (i.e., genotype) influence the numbers and how does the number of animals influence the kind, i.e., the genetical composition of the population? These questions are as ecological as they are genetical.
>
> L. C. Birch, 1960

This chapter will address studies bridging ecology and genetics, that is, relating issues discussed in the previous chapter to those discussed in chapters 2, 3, and 4. Much of the work discussed in chapters 2 and 3 regarding levels of genetic variation in populations was done with little or no regard to ecological context. The populations surveyed simply came from nature with little concern about the actual ecological setting. Chapter 4 did touch on some ecological issues (e.g., habitat choice in the laboratory and competition), but most of the work discussed in this chapter deals with wild populations, although when relevant, laboratory work will be mentioned.

Habitat Choice

Virtually all habitats are heterogeneous, both in nature and in the laboratory. Thus inevitably populations are faced with the problem of adaptation to more than a single environment. One possible response to this challenge is to maintain more than one genotype (Birch's "kind" in the opening quote) with alternative genotypes more or less well adapted to different aspects of the habitat. In chapter 4 we discussed studies in the laboratory indicating that genotype-based habitat selection can occur in those settings. Here we discuss evidence that it also occurs in nature, although the situation is much more complex than in the laboratory.

The Obscura Group at Mather

I will first describe a series of experiments on *D. pseudoobscura* and *D. persimilis* in which I participated (reviewed in Powell and Taylor, 1979; Powell et al., 1984; Taylor, 1987); their presentation here is not intended to imply this is the best or most important work in this field, simply that I am most familiar with it. Most of our work was performed at Mather, California so some familiarity with the locale will be useful.

It is located at about 1,400 meters elevation on the western slopes of the Sierra Nevada Mountains on the west boundary of Yosemite National Park. It is mostly primary and secondary forest with native oaks, pines, and cedars. It is maintained as a field station by the Carnegie Institution Washington's Plant Research Laboratory at Stanford; because of this, the area is relatively pristine, having little or no human-caused disturbance such as garbage that could influence the local *Drosophila* fauna. No human commensal *Drosophila* can be found at Mather, except that late in the summer sometimes a few show up in traps, presumably due to hikers and campers. The environment is highly seasonal with summer temperatures reaching as high as 33°C and with several feet of snow during the winter. *Drosophila* breed there for only about five months. The local populations have been studied for inversion frequencies for many years (table 3-10) as well as being studied for their dispersal behavior (chapter 5). Figure 6-1 is a schematic map of Mather.

One prediction of genotype-based habitat choice is that the frequencies of genotypes may be different in different parts of the environment or microhabitats. We

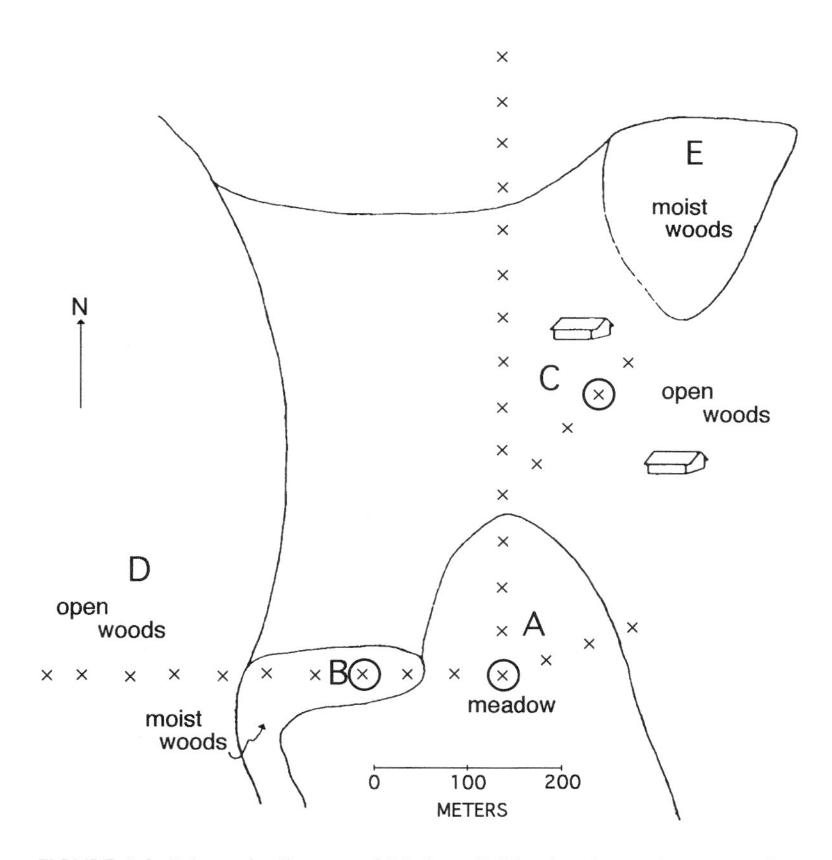

FIGURE 6-1 Schematic diagram of Mather, California where release experiments were performed. An x indicates placement of a trap, and a circled x is a release site. From Powell and Taylor (1979).

collected flies from the five areas indicated by letters A to E in figure 6-1. Area A was designated "open meadow." This is a low-lying meadow relatively dry and hot during the day, although a vernal pond exists in one corner. The area corresponds very well to Ornduff's (1974) montane meadow community with many species of herbaceous plants: *Oenothera, Iris, Achillaea, Sidalcea, Prunella, Potentilla*, and sedges. Area B is dense, well-shaded woods. The closely packed trees consist of *Pinus ponderosa* (Ponderosa pine), *Quercus kellogi* (Kellog oak), and *Calocedrus decurrens* (incense cedar). The ground is thickly covered with pine needles with remarkably little herbaceous ground cover. Areas C and D are similar with respect to tree species and are typical woodland/chaparral. These areas are open and dry with trees the same as in Area B but more widely spaced, along with considerable *Arctostaphylos mariposa* and *Ceanothus integerrimus*. Part of this area to the northwest of A is a quite xeric south-facing slope. The main difference between Areas C and D is that the moist ecologically distinct Area B is between D and A. Area E is dense, very wet woods. A wide stream trickles through the area for most of the summer. The cedars and Ponderosa pines are more closely packed than in any other area and the forest floor is thick with herbaceous cover and fungi. This area differs from B in having more dense and varied vegetation. While humans perceive these as ecologically distinct areas, *D. pseudoobscura* and *D. persimilis* also perceive differences among these areas, as demonstrated by their very different dispersal behavior in each (Dobzhansky et al., 1979). Thus, designation of distinct areas based on vegetation, moisture, and light is not simply a human-biased perception.

Table 6-1 shows some results. During the particular season these studies were carried out, *D persimilis* was much more common than *D. pseudoobscura*, so the sample size for the former species was much greater. Both the inversion polymorphism of the third chromosome and an X-linked allozyme marker showed significant heterogeneity in frequencies among the five areas. Our earlier study revealed that these species in this same area travel about 170 meters per day (Powell et al., 1976). In the previous chapter we noted these species should have little population structure,

TABLE 6-1 Third chromosome inversion frequencies and *Pgm* allozyme frequencies in *D. persimilis* collected in the five habitats indicated in figure 6-1.

	Third Chromosome Gene Arrangements (%)					*Pgm* Alleles (%)		
Area	WH	KL	ST	MD + SE	*N*	Slow	Fast	*N*
A	75	16	4	4	226	41	59	287
B	71	15	8	7	348	50	50	247
C	68	14	15	5	362	46	54	239
D	73	10	10	6	194	37	63	216
E	60	23	13	4	70	26	74	89
Chi-squared:	27.6, 12 df, $p < 0.01$					19.9, 4 df, $p < 0.001$		

Note: N is the number of genomes sampled.

From Taylor and Powell (1977).

given such high rates of dispersal, and that genetic differences among populations (for neutral, or nearly neutral polymorphisms) should be small. One way to reconcile these seemingly conflicting pictures, that is, high mobility with genetic uniformity over large geographic areas and microscale genetic differentiation, is to invoke behavioral differences among genotypes such that they do not distribute themselves randomly across microscale habitats. Models without some type of behaviorally based microhabitat choice could adequately explain the data only with highly unlikely assumptions (Taylor and Powell, 1977).

Given these results, we tested for the existence of habitat choice in another manner, namely by asking if flies exhibit microhabitat loyalty. Figure 6-1 also shows the schematic of the design for this experiment. Flies were collected in each habitat, marked with fluorescent dusts (a different color for each habitat), and released. Three release points were used, one each in the three most contrasting microhabitats (circled x's in figure 6-1). Table 6-2 shows some results. As predicted by habitat choice, flies had a significant tendency to return to the microhabitat in which they were initially captured. We found habitat fidelity even when we used different release sites varying in their distances from the recapture areas. This will be important when discussing critiques of this study.

To test whether this was simply a return to the geographical location, or return to an ecological habitat, we noted the behavior of flies collected in Area D (figure 6-1). Flies from areas B and D were released at the release site in Area A. If the D flies were returning to a geographic area, they would pass through Area B and tend to be found there. On the other hand, if habitat is important, they might go more directly to the ecologically nearly identical Area C. The latter was found (Taylor and Powell, 1978).

Of the 11 release experiments performed, 10 were done when the weather was normal, that is, sunny and warm. One release was performed during a rainy, cool day. It was the only one that did not show habitat loyalty. Presumably light and/or moisture are important cues for flies to distinguish habitats and on this rainy day, the distinction among habitats was much less distinct or absent.

TABLE 6-2 Recapture experiments of flies indicating habitat fidelity.

Release Site	Recapture Area	Original Source B	Original Source C	Contingency Chi-Squared (1 df)
A	B	53 (39)	63 (76)	16.00[a]
	C	15 (28)	70 (56)	
B	B	119 (110)	42 (51)	19.23[a]
	C	5 (14)	16 (7)	
C	B	29 (23)	12 (18)	5.48[b]
	C	18 (24)	24 (18)	

[a] $p < 0.01$
[b] $p < 0.05$

Note: Flies were originally captured in a source area, released in one of three localities (figure 6-1) and recaptured in areas B and C. Numbers in parentheses are those expected if movement were random.

From Taylor and Powell (1978).

One other negative result was obtained during these experiments, and that was when we attempted to determine the degree of additive genetic variance for the observed behavior. We collected flies in the different environments, placed them in bottles with standard laboratory medium, reared them at ambient Mather temperature, and released the resulting F_1 offspring. They did not show any nonrandom tendency to return to the microhabitat from which their parents came. Several interpretations of this result are possible. The additive genetic variance may be too small to detect with our sample sizes. Alternatively, flies reared on laboratory food, naive of any experience in nature, behave very differently from flies taken directly from nature. This explanation is consistent with the difference in dispersal behavior of the laboratory-reared flies in the Dobzhansky and Wright (1947) experiments compared to those using flies from nature (Powell et al., 1976). The offspring were recaptured only one and two days after release and perhaps longer time is required for them to get to know their environment. Finally, the negative results could be due to seasonal changes in the environment during the three weeks between collection of parents and release of offspring. Given the observation of seasonal cycling of inversions, this can be long enough for environmental changes to affect the genetics of populations.

The final experiment in this series was designed to test specifically for habitat preferences based on different inversion karyotypes in the third chromosome (Klaczko et al., 1986). Strains collected one season were established in the laboratory and homokaryotypic lines produced by one or two single-pair crosses. Homokaryotypic lines with the same arrangement were then intercrossed to restore genetic heterogeneity typical for the rest of the genome of flies in nature. *D. pseudoobscura* homokaryotypic for AR, CH, and ST and *D. persimilis* homokaryotypic for KL and WT were used. In addition to studying the distribution of these flies with regard to the five areas in figure 6-1, we added another variable, namely, seeding our baits with two different kinds of yeast to see if different frequencies of karyotypes were differentially attracted to different yeast. Two yeasts were used, commercial *S. cerevisiae* and a strain of *Kloekera apiculata* originally isolated from a *D. pseudoobscura* crop. Two or three different homokaryotypes were released in each of five different experiments. In four of five cases the karyotypes were significantly nonrandomly distributed with respect to the five areas; in two of the five releases karyotypes behaved significantly differently with respect to the different yeasts. In no release did the different genotypes sort themselves randomly. Even for these laboratory-reared flies, then, there was nonrandom behavior with respect to the microhabitats.

Other Studies

Considering all our experiments at Mather, we can only conclude there is good evidence that indeed habitat choice is occurring on a fine scale and that it is likely genotype-dependent. Observations on a number of other *Drosophila* species lend support to the generality of this conclusion. In a study of *D. melanogaster* in an orchard in Texas, Stalker (1976) found significantly different frequencies of inversions from flies at oranges versus those at grapefruits, although these fruits were only a few meters apart. Krimbas and Alevizos (1973) found inversion frequencies in *D. subobscura* in a natural setting to be temporally stable at traps when collected over several days, but traps 50 meters apart showed consistent frequency differences. Valente and

Araújo (1985) showed that different second and third chromosome karyotypes of *D. willistoni* were differentially attracted to different fruits in Brazil; furthermore, they found different proportions of karyotypes eclosing from the fruits. Cabrera et al. (1985) found microhabitat genetic differentiation for allozymes of *D. subobscura*. Richmond (1978) found significant microspatial heterogeneity in a 50 × 75 meter grid for allozymes of *D. affinis*; this study was carried out in a natural area in Indiana where the species is native. Richmond (1978) reviews other studies of microgeographic genetic differentiation in *Drosophila*.

There are several other studies of habitat fidelity based on mark-recapture experiments. Kekic et al. (1980) found evidence for *D. subobscura* in Yugoslavia returning to their original place of capture. Atkinson and Miller (1980) studied the same species in England and obtained negative results. Shorrocks and Nigro (1981), using the same site, did find habitat fidelity; the difference between this study and Atkinson and Miller was the particular microhabitats chosen for collecting and recapture. Shorrocks and Nigro (1981) also provide evidence that, as with the Mather studies, flies were returning to ecologically similar territories, and not simply to the geographic area where first captured. Hey and Houle (1987) studied three species of the *affinis* subgroup in New York. They found evidence for habitat fidelity for males of *D. affinis* but not for females, nor for either sex of *D. athabasca* and *D. algonquin*.

One study of habitat fidelity that gave negative results deserves special comment. Turelli et al. (1984) studied four species or species pairs in an orchard area in the Central Valley of California. The orchards produced two types of citrus (oranges and grapefruit), figs, and almonds; none of these fruits are native to California or, indeed, America. The four *Drosophila* studied were *D. immigrans, D. hydei*, and the pairs *D. melanogaster-simulans* and *D. pseudoobscura-persimilis*, the latter two pairs being indistinguishable morphologically (or nearly so). A study of a sample of the *D. pseudoobscura-persimilis* flies indicated they were about 99% *D. pseudoobscura*. With the exception of *D. melanogaster-simulans*, Turelli et al. found no indication of habitat fidelity with respect to these four *Drosophila* and the heterogeneous orchard resources. That the flies were using oranges, grapefruit, and figs as resources was confirmed by finding larvae in them.

How can the Turelli et al. (1984) results be reconciled with our observations with *D. pseudoobscura-persimilis* at Mather? Three factors seem important. First, the *D. pseudoobscura* population studied by Turelli et al. is not a natural one and may well be low in genetic variability, at least in the variability involved in differential habitat choice. When this species becomes a human commensal, it may go through selection and bottlenecks in adapting to human habitats, with perhaps a single behavioral genotype adapting. Second, the study of Vacek et al. (1979) on a similar commensal population of *D. pseudoobscura* in a citrus grove in Arizona indicated the yeast species harbored by the flies matched the species on the citrus. This is in stark contrast to the situation in more natural settings where the yeasts in adult guts are extremely variable and do not match well at all the yeasts in potential larval breeding substrates (see previous chapter). This shows that the behavior of this species in natural settings and in human habitats is very different. Third, an explanation favored by Turelli et al. is that the resources in the artificial orchard setting are much more abundant and evenly distributed than in natural settings, and the flies may not perceive much habitat heterogeneity.

Considerable attention has been given to the possibility that intraspecific variation in microhabitat choice is due to learning or experience rather than genetically based variation. It is well known that exposure to various environments (odors, media, temperatures, etc.) can subsequently affect adult behavior. A study of three species, *D. melanogaster, D. immigrans*, and *D. recens*, by Jaenike (1982) demonstrated that, generally, there was a tendency for a positive correlation between adult exposure to a chemical and females subsequently being attracted to it for oviposition. No effect of larval rearing was evident for subsequent female oviposition. But, as Jaenike points out, when females eclose, their first experience with a microenvironment will be their larval site, thus perhaps inducing a significant tendency to return. Hoffmann (1985) obtained ambiguous results with similar experiments.

Taylor (1986) studied *D. pseudoobscura* for subsequent behavioral choices after exposure to different temperatures, light levels, and foods. He found significant positive effect for temperature, a negative effect for light, and no effect for food. Jaenike (1985) exposed two strains of *D. tripunctata* to either tomatoes or mushrooms for seven days. After release into the field, he found a significant tendency for females of one strain to return to the substrate to which they had been exposed.

Hoffmann and Turelli (1985) studied the effect of stress (starvation) on the dispersal of *D. melanogaster* in nut and fruit orchards in California. They found that starving flies were more likely to choose a poor substrate. They propose a common sense explanation that starved flies tend to choose any substrate, whereas unstressed flies tend to be more choosy. They emphasize the idea that the physiological state of flies may be more important than genetic variation or learning.

Hoffmann and Turelli (1985) further argue from their results that they can explain the observations of *D. pseudoobscura* habitat preference in the Mather experiments. They base this on two points. The first is a misinterpretation of our work. They state that the habitats used in our microgeographic genetic differentiation (Taylor and Powell, 1977) were not the same as those used for the habitat fidelity studies (Taylor and Powell, 1978). In fact, while we did not use Area E in the habitat fidelity studies, all the other areas are identical to those used for the genetic microgeographic studies. The second claim is that the flies from the more stressful environment (which they assumed to be Areas C and D, figure 6-1) simply went to the nearest suitable habitat without making a choice. Again, this is mistaken. As can be seen in figure 6-1, Area B (the presumed most favorable site for these species based on the densities found there) is closer to the release site in the meadow, yet more stressed flies from Area C returned to Area C rather than move to Area B (table 6-2). Furthermore, more Area C flies moved through Area B to the traps in Area D than did Area B flies (Taylor and Powell, 1978); in other words, the stressed flies went further to find a habitat similar to that in which they were first captured than did the less stressed flies. Finally, when the presumed most favorable habitat (B) was used for the release site, more stressed Area C flies moved out of it to the presumed unfavorable Area C than did Area B flies. Overall, the evidence is that the phenomenon described by Hoffmann and Turelli (1985) is not relevant to the Mather studies.

General Comments

Is there genetically based microhabitat choice in natural populations of *Drosophila*, or is the evidence due to one or more sources of artifacts? It is true that adult behavior in the field is a complex phenomenon affected by genes, previous experiences, and physiological state. However, the evidence that some of the nonrandom distribution of flies in nature has at least a partial genetic basis that can account for microspatial genetic differentiation is strong. This is especially true for species studied in their native habitats. When one further adds the results from laboratory studies on genotype-dependent response to microhabitat variation (chapter 4), the support for the notion is strengthened.

Studies that provided no evidence for habitat choice may simply be a reflection of our lack of knowledge of the relevant ecological variables for flies in nature. As Barker (1990) has stated:

> The habitats of most species certainly are heterogeneous, but they are heterogeneous in many dimensions. Therefore critical experimental analysis of the heterogeneous environments–genetic polymorphism relationship would be greatly facilitated by identification of the relevant dimensions. This would seem to be fundamental, yet it apparently has not been seriously addressed.

The contrast in the studies of Atkinson and Miller (1980) and Shorrocks and Nigro (1981) on exactly the same species in the same area, but using slightly different habitats, highlights the importance of using ecological factors that are perceived by the species being studied. Further, it would seem that studying human commensals whose associations with human habitats must be very short in evolutionary time is not the optimal situation for understanding how most species of *Drosophila* are genetically adapted to environmental variation. There is a greater tendency for the experiments on natural populations in their native habitats to yield positive evidence for habitat choice compared to those dealing with commensals in human-disturbed habitats.

While my own judgment is that the current evidence is sufficient to warrant the conclusions reached here, the skepticism expressed by some authors (e.g., Barker [1990] and Hedrick [1990]) indicates that genotype-based habitat choice in *Drosophila* requires further investigation.

What kind of behavioral mechanism might be operating to affect habitat choice based on genotype? It might seem one would need to invoke some kind of genotype by environmental interactions of a complex nature such that a fly would "know" where its particular genotype has highest fitness. However, this is not really necessary and a very simple behavioral model would suffice. Assume that *Drosophila* exhibit a more or less random, Browning-type movement. Next assume that flies move relatively rapidly when they are stressed or "uncomfortable" and slow down when they are in a favorable "comfortable" habitat. The tendency will be to spend more time in the habitat where the flies feel comfortable. This type of behavior is reasonable and most insect ecologists would agree with some version of it. The only further assumption one needs for genotype-based habitat choice is that different genotypes feel differentially comfortable in different parts of the environment. If so, there will be a tendency for different genotypes to spend more time in one habitat than another, leading to microspatial genetic differentiation. If there is a correlation between the perception of comfort and fitness, then the conditions are met for stably maintaining such genetic polymorphisms. (I am acutely aware that the terms "comfortable" and "uncomfortable" in the preceding arguments are highly anthropocentric terms. However, I believe they are understandable in the context as referring to the flies' physiological reactions to variables in the environment, i.e., the light level, moisture, odors, presence of predators, etc.)

The late Herman Spieth once quipped to me that he had known about genotype-dependent habitat choice since he was a child. He noticed early on in his life in the midwestern United States that during a hot summer day, black cows stood in the shade more frequently than did white cows. This is the essence of the argument! In this regard, flies are as clever as cows.

Ephemeral and Newly Founded Populations

In the previous chapter, Jones et al.'s (1981) study on dispersal of *D. pseudoobscura* in the desert habitat of Death Valley was mentioned as evidence for very long distance dispersal in this species. Other aspects of these populations with more of a genetical theme were also examined. They have provided insight into the ephemerality of such populations, which may have general relevance for *Drosophila* in harsh and temporally varying environments.

No *Drosophila* can be collected in Death Valley during the summer; the average temperature is 44°C in August. From about November to May, however, *D. pseudoobscura* are found at oases, both naturally occurring ones and those formed by human activities such as cultivation of date palms at Furnace Creek. The nearest likely habitat to support continuous populations of *D. pseudoobscura* are on the slopes of the Panamint Mountains, some 25 kilometers from Furnace Creek. Bryant (1976) studied the recessive lethals in Furnace Creek in samples taken in April of two successive years. The level of allelism within a sample was higher than usually observed for this species, indicating relative genetic homogeneity consistent with a founder effect. Furthermore, the two yearly samples showed significantly less allelism between one another than within the samples. In fact, the allelism between yearly samples was about the same as between any two widely separated populations. The implication is that the population is ephemeral, being reestablished each fall by migrants coming from at least 25 km away.

In order to study the possible ephemerality of natural oases populations, Jones et al. (1981) performed a release experiment with laboratory-reared flies of known allozyme genotypes. The esterase allele $Est\text{-}5^{0.85}$ is relatively rare in *D. pseudoobscura* populations, being about 2% in the oasis populations of Death Valley. About 30,000 flies homozygous for this allele were released in each of two oases; almost all flies disappeared from one of the oases, but persisted in the other, called Salt Creek. Table 6-3 shows the frequencies of $Est\text{-}5^{0.85}$ for the collections taken after release. Obviously the addition of 30,000 flies into this population overwhelmed the natural populations, as indicated by a rise from 2 to 99% of this rare allele. In fact only one out of a sample of 77 individuals was not homozygous $Est\text{-}5^{0.85}$ one day after release. The frequency went down for about three weeks, presumably as the released flies died off. Then between 33 and 37 days, it rose again due to the eclosion of offspring of the released flies, some of whom were heterozygous, indicating they had mated successfully with natural flies. This gives an idea of the generation time under these conditions. The frequency again declined to an observed 6% on day 53; this was based on a sample of only 13 flies so is not very accurate, nor does it account for the frequency

TABLE 6-3 Results of release experiments with *D. pseudoobscura* homozygous for the rare *Est-5*$^{0.85}$ allele into an oasis population in Death Valley.

Day After Release	*N*	Frequency of *Est-5*$^{0.85}$
1	122	0.992
7	112	0.911
16	116	0.802
22	84	0.107
27	116	0.112
33	112	0.338
37	128	0.780
44	67	0.431
53	16	0.06

Note: The allele had a frequency of about 2% before release. *N* is the number of genomes sampled (the locus is X-linked).

From Jones et al. (1981).

in immature stages where it is presumably higher. After this, no more flies could be collected as the temperature became too high.

One year later, only 45 flies could be collected at Salt Creek in three days of effort. In this sample of 73 genes (the locus is sex-linked), not a single *Est-5*$^{0.85}$ allele was found. This is consistent with the population being ephemeral with no estivation or other mechanism for surviving the summer heat in situ. The nearest source for these oases is some 15 km away.

This perturbation study in Death Valley contrasts with that of Dobzhansky and Wright (1947) at Mather. Evidently *D. pseudoobscura* overwinters in situ at Mather, as indicated by the presence of the orange eye mutant 10 months after the release. The frequency was highest at the release point and decreased with distance (table 6-4).

Where do the flies come from that reestablish oasis populations in Death Valley? Interestingly, they may not be coming from the nearest possible source. Moore and Moore (1984) studied the inversions of flies at Furnace Creek and those at a likely source in the Panamint Range, Wildrose, more than 2000 meters above Death Valley

TABLE 6-4 Frequency of *D. pseudoobscura* orange eye mutant at Mather 10 months after release.

Distance	Chromosomes Tested	Frequency of Orange Eye
Point of release	646	0.0287
500 meters north	746	0.0177
500 meters south	698	0.0175
1000 meters north	334	0.0060
1000 meters south	312	0.0047

From Dobzhansky and Wright (1947).

and to the west. The populations of flies at Wildrose are summertime breeders as snow covers the ground in winter. Moore and Moore showed that the presence of *Drosophila* at Wildrose and Furnace Creek were temporally almost nonoverlapping. Furthermore, the inversion frequencies at the two sites were very different. Wildrose has a fairly even distribution of three common inversions, ST, AR, and CH, while Furnace Creek is about 80% AR. The high frequency of AR at Furnace Creek was noted in two different years, so if the population is extinct in the summer, it is being reestablished by very similar inversion types each fall. The high frequency of AR is typical of populations to the south and east of Death Valley (table 3-2 and figure 3-4). Thus, either one needs to postulate that some selection for an increase in AR occurs among migrants from closer populations of the Panamint Range, or the reestablishment does not come from the geographically closest area. Alternatively, the Furnace Creek population may not go extinct in the summer and the stable inversion frequencies represent the residents, although this is at odds with the previously discussed data.

The yearly reestablishment of the oases populations is comparable to the observations of Moore et al. (1979) on an area of Mt. San Jacinto that had been burned by a forest fire. About 70 square km had been burned with virtually all vegetation and above-ground life destroyed. Twelve days after the fire, while the burned area was still smoldering, *Drosophila* were captured 300 meters inside the burned area. Forty *D. pseudoobscura-persimilis*, 4 *D. melanogaster-simulans*, and 1 *D. immigrans* were found. (The U.S. Forest Service would not allow further penetration into the burned area at that time.) The next spring a collecting transect was set up inside the burned area one kilometer from the nearest edge. Over a thousand *Drosophila* were collected at this site over a six-month period, 85% being *D. pseudoobscura-persimilis*. The species composition inside the burn and adjacent to it were very similar.

The inversion frequencies in the burned area were also determined (Moore et al., 1979). Eight different third-chromosome gene arrangements were found, again in frequencies very similar to those found in adjacent nonburned areas, and indeed, very similar to those found for the previous 40 years of sampling this region. Furthermore, this burned site showed seasonal changes in inversions very similar to those noted by Dobzhansky in the 1940s. In other words, in less than a year, a completely devastated region of 70 square km had been recolonized by *D. pseudoobscura* and established the same inversion polymorphism that had been observed on the mountain for a long time. This is particularly interesting in light of the recent expansion into this region of two close relatives, *D. persimilis* and *D. miranda*. One might think that when species composition changes, especially if the species are closely related and likely to be competitors, it might have some impact on the genetic constitution of a species. This seems not to be the case with these species.

These studies have been revealing in regard to the dynamics of these populations and the large distances *D. pseudoobscura* can easily travel in invading habitats that vary over time in suitability for breeding. The indications are that the great mobility observed in the desert environment might well be a general one given the similar observations in the relatively benign conditions on Mount San Jacinto. That is, the behavior noted by Jones et al. (1981) is not a peculiar adaptation of the species to that particular environment. Forest fires are a natural part of the environment in California, so the behavior of the flies with respect to the burned area is also perhaps not so surprising. However, it would be of great interest to test whether this colonizing ability is geographically variable. The temporally stable highlands of Mexico provide perhaps the starkest contrast in ecology to Death Valley and Southern California. Studies of the dispersal behavior of the continuously-breeding populations of *D. pseudoobscura* in these highlands would be a good test of whether there is genetic variation in dispersal behavior depending on ecological context. The results of Dobzhansky and Wright (1947) at Mather indicate relative stability and less movement of flies such that homogenization of allele frequencies did not occur over one kilometer in 10 months (table 6-4).

Cactus-Breeding *Drosophila*

In the previous chapter, some aspects of the ecology of cactus-breeding *Drosophila* were mentioned. Here we take up genetic studies of these species, specifically in relationship to their well-studied ecologies. Several groups of researchers have carried out studies in different regions of the world and perhaps the easiest way to present the work is by region.

Sonoran Desert

The single best reference to ecological genetics of the Sonoran Desert species is Heed (1978), which cites much previously unpublished data; Etges et al. (1995) reviews the inversion data. There are only four endemic species in the ecosystem, all of which breed in cactus or in the soil soaked by cactus drippings (table 5-3). Three of the four species will shift host plant depending on the cacti in the given area.

D. pachea is the one species that does not change host; its host, senita cactus, exists throughout its range. Allozyme variation in this species is remarkably constant geographically (Rockwood-Sluss et al., 1973). The single inversion polymorphism for this species is much more interesting. It exhibits an extraordinarily regular north-south cline (figure 6-2). Several variables are correlated with this cline, such as temperature, moisture, and shape of the cactus; it isn't clear what the selective agent might be, but the authors favor an abiotic explanation.

D. nigrospiracula shows a shift in host plant, but no detectable genetic changes have accompanied the shift. There are no known inversion polymorphisms in this species (Cooper, 1964, cited in Heed, 1978) and allozymes are geographically constant (Sluss, 1975, cited in Heed, 1978). This species is capable of long distance dispersal (Johnston and Heed, 1976), and migrant males exhibit increased mating success over residents (Fontdevila et al., 1977). Together these phenomena promote extensive gene flow and can easily account for homogeneous allozyme frequencies.

D. mojavensis is probably the most interesting of these species in regard to ecological genetics, as considerable genetic differentiation does seem to have accompanied the host shift from agria cactus in Baja to organ pipe on the mainland. The exception is one small area of the mainland called Desemboque, where agria coexists with organ pipe; in this area *D. mojavensis* breeds almost exclusively in agria. The inversions in this species are clearly distributed with regard to host plant. Johnson (1973, cited in Heed, 1978; 1981; and Johnson, 1980, cited in Etges, 1990) found the Baja populations to be quite polymorphic for inversions of two chromosomes. The

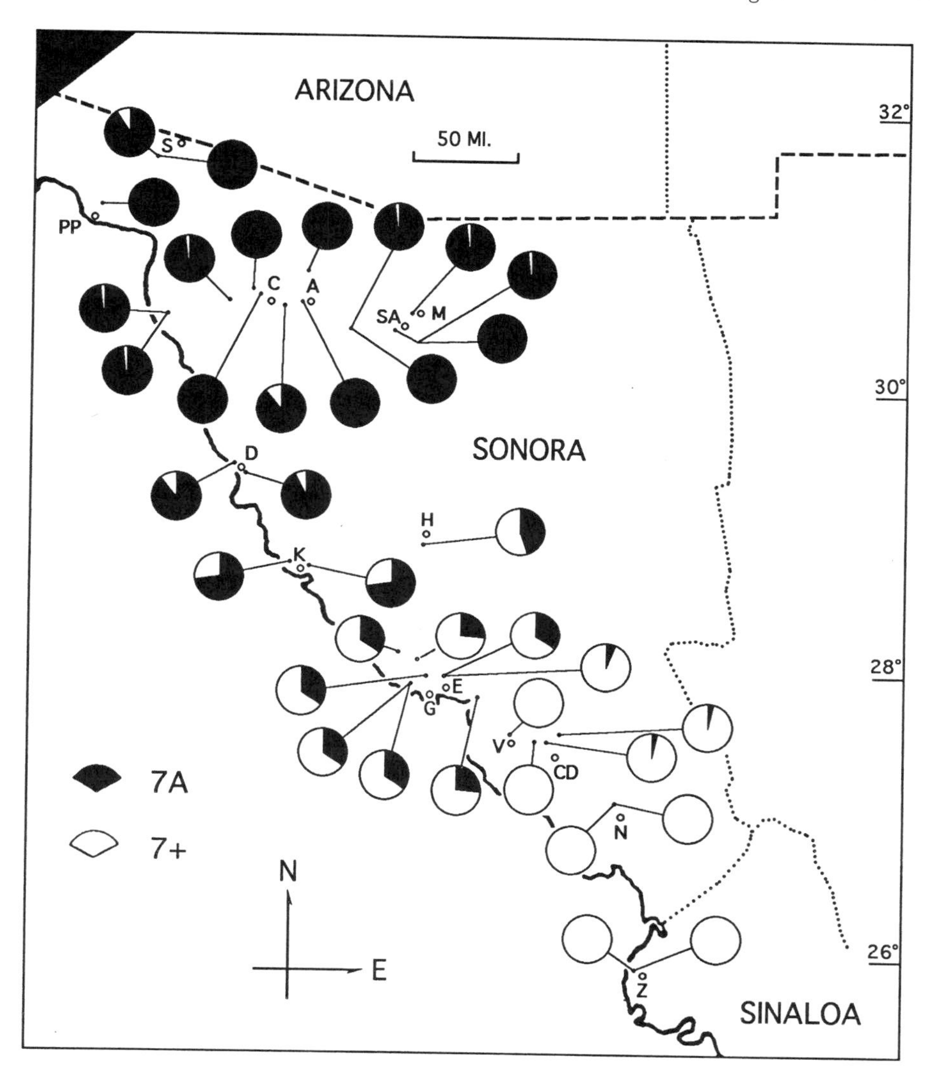

FIGURE 6-2 Graphic representation of frequencies of *D. pachea* inversions. From Ward et al. (1975).

polymorphism extends to the islands in the Gulf of Mexico and to the small isolated agria area of the mainland (figure 6-3), where the variation is present but much reduced. Elsewhere on the mainland where agria does not exist, the species is monomorphic for inversions. Curiously, where the species extends into Southern California it shifts host plants again, this time to barrel cactus (*Ferocactus acanthodes*), but has become fixed for an alternative inversion type (location 51 in figure 6-3).

Heed (1981) discusses the inversion polymorphism in *D. mojavensis* in relation to the "central/marginal" pattern and the stability and spatial distribution of resources.

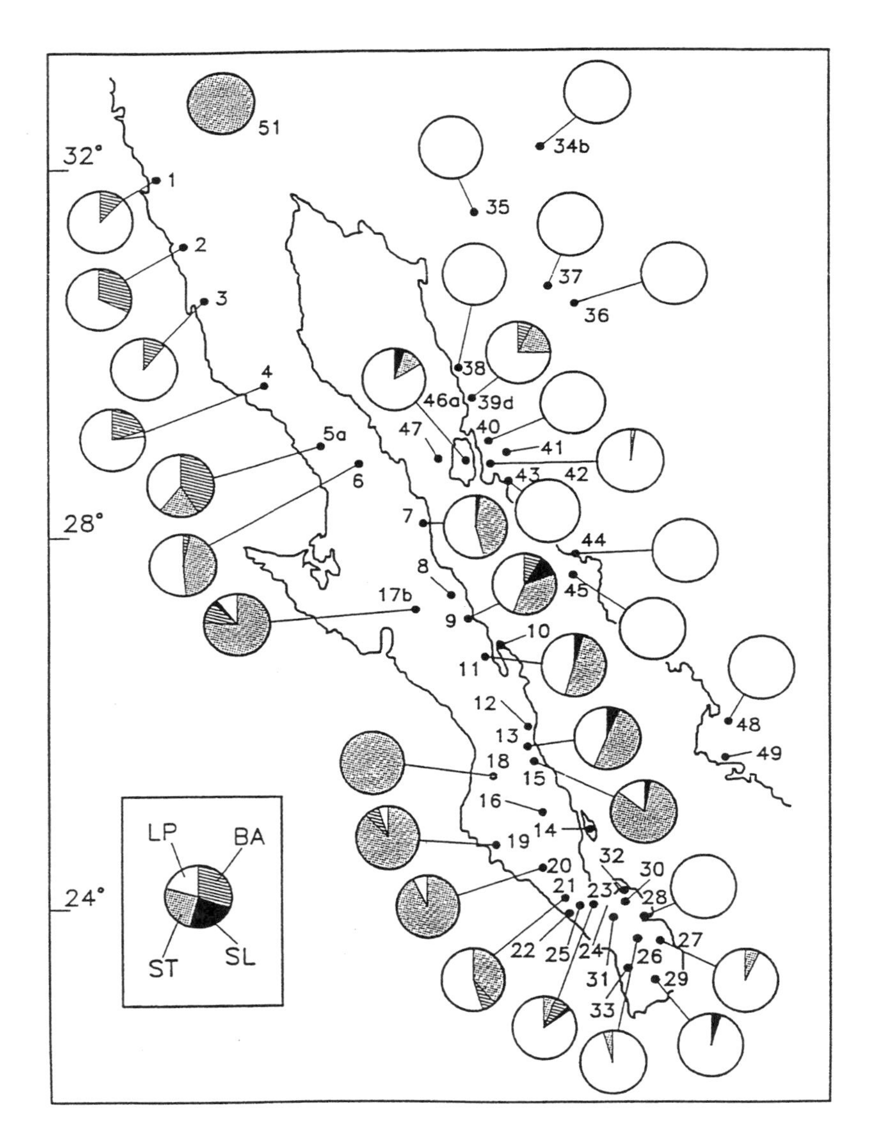

FIGURE 6-3 Graphic representation of frequencies of *D. mojavensis* inversions. This species breeds in agria cactus in the Baja Peninsula and organ pipe cactus on the mainland. Locality 39d is Desemboque, the only area where the two cacti coexist on the mainland. Locality 51 is in California, where a third cactus serves as the breeding site. From Etges et al. (1995).

The fact that almost all the inversion heterozygosity in this species is associated with breeding in agria cactus indicates this resource is "central" and the adaptation to breeding in organ pipe is "marginal." Heed goes on to cite data indicating the resource distribution of agria is much denser and temporally more stable than organ pipe. The trophic resource predictability, based on number of cacti, number with suitable rots, yeast species found in the rots, and so on all indicate the central populations on agria exist in a more predictable environment than the marginal ones in organ pipe. This is in agreement with those theories discussed in chapter 3 that emphasize the importance of predictability (both temporally and spatially) of resources as important in understanding the decrease in inversion heterozygosity toward the margins of species distributions (either geographical or ecological margins).

At least one allozyme locus also follows the Baja/mainland pattern, *Adh*. Figure 6-4 shows the pattern. In Baja the average ratio of *F* : *S* alleles is 97 : 3%, shifting to 15 : 85% on the mainland including the Desemboque region, where a pocket of agria is found. This correlation of an enzyme locus to a shift in host plant led to further investigations. A natural question stemming from this observation is whether agria and organ pipe cacti have different levels of alcohols. Table 6-5 shows they do. Ethanol and to a lesser degree 1-propanol are particularly high in organ pipe and 2-propanol and acetone in agria. The next observation is that *D. mojavensis* can use ethanol vapor as a nutritional resource (Starmer et al., 1977). Adults held over water with 1 to 4% ethanol survive longer than flies held over water alone. Flies from the mainland (high *S* allele frequency) utilize ethanol better than Baja flies; that is, the increase in longevity was greater in mainland flies than Baja flies. Sluss (cited in Heed, 1978) showed further that flies from the same population made homozygous for *S* or *F* also differed in survival in alcohol vapors (table 6-6). *S/S* individuals displayed a greater increased longevity over ethanol and 1-propanol than did *F/F* flies. *F/F* females did better over 2-propanol and were less susceptible to the toxic effects of methanol. Considering the distribution of alleles (figure 6-4), the alcohol contents of the substrates (table 6-5), and survival of genotypes over alcohol vapors (table 6-6), a strong case can be made that the allozyme shifts associated with the host switch is a selected adaptation. There is yet one further bit of evidence consistent with this conclusion and that is the biochemical properties of the proteins produced by the two alleles (Starmer et al., 1977). The product of the *Adh-S* gene is relatively heat and pH tolerant compared to the product of *Adh-F*. This again matches the host plants, organ pipe exhibiting greater variation in temperature and pH than does agria. Rarely has a more convincing story unfolded.

The Baja and mainland populations of *D. mojavensis* have diverged in other important ways. One is the beginning of premating reproductive isolation (Wasserman and Koepfer, 1977; Zouros and D'Entremont, 1980; Markow et al., 1983), a subject we return to in the next chapter. Etges (1990) has also studied the life history characteristics of the two groups and found them to differ. However, crosses among them yielded no evidence of coadaptation differences for life history traits; neither F_1 overdominance nor F_2 breakdown was noted. Despite the evidence for significant genetic divergence in inversions, allozymes, and mating behavior, no evidence for the buildup of epistatic blocks of genes affecting life histories was noted.

Lofdahl (1986) studied a population of *D. mojavensis* from Santa Catalina Island off the coast of California, where the species breeds in *Opuntia*. She was interested

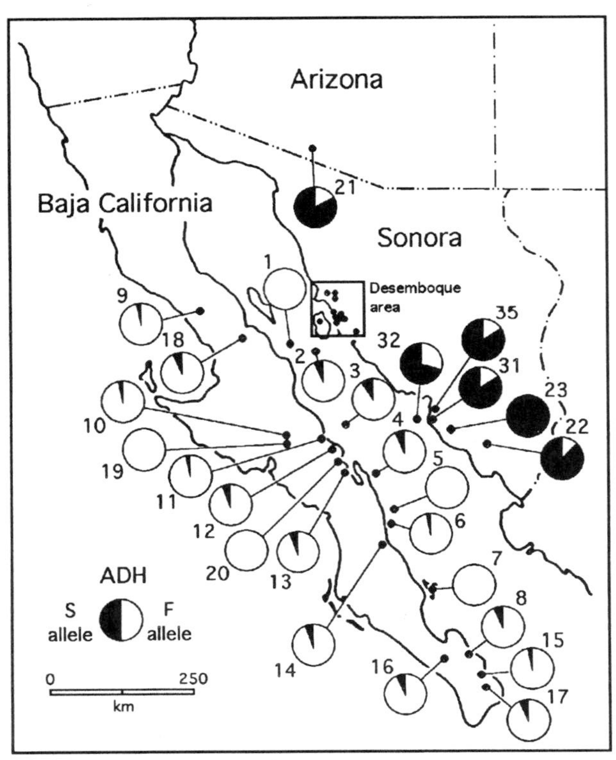

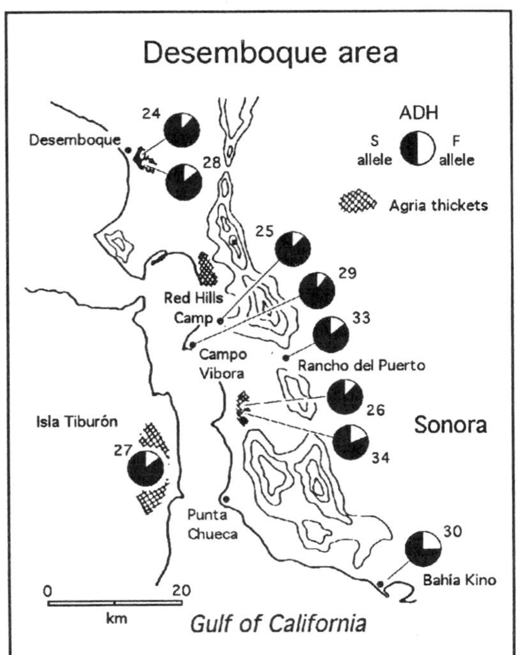

FIGURE 6-4 Graphic representation of *Adh* allele frequencies in *D. mojavensis*. From Heed (1978).

TABLE 6-5 Mean and maximum alcohol and acetone contents in the rot pockets of agria and organ pipe cacti, the host plants for *D. mojavensis*.

	Agria (12 samples)		Organ Pipe (9 samples)	
	Mean	Maximum	Mean	Maximum
Methanol	88	262	105	212
Ethanol**	8	31	142	361
Acetone**	285	1548	15	39
1-propanol	6	30	71	200
2-propanol*	403	2510	17	56

$^{*}p < 0.056$; $^{**}p < 0.01$.

Note: The figures are percent volume × 10^4. Significantly different means between cacti using a Mann-Whitney U test are indicated by asterisks.

From Heed (1978).

TABLE 6-6 Mean days survival (±S.D.) of *D. mojavensis* adults homozygous for alternative *Adh* alleles when held over vapors of various alcohols, with no other food.

	Females		Males	
	Adh^{SS}	Adh^{FF}	Adh^{SS}	Adh^{FF}
Water	11.2 ± 5.4	11.5 ± 2.9	10.8 ± 2.0	9.1 ± 1.6
2% Ethanol	15.3 ± 5.9**	12.9 ± 5.9*	13.5 ± 5.3**	12.4 ± 1.7**
4% Ethanol	18.0 ± 2.0**	15.4 ± 4.2**	14.1 ± 5.2**	11.9 ± 1.9**
8% Ethanol	11.8 ± 6.0	10.3 ± 2.1*	9.6 ± 3.1	9.3 ± 0.5
2% 1-propanol	14.1 ± 10.4*	11.1 ± 2.9	11.2 ± 3.4	10.4 ± 4.0*
2% 2-propanol	12.8 ± 2.0**	13.0 ± 1.5**	12.6 ± 8.1	8.5 ± 2.7
2% Methanol	10.8 ± 2.1	9.3 ± 1.2**	8.3 ± 2.2**	8.4 ± 0.5*

$^{*}p < 0.05$; $^{**}p < 0.01$

Note: All are from a single population in Arizona. Asterisks indicate significant differences from the control over water.

From Heed (1978).

in the amount of genetic variation for oviposition on agria cactus, which the population naturally never sees. She found additive genetic variance amounting to some 10 to 20% of the total variance. The conclusion was that even though this population was monophagous on *Opuntia*, it still harbored selectable genetic variation for acceptance of the presumed original host, agria. The colonization of this island and new cactus did not eliminate the species' genetic predisposition to oviposit on cacti used by other populations.

Hawaii

A. Templeton, J. S. Johnston, and their collaborators have been carrying out extensive studies of *D. mercatorum* (and to a lesser extent *D. hydei*) breeding in *Opuntia megacantha* cactus on a west-facing slope of the volcano Kohala on the island of Hawaii (Templeton et al., 1990a, b, are two reviews of this work). Figure 6-5 is a schematic of the site near Kamuela with the collecting localities labeled; all sites are within three kilometers of one another. In 1975 an unusual phenotype was found in *D. mercatorum* that was called abnormal abdomen, or aa. Abnormal abdomen adults retain juvenile

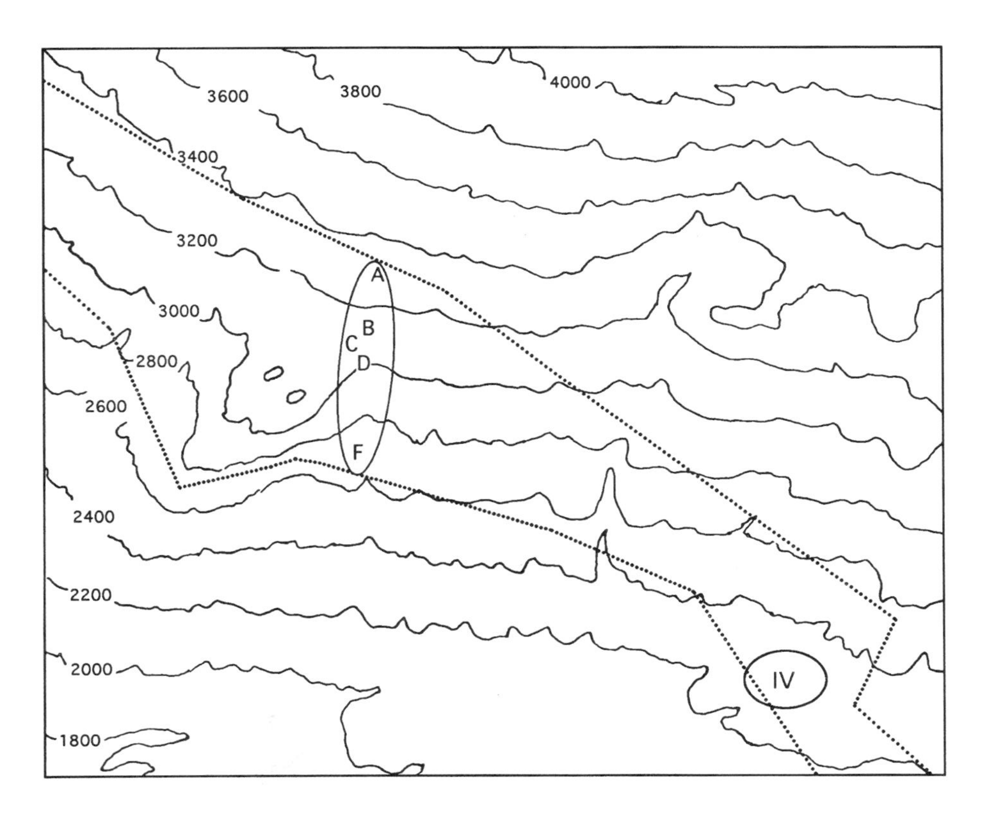

FIGURE 6-5 Schematic of the collecting sites near Kamuela, Hawaii used in the studies of *D. mercatorum* and *D. hydei*. From Templeton et al. (1990a).

cuticle due to improper juvenile hormone expression (Templeton and Rankin, 1978). Other aspects of the phenotype are a slowdown in larval development, but an earlier onset of female fecundity (Templeton, 1983). The expression of aa is due to two X-linked loci (about a half map unit apart near the heterochromatin) in females and an X-linked and a Y-linked locus in males (Templeton et al., 1985).

Several of the properties of aa resemble those of the bobbed phenotype in *D. melanogaster*, which is known to be due to ribosomal RNA (rRNA) deficiency. Subsequently, it was shown that up to 90% of rRNA genes in aa mutants of *D. mercatorum* have an insertion of a 5-kb sequence into the 28s rRNA genes (DeSalle et al., 1986a). However, the inserts alone did not produce the phenotype; rather, the copies of 28s genes with inserts were selectively being under-replicated in polytene tissues. Only when an allele at the linked locus was present would the inserted copies not be under-replicated and then give rise to the juvenilized cuticle (DeSalle and Templeton, 1986). This latter locus was designated aa in the early literature, but in recent publications *ur* (under-replication) is used with superscripts, ur^{+} designating the allele that selectively under-replicates; ur^{aa} is the recessive allele that does not selectively under-replicate inserted copies and leads to the phenotype. However, even in the absence of the expression of the cuticle phenotype (which is temperature-dependent and has low penetrance), the life history phenotype, that is, the longer development time and early onset of fecundity, is still expressed.

The fitness effects of ur^{aa} were studied in the natural population by means of a bagged experiment, that is, by placing bags over rots and collecting eclosing adults (Templeton et al., 1990a, b, 1993). On average, individuals carrying one or two copies of ur^{aa} had a development time 0.75 days longer than flies homozygous for ur^{+}. On the other hand, by comparing the frequency of ur^{aa} in adult females present in a local area (presumably those laying the eggs there) with the frequency of ur^{aa} in males eclosing in the bags, one could estimate the fecundity difference. The estimate is that females with the ur^{aa} allele have an early fecundity advantage of 71% over the population mean. Thus there is a trade-off between development time and fecundity.

Templeton and Johnston (1982) developed life-history theory to deal specifically with the abnormal abdomen syndrome in *D. mercatorum*. The predictions were that when the age structure of the population is young, early fecundity becomes the dominating selection component and ur^{aa} should rise. In older age structured populations the early fecundity advantage decreases and the other components become more important; ur^{aa} would be predicted to fall in frequency. The age structure of *D. mercatorum* in this site is largely controlled by humidity, the drier areas having a younger age structure (i.e., flies die earlier) than moister areas. As predicted, the frequency of ur^{aa} does track quite well the moisture and age structure of the different collecting sites demarcated in figure 6-5, that is, ur^{aa} is more frequent in drier areas, which in general means lower locations in the transect (table 6-7). That this is due to the demographics of the age structure and not directly to humidity levels was demonstrated in 1986–87 collections that occurred during unusually wet conditions. Two storms occurred just prior to two collections; these led to a great increase in appropriate rots for the flies, which led to a population explosion with a young age structure at the time of collecting. The effect was greater in the upper parts of the transect with large cacti, areas A–D in figure 6-5. The age structure was reversed from normal, be-

TABLE 6-7 Altitudinal distribution of frequency of abnormal abdomen (aa) in *D. mercatorum* in Hawaii and the bobbed syndrome in *D. hydei*.

Site	Altitude (m)	Year and Weather 1984 Normal	1985 Normal	1986 Rains	1987 Rains	Bobbed in *D. hydei*
A	1036	0.111	0.222	—	0.33	0.28
B	950	0.291	0.265	0.43	0.47	0.41
C	930	0.334	0.429	0.43	0.47	0.37
D	880	0.334	0.429	0.43	0.33	0.42
F	795	0.540	0.367	0.38	0.26	—
IV	670	0.360	0.276	0.32	0.34	—

Data in the first two columns are from Templeton and Johnston (1988) and in the last three from Templeton et al. (1989).

ing older in the lower sites in 1986–87, while generally being younger in previous years (cf. Johnston and Templeton, 1982). The frequency of ur^{aa} was also reversed (table 6-7).

D. hydei coexists with *D. mercatorum* in the upper sites in figure 6-5 and it too has a bobbed syndrome due to inserts into 28S rRNA (Franz and Kunz, 1981). Apparently it does not have the modifier *ur* locus; at least, these populations are not polymorphic at this locus. So the expression of bobbed in this species is due to 40% or more of the copies having insertions. Curiously, the altitudinal cline in this species parallels the normal cline found for *D. mercatorum*, being lower at the upper sites and higher at the lower sites (table 6-7). It was also found that *D. hydei* differed in other ways from *D. mercatorum* (Templeton et al., 1990a); for example, as noted in the previous chapter, *D. hydei* has a higher rate of dispersal at this site than does *D. mercatorum.* Thus these species present an attractive comparative study of two species with different life histories occupying the same environment with at least one similar genetic polymorphism.

The information on dispersal rates together with the clinal frequency distribution of ur^{aa} allowed calculation of the necessary selection coefficients needed to maintain the differences. Migration rates are high enough that, in the absence of selection, no differences should be seen. Among other pieces of evidence is the work from DeSalle et al. (1987a) on molecular variation in nuclear and mtDNA, which indicates *Nm* values of 4 to 8 (*Nm* being the product of effective population size and migration rate; values over 1 for neutral alleles should homogenize gene frequencies). Johnston and Templeton (1982) calculated that selection need be only about 6×10^{-4} between sites to maintain the differences.

Australia

Murray (1982) lists 32 species of cacti that have been introduced into Australia, most in the 19th century. Among them are 18 species of *Opuntia*, many of which became so widespread they were considered pests. Between 1925 and 1935, the cactus-breeding moth *Cactoblastis cactorum* was introduced to control *Opuntia*, a program that worked reasonably well. Two species of *Opuntia*-breeding *Drosophila* were also carried in with the *Cactoblastis: D. buzzatii* and *D. aldrichi* (Barker and Mulley, 1976). *Opuntia* now have a narrower, more island-like distribution, so the initially large population flush of *Drosophila* has contracted to much more spatially isolated populations. J. S. F. Barker and colleagues have been investigating this relatively simple *Drosophila*–cactus ecosystem, focusing on allozymes.

Barker and Mulley (1976) present considerable allozyme data that indicate *D. buzzatii* in Australia is relatively depauperate in this kind of variation. These authors do not attribute this to a founder effect but to the narrow niche of the species. On the other hand, the more recent data of Halliburton and Barker (1993), indicating no mtDNA variation in this species in Australia, makes it more difficult to ignore founder effects. Barker and Mulley (1976) also noted a general deficiency of heterozygotes over that predicted by Hardy-Weinberg expectations, an observation they attributed to a Wahlund effect. The small breeding populations associated with individual rotting cladodes was further examined by Prout and Barker (1993), who calculated population sizes from F statistics. They estimate N_e to be between 10 and 50 individuals. This estimate is lower for the presumed (nearly) neutral allozymes than for the other genetic variation studied, additive genetic variation in body size, presumably because this latter variation is subject to stronger selection.

Barker et al. (1986) present detailed analysis of temporal and microgeographic variation in allozyme frequencies over a four-year study. They conclude significant frequency changes occur over time and space, with individual rots representing microspatial variation of relevance to the species.

One of the more interesting developments from the Australian allozyme work has been the application of spatial autocorrelation analysis (Sokal et al., 1987). These authors studied the variation in frequencies of 12 allozyme alleles (from 5 loci) in 57 populations of *D. buzzatii*. The beauty of this type of analysis is that much more information can be gained compared to the F statistics discussed in the previous chapter. Not only can structure be detected, but the kind of structure with regard to distances and directions can be ascertained. From this data set, Sokal et al. concluded that at least 6 out of 12 alleles showed significant structure, but it is more intriguing that they tended to show independent structure. That is, some alleles were structured over different distances and some in different directions. Thus, no simple explanation based on drift, migration, or selection gradients along a single factor could explain the data. More than one environmental variable is likely affecting the different alleles that operate on different spatial scales. This type of analysis has often been used for other species, especially human data, but rarely for *Drosophila*. It would seem to hold great promise as indicated by this single example, yet is underutilized at least by *Drosophila* workers.

In considering what kinds of ecological factors could be operating in the Australian system, Barker (1990) has made a good case for the importance of yeasts as a source of environmental heterogeneity of evolutionary importance to *D. buzzatii* and presumably to *D. aldrichi*, although the latter species has been less thoroughly studied. There are 10 common yeast species found in Australian *Opuntia* rotting cladodes; they vary spatially and temporally (Barker et al., 1987). Barker et al. (1981a, b) further showed that adults can distinguish the different yeasts and respond differently to them.

That genotypic intraspecific variation may exist for yeast choice was indicated by nonrandom distribution of genotypes at the *Est-2* allozyme locus. Furthermore, laboratory studies indicate that feeding sites and oviposition choice are highly correlated (Vacek et al., 1985) so the attraction of flies to different yeast in the field may represent variation in oviposition preferences.

These results set the stage for further detailed studies. Unlike some other studies of habitat choice and environmental heterogeneity (our own included), Barker has identified a priori specific ecological factors known to be perceived by the flies and important to their fitness. The next step was to see if there was evidence for intraspecific genetic variation in choice of different species of yeasts. The studies have so far only been carried out in a controlled laboratory setting, but the results are encouraging. Barker (1992) studied isofemale lines of both *D. buzzatii* and *D. aldrichi*, giving females a choice to oviposit on *Opuntia* medium seeded with different yeast species. Three of the five species of yeast were differentially chosen by different isofemale lines. Thus the evidence is that yeast choice is a genotype-dependent trait although the heritability is low, around 9% or even less.

South America and Elsewhere

South and Central America are almost certainly the origin of the cactus-breeding species discussed above. Unfortunately, they have been little studied in this area of original diversification. Fontdevila et al. (1982) have studied the inversion polymorphisms of *D. buzzatii* in Argentina and compared them to those in populations that colonized parts of the Old World, especially the Iberian peninsula. They conclude there is little indication of a decrease in levels of inversion polymorphism that might have been attributed to founder effects. Some significant frequency changes have occurred during colonization, although their causes are not clear.

Fontdevila and his colleagues have also been studying *D. buzzatii* in Europe, where *Opuntia* has been introduced. We already mentioned their observations on seasonal cycling of linkage disequilibria between allozymes and inversions (figure 3-21; Fontdevila et al., 1983). A recent reference to this work is Santos et al. (1992), wherein references to earlier work can be found.

Flower-Breeders in South America

Like the cactus–*Drosophila* associations, flower-breeding flies also present a relatively simplified ecological interaction, making detailed ecological genetics possible. *D. flavopilosa* in South America, primarily Chile, is one such case, which has been exploited by D. Brncic and colleagues (major papers and reviews are Brncic, 1966, 1970, 1972, 1983).

It is difficult to imagine a simpler *Drosophila* ecology than that of *D. flavopilosa.* The Solanaceous plant *Cestrum* is the only larval breeding site. A single species, *C. parqui*, is used in Chile and a second species, *C. euanthes*, is also used in Argentina. These plants are thought to be toxic to most animals, so there are relatively few other insects in them and no other *Drosophila*. When liquid extract of the plants is added to standard laboratory medium every other *Drosophila* tested dies on the food; these included *D. melanogaster, D. immigrans, D. simulans, D. funebris, D. hydei, D. willistoni*, and others. Female *D. flavopilosa* have reduced ovaries with only one or a few mature eggs. Only a single egg is laid per flower and no egg is laid if an egg or larva is already present. Thus there is a single larva per flower and no competitors. The host plant is very common in Chile, often growing as large shrubs with a great number of flowers. Thus, despite the one larva per flower, the densities of flies may become extremely high, estimated by Brncic (1966) at 37,000 per 100 square meters! The only real biotic problem for the species is the wasp parasitism illustrated in figure 5-10. Two hymenopteran species parasitize *D. flavopilosa* from a level of 0 to over 90%. This, together with abiotic conditions such as weather, are about all the fly needs to contend with. The only unfavorable aspect of this system is *D. flavopilosa* cannot be bred on standard medium in the laboratory. All the sampling about to be discussed was done on larvae or adults taken directly from the natural populations.

Inversion polymorphism is common in *D. flavopilosa* and displays many of the features of a flexible system such as that of *D. pseudoobscura*. The geographic and temporal variation is somewhat simpler. On the macrogeographic scale, there is a distinct north-south cline with the B inversion in the south and the A in the north (Brncic, 1976). Altitudinally, there is likewise a clear cline with A decreasing and B increasing from lower to high altitudes (figure 6-6). Finally, the seasonal variation (figure 6-7) follows the pattern of A rising in the summer while B decreases, and vice versa in the winter. Clearly, the implication is that A is a warm-adapted inversion and B a cold-adapted one. When flowers were brought into the laboratory and kept at different temperatures, Brncic (1968) found A to have the advantage at 25°C and B at 16°C.

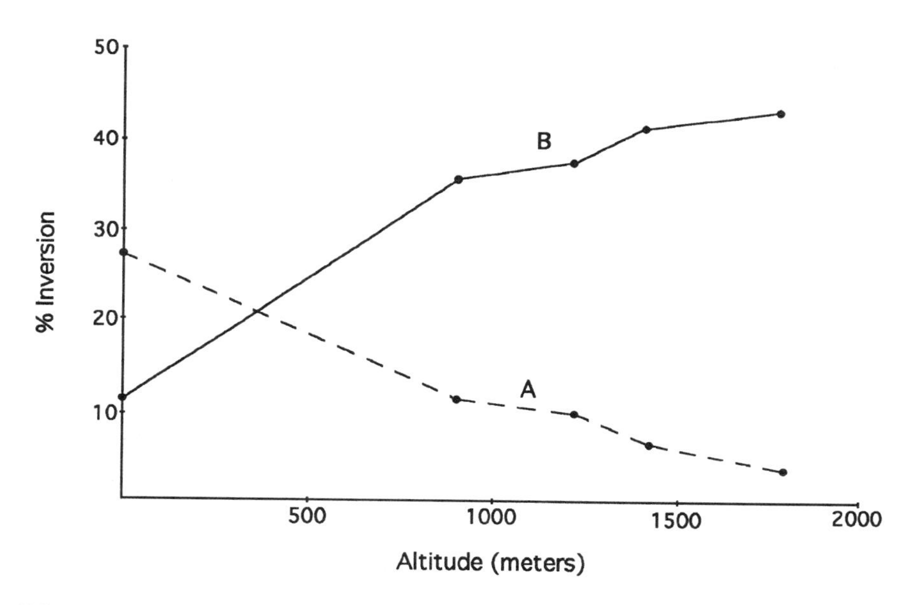

FIGURE 6-6 Altitudinal cline in inversion frequencies of *D. flavopilosa*. From Brncic (1966).

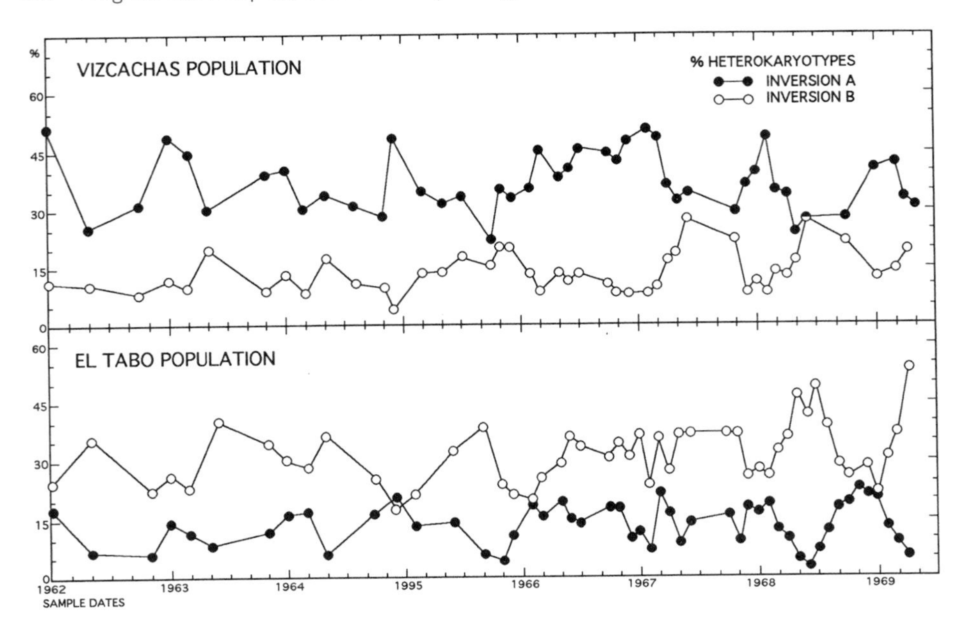

FIGURE 6-7 Seasonal fluctuations in frequencies of *D. flavopilosa* inversions. From Brncic (1972).

Given these results and the simplicity of the system, there are only two possible causes of the selection on inversions: weather and the wasp parasitoid. While the wasp parasitism rate also varies seasonally, this could well be due to weather and not have direct selection on the flies. There is no evidence of different karyotypes being differentially parasitized and all populations have the parasites, so it is hard to see how this could be an important factor. Density of flies is another possibility, but this could only be a factor at the adult stage, which seems unlikely. This leaves the abiotic conditions, and temperature in particular, as the direct cause of genetic variation in this species (Brncic, 1983 and references therein). It is rare in ecological genetics that one can make such a bold statement with such confidence.

Some allozyme work has been done on *D. flavopilosa* (Napp and Brncic, 1978). Perhaps the greatest surprise from these studies is the level of variation found. About 50% of all loci were polymorphic with heterozygosities typical of the neotropical *willistoni* group. The surprise is that one might not expect a species with such a restricted niche to be genetically so polymorphic. Whatever the cause of the maintenance of high levels of allozyme variation, in *D. flavopilosa* it is hard to evoke environmental heterogeneity. More likely large and stable population sizes are the important factors.

Mycophagous *Drosophila*

Fungus-breeding species also represent relatively simplified ecological settings. Relatively little genetic work has been done on these species, although work on the genetic basis of host preference has been done. The studies of J. Jaenike on species in the eastern United States have been particularly interesting (reviewed in Jaenike, 1990).

In the previous chapter some discussion of the fact that *D. tripunctata* is a broadly adapted species was presented. This species breeds on a large number of fruits and fungi. Jaenike's studies have been centered around two model breeding sites, tomatoes and commercial mushrooms. Strains of *D. tripunctata* vary greatly in both their settling choice and oviposition preferences, which may be negatively correlated (Jaenike, 1985, 1987). For example, two different strains were released into a native habitat and subsequently recaptured at tomato baits and mushroom baits; this provided an indication of what Jaenike calls settling behavior. The ratio of numbers of individuals on mushrooms to number of individuals on tomatoes was 62 : 38 and 19 : 81 for the two strains. Oviposition preference also varies considerably among isofemale lines, from about 23 to 97% preference for mushroom oviposition. This variation is found in all populations studied, with about the same total amount of variation everywhere, that is, there is no geographic pattern to genetic variation in host preference. The question of whether this genetic variation in host preference was correlated with performance (fitness) on the host was posed, with a fairly unambiguous negative conclusion, at least in the laboratory (Jaenike, 1989); no significant correlation was noted between a female's choice of mushroom or tomato and the development time of larvae on the two substrates. Whether this is true in nature is not clear.

Jaenike (1990) considered the patterns of genetic variation for host and settling behavior of *D. tripunctata* and proposed two possible mechanisms of maintenance of the variation. The first is density-dependent selection, that is, when one particular host resource is overcrowded with larvae, selection would favor genotypes preferring oviposition on alternative hosts. Larval competition for food is likely for these species given what is known of their ecology (Jaenike, 1990). An alternative mechanism of maintenance is more novel, namely, a mutation-selection balance. The hypothesis is that oviposition behavior is controlled to a large extent by the "motivation" of the female. When oviposition sites are rare, females would tend to harbor several mature eggs at any given time; when any suitable larval substrate is encountered, they will readily oviposit. At this time, genetic variation for differential preference in oviposition site would be neutral, that is, irrelevant. Only when oviposition sites are abundant (a condition hypothesized to be rare for this species) would the genetic variation in oviposition preference become manifest. Females would be more choosy and selection could act on the genetic variance for oviposition preference. This would lead to a form of density-dependent frequency-dependent selection. This motivational model is similar to the stress hypothesis of Hoffmann and Turelli (1985; Turelli and Hoffmann, 1988); the physiological state of the flies is important in expression of underlying genetic variation in behavioral choices.

Courtney and Chen (1988) also studied genetic variation in oviposition preferences in a fungus-breeding fly native to the west coast of the United States, *D. suboccidentalis*. They used a quantitative genetics, half-sib mating scheme to determine levels of heritability of the traits. They found narrow-sense (additive) heritability for ovipositing on a particular mushroom species to be about 0.189. For the number of eggs laid, the heritability was 0.45.

Male Mating Success

The measurement of selection, and even a precise component of selection, in natural populations is very difficult. The study of Templeton, Johnston, and colleagues on *D. mercatorum* is one such case. Another is that of W. Anderson and colleagues on male mating success in a natural population of *D. pseudoobscura* (Anderson et al., 1979; Salceda and Anderson, 1988). The Mexican populations they studied harbor high levels of inversion polymorphism. The strategy behind this study is as follows. Generally when flies are captured in nature, individual females are placed in bottles with considerable medium and allowed to oviposit. Larvae are harvested at the optimal time and polytene chromosome preparations made. Thus the chromosomes scored come from the female captured and the male or males whose sperm she was carrying at the time of capture. If one knew the frequency of karyotypes of the females themselves, karyotype frequencies of males in the population at the time of capture, and the frequencies in offspring, then one could determine the relative mating success of the males. This is a laborious task and the largest such study was that of Salceda and Anderson (1988), which will be described here.

Flies captured in nature were rapidly anesthetized and males and females separated in order to prevent mating in the collecting vials. Individual females were placed individually into bottles and eight larvae per female examined for karyotypes. After the females had exhausted their sperm obtained in nature, they were mated to a laboratory tester strain of known karyotype and larvae from this mating were examined to determine the karyotype of the female. Likewise, males captured in the natural population at the same time as the females were mated to the laboratory tester strain to determine male karyotypes. In this population at Amecameca there were three relatively frequent gene arrangements: CU (Cuernavaca), TL (Tree Line), and EP (Estes Park); there were also six rare inversions, which were lumped into a category Others, OT.

Table 6-8 summarizes the results of Salceda and Anderson (1988). 305 females and 156 males from the natural population were tested. As the final line in this table indicates, the relative mating success of males was inversely proportional to their frequency in the natural population. The commonest type, CU, suffered a mating deficit relative to its frequency in the natural population of about 16%; males carrying the other arrangements mated more frequently than found in nature, up to nearly 50% greater for the rare OT category. These differences are significant at $p < 0.025$. Very similar findings were reported in a smaller earlier study (Anderson et al., 1979). Thus the rare male mating advantage so well-studied in the laboratory (see chapter 4) appears to occur in nature. Whether this phenomenon is truly frequency dependent would require sampling the natural population when the karyotypes are at different frequencies, as occurs in seasonally cyclical populations. This heroic task has not yet been attempted, although in light of the paucity of studies on selection in nature and the demonstration that the above approach can detect a component of selection, such a study would be worthwhile.

Population Growth and Size

The final subject covered in this chapter is the effect of genetic variation on ecological parameters of populations such as innate rate of increase (r_m) and carrying capacity or

TABLE 6-8 Illustration of the data and calculations to determine relative mating success in a natural population of *D. pseudoobscura* from Mexico.

	CU	TL	EP	OT	Total %
1. Frequencies in 8 offspring from each of 305 females	59.45	29.94	8.30	2.32	100
2. Contribution of female parents to offspring	33.20	12.87	3.20	0.74	50
3. Contribution of male parents to offspring = (row 1 − row 2)	26.25	17.07	5.10	1.58	50
4. Male contribution scaled to 100% = frequencies in sperm stored in females (2 × row 3)	52.50	34.14	10.20	3.16	100
5. Frequencies in 156 adult males from nature	60.90	30.45	7.05	1.60	100
6. Δp_m due to differences in male mating success = (row 4 − row 5)	−8.40	3.69	3.15	1.56	0
7. % change in frequency = 100 × (row 6/row 4)	−16.0	10.8	30.9	49.4	

Note: All 305 females from nature contributed equally (8 larvae) to row one, so differences in fecundity among females was eliminated. Frequencies are given in percentages.

From Salceda and Anderson (1988).

population size (K). These parameters play an important role in ecological theory and are affected by genetic variation. Unfortunately, for this subject virtually nothing is known from nature, so laboratory work will need to suffice as a guide. The overall conclusion from studies of the relationship between genetic variation and the ecological parameters r_m and K is that they are positively correlated. That is, populations harboring more genetic variation tend to have higher intrinsic rates of growth and maintain larger population sizes and/or larger biomass.

We already mentioned the studies of Carson (1961b, see chapter 4), who found that population sizes underwent a dramatic increase when a single haploid genome was migrated into a standing laboratory population. Dobzhansky and colleagues carried out a series of experiments designed to determine whether populations of *D. pseudoobscura* that were monomorphic for gene arrangements differed in these parameters compared to populations polymorphic for inversions. The conclusion was that populations polymorphic for AR and CH compared to monomorphic ones for the same chromosomes produced more biomass per unit of food as well as having higher r_ms (Beardmore et al., 1960; Battaglia and Smith, 1961; Dobzhansky et al., 1964; Ohba, 1967). Thus the adaptedness or population fitness of genetically polymorphic populations is greater than that of monomorphic ones. Whether this is due to facilitation (which is sometimes seen and sometimes not in *Drosophila*, e.g., Lewontin, 1955) or some inherent superiority of heterokaryotypic flies is not clear.

As an aside, Dobzhansky (1968) used these experiments to distinguish between what he called fitness and adaptedness. He reserved the term fitness to describe intraspecific differences among genotypes and as a measure of the relative contribution of an individual genotype to future gene pools. The term adaptedness was used to describe populational phenomena such as rates of increase and population size, which may be absolute measures. Thus the term often used, population fitness, would be meaningless in this terminology.

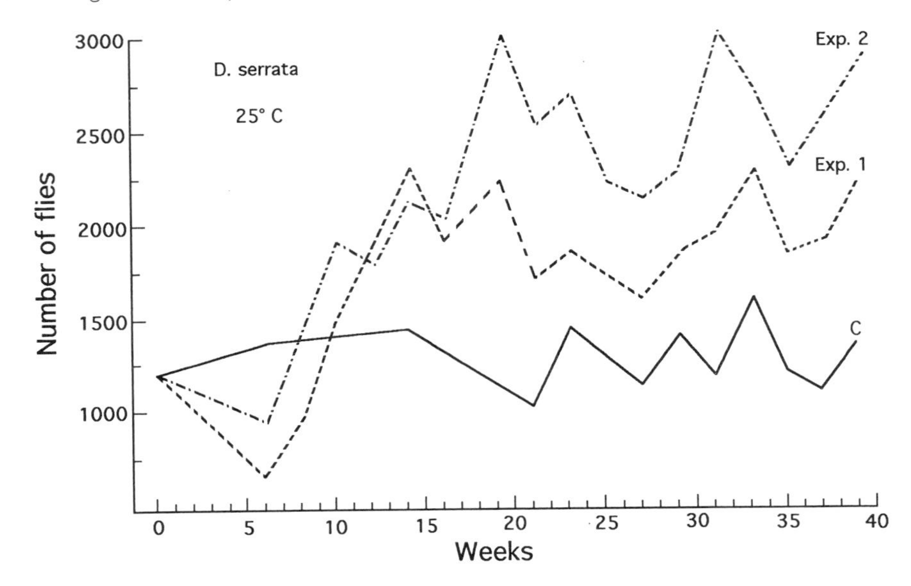

FIGURE 6-8 Numbers of *D. serrata* flies in three populations maintained by the serial transfer method. The two dashed lines are irradiated replicate populations of the nonirradiated control (c). Males were irradiated with 1,000 roentgens of x-rays for three consecutive generations and then mated to nonirradiated females and the offspring used to initiate experimental populations. From Ayala (1968b).

For *D. hydei*, Lopez-Saurez et al. (1993) found that more genetically heterogeneous strains had higher viability under competitive conditions than did genetically homogeneous strains.

Ayala (1968b) provided related observations on the relationship between genetic variation and population growth and size. He studied a series of strains of the Australian species *D. serrata* and *D. birchii*. He first noted considerable differences (presumably genetic) among strains in their productivity (defined as number of adults produced per food unit) and equilibrium population sizes in the serial transfer method. Hybrid populations made by crossing two strains of the same species outperformed, in general, both parental strains, thus indicating that increased genetic variation due to hybridization increased the adaptedness (*sensu* Dobzhansky) of the population. Similarly, he irradiated populations and determined what the effect of the added input of mutations would be on the parameters. Figure 6-8 illustrates his results. At first the irradiated populations suffered a decrease in population size relative to the nonirradiated control; presumably this was due to elimination of deleterious mutants caused by the radiation. After about 10 weeks both irradiated populations began to outperform the controls and maintained their higher population sizes for several weeks until termination of the experiments. This is a remarkable result: the random input of mutations can rapidly affect, and in a positive manner, the population's carrying capacity.

7

Speciation

> Of course I allude to that mystery of mysteries the replacement of extinct species by others. Many will doubtless think your speculations too bold—but it is as well to face the difficulty at once.
>
> John Herschel, 1836 (see Cannon, 1961)

> [W]e know virtually nothing about the genetic changes that occur in species formation.
>
> R. C. Lewontin, 1974

The foregoing chapters have all dealt with evolution within species, or microevolution, and is largely the purview of population genetics or ecological genetics. We now turn attention to the second major aspect of biological evolution, namely, the actual splitting of one species into two or more species. The next chapter will deal with even higher levels of biological organization, the relationship of species and species groups, concerns often designated macroevolution. Thus the present subject represents a bridge between microevolutionary concerns and macroevolution. We might then call studies of speciation *mesoevolution.*

The origin of species is obviously of primary concern to evolutionists as witnessed by the fact that the very phrase "origin of species" is used in the title of some of the most important books in evolution, starting with Darwin. It is an issue at the very heart of evolutionary thinking and crucial to understanding the generation of organic diversity. Yet, it is ironic that some of the best thinkers in evolutionary biology feel we have very little knowledge of the process, as indicated by R. C. Lewontin's quote at the head of this chapter. Equally ironic is the fact that natural selection may play a role in speciation only indirectly (in some cases) and thus Darwin's major contribution in his book *Origin of Species*, the documentation of natural selection as a force molding adaptations, may not help much in accounting for the origin of species! Natural selection can go on perfectly happily molding adaptations to environmental challenges while never causing a split in the lineage, what G. G. Simpson called *anagenesis*, as opposed to the generation of new clades, *cladogenesis.*

In the next several pages we will explore what is known of the process of speciation in *Drosophila*. If Lewontin's quote was true 20 years ago (which I doubted then), it certainly is not today. The study of mechanisms of speciation is one of the most active areas of evolutionary biology today, and *Drosophila* are playing a central role once again.

The Taxonomy of Species and Speciation

In discussing the process of generation of species, we must first be clear on what a species is. Various species concepts and/or definitions have been proposed with their own taxonomy: the biological species, cohesion species, phylogenetic species, and so forth (see Otte and Endler, 1989, for a compilation and discussion). Fortunately for *Drosophila* biologists, the species problem has been minimal; the usual biological species concept (BSC) works very well for *Drosophila*. This is perhaps not surprising, since it was developed by a drosophilist, Th. Dobzhansky. Species are defined as groups of organisms within which genes are exchanged freely by sexual reproduction, but between which genetic (reproductive) isolation exists and genes are not (or very rarely) exchanged. One of the beauties of this definition for genetic studies of the process is that clearly defined phenotypes are associated with the process, that is, sterility, inviability, and/or lack of mating. Thus much work on speciation has been on the genetic factors leading to those phenotypic manifestations and on the populational processes that bring about fixation of those genetic factors in different species.

The speciation process has also been defined in various ways. Some authors stress the geography of speciation, stressing sympatric, allopatric, parapatric, or other forms of speciation (e.g., Mayr, 1963). Others emphasize populational processes such as founder effects (e.g., Carson, 1982) or concentrate on genetic processes such as chromosomal changes (e.g., White, 1978). I have tried to clarify these discussions by explicitly separating the various factors (table 7-1). Some of my friends have called this the Chinese menu classification—one chooses one item from each column. The point is that one must keep the different levels of the process distinct and try to understand the connections between them. Not all paths through this menu are equally likely (or digestible?) and our goal is to decide which are the most common in our group of flies.

Already we can eliminate, or at least ignore as very unlikely, some of the possible paths through table 7-1. For example, little serious discussion of sympatric speciation has been evident in the literature. Allopatric speciation is almost certainly the rule in *Drosophila*, although some laboratory studies verge on sympatric speciation (see

TABLE 7-1 A categorization of speciation processes.

Ecological Context	Genetic Change	Cause	Result
Allopatric (large)	Polygenic	Selection	Hybrid sterility
Allopatric (small) Founder effects	Coadapted complexes	Recombination	Hybrid inviability
Parapatric	Chromosomal	Genetic drift	Habitat preference
Clinal	Polyploidy	Mutation	Temporal isolation
Sympatric	Single (few) major gene	Migration	Sperm rejection
Etc.			
	None	Combination of above	Behavioral (sexual) isolation
	Etc.		
		Symbiosis	Etc.
		Etc.	

From Powell (1982).

below). Polyploidy is unknown in *Drosophila* and can be ignored. Chromosomal changes are also unlikely to be of general concern as witnessed by the numerous examples of homosequential species, species whose polytene chromosome banding patterns are identical. At least to the level of resolution provided by polytene chromosomes, no chromosomal alterations can be detected between reproductively isolated species. Thus in what follows, we will limit our attention to the more likely mechanisms of speciation in *Drosophila*, and, of necessity, those processes that have been studied. This is not meant to imply other mechanisms of speciation are not operating in the genus; they certainly may be. However, it is difficult to discuss or assess processes that are rare and unstudied.

Three Caveats

A crucial and very difficult problem in studying speciation is to distinguish between cause and effect. Any species pair, after having completed the speciation process, will continue to accumulate genetic differences. There are many examples of pairs of *Drosophila* species that are morphologically identical or nearly so (sibling species), may be homosequential in their chromosomes, and are able to form viable hybrids, yet they have been reproductively isolated from one another for a million years or more, which may mean 10 million or more generations. An example is the classic pair *D. melanogaster* and *D. simulans*, which are thought to be on the order of 2.5 million years since last common ancestor (Lachaise et al., 1988). *D. pseudoobscura* and *D. persimilis* may also have diverged as long ago (Goddard et al., 1990; Aquadro et al., 1991).

The second caveat is perhaps obvious, but again often ignored: the process of speciation takes time. It is not an instantaneous event such that at one point in time it is clear that a lineage is a single species and immediately after, speciation has created two lineages. The continuum between one and two lineages may last a long time and at various points along this continuum, from being a freely interbreeding single species to reproductive isolation, the process could be halted and even reversed, start up again or even go back to the original state. It is often assumed, again without explicit statements, that once speciation has started down the path it cannot go back; Dobzhansky's term *in statu nascendi* describing the semispecies *D. paulistorum* would seem to imply some kind of inevitable "birthing" process is taking place.

The third caveat we need to make is that while the definition of species imposed by the BSC is considered to be a genome-wide definition, recent evidence from molecular studies indicates that there are exceptions. That is, as traditionally taken, it is assumed that once reproductive isolation has been completed, all genetic material ceases to flow between species. This may not always be so. For example, mitochondrial DNA may more easily cross species boundaries than does nuclear DNA, especially between species pairs where only hybrid females are fertile (e.g., Powell, 1983). Passage across even wider phylogenetic distance is indicated by P elements, mentioned in chapter 5 and discussed again in chapter 9. These observations are relevant to the issue of gene trees and species trees. The time when two genes (alleles) diverged from one another may not coincide with the attainment of reproductive isolation between species. In addition to problems of introgression, ancestral polymorphisms and lineage sorting (e.g., Avise, 1986; Pamilo and Nei, 1988) may lead to dissociation between gene trees and species trees. Examples of these processes in *Drosophila* are presented at the end of chapter 8.

The First Effects

In the following several sections, I will use terminology that is not universally in vogue. In particular, the term "isolation" and its related "isolating mechanisms" have been criticized as being misleading. While to some extent I am sympathetic to these critiques, the terms are still highly useful in communication, especially in regard to the BSC, which will be the orientation here. I will return at the end of this chapter to the "isolation" problem, but hold this discussion in abeyance while the data are presented.

Generally, reproductive isolation is divided into two major categories: premating isolation and postmating isolation. In the former case, due primarily to behavioral divergence, species do not exchange genes because they do not mate with one another, regardless of whether fertile viable offspring could be produced. Postmating isolation occurs after mating takes place but inhibits gene exchange because the offspring of interspecific matings are not viable and/or fertile. Thus one of the first questions we might pose about speciation in *Drosophila* is, which type of isolation occurs first? When two populations become allopatric and diverge sufficiently to begin speciation, is it more likely that mating behavior first diverges or that some kind of developmental incompatibility occurs such that gametes from the two populations cannot form a normal zygote? Note that I have explicitly stated this question not as what are the first *causes* of reproductive isolation, but what are the *effects*. The causes are some kind of genetic divergence, and the effect is either premating or postmating isolation.

Fortunately we can answer this basic question unequivocally: almost certainly both premating and postmating isolation have occurred first during speciation in *Drosophila*. There are cases where strong arguments can be made for both.

Premating

Premating isolation in the absence of (detectable) postmating isolation has been documented in a number of cases. One case is *D. mojavensis*. We already discussed (chapter 6) this species with regard to ecology and its host shift from breeding on agria cactus in the Baja Peninsula to organ pipe cactus in the mainland (Arizona and northern Mexico). In crosses between strains from the two areas, no indication of hybrid loss of vigor either in viability or fertility has been detected. However, strains from the two areas do exhibit premating isolation (Zouros and d'Entremont, 1980), an example of which is shown in table 7-2. Koepfer (1987a, 1987b), Krebs (1990), and Krebs and Markow (1989) have followed up these original observations and confirmed them. Interestingly, the other Sonoran species that make a similar host shift, *D. nigrospiracula* and *D. mettleri* (see table 5-3), do not exhibit any pre- or postmating isolation (Markow et al., 1983). Markow (1991) showed that it was not host shift per se that caused premating isolation in a strain from the mainland area of Desemboque (figure 6-3). The strain, reared from agria, exhibited the same premating isolation as mainland populations reared from organ pipe.

TABLE 7-2 Evidence for premating isolation between strains of *D. mojavensis* collected from Baja California and from the mainland (northern Mexico and Arizona).

Mainland Strain 1		Baja Strain 2	n_{11}	n_{12}	n_{21}	n_{22}	I_{total}	I_1	I_2
OPNM	×	A421	64	35	43	57	0.22 (.069)[b]	0.29 (0.96)[b]	0.14 (.099)
OPNM	×	A427	49	30	46	42	0.09 (.074)	0.24 (.109)[a]	−0.04 (.106)
OPNM	×	LP	82	30	78	118	0.30 (.054)[b]	0.46 (.084)[b]	0.20 (.070)
OPNM	×	A433	75	21	54	39	0.21 (.071)[a]	0.56 (.084)[b]	−0.16 (.102)
A514	×	LP	60	40	47	55	0.14 (.070)[a]	0.20 (.098)[a]	0.08 (.99)
A509	×	LP	75	42	59	47	0.09 (.067)	0.28 (.087)[b]	−0.11 (.096)
A559	×	LP	109	34	50	55	0.32 (.060)[b]	0.52 (.071)[b]	0.05 (.097)

Note: The n_{ij} is the number of matings between females of strain *i* and males of strain *j*. The isolation index is presented for the total data and for the effect of females (I_1) and males (I_2) separately. Clearly the discrimination is occurring because Mainland females prefer not to mate with Baja males. Standard errors are in parentheses with the [a]5% or [b]1% level of significance noted.

From Zouros and d'Entremont (1980).

Other examples of premating isolation in the absence of postmating isolation come from Hawaiian *Drosophila*. Ahearn et al. (1974) present the example of *D. heteroneura* and *D. silvestris* on the island of Hawaii. Fertile F_1 hybrids of both sexes are produced in both reciprocal crosses (e.g., Val, 1977), yet they remain distinct in areas of sympatry due to premating isolation. Carson et al. (1989 and references therein) do report the existence of hybrids at a low frequency in nature (somewhat less than 2%), which do not, however, obliterate the distinctness of the species. Evidently, the two mating systems of these species are sufficiently distinct that even in the face of potential gene flow via fertile hybrids, the hybrids quickly die out or evolve to one or the other mating types through backcrosses. Thus there may be some gene flow occurring between the species, yet they remain distinct. As we will present later, this can be thought of in Wright's adaptive landscape imagery. The two mate recognition systems of *D. silvestris* and *D. heteroneura* occupy two separate peaks. Hybrids between them fall into a fitness valley and are either eliminated or quickly get pushed back up to one peak or the other.

Wu et al. (1995) studied premating isolation between African and non-African *D. melanogaster* populations. The African populations from Zimbabwe represent the ancestral state, and the others represent the human commensal populations spread around the world by human activity. There was no indication of any postmating isolation between the Zimbabwe and non-African flies, but quite strong premating isolation. If this represents an early stage of speciation, it is another example of premating preceding postmating isolation.

Postmating

Postmating isolation in the absence of premating isolation is also known. The best example is perhaps the isolated population of *D. pseudoobscura* in the vicinity of Bogota, Colombia (figure 3-2). Prakash (1972; see also Dobzhansky, 1974) found that crosses between these populations and those from anywhere in North America yielded sterile male offspring when the female parent was from Bogota and the male from North America; the reciprocal produces fertile hybrids of both sexes. No premating isolation was detected by Prakash (1972); I confirmed these results using more recently collected strains (unpublished observations). Perhaps this result is not surprising given the complete allopatry of the two subspecies (as they have been designated by Ayala and Dobzhansky [1974]). There is evidence (discussed below) that sympatry promotes premating isolation.

Another example of the same pattern is *D. planitibia* on the island of Maui and its two close relatives on the island of Hawaii (Ahearn et al., 1974). Crosses between *D. planitibia* and either of the other two species produce sterile hybrid males, but little or no premating isolation exists between the allopatric species. Another example of this pattern occurs in the *willistoni* group. Ayala et al. (1974a) found that crosses of *D. equinoxialis* from the Caribbean Islands and mainland South America yielded fertile F_1 females but sterile males. Yet no premating isolation was found. The final example is from the *nasuta* subgroup of the *immigrans* group. Chang et al. (1989) could detect no premating isolation among strains of *D. nasuta* and *D. albomicans*, which are allopatric species. Postmating isolation occurs due to F_2 and F_3 hybrid breakdown, exhibited as a decrease in progeny numbers, reduced fertility of both sexes, and abnormal sex ratios.

As with observations of premating isolation taking temporal precedence over postmating, there are multiple examples of the opposite. This is consistent with Coyne and Orr's (1989a) observations that when many pairs of *Drosophila* species are compared, on average premating and postmating isolation occur at about the same rate. Assuming one or the other must occur first (which may not be a good assumption), this finding implies that about half the time each type of isolation has temporal precedence. However, we need to inject a word of caution in this conclusion. The manner in which the two types of isolation are measured and recorded could be biasing our perception. One problem is that postmating isolation is usually recorded as one of four discrete steps: viability of hybrid males and/or females, and fertility of hybrid males and/or females. The designation viable or fertile is taken as all or none in most studies, whereas quantitative decrease in the relative viability and/or fertility is rarely recorded, which of course is because it is much more laborious to quantify. A second problem is that premating tests probably systematically underestimate the true degree of isolation in nature. When two species are artificially placed in the confines of a mating chamber, interspecific hybridization occurs more frequently than it would in nature. For example, *D. pseudoobscura* and *D. persimilis* occur in sympatry at Mather. Dobzhansky (1951) collected 395 pairs of flies copulating on the collecting cups and determined the species of each sex; every case was an intraspecific couple. However, when flies in the same area were collected, left in the collecting vials, and then observed for copulations, 4 out of 36 were interspecific couples. Likewise for *D. heteroneura* and *D. silvestris*: in laboratory mating tests incorrect (i.e., interspecific) matings occurred at about a 15% rate (Ahearn et al., 1974), whereas in nature they exist at about 1/10 this frequency, 1.7% (Carson et al., 1989). My guess is that underestimating premating isolation is more serious than underestimating postmating. I would conjecture that premating isolation is more often the first effect of genetic divergence leading to speciation in *Drosophila*, but this remains conjecture until better means of ascertaining both types of isolation are developed.

Thus there is no general rule for the initial effects of genetic divergence during the first stages of speciation. Cases of both premating and postmating isolation occurring in the absence of the other have been found. We now explore a controversial connection between the two forms of isolation, namely, can one influence or even cause the evolution of the other?

Reinforcement

Early in the history of speciation genetics, the theory of speciation by reinforcement was put forth. Beginning in 1937, Dobzhansky presented the following reasoning: when two populations have been allopatric for a considerable time, genetic divergence, either by chance or adaptation to environmental changes, can cause the populations to be so divergent that gametes from the two cannot form normally fit hybrids. This interpopulational incompatibility clearly could not have been selected for, but must be a byproduct of the genetic divergence for other reasons. If the two populations come back into contact and if premating isolation has not also been a byproduct of the divergence, hybrids will be formed that are at a selective disadvantage. Selection will favor matings among individuals of the same population by partial elimination of offspring of interpopulational matings. In other words, the partial reproductive isolation due to postmating factors promotes the evolution of premating isolation after sympatry is reestablished. The initial isolation is reinforced by selection now acting specifically to effect more efficient premating barriers.

There has been considerable controversy over both the nature of reinforcement and its reality. One major objection has been that it is not really a theory of speciation. If two populations are so diverged as to produce inviable or infertile offspring, then speciation has already occurred and the subsequent selection for premating isolation has nothing to do with the speciation event itself. Rather, some authors claim, this would simply be a case of reproductive character displacement. For such selection to play a role in the actual attainment of reproductive isolation, the initial postmating isolation must be partial and the reinforcement complete it. But this brings its own problems. When two partially reproductively isolated populations reestablish contact, it is analogous to a polymorphic population with underdominance, an unstable situation in which the smaller population should be eliminated. Critiques of reinforcement include Spencer et al. (1986) and Butlin (1989); theoretical models indicating reinforcement is feasible include Felsenstein (1981), Sawyer and Hartl (1981), and Sved (1981a, b). Rather than getting bogged down in this controversy, we will confine ourselves to the empirical data from *Drosophila.*

Does premating reinforcement occur in *Drosophila*? The data relevant to this are skimpy at best, although they do provide some support. One of the original observations came from *D. paulistorum*, a complex species thought to comprise six semispecies that are given the non-Linnean names Andean, Amazonian, Orinocan, and so on, indicating their approximate centers of distribution in South America (see Ehrman and Powell, 1982). Most crosses among the semispecies yield fertile F_1 females but sterile males. Ordinarily this might be cause for simply raising them to species status were it not for some strains called Transitional, which are perfectly interfertile with more than one semispecies. Because of this, it is thought that the semispecies are only partly reproductively isolated and that they represent an early stage of the speciation process, the group for which Dobzhansky coined the term *in statu nascendi.* Whether this is the correct interpretation is moot for the present discussion. There can be no doubt they are a group of very closely related taxa (among other things this was confirmed by allozyme analysis [Richmond, 1972]) and most importantly for the present discussion, the distributions of the semispecies make them amenable to testing the reinforcement theory. Pairs of semispecies exist both sympatrically and allopatrically and exhibit incomplete premating isolation, at least in the laboratory.

Ehrman (1965) studied pairs of semispecies of *D. paulistorum* using strains collected where the taxa are sympatric and where they are allopatric. Table 7-3 illustrates her results. The figures given here are the isolation index, calculated as the number of homogametic (same strain) matings minus the number of heterogametic matings divided by the total number of matings, when the two types are placed in equal numbers in the mating chamber. Thus, this isolation should be zero if the strains mate randomly, negative if they prefer to mate with the other strain, and positive if some premating isolation exists. In seven out of eight of the combinations tested, the strains coming from the region where the semispecies coexist exhibited greater premating isolation than when strains were used that came from areas where the semispecies have not been in contact, that is, allopatric. This pattern is completely consistent with the expectations of reinforcement.

A similar situation has been found for *D. pseudoobscura* and *D. persimilis* (Noor, 1995). Recall that these species are sympatric over much of their distributions, al-

TABLE 7-3 Results of tests between allopatric and sympatric strains of semispecies of the *D. paulistorum* complex.

Semispecies Pair	Origin	No. Matings	*I*
Amazonian × Andean	Sympatric	108	0.86 ± 0.049
	Allopatric	100	0.66 ± 0.074
Amazonian × Guianan	Sympatric	104	0.94 ± 0.033
	Allopatric	109	0.76 ± 0.061
Amazonian × Orinocan	Sympatric	106	0.75 ± 0.065
	Allopatric	124	0.61 ± 0.070
Andean × Guianan	Sympatric	109	0.96 ± 0.033
	Allopatric	102	0.74 ± 0.066
Orinocan × Andean	Sympatric	100	0.94 ± 0.033
	Allopatric	111	0.46 ± 0.084
Orinocan × Guianan	Sympatric	104	0.85 ± 0.053
	Allopatric	100	0.72 ± 0.069
Centro-American × Amazonian	Sympatric	102	0.68 ± 0.072
	Allopatric	103	0.71 ± 0.070
Centro-American × Orinocan	Sympatric	110	0.85 ± 0.052
	Allopatric	103	0.73 ± 0.069
		Average Sympatric = 0.85	
		Average Allopatric = 0.67	

Note: The isolation index, *I*, is a direct measure of the degree of isolation and is the same as in table 7-2.

From Ehrman (1965).

though *D. pseudoobscura* has a much larger distribution (figure 3-2). Therefore, it is possible to find both sympatric and allopatric populations of *D. pseudoobscura*, while all populations of *D. persimilis* are sympatric with *D. pseudoobscura.* Table 7-4 presents Noor's results using *D. persimilis* males from two localities and *D. pseudoobscura* from the same two localities as well as from three regions where only *D. pseudoobscura* exists. Clearly, there is more premating isolation between sympatric populations than when the females in the tests originated from allopatric populations.

A third case for reinforcement has been the example of *D. mojavensis* and *D. arizonae.* (The latter species presents some nomenclature difficulties; originally it was called *D. arizonensis*, but revision according to finer points of the nomenclature rules led to the name change in 1990. The reader need only be aware that the two names refer to the same species. We adopt the newer terminology here.) Wasserman and Koepfer (1977) studied these species from areas of both sympatry and allopatry and their data are in table 7-5. Again, there is more isolation in areas of sympatry than in areas of allopatry. Wasserman and Koefper referred to this as reproductive character displacement. However, one could make an argument that it is reinforcement *sensu* Dobzhansky because the postmating isolation between these species is only partial: hybrid males are sterile, but females are fertile.

Recall that within *D. mojavensis* there exists prereproductive isolation in the absence of postmating isolation (table 7-2). Zouros and d'Entremont (1980) point out that the discrimination by mainland *D. mojavensis* females against Baja males may have been induced by the mainland populations having been sympatric with *D. arizo-*

TABLE 7-4 Results of mating tests between *D. persimilis* males from two populations, Mather and Mt. St. Helena, and females of *D. pseudoobscura* from the same population (designated S, sympatric) or from an allopatric *D. pseudoobscura* population (designated A, allopatric).

	Total Observed	Mated	Probability
Mather *persimilis* male			
(S) Mather *pseudoobscura* female	200	73	0.5
(A) Tempe *pseudoobscura* female	200	72	
(S) Mather *pseudoobscura* female	102	16	<0.001
(A) Flagstaff *pseudoobscura* female	102	63	
(S) Mather *pseudoobscura* female	100	18	<0.001
(A) Provo *pseudoobscura* female	100	49	
Mt. St. Helena *persimilis* male			
(S) Mt. St. Helena *pseudoobscura female*	147	23	<0.001
(A) Tempe *pseudoobscura* female	147	51	
(S) Mt. St. Helena *pseudoobscura* female	107	28	<0.001
(A) Flagstaff *pseudoobscura* female	107	56	
(S) Mt. St. Helena *pseudoobscura* female	100	33	<0.001
(A) Provo *pseudoobscura* female	100	56	

Note: Single pairs of males and females were placed in a vial and the females' spermathecae checked for presence of sperm after 24 hours; number with sperm are number mated. The difference between S and A females was tested with a one-tailed Fisher's exact test.

From Noor (1995).

TABLE 7-5 Results of mating tests between *D. mojavensis* and *D. arizonae* with strains from sympatric and allopatric species.

	Number of Matings		
Strain Origins	Homogametic	Heterogametic	$I \pm$ S.E.
Allopatric × Allopatric	354	119	0.497 ± 0.040
Allopatric × Sympatric	780	138	0.699 ± 0.024
Sympatric × Sympatric	363	14	0.936 ± 0.019

Note: Isolation index, $I \pm$ S.E., is the same as in the previous tables.

From Wasserman and Koepfer (1977).

nae. The selection on mainland *D. mojavensis* to increase premating isolation with *D. arizonae* has produced it within *D. mojavensis*. If this is the case, then it may truly be an example of reinforcement (or character displacement) actually initiating a speciation event, assuming that the partial isolation within *D. mojavensis* could lead to a new species.

Another line of evidence in favor of reinforcement came from Coyne and Orr's (1989a) extensive review of *Drosophila* speciation. Their approach was to use all species pairs for which knowledge of the degree of reproductive isolation was available along with allozyme estimates of genetic divergence. To appreciate this discussion, we need to review some aspects of allozyme data. Generally, two or more closely related species were studied for differences in the electrophoretic mobility of proteins produced by a number of genes. Identical mobility was assumed to indicate the same allele, while differences in mobility indicated allelic differences. Genes could fall into one of three categories: they have the same allele or alleles in the same frequency in the two taxa, they are completely differentiated in sharing no alleles, or they display partial differentiation in that the taxa share some alleles but in different frequencies. Nei (1972) devised what was to become the most popular statistic to describe the overall genetic "distance" based on such data sets. This became known as Nei's *D* or simply *D*. In its original derivation it is simply interpreted as the average number of allelic substitutions per locus. For example, a *D* of 0.25 indicates 25 out of 100 loci are different between the taxa compared; this is due to a combination of complete fixation of alternative alleles and weighting loci with partial differentiation. This statistic has been widely discussed, criticized, and so on, but we need not be too concerned about these details. Suffice it to say, it is a standard measurement of overall genetic divergence between taxa, and it has a reasonably good proportionality to time since divergence (Nei, 1975, 1987).

The second index Coyne and Orr (1989a) needed was one of reproductive isolation. Following Zouros (1973), they considered four postmating isolating factors: viability of both sexes and fertility of both sexes. They counted the number of sexes in both reciprocal crosses that were either sterile or inviable and divided by four. The measurement can range from 0 when both sexes are fertile in reciprocal crosses to 1 when both sexes are inviable or sterile in reciprocal crosses. Thus this measure takes on discrete values of 1/4, 1/2, 3/4, and 1. The premating isolation index used was simply

1 − (number of heterogametic matings)/(number of homogametic matings).

If the numbers of the two taxa being tested are equal, then this index ranges from 0 when random mating occurs to 1 when complete assortative mating (isolation) occurs.

Figure 7-1 illustrates one result of Coyne and Orr (1989a). To summarize, the degree of premating isolation for sympatric pairs is greater than for allopatric pairs at the same time level of divergence (D). These differences are statistically significant. While these workers found that premating and postmating isolation occurred first about equally frequently during these early stages of divergence, when they broke the data into allopatric and sympatric pairs there is a clear pattern of sympatry being associated with premating isolation. As Coyne and Orr (1989a, p. 376) state: "It is remarkable that in our corrected data, *every* case of strong prezygotic isolation at low genetic distances occurs in a pair of species that is sympatric." (The correction of the data was for phylogenetic correlations.) This is one of the stronger analyses favoring reinforcement.

In a critique of reinforcement, Templeton (1981) pointed out that when two divergent allopatric populations come back into contact, fusion of the gene pools or elimination of one of them occurs, except in those cases where isolation is strong already. Thus the sample of sympatric cases is a biased sample of only those cases where they remained distinct. However, as Coyne and Orr (1989a, b) argue, this should be the case for both pre- and postmating isolation, and it is not (figure 7-1).

Coyne and Orr's results, taken together with the three previous examples in *D. paulistorum*, *D. pseudoobscura-persimilis*, and *D. mojavensis-arizonae* make it hard to avoid the conclusion that, at least sometimes, reinforcement of reproductive isolation can occur in sympatry. The stickier question is whether this should be considered part of the speciation process or simply character displacement after speciation has occurred. The very term reinforcement indicates that it was never intended as a scenario to account for the initiation of reproductive isolation. Just when one should consider the speciation process over is at least partly a matter of opinion or semantics. Yet this seems to be the crux of the problem in deciding if selection for increased mating discrimination is part of, or subsequent to, speciation.

Genetic Divergence

Traditional genetics requires that one must be able to perform controlled crosses and quantify phenotypes in progeny. The very definition of species as units between which genes cannot be exchanged precludes, for the most part, measuring genetic differences between species. (Exceptions to this rule are cases where at least one of the F_1 hybrid sexes is fertile, cases discussed in some detail in the next two sections.) Thus when molecular and biochemical techniques began to play a prominent role in evolutionary genetics beginning with allozyme studies in the mid-1960s, there was considerable excitement in being able to quantify genetic differences among taxa. Before that time, it was impossible to really ascertain the amount of genetic difference between subspecies, semispecies, sibling species, and so forth. If one could, perhaps some insight into speciation processes would emerge. This hope has only been partly realized, but some discussion of the attempts to quantify the level of genetic divergence at the early stages of species formation provides some insight. In fact, the use of protein

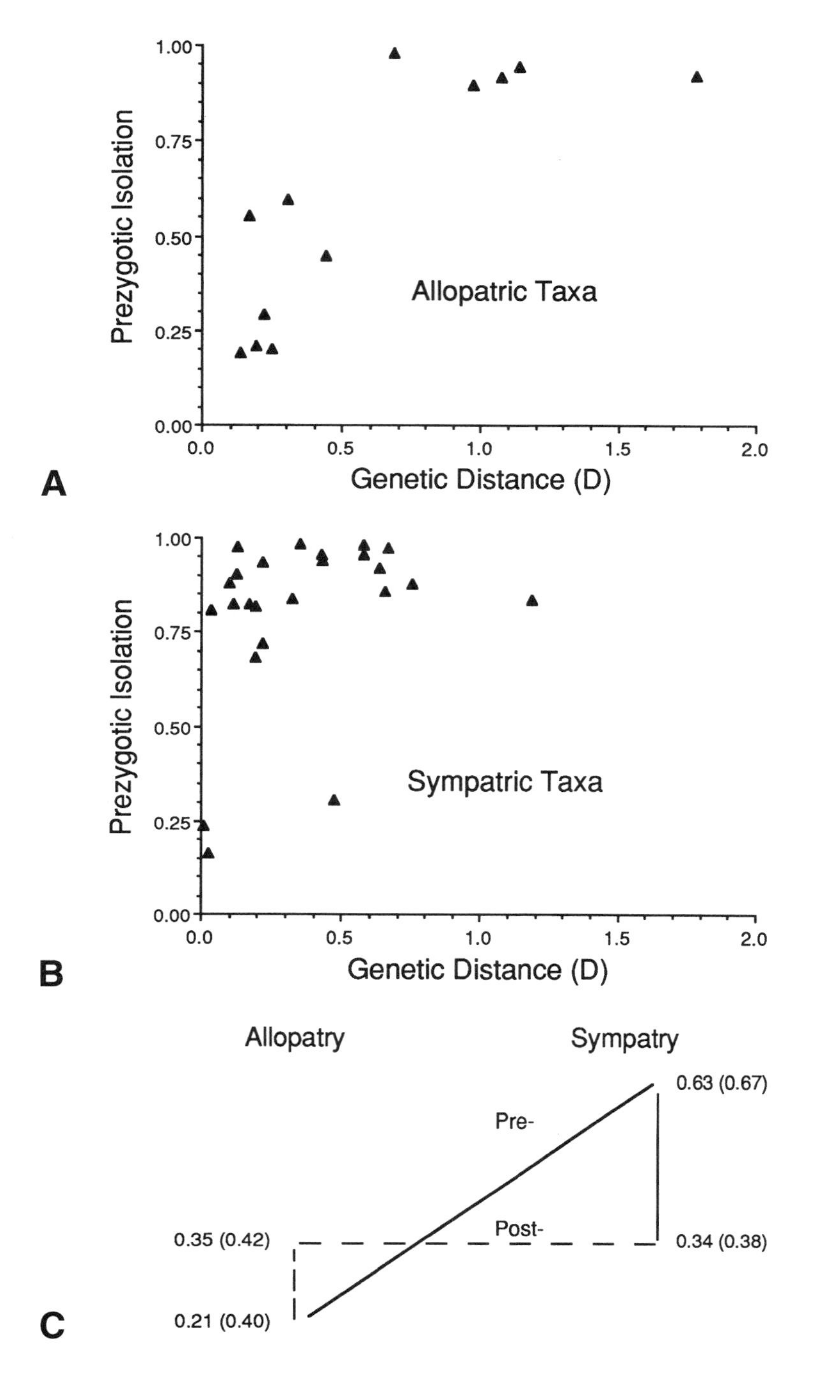

FIGURE 7-1 Scatter diagrams relating Nei's genetic distance (*D*) to degree of reproductive isolation in many pairs of *Drosophila* species. In the upper two graphs all pairs are shown; in the bottom graph only pairs with $D \leq 0.5$ are included. In this lower graph, values connected by a solid line are statistically different by a Mann-Whitney U test at $p < 0.05$; those connected by a broken line are not significantly different. From Coyne and Orr (1989a).

electrophoresis in *Drosophila* was first proposed to be a tool for systematics study (Powell, 1994). J. Hubby and L. Throckmorton had published on this use before it was used in population genetics (e.g., Hubby and Throckmorton, 1965).

Allozymes

One of the more thorough attempts to gain insights into speciation from this approach was that of F. Ayala and colleagues, working with the *D. willistoni* group (summarized in Ayala, 1975). Because this group contains taxa that display varying degrees of reproductive isolation, it was thought to represent the speciation process at different times. The taxa compared, in order of degree of expected divergence, are populations within a species, subspecies, semispecies, sibling species, and morphologically distinct species. This group has taxa falling into each category. We can ask how much divergence is detected at each level of the speciation process, assuming these represent such levels. Table 7-6 summarizes the results. Local populations of species are genetically very similar for allozymes as emphasized in chapter 2 (table 2-12). In the two cases of subspecies in this group, the D takes a jump to 0.23, indicating about 23% of the loci are diverged; the semispecies of the *paulistorum* complex are at divergence virtually identical to subspecies. Ayala interpreted these cases as follows: subspecies represent the first stage of speciation, being the beginning of genetic divergence in allopatry with only partial reproductive isolation. The semispecies of *paulistorum* being often sympatric, yet genetically distinct, represent a further stage in the process with a greater degree of reproductive isolation. This jump from partial to nearly complete isolation was accompanied by very little, in fact undetected, increase in allozyme divergence.

Perhaps more interesting was the pattern of distribution of divergence when individual loci are considered. The four histograms in figure 7-2 show the pattern; I in these histograms, genetic identity, is simply the converse of D. The indications are that loci tend to fall into one of two classes: completely differentiated or remaining identical; the semispecies have perhaps the most intermediate cases. It is as if during speciation, either a locus is completely free to change and given time will become fixed for alternative alleles, or loci are severely constrained between species (presumably by selection) and no divergence is tolerated.

TABLE 7-6 Allozyme divergence between taxa at different taxonomic levels in the *D. willistoni* group.

Taxonomic Level	D (average) ± S.E.
Local populations within a species	0.031 ± 0.007
Between subspecies	0.230 ± 0.016
Between semispecies	0.226 ± 0.033
Between sibling species	0.581 ± 0.039
Morphologically distinct species	1.056 ± 0.068

Note: Values given are Nei's D ± S.E.

From Ayala (1975).

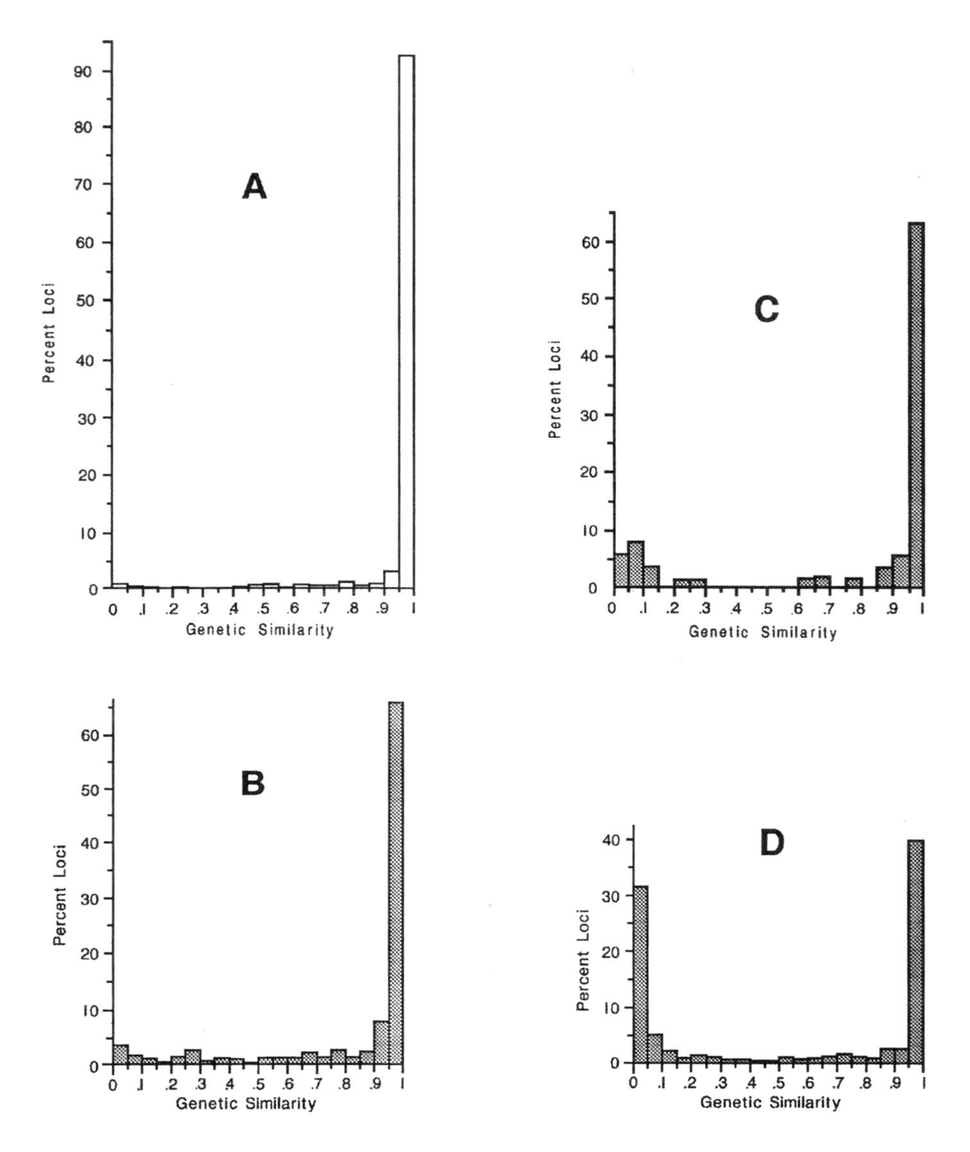

FIGURE 7-2 Genetic differentiation at different levels of taxonomic status in the *D. willistoni* group. A is between populations within a species, B is between semispecies of the *paulistorum* complex, C is between subspecies, and D is between sibling species. From Ayala (1975).

A broader picture of allozyme divergence and degree of reproductive isolation was presented by Coyne and Orr (1989a) in that they collected all species pairs where the information was available. Some of the data were presented in figure 7-1, to which we now add figure 7-3. While there is considerable scatter to these data, by and large once a D of 0.5 is reached, complete reproductive isolation has occurred. Nevertheless, in figure 7-3 there are cases where virtual allozyme identity (D of 0) is associated with some isolation (upper left part in figure 7-3), and cases of D greater than 1 where isolation is not complete. Considering this spread, and assuming D is reasonably correlated with time, we must conclude that different lineages require different amounts of time to evolve reproductive isolation. Ayala's *willistoni* group data would appear to be fairly intermediate in this regard with partial isolation having been reached at a D of 0.25 and complete isolation at a D of 0.58.

DNA

Better resolution on this problem should come from DNA studies. Direct sequence data is not sufficient to provide an overall measure of DNA divergence among several taxa. However, DNA–DNA hybridization does provide this kind of data and has been applied to several groups of *Drosophila*. This technique relies on the fact that the thermal stability of DNA duplexes is dependent on the fidelity of base-pair matching, A with T and G with C. If a duplex is artificially prepared in the laboratory in which each complementary strand comes from a different taxon, then the decrease in the thermal stability compared to perfectly match duplex should provide an overall measure of base-pair differences between taxa. In fact the change in thermal stability, usually measured by ΔT_m (the change in median melting temperature), is remarkably

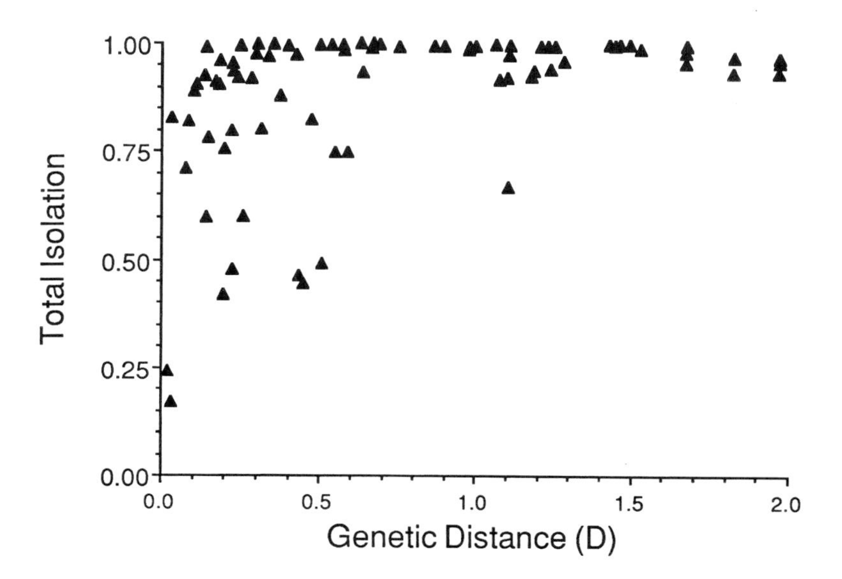

FIGURE 7-3 Same as figure 7-1, except total isolation is shown, which is premating plus postmating. From Coyne and Orr (1989a).

linear with degree of base-pair matching and can be converted into percentage of base-pair mismatch (Caccone et al., 1988a; Springer et al., 1992). A ΔT_m of 1°C is caused by about 1.5% base-pair mismatch. It is also important to realize these studies are usually done on the DNA that have had the repetitive fraction removed, that is, the results are for single-copy DNA or scDNA. We will have reason to discuss this technique in more detail in the next two chapters, but the above will suffice for now.

Figure 7-4 presents the results of application of DNA–DNA hybridization to several groups of closely related species often used in speciation studies. The interesting

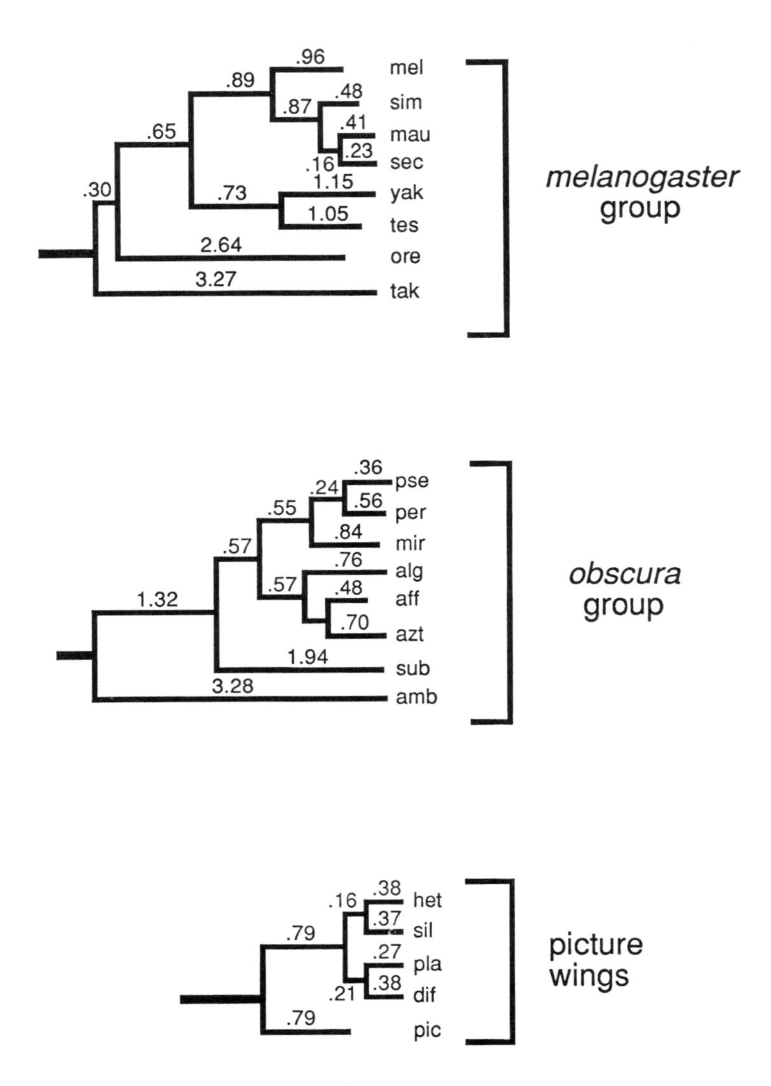

FIGURE 7-4 Degree of DNA differentiation among various taxa as measured by DNA-DNA hybridization. Lengths of branches are proportional to DNA divergence as determined by Neighbor-Joining. From Powell and DeSalle (1995).

point here is that on the DNA level the *melanogaster* subgroup species such as *D. melanogaster, D. simulans, D. sechellia*, and *D. mauritiana* are about as diverged as the *obscura* group species *D. pseudoobscura, D. persimilis*, and *D. miranda*. The Hawaiian picture-wing species illustrated here form a tighter cluster, presumably being younger than the other groups, a conclusion consistent with biogeographic considerations. The splits in the former two groups are generally considered to be a million years or more, while the picture-wing species illustrated here are thought to have differentiated after the origin of the island of Hawaii some 700,000 years ago.

What can be learned from considering the overall degree of genetic divergence between taxa at different stages of the speciation process? Perhaps the most important feature is that some idea of the time span needed to evolve different types of reproductive isolation can be gleaned from these data. This is assuming that these measures of genetic distance are related to time since divergence. The insights provided by Coyne and Orr (1989a) of the timing of appearance of different types of isolation, especially with respect to sympatry and allopatry, have been substantial. Ayala's (1975) conclusion that perhaps very little genetic change accompanies the step from partial to nearly full isolation is also important. Comparative studies like those in figure 7-4 can address questions such as how much genetic differentiation occurs between closely related species in different species groups. However, such studies do not begin to identify actual numbers of genes directly involved in speciation—we are dealing with the problem of cause and effect here: a single gene could cause the isolation and once established, a whole host of genes could also differentiate.

One intriguing anomaly was identified in the study of the *obscura* group species (Goddard et al., 1990). Using DNA–DNA hybridization to study the overall DNA divergence of the scDNA, they obtained the results in table 7-7. In this table the percentage of the scDNA that hybridized between species is presented along with the ΔT_m of that fraction that did hybridize. The hybridization conditions used require that about 75% of the base pairs match in order to form a stable duplex (this assures

TABLE 7-7 Degree of DNA differences among species of the *D. obscura* group and their ability to form interspecific hybrids.

Species Pair	Percentage scDNA Hybridized	ΔT_m	Hybrids
N. American × Bogota *pseudoobscura*	100	0.68	Yes, sterile males, one direction only
pseudoobscura-persimilis	98	0.92	Yes, sterile males, both directions
pseudoobscura-miranda	88	1.55	Yes/No, sterile males, both directions
pseudoobscura-affinis subgroup	75	2.49	No
pseudoobscura-subobscura	65	3.82	No
pseudoobscura-ambigua	50	5.68	Yes, development to pupal stage, adults?

Note: The Yes/No for crosses to *D. miranda* indicates some reports are positive, some negative, probably depending on geographic origin of the strains. Most studies with *D. ambigua* report hybrid development to the pupal stage. One publication reported adults.

From Goddard et al. (1990).

avoiding random reassociations). As expected when very closely related species are crossed, often viable progeny with one fertile sex are produced. As phylogenetic distance and DNA divergence increases, the ability to form hybrid progeny ceases. However, unexpectedly, at even greater divergence, hybrid progeny can be formed! The sobering thing about these data is that even with the great DNA divergence between *D. pseudoobscura* and *D. ambigua*, they are able to form viable zygotes that go through development at least to the pupal stage. About half of the single-copy genome will not even cross-hybridize, and that which does is about 10% diverged. Clearly, the overall DNA divergence between taxa can sometimes be a poor indicator of reproductive isolation.

General Proteins

R. Singh and colleagues (Thomas and Singh, 1992; Zeng and Singh, 1993; Civetta and Singh, 1995) have taken a very different approach to studying species divergence. They used two-dimensional electrophoresis to resolve about 1,000 proteins from testes. Comparing *D. simulans* with *D. sechellia*, they noted that about 8% of the protein spots differed. In backcrosses they found that most of the protein differences segregated as autosomal traits. Furthermore, because some backcross males are sterile, some fertile, they found "a few *D. sechellia*-specific proteins" were consistently associated with the sterile males. These results indicate that perhaps relatively few, mostly autosomal, protein-coding genes are associated with sterility. Furthermore, in comparing the divergence of testes-specific proteins with proteins from other tissues, they find the testes-specific proteins are more often diverged, suggesting faster evolution of these genes.

Genetics of Premating Isolation

Rather than assessing overall genetic divergence between taxa at various points in speciation, an alternative approach is to try to study the genetic basis of the isolating factors directly. In this and the next section we take up what is known of the genetic basis for the two major categories of isolation, premating and postmating. These studies have generally taken two forms. One is to artificially select for isolation and try to determine the genetic response to selection. Another set of studies concerns those cases for which fertile hybrids can be formed between taxa. Through use of mutant markers and backcrosses, genetic analysis is possible.

Behavior

First we examine the genetic basis of mating behavior differences among taxa. One of the major tenets of the Neodarwinian view of evolution is that the kinds of genetic differences that characterize differences between species is the same kind of variation that exists within a species. Mating behavior in *Drosophila* certainly conforms to this tenet. Anderson and Ehrman (1969) surveyed the then-available literature on tests of mating isolation among strains within species (table 7-8). They found 21 studies among which 19 showed some combinations of strains that exhibited small degrees of mating isolation; the two negative tests are cases where only two strains were

TABLE 7-8 Tests of mating preference between geographically separated strains of *Drosophila*.

Species	Number of Strains Tested	Pairs of Strains with Preferences
americana	4	5/6
arizonensis	5	4/10
athabasca	6	6/6
auraria	14	11/27
birchii	6	4/6
crocina	3	2/3
gasici	4	3/5
miranda	2	1/1
montana	4	3/3
nebulosa	2	1/1
paulistorum	24	29/46
pavani	2	0/1
peninsularis	2	1/1
prosaltans	7	18/21
repleta	6	6/15
serrata	15	13/80
sturtevanti	5	7/10
texana	4	2/3
virilis	4	3/3
willistoni	5	4/8
Totals:	126	123/257 = 0.479

From Anderson and Ehrman (1969), which contains the original reference to each study.

tested. In fact, their own study reported in that publication was on *D. pseudoobscura*, which fell into the minority category in showing no premating isolation among five geographically widespread strains; however, the previously discussed studies of Noor (1995) do indicate populations of this species differ in their propensity to mate with *D. persimilis*. In Spieth and Ringo's (1983) review, some 30 citations are given to studies revealing intraspecific variation in mating behavior leading to nonrandom mating.

Given that intraspecific genetic variation in mating behavior is the rule in *Drosophila* rather than the exception, it should be possible to carry out artificial selection experiments to study this trait. This has been done numerous times, almost always successfully (table 7-9, category C). We illustrate typical results with the longest such study of which I am aware. Dobzhansky and Pavlovsky (1971) observed an unusual phenomenon with some strains of *D. paulistorum*. A strain was collected in the Llanos of Colombia in 1958; it was tested with various tester strains of the semispecies and proved to be interfertile with Orinocan and only Orinocan, and so was classified as Orinocan. In 1963 the strain was again tested and it now proved intersterile with Orinocan, but interfertile with a newly discovered semispecies, Interior (not known in 1958). Given this spontaneous generation of postmating isolation, the project was undertaken to impose premating isolation by reinforcement. Recessive mutants were generated in the New Llanos strain and the Orinocan strain it had been interfertile

TABLE 7-9 Studies on laboratory populations of *Drosophila* for which premating isolation was tested after various types of selection.

Study	Premating Isolation
A. Divergent selection in allopatry	
Koref-Santibanez and Waddington (1958)	No
Ehrman (1964, 1969b)	Yes/No, inconsistent
del Solar (1966)	Yes
Kessler (1966)	Yes, asymmetrical
Barker and Cummins (1969)	No
Grant and Mettler (1969)	Yes
Burnet and Connolly (1974)	Yes
van Dijken and Scharloo (1979)	No
de Oliveira and Cordeiro (1980)	Yes
Kilias et al. (1980)	Yes
Markow (1981b)	Yes
Koepfer (1987a)	Yes, asymmetrical
Dood (1989)	Yes
Ehrman et al. (1991)	No
B. Parallel selection in allopatry	
Kilias et al. (1980)	No
Dodd (1989)	No
C. Divergent selection in sympatry with destruction of hybrids	
Koopman (1950)	Yes
Wallace (1953)	Yes, transient
Knight et al. (1956)	Yes
Kessler (1966)	Yes
Ehrman (1971, 1973, 1979)	Yes, but complex
Barker and Karlsson (1974)	Yes
Crossley (1974)	Yes
Dobzhansky et al. (1976)	Yes
D. Divergent selection in sympatry without destruction of hybrids	
Thoday and Gibson (1962)	Yes
Grant and Mettler (1969)	No
Eighteen experiments cited in	No, 18/18
Thoday and Gibson (1970)	
Scharloo (1971)	
Spiess and Wilke (1984)	No
E. Divergent selection in sympatry and with isolation via pleiotrophy	
Coyne and Grant (1972)	Yes, 1/2 replicates
Rice (1985)	Yes
Rice and Salt (1988, 1990)	Yes

From Rice and Hostert (1993) with modification.

with; at the start of the experiments, the two strains showed no statistically significant deviation from random mating. The mutants allowed artificial selection against hybrids by the removal of all wild type from the breeding population each generation. Table 7-10 shows the results over 132 generations. Partial premating isolation was built up rather rapidly, reaching a maximum of about 0.8 after 50 generations, then seeming to decrease slightly. Such a pattern is typical of the expectations of a character with multigenic inheritance.

Unfortunately, attempts to delve much deeper into the genetic basis of such behavioral differences are extremely difficult as with all multigenic characters, but especially for one in which the phenotype is so difficult to measure. As just one example, we take the *D. paulistorum* complex, which should represent early stages of speciation and which exhibit the classic allopatric/sympatric differences in premating isolation (table 7-2). Koref-Santibañez (1972a, b) performed painstakingly detailed analysis of the behavior of these semispecies; part of her data are in table 7-11. The courtship behavior of the various semispecies differed only quantitatively, that is, there are small differences in number of times an element is performed and in length of time spent performing the elements. Considering that these different individual elements of courtship are control by multiple loci (as well as environmental interactions), the extreme difficulty of precise genetic characterization of premating isolation should be obvious. Ehrman's (1961) pioneering studies on the genetics of premating isolation in this group are consistent with this complex view of the genetics of these traits.

Nevertheless, some studies have shed light on the genetic basis of premating isolation. For example, Ewing (1969) studied the courtship songs (pattern of male wing vibrations during courting) of *D. pseudoobscura* and *D. persimilis*. *D. pseudoobscura* males exhibited a two-part song including a low repetition rate song with pulses at 6 per second and a high repetition rate song with pulses at 24 per second. *D. persimilis* had only the high repetition rate song, but somewhat slower at 15 pulses per second. By performing interspecific crosses and backcrosses he showed that the low repetition song and the pulse of the high repetition song were controlled by X-

TABLE 7-10 Results of selection for increased premating isolation between strains of *D. paulistorum*.

Generation	$I \pm$ S.E.
0	0.12 ± 0.10
12	0.52 ± 0.08
27	0.59 ± 0.07
50	0.84 ± 0.05
64	0.82 ± 0.05
90	0.74 ± 0.06
100	0.65 ± 0.05
118	0.63 ± 0.07
132	0.63 ± 0.07

Note: I is as in previous tables.

From Dobzhansky et al. (1976).

TABLE 7-11 Analysis of courtship differences among semispecies of *D. paulistorum*.

	Orinocan	Centro-American	Amazonian
Orientation	5.2 ± 0.4	4.5 ± 0.4	5.7 ± 0.5
Tapping	21.9 ± 2.7	17.2 ± 1.9	15.8 ± 1.6
Vibration	14.1 ± 1.5	21.8 ± 2.4	13.1 ± 1.6
Scissoring	14.0 ± 2.9	10.4 ± 1.6	4.3 ± 1.0
Circling	5.7 ± 1.2	1.5 ± 0.3	2.0 ± 0.3
Licking	10.4 ± 1.4	9.6 ± 1.2	8.0 ± 0.9
Running	39.8 ± 5.2	59.8 ± 6.8	26.2 ± 3.5
Standing	17.3 ± 2.3	18.0 ± 2.0	13.9 ± 1.5
Rubbing	8.8 ± 1.3	15.9 ± 2.5	8.3 ± 1.2
Extruding	5.9 ± 0.8	5.0 ± 0.6	4.3 ± 0.7
Time courting	226 ± 22	241 ± 24	163 ± 17

Note: Numbers are the mean and standard error of courtship elements performed by males and females, and the courtship time in seconds. The data for the other three semispecies are similar.

From Koref-Santibañez (1972a).

linked factors, whereas the pulse interval rate of the high repetition song (15 versus 24 per second) was inherited independently, that is, was autosomal.

We have already discussed the premating isolation between Baja and mainland populations of *D. mojavensis* (table 7-5), in which the primary isolation is due to mainland females rejecting Baja males. Koepfer (1987a, b) performed selection experiments on these populations and found a distinct asymmetrical response; she could increase the isolation between Baja males and mainland females, but despite selection, no induction of isolation occurred between mainland males and Baja females. Thus the mainland populations that had been hypothesized to have shifted their mating behavior in response to the presence of another species (*D. arizonae*) still harbored more genetic variance than the presumed ancestral Baja populations.

Krebs (1990), Markow (1981a), and Krebs and Markow (1989) studied the genetics of the courtship behavior differences between Baja and mainland *D. mojavensis.* They found that male courtship success was dominantly inherited and due to autosomes; F_1 males of either reciprocal cross behaved like mainland males. Female hybrids, on the other hand, were intermediate in behavior, the relative receptivity of which appeared additively inherited. Given these two different modes of inheritance in the different sexes, Krebs (1990) concludes it is likely that male and female mating behaviors are controlled by different sets of genes.

Zouros (1981) studied the genetics of premating isolation between *D. mojavensis* and *D. arizonae*. He used isozyme markers to follow chromosomes. Male mating behavior was linked to the PGM locus, which is assumed to be X-linked (by chromosome homology). Female behavior was linked to chromosomes marked by two other presumed autosomal loci. Here again, the evidence is for independent genetic control of male and female mating.

Ahearn and Templeton (1989) studied the Hawaiian *planitibia* picture-wing group. We already mentioned this as an example of premating isolation in the absence of postmating isolation and it will also feature in discussions later. For the present, however, we point out that the genetic control of premating isolation, between *D. silvestris* populations intraspecifically and with *D. heteroneura* interspecifically, were shown to be under control of epistatically interacting blocks of genes, that is, coadaptation was detected. The strength of the coadaptation was greater between forms of *D. silvestris* than between species.

As perhaps expected, *D. melanogaster* and its close relatives have been the subject of a large number of studies on the genetic basis of premating isolation. As with all other studies, polygenic inheritance is the prevalent mode of inheritance. However, contrary to what has been found with postmating isolation in this group (discussed in the next section), most or nearly all of the control appears autosomal with little X-linked effect (e.g., Coyne, 1989; Welbergen et al., 1992). Coyne (1992b) found that at least two autosomal genes are involved in behavioral isolation between *D. simulans* and *D. sechellia* and three autosomal genes between *D. simulans* and *D. mauritiana*, with little or no effect of the X. Further, he argued the genes are likely not the same in the two species pairs. This finding is somewhat enigmatic in light of the following two observations on sex-linked genes in these species.

Coyne (1993) also studied the genetic control of the length of copulations between *D. simulans* and *D. mauritiana*, which are short compared to intraspecific copulations, meaning that little sperm is transferred. He found every chromosome had factors affecting this trait, although autosomes had a larger effect than the X. Male genotype had a greater effect than female genotype. Evidently, the female can detect the genetic differences and discriminate against males of the "wrong" genotype.

While all of the above evidence strongly supports the multigenic nature of premating isolation, there is evidence that sometimes single genes can have major effects. Bastock (1956) carried out an elegant and thorough study of the effect of the single-gene mutant *yellow* (*y*) on various aspects of male mating behavior. This mutant was not a random choice, but chosen because previous studies had shown it to suffer a mating disadvantage. Bastock showed the major effect was to change the courtship song. The morphological change, yellow body, is independent of the courtship change.

By far the most intriguing evidence for a single-gene major effect on premating isolation comes from studies of the *period* (*per*) gene. This sex-linked gene was isolated in a screen for circadian rhythm mutants and was first characterized as affecting the rhythm of eclosion and activity (Konopka and Benzer, 1971). Three types of mutants were isolated: those producing rhythms shorter than 24 hr (per^s mutants), longer than 24 hr (per^L), and arrhythmic flies (per^o). This gene was also found to affect the cycle of the interpulse interval (IPI) of the courtship song of male *D. melanogaster* (Kyriacou and Hall, 1980). Curiously, the long-day circadian rhythm alleles have longer IPI oscillations than wild type, and short-day alleles cause shorter IPI oscillations. At least within the *melanogaster* subgroup, there are species-specific IPIs; *D. melanogaster* has an IPI of about 55 sec and *D. simulans* has an IPI of 35 to 40 sec. Kyriacou and Hall (1982, 1986) showed that females mated more readily with males with conspecific courtship songs. They also found that hybrid females preferred the song of hybrid males; this is unexpected if *per* were the only genetic factor involved in male song, as hybrid males carry a single *per* allele and thus should produce species-pure song. It has been hypothesized, and sometimes claimed to have been observed, that the same genes affecting male courtship behavior control female receptivity via some kind of genetic coupling (see Butlin and Ritchie, 1989, for a critical review). Greenacre et al. (1993) addressed more directly the question of whether there was genetic coupling between male song production and female response. This was ruled out as per^s females still responded preferentially to 55 sec songs. However, Greenacre et al. (1993) also observed that females could change their preference of IPI. Females from a strain of per^s maintained for 15 years in culture in the laboratory had changed their mating preference to the short IPIs of the males in the culture.

Molecular analysis done by either in vitro mutagenesis (Yu et al., 1987) or differential cutting and splicing (Wheeler et al., 1991) followed by transformation of the chimeric genes into a *per* null mutant allowed identification of the specific part of the gene responsible for the song difference between *D. melanogaster* and *D. simulans*. The difference in IPI was mapped to a 700-bp region near a Thr-Gly repeat unit. The sixty amino acids 5′ to the repeat are identical in the two species and the Thr-Gly repeat is highly polymorphic within both species. So the implication is that the species specificity resides in the 122 amino acids 3′ to the Thr-Gly repeat; there are 8 amino acid substitutions between species in this region. (For more information on *per*, consult the review by Kyriacou and Hall, 1994.)

At least two other species groups have been studied for male courtship songs with some indication of the importance of the IPI interval; these are *D. ananassae-pallidosa* (Oguma, 1993) and the *bipectinata* complex (Crossley, 1986). In neither case has the *per* locus been implicated, although this has not been investigated in any detail.

The studies of *per* are the first to molecularly characterize any gene directly involved in reproductive isolation in *Drosophila* and would appear to be very important as clues for studies of other species. However, the enthusiasm for *per* as a general "speciation gene" needs to be tempered. First, other evidence indicates that premating isolation in this species group is not strongly affected by sex-linked genes (see above), yet *per* is sex linked. That the X chromosome is important in male courtship song was also confirmed by von Schilcher and Manning (1975) and Kyriacou and Hall (1986). Thus, we might conclude that male courtship song is not really the most important element in premating isolation in all groups of *Drosophila*, although it clearly can affect it. Second, the role of male courtship song in other species groups is not at all well established, although there is reason to believe it may not be the crucial factor controlling premating isolation. In many Hawaiian flies, visual stimuli are the predominant feature of courtship. In the *repleta* group there is remarkably little variation in visually detectable courtship behavior, including male courtship song (e.g., Ewing and Miyan, 1986); the effort on pheromones in this group was undertaken in light of these findings (Markow and Toolson, 1990; discussed below). In the *willistoni* group the variation in song among species is considerable, greater than that found among members of the *melanogaster* subgroup (Ritchie and Gleason, 1995), yet the region of the *per* locus implicated in controlling the isolation between members of the *melanogaster* subgroup is remarkably constant both within species and across species of the *willistoni* group (J. Gleason, unpublished data). Finally, Ford et al. (1994) failed to find any association between *per* variation and the isolation among subspecies of *D. athabasca*.

Morphology

Premating isolation can be attributed to morphological changes as well as solely behavioral changes, although the two may sometimes be confounded. A great advantage of studying the genetics of morphological characters is that they have easily ascertained phenotypes. Three examples of morphological change associated with premating isolation deserve mention.

The premating isolation between *D. silvestris* and *D. heteroneura* must be at least partly due to the very marked differences in morphology of males, especially the head (figure 7-5); females are virtually identical morphologically, which is a typical example of strong sexual dimorphism in Hawaiian *Drosophila*. Visual cues in mating are crucial in these flies, as no courtship occurs in the dark, and these head shapes and pigment patterning are crucial to mating success in these species (Spieth, 1981). Val (1977) carried out detailed genetic studies of these morphological differences and Templeton (1977) analyzed the results. The conclusion of Val was that 15 to 20 loci, most autosomal, and at least one major X-linked gene could account for the morphological differences. Templeton's analysis confirmed the polygenic nature involving largely additive autosomal genes and a major X-linked gene or genes with sex-limited expression. Thus the genetic architecture controlling this premating isolating factor is not extremely complex.

Within *D. silvestris* there is also a morphological change associated with courtship. During courtship, males use the foreleg tibia to vibrate against the female abdomen (Spieth, 1978). *D. silvestris* from the south and west sides of the island of Hawaii, the Kona side, have a typical number of cilia arranged in two rows on the foreleg

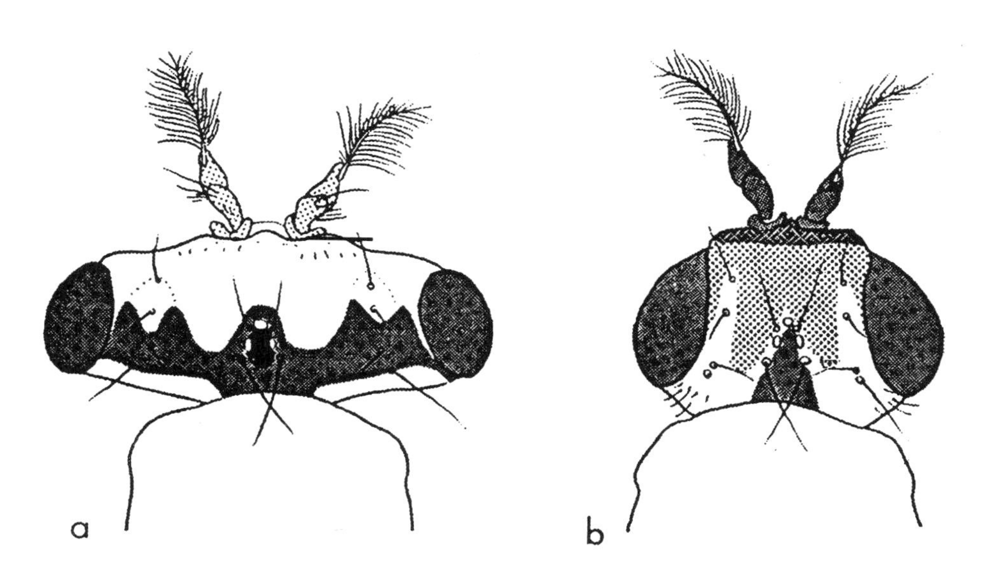

FIGURE 7-5 Head morphologies of males of two species of Hawaiian *Drosophila*; (a) *D. heteroneura*, and (b) *D. silvestris*. From Val (1977).

tibia; this is the pattern for the other species in this group. Populations of *D. silvestris* to the north and east, the Hilo side, have an extra row of cilia, adding some 25 extra bristles (Carson and Bryant, 1979; Carson, 1982). Evidently the presence or absence of the bristles can affect the mating success, with females from the Kona side discriminating against males from the Hilo side (Kaneshiro and Kurihara, 1982; Carson, 1982). Unfortunately, the genetic control of the two-row versus three-row phenotype is complex with typical highly multigenic inheritance (Carson and Lande, 1984).

Carson et al. (1994) adduced evidence that the two morphological differences between *D. silvestris* and *D. heteroneura*, the head shape and cilia number on the tibia, evolved independently. They studied two hybrid populations designated HS and SH, the first letter indicating the female parent of the hybrid population. After 14 generations, SH males had evolved toward a *heteroneura*-like head, but the tibia cilia evolved to be *silvestris*-like. In the HS population, the head did not change, but the cilia evolved to be *heteroneura*-like. Evidently the hybrid populations had evolved changed mate recognition systems that differed depending on the parental makeup. The two most obvious morphological traits showed no correlation in direction of change.

Differences in male genitalia are often considered premating isolating factors, the so-called lock and key concept. It is certainly true that in many closely related *Drosophila* the only morphological differences discernible to a human peering through a microscope are changes in male genitalia. Thus the genetic control of differences in male genitalia should be of interest in understanding speciation in this genus. Coyne (1983; 1985a; Coyne and Kreitman, 1986) has carried out genetic studies of the differences in male genitalia of four members of the *D. melanogaster* subgroup, *D. melanogaster, D. simulans*, *D. sechellia*, and *D. mauritiana.* The first two species are relatively distinct phylogenetically, while the latter three are extremely close and almost certainly represent species close to the splitting event (chapter 8). Table 7-12 summarizes the results for these latter three species. Because of the design of his study, only the effect of whole chromosomes could be detected, so the maximum number of single genes that could unambiguously be detected is five, one for each chromosome arm. This maximum was found. Thus the results are not distinguishable from a highly

TABLE 7-12 Effects of different chromosomes on male genitalic differences among very closely related species of the *D. melanogaster* subgroup.

		Proportionate Contribution		
Character	Species Pair	X	2	3
Sex comb teeth	*mauritiana/ simulans*	0.11	0.39	0.50
Genital size	*mauritiana/ simulans*	0.08	0.39	0.56
Genital size	*sechellia/simulans*	0.30	0.52	0.18

Data from Coyne (1983, 1985a), Coyne and Kreitman (1986); table from Coyne and Orr (1989b).

polygenic trait controlled by genes spread all over the genome. Of particular note is no disproportionate influence of the X chromosome; indeed, in one case (genital size in *mauritiana/simulans*) the X has a disproportionately low contribution given its relative size in the genome. Coyne (1985b) found very similar kinds of patterns for sex-comb differences and testes color. Liu et al. (1996) used molecularly marked regions of the genome to study the genetic control of differences in the posterior lobe of the male genital arch of *D. simulans* and *D. mauritiana*; they found that 9 out of the 15 regions marked had a significant effect.

Pheromones

Pheromones are known to be important in *Drosophila* mating and could influence premating isolation. Relatively little work on this factor has been done in relationship to the speciation process in *Drosophila.* The cuticular hydrocarbon (CHC) profiles have been studied in a number of species of the *melanogaster* group (Jallon and David, 1987; Cobb and Jallon, 1990) and the *virilis* group (Bartelt et al., 1986). One possible manner of identifying CHCs involved in mate recognition is looking for sexual dimorphism. The results are somewhat confusing in that sometimes sexual dimorphism is observed, sometimes not. The degree of CHC similarity and difference was also not related to phylogenetic relationships. One problem with these studies is the great changes in CHC that can come about by laboratory rearing (Toolson and Kuper-Simpron, 1989). Nevertheless, Cobb and Jallon (1990, and references therein) make a good case that pheromones are involved in mate recognition in the *melanogaster* subgroup.

Coyne et al. (1994) provide more direct evidence that pheromonal differences between *D. simulans* and *D. sechellia* may be the direct causal factor in the premating isolation between these two species. These two species display asymmetrical premating isolation: *D. sechellia* males will court and mate with either type of females, while *D. simulans* males do not court *D. sechellia* females. This was postulated to be due to the fact that *D. sechellia* is sexually dimorphic for CHCs, whereas *D. simulans* is not (Cobb and Jallon, 1990). By confining *D. simulans* females with an excess of *D. sechellia* females, the CHC became transferred and the *D. simulans* females acquired a CHC profile more similar to *D. sechellia.* These "perfumed" females were no longer courted by their conspecific males (Coyne et al., 1994). The converse experiment also worked, that is, *D. sechellia* females perfumed by confinement with *D. simulans* females were now courted by *D. simulans* males. Genetic crosses were performed that indicated one chromosome (the third) had a major effect in the crucial CHC profile, although the genetic control could still be highly multigenic.

Another intriguing study of the possible role of pheromones in premating isolation is that of Markow and Toolson (1990). They studied the CHC profiles of the Sonoran Desert *repleta* group flies and found they have relatively high concentrations of longer-chained molecules, as do most insects adapted to dry hot climates. They also found that the Baja and mainland populations of *D. mojavensis* differed significantly in their CHC profiles. Laboratory studies had indicated that mating success in this species was correlated with the ratio of $C_{35:2}/C_{37:2}$ dienes: females with a relatively low ratio were courted more and males with a relatively high ratio were more successful in courtship. Table 7-13 summarizes the results. Mainland males have a higher

TABLE 7-13 Ratio of two cuticular hydrocarbons found in *D. mojavensis* strains from different geographic localities.

Locality	Sex	Mean Ratio $C_{35:2}/C_{37:2}$	S.E.	Range
Baja	F	1.83	0.458	1.38–2.28
	M	2.83	0.509	2.36–3.28
Mainland	F	2.56	0.633	2.02–3.47
	M	4.31	0.801	3.26–5.67

From Markow and Toolson (1990).

ratio, so in general their mating success is high. Mainland females have a higher ratio too, so they would be courted less. This is in agreement with the asymmetrical isolation between these populations noted in table 7-2.

Another sort of mating-related pheromone exists in the *adiastola* subgroup of Hawaiian *Drosophila.* During courtship, males of this group engage in a unique behavior: they raise their abdomens over their heads and emit anal droplets. Tompkins et al. (1993) studied the hydrocarbons in the male's droplets and found them to be species specific and implicated them in mate recognition. They also studied the cuticular hydrocarbons of both males and females and found them to be strongly sexually dimorphic and species specific. The two types of evidence for the involvement of these pheromones in mating behavior are: unlike many Hawaiian flies, these species will court in light conditions in which they cannot see, that is, they do not require visual cues; and "anointing" flies with a solvent of extracts induces courtship, including homosexual courtship when males are anointed with female extracts.

Asymmetry in Premating Isolation

It is often observed that when two closely related taxa exhibit premating isolation, the pattern is not symmetrical, as already noted in some of the examples above. During the 1970s two theories were put forward to explain asymmetry in behavioral isolation in *Drosophila*, both based on the direction of evolution.

Kaneshiro (1976) noted that among Hawaiian *Drosophila* asymmetrical premating isolation was often correlated with direction of evolution, which could be inferred from the age of the Hawaiian Islands. Females from the older islands often showed more discrimination against males from the newer islands than did females from newer islands against males from older islands. He put forward the hypothesis that when a new island is colonized by one or a few individuals (see below on the founder effect and speciation), the derived population will evolve a less discriminating mating behavior. This may be due to a loss of courtship elements by males due to drift (as initially hypothesized by Kaneshiro) or to sexual selection for a simplified mate recognition system (proposed in more recent writings, e.g., Kaneshiro [1989]). Ancestral populations would retain the more elaborate mate recognition system and discriminate against males with the simplified system, whereas females with the simplified requirements would accept equally readily both derived and ancestral males.

At about the same time, Watanabe and Kawanishi (1979) put forth an alternative theory that makes exactly the opposite prediction: ancestral species show less premating isolation from derived species than vice versa. Their arguments were as follows: generally, the newly formed derived species will be smaller than the ancestral species. When sympatry is reestablished, if hybrids formed between the two incipient species are at a selective disadvantage, then one has the equivalent of an underdominant population genetics model that is unstable and would lead to the elimination of the smaller population, in this case the derived one. Thus there would be pressure on the derived population to evolve mate discrimination against the ancestral if they are to survive. This sounds a bit teleological, but another way to look at it is that the only species we see today are those that did not get swamped by the larger ancestral populations and thus were the ones that had evolved premating discrimination.

These models have been discussed extensively and the interested reader is referred to three successive articles in a volume of *Evolutionary Biology*: Ehrman and Wasserman (1987), DeSalle and Templeton (1987), and Kaneshiro and Giddings (1987). Suffice it to say, the evidence is such that using asymmetrical premating isolation to infer the direction of evolution in the absence of any independent corroborating evidence is not wise. The instances where such models may be helpful would seem to be too restrictive to be of general use. Furthermore, we must bear in mind any such asymmetry may be temporary and depend on where the taxa are in the speciation process. For example, the idea of simplifying mate recognition systems in derived populations cannot go on indefinitely; at some point more elaborate systems must get rebuilt; otherwise evolution would have led to virtually no mate recognition system.

Genetics of Postmating Isolation

Given the fact that the phenotypes characterizing postmating isolation, infertility or inviability, are much more clearly defined than premating isolation phenotypes, genetic studies are likewise better defined. While some very serious issues remain, some clear generalizations are emerging from postmating isolation genetics of *Drosophila*. Our discussion will refer to two aspects of postmating isolation, the two best characterized: infertility and inviability. There are two other phenomena that come into play in *Drosophila* that come under the heading postmating. One is the reaction sometimes seen in females inseminated by non-conspecific males; sperm is expelled from the spermetheca. This reaction has not received much attention and little is known of the genetic basis of it, and therefore we will not discuss it further here. A thorough discussion of the insemination reaction is chapter 8 in Patterson and Stone (1952), and a more up-to-date discussion can be found in Alonso-Pimentel et al. (1994). Likewise, F_2, or other post-F_1, breakdown is known in *Drosophila* but has not been well-studied. However, these phenomena highlight the fact that sometimes the simplification of focusing on one or a few factors can be distorting. One must bear in mind reproductive isolation may come about by a combination of many separate phenomena.

Haldane's Rule

In 1922 J. B. S. Haldane pointed out a phenomenon that came to bear his name, namely that in interspecific hybrids if only one sex is sterile or inviable, it is the heterogametic sex. This rule has stood the test of time remarkably well and is one of the few generalizations in all of evolutionary biology. In *Drosophila* this is certainly the case. In Bock's (1984) extensive and highly useful compilation of interspecific hybridizations in *Drosophila*, out of 145 cases where only one sex is sterile or inviable, it is the males in 141 of them. Equally interesting is the pattern observed by Coyne and Orr (1989a) concerning the genetic distances characterized by the different levels of postmating isolation (table 7-14). Recall that the *D* in this table is indicative of time. Isolation up to 0.5 (which is the case when only one sex is sterile/inviable in reciprocal crosses, which means almost always males) occurs relatively rapidly. There is then a stalling of the process at males-only sterile/inviable and only with considerably more time do genetic differences accumulate to cause female hybrids to become sterile/inviable (isolation index of 0.75 or 1.00). In *Drosophila*, then, male infertility or sterility is generally the first form of postmating isolation to evolve.

What is the cause of Haldane's rule? Considerable discussion abounds on this question, and we will return to it at the end of this section after we have discussed what is known about the genetics of infertility/inviability.

Male Sterility

Just as it is generally true that males are the first sex to be affected, it is also a valid generalization that the fertility of hybrid males is affected before viability. Wu and Davis (1993) tabulated from Coyne and Orr's (1989a) compilation that in 19 cases of species hybrids separated by a $D < 0.3$ (earliest times of divergence), the males are sterile, but viable (and females are viable and fertile). It seems a safe generalization that in the first stages of the evolution of postmating isolation, male hybrid fertility is the first factor affected. Thus the genetic basis of male infertility in hybrids is of considerable interest.

Dobzhansky (1936) was the first to undertake a systematic study of the genetic basis of hybrid male sterility in *Drosophila*. Figure 7-6 is an example of his data from *D. pseudoobscura* and *D. persimilis*. Until quite recently, virtually all subsequent studies were some variation on this theme. One point we need emphasize is the manner in which male fertility-sterility is measured. Dobzhansky used two criteria: actual

TABLE 7-14 Evidence for the "stalling" at postmating isolation index 0.5, which represents cases where male hybrids are sterile.

Isolation Index	Mean Genetic Distance ± S.E. (*N*)
0.00	0.138 ± 0.058 (8)
0.25	0.251 ± 0.083 (5)
0.50	0.249 ± 0.032 (16)
0.75	0.722 ± 0.198 (5)
1.00	0.991 ± 0.127 (8)

Note: Nei's genetic distances, meant to indicate relative time, are given. *N* is number of pairs of species.

From Coyne and Orr (1989a).

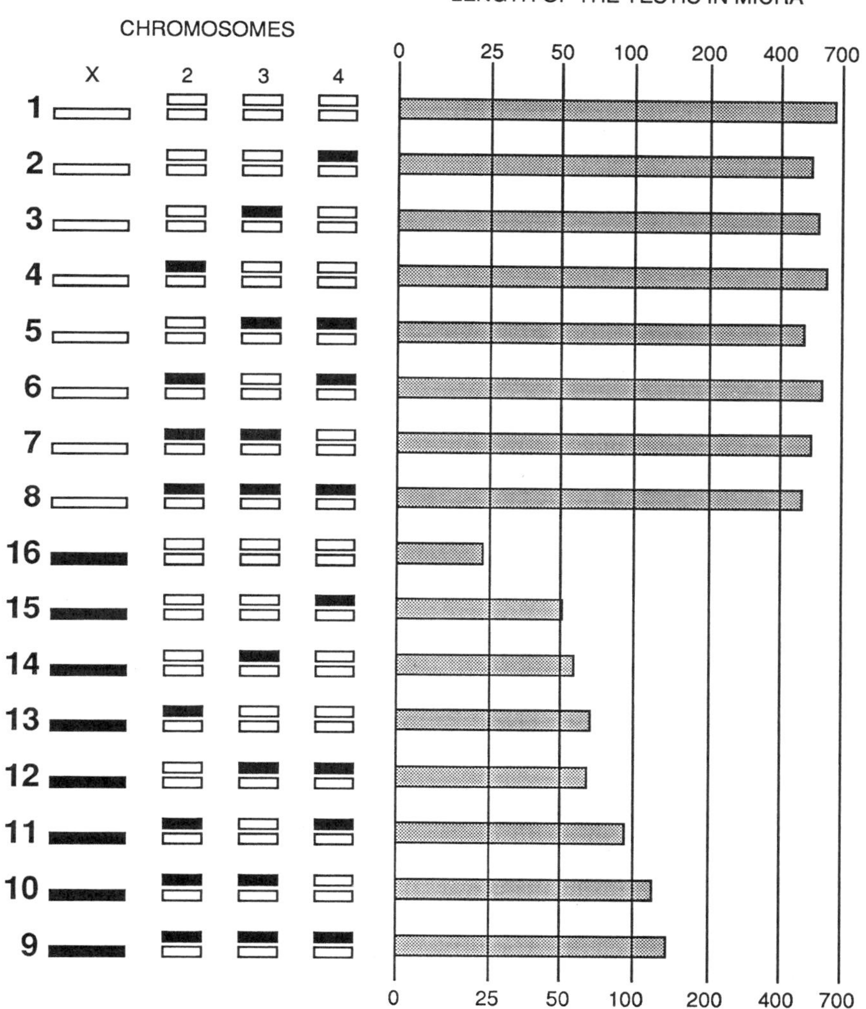

FIGURE 7-6 Results of genetic analysis of male sterility in backcross hybrids between *D. pseudoobscura* and *D. persimilis*. White chromosomes are from *D. pseudoobscura* and the black from *D. persimilis*. The length of the testis is taken as a measure of sterility. From Dobzhansky (1936).

sterility as measured by failure to produce offspring, and size of the testes in the hybrid male, which is clearly correlated with sterility. Other studies use microscopic examination of dissected males to determine the presence or absence of motile sperm; any motile sperm is taken as evidence of fertility. In any case, Dobzhansky's major conclusion was that genes affecting male fertility are on all chromosomes and thus the trait is multigenic. This multigenic nature of species difference was taken more or less as dogma for a considerable time.

In the 1980s this field had a rejuvenation. An example of the same species in the newer studies is in figure 7-7. In these studies as compared to Dobzhansky's (figure 7-6), the motility of sperm is the criterion for fertility. The results are somewhat different, although factors spread around the genome seem to be involved. However, in many of these newer tests, a claim has been made for a major effect of the X chromosome in controlling male hybrid sterility-fertility (Coyne and Orr, 1989b; Coyne, 1992a). The primary evidence is the kind of data in figure 7-7, wherein the presence of the "wrong" X seems to be the dominating factor regardless of the autosomal constitution. In addition to Haldane's rule, Coyne and Orr have dubbed this the "second rule of speciation," namely that male sterility factors are primarily X-linked.

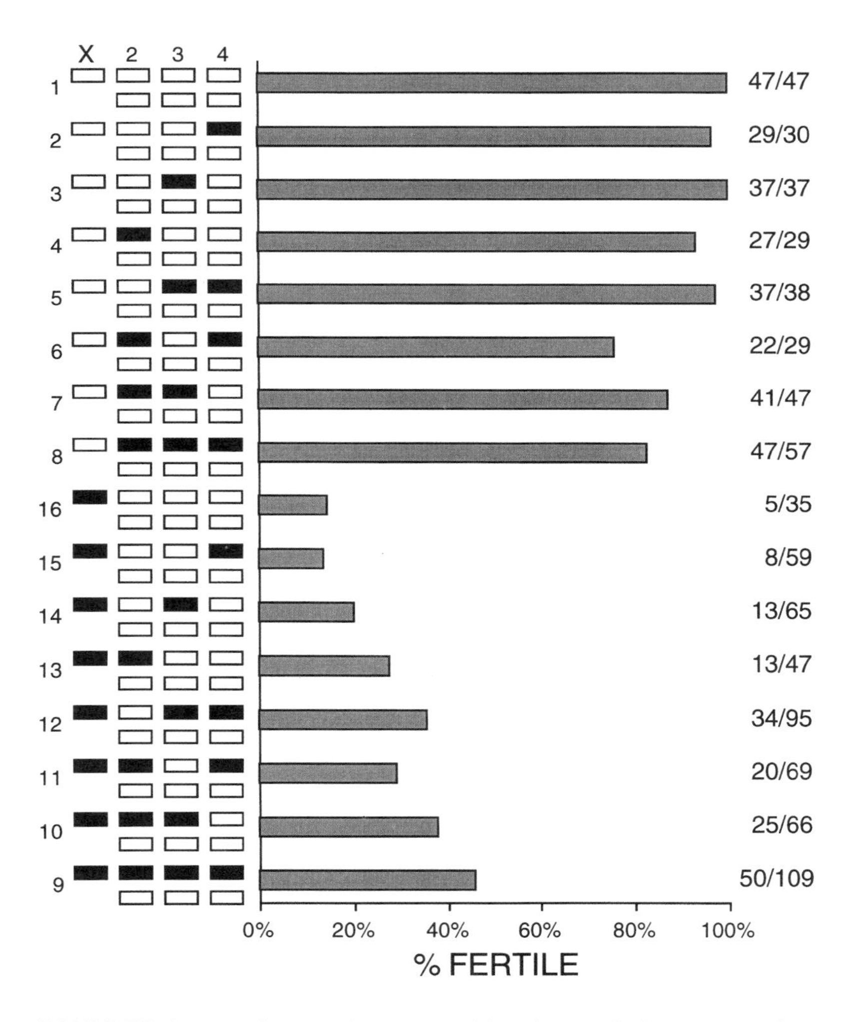

FIGURE 7-7 Same as figure 7-6, except motility of sperm is the measure of male sterility/fertility. From Orr (1987).

Similar evidence exists for the *melanogaster* subgroup (Coyne, 1984; Coyne and Charlesworth, 1986, 1989) and the *virilis* species group (Orr and Coyne, 1989).

However, there is likewise evidence for autosomal genes. Pantazidis and Zouros (1988) and Pantazidis et al. (1993) have clear-cut evidence that an autosomal factor causes male sterility in crosses between *D. mojavensis* and *D. arizonae*. Orr (1992) has evidence that the dot fourth chromosome of *D. simulans* causes male sterility when introgressed into *D. melanogaster*.

Wu and Davis (1993) have criticized the X-chromosome effect on technical grounds. They point out that in studies such as that illustrated in figures 7-6 and 7-7, the autosomes are always tested in the heterozygous state, whereas the hemizygous effect of the X is expressed. If many or most genes affecting male sterility (or indeed any other kind of reproductive isolation) are mostly recessive, then these studies would bias the results to indicate a strong X effect. In fact, the results of Pantazidis and Zouros (1988) show just such an effect; the homozygosity of the autosomal fourth in *D. mojavensis-arizonae* is necessary in order to detect the sterility action of this chromosome. Hennig (1977) presents similar data for *D. hydei-eohydei*.

A second method of studying the genetics of postmating isolation has been by introgression. Naveira and Fontdevila (1986, 1991) have studied the effect of introgression between the *repleta* group species *D. buzzatii* and *D. koepferae* (formerly *D. serido*). They used a technique of tracking introgression by determining asynapsis of chromosome regions in backcross generations. They conclude that sterility factors are all over the genome and act in an additive manner. Any X-chromosome fragment introgressed from *D. koepferae* causes male sterility, whereas for the three autosomes there is a threshold of around 30% of each chromosome. That is, when 30% or more of any of the *D. koepferae* three autosomes is introgressed (in the heterozygous state) into *D. buzzatii*, male sterility results. Unfortunately, this technique of detecting introgression by asynapsis cannot detect homozygous introgression, but presumably the effect of homozygosity of autosomal regions would be even greater.

How many genes cause male sterility? It has been estimated that about 10% of all genes in *D. melanogaster* can mutate to a male-sterile phenotype (Lindsley and Tokuyasu, 1980) and they are found in autosomes and X chromosomes in proportion to their euchromatic lengths (Lindsley and Lifschytz, 1972). If the genome has about 10,000 genes, then about 1,000 can cause male sterility. One mutation in any of about a thousand genes is sufficient to induce male sterility, but the more important question is, what actually happens in nature? In a similar vein, while the evidence is that factors affecting male sterility are spread all around the genome, are there any genes with major effects that can be isolated by genetic techniques? The genetic architecture of reproductive isolation genetics will play an important role in trying to determine the kinds of populational phenomena one might envision in inducing speciation.

There have been attempts, some successful, to isolate genes with major effects on male sterility in interspecific hybrids. Coyne and Charlesworth (1986) have actually mapped such a factor in *D. melanogaster-mauritiana* to a resolution of about 2 centimorgans. Orr (1992) has mapped the major fourth chromosome element to five polytene bands, and it may in fact be a single gene. Pantazidis and Zouros (1988) and Pantazidis et al. (1993) have isolated their autosomal factor to a single Mendelian location in *D. mojavensis-arizonae*. There is evidence that not all male sterility-inducing genes are of the polygenic sort, each with a very small effect, and thus virtually unanalyzable by standard genetic procedures.

C.-I. Wu and colleagues (e.g., Wu et al., 1993; Perez et al., 1993; Wu and Davis, 1993; Johnson and Wu, 1993) have taken a variation on the introgression theme in studies of isolation among *D. simulans-mauritiana-sechellia*. They have taken advantage of molecular markers, primarily RFLP, to detect introgressed parts of the X chromosomes among backcrosses. The beauty of this scheme is that by taking the introgressed lines through many backcross generations (in their case usually 20 or more) the introgressed fragment can be "cleaned" of any other interacting fragments. That is, the whole genome is one species, except for the single molecularly marked introgressed fragment. This also allows an entrée into the genome for molecular isolation. They have shown that at least six different X-chromosome regions can cause male sterility among these species. They have even mapped one they call *Odysseus* to cytological interval 16D, which, when introgressed from *D. mauritiana* to *D. simulans*, causes male sterility, but when the same locus is taken from *D. sechellia* into *D. simulans*, it does not cause sterility. Other mapping using this strategy has indicated many more sterility-causing genes on the X chromosome in these species (Cabot et al., 1994) and these authors call into question the generality of genes with major effects. Palopoli and Wu (1994) estimate there are at least 40 loci on the X chromosome that affect male sterility in crosses between *D. simulans* and *D. mauritiana*. Naveira (1992), using more traditional crossing, also argues for many genes of minor effect causing male sterility in this species group.

The multigenic control of male sterility in *D. simulans-mauritiana* hybrids has been confirmed by Hollocher and Wu (1996) and True et al. (1996). The former study used mutant markers to study the density of sterility-inducing factors. They conclude that there is about one sterility factor per 1% of autosomes (euchromatin) and about two factors per 1% of the X, for a total of about 120 genes. True et al. (1996) marked 87 positions in the *D. mauritiana* genome with P elements and by 15 generations of backcrossing, placed approximately 7% of the *mauritiana* genome in an otherwise homozygous *simulans* background. In the heterozygous state virtually all the introgressed segments were viable and fertile but when homozygous, male sterility was often induced, much more frequently than female sterility or inviability. As with the previous study, a somewhat greater effect for introgression of X chromosomal material was noted.

Thus in regard to fertility of hybrid males, the bulk of the evidence would indicate that many genes, often with substantial epistatic interactions, are the underlying architecture of this form of postmating isolation. Single genes of relatively major effect can be identified and indeed mapped, but these isolated major effect genes are likely to be part of a much larger set of polygenes. It is very difficult to state unequivocally that the major effect genes are not just a function of the manner in which the studies were done, that is, if all the autosomes are heterozygous, major factors on the hemizygous X have their effects magnified.

The Y Chromosome

Since the Y chromosome is known to contain male fertility factors, it is not unreasonable to suppose it has an effect on inducing male sterility in interspecific hybrids. This is sometimes observed, and sometimes not. For example, when the Y chromosome of *D. sechellia* is introgressed into a genetic background consisting solely of *D. simulans* X and autosomes, males are fertile. When the reciprocal is done (*D. simulans* Y in an otherwise pure *D. sechellia* genotype), the males are sterile (Johnson et al., 1992; 1993).

A more controversial issue is whether the Y chromosome, when it does have an effect on hybrid male sterility, interacts with X-linked or autosomal factors. Obviously, by itself the Y does not cause sterility or it would not exist; it is only when a Y chromosome of one species is placed in a cell with the X and/or autosomes of a second species that sterility is induced. Arguments were made that in the case of the triad *D. simulans-mauritiana-sechellia* that it was an X-Y interaction that had a large effect on sterility (Coyne, 1984; 1985a). Likewise for the *D. pseudoobscura-persimilis* pair (Orr, 1987), against Dobzhansky's (1936) original findings, which Orr attributes to the difference in using testes size (Dobzhansky) versus motility of sperm (Orr) as the fertility measure.

On the other hand, there is also good evidence that Y-autosome interactions can also occur. We have already mentioned the case of *D. mojavensis-arizonae*, studied by Pantanzidis and Zouros (1988; Pantanzidis et al., 1993). When a *D. arizonae* Y chromosome is introgressed into an otherwise pure *D. mojavensis* genetic background, male sterility occurs. If a single fourth chromosome (i.e., in the heterozygous state) of *D. arizonae* is also introgressed, fertility is restored.

Finally, the elegant study of Johnson et al. (1992) indicates that some methods may be misleading in indicating X-Y interaction. We already mentioned the strategy of the Wu group in introgressing parts of the *D. sechellia* X chromosome into an otherwise pure *D. simulans* background. Figure 7-8 shows one such construct involving a region near the tip of the X marked by the mutant *y*. The important point in this study is the comparison of genotypes B and C in this figure. If the male sterility factor linked to *yellow* interacted with the X in causing the sterility, genotype B should be sterile and C fertile; in fact C is also sterile. Therefore, the sterility factor close to the *y* mutation must interact with *D. simulans* autosomes in inducing sterility. Interestingly, genotype D in figure 7-8, while also being sterile, has spermatogenesis much closer to normal compared to genotypes B and C. Even one haploid set of autosomes compatible (i.e., conspecific) with the X-linked sterility factor almost rescued the effect. The argument made by Johnson et al. (1992) is that previous studies had only tested the effect of intact X chromosomes. If any (of the presumed many) X-linked sterility factors interacted with the Y, they would be detected and attributed to the X.

Female Sterility

For reasons already mentioned, the genetics of male hybrid sterility have received the bulk of attention by workers in the field. However, it is also of interest to consider the genetic basis of female sterility in cases where this sex is also affected.

While F_1 female hybrids between *D. pseudoobscura* and *D. persimilis* are normally fertile, some backcross female progeny show varying degrees of infertility. Orr (1987) seems to be the first to have made a systematic study of this situation, the results of which are shown in figure 7-9. As with male sterility, Orr argues for a large X-chromosome effect, which is obvious from genotype 16 in this figure: in an other-

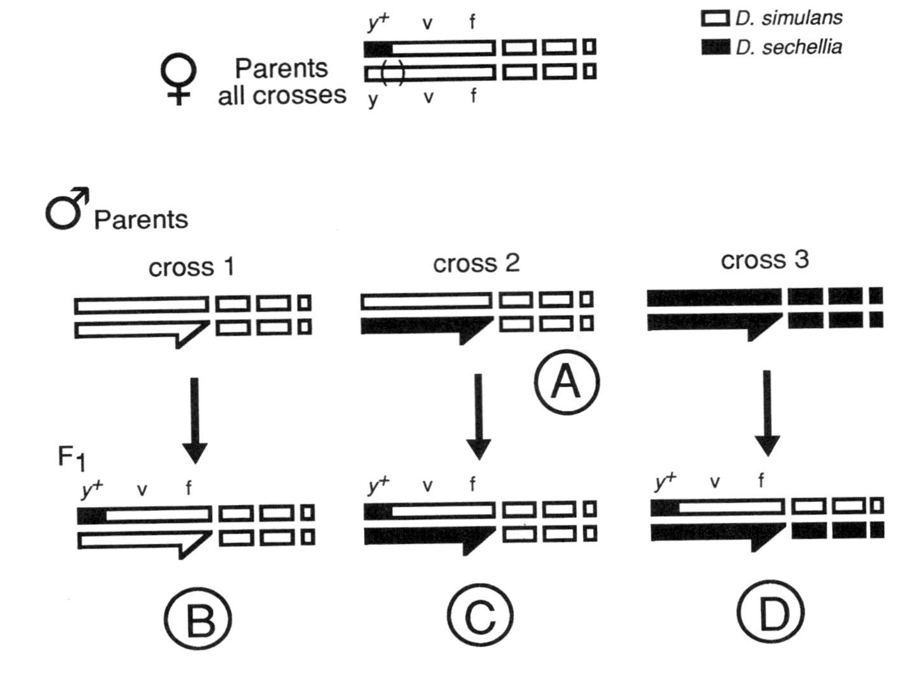

FIGURE 7-8 Interspecific hybrids between *D. simulans* and *D. sechellia*. The letters on the chromosome are morphological mutants and the parentheses indicate an inversion. The female in all crosses is at the top. The sex chromosomes are drawn out of proportion to the autosomes. Male A is fertile, while males B, C, and D are sterile. From Johnson et al. (1993).

wise *D. pseudoobscura* autosomal genotype, the heterozygosity for the X was sufficient to cause almost complete female sterility. However, clearly there are interactions with the autosomes, as semifertility is restored when one set of autosomes of *D. persimilis* is added (genotype 9).

A stronger case for the importance of X-linked factors controlling female hybrid sterility-fertility comes from the *virilis* group study of Orr and Coyne (1989). There, in backcrosses among all seven species of the group tested, backcross females with conspecific X's were inevitably fertile, while those heterozygous were variably sterile. Furthermore, in this study there was evidence that the genes conferring female sterility were different from those conferring male sterility.

Davis et al. (1994) studied the genetic basis of female sterility in the *D. simulans-mauritiana-sechellia* triad. Using a strain of *D. simulans* with a compound X chromosome, they could construct backcross female progeny that were homozygous for the *simulans* X, but carrying various combinations of *D. sechellia* or *D. mauritiana* autosomes. They found autosomal effects, including strong evidence for epistatic interactions.

Johnson and Wu (1993) studied the effects of two of the six X-chromosome male sterility regions identified between *D. simulans-mauritiana* on the fertility of females.

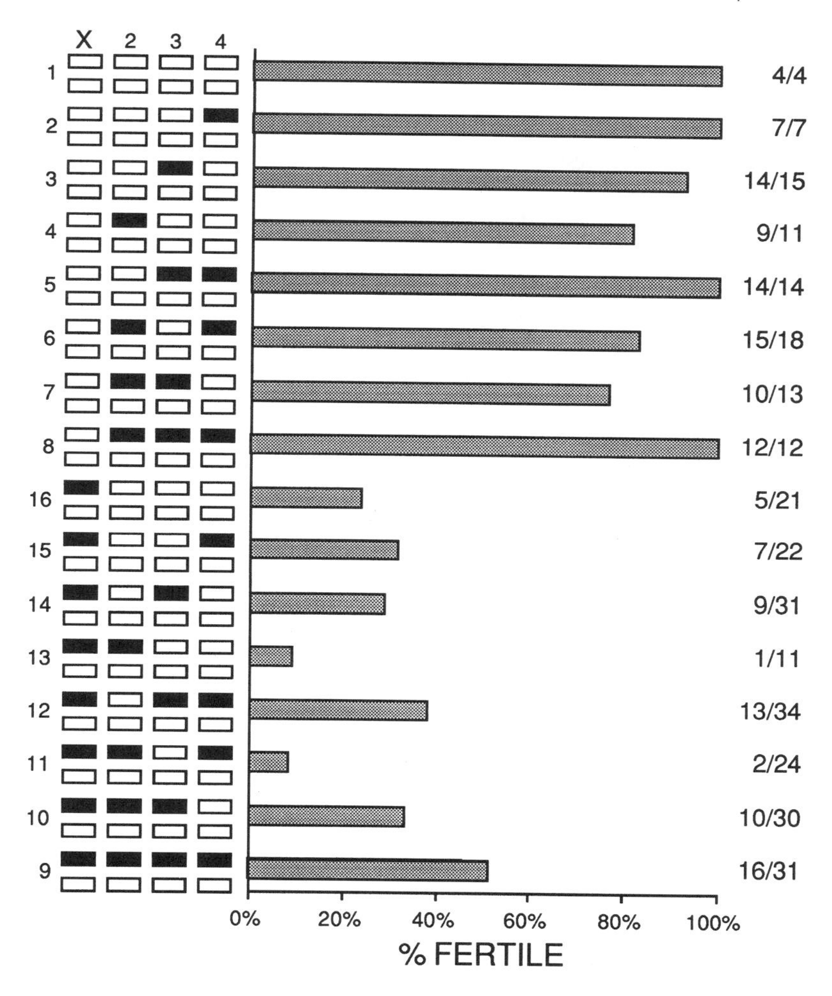

FIGURE 7-9 Similar to figures 7-6 and 7-7, except backcross females are analyzed. Numbers on the right are the number of fertile females over total inseminated; length of bar is proportional to this fraction. From Orr (1987).

They showed that each by itself had a significant effect in lowering fertility in females heterozygous for these factors. Other regions of the X that did not have an effect on male fertility did not affect female fertility. They conclude that the same genes (or genes tightly linked) that can cause sterility in males also affect female fertility.

Thus as with male sterility-fertility studies, different species groups and/or different experimental approaches seem to yield conflicting results. The large X chromosome effect has been seen, but autosomal effects are also often observed. The genes affecting male and female sterility may be different in some cases, but may be the same in other cases. Generalizations are hard to come by.

Inviability Genetics

Because hybrid male sterility, as a rule, precedes inviability of hybrids, this latter subject has received less attention, as most workers are preoccupied with the first steps in speciation. Nevertheless, some work has been done on the genetic basis of inviability.

In crosses between *D. melanogaster* females and *D. simulans* males, F_1 females are viable, but males are not. In the reciprocal cross the F_1 males are viable and the females are not. Pontecorvo (1943) carried out studies of the genetics of this phenomenon, using a triploid strain and irradiation of sperm to produce aneuploids. He could demonstrate and map at least nine factors involved in the inviability. Two caused inviability when introgressed from *D. melanogaster* to *D. simulans*, and two caused inviability in the reciprocal direction. The five remaining factors suppressed the inviability effects when introgressed in either direction. Thus the inviability genetics using this technique appeared to consist of relatively few factors (genes or gene blocks), certainly simpler than the sterility genetics.

The strongest evidence that the genetics of inviability may have a simpler basis, that is, fewer loci with large individual effects, comes from studies of hybrid rescue mutants. Watanabe (1979) found a single gene mutation in *D. simulans* that rescued the inviability of hybrids when crossed to *D. melanogaster*. In a screen of several *D. melanogaster* strains, Hutter and Ashburner (1987) found a similar mutation in this species. This mutant, called *Hybrid male rescue* (*Hmr*), remarkably enough, also rescues the inviable male hybrids normally formed between crosses of *D. melanogaster* and *D. sechellia* or *D. mauritiana*. Further work indicated its map location and recessive action (Hutter et al., 1990). Hutter et al. (1990) found another hybrid rescue gene and Sawamura et al. (1993a, b) have found two others in *D. simulans*. Similar variation among strains in their ability to produce viable interspecific hybrids exists for *D. mulleri-aldrichi* (Crow, 1942) and the *virilis* group (Patterson and Griffen, 1944; Mitrofanov and Sidorova, 1981).

The important point to be emphasized is that single loci have major effects on inviability and that the number of such factors seems small. Secondly, *in no cases of rescue of viability is the rescued sex fertile.* In other words, the genetic bases of fertility and viability would seem to be independent.

Meiotic Drive

Hurst and Pomiankowski (1991) and Frank (1991) independently proposed that meiotic drive could account for the evolution of hybrid sterility, in particular due to genes on the X chromosome. The argument is that X chromosomes are particularly likely to become meiotic drivers causing sex-ratio distortion (see chapter 4). Species will evolve suppressers of the driver, often on the Y. Thus, within each species the set of X-driver genes and their Y suppressers will be in balance. When a hybrid is formed there will be a lack of the proper suppression in hybrid males and the driven X will cause inviability or sterility in males.

In support of this hypothesis, Hurst (1992) sites the *Stellate* (*Ste*) locus in the *melanogaster* subgroup. This gene is X linked and can cause disruption of normal spermatogenesis unless it is suppressed by a Y-linked *Suppresser of Stellate*, *Su*(*Ste*).

In X/O males lacking the suppresser, the *Ste* gene produces crystals in the sperm and renders them nonfunctional, causing sterility. *Ste* sequences are found in *D. melanogaster, D. simulans*, and *D. mauritiana*, but not in *D. erecta, D. teissieri*, or *D. yakuba*. The lack of *Ste* in the last three species is taken to be evidence that the product of *Ste* is not required for normal spermatogenesis. Hurst poses the conundrum: "How could a gene which is possibly not required for spermatogenesis, and which can render the host sterile unless suppressed ever have evolved?" He contends this is best understood if *Ste* is a driver and *Su*(*Ste*) the suppresser; thus, the system conforms to Haldane's rule.

Another system conforming to the prediction of the Hurst/Pomiankowski/Frank theory is a sex-ratio inversion in *D. subobscura*. Hauschteck-Jungen (1990) has found an X-chromosome gene arrangement (designated $A_{2+3+5+7}$) in a Tunisian population; males from Tunisia carrying this chromosome produce 90%+ females when mated to laboratory strains from Europe. When the F_1 females are backcrossed to the European strain, the resulting males are sterile. The backcross sterile males could be rendered fertile by replacing the European autosomes with Tunisian autosomes. In other words, the Tunisian autosomes were coadapted to the $A_{2+3+5+7}$-driven X, whereas autosomes from elsewhere were not. This intraspecific sterility appears then to be due to an imbalance between a meiotic drive chromosome and coevolved suppressers of its effects.

Another example is in *D. mediopunctata*, where both autosomal and Y-linked suppressers of a sex-ratio X chromosome have been found (Carvalho and Klaczko (1993; 1994). The final example is in *D. simulans* (Mercot et al., 1995). Two strains, one from New Caledonia and one from the Seychelles, each produced sex ratio when outcrossed to other populations, but their native populations had suppressors of the driven X chromosome; the study was not sufficiently detailed to determine whether the suppressors were autosomal or Y linked.

The theory that meiotic drive can account for Haldane's rule has not been without its critiques. Coyne et al. (1991a) raise six specific problems with the theory. The most devastating objection is that there is little evidence for any meiotic drive in F_1 interspecific hybrids (e.g., Coyne, 1986). In hybridizations where F_1 males are partially fertile, the offspring produced show a 1 : 1 sex ratio (Patterson et al., 1942; Patterson and Stone, 1949). While it is true that there are examples of X chromosome meiotic drive with suppressors, their role in speciation remains in doubt until further evidence is accumulated.

Haldane's Rule Revisited

Now that we have discussed some of the data pertaining to the genetic basis of postmating isolation, we return to the issue that opened this section, namely Haldane's rule. Have the genetic studies shed light on this nearly universal phenomenon such that we can provide an explanation for it? The question can be broken into several subissues.

First is the question of whether the phenomenon that comes under the heading of Haldane's rule is a unitary one. Is hybrid male sterility/inviability a single thing with a single genetic explanation? Wu and Davis (1993) have made quite compelling arguments that this is not the case. In particular, they present evidence that infertility of males and inviability are genetically two different phenomena and that different explanations are needed. Part of their argument is illustrated in figure 7-10. Two related explanations for the sterility/inviability of the heterogametic sex were put forward by H. J. Muller. One is that it is due to dominance: X-linked recessive genes will have their effects expressed in male hybrids but not females. Second, the matching or balance of X chromosomes to sets of autosomes may be important; in hybrid males only one haploid set of autosomes is balanced by the X of the same species, while in hybrid females, both sets of autosomes are balanced by an X. These hypotheses have been dismissed at least for sterility in the case of *D. simulans-mauritiana* by Coyne (1985b), the relevant observation being cross 1 in figure 7-10. By using attached X's, Coyne generated females with two X chromosomes from the same species, so recessive effects should have been expressed and such females are fertile (column 3, cross 1). Both males and attached-X females have similarly imbalanced sets of chromosomes and only males are sterile. On the other hand, making the balance of X: autosomes the same in males and females in hybrids between *D. melanogaster* and

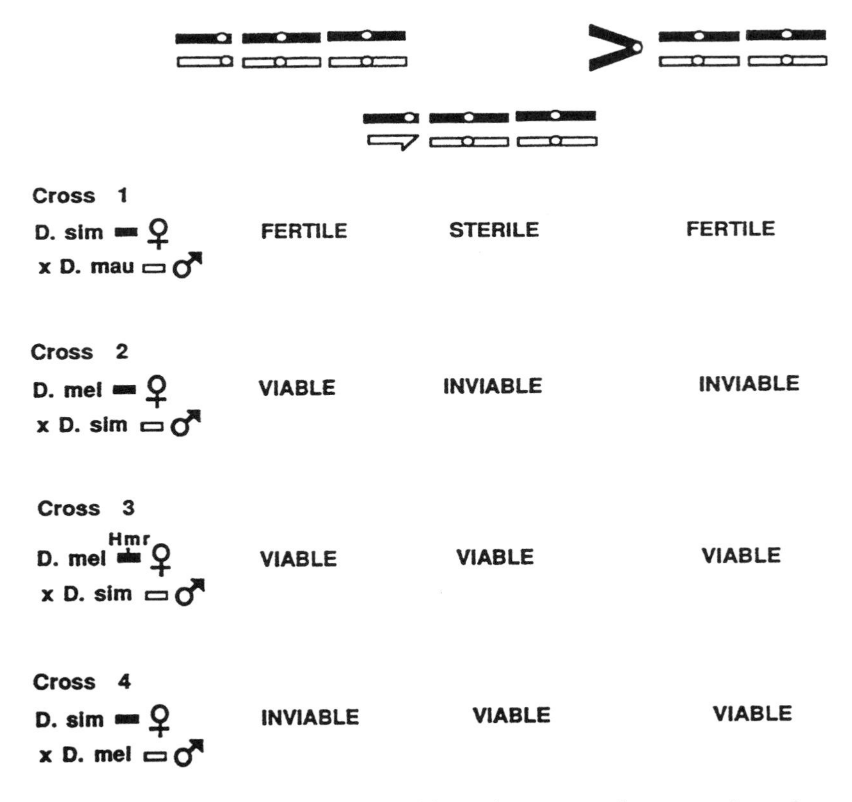

FIGURE 7-10 Hybrids among species of the *melanogaster* subgroup: mel = *melanogaster*; sim = *simulans*; mau = *mauritiana*. The crosses are on the left and the resulting genotypes on the top. In the first two columns, the female parent had two free Xs and in the third column she had an attached X. In cross 3, the female X chromosome carries the hybrid rescue mutant, *Hmr*. From Wu and Davis (1993).

D. simulans produces the same pattern in the sexes with regard to viability, crosses 2 and 4 in figure 7-10. Finally in this figure, cross 3 illustrates that the *Hmr* mutant of Hutter et al. (1990) rescues both sexes when the X : autosome balance is the same. It would seem that Muller's balance theory holds for viability but not fertility and thus the underlying genetic mechanisms must be different. Turelli and Orr (1995) provide a somewhat modified dominance theory which may overcome some of the objections just presented.

A second argument against inviability and infertility being caused by the same kinds of genetic changes was put forward by Wu and Davis (1993): the numbers of genes that are usually detected causing inviability and infertility in hybrids are very different. As we have seen in the above discussions, genes affecting infertility in hybrid males are usually very many and spread all around the genome, while the number affecting inviability are much fewer with major effects. Thus the overall architecture of the two genetic systems is different. The strength of Haldane's rule with regard to these two phenomena is also different. Wu and Davis (1993) show from Bock's (1984) compilation that when only one sex in the hybrids is sterile and one fertile, it is males sterile 199 times to 3 times for females. On the other hand, in regard to inviability, the ratio is 14 : 9; male inviability is only marginally more common in hybrids than is female inviability in cases where only one sex is inviable. Thus genes affecting viability act more or less equally in the sexes, whereas those affecting fertility act primarily in males.

A second major question that needs to be addressed is whether the sex chromosomes have a disproportionate effect on hybrid sterility/infertility, the so-called second rule of speciation genetics (Coyne and Orr, 1989b). This is a more difficult issue to resolve. As pointed out, the data are somewhat conflicting on this issue. Clearly the X chromosome often has major effects, but autosomes can also play a large role, as illustrated by the *D. mojavensis-arizonae* work. The solely technical difficulty of being able to easily assay the homozygous effects of autosomal regions versus the ease of testing the hemizygous effect of the X may be biasing our perception of the relative importance of the sex chromosomes versus autosomes. In the two studies that did make regions homozygous in backcrossed hybrids, one study (Hollocher and Wu, 1996) found that X-linked sterility factors were about twice as dense as autosomal factors; the other (True et al., 1996) found about 50% higher probability that X chromosome fragments produced male sterility. On the other hand, if the autosomal effects are only expressed in the homozygous state, and if backcrosses are never formed in natural populations, these genes are irrelevant, as they will never be homozygous in nature. It is not clear, especially in the earliest stages of speciation, whether such backcrosses are commonly formed.

Assuming that the X chromosome does play a disproportionate role in the evolution of postmating isolation (a tentative conclusion at this time), why might this be so? Charlesworth et al. (1987; with an elaboration by Coyne and Orr, 1989b) have put forth the theory that the X evolves faster with respect to selectively advantageous mutants, especially those with at least partial recessivity. This is simply due to the fact that X-linked recessive genes get exposed to selection because of male hemizygosity. So that in general, if the genes that affect hybrid male fertility/inviability evolved within each species because of some favorable effects and, if they are at least partially recessive, this could explain the disproportionate effect of the X, and as such it could also account for Haldane's rule. At least in regard to the viability component of isolation, this theory has several problems (e.g., Wu and Davis, 1993). First, there is much less evidence of a large X effect for this component of isolation. Second, this theory also requires that the genes have a sex-specific expression, a phenomenon not usually seen with mutations that affect viability. Third, as noted before, for viability the X : autosome ratio explanation seems to work, so there is no need to postulate a faster evolution of X-linked genes.

How about the sterility component of isolation: can the faster evolution of male sterility genes on the X be supported? Two objections can be made. First, the theory postulates that there are genes that are favorable in males in one species but cause male sterility in hybrids. It is difficult to imagine such genes are very common. Second, the explanation holds for all favorable recessive X-linked genes, so one might expect the X to evolve more rapidly in general, and not just with regard to male sterility-inducing genes. There is no evidence for this, and in fact Charlesworth et al. (1987) have made arguments that for other traits the X does not play a disproportionate role (see table 7-12). This led them to postulate that male sterility genes are a special class of genes unlike those affecting morphology and so forth. Finally, there is one other testable prediction of the faster evolution of the X theory. If, as seems to be the case, the male sterility genes are spread fairly widely and evenly along the X, then their faster evolution should affect neutral (or nearly neutral) linked DNA variation. Thus one might expect DNA sequences in the X to be more diverged between species than autosomal sequences; this is not the case (E. N. Moriyama and J. R. Powell, unpublished data).

We have already mentioned the meiotic drive theory to explain Haldane's rule and need not reiterate it here. There is one other theory based solely on mechanical developmental phenomena. Jablonka and Lamb (1991) have postulated that sterility in the heterogametic sex can be explained by the conformational changes that take place in the chromosomes during gametogenesis. Such changes seem to be more pronounced in sex chromosomes, so in the heterogametic sex, divergence between the sex chromosomes may often result in abnormal condensation-decondensation, pairing, and so forth, leading to sterility or inviability. This explanation suffers in two regards. First, most male sterility in hybrids occurs postmeiotically. That is, meiosis seems normal, but the next stages of spermatogenesis, the actual formation of a functional sperm, is abnormal. Second, this explanation would place all the effect on the sex chromosomes and is difficult to reconcile with the evidence of the involvement of autosomal genes in hybrid male sterility. The ubiquity of interactions of different chromosomes would seem to be at odds with this explanation.

Finally, Wu and Davis (1993) have suggested that the faster evolution of genes causing male sterility may be due to sexual selection. Perhaps sperm competition is driving the divergence. The results of Thomas and Singh (1992) and Civetta and Singh (1995) showing that testes proteins tend to be more diverged than proteins from other tissues may be relevant. However, at this point, the role of sexual selection in generating Haldane's rule must remain speculative. It would seem not a very promising hypothesis in cases where the female is the heterogametic sex.

My own judgment is that we have yet to reach a satisfactory explanation for Haldane's rule. Some of the theories put forward are promising, but all still seem to lack the sense of satisfactory finality. It may well be there is no single overriding explanation, although given the ubiquity of Haldane's rule, one would think there is some unifying underlying mechanism. Studies of the genetics and mechanisms of spermatogenesis would seem to be an avenue of research that might lend more insights into why this process seems particularly sensitive to disruption when populations begin to genetically diverge. Another avenue would be to heed the geneticist's dictum, value the exceptional. Examination of those very few cases in *Drosophila* that do not conform to Haldane's rule might lead to insights about why the majority do.

Symbionts

In chapter 5 we discussed the many symbionts and parasites known to associate with *Drosophila*. In the present context, the most interesting ones are those that can cause reproductive isolation, at least two cases of which are known in these flies. This type of reproductive isolation is very different from those discussed previously, as it may involve no genetic changes in the flies themselves, what I have called a "non-genetic mechanism of speciation" (Powell, 1982).

D. paulistorum

The first example is from *D. paulistorum*. L. Ehrman, in a long series of experiments over 30 years, has documented the role of a bacterial symbiont as the causative agent in hybrid male sterility between semispecies of *D. paulistorum* (reviewed in Ehrman et al., 1986). Each of the semispecies is infected with a cell-wall deficient L-form strain of bacteria, probably *Streptococcus faecalis.* In its native host semispecies, the bacteria is benign and is transmitted cytoplasmically through the egg. However, as the testes of hybrids between semispecies develop, the bacteria grow out of control, taking over the tissue and virtually destroying the testes. The evidence for the direct involvement of the bacteria comes from a number of sources. First, antibiotic treatment can partly alleviate F_1 male sterility (Kernaghan and Ehrman, 1970), although apparently the bacteria is required for survival of the flies as no true bacteria-free strain can be produced; prolonged treatment with the antibiotic leads to the extinction of the strain. Second, the sterility agent can be injected (Williamson et al., 1971). Taking the homogenate of one semispecies and injecting it into females of a second semispecies will induce sterility in sons of these females even when mated to males of their own strain; table 7-15 shows some of the data. Note in this table that it is the foreign cytoplasm that causes the sterility when hybrid material is injected. Finally, when cultured isolates of a foreign semispecies are injected into females, male sterility is induced (Ehrman et al., 1986).

We already mentioned the curious observation that during laboratory culturing of *D. paulistorum* semispecies, the patterns of intersterility may change. This was first observed by Dobzhansky and Pavlovsky (1967, 1971) when a strain that had initially been characterized as the Orinocan semispecies became intersterile with the Orinocan tester strain a few years later. That this is a repeatable observation was shown by Dobzhansky and Pavlovsky (1975) when 6 out of 10 strains collected from the same region also exhibited changes during laboratory culture. Interestingly, this time the change was the opposite: strains originally interfertile with Interior then changed to being interfertile with Orinocan. These otherwise mysterious changes are explicable

TABLE 7-15 Effects of injection of homogenates into females of *D. paulistorum.*

Recipient Strain	Material Injected	Sterility of Males (%)
M	Fertile M	3.3
M	Fertile S	80.0
M	Sterile F_1 (M × S)	6.7
M	Sterile F_1 (S × M)	60.0
S	Fertile S	6.7
S	Fertile M	53.3
S	Sterile F_1 (S × M)	6.7
S	Sterile F_1 (M × S)	60.0

Note: The strains used are M (Mesitas = Andean-Brazilian semispecies) and S (Santa Marta = Transitional semispecies). In all cases the recipient female was mated to a male of her own strain and the sterility/fertility of her sons monitored.

From Williamson et al. (1971).

in light of the microorganismal basis of the sterility between semispecies. Presumably some changes take place in the microbial flora harbored by the flies during laboratory culturing. Strains may lose bacteria and/or become infected with other bacteria, which then changes their mating compatibilities.

The above experiments leave little doubt that bacteria are causing hybrid male sterility in *D. paulistorum.* However, the question of whether the bacteria are the cause of the incipient speciation in this group is harder to answer. In Ehrman's studies of the genetics of reproductive isolation among these semispecies (e.g., Ehrman, 1960), she showed that the male sterility was due to complex interaction of chromosomal factors and the cytoplasm, this latter finding being what led her to pursue the symbiont hypothesis. But was the divergence already underway when the infection occurred and would incipient speciation have proceeded in the absence of the bacteria? One piece of evidence is the fact that the very closely related species *D. pavlovskiana* does not have the bacteria (Ehrman et al., 1989) nor is there any evidence it exists in any other members of the *willistoni* group. Thus the most parsimonious explanation would be that the ancestral *D. paulistorum* lineage became infected. As the semispecies diverged, presumably in allopatry in the rainforest refugia during cool periods, the bacteria in each isolated population coevolved with the flies. Each strain of bacteria has remained benign in its native host. It is only in hybrids that its growth becomes unchecked, especially in testes. It becomes a bit of a semantic problem to say whether bacteria or genetic divergence of the hosts was the "cause" of the evolution of the partial reproductive isolation. As a final comment on this case, it is curious that it is once again the heterogametic sex that is affected, although clearly the cause of the concordance with Haldane's rule in this case must be very different from those cases discussed in the previous section.

Wolbachia

The second case of an endosymbiont producing something approaching incipient speciation involves *D. simulans* and the endosymbiont *Wolbachia pipientis.* This association was first detected in California by Hoffmann et al. (1986). Males from a strain with the symbiont, when crossed to a strain without it, show a large reduction in fertility, that is, few viable eggs are formed. While the sperm do not directly carry the bacteria, presumably by developing in an environment with the bacteria they pick up some factor that makes them unable to fertilize eggs not harboring it. The intriguing aspect of this infection is the rapidity with which it is spreading. Originally it was found only in the Los Angeles basin, but in 6 years of monitoring, it has spread some 600 km north (Turelli and Hoffmann, 1991). Some populations that were previously uninfected were seen to become 80% infected in this short time. The spread of the bacteria is due to the fact that it confers an advantage on its host. Females that have the bacteria produce offspring from matings with both infected and uninfected males, while uninfected females cannot produce offspring with infected males. Thus it pays to be infected. Turelli and Hoffmann (1991) present a model compatible with this rapid spread. MtDNA is another maternally inherited factor and Turelli et al. (1992) also monitored variation in mtDNA in the California populations. They showed that all infected individuals had the same mtDNA, whereas uninfected individuals were polymorphic. Thus the spread seems to be due to a single type of cytoplasm; the results are consistent with the hypothesis that the infection began in a single individual, a *D. simulans* Eve.

Bidirectional intersterility is also sometimes seen, what Montchamp-Moreau et al. (1991) have called S strains. This was first found by O'Neill and Karr (1990) for a *D. simulans* strain from Hawaii. S strains are bidirectionally incompatible with R strains, but unidirectionally incompatible with W strains. While no morphological distinctions can be observed, evidently the *Wolbachia* symbionts in W and S strains are different. Curiously, S strains are confined to islands and associated with a particular mtDNA haplotype, the SiI type (Montchamp-Moreau et al., 1991). R and W strains are found on both mainland and islands and are associated with the worldwide mitochondrial type SiII.

While for the most part, maternal transmission of the *Wolbachia* is thought to be the primary mode, there is evidence for some paternal transmission in *D. simulans* (Nigro and Prout, 1990).

D. melanogaster in Australia have *Wolbachia* infections and populations are polymorphic for it (Hoffmann et al., 1994). However, in this species the reduction in fertility is much less than in *D. simulans*.

Could *Wolbachia* infection, known to induce incompatible crosses in a number of insect orders (Stevens and Wade, 1990) be a cause for speciation? This is not at all clear. However, if it can, then this is truly a case of nongenetic speciation. There need not be any genetic change in the host, since simply being infected or uninfected confers the speciation status. Such a scenario is still speculative in the absence of more data and relevant theoretical models.

Laboratory Studies

Generally, the process of speciation is thought to require a time frame that makes it unamenable to laboratory study, that is, it takes too long to evolve reproductive isolation. All of the above studies were done with species, semispecies, incipient species, or whatever, which had evolved to their state in natural conditions. Cases exist where laboratory manipulation of *Drosophila* has resulted in the beginnings of reproductive isolation. We already mentioned cases where direct artificial selection for increased premating isolation were carried out. The cases we now discuss fall into two types: reproductive isolation as a byproduct of adaptation to different environments, and attempts to induce reproductive isolation by founder effects.

Isolation as a Byproduct of Adaptation

A number of cases exist where replicate populations were maintained in different environments or were selected for different behaviors and then tested positive for premating isolation; these are listed in table 7-9, part A. One case was that of Dodd (1989) whose experiments were also discussed in chapter 4 (table 4-16) in regard to habitat choice. Her populations were four replicates that had been maintained on medium where the only carbohydrate source was starch and four on medium in which the only carbohydrate source was maltose. The populations all began from the same base population and after two years they had diverged with respect to their fitnesses on the two different media as well as in larval behavior with respect to the media. Presumably as a byproduct of the selection for adaptation to the different media, they also evolved some premating isolation. Table 7-16 illustrates the results. The most remarkable aspect of this study is that not only had mating preferences diverged, but they did so in a nonrandom manner. Starch-adapted and maltose-adapted flies showed isolation between populations, but no isolation was detected between populations adapted to the same medium. Evidently, the four replicate populations on each medium underwent similar genetic changes. The genes for mating preferences must have been linked to the genes for adaptation to different carbohydrate sources.

The other studies listed in table 7-9, part A involve selection for a number of traits, such as adaptation to temperature, phototaxis, geotaxis, oviposition behavior, and so forth. Altogether 10 out of 14 studies gave at least some sign of premating isolation as a byproduct of adaptation to an environmental variable. That such changes in mating preferences do not always take place when *Drosophila* populations are subjected to selection is illustrated by the long-term study of Ehrman et al. (1991). They selected replicate populations for adaptation to salts for more than 10 years. No

TABLE 7-16 Summary of mating tests on *D. pseudoobscura* strains adapted to starch (St) or maltose (Ma).

Population		Matings				
A	B	A × A	B × B	A × B	B × A	*I*
St	Ma	290	312	149	153	+0.33
St	St	87	84	70	85	+0.05
Ma	Ma	77	85	81	84	−0.01

Note: There were four populations of each type and the figures given are the sum of all pairwise mating tests. The isolation index is shown, but no standard error or chi-squared as these are composite data. Individual tests and each pairwise test are in Dodd (1989).

indication of premating isolation could be detected. Doubtless there are many more examples of such negative results that have never been published; I have some in my notebooks. The positive results are almost certainly a biased sample of all such experiments attempted. However, this does not detract from the fact that premating isolation can be observed to evolve in the laboratory when populations are subjected to selection for something else. To my knowledge, with the exception of the suspected endosymbiont cases, no experiments have ever detected any evolution of postmating isolation evolving in the laboratory.

All of the studies just discussed were carried out using replicate populations kept completely separate, that is, no gene flow was allowed between the allopatric populations adapting to different environments. Studies have been done in which gene flow is still allowed to exist (table 7-9, parts D and E). One of the more famous cases was that of Thoday and Gibson (1962), who carried out a disruptive selection program for bristle number in a population of *D. melanogaster*. In each generation they destroyed individuals with intermediate numbers of bristles and kept equal numbers of flies with extremely low and high numbers of bristles and mixed them in the same cage to start the next generation. After 30 to 40 generations, a bimodal distribution of bristle numbers was found in what had been a unimodal distribution. But most surprisingly, they also found almost complete premating isolation between the two types! Quite naturally, this led to a number of attempts to replicate this observation, but as indicated in table 7-9, 19 out of 19 attempts failed. The Thoday and Gibson (1962) result remains one of the mysteries in the *Drosophila* literature.

Rice and Salt (1988, 1990) performed experiments more or less intermediate between true allopatry and sympatry. They constructed a complex maze in which flies could choose different habitats that varied by light level, odor, geotaxis (whether flies had to move up or down to get to the habitat), and temporally by collecting emerging flies at different times (i.e., selection on developmental time). Flies from the two most extreme habitats were chosen each generation to begin the next generation. After about 30 generations, virtually complete premating isolation had evolved because flies returned to the habitat in which their parents were found. The almost complete habitat selection meant the differently-adapted flies never met to mate. When placed together in mating vials, however, the two groups of flies displayed no assortative mating. Evidently no changes had occurred with respect to the mate recognition system.

Founder Flush in the Laboratory

The second set of experiments relate to a theory of speciation that has come to be called the founder-flush theory, developed by H. Carson, specifically in regard to Hawaiian *Drosophila* (e.g., Carson, 1971b; 1975). Carson postulated that populations that go through small bottlenecks and are then allowed to increase rapidly (the flush) are particularly prone to evolving reproductive isolation and hence to speciate. The genetic divergence is caused first, by the founder event, which increases homozygosity (including for inversions) so recombination would be enhanced; and, second, during the time of rapid growth in population size, selection is relaxed so that novel genotypes (produced mostly through recombination) that might never have survived in a dense population subjected to strong selection, now have a chance. Once the new population reaches carrying capacity, selection is increased and it is not unlikely that the set of novel genotypes on which it acts will come to a different adaptive peak than that occupied by the ancestral population. This clearly is related to the asymmetrical isolation theories mentioned previously.

Carson (1971b) postulated that "if single founders are involved in some way in the formation of species, it should be possible through laboratory manipulations to bring the process under close observational and experimental scrutiny." In 1972 I began such laboratory manipulations on populations of *D. pseudoobscura* (Powell, 1978; 1989; Dodd and Powell, 1985). The course of the experiment is outlined in figure 7-11. The populations I worked with were begun by crossing flies from four

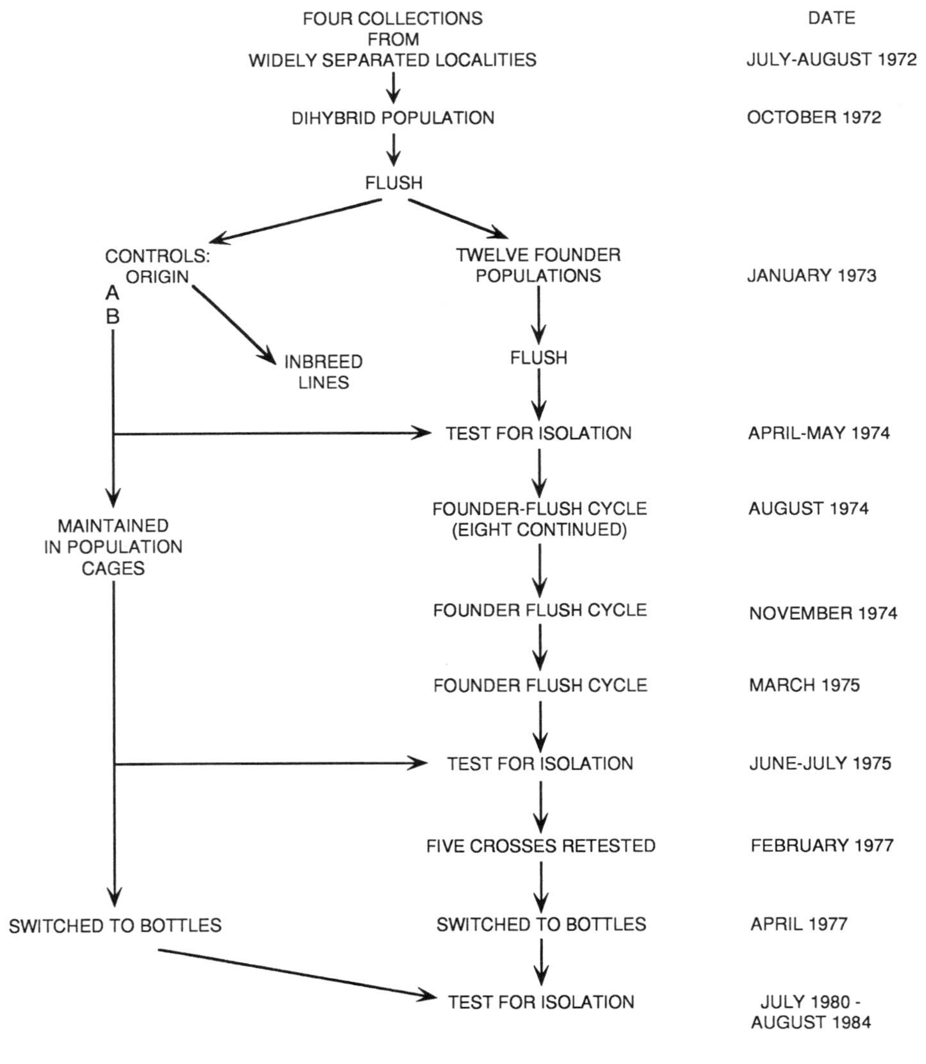

FIGURE 7-11 Schematic representation of laboratory experiments on the founder-flush speciation theory using *D. pseudoobscura*. From Powell (1989).

different natural populations. The purpose of using dihybrid populations was to maximize the probability of destabilizing coadapted complexes. Carson had postulated that the founding of a new population may occur when the ancestral population has recently gone through a populational flush, which induces increased migration out of the ancestral territory. Based on the experiments indicating that immigrants can cause a population to expand rapidly (Carson, 1961b, discussed in chapter 4), this could be one cause of increased emigration. Thus, using as the basis for my studies a hybrid population was meant to mimic to some degree a population that had recently undergone a flush due to migration. This aspect of these experiments has been criticized as not being realistic in nature, but that seems to me a moot point. The idea was to manipulate the populations to maximize the chances for destabilization of genetic systems to see if one could induce changes indicative of speciation.

Twelve populations founded by a single pair were begun from this Origin population after it had rapidly expanded; the Origin was maintained as three separate large populations to act as the controls. Forty single-pair matings were made and their larvae studied for inversions; the 12 most homokaryotypic were chosen, with only one founder line, designated 1, being completely homokaryotypic. The single-pair founder populations were allowed to expand rapidly before inducing selection again. They were tested for both postmating and premating isolation for the first time after about 15 months. Eight populations were continued and put through three more rounds of founder-flush cycles and were tested again three to four months later, and again five to nine years later (figure 7-11). Table 7-17 summarizes the results.

In none of the tests could any indication of postmating isolation be detected. Premating isolation was detected between some pairs of derived founder populations, as well as between pairs of derived and the controls; the three replicate controls never showed isolation. One interesting aspect of these results is that one population, that designated 1, consistently showed the greatest degree of isolation. It was the only one showing some isolation after just one founder-flush cycle and was involved in 11 out of 19 significant isolation in subsequent tests. Some combinations that tested positive in early tests did not show isolation in subsequent tests, but those involved with population 1 tended to be stable. This was the only population that was homokaryotypic from the start, it was the one that showed the greatest isolation, and it was the most stable. This is consistent with the idea that free recombination is a crucial aspect of the founder-flush theory, but clearly one cannot make much of a sample of one.

One effect of founder populations is to induce inbreeding, which may confound the effects of the founder-flush cycles. Controls were done where we inbred lines from the Origin population. No sign of premating isolation was detectable among inbred lines (Powell and Morton, 1979).

Thus we were able to conclude that destabilizing genetic systems through the kinds of manipulations involved in founder-flush cycles could induce some degree of premating isolation. Asymmetry in the isolation was often detected and in virtually every case it was in the direction predicted by the Kaneshiro hypothesis. Females from the Origin tended to discriminate against males from the founder populations more than founder females discriminated against Origin males (Powell, 1989).

This type of experiment has been repeated twice by others. Ringo (1986) worked with *D. simulans* and began his populations by intercrossing strains from many localities. He found little indication of premating isolation, although one population out of

TABLE 7-17 Results of mating tests between ancestral control populations and derived founder populations.

Year	No. Tests	No. with Significant Chi-Squared Tests $0.05 > p > 0.01$	$p < 0.01$	No. Involving Population 1
1974				
control : control	3	0	0	
control : derived	12	0	0	
derived : derived	45	0	1	1
1975				
control : control	3	0	0	
control : derived	8	1	0	1
derived : derived	28	4	6	4
1980s				
control : control	3	0	0	
control : derived	18	2	2	3
derived : derived	20	3	1	3

Isolation Index not Including Control : Control		
Positive	Zero	Negative
58	4	12

Note: All significant mating tests were in the direction of positive assortative mating.

From Powell (1989).

eight did show reasonably consistent positive assortative mating. Overall, however, the results were much less striking than those discussed above. One possible problem with this study is that *D. simulans* is a human commensal and may have a genetic architecture much different from *D. pseudoobscura*, a subject addressed by molecular studies discussed in chapter 10. Most of the strains used to start Ringo's experiment were from human-associated populations, although he did use strains from Africa, but it is not clear if they were from truly ancestral native populations.

Galiana et al. (1993) studied *D. pseudoobscura* in such experiments, but their studies differed in two ways from our own. First, they did not start with a hybrid population but with two different populations, one relatively monomorphic for inversions (Bryce Canyon) and one very polymorphic (Zirahuen, Mexico). They also studied the effects of different numbers of founders, 1 to 9 single pairs. They subjected 45 experimental lines to seven founder-flush cycles. Their results paralleled our own, although the indications of positive assortative mating were not as strong. As with our results, some combinations that showed isolation in one test were unstable in subsequent tests, but one combination exhibited quite stable isolation over the course of the experiment (50 generations). Another result was that populations with fewer founding pairs (1 or 3) showed greater positive assortative mating than did populations founded by 5, 7, or 9 pairs.

While Galiana et al. (1993) obtained results similar to our study on the same species, it is of interest to inquire why their results were less striking than ours. Two factors may be involved. One was not using a hybrid population as the origin of the founder lines; as mentioned above, this would have the effect of destabilizing coadapted complexes and thus increasing the effects of founder events. A second difference is that they incorporated eye mutants into their stocks through backcrosses. This was done to guard against accidental contamination during the experiments. However, eye mutants are often known to cause changes in mating behavior, at least in some species of *Drosophila*. While I know of no study that indicates these two mutants, *orange* and *sepia*, in fact affect mating behavior in *D. pseudoobscura*, their incorporation was one added variable whose possible effects we do not know.

While not completely comparable to the above studies, the work of Templeton (e.g., 1979) on *D. mercatorum* is also relevant to discussion of founder effects on reproductive isolation. *D. mercatorum* is a species that can sometimes reproduce parthenogenetically (for a review of parthenogenesis in *Drosophila*, see Templeton, 1983). When this species adopts this mode of reproduction, it is due to a haploid egg spontaneously doubling its genome and producing a female, which gives rise to an all-female parthenogenetic strain. This is the ultimate founder effect: a single genome! Templeton (1979) finds that such strains exhibit signs of incipient reproductive isolation.

Population Processes

Thus far this chapter has been primarily concerned with genetic aspects of speciation in *Drosophila*. The kinds of populational phenomena that might lead to the splitting of lineages depends on the genetic architecture of the genetic systems involved in reproductive compatibility, that is, mate recognition and developmental compatibility of gametes to form normal zygotes. This is one reason we have spent so much time discussing the genetic evidence. Templeton (1981) defined for convenience three basic types of genetic architecture. In type 1, a trait is controlled by a large number of loci, each with a small effect, usually acting additively; this is the usual kind of picture one assumes for polygenes *sensu* Kenneth Mather. In type 2, one or a few (say fewer than 10) genes have major effects on a trait, but there are also minor epistatically interacting modifiers. Finally, in type 3, complementary pairs of loci control a trait. Obviously, in real biological systems these are not discrete categories, but rather there is a continuum among these architectures. Nevertheless, we adopt this categorization for heuristic reasons.

There are two questions, both of which are still open, about this issue. The first is, which of the above architectures is most likely the underlying basis of traits involved in speciation? We presented conflicting evidence for both type 1 and type 2 architectures. This issue still needs to be clarified. It may well be the case that both actually occur and that different species and aspects of isolation may have different underlying genetic architectures. For example, F_1 male sterility may have a type 2 architecture and behavioral isolation a type 1 in one species, and vice versa in another. This issue can only be clarified by the kinds of studies discussed above; the progress in this area is certainly encouraging and this aspect of speciation may well be clarified in the not-too-distant future.

The second question about these architectures is whether one will lead to speciation faster than another. Templeton (1980) has argued that the two different architectures make very different predictions about what populational processes will cause speciation. To simplify his analysis, if type 1 genetic architecture underlies most traits important in speciation, then one would expect relatively slow divergence of isolated populations, with many genes with small effect finally building up enough genetic differentiation to cause reproductive isolation. On the other hand, type 2 architectures can change much more rapidly, for example by populational processes such as founder effects. Such rapid changes are what Templeton calls a genetic "transilience."

These two views of speciation can be thought of as the kind of different views of population genetics represented by R. A. Fisher and S. Wright (as discussed by Wright, 1980, 1982). Fisher was a proponent of mass selection, that is, he thought most populations were large, most important traits had a type 1 architecture, and that evolution proceeded by small steps, imperceptible to an observer because of the protracted timescale envisioned by Darwin. Wright introduced the idea that population structure and genetic drift can play an important role. His adaptive landscape model provided a means whereby evolution could proceed much faster than in the mass selection view. Chance events due to small population size could suddenly change the genetic environment and natural selection would mold adaptations from the new set of genotypes. The adaptive landscape was introduced as a metaphor: natural selection keeps populations on adaptive peaks until or unless something, usually a small bottleneck, knocks it off the peak and into a saddle or onto the side of another peak. Obviously, the kind of founder-flush speciation scenario envisioned by Carson for the Hawaiian *Drosophila* owes a lot to the Wrightian view of evolution, as well as Mayr's (1954) invocation of small peripheral isolates as prime settings for speciation.

Which is the best model for speciation in *Drosophila*? This remains an open and controversial question. The back-to-back articles by Carson and Templeton (1984) and Barton and Charlesworth (1984) present two contrasting sets of arguments. My own conclusion from this is that one needs to take an ecumenical view of speciation, even for a single genus like *Drosophila*. The Hawaiian archipelago and its magnificent drosophilid fauna would certainly be a situation ripe for a Wrightian kind of process, and much of the work by Carson and colleagues on these flies has led them to adopt these views. The founder-flush speciation theory was not developed out of thin air; there are good empirical reasons for adopting it (Carson and Templeton, 1984). On the other hand, much of the empirical data on other species would indicate that speciation is a much slower process and that when studied in detail, a type 1 genetic architecture is often found for the isolating traits. I find no compelling reason at this point not to admit both kinds of population genetic processes have been important in speciation in *Drosophila*.

Finally, I think it must be pointed out that some authors have maintained that the slow speciation process based on many genes of small effect is the Neodarwinian view. I find nothing non-Neodarwinian about the alternative view. Much of the theory stems directly from S. Wright, one of the "holy trinity" credited with developing Neodarwinism, as well as from E. Mayr. Both processes rely on natural selection and both have perfectly good population genetics bases. A major difference is simply time scale, but this by itself should not define what is meant by Neodarwinism. After all, Neodarwinism had no problem subsuming such processes as almost-instantaneous speciation due to polyploidy in plants, witness G. L. Stebbins.

Sexual Selection and Isolation

Throughout this chapter I have used the term *reproductive isolation* or simply *isolation* rather freely. This could be considered a traditional manner of thinking about species and speciation. Several authors, such as Paterson (1978), have promoted what they consider an alternative view. They place much more emphasis on the flip side of the coin of isolation, mate recognition. The argument is that natural selection cannot select for isolation, but that natural selection is for proper mate recognition. Specific mate recognition systems evolve; this has become established enough that the initials SMRS are even used. A related phenomenon is sexual selection, that is, the competition among members of the same sex for mates. The importance of sexual selection specifically as related to speciation in *Drosophila* is nicely laid out by Carson (1978; 1986), what he has called "the mutual adjustment of the sexes to what may be called the intraspecific sexual environments." While traditional evolutionary thinking is usually focused on the external environment to which species need adapt, the emphasis in this Carson view is that there is also the sexual environment, which is more important during the speciation process.

I have no objection to this somewhat altered view of speciation, one that emphasizes the positive side of proper mate recognition and the role sexual selection plays. In fact, I proposed just such a scenario to explain the results from the founder-flush laboratory experiments (Powell, 1989). Obviously, the induction of premating isolation in those experiments could not have been selected for by some kind of process such as reinforcement. Rather, the founder-flush process knocked the populations off an adaptive peak of mate recognition. When sexual selection was reintroduced over a period of time, a somewhat altered mate recognition evolved in some of the founder populations. The change was induced by the stochastic events of founder effect and free recombination during the flush, but the adoption of an altered mating system was caused by sexual selection.

This brings me to the last point for this chapter: should speciation, in particular in *Drosophila*, be viewed as a product of natural selection? Darwin was looking for a mechanism for the origin of species, but did he find it in natural selection? Probably the most ardent supporter of the adaptive nature of species was the drosophilist Th. Dobzhansky (1937, 1970). The whole terminology of isolating mechanisms implies selection for mechanisms to protect the integrity of coadapted gene pools. In the above, while using the word "isolation" freely, I have avoided using "mechanism." With the exception of reinforcement, the existence of which is doubted by some, there is little theoretical framework on which one can hang natural selection as the direct cause of evolution of mechanisms of isolation. When two populations begin to evolve allopatrically, there is little reason to think natural selection acts directly to evolve pre- or postmating isolation. Rather, the divergence of the populations is due to other factors, such as drift and adaptation to environmental differences. Sexual selection will be acting to assure proper mate recognition, but not isolation. Isolation is the result, but not the primary target of selection.

In referring this discussion specifically to *Drosophila*, I think we need to reemphasize that in the very earliest stages of speciation, when reproductive isolation first begins to evolve to a partial state, there are two processes that are affected in most cases. One is mating behavior, as demonstrated by all those cases discussed above where there is premating isolation in the absence of postmating isolation. Here the emphasis is correctly placed, I think, on the role of sexual selection, as clearly enunciated by Carson (1978; 1986). The other trait that evolves first in many cases is that the genetic control of spermatogenesis is changed. This is a phenomenon that seems to me to be more mysterious in that it is difficult to think of a reason why this should be the case. Whether sexual selection can also account for this remains speculative.

8

Phylogenetics

> Naturalists try to arrange the species, genera and families in each class on what is called the Natural System. But what is meant by this system? Some authors look at it merely as a scheme for arranging together those living objects which are most alike, and for separating those which are most unlike. . . . I believe that something more is included than mere resemblance; and that propinquity of descent—the only known cause of the similarity of organic beings—is the bond, hidden as it is by various degrees of modification.
>
> C. Darwin, 1859

> But the orderly sequence, historically viewed, appears to present from time to time something genuinely new.
>
> C. Lloyd Morgan, 1927

The topics covered in this chapter traditionally fall into three disciplines: taxonomy, systematics, and phylogenetics. Taxonomy is the science of accurately naming, defining, and identifying various taxa. Systematics is concerned with arranging the named taxa into a logical hierarchical framework to make sense of the many millions of named taxa. Phylogenetics aims to understand the historical ancestral-descendent relationships of the taxa, both extant and extinct. Clearly the three disciplines are tightly integrated in modern biology, although this has not always been the case. For example, today it is quite widely accepted that systematics should, as far as practical, reflect phylogeny, a proposition of considerable controversy not so long ago. Because this book is primarily concerned with evolutionary issues, phylogenetics will be the primary focus here, although we can hardly address these issues without frequent reference to systematic and taxonomic issues as well.

While the family Drosophilidae, and the genus *Drosophila* (Fallen, 1823) in particular, have long been recognized, it wasn't until the early 20th century, when geneticists began using the genus so extensively, that more intensive taxonomic and systematic work was undertaken. A. H. Sturtevant, the best systematist among the founders of *Drosophila* genetics, was the first to closely examine the diversity of this genus, and began to define and systematize the group; his 1921 publication was particularly influential in setting the foundation for virtually all subsequent work. Until recently, *Drosophila* systematics did not attract very many workers and one can fairly state that in the first three-fourths of this century, along with Sturtevant, Marshall Wheeler and Lynn Throckmorton were the two who devoted entire careers to systematizing the genus. Today the field is much more active with many workers applying the latest technologies and conceptual frameworks in gaining an understanding of the group.

Background

As mentioned in the first chapter, today over 3,200 valid binomial Latin names have been assigned in the family Drosophilidae, with about 1,600 of these species belonging to the genus *Drosophila.* Catalogs of these names include Wheeler's (1981, 1986) and, most recently, G. Bächli's (1994) compilation, available by computer network. The systematics of this group is hardly complete, since many new species are described every year; in fact, Ashburner (1989) states that about 15% of all Drosophilidae have been added since Wheeler's 1981 compilation.

Beginning with Sturtevant, and especially in Patterson and Stone (1952), the tradition in systematizing this large number of taxa has followed the scheme outlined in table 8-1. In addition to the formally recognized Family, Genus, and Species, Drosophilists have found it convenient to recognize "groups," "subgroups," and "complexes," none of which has formal recognition by the International Code for Zoological Nomenclature. These lower categories have been especially convenient in subdividing what would have been extremely huge groups by most standards. One could question whether the great diversity and numbers of species really should be a single genus; a genus with 1,600 named species (and probably at least 2,000 total) is seldom found in any other group. For example, genera could be raised to subfamilies, subgenera to genera, and so forth. However, such radical taxonomic revision is not advisable at this time, as the literature and traditions are so well established that any such formal reassessment would not be worth the confusion engendered. Drosophila workers, by and large, are quite comfortable with the informal subdivisions traditionally used, and it would seem advisable to keep them.

The first attempts to place this large number of taxa into an evolutionary framework (i.e., draw phylogenies rather than simply systematize) were made by Sturtevant (1921) and Patterson and Stone (1952). Their scheme is shown in figure 8-1. Such trees need to be understood in the context of the times whence they came as they do not resemble the kinds of bifurcating diagrams we have become accustomed to in modern phylogenetics. Rather, they are attempts to indicate which groups are more closely related, which gave rise to others, and so on. Extant species are placed on

TABLE 8-1 The nomenclature used to describe taxonomic ranks in Drosophilidae.

- Family (Drosophilidae)
  - Subfamilies (Drosophilinae and Steganinae)
    - Genus (e.g., *Drosophila*)
      - Subgenus (e.g., *Sophophora* and *Drosophila*)
        - Groups (e.g., *melanogaster, obscura, willistoni*)
          - Subgroups (e.g., *affinis, hydei*)
            - Complexes (e.g., *ananassae*)
              - Species (e.g., *Drosophila pseudoobscura*)
                - Subspecies (*D. pseudoobscura bogotana*)

Note: Some (e.g., family, genus, species) are formally recognized while others (groups, subgroups, and complexes) are used solely by tradition and convenience.

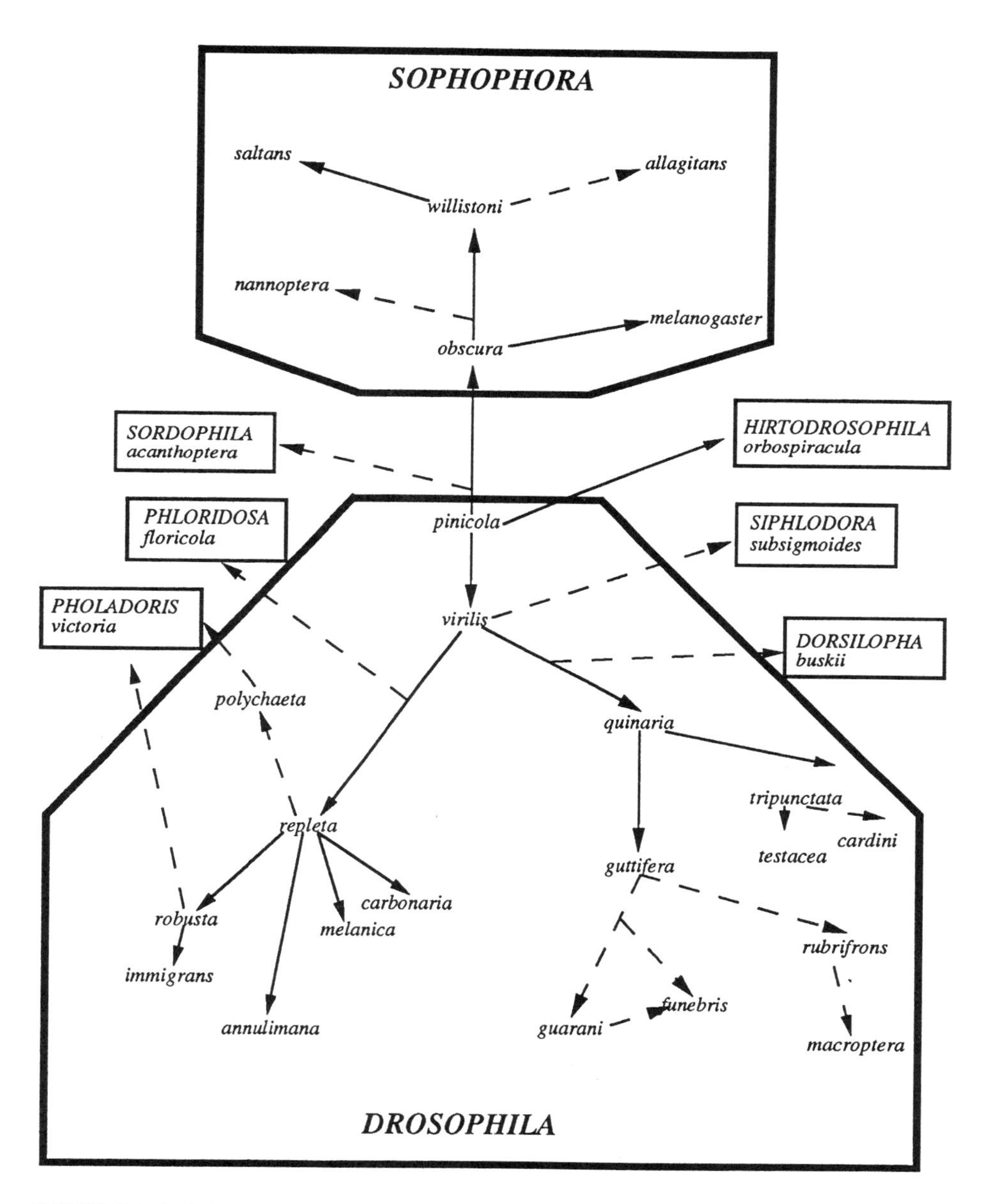

FIGURE 8-1 Relationships of Drosophilidae as deduced by Sturtevant (1921) and Patterson and Stone (1952). From Powell and DeSalle (1995) after Patterson and Stone (1952).

nodes and polyphyletic origins seem acceptable. Throckmorton's more recent attempts at phylogenetic trees are somewhat better, although strict interpretation remains difficult (figure 8-2). The emphasis in Throckmorton's scheme is on various radiations of groups rather than phylogenetic definition. The presentation of these early attempts is not meant in any way to ridicule early workers. Rather, as we will see, the phylogenetics of *Drosophila* has greatly advanced in recent years and the presentation of such early trees is helpful in understanding how the knowledge has evolved. The more modern data (e.g., DNA sequences) and analytical techniques (e.g., cladistics) have served to help define more explicitly these vague relationships, rather than to fundamentally overthrow the conclusions reached by these pioneers.

Basics of Phylogenetics

Since many readers of this book may be unfamiliar with the technologies and terminologies used to reconstruct phylogenetic relationships, a short digression is in order. Those reasonably versed in the subject can easily skip this section.

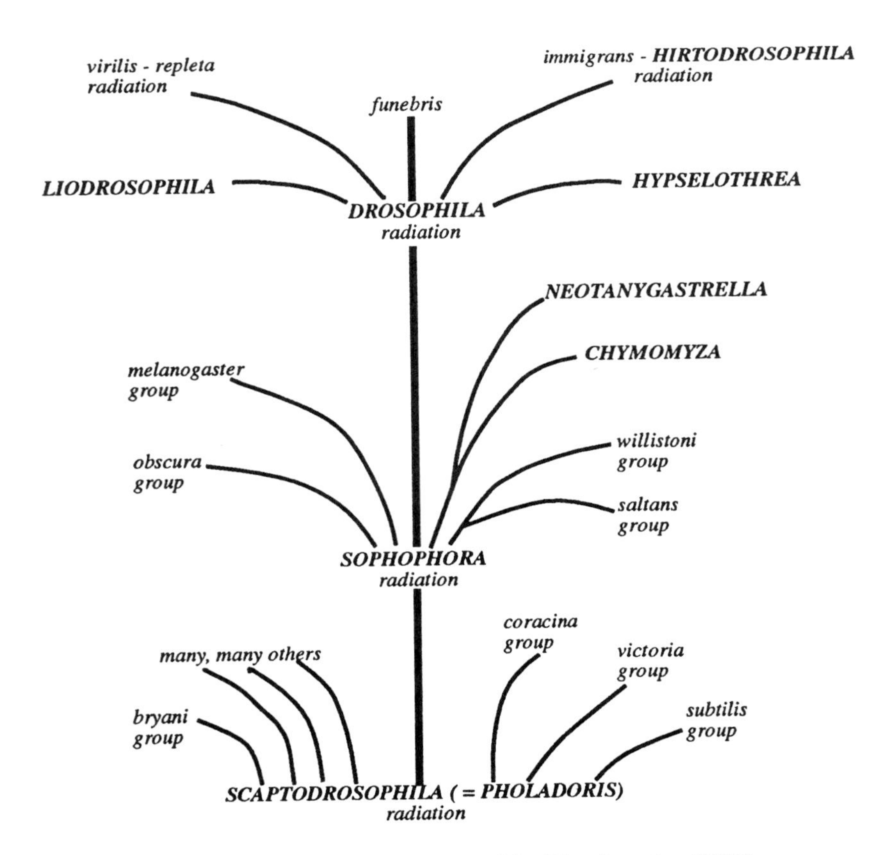

FIGURE 8-2 Radiations as proposed by Throckmorton (1975).

Except where one has an exceptionally complete set of fossils, the problem of phylogenetics is to take existing taxa, look at a set of their characteristics, and then make a plausible historical scenario to describe how they came about from a single common ancestor. Traditionally this was done by experts who studied groups in great detail until they knew them very well, to such an extent that they thought they understood how evolution had brought about the different forms. Rarely, if ever, were explicit rules set out as to how one might rigorously and quantitatively construct and assess any given phylogenetic hypothesis. In recent years, a renaissance in phylogenetics has occurred with the institution of techniques that allow objective and often quantitative analysis and evaluation of phylogenetic hypotheses. Note that we use the term hypothesis to describe any phylogenetic tree; as with any historical science, one can never know with certitude exactly what happened. Rather, given a set of data, and some assumptions about how evolution works, we can reconstruct an historical hypothesis and determine how strongly the existing data support that hypothesis relative to other possibilities.

There are two very basic ways in which one can reconstruct historical relationships. One is to base them on overall similarity or differences (reviewed in Sokal, 1986). The assumption is that organisms that are more similar to one another shared a common ancestor more recently than did organisms less similar. The overriding assumption of using similarity in creating a phylogeny is that changes accumulate directly proportional to time, and in fact, some methodologies require that the changes measured are linear with time. The usual method for inferring phylogenies is to create some kind of "distance" measure, which is a quantitative assessment of similarity and differences. The resulting tree is often called a phenogram because originally the data came from measurements of phenotypes. Today, genotypes (e.g., DNA sequences) are often the source of data, but the term phenogram seems to have stuck.

A second major methodology is to use individual character analysis; the most widely used procedure of this sort is cladistics. This involves deducing, by a set of explicit rules, which characters or character states are ancestral and which are derived. Taxa are grouped based on shared derived states, or synapomorphies. Ancestry is determined most often by the use of an outgroup, a taxon or taxa which, for reasons independent of the data being analyzed, are known to be phylogenetically outside the group one is trying to resolve into a phylogeny. The phylogenetic trees derived from a cladistic analysis are called cladograms and have extant taxa only at the tips of branches and not at nodes as did some older ways of indicating relationships (e.g., figures 8-1 and 8-2).

Another important assumption in almost all phylogenetic analyses is that the best hypothesis is a parsimonious one. This is the explicit criterion of the cladistic methodology: the most acceptable tree is that which requires the smallest number of steps, that is, the smallest number of character state changes. While in reality evolution may not necessarily proceed in the "most parsimonious" manner, in the absence of evidence to the contrary, the assumption of parsimony is generally acceptable.

The type of data used in phylogeny reconstruction is also highly variable. Traditionally, morphological, physiological, and behavioral data were used. Genetic data such as chromosomal analysis, especially inversions, have proven particularly useful in some groups of *Drosophila*. In more recent years, molecular data have become more widely used. Molecular data have been especially helpful for *Drosophila*, as the number of species is very great and the small size of the organisms has meant that traditional methodologies had few characters with which to work. Another attractive aspect of molecular data is that they are often less subject to selection than are morphology, physiology, and so forth. Selection can be a cause of incorrect phylogenetic analysis as it may induce convergences, parallelism, or other problems. The "neutral" theory of molecular evolution, if correct or even nearly correct, would argue that molecular data are ideal for phylogenetic reconstructions. This, along with the potentially large number of characters (nucleotide positions), makes these data extremely rich and powerful for phylogenetics.

Regardless of which methodology and what kind of data one uses, there always remain problems in weighting data. One could simply choose not to weight character state changes and thus be more objective or assumption-free (although one could argue no weighting is itself an assumption). However, there is often good reason to weight some changes more heavily than others. For example, in deciding that dolphins are mammals rather than sharks, one would not put much weight on overall body shape. For DNA sequence data, it is known that transitions (purine for purine or pyrimidine for pyrimidine) often occur more frequently than do transversions (change from purine to pyrimidine or vice versa). Thus there is a much greater chance that two taxa have independently acquired the same nucleotide at a position if it only required a transition from the ancestral state rather than a transversion. Homoplasy, the sharing of a derived character state not through common ancestry, is more likely for frequent events, and thus to minimize homoplasy, transitions should be weighted less than transversions. Also for molecular data there is often the problem of weighting base changes relative to insertions and deletions. Further discussion of weighting DNA sequence data can be found in Brower and DeSalle (1994) and Simon et al. (1994).

One problem with modern phylogenetic analysis is that given a set of data such as DNA sequences, and given the software packages readily available, it is very easy to have a computer generate a tree. What becomes important is the evaluation of the robustness of the tree relative to other possible trees. How well is a tree supported by the data on hand? This is a difficult statistical problem and only a few methodologies are available. For some algorithms such as maximum likelihood, there are explicit statistical properties of the trees so that the probability of one tree versus another can be evaluated in a fairly straightforward manner (although with subtle problems). For other algorithms such as the cladistic maximum parsimony approach, one ends up with the tree that requires the minimum number of changes, say 100, but is that really better from a statistical standpoint than one requiring 101 steps, or three other topologies requiring 102 steps each? The most commonly used methodology for obtaining some idea of the robustness of maximum parsimony trees has been to "bootstrap" the data (Felsenstein, 1985). Bootstrap is a resampling technique which allows one to evaluate how well a given data set supports a given node in a tree.

One further point that must be explicitly acknowledged when using molecular data such as DNA sequences is that the derived tree is the one that best describes the evolutionary history of that particular sequence of nucleotide bases. Most often what is desired is the phylogenetic history of the taxa from which the data have been derived. This has come to be known as the gene tree-species tree problem and will be illustrated with examples from *Drosophila* at the end of this chapter.

From the above it should be clear that the modern methodologies, both in data gathering (e.g., DNA sequencing) and analysis (by computer programs with explicit rules), are more objective and quantitative than older traditional procedures. However, correct implementation of the procedures requires considerable knowledge and skill. Sound judgments are required for weighting character state changes, finding appropriate outgroups, choosing an appropriate algorithm for deriving the tree(s), choosing and interpreting the statistical tests that evaluate trees, and deciding when a gene tree accurately reflects the history of the taxa itself. The above is just a very brief introduction to allow the naive reader to appreciate what follows. The issues are really much more complex and subtle than implied by the above and the reader is referred to a number of more thorough treatments of these issues in greater detail: Felsenstein (1988), Swofford and Olsen (1990), Miyamoto and Cracraft (1991), and Hillis et al. (1993).

The Larger Context

Drosophilidae within Diptera

Table 8-2 presents a systematic treatment that places the family Drosophilidae in the context of Diptera. This family is considered a member of the higher Diptera, compared to lower Diptera such as mosquitoes, midges, gnats, and crane flies. Within the higher Diptera, Drosophilidae belongs to the Cyclorrhapha, a taxon variously considered as a suborder and as a part of the suborder Brachycera. The main defining characteristic of the Cyclorrhapha is the rotation of the male hypogonium. A major group

TABLE 8-2 The systematics and taxonomy of Diptera with special reference to Drosophilidae.

Order:	Diptera
Suborders:	"Nematocera": Lower Diptera: mosquitoes, midges, gnats, crane flies
	Brachycera: Higher Diptera: horse flies, deer flies, robber flies
	Cyclorrhapha: with two Divisions:
Divisions:	"Aschiza": humpback flies, flower flies
	Schizophora: with two Sections:
Sections:	Calyptratae: house flies, blow flies, tsetse, bot flies
	"Acalyptratae": true fruit flies, stalk-eyed flies
Superfamily:	Ephydroidea with seven families:
Families:	Curtonotidae
	Campichoetidae
	Diastatidae
	Camillidae
	Risidae
	Ephydridae
	Drosophilidae with two Subfamilies
Subfamilies:	Steganinae
	Drosophilinae with about 40 genera

Note: One controversial point is whether two or three suborders are recognized; Cyclorrhapha is sometimes included in Brachycera and sometimes given subordinal status. Taxa in quotes indicate probable paraphyletic groups, while the rest, so far as known, are monophyletic.

of Cyclorrhapha is the acalypterate flies, those with a highly reduced single pair of calypers. These include, in addition to Drosophilidae, the true fruit flies (Tephritidae) and stalk-eyed flies (Diopsidae). A superfamily of acalypterates is Ephydroidea, which contains seven families. Grimaldi (1990) presents strong evidence that Drosophilidae belongs in this superfamily, but just how the seven families are related to one another is not clear. Three cladograms of equal length were found by Grimaldi using 30 morphological characters; Grimaldi favored Curtonotidae as the sister family to Drosophilidae ("sister taxa" is the term used to describe the two closest taxa being considered). The two subfamilies of Drosophilidae, Drosophilinae and Steganinae, always formed sister taxa in Grimaldi's analysis. Thus there is good reason to believe the family Drosophilidae is a good monophyletic group.

From an evolutionary perspective, it is of some interest to examine the biology of the closest relatives of Drosophilidae. Except for the family Ephydridae, the other close families are not well known. Ephydridae are called shore flies because they are most often found along shores of oceans, ponds, and streams as the larvae are aquatic, breeding in both fresh water and brackish water; the major sources of food are microorganisms, especially algae. However, the family has members whose larvae breed in a large variety of substrates, including leaves of cereals and sugar beet, dead snails and other small carrion, urinals, feces, and egg sacs of spiders and frogs (Ferrar, 1987). Thus, as with Drosophilidae, while one can define one or a few major breeding sites, the range of substrates used by Ephydridae larvae is very large.

The proposed sister family to Drosophilidae, Curtonotidae (Grimaldi, 1990), is a very small family of just three genera. Almost all species are known only from the African tropics; only a single species is known from the Nearctic. The only confirmed larval breeding site is the egg pods of the desert locust. These flies are often observed in burrows of small animals, although it is not known if this is a breeding site (Ferrar, 1987). Thus, other than the tropical distribution of Curtonotidae, what little is known of the ecology of this family sheds little light on the origin of Drosophilidae.

Steganinae

Just as Drosophilidae is a good monophyletic group, all evidence is that the two subfamilies within Drosophilidae, Drosophilinae and Steganinae, are also monophyletic taxa. The breeding habits and biology of some members of the Steganinae are well known. Table 8-3 is a summary of the larval breeding sites of the 11 most common genera of this subfamiliy. Although this subfamily (having about 400 species) is relatively small compared to Drosophilinae, it occupies almost as diverse a set of larval breeding substrates: rotting vegetation, slime fluxes, fungi, flowers, other insects, and insect spittle. Perhaps most significant is that there is one major breeding site for *Drosophila* that has not been exploited by any known member of the Steganinae, namely rotting fleshy fruit. It is tempting to speculate that perhaps the exploitation of rotting fruit was the ecological expansion that led to the origin of the Drosophilinae, as neither its sister subfamily nor any of the other Ephydroidea exploit this niche. In this sense *Drosophila* perhaps should be called fruit flies.

Drosophilinae

Some 35 to 40 genera have been assigned to this subfamily. Grimaldi's (1990) monograph is the best recent reference for their relationships based on morphological data.

TABLE 8-3 Larval breeding sites of members of subfamily Steganinae.

Genus	Larval Habitat									
	1	2	3	4	5	6	7	8	9	10
Acletoxenus										X
Amiota		X	X				X			X
Apenthecia				X						
Cacoxenus			X						X	X
Gitona	X				X	X				
Leucophenga	?	X	X					X		
Pseudiastata										X
Rhinoleucophenga	?									X
Stegana		?		X						

Note: An X indicates a member of that genus is known to use that larval habitat.

1, General rotting vegetable matter including leaves; 2, tree sap/slime fluxes; 3, fungi; 4, living flowers; 5, leaf miners; 6, stem borers; 7, living in beetle tunnels; 8, commensal in spittle of Cercopidae; 9, commensal/predacious in nests of Apoidea; 10, predacious on other insects.

Modified from Ferrar (1987).

Unfortunately the literature is often quite confusing in regard to the systematics of this group, as some of the presently accepted genera have variously been called subgenera within genus *Drosophila*, while others are almost certainly not monophyletic groups. Most of the genera are small and have not been of great interest to experimental biologists. However, for comparative purposes it is often useful to have an outgroup or outgroups for genus *Drosophila*, so the better defined, and easily available, close relatives of *Drosophila* deserve discussion.

Figure 8-3 is a cladogram including three potential outgroups, the branching pattern of which is consistent with both molecules and morphology. Thus we can accept this phylogenetic hypothesis with some confidence (Grimaldi, 1990; DeSalle, 1992a; DeSalle and Grimaldi, 1991, 1992). *Zaprionus* would appear to be a good choice for a taxon close to genus *Drosophila*, with *Chymomyza* a bit more distantly related, followed by *Scaptodrosophila* (see, however, Kwiatowski et al., 1994, for an alternative placement of *Chymomyza*). These taxa are highlighted not only because they are placed with some confidence, but also because they are available. The National Drosophila Species Resource Center at Bowling Green University lists five species of *Zaprionus*, two species of *Chymomyza*, and seven *Scaptodrosophila*. This last genus was sometimes considered a subgenus of *Drosophila* and thus some of the species, *S. lebanonensis* being the best known, have also been called *Drosophila* in the literature.

In addition to providing a firm phylogenetic framework for comparative studies, the work of Grimaldi and DeSalle has also dispelled what appears to be a misconception of earlier phylogenetic hypotheses. This has to do with the relative ancestral or derived status of *Drosophila*, that is, was it an early existing lineage from which other major lineages were formed, or is the genus *Drosophila* more derived, appearing rather late? The early hypotheses such as those in figures 8-1 and 8-2 imply that several genera are derived from *Drosophila*, whereas the revised phylogeny (figure 8-3) implies *Drosophila* is itself a relatively recently derived genus.

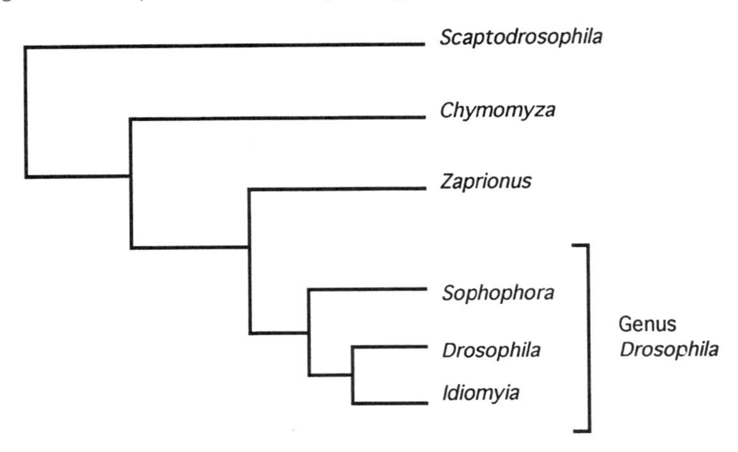

FIGURE 8-3 Relationships of three genera with respect to *Drosophila*. This relationship is supported by both morphological and molecular data (DeSalle and Grimaldi, 1992). The three top genera are all available from the U.S. National Stock Center and can serve as outgroups for the genus *Drosophila*.

A few other genera of Drosophilinae are also available from the National Stock Center, which may be useful for comparative studies, although their phylogenetic affinities are not as certain as those presented in figure 8-3. *Hirtodrosophila* is placed on the branch between *Chymomyza* and *Scaptodrosophila* by DeSalle's (1992a) mtDNA sequence data and between *Chymomyza* and *Zaprionus* by Grimaldi's (1990) morphological data. *Liodrosophila* appears on the branch leading to *Zaprionus* in Grimaldi's cladogram, whereas the mtDNA data place it between *Sophophora* and subgenus *Drosophila*. *Samoaia* is also placed by Grimaldi very close to *Zaprionus* but was not studied by DeSalle for mtDNA.

The genus or subgenus *Scaptomyza* may or may not be a monophyletic taxon. Species placed into this taxon are found endemic to Hawaii as well as endemic to other continents. As we will see later, all the species endemic to Hawaii, including *Scaptomyza*, fall into a monophyletic lineage which clearly falls into genus *Drosophila*. It is possible that continental *Scaptomyza* belong to this same lineage and represent a migration to the continents from Hawaii, which as far as known, would be a unique case. Alternatively, continental and Hawaiian *Scaptomyza* are a case of convergence and really belong to different lineages. This issue should be resolvable with further DNA sequence data.

Genus *Drosophila*

The rest of this chapter will be concerned with relationships within the genus *Drosophila*, with an emphasis on those used in experimental work. Much of the work in this section will be concerned with molecular phylogenies of the genus, as they have provided the firmest definition of relationships; this work was reviewed in Powell and DeSalle (1995).

DNA–DNA Hybridization

The technique of DNA–DNA hybridization provides an overall measure of the DNA differences between two taxa. This is done by measuring the decrease in melting temperature (the temperature at which DNA breaks from a double-stranded duplex into single strands) due to base-pair mismatch in heteroduplexes (duplexes in which each strand comes from a different taxon) compared to homoduplexes (duplexes formed by strands from the same taxon). The technique, with particular emphasis on its use in phylogenetics, is reviewed in Werman et al. (1990a) and Powell and Caccone (1991).

In regard to the phylogenetic relationships of the major groups of *Drosophila*, it was observed that when stringency conditions are used that require about 75% base pairing to form stable duplexes, very little of the total single-copy DNA cross-hybridized, on the order of 20% or less. This phenomenon of a very rapidly evolving fraction of the genome will be discussed in some detail in the next chapter. For the present concern, phylogenies, the importance of this observation is that it is difficult to perform DNA–DNA hybridizations across the major lineages using the total single copy genome. Therefore, we devised a strategy to hybridize a relatively conserved fraction of the genome, that coding for amino acid sequences in proteins. This could be fairly readily done by using reverse transcriptase to copy DNA from poly-A^+ RNA. The resulting cDNA was then used to cross hybridize across the major lineages of *Drosophila.* Details of these experiments are in Caccone et al. (1992) with the major conclusions illustrated in figure 8-4.

The relationships in figure 8-4 are more or less typical of all attempts to construct phylogenies of the major groups of *Drosophila.* Subgenus *Sophophora* contains three major groups whose relationships are very well supported. The *melanogaster* group is closest to the *obscura* group, with the *willistoni* group being the oldest lineage. This is consistent with the biogeography discussed in the first chapter. The major *Sophophora* groups evolved in the tropics, the *melanogaster* lineage in the Old World (Asia and Africa) and the *willistoni* lineage in the New World tropics. The *obscura* group likely originated in Africa where a few species still exist in temperate high elevations.

The relationships of members of subgenus *Drosophila* have proven much more difficult to define. As seen in figure 8-4, except for fairly strong evidence indicating that the *virilis* and *repleta* groups belong to the same lineage, the cDNA-DNA hybridization data could provide little resolution for this subgenus. Some better resolution is afforded by DNA sequence data.

DNA Sequences

Four DNA sequence data sets are relevant to the placement of the major groups of *Sophophora: Adh* (Anderson et al., 1993; Russo et al., 1995), nuclear 28S rDNA (Pelandakis and Solignac, 1993), mtDNA (DeSalle, 1992a, b), and *Sod* (Kwiatowski et al., 1994). Figure 8-5 presents the parsimony trees for these four data sets. The *Adh, Sod*, and mtDNA sequence data are consistent with the cDNA–DNA hybridization result in indicating that *Sophophora* is a monophyletic taxon with the three major species groups related as before. Only the 28S rDNA data are at odds with all the

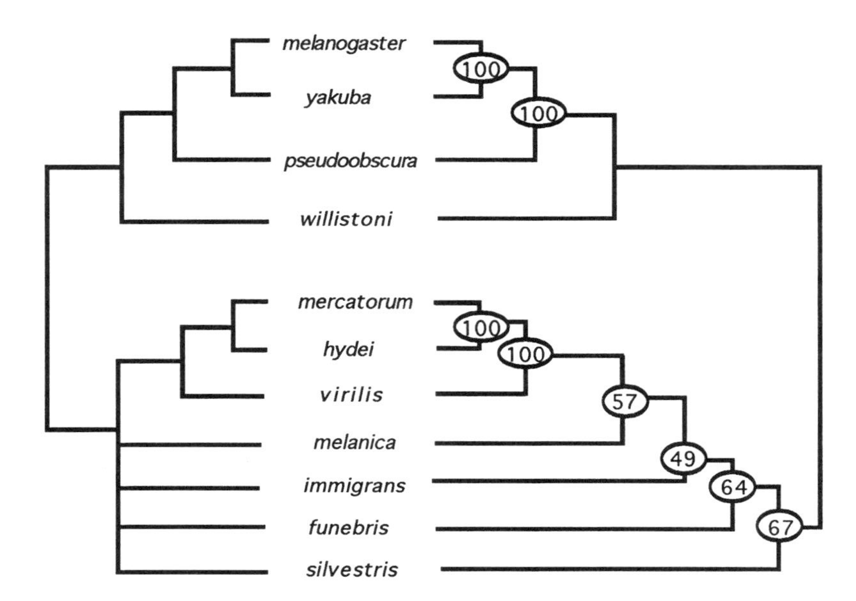

FIGURE 8-4 Trees deduced from cDNA–DNA hybridization studies. The tree on the left is the Jackknife consensus tree. The right tree was generated by the FITCH algorithm of PHYLIP with the bootstrap support (out of 100 trials) at some nodes; nodes without numbers had bootstrap values of 100%. From Caccone et al. (1992).

other information, including morphology and behavior. Thus we are inclined to discount this data set.

However, there is a manner in which data sets can be combined when individual data sets disagree, and that is to use total evidence, that is, simply combine the data sets so those with the strongest phylogenetic information tend to dominate. Figure 8-6 shows the resulting trees when the 28S rDNA and mtDNA sequence data are combined with the morphological and behavioral data (see Powell and DeSalle, 1995, for more details). The important point is that when combined, the conflicting 28S rDNA phylogenetic signal is not as strong as the other data sets. The combined data set supports the same relationship within the *Sophophora* as obtained by most data sets, as well as supporting the monophyly of this subgenus. Overall, the bulk of the data is quite convincing for this subgenus.

Hawaiian Drosophilids

A major concern with recent Drosophilidae phylogenetics has been the placement of the large group of endemic Hawaiian drosophilids. These species have been placed in a number of genera and subgenera based primarily on morphology. The morphological diversity of this group is great, greater than all other species combined. Despite the amazing diversity, Throckmorton (1966), working with morphological characters, and Spieth (1966; 1984), working with behavioral characters, both concluded there was

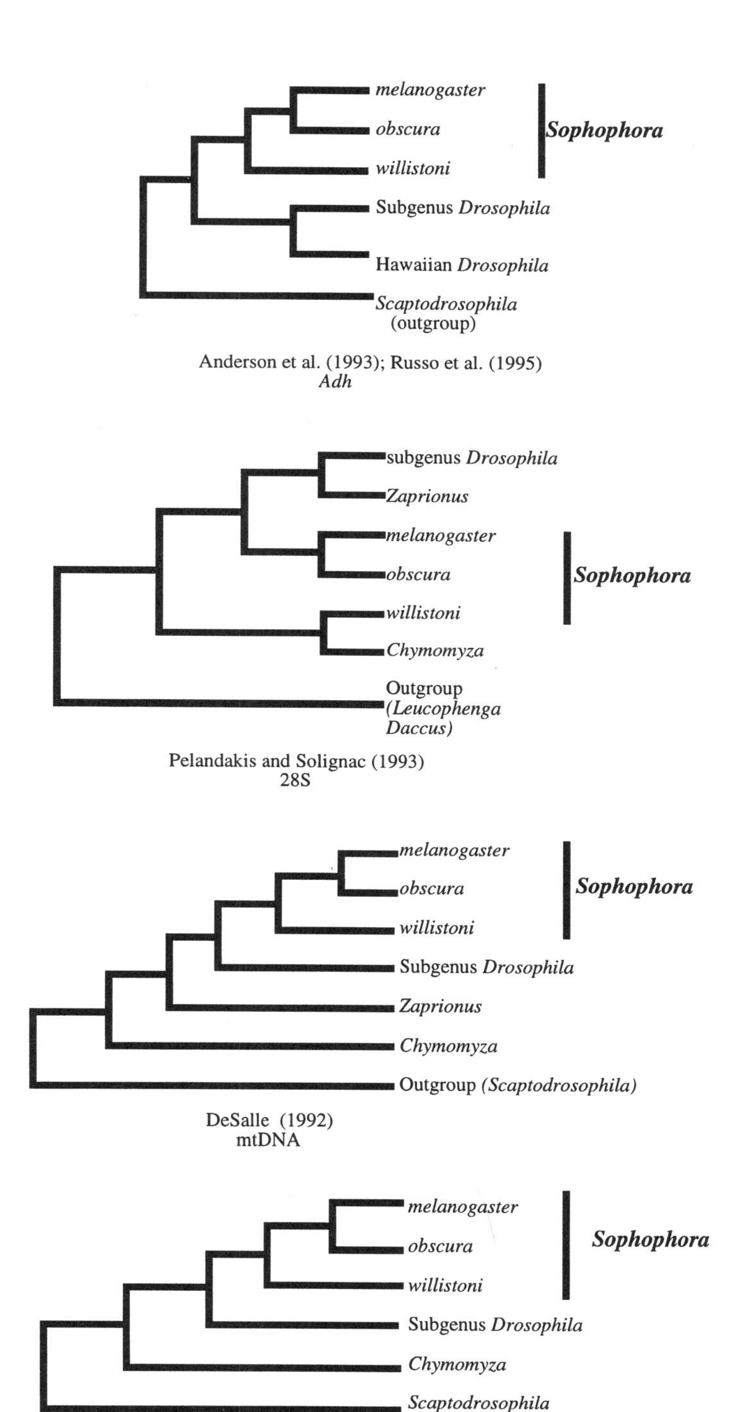

FIGURE 8-5 Four hypotheses deduced from four DNA sequence data sets.

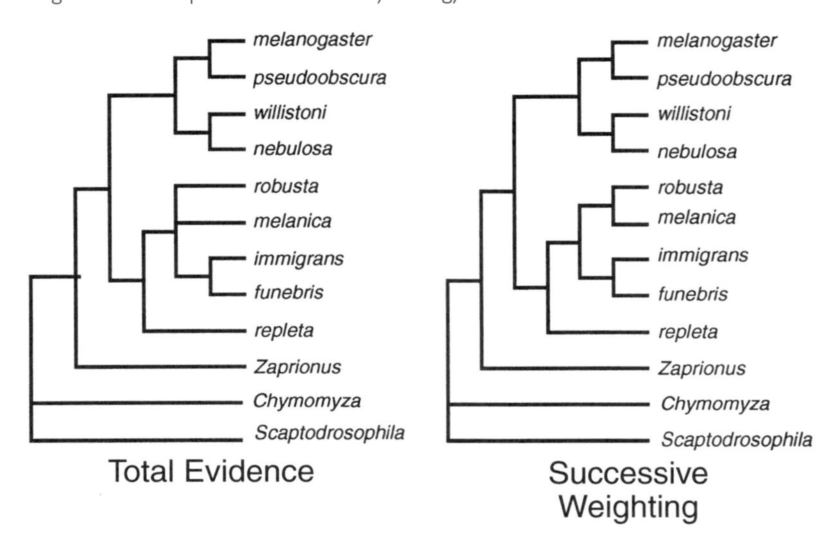

FIGURE 8-6 Trees generated by combining the data sets, including the data set from Figure 8-5 that showed the *Sophophora* not to be a monophyletic group. From Powell and DeSalle (1995).

good evidence for a single founder (or perhaps two) giving rise to the entire Hawaiian fauna. Classically, this group had been considered part of the subgenus *Drosophila.* However, Grimaldi's (1990) reanalysis of morphological characters placed this lineage outside the genus *Drosophila* and as a close relative of *Hirtodrosophila.* In fact, Grimaldi suggested erecting a new genus, *Idiomyia*, for the Hawaiian drosophilids.

Thomas and Hunt (1991, 1993) and DeSalle (1992b) have shed light on this problem using, respectively, *Adh* and mtDNA sequence data. Figure 8-7 displays the relationships supported by both data sets. It is important to stress that the 11 species studied by these workers represent what should be the most diverse groups including members of the three "genera" of Hawaiian flies. Two very important results are clear. First, all the evidence indicates that despite the great diversity, the Hawaiian fauna is a monophyletic group. Second, the group is an early offshoot of subgenus *Drosophila*, possibly occurring before the diversification of this subgenus, and definitely occurring after the *Sophophora* lineage split. Thus from a phylogenetic standpoint, the Hawaiian monophyletic lineage belongs in genus *Drosophila*, and as a sister group to the classically defined members of subgenus *Drosophila.* This result has made moot the question of which species is the closest relative to the Hawaiians, a favorite source of much speculation. All species of subgenus *Drosophila* are equally distantly related to the Hawaiians, as they were an early independent lineage.

Given that the Hawaiian lineage falls clearly within genus *Drosophila*, it makes little sense to maintain "*Scaptomyza*" and "*Engiscaptomyza*" as genera (at least not for Hawaiian species). On the subgeneric level, one could either simply include the Hawaiians as members of subgenus *Drosophila* as has been traditional or, as we suggest (Powell and DeSalle, 1995), perhaps their early origin and uniqueness deserve

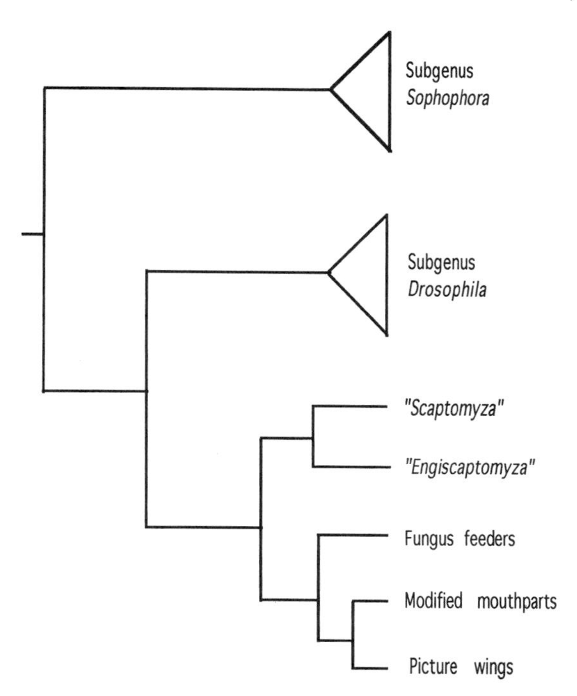

FIGURE 8-7 Deduced trees from mtDNA and *Adh* sequence data for Hawaiian *Drosophila*. Taxa are placed in quotes to indicate they should probably not be considered separate genera or subgenera. Data supporting this are from Thomas and Hunt (1991, 1993) and DeSalle (1992b).

the erection of a third major subgenus of *Drosophila*, subgenus *Idiomyia*, as the sister to subgenus *Drosophila.*

Recent evidence from DNA sequences indicate that it is possible not all members of subgenus *Drosophila* originated after the split from the Hawaiian lineage (R. DeSalle, personal communication). If this is confirmed, then raising the Hawaiian lineage to subgenus status would render subgenus *Drosophila* nonmonophyletic. Once these issues are clarified, clearly some nomenclature changes would be in order.

Subgenus *Drosophila*

While DNA data provide good evidence for the phylogenetic placement of the Hawaiian lineage, the phylogenetic relationships of groups within subgenus *Drosophila* remain poorly defined. Both the DNA–DNA hybridization data (figure 8-4) and the available sequence data provide low statistical confidence for the relationships in this subgenus. Figure 8-8 summarizes what can be defined in subgenus *Drosophila* with reasonable confidence. First, all indications are that the subgenus is a monophyletic taxon with the traditionally placed species and species groups. Within the subgenus,

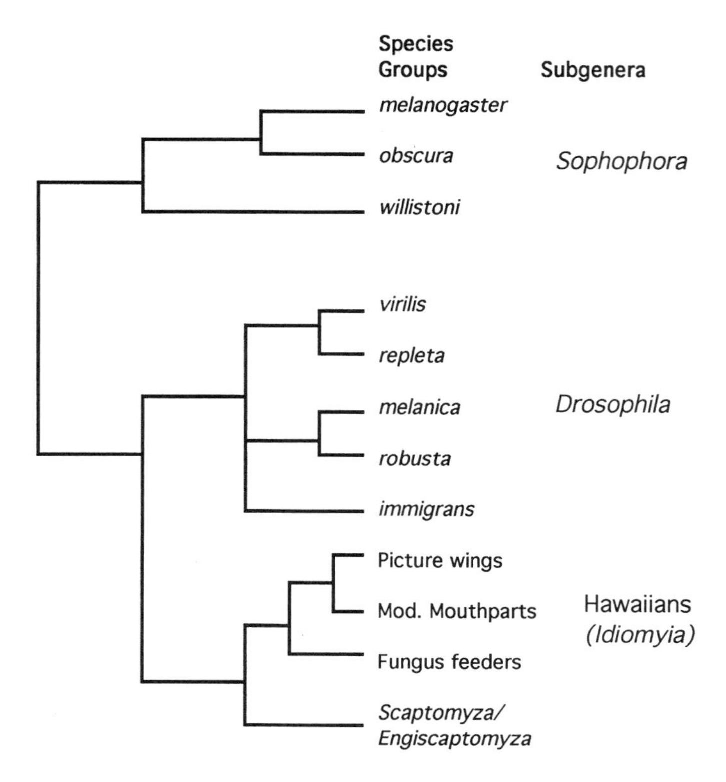

FIGURE 8-8 Consensus relationships supported by several data sets. The lack of resolution of subgenus *Drosophila* is apparent.

the *virilis* and *repleta* groups are almost certainly sister taxa, as are the *melanica* and *robusta* groups. Other than this, the relationships of the major groups remain largely unknown.

Absolute Times

Branching patterns are one form of phylogenetic information, but it is also of interest to have estimates of the actual times of the major events in a phylogeny. In attempting to estimate absolute times there are three sources of data: fossils, biogeography, and molecular clocks. To use molecular clocks requires independent estimates from one of the two former sources to calibrate the clock. There is a reasonable amount of information from these three sources, which are often consistent among themselves, which allows us to place some degree of confidence in absolute times of major events. Table 8-4 is a summary of this information for Drosophilidae and some related Diptera; figure 8-9 illustrates the information graphically.

In trying to estimate absolute times from the information in table 8-4 and figure 8-8, it is important to note the range of suggested dates. Obviously a fossil is a minimum estimate of the time of origin of a lineage. Also, basing estimates of time solely on molecular clock estimates is risky, as clocks are often inaccurate and lin-

TABLE 8-4 Absolute time estimates for Drosophilidae evolution based on three sources of data: fossils, biogeography, and molecular clocks.

Fossils Major Taxon		Age	Reference
Lower Diptera	Nematoceran and Brachyceran	Jurassic (135 my)	Hennig (1973)
Cyclorrhapha	Phoroidea	Cretaceous (70 my)	MacAlpine (1970)
Muscoidea	Calliphoridae	Cretaceous (70 my)	MacAlpine (1970)
Steganinae	Electrophoretica succini	Eocene (40 my)	Hennig (1965)
Chymomyza	C. primaeva	Oligocene (30 my)	Grimaldi (1987)
Protochymomyza	P. miocena	Oligocene (30 my)	Grimaldi (1987)
Drosophila	D. succini	Oligocene (30 my)	Grimaldi (1987)
Drosophila	D. poinari	Oligocene (30 my)	Grimaldi (1987)
Hirtodrosophila	H. paleothoracis	Oligocene (30 my)	Grimaldi (1987)
Myiomia	M. io	Oligocene (30 my)	Grimaldi (1987)
Neotanygastrella	N. wheeleri	Oligocene (30 my)	Grimaldi (1987)
Scaptomyza	S. dominicana	Oligocene (30 my)	Grimaldi (1987)
Drosophilinae	D. spA	Oligocene (30 my)	Grimaldi (1987)

Biogeography Divergence event	Time	Reference
Drosophilidae-Ephydroidea	80 my	Beverley and Wilson (1984); Hennig (1960)
Steganinae-Drosophilidae Mayagueza	80 my	Grimaldi (1988)
Scaptodrosophila divergence	36 my	Throckmorton (1975)
Obscura Sophophora	36 my	Throckmorton (1975)
virilis-melanica-robusta-immigrans species group divergence	36 my	Throckmorton (1975)
Hirtodrosophila divergence	36 my	Throckmorton (1975)
melanogaster subgroup	17 my	Lachaise et al. (1988)
melanogaster-orena	6–15 my	Lachaise et al. (1988)
orena-erecta	6–7 my	Lachaise et al. (1988)
melanogaster-simulans	2.5 my	Lachaise et al. (1988)
Molecular Clock		
Sophopora-subgenus Drosophila	61–65 my	Beverley and Wilson (1984)
obscura-melanogaster	46 my	Beverley and Wilson (1984)
melanogaster-willistoni	53 my	Beverley and Wilson (1984)
melanogaster-orena	5.5 my	Moriyama (1987)
	6–15 my	Lachaise et al. (1988)
melanoster-simulans	2.5 my	Lachaise et al. (1988)
	2.3 my	Moriyama (1987)
repleta-robusta	35 my	Beverley and Wilson (1984)
Hawaiian Drosophilidae	40 my	Beverley and Wilson (1984)
	23 my	Thomas and Hunt (1991)

From Powell and DeSalle (1995).

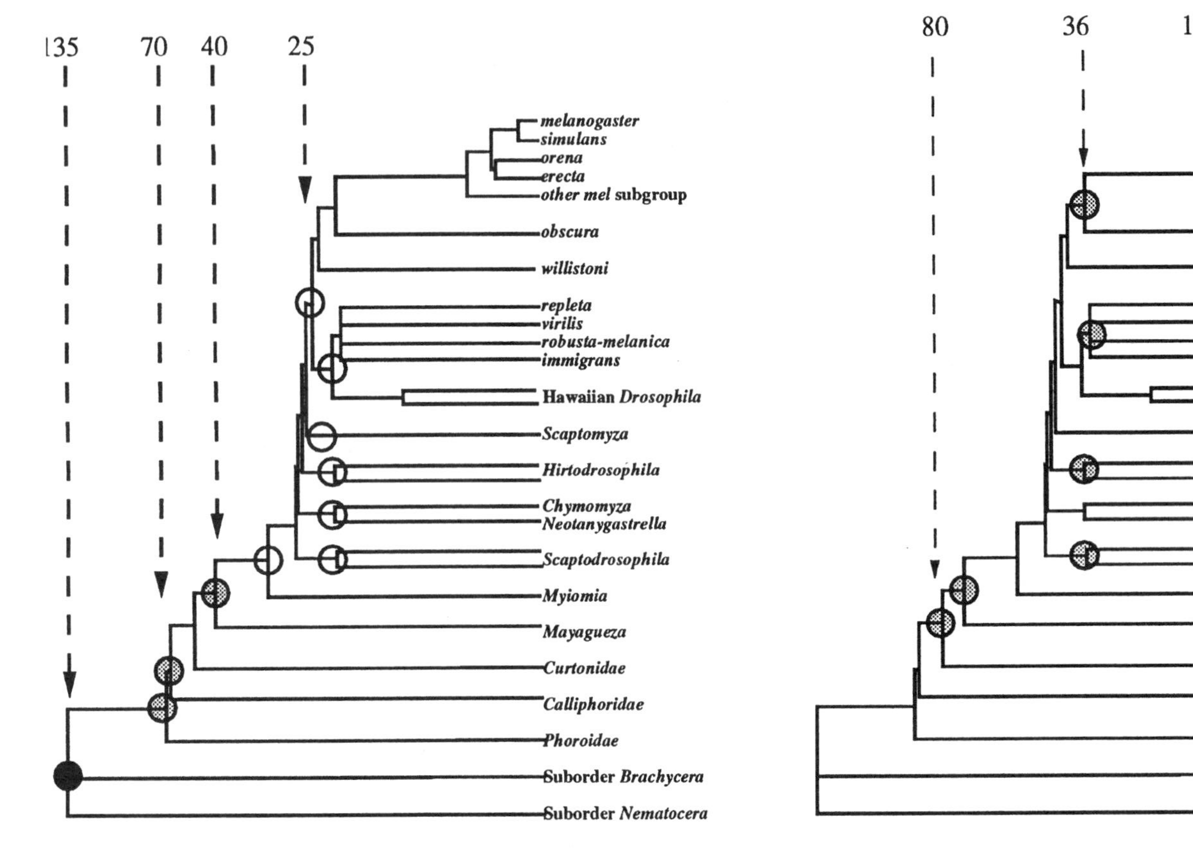

FIGURE 8-9 Attempts to place absolute time estimates on Drosophilid cladogenic events. Numbers at top are in millions of years. References are in table 8-4. From Powell and DeSalle (1995).

eage-specific. However, even given all these uncertainties, we can place approximate ranges on the time of certain major events in the phylogeny and we have done so in figure 1-2. As can be seen in that figure, the major lineages of *Drosophila* are really quite old. The two major subgenera split no more recently than 30 million years ago (mya) and perhaps as long ago as 60 mya; by comparison, all the major orders of mammals diversified in this time. The *melanogaster* group emerged about the same time as the Great Apes. Likewise, the major splits in subgenus *Drosophila* occurred early, having been established no less than 20 mya. Thus we can conclude that *Drosophila*, like many other insects, tended to diversify early and then remain relatively stable for a long time. A modern day biologist transported back some 15 million years would find the drosophilid fauna very much like it is today (with the possible exception of Hawaiian flies).

The absolute time estimate on the origin of Hawaiian flies deserves special comment. The presence of a fossil of a species in subgenus *Drosophila* (the sister clade to the Hawaiians) dated at about 30 my (Grimaldi, 1987) would argue for this as a minimum age of this lineage; this estimate agrees well with molecular clock estimates (Beverley and Wilson, 1985; Thomas and Hunt, 1991) all of which indicate a minimum age of about 25 my. This brings up the intriguing problem of how this lineage formed when the oldest extant Hawaiian island (Kauai) is only about 5.5 my old. Beverley and Wilson (1985) were the first to hypothesize that the Hawaiian lineage was founded on islands now submerged. The Hawaiian archipelago is being formed by a hot spot of volcanic activity now located in the extreme southeast of the youngest island, Hawaii. However, the tectonic plate has been moving northwestward over this spot for 50 my or more forming a long line of volcanoes, some of which rose above sea level and have since eroded to below the surface. (See Rotondo et al. [1981] and Carson and Clague [1995] for a discussion of the biogeography of this case.) The question of whether these presently submerged islands had the type of rain forest or vegetation to support a drosophilid fauna has not been established in all cases, although some certainly did have vegetation. Given the present information, the submerged island hypothesis would seem to be the best explanation. The alternative is that the group started evolving somewhere else and then moved *en masse* to the Hawaiian archipelago. A final alternative is that both the fossil is misinterpreted and all the molecular clocks are wrong; this seems highly unlikely.

Individual Groups

The *melanogaster* Group

Some 150 species have been placed into 10 subgroups of the *melanogaster* group (Lemeunier et al., 1986; Ashburner, 1989). Most groups have an Oriental center of distribution, suggesting the group originated in Asia; however, some subgroups are in Africa, Australia, and South America. Except for the *melanogaster* subgroup, by far the most studied species in this group belong to the *ananassae* subgroup (Tobari, 1993a). Based on banding patterns of the polytene chromosomes, Lemeunier et al. (1986) suggest the group's subgroups can be divided into three major lineages: *ananassae, montium*, and *melanogaster-takahashii-suzukii-eugracilis-ficusphila-elegans*.

Other than this suggestion, little further information is available on the relationship among subgroups within the *melanogaster* group.

The *melanogaster* subgroup, being the most intensely studied group of *Drosophila*, has had its phylogeny examined from a variety of standpoints. The data from biogeographical considerations (Lachaise et al., 1988), polytene banding patterns of chromosomes (Lemeunier and Ashburner, 1976; 1984), DNA–DNA hybridization (Caccone et al., 1988b) and *Adh* DNA sequences (Jeffs et al., 1994; Caccone et al., 1996) all converge on the same topology, indicated in figure 8-10; also in this figure, some absolute times of cladogenic events are indicated based on biogeographical considerations (Lachaise et al., 1988). The biggest problem in this phylogeny concerns three very closely related species; *D. simulans* and the two island endemics, *D. sechellia* and *D. melanogaster*. Neither morphology nor chromosomes could resolve the relationship of the members of this triad (Lachaise et al., 1988) nor could the DNA sequence from any single gene. By combining information from 12 genes, the rela-

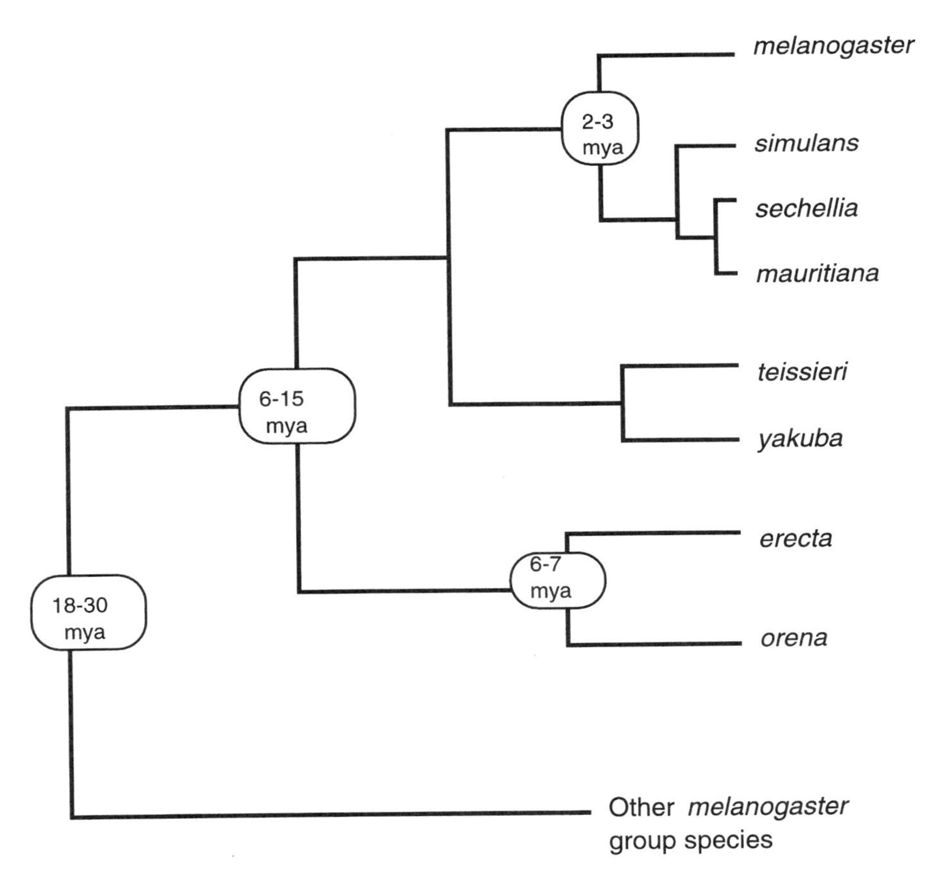

FIGURE 8-10 Relationships within the *D. melanogaster* subgroup. Data of several sorts support this tree (see text). Dates are from biogeographic considerations, largely from Lachaise et al. (1988).

tionship indicated in figure 8-10 was very strongly supported (Caccone et al., 1996), which is in agreement with DNA–DNA hybridization (Caccone et al., 1988b). Therefore, until contrary evidence arises, we can take the relationships in figure 8-10 as being very well supported.

The *obscura* Group

While the placement of the *obscura* group as a sister to the *melanogaster* group is quite secure, the relationships within this group have proven difficult to resolve. Classically it was divided into two subgroups based on the number of sex combs on male forelegs, testes shape, and numbers of acrostichal hairs; these are the *obscura* subgroup and *affinis* subgroup (Sturtevant, 1942; Patterson and Stone, 1952). The former subgroup has more recently been split into two subgroups, the Old World *obscura* subgroup and the New World *pseudoobscura* subgroup (Lakovaara and Saura, 1982), a classification which fits better the phylogenetic relationships. With one ambiguous exception (*D. helvetica*), the *affinis* subgroup is also confined to the New World. Most species today are temperate and subtropical in the Northern Hemisphere, although the likely origin of the group is Africa, where some relictual species still exist.

The phylogenetic relationships in the New World *obscura* group are summarized in figure 8-11. The phylogeny here is a consensus of the DNA–DNA hybridization results of Goddard et al. (1990), RFLP mtDNA data of Latorre et al. (1988) and Gonzalez et al. (1990), and DNA sequence data (Beckenbach et al., 1993; Barrio et al. 1994; Gleason et al., 1997), as well as with older chromosomal, allozyme, and morphological data. All these data sets agree on the essentials of what is shown in

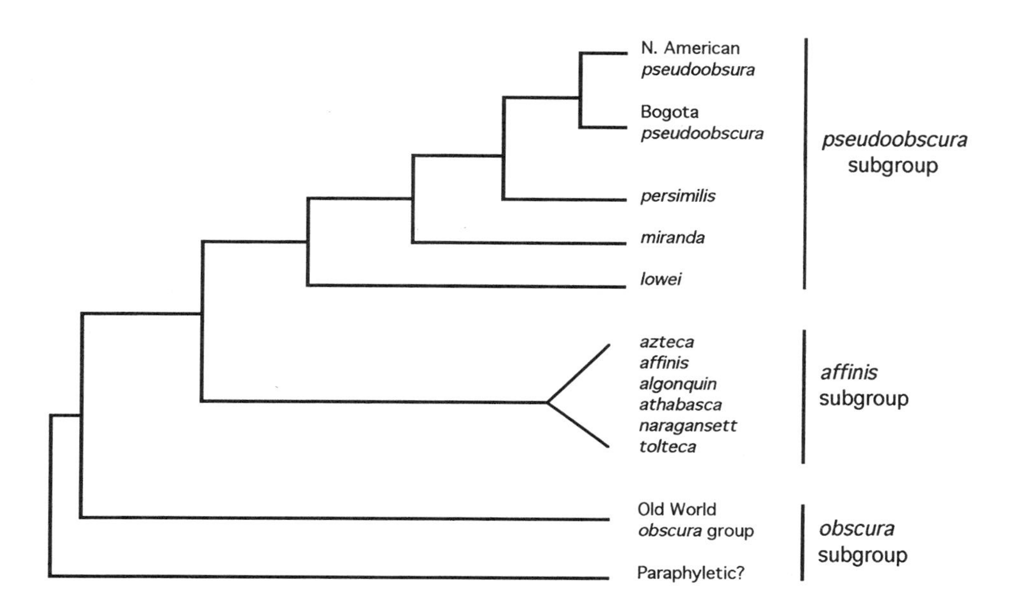

FIGURE 8-11 Relationships among New World *obscura* group species. Several data sets support this tree. Little resolution of the *affinis* subgroup has been obtained.

this figure and can be summed up as follows. The subspecies of *D. pseudoobscura* are the closest taxa, followed by *D. persimilis, D. miranda*, and *D. lowei.* This last species has not been discussed in this book as it is a somewhat rare, narrowly distributed species that does not breed well in the laboratory, although it is clearly a member of the *pseudoobscura* subgroup as already hypothesized by the discoverers of this species (Heed et al., 1969). While the *affinis* subgroup is monophyletic (at least for those species studied), the relationships within this subgroup are not clear; none of the data sets resolves them with any certainty. Perhaps the most interesting thing about this subgroup is its recency. Both the hybridization data (Goddard et al., 1990) and DNA sequence data (Beckenbach et al., 1993) indicate the deepest branches in this subgroup are only about as deep as the node leading to *D. miranda.* Assuming a molecular clock, this would indicate that this whole subgroup radiated since this time, forming ten or so extant species. Beckenbach et al. (1993) make some time estimates for the *obscura* group lineage splitting based on molecular clocks, and conclude that the *affinis* subgroup is at most 3.5 million years old.

The Old World *obscura* species present a more complex picture. Figure 8-12 shows the conclusions from four separate molecular data sets, all indicating this set of species is paraphyletic. Paraphyly of the Old World subgroup is not surprising in light of the biogeography of the group. Africa is the likely origin of the *obscura* group, where there are still a few members; the mtDNA data of Ruttkey et al. (1992) is consistent with this being the oldest lineage in the group. Two African species have been studied, *D. kitumensis* and *D. microlabis*, and their position in the proposed tree (upper right tree in figure 8-12) indicates they are the oldest lineages yet found. Members of this group became adapted to temperate regions and moved north into Europe and Asia. Sometime later, the New World was colonized by the group. Throckmorton (1975) suggests this had occurred by the mid-Miocene period, some 15 million years ago, when temperate forest still connected North America and Europe. On the other hand, assuming a molecular clock, Goddard et al. (1990) pointed out this time is perhaps too early for the split between the New World and Old World lineages and that perhaps an introduction into the New World via the Bering Strait and Aleutian Islands is more likely, where connections existed up to 3 million years ago. In any case, the evidence is that the introduction was likely a single lineage after the Old World species had already begun to diversify. Thus the paraphyly of the Old World subgroup relative to the New World subgroups is the expected pattern.

While we have stated that the phylogenetic relationships within the Old World *obscura* subgroup are ambiguous, it is only fair to point out there is considerable data that produce some degree of resolution. Four different groups have performed allozyme studies on this subgroup, which included proposed phylogenies: Cabrera et al. (1983), Lakovaara et al. (1972; 1976), Loukas et al. (1984), and Marinkovic et al. (1978). Unfortunately different groups used different sets of loci and different methods of analysis. They all come up with different phylogenies. Lakovaara and Saura (1982) go into more detail in discussing various attempts to resolve relationships within this subgroup. Steinemann et al. (1984) and Felger and Pinsker (1987) also present phylogenies based on chromosomal evolution, which provides modest resolution, although it is not totally congruent with some of the other data. The DNA-based data sets illustrated in figure 8-12 show that this source of information has not yet clarified matters to any great degree. The largest DNA data set is that for four mtDNA

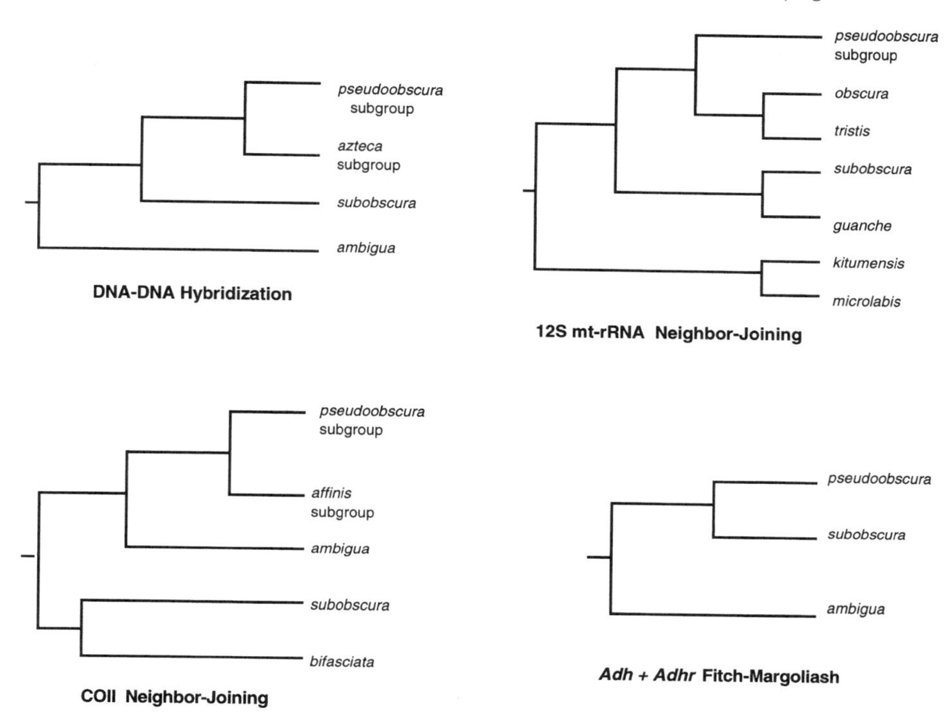

FIGURE 8-12 Four hypotheses derived from four different molecular data sets. In all cases the paraphyly of the Old World *obscura* species is apparent. Except for those taxa called subgroups, all the species on the trees are Old World. *D. kitumensis* and *D. microlabis* in the lower right tree are from Africa. DNA–DNA hybridization data from Goddard et al. (1990), COII data from Beckenbach et al. (1993), 12S data from Ruttkey et al. (1992), and *Adh* data from Marfamy and Gonzàlez-Duarte (1992b).

genes that show good support for two major lineages in the Old World, one containing *subobscura/guanche/madeirensis*, and one containing *obscura/ambigua/tristis* (Barrio et al., 1994); however, even with this data set, the relationship of these two major lineages to the New World lineage was ambiguous, with paraphyly and even polyphyly sometimes supported.

The *willistoni* Group

Until recently, the relationships among the sibling species of this group were quite obscure. Spassky et al. (1971) presented a diagram of relationships based on biogeography, ability to hybridize, and chromosomes. Ayala et al. (1974c) presented a tree based on allozyme distances. The two sets of relationships were not congruent, with the exception that *D. nebulosa* (a nonsibling member) was the outgroup and *D. insularis* was the oldest lineage. With the addition of DNA sequence data, in particular from *Adh* (Carew, 1993), *Per* (Gleason, 1996), and the COI gene in the mtDNA (unpublished data), a firm hypothesis has emerged. Figure 8-13 indicates the relationships supported by these data and the bootstrap confidence. In agreement with the previous studies, *nebulosa* is the outgroup, with *insularis* the oldest lineage. *D. equinoxialis* and *D. paulistorum* are the two closest relatives followed by *willistoni*, then *tropicalis.*

The *virilis* Group

We now shift attention to subgenus *Drosophila.* The *virilis* group of this subgenus consists of 13 (or perhaps 14) species and has been the subject of considerable work including phylogenetic studies (reviewed in Throckmorton, 1982b). Spicer (1992) analyzed all data relevant to phylogenies in this group, which includes 10 molecular studies (almost all based on proteins rather than DNA), 7 behavioral studies, 6 chromosomal studies, 4 morphological studies, and 1 developmental study. After careful consideration of all the information and critical evaluation, Spicer (1992; and personal communication, 1994) produced the consensus tree shown in figure 8-14. This tree, including the root, is congruent with one produced independently based on *Adh* sequences (Numinsky et al., 1996). The group can be subdivided into four monophyletic lineages (what Spicer calls subphylads). One contains five species: *virilis, lummei, novamexicana, americana*, and *texana.* A second lineage contains four species: *mon-*

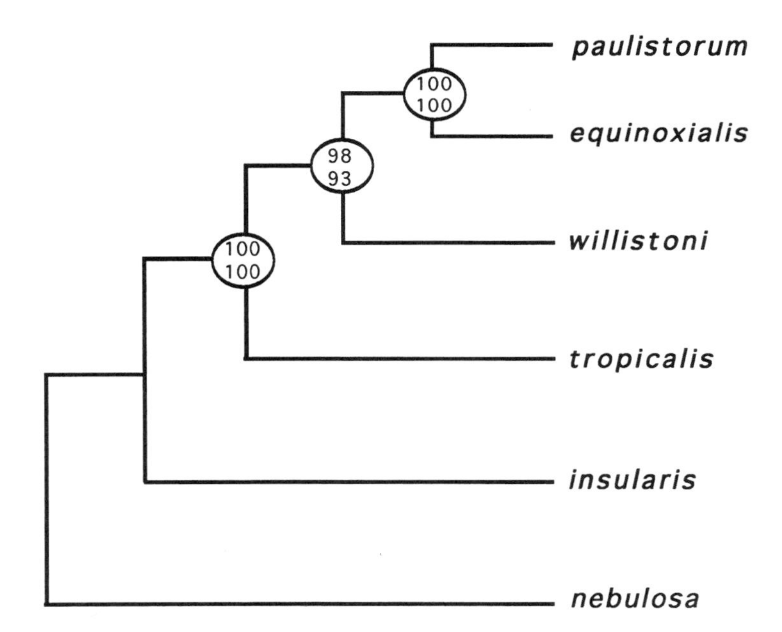

FIGURE 8-13 Results from combined DNA sequence data for *Adh* (Carew, 1993), *Per* (Gleason, 1995), and mtDNA COI (unpublished) from the *willistoni* group. An identical topology was obtained for Maximum Parsimony and Neighbor-Joining; bootstrap values are noted at the nodes, the upper figure for parsimony, the lower for NJ. From Gleason (1995).

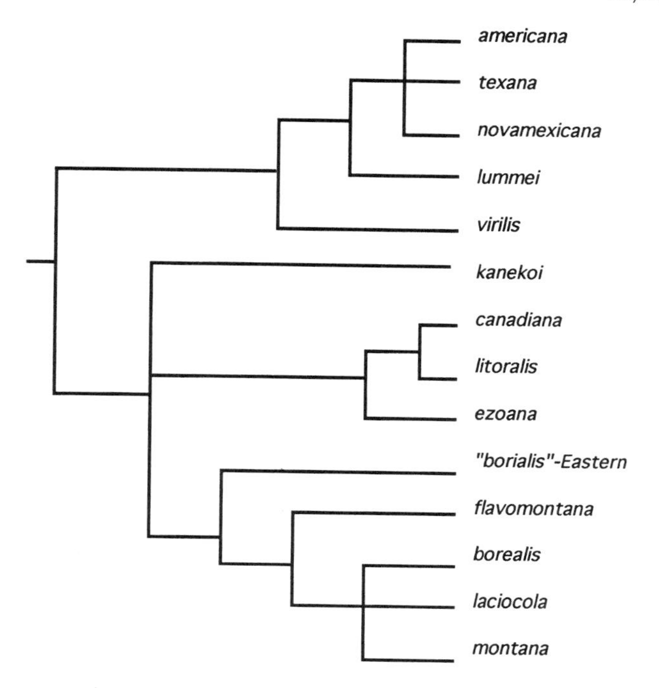

FIGURE 8-14 *Virilis* group phylogenetic relationships based on several sources of data. From Spicer (1992) and G. Spicer (personal communication).

tana, lacicola, borealis, and *flavomontana*. The third lineage contains *littoralis* and *ezoana* and the fourth is a monotypic lineage with *kanekoi*. Spicer concludes that there is some resolution of relationships within the two larger lineages, as indicated in figure 8-14. The root of the tree is between the *virilis*-containing subphylad on one side, and the other three on the other side. Very recent evidence indicates that *D. borealis* may be two species with the flies from the eastern part of the distribution being genetically distinct and related to the other species as indicated in figure 8-14 (G. Spicer, personal communication).

The *repleta* Group

This large group has been studied in detail by M. Wasserman with regard to chromosomal phylogenies (Wasserman, 1982, 1992). There are five major subgroups of the *repleta* group, the *repleta, fasciola, mulleri, mercatorum*, and *hydei* subgroups. Figure 8-15 is a cladogenic interpretation of Wasserman's (1992) chromosomal phylogeny. Among these, only the *mulleri* subgroup has been studied for DNA variation by Baker et al. (1994; figure 8-16). This tree is based on a combination of a nuclear gene (acetylcholinesterase), two mtDNA genes (COII and 16s rRNA), plus chromosomal inversions. The groupings in this figure coincide quite well with Wasserman's hypotheses based on inversions alone, that is, the molecular and inversion trees are consistent with one another.

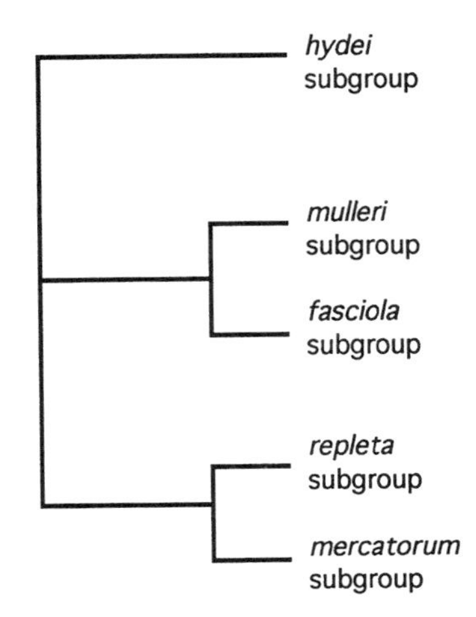

FIGURE 8-15 Relationship of the five major subgroups of the *repleta* group. This is an interpretation of the network in Wasserman (1992).

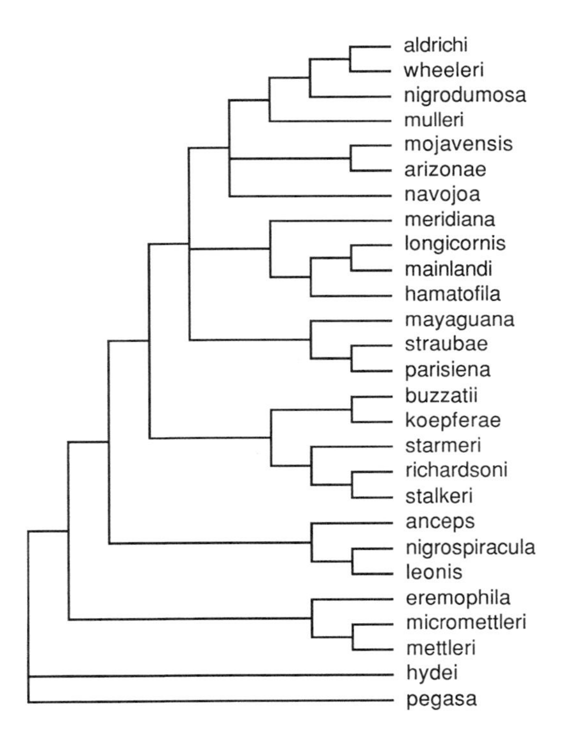

FIGURE 8-16 Maximum parsimony tree from a combined analysis of mtDNA and nuclear DNA sequences along with chromosomal inversions for species of the *mulleri* subgroup of the *repleta* group. From Baker et al. (1994) and Baker (personal communication).

The Hawaiian *Drosophila*

We have already emphasized that the molecular data all support the monophyly of the Hawaiian flies. In addition, some further resolution within the fauna can be seen in the data of DeSalle (1992b) and Russo et al. (1995); these are depicted in figures 8-7, 8-8, and 10-20. The most consistent conclusion from these data is that the Hawaiian *Scaptomyza* and *Engiscaptomyza* are a monophyletic lineage separate from the Picture Wing, Modified Mouthparts, and White-tip Scutellum lineage; the fungus feeders form a third lineage. DeSalle (1995) has elaborated on these species phylogenies in the context of biogeographic considerations (area cladograms).

By far the best phylogenies of the Hawaiian fauna come from the work of Carson and colleagues on the Picture Wing group using inversions in polytene chromosomes. Figure 8-17 is a depiction of a network of 106 species! This is the most spectacular inversion phylogeny ever produced, at least in terms of numbers of species. Not all of the individual species are resolved in this phylogeny as some are homosequential, that is, having identical polytene banding patterns. Relatively little independent molecular work has been done on these species, although what little has been done has been congruent with the inversion phylogeny. The best example is the *planitibia/differens/silvestris/heteroneura* quartet, which has been discussed in the previous chapters. The DNA–DNA hybridization results of Hunt and Carson (1983) are depicted in figure 7-4 and this deduced topology is consistent with the topology in figure 8-17. Kaneshiro et al. (1995) have analyzed these same inversion data using cladistic techniques and produce results concordant with the more classical method used to generate the network in figure 8-17.

Gene Trees–Species Trees

Many, though not all, of the phylogenetic hypotheses presented in this chapter are based on molecular data of one sort or another. We should be explicit, although seldom are, that in fact when one analyzes molecular data, the resulting trees are the best representation of the data given some framework or optimizing principle. For example, given a set of sequences, one can ask computer programs to construct a network connecting the set of sequences in a way that requires the minimum number of changes to derive one from the other, the maximum parsimony criterion. Or under maximum likelihood, one constructs the network that connects the sequences in the most probable manner given a certain probability of changing each nucleotide for another. Such networks are of rather limited interest of and by themselves; they are produced mechanically by feeding a set of A's, T's, C's and G's into a computer program written by someone else. The networks become phylogenetic trees of the taxa involved if the network accurately reflects the history of the taxa from which the sequences were taken. This has become known as the *gene tree–species tree problem*: can one assume that the tree generated for a set of sequences (regardless of the specific gene or method used to construct the tree) accurately reflects the species phylogeny? In *Drosophila* there are two empirical examples of the complexity of this question. The first has to do with complexity introduced by the ubiquity of inversions in *Drosophila* and their effects on DNA sequence evolution, and the other has to do with

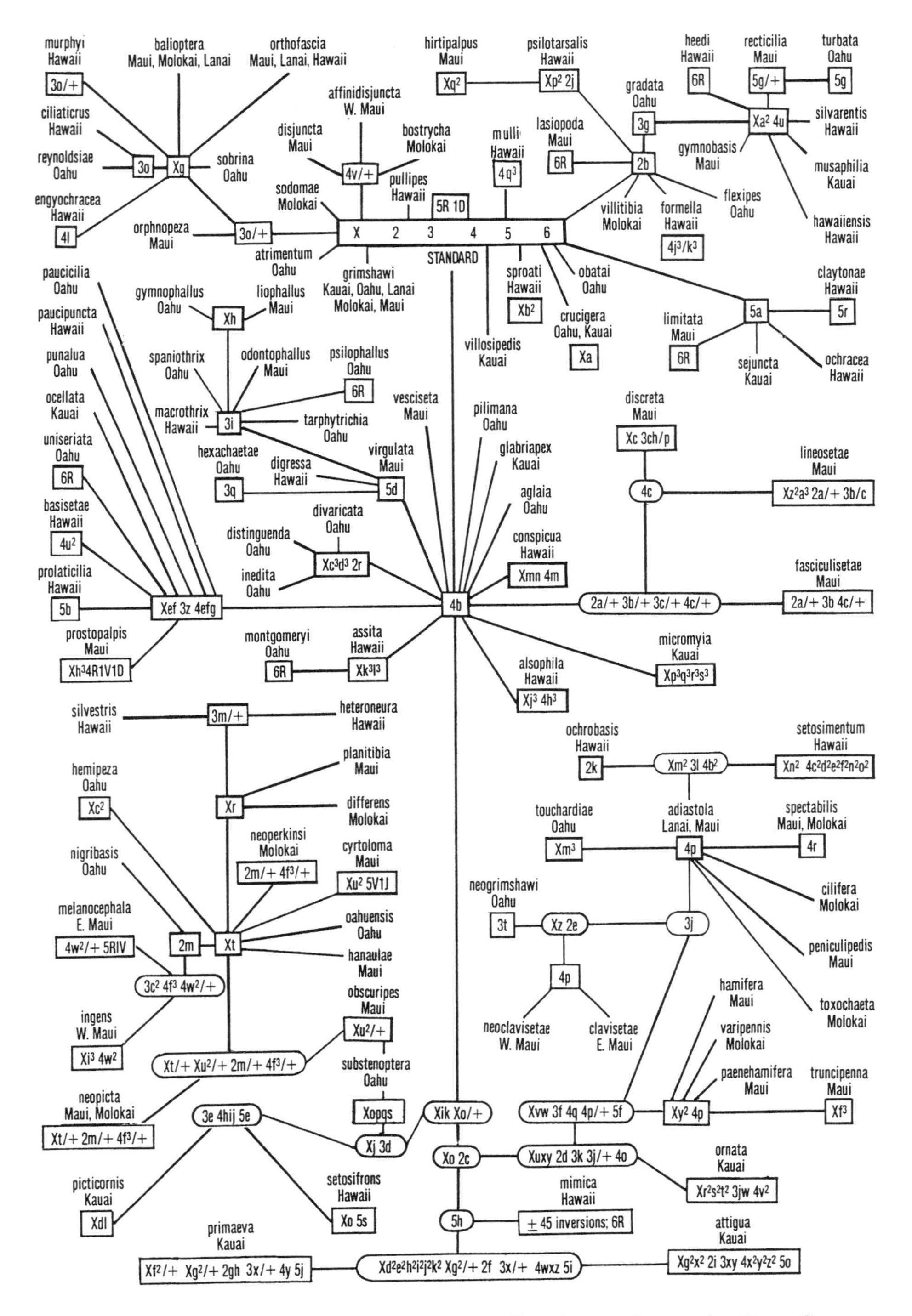

FIGURE 8-17 Chromosomal inversion tree for Hawaiian picture wing species. From Carson (1992).

studying extremely closely related taxa, taxa so recently separated that not enough time has elapsed for ancestral polymorphisms to decay.

Inversion, Gene, and Species Trees

The first example is taken from the American *pseudoobscura* subgroup. In figure 8-11 we presented a phylogeny for this group and stated it was very well-supported by a diverse set of data: molecular, chromosomal, morphological, behavioral, and degree of reproductive isolation. It almost certainly reflects the history of the taxa involved (see figure 8-18, part A). We have also discussed in detail the inversion polymorphisms in these taxa, especially in *D. pseudoobscura* and *D. persimilis* (chapters 3 and 4). In figure 3-3 we presented the phylogeny deduced for the inversions. In the Goddard et al. (1990) study, two strains of *D. pseudoobscura* from North America were used, one fixed for TL and the other polymorphic for PP and ST; these represent the TL and ST phylads respectively. The Bogotá strain of the subspecies *D. pseudoobscura bogotana* was fixed for TL; the Bogotá populations have only TL and SC inversions. The one *D. persimilis* strain studied was polymorphic for KL and WT.

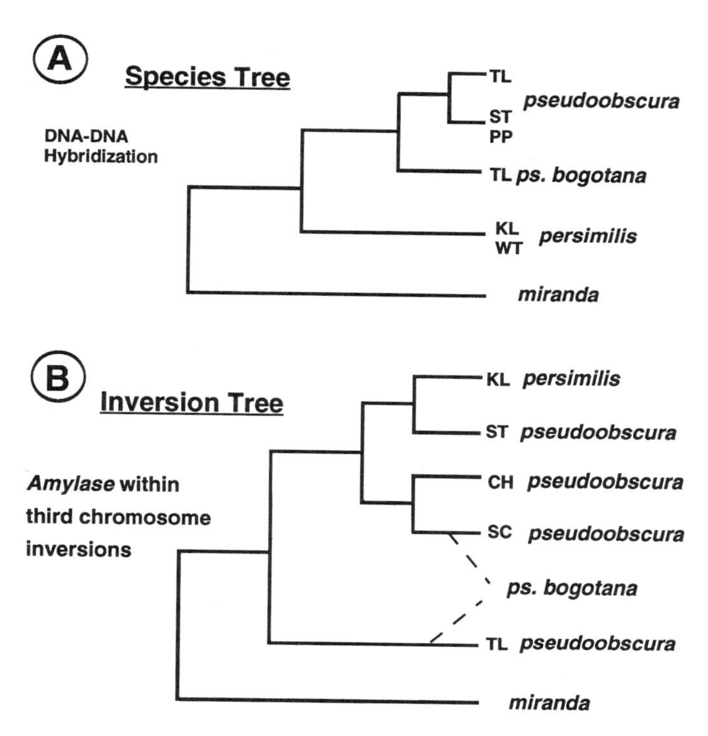

FIGURE 8-18 Two phylogenies of the New World *pseudoobscura* subgroup. (A) is supported by a number of data sets. (B) is for the amylase gene located within the third chromosome inversions (Aquadro et al., 1991). Letters at tips of branches refer to the inversions. From Powell (1991).

Aquadro et al. (1991) studied molecular divergence (by means of RFLP analysis) of a gene region within the breakpoints of most of these inversions on the third chromosome, the region around the amylase locus. From these data they reconstructed a phylogenetic tree for this region; this is presented in figure 3-23. As we noted in chapter 3, the resulting topology is consistent with the inversion topology, which is what one would expect if inversions were monophyletic and suppressed recombination within the breakpoints. We have summarized this phylogeny in part B of figure 8-18. For these inversions then, *D. pseudoobscura* would be considered a paraphyletic taxon with *D. persimilis* a derivative of a subset of the polymorphism within *D. pseudoobscura* (that of the ST phylad). Yet we see in part A of figure 8-18, there is considerable evidence indicating that the *D. pseudoobscura* ST phylad is genetically closer to the *D. pseudoobscura* TL phylad than it is to *D. persimilis.* Furthermore, if one considers the subspecies *bogotana*, note that it is composed of inversions from two different parts of the inversion phylogeny (the TL and SC phylads), so it would be considered a polyphyletic taxon.

The important point from this example is that, depending on the data set used, one could conclude that these taxa are either simple monophyletic groups, as indicated in part A of figure 8-18, or display the complexity in part B of this figure. Gene sequences found within inversions that do not undergo recombination would presumably all display the complex phylogeny observed for the amylase region. This is almost certainly a case of the inversion polymorphism predating the speciation events, so the assortment of alleles at speciation can cause confusion. In this case, *D. persimilis* was derived from ST, which itself is a derived gene arrangement. Because inversion polymorphisms in *Drosophila* are most often maintained by balancing selection (chapters 3 and 4), they might be expected to persist in lineages longer than neutral polymorphisms. Thus in groups like *Drosophila* that are generally replete with inversions, it is important to know the chromosomal location of any genes used for phylogenetic work and to be able to evaluate whether the gene is confusing species histories and inversion histories. We will have more to say about the influence of inversions on sequence evolution in chapter 10.

The *simulans/mauritiana/sechellia* Triad

The second example of confusion between gene and species trees concerns the three very closely related members of the *melanogaster* subgroup, *D. simulans, D. mauritiana*, and *D. sechellia.* In the chapter on speciation (chapter 7) these species played a large role, as presumably they have very recently speciated and thus reflect some of the early genetic changes occurring during speciation. The species are interfertile with one another, with F_1 females fertile, males sterile. Chromosomally, the polytene banding patterns are identical with no common polymorphisms, so the complications in the last example do not affect this triad. That these three species are extremely closely related is confirmed by allozyme data and DNA-based data.

J. Hey and colleagues have studied this triad by sequencing several alleles from each species for three genes, *period, Yolk protein2*, and *zeste* (Hey and Kliman, 1993; Kliman and Hey, 1993a; Hey, 1994). Studying multiple copies of genes (alleles) from each taxa, as well as several independent genes, becomes extremely important for such closely related taxa. Hey's program has been one of the few to address this in explicit detail. Figure 8-19 is an example of the results for *zeste*. *D. melanogaster* is the sister taxon to the group and thus is a good outgroup for the triad (figure 8-10). However, within the triad, relationships are not at all clear. Different copies of the gene (alleles) seem to indicate different relationships. This has also been true for a number of other loci studied for these species as well as for mtDNA. Presumably, many of these nucleotide polymorphisms predate the recent speciation event, thus obscuring species relationships. These polymorphisms are very likely closer to being neutral than the inversion polymorphisms, and yet for such closely related species, not enough time has elapsed to have fixed a monophyletic gene lineage in each species lineage.

How could one overcome this problem? The usual way is to add more independent (not tightly linked) genes rather than more alleles of a few genes. When this is done for the *simulans/mauritiana/sechellia* triad, the results are congruent with the total single-copy genome as measured by DNA–DNA hybridization (Caccone et al., 1988b; Jeffs et al., 1994; Caccone et al., 1996). The results are as indicated in figure 8-10, showing that the two island species, *D. mauritiana* and *D. sechellia*, are the two most closely related, followed by *D. simulans*. This triad remains one of the best examples of the gene tree-species tree problem and serves as a model for the complexity of trying to resolve the relationship of extremely closely related species using DNA data.

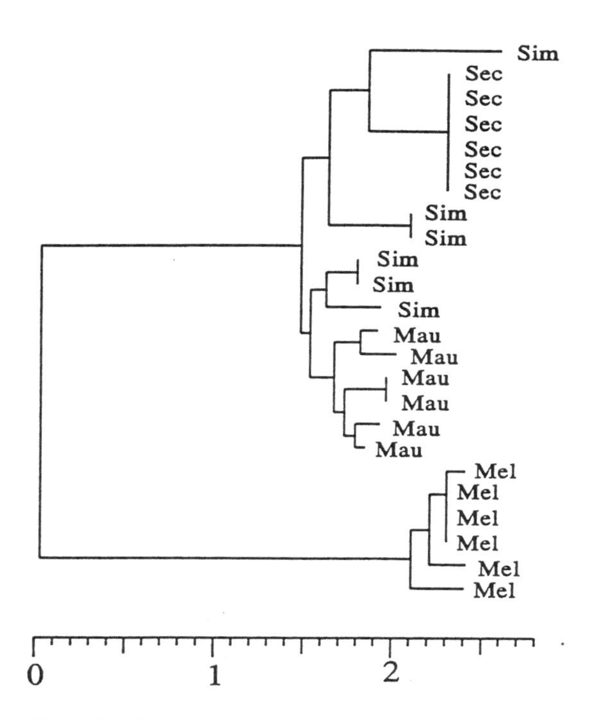

FIGURE 8-19 Gene tree for *zeste* from *D. simulans* (Sim), *D. sechellia* (Sec), *D. mauritiana* (Mau), and *D. melanogaster* (Mel). Each branch represents a different allele. From Hey (1994).

Conclusions

The knowledge of the phylogenetic relationships of members of the Drosophilidae, and in particular *Drosophila*, has undergone a recent saltational increase. It is now possible to define with some confidence and rigor many of the major cladogenetic events in this group and even to begin to date events with some degree of confidence. These phylogenies in and of themselves are interesting, but as we will illustrate in the following chapters, they can also be used to analyze evolutionary events of all sorts (see also Powell and DeSalle, 1995). The comparative method of evolutionary biology requires sound phylogenies such as most of those presented in this chapter.

Despite this progress, clearly much remains to be done in the field of *Drosophila* phylogenetics. Major questions remain, such as the relationships of the Old World *obscura* subgroup, and among major groups of subgenus *Drosophila.* Having more phylogenetic information can only enhance the usefulness of this model for all sorts of biological work. Unfortunately, as with so many other groups of organisms, the environmental changes brought on by human activity have and will continue to threaten many species. Perhaps the most critical situation is in Hawaii, where the magnificent drosophilid fauna is precariously poised and where no doubt extinctions are taking place. The fact that most of these species cannot be easily maintained in laboratory culture means their existence is totally dependent on natural habitats.

9

Genome Evolution

> In many ways we are like children in an enchanted forest, wandering almost aimlessly from discovery to discovery. For the moment, at least, that should be sufficient. At some point we will inevitably emerge into a clearing where principles and patterns in the organization and evolution of the genome are evident. Until then, let us be thankful that the pleasures of the forest are so numerous and diverse.
>
> Ross J. MacIntyre, 1985

This is the first of two chapters that focus on how the genetic material, DNA, and its primary products (e.g., proteins) evolve. The subject is somewhat arbitrarily divided into two views. The "macro" view, which will be the subject of this chapter, considers issues such as the evolution of chromosome number, genome size, and organization of the genome. The "micro" level, the focus of the next chapter, is concerned with the specifics of DNA and protein evolution and the mechanisms involved. Although these two levels of analysis are really a continuum, there is some conceptual justification for separating them.

Drosophila Genome Structure

Sizes and Patterns

Before proceeding to issues of evolution of genomes, we need to present what is known about the structure and organization of the *Drosophila* genome. Obviously, to speak of "the" genome is inaccurate, as there are about 2,000 different genomes in the genus. What we will present here is the *D. melanogaster* genome, as it is by far the best known. It will serve, as so often has been the case, as the prototype.

The *D. melanogaster* haploid genome is 0.18 to 0.21 picograms (10^{-12} grams) of DNA, which corresponds to about 1.65 to 2.0×10^8 base pairs (bp). A range is presented because different methods of measurement have given somewhat different estimates and there is also some indication that different strains and sexes may differ in genome size (see Laird, 1973; Rasch et al., 1971; Dawley et al., unpublished). This is a small genome compared to other multicellular eukaryotes. For example, the typical mammalian haploid genome is some 3×10^9 bp, which means a *Drosophila* genome is about 5% that of a typical mammal's. However, among Diptera, *Drosophila* genome size is more or less typical, especially if one considers the variation in genome sizes found among species of *Drosophila* (discussed in more detail later). Table 9-1 shows estimates of genome size for several Diptera in order to see how *Drosophila* compare.

All evidence is that the genome of *Drosophila*, as well as most eukaryotes, consists of single long strands of DNA that are uninterrupted along the whole length of each chromosome. However, the genome is far from a homogeneous structure. The first indication of this came from the early observation that use of classical stains in cytological preparations produces two very distinct parts of chromosomes, parts that stain relatively weakly (euchromatin) and parts that stain very darkly (heterochromatin) (Heitz, 1933, 1934). Figure 9-1 is a diagram of the distribution of euchromatin and heterochromatin in the *D. melanogaster* genome in terms of approximate megabases (millions of base pairs). Autosomal heterochromatin is mostly confined to the centromeric regions. About half of the X chromosome is heterochromatic, while the entire Y chromosome is considered heterochromatic. Altogether, heterochromatin composes approximately 30% of the genome. Initially, it was thought that heterochromatin is genetically inert, that is, does not have functional genes. However, several functional genes, including the rRNA-encoding regions, are localized to heterochromatin (reviewed in Gatti and Pimpinelli, 1992), although compared to euchromatin, the density of (known) functional units is much less in heterochromatin.

From a molecular perspective, the *Drosophila* genome is also highly heterogeneous. As with most eukaryotes, the degree of repetition of DNA sequences varies.

TABLE 9-1 Estimates of genome size for several Diptera.

Species	Size	Reference
Prodiamesa olivacea (chironomid)	0.13	Zacharias, 1979
Drosophila simulans	0.13	Laird & McCarthy, 1969
Drosophila melanogaster	0.18–0.21	Laird, 1973
		Rasch et al., 1971
		Dawley et al., unpublished
Drosophila neohydei	0.19	Zacharias et al., 1982
Drosophila pseudoobscura	0.18–0.21	Dawley et al., unpublished
Chironomus tentans	0.21	
Drosophila affinis	0.22	Dawley et al., unpublished
Drosophila persimilis	0.22	Dawley et al., unpublished
Drosophila miranda	0.23	Dawley et al., unpublished
Drosophila eohydei	0.23	Zacharias et al., 1982
Drosophila algonquin	0.23	Dawley et al., unpublished
Drosophila hydei	0.23–0.24	Laird, 1972
		Dawley et al., unpublished
Drosophila willistoni	0.24	Dawley et al., unpublished
Drosophila funebris	0.26	Laird & McCarthy, 1982
Anopheles gambiae	0.26	Besansky & Powell, 1992
Drosophila virilis	0.34–0.38	Laird, 1972
		Dawley et al., unpublished
Sarcaphaga bullata	0.61	Samols & Swift, 1979
Drosophila nasutoides	0.79	Zacharias, 1986
Culex pipiens	0.87	Jost and Mameli, 1972

Note: Values are in picograms per haploid genome. If the sexes were reported separately, the mean is shown. Nondrosophilids shown in bold type.

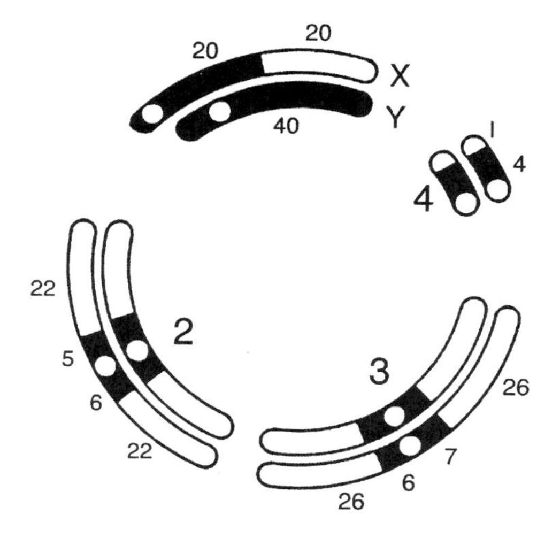

FIGURE 9-1 Schematic diagram indicating relative proportions of heterochromatin (black) and euchromatin (open) regions in the *D. melanogaster* genome. Numbers on the outside indicate Mb, or millions of base pairs, in that part of the genome. Large numbers on the inside identify the chromosome numbers. From John and Miklos (1988).

Dividing the genome, somewhat arbitrarily, into highly repetitive, moderately repetitive, and unique sequences, the fractions of the total DNA of *D. melanogaster* that fall into these categories are 25%, 15%, and 60% respectively. Most of the highly repetitive DNA consists of simple sequences repeated many thousands or millions of times, located in the heterochromatin, often called satellite DNA. The bulk of the moderately repetitive fraction, with ten to a few hundred copies, is generally thought to be largely transposable elements, either active ones or inactivated copies. These sequences exist both in euchromatin and heterochromatin. The unique or single-copy class is found almost exclusively in euchromatin and is thought to comprise the real functional stuff of the genome with active genes and sequences needed to regulate them. An important caveat concerning this class of DNA is the rather loose use of the word "unique." Many of the sequences in this class of DNA have related sequences in the genome, for example, pseudogenes and evolutionarily related duplicates (paralogous sequences). In fact, if genes evolve from preexisting genes via duplication and divergence, then all genes must be traceable to a single "Eve gene." There is evidence that the primordial world of exons was a fairly limited number of sequences (Dorit et al., 1990). The point is, the term "unique" is a misnomer as it is being used in a relative sense: relative to the other two components, these sequences are in fewer copies and depending on how diverged they are from evolutionarily related copies, they are "relatively unique."

The interspersion of repetitive sequences and single-copy sequences in the euchromatin of *Drosophila* displays a pattern called the long period pattern. This consists of a few to several kb of single-copy sequence followed by a few kb of middle repetitive sequence. This contrasts with most mammalian genomes, where the single-copy portion is more frequently interrupted by short repetitive sequences of one to a few hundred bp, the most well-known of which are the Alu-sequences. The phylogenetic distribution of the long and short period patterns is complex, as there are some mammals with the long period pattern (e.g., the Syrian hamster) and some insects (e.g., houseflies and butterflies) with the short period pattern (John and Miklos, 1988).

TABLE 9-2 Base composition of the nuclear genomes of several species.

Species	A+T Content	Reference
D. melanogaster	57%	Laird & McCarthy, 1969
D. virilis	58%	Gall et al., 1971
D. miranda	54%	Steinemann, 1982b
D. funebris	61%	Laird & McCarthy, 1969

Thus, it is not at all clear whether these two very different patterns of interspersion of unique and middle repetitive DNA have any adaptive significance.

The base composition of *Drosophila* genomes is relatively A+T-rich. Table 9-2 gives estimates for four species. The mitochondrial genome is even more A+T-rich, being about 79% A+T in *D. yakuba* (Wolstenholme and Clary, 1985) and *D. melanogaster* (Garesse, 1988).

Base composition is not uniform over the *Drosophila* nuclear genome. Pardue et al. (1987) and Lowenhaupt et al. (1989) hybridized simple repeats of mono- and dinucleotides and found their distributions to be nonrandom. While there is this nonrandom distribution of simple sequences, it needs to be emphasized that, unlike warm-blooded vertebrates (Bernardi, 1993), *Drosophila* do not have long stretches of A+T-rich and G+C-rich isochores. Another important difference from mammalian genomes is that, as far as can be detected, there are no methylated bases in *Drosophila* genomes (Bird and Taggart, 1980; Urieli-Shoval et al., 1982).

Function

In regard to the function of various components of the genome, John and Miklos (1988) present the diagrammatic view in figure 9-2. They argue that only about 20%

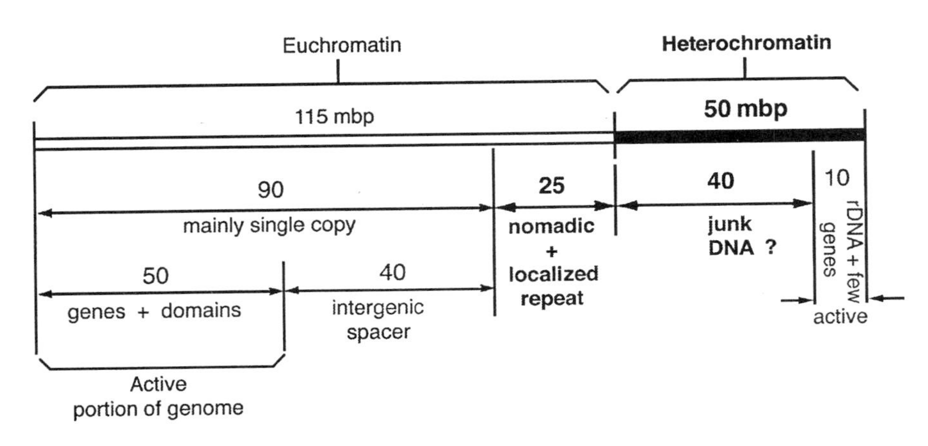

FIGURE 9-2 Schematic representation partitioning the *D. melanogaster* genome into different functional classes. Numbers indicate approximate mbp (million base pairs) in each category. From John and Miklos (1988).

of the heterochromatin (or 10 megabases out of 50) accounts for the functional portion of this component and the rest may be "junk." Trying to quantitatively divide up the euchromatin into functional units and spacer DNA is difficult given our somewhat vague notion of what a functional unit is. Nevertheless, John and Miklos (1988) use a reasonable line of argument based on experimental data that not more than about 50 million bp are needed to form the estimated 10–12,000 genes and their regulation. This conclusion was based on the average sizes of transcripts, the number of transcripts per polytene band, the number of polytene bands in the genome, and the estimated size of controlling regions. Probably the area of most doubt is the role(s) of what John and Miklos call the inactive part of the genome, the intergenic spacer DNA, and other "junk" categories. Whether this is really inactive in the sense of nonfunctional remains to be determined. It does not code for genes, and it may not be concerned with control of transcriptional units, but it could still have an important role in the biology of the organism.

We will return to some of these issues later in this chapter and discuss possible roles for heterochromatin, including how the functional heterogeneity of the genome leads to a great heterogeneity in evolutionary rates. However, first we consider the issue of changes in size of the genome in drosophilids.

Genome Size Evolution

We have already seen in table 9-1 that different species of *Drosophila* vary in the overall size of the genome. In this table the total known range of genome size is shown, with the genome of *D. simulans* being the smallest and that of *D. nasutoides* the largest. This latter species is very unusual in having an extreme amount of heterochromatin, as we will illustrate later. Of normal genomes, *D. virilis* has the largest known, being about twice that of *D. melanogaster.* Why this range of size exists is not known. One of the major adaptive hypotheses about genome size is that it is related to rate of cell division and thus development time; furthermore, larger genomes are associated with larger cell volume, which often increases the size of the organism (Cavalier-Smith, 1985). This explanation does not seem to hold for *Drosophila.* Species with relatively large genomes like *D. virilis* and *D. funebris* are quite large and slowly developing, but *D. willistoni*, with genome size nearly identical to *D. funebris*, is a rapidly developing small species. *D. hydei* and its close relatives including *D. neohydei*, are nearly as large as *D. virilis.* Finally, *D. nasutoides* is not particularly large or slowly developing, yet has a huge genome relative to other drosophilids.

The most thorough study of genome size variation within a group is by Dawley et al. (unpublished) on the *obscura* group. Figure 9-3 summarizes some of their results. It is difficult to detect any phylogenetic trend in these data. Perhaps the most interesting observation is that, within a species, female genomes tend to be larger than male genomes. The exception to this pattern is *D. miranda*, which has an unusual karyotype: one of the autosomes has become attached to the Y chromosome to become a neo-X. This situation will be discussed in more detail later.

Dawley et al. (unpublished) have also studied intraspecific variation in genome size of *D. pseudoobscura*, figure 9-4. Again, in this figure the larger size of female genomes is evident. There are also significant differences among strains. However, it is difficult to discern any trends such as geographic clines. In chapter 4, we discussed

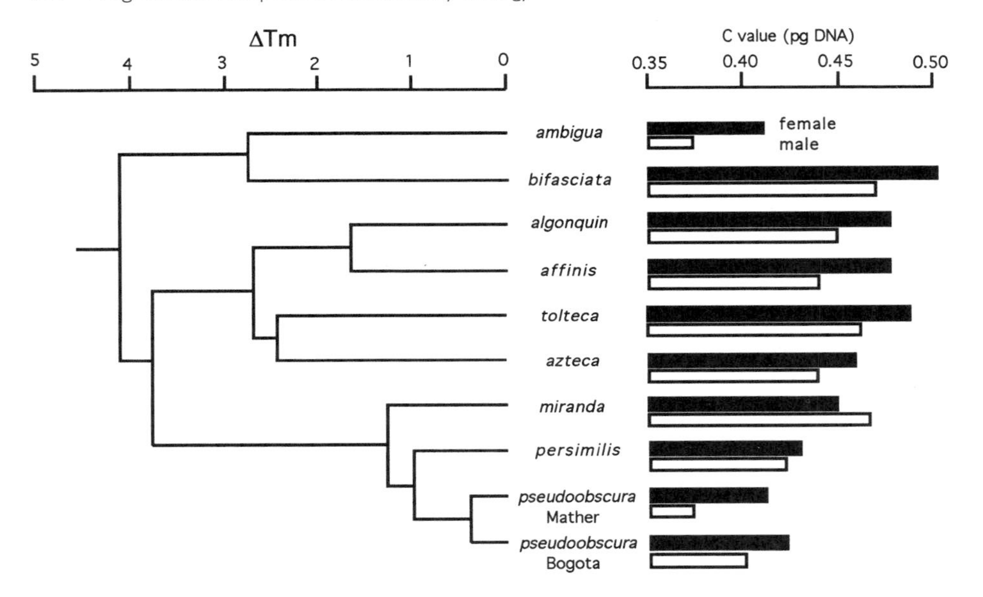

FIGURE 9-3 Haploid size of the nuclear genome of several species of the *Drosophila obscura* group, mapped onto a phylogeny. The phylogeny is based on the DNA–DNA hybridization results from Goddard et al. (1990). Genome sizes were determined by R. Dawley et al. (unpublished data).

studies indicating that populations adapted to cooler temperatures tend to be larger than conspecific populations adapted to warmer temperatures. If organism size and genome size are positively correlated, one might expect to see a trend of smaller genomes from warmer regions. This is not evident in figure 9-4. We can only conclude from these studies that an adaptive basis of variation in genome size in *Drosophila* remains to be documented.

Karyotype Evolution

We now turn to the issue of how the *Drosophila* genome is packaged and how the packaging changes through evolution, that is, karyotype evolution. It is important to emphasize that chromosome evolution in *Drosophila* has been studied using two different types of chromosomes for observation. To determine chromosome number and morphology in a species, mitotic or meiotic metaphase chromosomes are examined, usually from brain or testes, respectively. Polytene chromosomes are formed by endoreplication (replication of DNA without nuclear or cell division) in some dipteran cell types, most often studied in larval salivary glands. Because only the euchromatic part of the genome undergoes polytenization, only this fraction of the genome can be studied in these terminally differentiated cells. The high resolution afforded by the banding patterns in polytene chromosomes makes this material extremely useful for following chromosomal rearrangements and determining homologies.

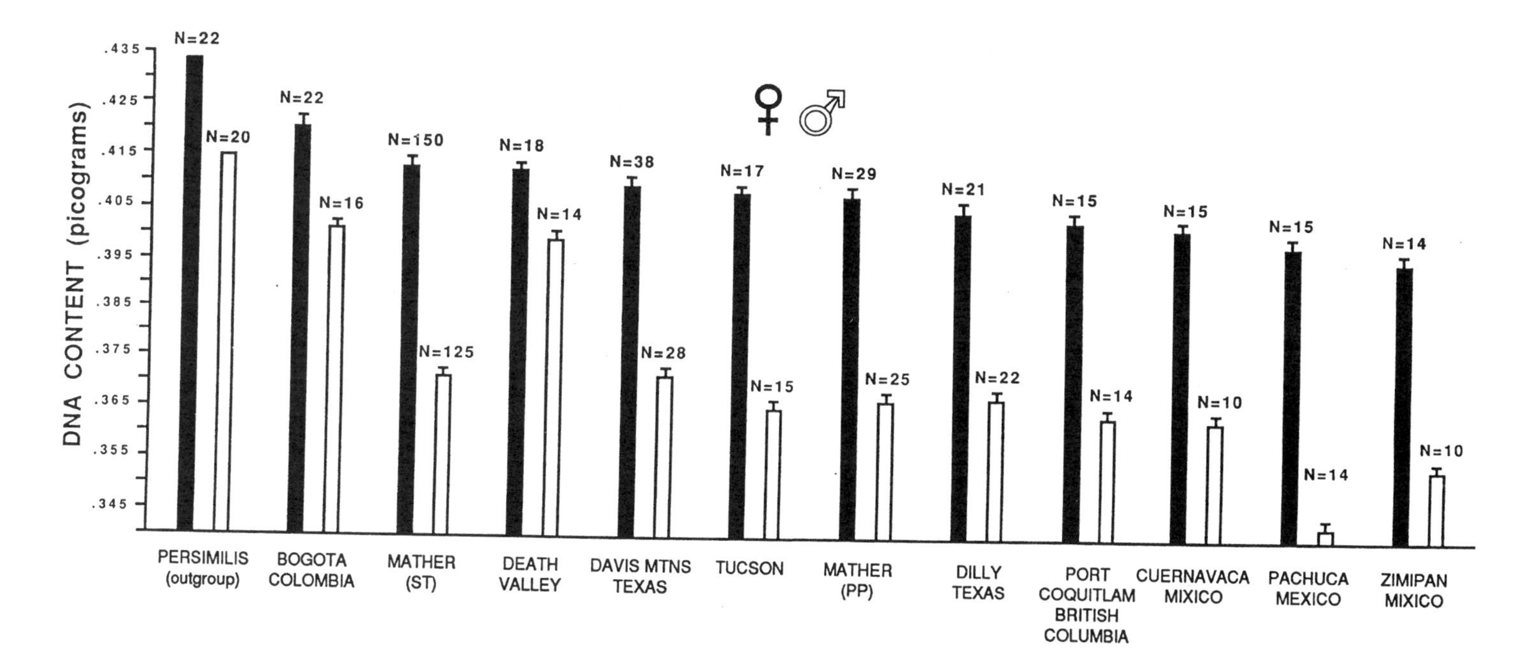

FIGURE 9-4 Intraspecific variation in genome size of *D. pseudoobscura*. Females are solid bars, males open bars; N is the number of cells counted in the flow cytometer. From Dawley et al. (unpublished data).

Chromosome Changes

The haploid number of chromosomes in *Drosophila* is between three and six (the exceptions are two species possessing an extra pair of heterochromatic dots). Clayton and Wheeler (1975) and Clayton and Guest (1986) catalog the chromosome numbers and configurations of some 600 species of *Drosophila*, as well as about 50 species of Drosophilidae outside this genus. Table 9-3 is a numerical summary of some of the groups highlighted in this book. Perhaps the most striking trend in these data is that a haploid number of six predominates in subgenus *Drosophila* and the Hawaiian *Idiomyia*, while lower chromosome numbers characterize members of subgenus *Sophophora.*

While chromosome number is variable within the genus (table 9-3), there is remarkable conservation of karyotype with respect to chromosomal arms. Virtually all *Drosophila* have five chromosomal arms and many have a dot (a very small chromosome). In *D. melanogaster*, these arms are arranged into two metacentric chromosomes (the second and third), one acrocentric X chromosome, and the dot fourth (figure 9-1). Most karyotypes can be derived from one another simply by rearranging the arms via centric fissions and fusions. The dot chromosome either remains distinct or, when missing, is thought to have been incorporated into one of the other larger chromosomes (for which there is evidence in a few species).

Muller (1940) and Sturtevant and Novitski (1941) were the first to note in comparative genetic studies that different species tend to have phenotypically similar mutants in the same linkage groups. They erected the system of lettering the chromosome arms as elements, with the A element being the *D. melanogaster* X chromosome, elements B and C the two arms of the second, D and E the two arms of the third, and element F the dot. By comparing genetic maps, polytene banding patterns, and localization by in situ hybridization, a number of species' chromosomal arms have been "homologized" to the Muller/Sturtevant/Novitski elements. Table 9-4 summarizes the available information.

TABLE 9-3 Numbers of species from selected genera and subgenera with different haploid chromosome numbers.

		Haploid Number of Chromosomes					
Genus	Subgenus	7	6	5	4	3	Total
Drosophila							
	Sophophora		5	24	63	49	141
	Drosophila						
	(non-Hawaiians)	2	146	52	69	9	278
	Hawaiian		136	2	3		141
Chymomyza					6		6
Zprionus			9		1		10
Scaptodrosophila			5	2	10	2	19
Steganinae							
Leucophenga			2				2

From Clayton and Guest (1986).

TABLE 9-4 Correspondence of chromosomal arms to the Muller/Sturtevant/Novitski elements.

	Element						
	A	B	C	D	E	F	Reference
Subgenus Sophophora							
melanogaster	X	2L	2R	3L	3R	4	Muller, 1940
simulans	X	2L	2R	3L	3R	4	Muller, 1940
ananassae	XL-XR	3R	3L	2R	2L	X	Kikkawa, 1938
willistoni	XL	2R	2L	XR	3		Lakovaara & Saura, 1972
subobscura	A	U	E	J	O	D	Steinemann et al., 1984 Lakovaara & Saura, 1982
madeirensis	A	U	E	J	O		Loukas & Kafatos, 1986 Krimbas & Loukas, 1984
pseudoobscura & *persimilis*	XL	4	3	XR	2	5	Steinemann et al., 1984 Lakovaara & Saura, 1982
miranda	X^1L	4	X^2	X^1R	2	5	Steinemann et al., 1984 Lakovaara & Saura, 1982
affinis	XL	4	3	XR	2	5	Sturtevant & Novitski, 1941
azteca and *algonquin*	XS	C	A	XL	B	D	Sturtevant & Novitski, 1941
athabasca	XS	B	C	XL	E	F	Paika & Miller, 1974
bifassciata	XL-XR	4L-4R	2L	2R	3L-3R	5	Felger & Pinsker, 1987
obscura	AL-AR	BL-BR	JL	JR	EL-ER	D	Steinemann et al., 1984 Lakovaara & Saura, 1982
ambigua	AL-AR	EL-ER	JL	JR	UL-UR	D	Steinemann et al., 1984 Lakovaara & Saura, 1982
subsilvestris	AL-AR	BL-BR	C	D	EL-ER	D	Steinemann et al., 1984
Subgenus Drosophila							
macrospina	X	4	3	5	2	6	Weinberg, 1954
hydei	X	4	3	5	2	6	Spencer, 1949 Hess, 1976 Wasserman, 1982 Loukas & Kafatos, 1986
peruviana and *repleta*	X	3	5	4	2	6	Wasserman, 1982
pavani	X	3	4L	4R	2	5	Wasserman, 1982
nigromelanica	XR	4	3	XL	2	5	Wasserman, 1982
robusta	XL	3	2R	XR	2L	4	Wasserman, 1982 Loukas & Kafatos, 1986
virilis	X	4	5	3	2	6	Throckmorton, 1982b Loukas & Kafatos, 1986 Whiting et al., 1989 Kress, 1993
grimshawi	X	3	2	5	4	6	Loukas & Kafatos, 1986 Jeffery et al., 1988
Scaptrodrosophila lebononensis	X	4L-4R	2R	2L	2L-3R		Papaceit & Juan, 1993

Note: L and R refer to "left" and "right" arms, although sometimes L and S are used to refer to long and short arms. Compilation taken from Ashburner (1989) with additions.

Two types of chromosomal changes have predominated in *Drosophila* karyotype evolution: fusions or fissions of centromeres (Robertsonian or centric fusions/fissions) and paracentric inversions (inversions not including the centromere). Two other types of possible chromosomal rearrangements, translocations and pericentric inversions, are rare. This is as it must be if the chromosomal arm homologies indicated in table 9-4 are real. The linkage groups as represented by arms remain, but the order of genes is scrambled from species to species via paracentric inversions.

The rarity of pericentric inversions relative to paracentric inversions is highlighted in Wasserman's (1992) thorough study of the *repleta* group. He reports a total of 296 inversions fixed or polymorphic among the 70 species studied. All are paracentric. Four centric fusions are also present in this group. No translocations were noted. This is more or less typical of *Drosophila* lineages.

Pericentric inversions do occur rarely. In table 9-4 the species *D ananassae, D. obscura, D. ambigua*, and *D. silvestris* would appear to have generated more chromosomal arms having three metacentric and two acrocentric chromosomes (plus a dot). In fact, the metacentric chromosomes have been generated by pericentric inversion in acrocentric ancestral chromosomes (Steinemann et al., 1984), which is also the case for *Scaptodrosophila lebanonensis* (Papaceit and Juan, 1993).

While the single-armed chromosomes in *Drosophila* are often called telocentric, there is some doubt that truly telocentric chromosomes, in which the centromere is really at the tip, exist; some chromosomal material likely extends from the centromere but is too small to be cytologically detected under most conditions of staining and power of magnification. The generation of metacentric chromosomes from "telocentric" ones (as in the cases of the species just noted) would argue for some material on each side of the centromere so that a pericentric inversion is possible. Otherwise, one would need to hypothesize the placement of the terminal centromere in the middle of the chromosomes, followed by the generation of a new telomere.

Another possible example of a pericentric inversion is between the Old World and New World *obscura* group species. Segarra and Aguadé (1992) mapped by in situ hybridization several molecular markers for the X chromosome. In the New World species, the metacentric X chromosome is composed of elements *A* (=XL) and *D* (=XR), while in Old World species it is acrocentric, element *A*. Of the eight genes studied by Segarra and Aguadé, all mapped to the X in the Old World species; six mapped to the XL of the New World species; the other two mapped to the XR. Thus, subsequent to the fusion of elements A and D, a pericentric inversion occurred.

Intraspecific polymorphisms for pericentric inversions were discussed in chapter 3. Such do exist, although much less frequently than polymorphisms for paracentric inversions. As for fixed inversions, Stone et al. (1960) estimated only 32 pericentric inversions have occurred in the genus. Clayton and Guest (1986) cite only four cases of translocation heterozygotes: *D. ananassae* (Dobzhansky and Dreyfus, 1943), *D. melanica* (Ward, 1952), *D. prosaltans* (Dobzhansky and Pavan, 1943), and *D. rubida* (Mather, 1962). There are three documented cases of fixed translocations: *D. ananassae* (Kaufmann, 1937), *D. repletoides* (Hsiang, 1949), and *D. miranda* (Dobzhansky and Tan, 1936). Totaling pericentric inversions and translocations, the genus *Drosophila* has experienced less than 40 of these chromosomal changes. In contrast, it has been estimated that between 20,000 and 50,000 paracentric inversions have occurred (Stone et al., 1960). It is clear, then, that paracentric inversions, along with an estimated 58 centric fusions/fissions (Stone et al., 1960), are the predominant features of *Drosophila* karyotype evolution.

The Ancestral State

It has long been the lore in *Drosophila* biology that the ancestral chromosomal configuration for the group was five rods (acrocentric chromosomes) and a dot. The many different karyotypic configurations now seen were produced by fusions of centromeres, with fissions being very rare. With the phylogenies presented in the previous chapter, it is now possible to test this contention using the comparative method. If one plots the chromosomal configurations onto a phylogeny, can fusions alone account for the observed karyotypic changes and is this the most parsimonious scenario? This will be a good example of the utility of robust phylogenies.

In testing this hypothesis, it is necessary to use examples for which chromosomal homologies have been established (i.e., the cases in table 9-4). When this is done, the pattern revealed is in figure 9-5. A most parsimonious explanation of karyotype evolu-

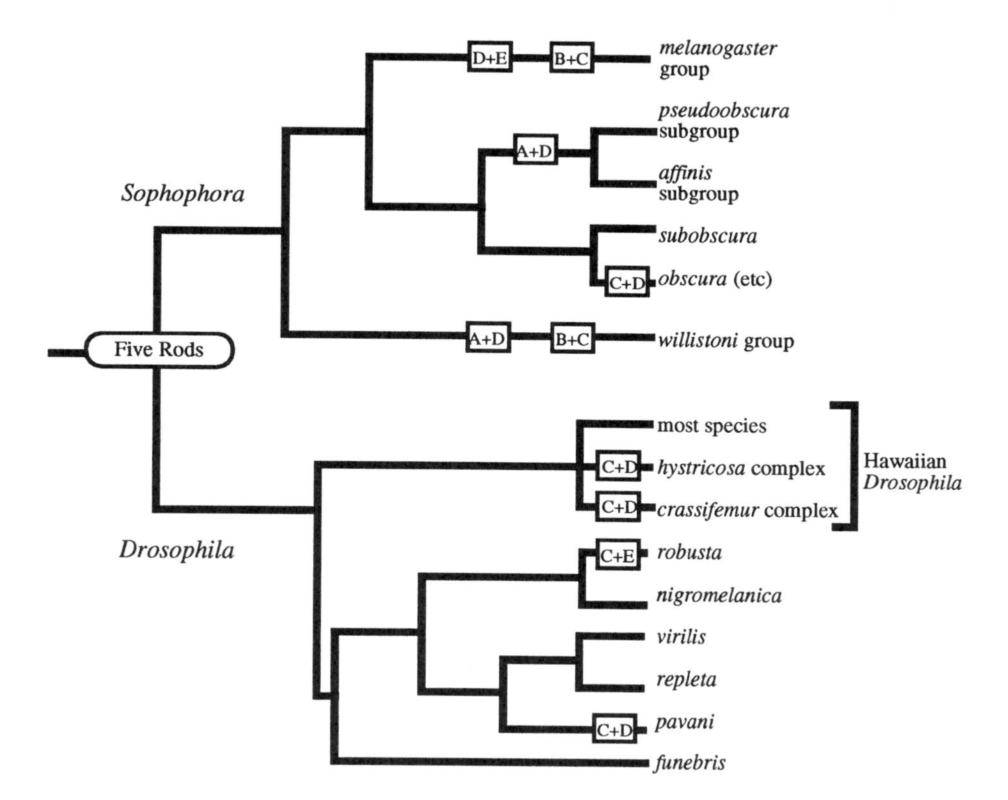

FIGURE 9-5 The distribution of chromosomal arm or element configurations mapped onto a well-established phylogeny (chapter 8). Assuming the ancestral state was five arms, a most parsimonious scenario requires only fusions of arms. The C+D fusion in Hawaiian flies is thought to have arisen twice independently (Carson, personal communication). The dot is not considered. From Powell and DeSalle (1995).

tion requires only fusions of arms (the dot is ignored). There is one equally parsimonious hypothesis, which would require a fusion of B+C on the ancestral branch of *Sophophora* followed by a fission on the branch leading to the *obscura* group. This phylogenetic analysis is consistent with five rods being the ancestral state with fusions being the predominant, if not exclusive, direction of change. Examining outgroups is another means of elucidating ancestral states. If we look at the genera in table 9-3, we see that most outgroup species have the hypothesized five rods and a dot (i.e., are under the 6 column). The only two cases examined in the most distant outgroup, members of the subfamily Steganinae, also have six chromosomes. Thus, outgroup analysis is also consistent with the ancestral state being six chromosomes, further supporting the predominance of fusions hypothesis.

Linkage

Conservation

Before the advent of comparative mapping using the physical means of in situ hybridization, the most thorough comparisons of linkage groups across species was Sturtevant and Novitski's classic 1941 paper, a truly remarkable study. They list a total of 66 morphological mutants of *D. melanogaster* for which homologous mutants had been found in other species. The use of the term homologous in describing the relationship among these mutants was something of an act of faith or insight in 1941; however, subsequent isolation and DNA sequencing has established that mutants presenting very similar or identical phenotypes in different species are evolutionarily related. This being the case, the use of the term homology to describe the conservation of chromosomal arms would also appear justified. Figure 9-6 is an example of the comparative positions of several X-linked morphological mutants in five species. Note that while the mutants do map to the same chromosomal arm, the recombination maps and gene order are different among species. The arms are about the same physical length, but clearly some species recombine at greater frequency than others.

With the introduction of allozyme mapping and use of in situ hybridization of cloned DNA segments to polytene chromosomes, comparative mapping became more common; references to such studies are in table 9-4. A good example of results of comparing linkage relationships using allozymes is the work of Pinsker and Sperlich (1984) on *D. melanogaster* and *D. subobscura* (figure 9-7). Probably the most thorough example of using in situ hybridization of cloned probes (as well as other markers) is between *D. virilis* and *D. melanogaster* (Whiting et al., 1989; Kress, 1993; Lozovskaya et al., 1993). Figure 9-8 shows the results for two chromosomes. Recall that these two species belong to subgenera *Drosophila* and *Sophophora*, respectively, so they last shared a common ancestor some 30 to 50 million years ago. While the gene order has become considerably scrambled, the arm positions remain conserved. Much of the scrambling in all these examples is almost certainly due to paracentric inversions, although the possibility of intra-arm transposition cannot be excluded.

Conservation over even longer evolutionary time is indicated by the ability to identify the Muller/Sturtevant/Novitski arms in *Scaptodrosophila lebanonensis*, often called *Drosophila lebanonensis* (Papaceit and Juan, 1993). Conservation of the X chromosome is particularly clear. As Papaceit and Juan point out, the same kinds of

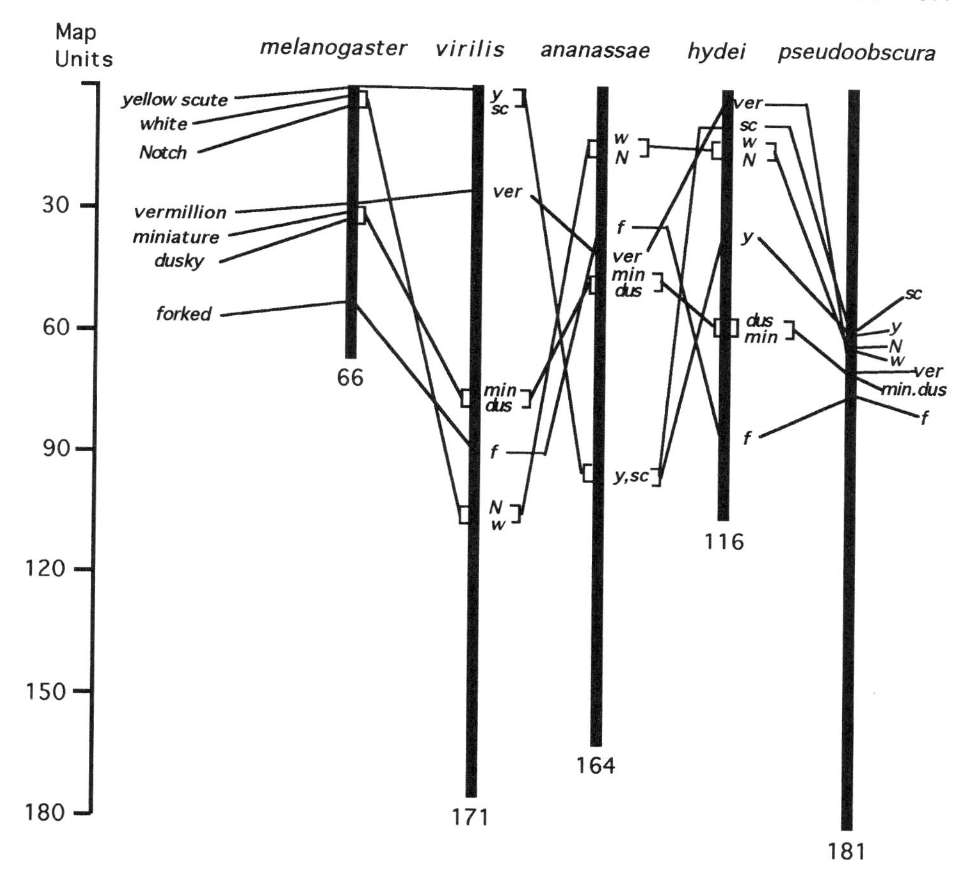

FIGURE 9-6 Comparative linkage maps of X chromosome morphological mutants in five species of *Drosophila*. Only one arm of the metacentric X, the XR, is shown for *D. pseudoobscura*. From John and Miklos (1988).

chromosomal rearrangements occurring within *Drosophila*, namely paracentric inversions and, more rarely, pericentric inversions of acrocentric chromosomes to produce metacentric chromosomes, have occurred between *Drosophila* and *Scaptodrosophila*. Matthews and Munstermann (1994) have extended the conservation of linkage groups of *Drosophila* to mosquitoes.

Nonconservation

Despite this overall conservation of linkage groups and chromosomal arms in the genus, there are exceptions. Tonzetich et al. (1990) studied seven tRNA's by in situ hybridization to polytene chromosomes from *D. melanogaster, D. pseudoobscura, D. hydei*, and *D. virilis*. Most probes hybridized to the predicted elements, but some did not. Duplications were also detected among species. The evidence is for a general

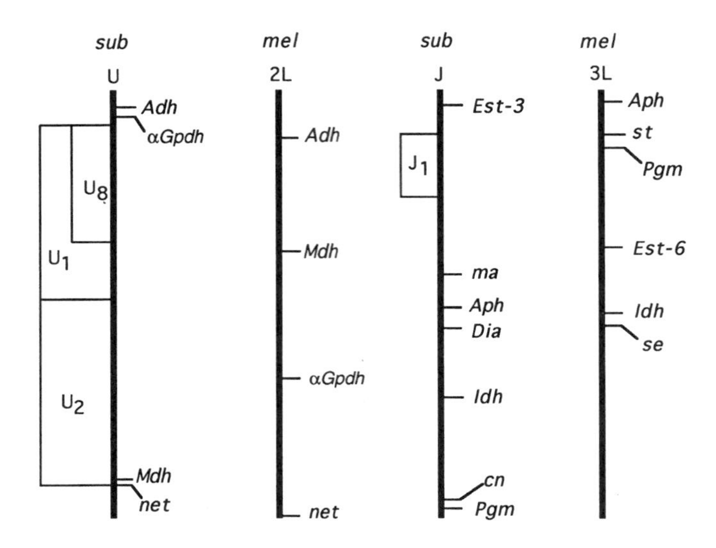

FIGURE 9-7 Comparative linkage maps for enzyme-coding loci for *D. subobscura* and *D. melanogaster*. Homologous chromosome arms are U = 2L and J = 3L. From Pinsker and Sperlich (1984).

conservation of linkage of these genes with a limited amount of exchange between elements.

Another class of exceptions are tandemly repeated clusters of genes, the best documented being genes encoding 5S rRNA and histones (Steinemann, 1982b; Felger and Pinsker, 1987). Both between the *melanogaster* and *obscura* groups, as well as within the *obscura* group, copies of these sequences appear to have transposed to nonhomologous chromosomal arms. Likewise, in comparing *D. virilis* and *D. melanogaster*, there is evidence that the repeated larval glue protein genes have transposed to nonhomologous arms (Kress, 1993). Why it should be the case that single-copy genes are more conserved in being maintained on the same chromosomal element than are tandemly repeated genes is not at all clear. As Kress (1993) points out, the functioning transposed 5S rRNA and histone clusters require multiple copies for proper expression. The multiplication of tandem clusters is generally thought to occur by unequal crossing-over, which requires an initial situation of at least two tandem copies of the gene. Because the transposed copies are functional repeat units, Kress concludes that at least two copies must have been transposed. Perhaps the mechanism of interarm transposition requires that multiple copies of the sequence be transposed. It is interesting to note that a single case of transposition between elements was postulated by Sturtevant and Novitski (1941). This was for the nucleolus organizer and *bobbed* locus in *D. ananassae*. It is now known that these two units are the tandemly repeated rRNA-coding regions.

Thus far we have been mostly concerned with the single-copy fraction of the genome, and, by and large, it is this fraction that remains conserved in arms. The fraction of middle repetitive DNA composed of transposable elements, by definition, is not conservative in linkage. We will address highly repetitive sequence evolution

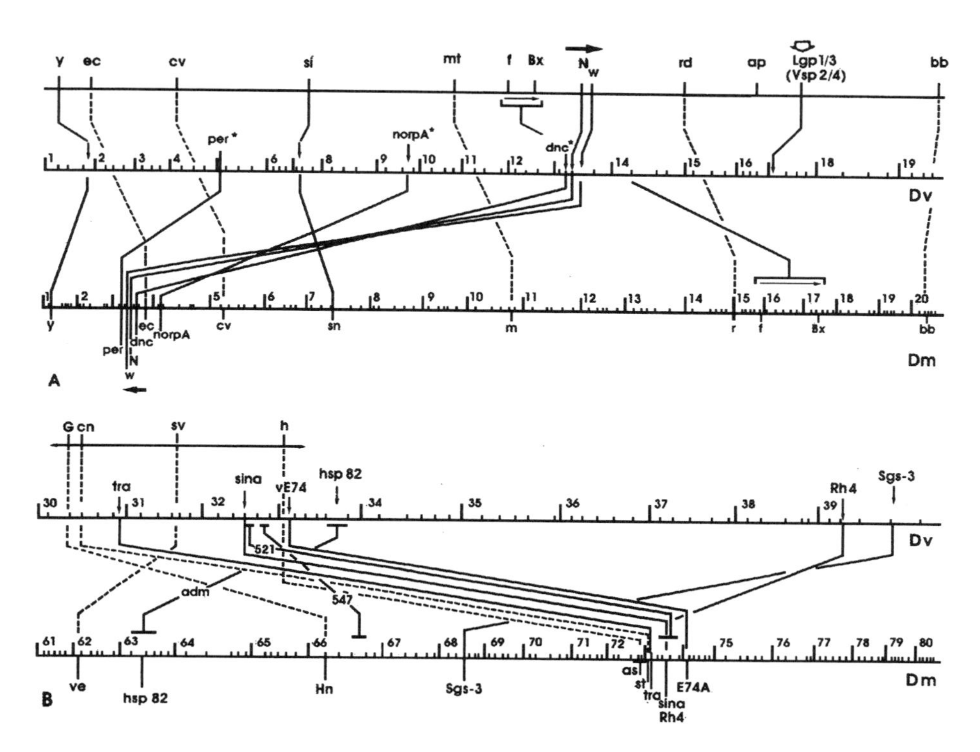

FIGURE 9-8 Comparative linkage maps for *D. virilis* (Dv) and *D. melanogaster* (Dm). Part A is for the X chromosome, part B for the third of *D. virilis* and 3L of *D. melanogaster*. Solid lines are determined by in situ hybridizations and broken lines are determined by genetic and other criteria. The middle line in part A is the genetic map scaled in crossover units. From Kress (1993).

in a later section of this chapter, but simply state at this point that highly repetitive sequences are not conserved in relation to the major chromosomal elements.

B Chromosomes

While supernumerary B chromosomes are fairly common in many insects, they are rare in *Drosophila* (their evolutionary significance is reviewed in Shaw and Hewitt, 1990). Such chromosomes have been reported only from two species of *Drosophila: D. nausuta albomicans* (Ramachandra and Ranganath, 1987; Hatsumi, 1987) and *D. subsilvestris* (Gutknecht et al., 1995). In the latter case, a species-specific satellite-DNA family has been localized to the one to five extra small B chromosomes. The presence of this satellite DNA in one of the acrocentric chromosomes suggests that the B chromosomes arose by spontaneous amplification of this satellite family.

Global View of scDNA Evolution

In keeping with the theme of this chapter, a general or global view of single-copy DNA (scDNA) evolution follows; the next chapter examines this at finer resolution.

The global data come from DNA–DNA hybridization studies in which average divergence of all scDNA is measured, although some studies have measured average divergences of different classes of scDNA. Consistent with the notion that the *Drosophila* genome is not a uniform entity, we present evidence that the scDNA portion of the genome is highly heterogeneous in rates of evolution.

The Nature of DNA–DNA Hybridization Experiments

To appreciate the conclusions drawn in this section, some details of the technique of DNA–DNA hybridization need to be kept in mind. Readers familiar with this technique can skip this discussion without loss of understanding.

In DNA–DNA hybridization experiments, the object is to determine two parameters that quantitatively measure the divergence of two sources of DNA. The first measure is the fraction of the genome that cross-hybridizes between taxa, and the second is the degree of base-pair divergence of that fraction that does hybridize, determined by the decrease in thermal stability due to base pair mismatch.

If two sources of DNA, A and B, were heated until they became single-stranded (denatured), then cooled and allowed to reassociate into duplexes, three kinds of duplexes are potentially possible: A/A, A/B, and B/B. To determine only the reassociation and thermal stability of A/B duplexes, one of the DNAs is radioactively labeled and is called the tracer. When setting up a reassociation reaction, the tracer DNA is added to a 1000-fold or greater excess of nonradioactive DNA; this nonradioactive DNA raises the concentration and "drives" the reassociation, and hence is called the driver DNA. When duplexes are formed under these conditions, many of them will be driver-driver, some driver-tracer, and virtually none will be tracer-tracer. Assuming random reassociation and a 1 to 1,000 ratio of tracer to driver, we expect more than 99.9% of the tracer (radioactivity) to be in tracer-driver hybrid molecules. Because it is the radioactivity that is "traced" in the experiments, it is the behavior of A/B duplexes that is followed.

To avert random reassociation of single-stranded DNA molecules, it is necessary to adjust the reassociation conditions (salt concentration and temperature) to require a certain level of fidelity of pairing, A with T and C with G. Standard conditions for such experiments are usually adjusted for a stringency requiring about 75% base-pair matching to form stable duplexes, well above the 25% expected of random sequences. It is assumed that the 75% level of stringency is sufficient to assure true homologous pairing, at least for the majority of sequences.

When DNA is made single-stranded and allowed to reassociate, 100% reassociation never occurs. Therefore, when measuring how much of two heterologous genomes cross-hybridize, it is necessary to normalize the measure for how much hybridization occurs under identical conditions for DNA from the same source. In performing DNA–DNA hybridizations then, the tracer is mixed in separate reactions to an excess of driver DNA from the same source and driver from the heterologous source. The normalized percent reassociation, NPR (also sometimes called normalized percent hybridization, NPH), is the degree of heteroduplex formation (with tracer and driver DNAs from different sources) divided by the degree of homoduplex formation (tracer and driver from the same source).

In the experiments to be discussed, scDNA was used by running a C_ot curve long enough for highly repetitive and middle-repetitive DNA to reassociate and then using the nonreassociated scDNA. In reality, this fraction does contain highly and moderately repetitive sequences, but now they are represented in the same molar ratio as the truly single-copy sequences. This is because as the reassociation proceeds, eventually repetitive sequences will be sufficiently removed (i.e., made double-stranded) to make the reassociation dynamics of the remaining single-stranded copies behave like sequences that are in single-copy in the genome.

After the reassociation mixture has come to equilibrium (i.e., further incubation produces no more double-stranded DNA), the single-stranded DNA is separated from the double-stranded by running the mixture over a HAP (hydroxyapatite) column, which specifically binds double-stranded DNA. The thermal stability of the reassociated duplexes can be determined by two methods. The temperature of the HAP column can be raised and the amount of radioactivity released by denaturation can be measured at each temperature. Alternatively, digestion with a single-stranded specific nuclease, S_1, can be used to determine the degree of denaturation at different temperatures. Further technical details of these two procedures can be found in Werman et al. (1990a) and Caccone and Powell (1991).

The measure of change in stability of heteroduplexes compared to homoduplexes is usually expressed as ΔT_m, the change in median melting temperature. The correspondence of ΔT_m to base-pair mismatch is remarkably linear and accurate. Quantitatively, the relationship has been empirically determined to be such that a ΔT_m of 1°C corresponds to 1.7% base-pair mismatch for the S_1 nuclease procedure (Caccone et al., 1988a) and to 1.5% base-pair mismatch for the HAP method (Springer et al., 1992).

Extreme Heterogeneity in Divergence

DNA–DNA hybridization studies with *Drosophila* total scDNA produced a surprising result: even when scDNA from quite closely related species (e.g., *D. melanogaster* and *D. simulans)* is cross hybridized, a fairly large fraction of the genome does not form stable duplexes when 75% base pairing is required, yet that fraction that does hybridize is less than 10% divergent. Figure 9-9 illustrates a typical example. Between the scDNA genomes of *D. melanogaster* and *D. simulans*, 35% of the sequences will not cross hybridize, yet that fraction that does is only about 3 to 4% divergent. For *melanogaster-orena*, fully 60% of the scDNA will not cross hybridize, yet that which does is 7 to 9% divergent. The important point to note in figure 9-9 is that the hybridizing and nonhybridizing portions of the genome are not a continuum of divergence. The hybrid molecules that are formed do not start melting until the temperature is raised several degrees above that used for reassociation. This indicates there is part of the single-copy genome that evolves very rapidly and a portion that remains quite conserved, and these two fractions are discrete. Figure 9-10 gives a pictorial view of this situation.

The above phenomenon has been noted for a number of different *Drosophila* groups: the *repleta* group (Schulze and Lee, 1986), the *melanogaster* subgroup (Caccone et al., 1988b), the *obscura* group (Goddard et al., 1990), and Hawaiian *Drosophila* (Hunt et al., 1981; Hunt and Carson, 1983). Even between haploid genomes from

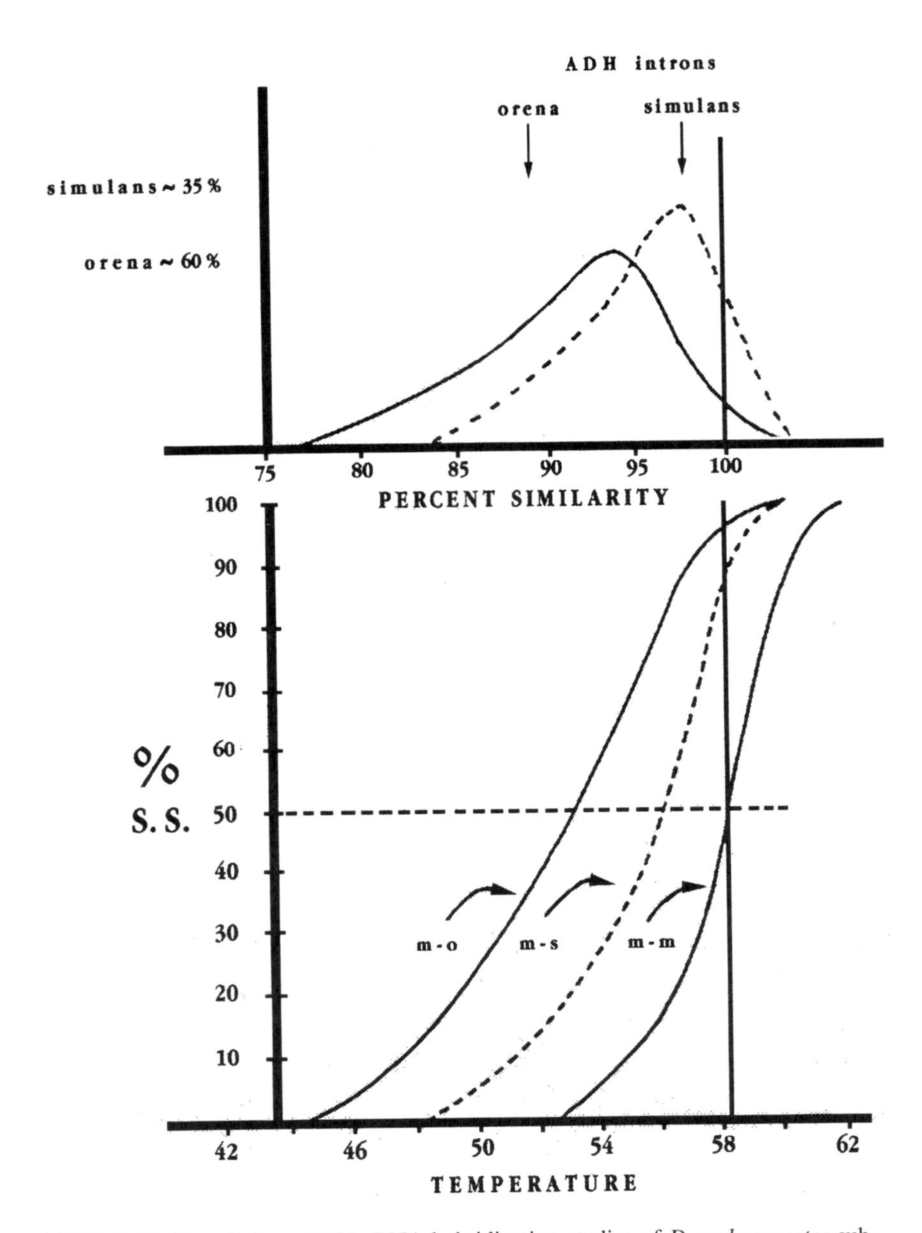

FIGURE 9-9 Results from DNA–DNA hybridization studies of *D. melanogaster* subgroup species. The lower graph depicts the melting curves of homoduplex (m-m) and heteroduplexes of *D. melanogaster* with *D. simulans* (m-s) or with *D. orena* (m-o); the cumulative proportion of DNA becoming single-stranded with increasing temperature is shown. The upper graph indicates absolute amount melting at each temperature. The heavy vertical line represents the conditions of stringency of hybridization requiring about 75% or greater base pairing to form stable hybrid molecules. The percentages at upper left represent the proportion of the single-copy genomes that do not hybridize to *D. melanogaster*. Arrows on the upper graph indicate the divergence of alcohol dehydrogenase introns. m = *D. melanogaster*, s = *D. simulans*, o = *D. orena*. From Powell and Caccone (1989).

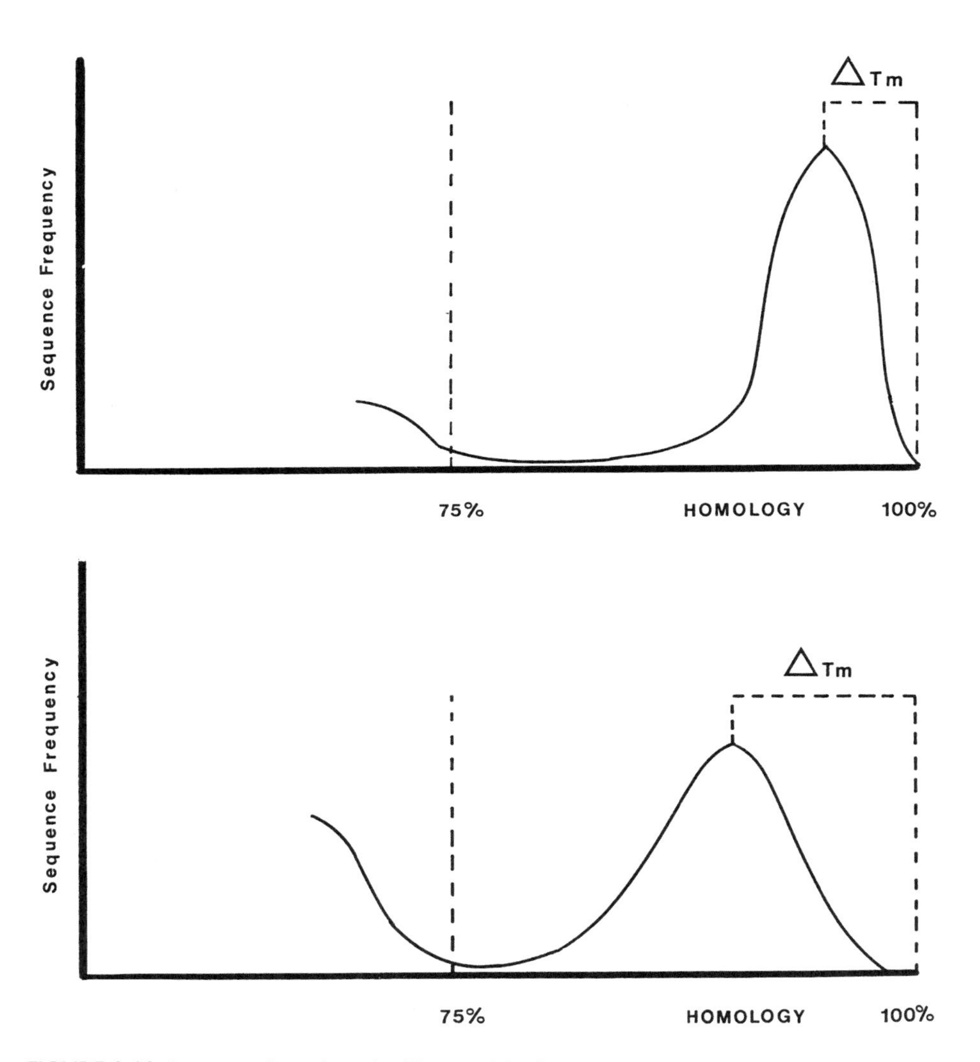

FIGURE 9-10 Interpretation of results illustrated in figure 9-9. The upper graph represents divergence of sequences of two closely related species, the lower graph more divergent species. The important point is there are two discrete fractions of the genome, one relatively conserved and represented by the right hump, and a fraction that quickly passes the 25% divergence threshold. The distribution of the fast-evolving fractions is left purposely ambiguous as there is no information relevant to how this fraction is distributed. From Caccone and Powell (1990).

parthenogenetic strains of *D. mercatorum*, this fast-evolving, nonhybridizing fraction can be detected (Caccone et al., 1987). DNA sequence data also consistent with this picture come from Martin and Meyerowitz's (1986) study of glue protein genes in the *melanogaster* subgroup. They observed a relatively conserved set of sequences followed by a highly divergent set; the transition from relatively conserved to highly divergent occurs over fewer than 50 base pairs.

What is this highly divergent fraction of the genome? We present data later to show that coding sequences are primarily in the conserved fraction, as might be expected. In figure 9-9 the divergences of introns (determined by DNA sequences) between the species compared are noted. At least for *Adh* introns, they too seem to fall into the divergence range of the conserved fraction.

In theory, the degree of nonhybridization of tracer sequences could be due to one of two phenomena: either nucleotide substitutions have occurred very rapidly so the sequences are more than 25% divergent, or insertions/deletions (indels) have occurred such that the tracer sequences do not have matching driver sequences. The evidence is that both may be occurring. Werman et al. (1990b) reasoned that if the nonhybridizing fraction was due to simple base-pair substitutions, then by lowering the stringency of reassociation conditions, NPR should increase; if indels were the explanation, lowering the stringency would have no effect. They observed an increase in hybridization when stringency was reduced. On the other hand, Caccone and Powell (1990) compared the NPR determined by two different methods, binding of double-stranded DNA (but not single-stranded) to hydroxyapatite (HAP) and S_1 nuclease digestion. Indel differences between tracer and driver would form loops. The loops formed by such events would bind to HAP due to the double-stranded adjacent sequences, and thus indels would be placed into the reassociated fraction. S_1 nuclease digestions would digest the loops and thus they would be considered in the nonreassociated fraction. For the same pair of *Drosophila* species, NPR determined by S_1 digestion consistently indicated about 10 to 20% less hybridization than did HAP (Caccone and Powell, 1990). This argues that some of the non-cross–hybridizing fraction is due to indels. Thus this rapidly evolving fraction is likely made up of both highly divergent sequences due to base pair substitution as well as due to indel differences, including transposable elements.

Divergence of Functional Subclasses

Because of this very rapid divergence of total scDNA, to the point of not cross-hybridizing, we performed DNA–DNA hybridizations using only coding sequences by using cDNA made from poly(A)$^+$ mRNA as a tracer (Caccone et al., 1992; Powell et al., 1993). We also performed experiments using the intergenic DNA (igDNA) as tracer. This was prepared by labeling total scDNA and then cross-hybridizing it with a 1000-fold excess of poly(A)$^+$ mRNA. The fraction of scDNA not binding to HAP (i.e., sequences that did not hybridize with the mRNA) was used as the igDNA tracer. Because HAP was used, we would expect that introns forming loops would be removed from the igDNA, as well as sequences immediately adjacent to coding regions due to over-hanging single strands.

Table 9-5 compares the divergence of three types of DNA, the total single-copy DNA, that coding for mRNA, and the intergenic DNA. As expected, cDNA is the

TABLE 9-5 Comparison of DNA divergence of different classes.

Species Hybridized	cDNA	igDNA	scDNA
melanogaster-simulans	1.25 ± 0.12	3.90 ± 0.31	2.17 ± 0.09
melanogaster-yakuba	2.24 ± 0.12	9.00 ± 0.27	4.02 ± 0.08
melanogaster-orena	3.31 ± 0.10	9.56 ± 0.27	5.11 ± 0.08
melanogaster-takahashii	5.18 ± 0.12	13.76 ± 0.28	6.16 ± 0.11

Note: cDNA represents coding sequences expressed in adult mRNA, igDNA is the intergenic DNA as described in the text, and scDNA is total single-copy DNA. Figures shown are the ΔT_ms for the hybridizing sequences ± S.E.

From Powell et al. (1993).

least divergent fraction and igDNA the most divergent; total scDNA, comprising both fractions, is intermediate. We will return to these experiments in chapter 11 when comparing the degree of divergence of messages expressed at different developmental stages.

One other result from these experiments is illustrated in figure 9-11. We have already mentioned that when total scDNA is cross-hybridized, even at quite low evo-

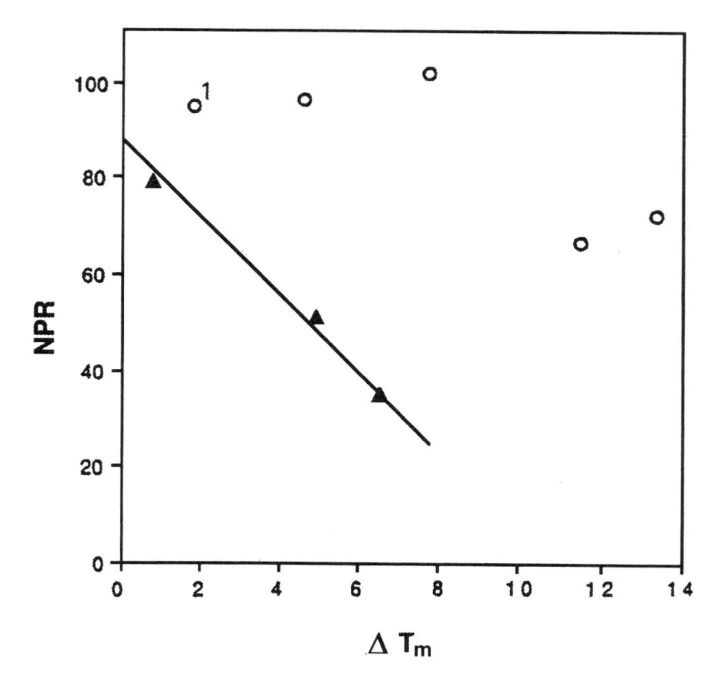

FIGURE 9-11 Relationship between normalized percent reassociation (NPR) and the divergence measured by ΔT_m as seen in DNA–DNA hybridization experiments. Triangles are for total single-copy studies and open circles are for cDNA studies that only involve coding DNA. Points represent averages for the intervals of ΔT_m units: 0–3, 3–6, 6–9, etc. From Powell et al. (1993).

lutionary divergence, a large fraction does not hybridize. The solid line in figure 9-11 shows this graphically; in fact, as has been shown by a number of workers, the relationship between NPR and the ΔT_m is remarkably linear (for reasons not at all clear). More importantly, in this figure it can be seen that the cDNA tracers do not exhibit this decrease in NPR. This suggests two things: First, the part of the genome represented by cDNA is not part of the rapidly evolving fraction of the genome defined by lack of hybridization. Second, the great decrease in NPR seen for total scDNA is not an artifact; when relatively homogeneously divergent sets of sequences are cross-hybridized, the dramatic fall-off in NPR does not occur. This is also true for mtDNA (Caccone et al., 1988b).

It should be emphasized that, with few exceptions, the genomes of most other organisms do not show this unusual behavior in DNA–DNA hybridization experiments. The extremely fast evolving fraction has not been detected in extensive studies of birds and mammals, for example. It was detected in cave crickets (Caccone and Powell, 1987). Relatively few insects other than *Drosophila* have been examined using these methods, so it is not clear if this phenomenon is generally true for the whole Class Insecta.

Heterochromatin

Molecularly, heterochromatin differs from euchromatin in containing exclusively repetitive DNA. Two classes of repetitive sequences are known. Highly repetitive "satellite" DNA was first noted in cesium chloride gradients as discrete bands with density different from the bulk DNA band. These were subsequently shown to be simple repeats of 5 to a few hundred nucleotides repeated thousands or millions of times; they are usually A+T-rich. Moderately repetitive DNA, usually longer sequences repeated fewer times than satellite DNA, is also in heterochromatin. Some of these sequences are copies or partial copies of known transposable elements. Also, the moderately repeated clusters of genes coding for rRNA usually reside in heterochromatin. Single-copy DNA sequences have not been detected in heterochromatin.

The evolutionary significance of heterochromatin in *Drosophila* is far from clear if not downright confusing. Some authors have concluded it is simply "junk" DNA, the "flotsam and jetsam" of the genome (Miklos, 1985). On the other hand, there is mounting evidence for a relatively large number of important gene functions associated with heterochromatin; the following discussion of functional units in *D. melanogaster* heterochromatin is from Gatti and Pimpinelli (1992), which can be consulted for details and original citations.

Functions

Figure 9-12 illustrates the genetic functions now known to be associated with heterochromatin in *D. melanogaster.* As X/O males are sterile, it is not surprising that fertility factors exist on the heterochromatic Y chromosome, six of which have been mapped. Because Y chromosomes do not recombine, genetic analysis of this chromosome has been primarily accomplished by deletion mapping and defining complementation groups. The six fertility factors on the Y are huge, each occupying about 10% of the chromosome. These regions form transcriptionally active lampbrush-like loops

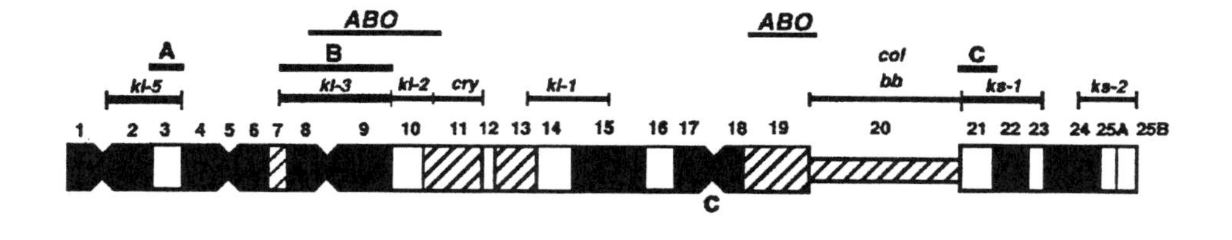

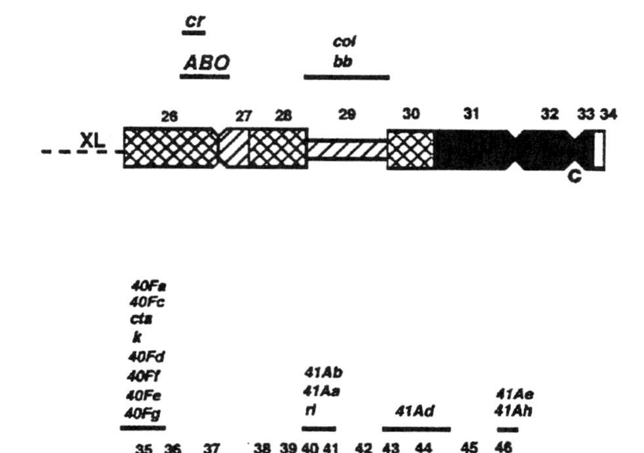

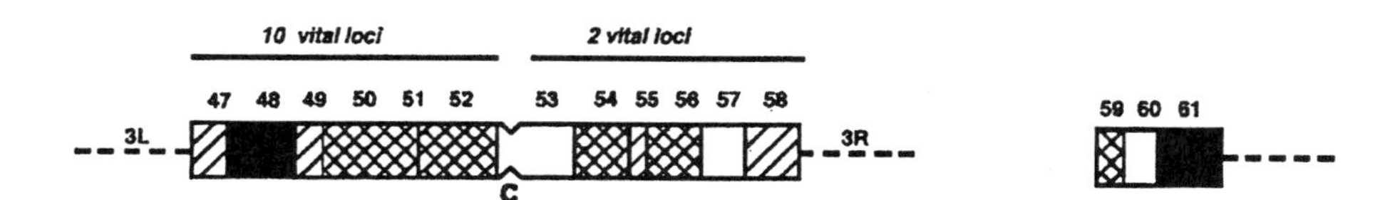

FIGURE 9-12 Cytological maps of heterochromatin from *D. melanogaster*. The upper chromosome is the Y, with the others identified at the ends. The C beneath indicates the position of the centromere. The shading indicates the relative brightness of fluorescence after staining with Hoechst 33258, filled being strongest and open the weakest. Various other symbols are defined in the text. From Gatti and Pimpinelli (1992).

in the spermatocytes; these loops only form in this tissue and only during the spermatocyte maturation period. In sterile X/O males, meiosis proceeds more or less normally and sterility is caused by a breakdown of the maturation process. *D. hydei* is the only species other than *D. melanogaster* where the Y chromosome has been studied in genetic and cytogenetic detail. The loop formation associated with the fertility factors is even more prominent than in *D. melanogaster* (reviewed in Hackstein, 1987). Russell and Kaiser (1993) have identified a set of germ line-specific transcripts that map to the Y and turn out to have similarity to an autosomal gene coding for a cAMP-dependent protein kinase. All the Y copies have the same intron/exon structure as the autosomal one, but also have stop codons in exons, so it is not clear if any of the Y chromosome copies encode a functional protein. No similar Y-linked copies of this gene were detected in the close relatives of *D. melanogaster*. Satellite sequences have also been shown to be located within both the loop-forming and non-loop-forming male fertility genes (Bonaccorsi and Lohe, 1991); each fertility region contains its own specific set of satellites. Considering the importance of male fertility/sterility in the early stages of speciation in *Drosophila* (chapter 7), the function and evolution of Y chromosome fertility factors is clearly a subject of considerable interest.

The *bobbed* (*bb*) region encodes rRNA, and in *D. melanogaster* there is one region on the Y chromosome and one on the X chromosome. These regions are also called nucleolar organizer regions (NORs) as they form the nucleolus. The number and location of NORs vary from species to species and may sometimes be autosomal as well as usually associated with sex chromosomes.

Another function associated with NORs is proper pairing of X and Y during meiosis to assure proper segregation. These regions promoting pairing are called collochores (the *col* in figure 9-12). It is a 240-bp tandem repeat located in the intergenic spacer of the rDNA cluster (McKee et al., 1992). The fidelity of pairing and subsequent proper segregation of the sex chromosomes is additively related to the number of copies of this repeat. A third function embedded in the NORs is one causing ribosomal exchange and deletions at a high rate. This is a maternal effect gene that also has suppressers in the NOR.

We already mentioned the *Stellate* gene associated with the crystal formation in X/O sterile male spermatocytes (chapter 7). *Suppressor of Stellate* (also called *crystal*) maps to a particular region of the Y (figure 9-12, bands 11–13). Recall also that not all species of *Drosophila* have the *Stellate* gene, so it is not a gene crucial to spermatogenesis.

Other genes mapped to heterochromatin are detected because of interactions with other euchromatic genes. *Abnormal oöcyte (abo*) is a recessive maternal effect gene that maps to the euchromatin of the second chromosome; eggs from homozygous *abo* mothers have a low probability of hatching. The probability depends on the amount of heterochromatin on the X, Y, and second chromosome noted in figure 9-12 as ABO. In chapter 4, it was noted that the effect of the euchromatic segregation distortion genes (*SD*) is dependent on the heterochromatin of the second chromosome, which contains two elements, *Enhancer of SD* [*E*(*SD*)], and *Responder* (*Rsp*). In addition, there are a number of autosomal vital genes associated with heterochromatin whose functions are not well known. They are indicated by numbers in figure 9-12 in cases where some mapping has been done.

That repetitive DNA in the heterochromatin can have a direct effect on fitness was demonstrated by Wu et al. (1989). Population cages had two kinds of second chromosomes, those with *Rsp* repeats and a chromosome where these repeats had been deleted. The chromosomes with *Rsp* repeats out-competed the deleted chromosomes. Surprisingly, the major effect seemed to be viability differences; it had been postulated that if heterochromatin had a fitness effect it might be most likely to occur in the germ line, as we discuss next.

Recent evidence of a general role of heterochromatin in normal cell function as a promoter of meiotic pairing is reviewed by Irick (1994) and is based on the work of Hawley et al. (1993). These workers show that the achiasmatic (nonrecombining) pairing and segregation of the dot fourth chromosome is dependent on the heterochromatin of this chromosome. Furthermore, Hawley et al. (1993) demonstrated a role of heterochromatin not associated with the NORs in proper pairing and segregation of X chromosomes in female meiosis. Irick (1994) also notes that heterochromatin "pairs" in polytene cells to form the chromocenter.

Two other aspects of heterochromatin are worth noting. One is that position-effect variegation is caused by translocation of a normally euchromatic gene to the vicinity of heterochromatin. The gene is inactivated only in some cells and the inactivated state is maintained in descendent cells. Finally, when transposable elements are hybridized to polytene cells, most of the time strong hybridization is noted in the chromocenter, which is composed of nonamplified heterochromatin. Many of these copies of transposable elements in heterochromatin are partial copies or otherwise inactivated.

Extreme Variation

Given all these functions for heterochromatin, one might think it should be a conservative structure through evolutionary time. Nothing is further from the case. As we already noted in the discussion of genome size, large amounts of heterochromatin appear and disappear through evolutionary time. The large sizes of the *D. virilis* and *D. nasutoides* genomes are due largely or exclusively to an accumulation of heterochromatin. In addition to amounts, the composition and chromosomal positions also change drastically. Figure 9-13 shows the distribution of heterochromatin in *D. nasu-*

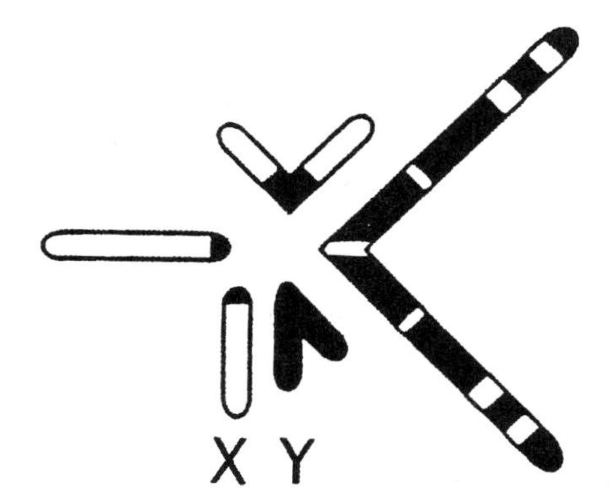

FIGURE 9-13 Haploid chromosome configuration of *D. nasutoides* with the euchromatin represented as open regions and heterochromatin as black regions. From John and Miklos (1988) with data from Wheeler and Altenberg (1977).

toides; an autosome (the element homologous to the dot fourth of *D. melanogaster*) has accumulated a large amount of heterochromatin with the euchromatin interspersed. So the confinement of heterochromatin to centromeric regions (e.g., figure 9-1) does not seem to be an absolute evolutionary requirement.

Closely related species, those that are morphologically, ecologically, and physiologically very similar, can also vary considerably in heterochromatin. For example, one of the sibling species of the *melanogaster* subgroup, *D. orena*, has acquired a second arm on the X chromosome, making it metacentric; all this material is added heterochromatin composed mostly or exclusively of satellite DNA (Lemeunier et al., 1978).

One of the more interesting and puzzling patterns of satellite DNA distribution occurs among very closely related species in the *virilis* group (see figure 8-14). *D. virilis* has a relatively large genome due to the accumulation of heterochromatin. About 42% of the genome is composed of four satellite families, each with a seven-nucleotide repeat unit. This pattern is similar in other members of this tightly-related group. However, at least two members of the group, *D. littoralis* and *D. ezoana*, have no detectable satellite DNA (Holmquist, 1975; Cohen and Bowman, 1979). The ability of these species to produce viable and/or fertile hybrids seems not at all related to heterochromatin. *D. virilis* produces fertile F_1 males and females when crossed to *D. littoralis*, but when crossed to *D. ezoana*, no F_1 develop at all (Throckmorton, 1982b). Thus whatever the role of satellite DNA is in this group, it does not seem crucial to reproductive isolation. In one case, F_1's heterozygous for one haploid set of chromosomes with relatively large amounts of satellite DNA and one with no detectable satellite DNA develop into fertile adults, while in another case, they do not even develop past embryogenesis. It should be admitted, however, that not detecting satellite DNA in the classical method of banding on cesium chloride gradients does not necessarily mean the species lack highly repetitive simple sequences. In fact, using *D. virilis* satellite probes, Cohen and Bowman (1979) could detect simple sequences in the two *virilis* group species that lacked satellites; some were in the chromocenter, some in euchromatin.

Another case of a large change in number and locality of satellite sequences (but with conservation of the exact nucleotide repeat unit) is in the *melanogaster* subgroup. Table 9-6 shows the distribution of the 11 major satellites of *D. melanogaster* in the two sibling species *D. simulans* and *D. erecta*. While most satellites are found in all species, the relative abundance varies considerably, with *D. erecta* having 350 to 19,000 copies of sequences present in hundreds of thousands or millions of copies in *D. melanogaster*. The locations can also change. For example, the 1.705 satellites of *D. melanogaster* (5.6% of the genome) are on all chromosomes but especially the second; this same sequence cannot be detected on the second chromosome of *D. simulans* but is primarily on the sex chromosomes in this species (Lohe and Roberts, 1988). While number and position varies, the sequence does not.

Other variations in satellites are equally inexplicable. In the huge genome of *D. nasutoides* all the satellite DNA, composed of four different satellites, is confined to the enlarged autosome that makes up a total of 57% of the genome (Cordiero et al., 1975; Wheeler et al., 1978). Three species of Hawaiian *Drosophila* (*D. gymnobasis, D. silvarentis*, and *D. grimshawi*) were found to contain about 40% of their genome in a single satellite composed of a 189-bp repeat unit (Miklos and Gill, 1981). Thus

TABLE 9-6 Satellite DNA distributed among three species of the *melanogaster* subgroup.

Sequence Repeat Unit, 5′ → 3′	CsCl Buoyant Density (g/cm^3)	Amount in Genome (%)		
		melanogaster	*simulans*	*erecta*
AATAT	1.672	3.1	1.9	0.0088
AATAG	1.693	0.23	2.4	0.041
AATAC	1.680	0.52	0.0065	0.0018
AAGAC	1.689/1.701	2.4	—	0.011
AAGAG	1.705	5.6	0.71	0.055
AACAA	1.663	0.06	—	—
AATAAAC	1.669	0.23	0.10	0.0015
AATAGAC	1.688	0.23	0.036	0.0016
AAGAGAG	—	1.5	0.074	0.0070
AATAACATAG	1.686	2.1	—	0.0091
359 bp	1.688	5.1	0.11	0.24

From Lohe and Roberts (1988).

there are species with more than half their genomes composed of four different simple sequence satellites (*D. nasutoides*), species with 40% of their genomes in a single relatively large satellite (the three Hawaiians), and some species (e.g., *D. melanogaster*) with much less of the total genome consisting of satellite DNA, but with at least 11 different kinds. Table 9-7 and figure 9-14 summarize the variation known to exist for satellite DNA in *Drosophila.*

There would seem to be a real enigma attached to the evolutionary significance of heterochromatin. This section started with an enumeration of known functions for this not insubstantial portion of the *Drosophila* genome, and ends with an elaboration of the extreme variation evolution has tolerated for this material. Perhaps a partial understanding of the enigmatic evolutionary significance of heterochromatin comes from two points emphasized by Gatti and Pimpinelli (1992). The first is that many of the functions of heterochromatin are cryptic in the wild-type genetic background, for

TABLE 9-7 Satellite DNA variation among species of *Drosophila.*

Species	Percentage of Heterochromatin in Males	Numbers of Satellites	Percentage of Genome in Satellites
melanogaster	35	11	21
simulans	30	2	22
virilis	52	3	42
texana	47	3	35
ezoana	30	no detectable satellite DNA	
hydei	30	2	17
nasutoides	>50	4	57
gymnobasis		1	40

From John and Miklos (1988) with modifications.

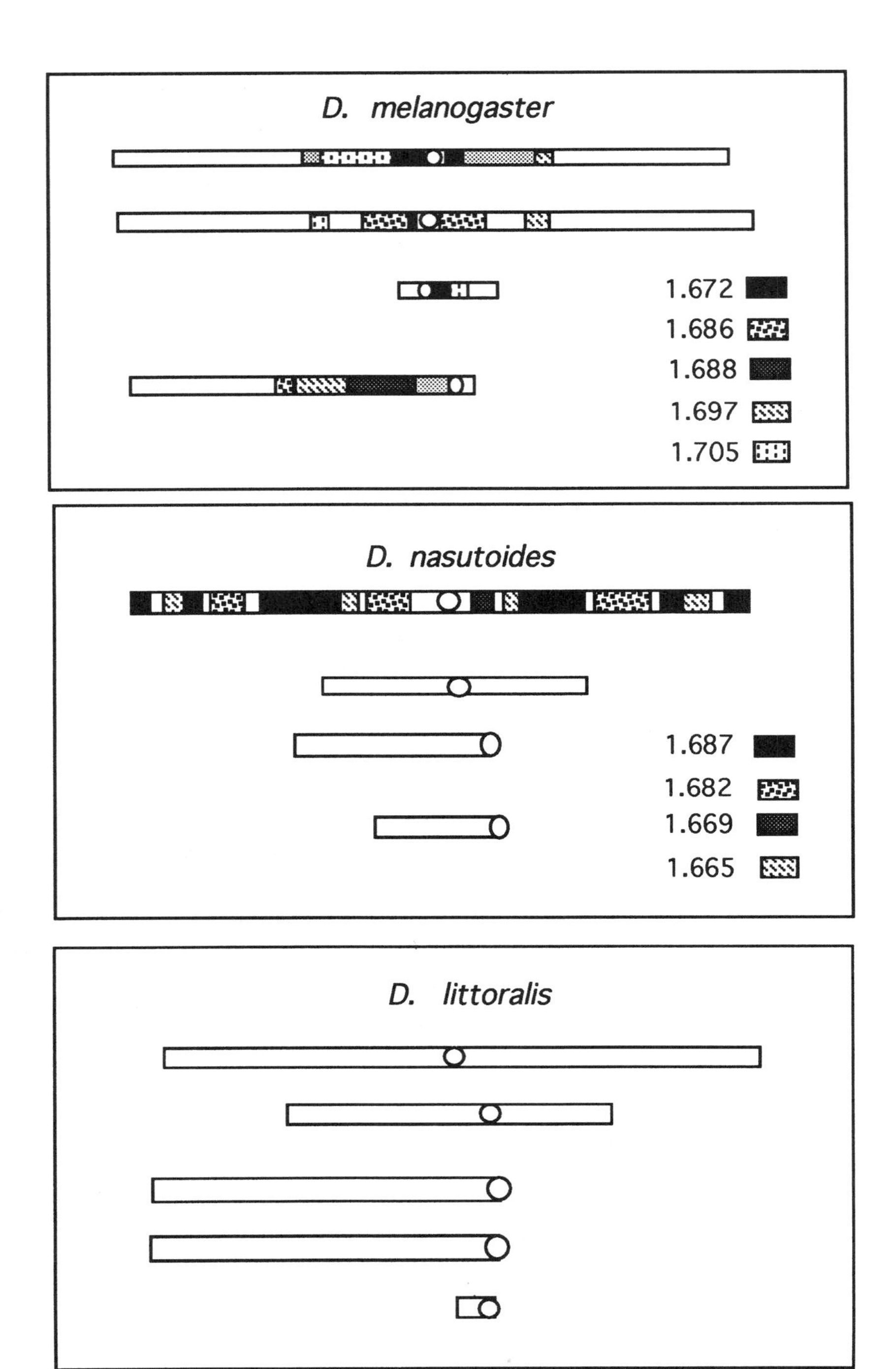

FIGURE 9-14 Comparison of satellite DNA in three species of *Drosophila*. Shaded areas represent different satellites with the buoyant densities indicated. From John and Miklos (1988).

example, ABO, *Rsp*, and *crystal*, although others (e.g., male fertility factors) are not. Second, almost all DNA sequences found associated with heterochromatic functions are also found elsewhere in the genome, often in the euchromatin. Gatti and Pimpinelli speculate that the origin of some of the heterochromatin functions came from the acquisition of a euchromatic sequence followed by amplification and then evolutionary divergence to new functions. While this may partially explain the evolution of this type of DNA, clearly much remains to be known about the basic biology of heterochromatin before a final understanding of its evolutionary significance can be achieved.

The Y Chromosome

The evolution of sex and sex chromosomes is a perennially intriguing problem, and *Drosophila* have offered insights into some aspects of the issue. The variation in the kinds of Y chromosomes in the genus has proven useful in testing some hypotheses about their origin.

Variation

Y chromosomes are known to be polymorphic within populations, the best examples being in *D. pseudoobscura* and *D. persimilis* (reviewed in Dobzhansky and Epling, 1944). The primary differences among the six or so different Y chromosomes is in the length of the shorter arm. The mitotic appearance of the Y varies from an almost equal armed V to a nearly acrocentric J. Presumably this is due to accumulation of heterochromatin of little function or consequence, much like the acquisition of heterochromatin in the several cases discussed in the previous section. This polymorphic case has not been investigated in much detail.

The majority of *Drosophila* species have the usual XX/XY sex-determination, although variation does exist. For example, there is a handful of species with XO males that are either known to be fertile from laboratory studies or are collected in nature and thus assumed to be fertile. Table 9-8 is a listing of these species. Clearly

TABLE 9-8 Species of *Drosophila* known to have fertile XO males or to have XO males in natural populations that are thus assumed to be fertile.

Species	Reference
D. auraria	Zhuohua & Fushan, 1986
D. affinis	Miller & Stone, 1962
D. longala	Patterson & Stone, 1952
D. annulimana	Dobzhansky & Pavan, 1943
D. thoracis	Clayton & Ward, 1954
D. pictiventris	Clayton & Ward, 1954
D. orbospiracula	Wharton, 1943
Samoaia spp.	Ellison, 1968

From Ashburner (1989).

these species would be of considerable interest to study with regard to all the functions associated with the Y chromosome in *D. melanogaster*, especially the fertility factors. Unfortunately, most of the species in this table have not received much attention. However, *D. affinis* is a common species that has been studied. In this species, both XO and XY males occur in nature and both types are fertile (Miller and Stone, 1962). All other closely related members of the *affinis* subgroup have only XY males with XO males being sterile.

Other modifications in sex chromosomes are illustrated in figure 9-15. *D. busckii* has an unusual situation; the dot fourth chromosomal element F is attached to the X and Y chromosomes (Krivshenko, 1959). This causes inviability in XO males due to the imbalance of element F genes. In *D. americana* (a member of the *virilis* group) a similar thing has happened in one of the subspecies (*D. a. americana*), except the translocation of the fourth dot chromosome is only to the X chromosome (Stalker,

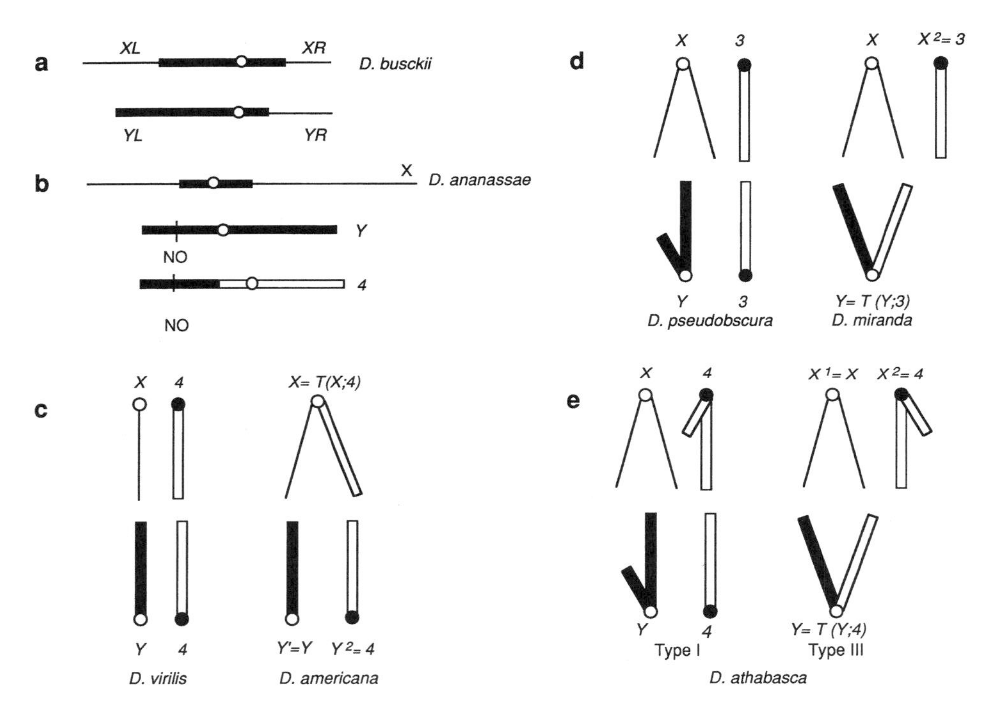

FIGURE 9-15 Variation in sex chromosomes among species of *Drosophila*. Dark areas represent heterochromatin, or in the case of *D. miranda*, an arm evolving to be heterochromatic. (a) In *D. busckii*, both the XR and YR are homologous to the dot fourth chromosome of *D. melanogaster*. (b) In *D. ananassae*, a nucleolar organizer has been translocated to the fourth chromosome. (c) Illustration of the translocation of the fourth chromosome to the X in *D. americana*, contrasted with the configuration in the closely related *D. virilis*. (d) Illustration of the translocation of the third chromosome to the Y in *D. miranda* compared to the configuration in the closely related *D. pseudoobscura*. (e) Two sex-chromosome configurations are known for *D. athabasca*, one normal and the other with a translocation of the fourth to the Y. From Ashburner (1989).

1942). This means that in males the fourth appears as a second Y chromosome since it is in a single copy in males only. The complement has occurred in *D. athabasca*: the fourth dot has translocated only to the Y chromosome, so its partner now acts as a second X (Miller and Roy, 1964; Paika and Miller, 1974).

The *D. miranda* neo-Y

The situation with respect to *D. miranda* has already been alluded to above and is illustrated in figure 9-15. In this case a much larger chromosome, the third of *D. pseudoobscura* (=element C = 2R of *D. melanogaster*) has been translocated onto the Y chromosome (Dobzhansky, 1935; MacKnight, 1939; Cooper, 1946). This leads to the remaining third chromosome being hemizygous in males. It is now called a second X chromosome, denoted as X^2 to distinguish it from the X of *D. pseudoobscura* and *D. persimilis*, which is now called X^1 in *D. miranda*. The new Y chromosome consisting of the previous Y plus the translocated third is now called the neo-Y. If, as is usually assumed (e.g., Charlesworth, 1991), sex chromosomes evolve from previous autosomal units, then *D. miranda* presents an ideal opportunity to study the transition of a chromosome from an autosome to a Y chromosome.

In *Drosophila* the two most notable differences between autosomes and sex chromosomes are: (1) dosage compensation for genes on the X; in males transcription of X-linked genes is about double that in females (Lucchesi, 1977; 1978); and (2) most of the genetic material on the Y chromosome is genetically inert in the classical sense. The genes on the new X^2 of *D. miranda* appear to be at an intermediate stage in the evolution of dosage compensation. Strobel et al. (1978) showed that the rate of transcription per chromosome of the terminal 10% of the X^2 was about the same in males and females, but that in the internal part of the X^2, transcription per chromosome in females (with two copies) was about half that in males (with one copy). Thus dosage compensation is incomplete, being observed only parts in of the X^2.

In addition, evidence of the second aspect of sex chromosome evolution, the relative genetic inertness of the Y chromosome, can also be detected in *D. miranda*. In polytene cells the autosomal arm of the neo-Y of *D. miranda* does undergo polytenization, although in an altered fashion. The polytene bands are not distinct and have heterochromatin interspersed (MacKnight, 1939; Steinemann, 1982b); the banding is sufficiently indistinct that identification of bands homologous to the third cannot be made, although doubtless this arm was a third chromosome at some time.

In a series of papers, M. Steinemann and colleagues have demonstrated that the neo-Y has accumulated a considerable amount of repetitive DNA in the new arm. The evidence comes from several observations including in situ hybridization to mitotic chromosomes of cRNA made from bulk DNA; under these conditions one would expect hybridization of highly repetitive elements to predominate. The entire neo-Y stained heavily while the X^2 exhibited no hybridization outside the centromere (Steinemann, 1982b). A new transposable element found on the neo-Y was called TRIM (Steinemann and Steinemann, 1991, 1993; Ganguly et al., 1992). In a study of four genes encoding larval cuticle proteins, Steinemann and Steinemann (1990; 1992; Steinemann et al., 1993) showed that while all four are transcribed from the X^2, three were silenced in the neo-Y and one showed very limited activity. The silenced copies were interrupted by transposable elements.

The conclusions from these studies are:

1. When a previously autosomal chromosome becomes a Y chromosome, its genes become inactive.
2. The cause of inactivation is the accumulation of transposable elements.
3. The genes on the new X (previously an autosome) become dosage compensated.

These processes are at an intermediate stage in *D. miranda* and thus are an excellent model for investigating the evolution of sex chromosomes.

In relation to the size of the genome, it seems likely that the accumulation of repetitive DNA in the neo-Y chromosome is the cause of the greater genome size of males than of females (see previous section on genome size and figure 9-3). In all other species females have slightly larger total genome size, presumably due to the DNA content of XX being somewhat greater than XY. While *D. miranda* males have one less chromosome than do females, the added DNA in the neo-Y more than compensates.

The accumulation of transposable elements in the Y chromosome conforms to a theory that part of the function of heterochromatin is to act as a "graveyard." That is, either deleterious and/or no longer functioning DNA sequences get trapped there. It is quite likely this is due to lack of recombination in the Y, since recombination is important in eliminating transposable elements (Charlesworth and Langley, 1986). As we will see near the end of this chapter, there is some evidence of higher accumulation of transposable elements elsewhere in the genome in positions of low recombination. Stephan (1989a) has shown that low recombination also promotes the generation of simple tandem repeats such as satellites. Thus recombination is an important factor in understanding how genomes evolve, and we take this up in the next section.

Recombination

Recombination is one of the most important genetic properties of genomes, without which, it is safe to speculate, life would not have evolved as it has. It is thus of some interest to ask what is known about the evolution of recombination in *Drosophila.* The theoretical literature in this area is immense and not within the purview of this book. Unfortunately, the experimental work in this regard is much more limited.

Variation

The first fact we can glean from a consideration of interspecific patterns is that different species recombine at different rates. Figure 9-6 shows the X chromosome homologous arms drawn to scale with recombination. While the physical size and gene content of these arms remain more or less constant, differing rates of recombination result in considerable variation in map lengths. Also, there is variation in the generality that only females undergo recombination in *Drosophila, D. ananassae* being the best studied exception (Moriwaki and Tobari, 1975; Matsuda et al., 1993).

Another fact of *Drosophila* recombination is that there is genetic variance in populations for increasing and decreasing recombination. Table 9-9 reports cases where selection for increased and/or decreased recombination has been attempted in

TABLE 9-9 Studies applying selection for increased and decreased recombination in *Drosophila melanogaster*.

Reference	Chromosome	Outcome
Detlefsen & Roberts, 1921	X	Decrease attained Increase failed
Parsons, 1958	2nd	Increase attained Decrease not attempted
Moyer, 1964	3rd	Decrease attained Increase failed
Acton, 1961	2nd	Decrease failed Increase not attempted
Chinnici, 1971a, b	X	Increase and decrease attained
Kidwell, 1972a, b	3rd	Increase attained Decrease varied
Charlesworth & Charlesworth, 1985a, b	3rd	1/16 lines increased Decrease not attained

normal *D. melanogaster* strains. The emphasis on normal variation here is because it is now known that certain "dysgenic" phenomena, such as the activation of transposable elements, can cause a great increase in recombination, including inducing it in males. As far as is known, all the studies reported in this table did not involve dysgenic crosses, so these studies give an idea of the degree of additive genetic variance that exists in populations of *D. melanogaster*. In addition, there is evidence that some of the classical morphological mutants (mutations of relatively drastic morphological and physiological effect, rare in natural populations; see chapter 2) can also affect recombination (e.g., Hinton, 1967). All of the studies concerning selection on natural populations for increased or decreased recombination have used *D. melanogaster*, so nothing is known of genetic variance for this character in any other species.

The genetic basis of the variance in recombination rate has been studied in a few cases. Chinnici (1971b) used chromosomal substitutions to demonstrate that the effect of increasing and decreasing recombination between *sc* and *cv* on the X chromosome was controlled by genes on all three major chromosomes and the effect was specific to that region of the X. For the low recombination line, the effects were approximately additive, while for the high recombination line, chromosome 3 acted epistatically. Kidwell (1972a, b) found that high recombination was almost completely recessive to low recombination and that during artificial selection for high recombination, natural selection retarded the rate of increase; she also found that genetic drift had a strong effect on recombination. Charlesworth and Charlesworth (1985a, b) also detected effects of all chromosomes in chromosome substitution experiments, but they found considerable complexity in interactions; they also produced evidence that the actual numbers of genes affecting differences in recombination may be relatively few (at least for the particular intervals of chromosomes they studied).

Sex Chromosomes

Empirical data regarding the importance of high and low recombination in *Drosophila* have been discussed in chapters 3 and 4 with regard to response to selection and the evolutionary importance of inversions (crossover suppressers). One study on the effect of recombination on the evolution of genes with sex-specific effects is especially relevant to the evolution of the Y chromosome and has not yet been discussed. Bull (1983) and Rice (1987) argued from theoretical considerations that one reason recombination may become inhibited in primitive sex chromosomes is the accumulation of alleles that affect the fitness of the sexes in opposite directions, what they termed sexually antagonistic alleles. For example, the primitive state might have been a single sex-determining locus with heterozygotes (M/m) being males and the homozygotes (m/m) females (which is the case, for example, in some mosquitoes). If there were loci linked to the sex-determining locus that had alleles that conferred higher fitness on males and lower fitness on females, there would be selection to increase the linkage of such alleles to the M allele at the sex-determining locus, that is, selection to decrease recombination of the chromosome. Remarkably, Rice (1992) was able to demonstrate the evolution of just such a situation in *Drosophila*. He devised a scheme with *D. melanogaster* whereby two previously autosomal loci were made to segregate as would a pair of sex-determining genes. He found that after 29 generations the sexually antagonistic genes accumulated in tight linkage to the new sex-determining loci. This demonstrated that such alleles are not infrequent and they will accumulate under the conditions of the model.

Transposable Elements

Some History

During genetics meetings and congresses in the 1950s and 60s, two individuals often found themselves talking to each other, as no one else seemed to understand or appreciate what they were talking about. They were Barbara McClintock and Melvin M. Green. McClintock, working on corn, and Green, working with *Drosophila*, had evidence that genetic elements sometimes spontaneously change positions in genomes. This was contrary to all prevailing notions of genetics, including the very basic principle that genes could be mapped on genomes, and, as is so often the case, such heretical ideas were ignored. The clearest statements and elegant deduction that the corn and *Drosophila* observations were the same phenomenon can be found in Green (1969a, b). The recognition that movable DNA segments, or transposable elements, are ubiquitous in living organisms was one of the most surprising and far-reaching discoveries in genetics in the last 50 years. For this work, McClintock was awarded the Nobel Prize in 1983.

Other historically important pieces of work deserve special mention in discussions of *Drosophila* transposable elements (TEs). Hiraizumi (1971) had found a strain of *D. melanogaster* from South Texas that displayed unusual male recombination. Such MR (male recombination) strains were found to be fairly common throughout the distribution of this species. Subsequently, Kidwell et al. (1977) began to define and document what they called hybrid dysgenesis, which occurred when strains from natural populations were crossed to old laboratory marker strains. In addition to hybrid dysgenesis inducing male recombination, a number of mutation-like phenomena were induced. It was soon found that hybrid dysgenesis was due to a TE that became highly mobile and mutagenic in certain hybrids (Bingham et al., 1982); it only occurred when the *P*aternal parent carried the element, so it became known as the *P* element. This *P* element was soon co-opted as a transformation vector for placing foreign DNA into an embryo and having it incorporated into the host genome; if under proper control, such genes are even expressed (Rubin and Spradling, 1982). This is arguably the most important advance in multicellular eukaryotic developmental biology in the last 50 years.

Much has been written on TEs in general and those in *Drosophila* in particular; we will confine this discussion to the highlights of the evolutionary biology of TEs in *Drosophila.* A good source for reviews on the subject is McDonald (1993a), while the population genetics theory, with special reference to *Drosophila*, is reviewed by Charlesworth and Langley (1989).

Types of *Drosophila* TEs

Altogether there are thought to be about 40 to 60 families of TEs in *Drosophila.* They can be divided into various types based primarily on their structures, which also implies something about their mode of transposition (figure 9-16). The first division is

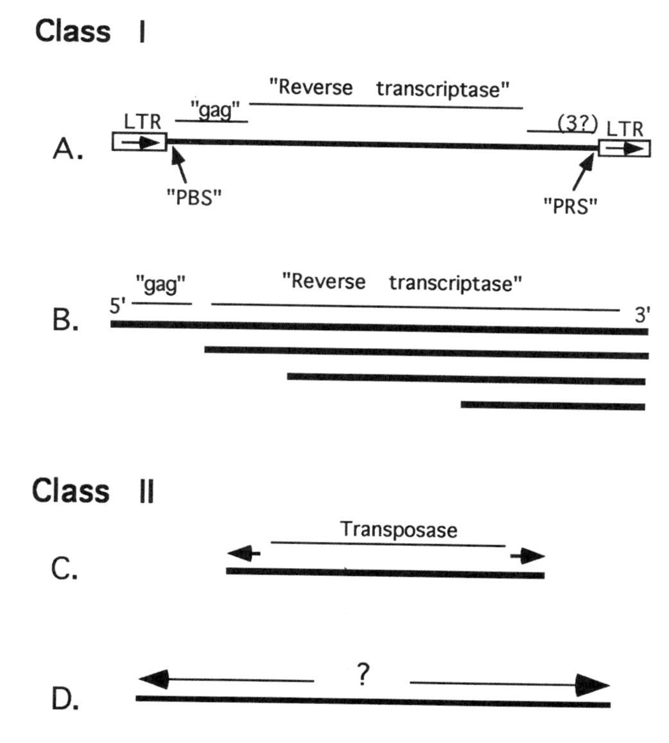

FIGURE 9-16 Transposable element taxonomy. Class I includes those that go through an RNA intermediate. Class II go directly from DNA to DNA. Arrows indicate repeat regions. LTR = long terminal repeat, PBS = primer-binding site, PRS = purine-rich site. From Finnegan, 1989.

into two major classes, based on whether the elements transpose through an RNA intermediate or whether they can go directly from one DNA copy to another. The former types require a reverse transcriptase to copy the RNA back into a DNA molecule capable of inserting into the host genome. Such a mechanism is reminiscent of many retroviruses, so these types have sometimes been called retroposons. They can be further subdivided by structure with regard to having long terminal repeats (LTRs) or not. When they do not, very often they are truncated on the 5′ end, category I-B in figure 9-16. The DNA directly to DNA TEs can also be divided into two types: those with short inverted repeats with a transposase gene between and those with long inverted repeats with no apparent gene function coded between.

Table 9-10 lists *Drosophila* TEs that have been identified and placed in the different categories; almost all of these were first found and characterized for *D. melanogaster.* Thirty-seven element families are listed in this table, implying that the majority of TEs in *D. melanogaster* have been identified. Most fall into Class I-A, those containing an LTR and transposing through an RNA intermediate; 28 out of 37 fall into Classes I-A and I-B, so the RNA intermediate path would seem to be the most common. The fold-back types, II-D, are the least common.

Population Genetics

One of the major issues in the evolutionary biology of TEs is to understand the dynamics of forces that control their frequencies in populations. As with the classical population genetics theory of mutations, two important parameters in understanding TE dynamics are the rate of transposition, μ, and the rate of excision, υ (analogs of the forward and reverse mutation rates, respectively). If, as seems to be the case, most transpositions occur replicatively (i.e., the original copy remains while an additional copy finds a new position) then as long as $\mu > \upsilon$, the element will go to fixation. By fixation, it is meant that all sites that are targets for insertion for that particular element will be occupied. The observations are that μ is indeed greater than υ, being approximately 10^{-4} and 10^{-5}, respectively (Charlesworth and Langley, 1989) and that not all sites are occupied in *Drosophila.* The latter is inferred from the fact that most sites of

TABLE 9-10 Names of *Drosophila* transposable elements falling into the categories noted in figure 9-16.

Class	Other Names	Examples
I-A	Long terminal repeats, LT	Copia, mdg, HMS Beagle, 17.6, 297, Gypsy (=mdg4), Springer, Flea, Opus, Hopper, Tom, 412, roo, B104, Blood, 1731, micro-copia, NEB, 3S18, BEL, Harvey?
I-B	Retroposon, RP	I, F (=jiminy, 101), D, G, 2161, Doc, Jockey (=Sancho), Wallaby
II-C	Inverted repeats, IR	P, Hobo, Pogo, 1360, HB, Mariner (*D. mauritiana*), Minos (*D. hydei*)
II-D	Fold back	FB, BS

Note: All are known from *D. melanogaster*, except where noted. If more than one name has been used, they are indicated in parentheses.

From Ashburner (1989), wherein references can be found.

insertion are polymorphic in populations (in sharp contrast to humans, where polymorphism in TEs at a site are rare). So the conclusion is that some factor is acting to keep TEs from fixation in *Drosophila.* Three opposing forces have been suggested:

1. Simple mutation-selection balance, with the idea that insertions most often result in a deleterious mutation.
2. Some sort of self-regulation occurs such that up to a certain number of copies, TEs are neutral for the host, but that too many begin to lower its fitness.
3. Recombination of ectopically paired regions due to the presence of TEs leads to chromosomal abnormalities and selects against high densities of TEs.

Fortunately, these models make predictions that empirical data, in theory, could falsify. Unfortunately, the data to date have proven insufficient to do so.

If most new transpositions result in a deleterious mutation and if most such deleterious mutations are recessive, then if hypothesis (1) above is correct, we would expect fewer TEs on the X chromosome than on autosomes due to more efficient selection in hemizygous males. Alternatively, if the neutral hypothesis (2) is correct, there is no reason to expect autosomes and X chromosomes to be any different. Biémont (1993) reviewed the empirical observations based on in situ hybridizations to polytene chromosomes; table 9-11 summarizes the results. Clearly no single hypothesis is supported for all TEs and it remains questionable whether data are sufficiently consistent to make a firm conclusion about any element. Either the data are insufficient to distinguish between hypotheses (1) and (2) or neither is generally true. Given the large number of studies and the variety of TEs examined, it seems unlikely there are insufficient data.

The third hypothesis, ectopic pairing and exchange, was first suggested by Langley et al. (1988b). In examining cells at meiosis, or in other cells where homologous chromosomes pair (such as polytene cells), it is often observed that nonhomologous regions show some degree of pairing; in polytene cells this is often associated with visible amounts of chromosomal material pulled from the normal chromosome to pair with nonhomologous regions. It can occur within a chromosome or between nonhomologues. This ectopic pairing is thought to happen because multiple copies of sufficiently similar sequences induce pairing based on local nucleotide sequence. If such ectopically paired regions recombine, then chromosomal abnormalities (translocations, inversions, etc.) can arise that are usually deleterious or lethal and thus rapidly eliminated. Accordingly, the rate of recombination is predicted to be positively correlated with the rate at which TEs are eliminated. Assuming ectopically paired TEs recombine at a rate similar to that of the chromosomal region in which they reside, then TE density should be negatively correlated with the recombination rate for their chromosomal location. It is well known that recombination in *Drosophila* tends to be highest in the middle of arms of chromosomes and decrease toward the centromere and telomere (see figure 10-1 and the accompanying discussion in chapter 10). Unfortunately, as with trying to decide if the empirical data support hypotheses (1) or (2), it is equally ambiguous as to whether the data conform to the predictions of hypothesis (3). Two recent studies can be cited. Biémont et al. (1994) studied five TEs and their chromosomal locations in males from natural populations of *D. melanogaster.* They found no support for variation in TE density correlated with position along a chromo-

TABLE 9-11 Tests of whether the proportion of TEs found on the X chromosome relative to autosomes is compatible with the deleterious selection hypothesis or neutral hypothesis.

Element	Mean no. per Genome	Proportion on X	Inference		Reference
			Test[a]	Test[b]	
mdg-1	383	0.16	neutral	both	Biémont, 1986
mdg-1	285	0.17	neutral	both	Biémont & Gautier, 1988
mdg-1	493	0.09	selected	neither	Belyaeva et al., 1984
mdg-3	220	0.23	neither	neutral	Belyaeva et al., 1984
copia	303	0.09	selected	neither	Biémont & Gautier, 1988
copia	105	0.12	both	selected	Biémont et al., 1994
copia	511	0.12	selected	selected	Yamaguchi et al., 1987
copia	124	0.10	selected	selected	Strobel et al., 1979
412	115	0.17	neutral	both	Strobel et al., 1979
412	411	0.13	selected	selected	Montgomery et al., 1987
297	92	0.24	neutral	neutral	Strobel et al., 1979
297	454	0.20	neutral	neutral	Montgomery et al., 1987
roo	1225	0.19	neutral	neutral	Montgomery et al, 1987
I	427	0.16	neutral	selected	Biémont, 1986
I	298	0.19	neutral	neutral	Biémont & Gautier, 1988
I	487	0.11	selected	neither	Ronsseray & Anxolabéhère, 1986
I	326	0.17	neutral	neutral	Ronsseray et al., 1989
FB-NOF	60	0.3	neither	neither	Harden & Ashburner, 1990
P	316	0.22	neither	neutral	Biémont & Gautier, 1988
P	721	0.24	neither	neither	Ronsseray & Anxobaléhère, 1986
P(P)	533	0.23	neither	neutral	Ronsseray et al., 1989
P(Q)	522	0.24	neither	neither	Ronsseray et al., 1989
P(M′)	523	0.22	neither	neutral	Ronsseray et al., 1989
P	144	0.23	neutral	neutral	Shrimpton et al., 1990

Note: Two chi-squared tests are presented, [a]that of Montgomery et al. (1987) and [b]that of Langley et al. (1988b), the latter correcting for the unequal number of Xs in the sexes. Under the Inference column, it is indicated whether the data are consistent with only one hypothesis, both hypotheses (i.e., neither rejected), or with neither (i.e., both rejected).

From Biémont (1993).

some (and thus recombination rate) and therefore the data reject the ectopic-pairing hypothesis. On the other hand, Sniegowski and Charlesworth (1994) studied the density of 10 families of TEs on inverted and standard chromosomes of *D. melanogaster.* Inverted chromosomes are relatively rare compared to the standard chromosome and therefore are in the heterokaryotypic state most of the time (where they cannot recombine), whereas the relatively frequent standard chromosomes are often homokaryotypic and thus free to recombine. Sniegowski and Charlesworth found that TEs were significantly more dense in the inverted chromosomes, thus supporting the ectopic-pairing hypothesis (see also Eanes et al., 1992).

Clearly, we do not yet have a satisfyingly comprehensive explanation of the population genetics of *Drosophila* TEs. However, sufficient data have been gathered that we can reach a tentative conclusion: none of the three hypotheses advanced thus far is the universal key to understanding the forces that regulate TE copy number in *Drosophila.* It may be that there is no single, all-comprehensive hypothesis that can account generally for all families of TEs. Each TE may need to be studied individually and once its molecular and evolutionary biology is sufficiently understood, how it is regulated will become obvious. This is not a very attractive thought as science much prefers general principles.

Dynamics of Spread

An issue related to that just discussed is the rate at which a newly arisen (or acquired) TE can spread through a population and species. Notably, there is information on this. The first hints that TEs can rapidly spread came from the hybrid dysgenesis studies. When newly collected males from natural populations of *D. melanogaster* were crossed with old laboratory stocks, dysgenesis was observed in the offspring; the reciprocal cross did not show dysgenesis, nor did some crosses of newly collected males to newly collected females. Once the cause of the dysgenesis was found to be the transposable *P* element (Bingham et al., 1982), it became clear that the newly collected strains often had *P* elements, while the older laboratory strains lacked them. Two explanations were possible. First, something about the long-term maintenance of laboratory cultures could have caused these strains to lose the *P* element. Second, the *P* element may have appeared in *D. melanogaster* very recently and become established in natural populations very rapidly.

The first possibility was tested by Kidwell et al. (1981), when they studied what happened to *P* elements in strains maintained under the usual husbandry condition Drosophila workers use to maintain stocks. In fact, they showed a tendency for *P* elements to increase in strains under laboratory culture conditions. When strains free of *P* elements had a low frequency of *P* introduced, most often the strains went to fixation for *P*. The results are a bit more complicated, as the populations into which *P* was introduced did not always maintain the *P* at fixation and there was also evolution of regulation of transposition; see Kidwell (1987) for more details. The overall conclusion, though, is that under laboratory maintenance regimes, *P* elements tend to increase in *D. melanogaster* populations, making the laboratory-loss hypothesis unlikely.

Kidwell (1983) made a systematic study of strains collected from the wild at various times and assayed them for hybrid dysgenesis involving the *P-M-Q* system (*M* is the symbol for strains lacking *P* elements; some strains that have *P* elements exhibit only some of the dysgenic syndrome and are called *Q* strains). She also studied another dysgenic system called the *I-R*, which is similar in having an *I*nducer parent and a *R*eactive strain, the *I* strains having the *I* element (some strains lacking *I* are neutral to the inducer strains and signified at *N*). Figure 9-17 shows the results of these surveys. In both cases the oldest strains, collected in the 1920s, lack the element. By the 1930s the *I* element appears and by 1980 almost all strains tested had *I*. The *P* element did not appear until the 1950s and is predominant by 1980. Clearly transposable elements can spread through populations and species extremely rapidly. There is a third dysgenesis system based on the *hobo* element, which also appears to have spread through *D. melanogaster* very recently, temporally somewhere between the *I* and *P* rises (Pascual and Periquet, 1991). Perhaps more remarkable than the rapid rise in these TEs is the fact that *three* such elements seem to have undergone this dramatic increase during the last 50 years, the time when humans have been most active in

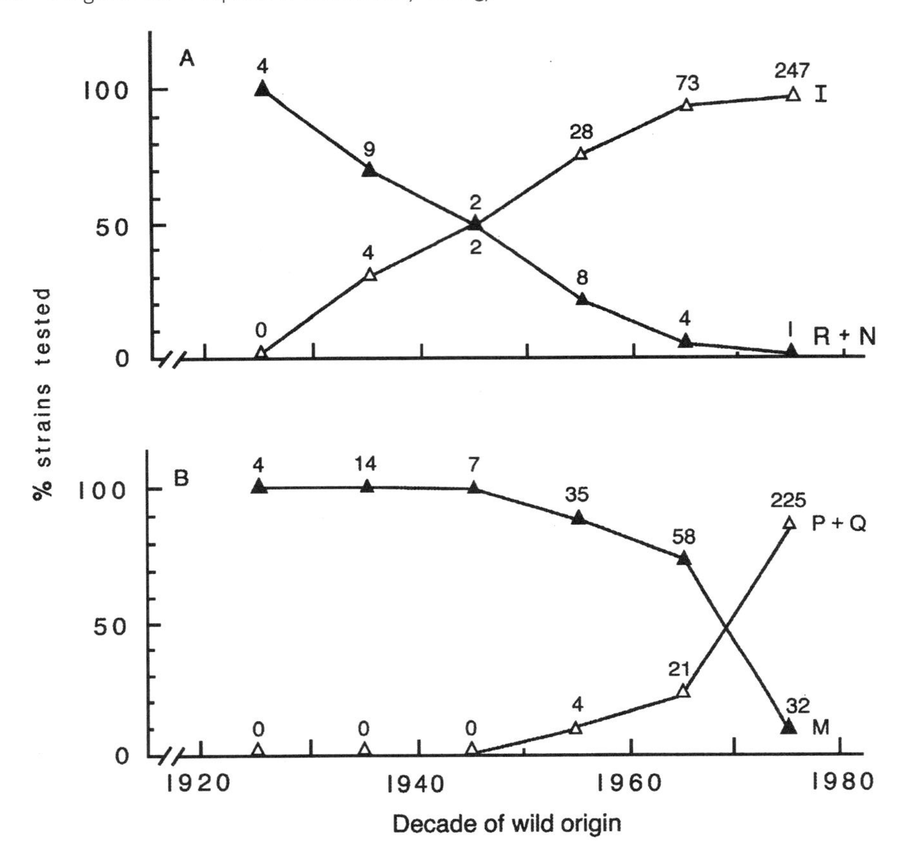

FIGURE 9-17 Graphs showing the frequencies of the transposable elements I (A) and P (B) in strains first brought into the laboratory at the time indicated on the X axis. Lines marked I and P + Q are those with the elements; lines marked R + N and M are those without. The number of lines tested in each case is indicated above the points. From Kidwell (1983).

studying this species. Either such sweeps occur very commonly, or it is a stroke of remarkable luck to have observed these events at this time. There is a third possibility: the recent invasion of *D. melanogaster* with TEs previously unknown in the species could be related to its relatively recent spread (via human transport) around the world (see chapter 10 and especially figure 10-12). Coming into contact with a variety of new species may have led to a relatively large number of horizontal transfers of TEs; in a later section the evidence for horizontal transfer among species will be discussed.

Codon Usage in TEs

Sharp and Lloyd (1993) and Shields and Sharp (1989) have reviewed the codon usage implied in the open reading frames of *Drosophila* TEs and conclude that, in general, the codon usage is quite different from that of its host, in this case almost exclusively

D. melanogaster. They also point out that codon usage varies from element to element, but remains fairly conserved among copies of the same family. In the cases where transposition occurs by reverse transcriptase, Sharp and Lloyd speculate that because most mutations occur during reverse transcription, it may be that the reverse transcriptase for each element differs in its mutation bias. Xiong and Eickbush (1990) have shown that the amino acid sequence of the reverse transcriptases of different *Drosophila* TEs varies considerably.

Horizontal Transfer

The possibility that TEs in *Drosophila* might be acquired in some manner other than simple parent-offspring inheritance (vertical transmission) was recognized early after the discovery of the *P* element. The pertinent observations were the sudden rise in *P* frequency from zero to being extremely common in 50 years, plus the fact that *P* elements are not found in any of *D. melanogaster's* close relatives. It is hard to imagine a *de novo* appearance of the fairly complex element, so it was speculated this species may have acquired *P* by some kind of horizontal transfer such as from a virus or bacteria. The mystery of the origin of *D. melanogaster P* was solved when it was found in the *D. willistoni* group (Daniels et al., 1984; Daniels and Strausbaugh, 1986). Most remarkably, the DNA sequence of *P* from *D. willistoni* and most of its relatives shows a 99% or greater nucleotide identity with the elements from *D. melanogaster*, a much greater degree of similarity than found with any other gene compared between these two species. Furthermore, the distinctive codon usage in the *D. willistoni* group matches very well the codon usage of *P* elements derived from *D. melanogaster*, and is very different from codon usage of any other *D. melanogaster* gene (Powell and Gleason, 1996; table 10-16 in next chapter). DNA sequences with sufficient similarity to *P* elements to indicate homology are found in other species in genus *Drosophila* (Daniels et al., 1990a; Hagemann et al., 1990) and progenitor *P* elements may be present outside Drosophilidae (Anxolabéhère and Periquet, 1987), but the similarities are far less than the 99% plus identity between *D. melanogaster* and *D. willistoni.*

Clark et al. (1994) and Powell and DeSalle (1995) have explicitly analyzed this situation from a phylogenetic approach (figure 9-18). Two points seem clear from this figure. First, the most parsimonious explanation is that *D. melanogaster* acquired its *P* elements from *D. willistoni*. Second, the loss of a transposable element is indicated by the complete lack of *P* in *D. insularis*; this was first noted by Daniels and Strausbaugh (1986) based on two old laboratory strains and later confirmed by analysis of 24 freshly collected strains (Powell et al., unpublished data). The most parsimonious explanation consistent with the distribution of *P* elements in this phylogeny, as well as other information, is that ancestral *P*-like sequences are old, certainly older than the genus *Drosophila.* Within *Drosophila*, the element was lost in the ancestral lineage leading to the *melanogaster* group but retained in the *obscura* and *willistoni* groups (at least in some species of these groups). *D. melanogaster* subsequently acquired the element from the *willistoni* group. In considering a broader phylogeny including *Scaptomyza* and a *Lucilia* (a calliphorid), Clark et al. (1994) found further support for at least two other likely instances of *P* element transfer.

Two other pieces of information are relevant to this scenario. First, Kidwell (1983) could show that *P* elements first appear in strains collected in the Americas,

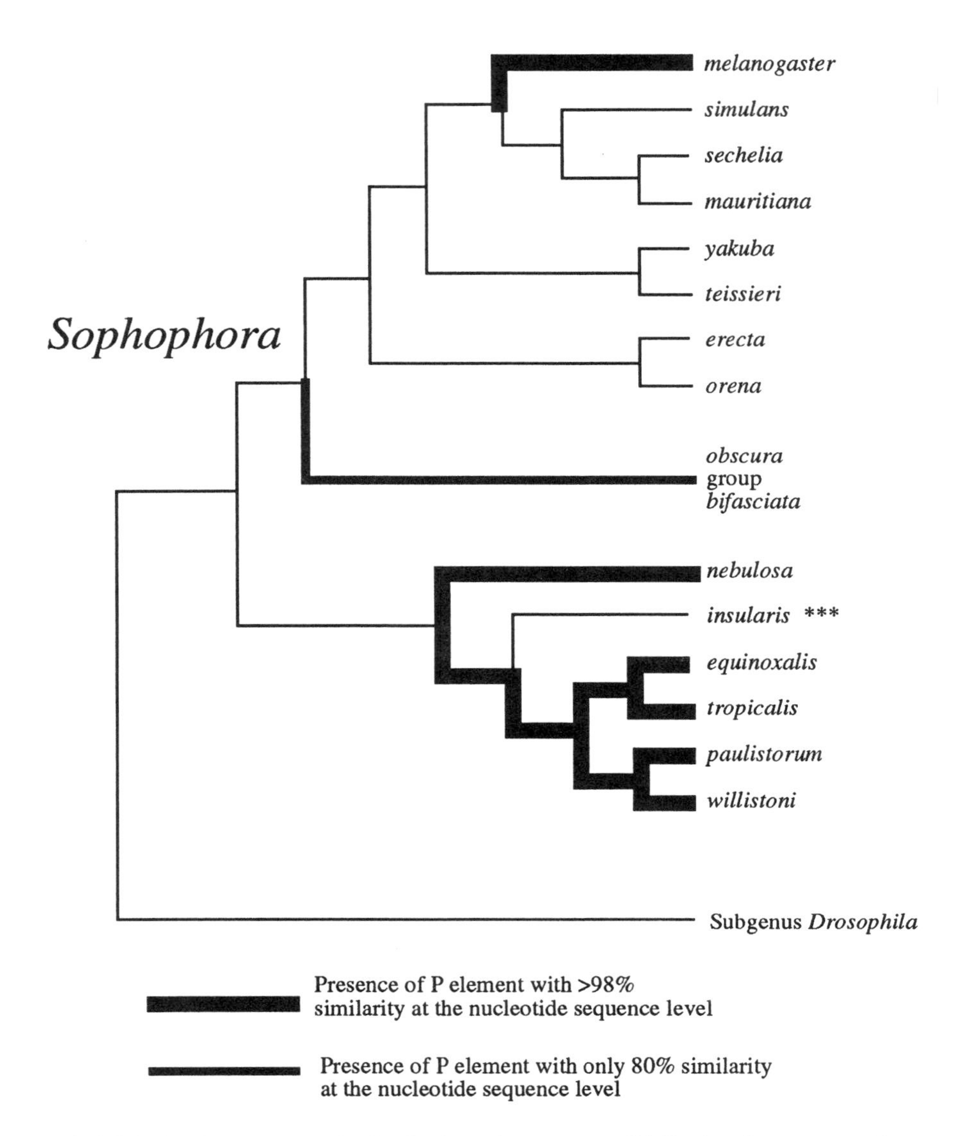

FIGURE 9-18 The phylogenetic distribution of *P* elements. While the *obscura* group has a *P*-like sequence, it shows much less similarity to those of *D. melanogaster* and the *willistoni* group than the latter two do to one another. The asterisks next to *D. insularis* emphasize this lineage lost the element. From Powell and DeSalle (1995).

an observation that predated the finding of *P* elements in the neotropical *willistoni* group. Second, a mechanism for horizontal transfer of *P* elements has been proposed by Houck et al. (1991), who found evidence that a species of mite, when co-cultured with *P*-containing flies, could acquire *P* element DNA. As mentioned in chapter 5, mites are a common pest in *Drosophila* cultures and are often found associated with flies in nature. It is curious that *P* first appears in *D. melanogaster* strains from the 1950s and Dobzhansky started working with *D. willistoni* in the 1950s, so the two species were cultured in the same laboratory beginning at that time. However, if the transfer occurred in the laboratory, it seems strange that *P* only found its way into strains captured since 1950 and not into older laboratory strains. The exact mechanism and site of the horizontal transfer remains unknown, but its occurrence seems inescapable. It is further conceivable that the other two elements documented to have risen in frequency in the last 50 years, *I* and *hobo*, were also acquired horizontally since the expansion of the range of *D. melanogaster*. In these two cases, however, the elements are not confined to the single species *D. melanogaster*, but exist in the other members of the subgroup, although not in most other groups; they are among the most restricted elements known (Stacey et al., 1986; Daniels et al., 1990b). The similarities among copies from different species of the subgroup are so slight as to imply horizontal transmission (Simmons, 1992). If *I* and *hobo* were acquired due to the expansion of the range of *D. melanogaster* and *D. simulans*, then somehow the element got back to the native region of sub-Saharan Africa, where it moved to its siblings—but this has not occurred with the *P* element.

Other cases of possible horizontal transfer in *Drosophila* (and other organisms) are reviewed in Kidwell (1993). *Drosophila* TEs implicated in horizontal transfer include *jockey* (Mizrokhi and Mazo, 1990), *copia* (Mount and Rubin, 1985; Stacey et al., 1986), *gypsy* (Alberola and de Frutos, 1993), and *mariner* (Maruyama and Hartl, 1991; Capy et al., 1992; Lohe et al., 1995). This last-mentioned element is among the more intriguing, as it has about the broadest range across insects of any known TE (Robertson, 1993) and has the potential to be developed as a transformation vector of general use in insects.

Other Species

Most of what has been discussed so far concerns TEs first isolated and characterized from *D. melanogaster*. Few other species have been studied in detail. In *D. virilis* a hybrid dysgenesis syndrome has been characterized that appears to differ from that in *D. melanogaster*. The major difference is that it appears that the TE present/absent in the parents of crosses not only mobilizes copies of itself in the hybrids, but also may mobilize other TEs as well (Lozovskaya et al., 1990; Scheinker et al., 1990; Evgen'ev et al., 1993). Dowsett and Young (1982) demonstrated that, compared to *D. melanogaster, D. simulans* has a much smaller proportion of the genome composed of middle repetitive DNA and, by inference, fewer TEs. Hey (1989) studied *D. algonquin* and *D. affinis* and found that the TEs in these species occur in relatively few families, each in high frequency, a pattern in contrast to that in *D. melanogaster*. Brezinsky et al. (1993) have studied a TE from the endemic Hawaiian *Drosophila*.

Effect of TEs on Their Hosts

It was pointed out above that there must be some factor(s) retarding the takeover by TEs of the fly's genome; in the "selfish DNA" view of TEs, the fly genome can be thought of as the host of the TE parasite. Generally we consider parasites to affect their hosts negatively and there is good evidence that the fitness of flies generally decreases due to transpositional activity (e.g., Mackay, 1986; Eanes et al., 1988; Woodruff, 1993). The majority of spontaneous visible mutants are due to TE insertions (Green, 1987), and in chapter 2 we reviewed the evidence that most visible mutations are detrimental. Often recessive lethals are due to TE insertions (e.g., Sheen et al., 1993; Woodruff, 1993); hybrid dysgenesis is an extreme form of fitness reduction. As a general rule, there is little doubt that TEs are detrimental to their *Drosophila* hosts.

But can TEs play a positive role in *Drosophila* evolution? In a trivial sense, the answer is yes. TEs are a source of mutation, and mutations, are the basis of all evolutionary change. In chapter 2, studies indicating populations with active transposable elements responded faster to selection for quantitative traits were reviewed. But this would simply be assigning a positive role for any agent that increases mutation rate, such as radiation, chemical mutagens, and so on. While in the ultimate view, mutations are necessary for evolution, more proximally, mutagenic agents do more harm than good, that is, they decrease population fitness. The maintenance of such factors by positive selection would require some kind of group selection.

Other than mutation, TEs have been invoked to perform at least two other functions in *Drosophila*. Biessmann et al. (1992) have evidence that a TE may be involved in the "healing" of broken chromosomes and thus may be involved in the formation of telomeres. Thompson-Stewart et al. (1994) have shown that *P* element excision may be accompanied by changes in the number of repeats near the *P* element site. They speculate that this is due to a double-strand break that accompanies excision; the repair of such breaks is accomplished by pairing to similar sequences. Such similar sequences may be close by or elsewhere in the genome, resulting in ectopic pairing and copying. This has the potential to be a mechanism of concerted evolution of not only tightly linked copies of duplicated genes, but also of unlinked copies.

Of perhaps more interest is the possibility that TEs are useful during times of stress when populations and species are required to evolve rapidly to overcome immediate new challenges. McClintock (e.g., 1978, 1984) long felt that movable elements were responsive to environmental stress and thus may be a mechanism for rapid evolution. It is known that transposition can be induced by stress such as heat. However, the ubiquity of transpositional activity increasing under stress has been critically assessed by Arnault and Dufournel (1994), who conclude it is not a general phenomenon. McDonald (1990; 1993b) has argued TEs may be particularly effective in inducing changes in gene regulation and development, which may be the most important aspects of phenotypic evolution.

All of these possibilities that TEs positively affect the fitness and evolution of species (including speciation) remain highly speculative. At this point it would seem more appropriate to take the selfish-DNA point of view and conclude that TEs have not evolved for the benefit of their host flies. When and if TEs play a role in increasing a host's fitness, it is probably best to view this as an exaptation (*sensu* Gould and Vrba, 1982) rather than an adaptation.

Genome Projects

The process of physically mapping and ultimately sequencing complete genomes of *Drosophila* is rapidly progressing. When such information is available for a few species, our knowledge of genome evolution will doubtless take a quantum leap forward. Preliminary reports and discussions of mapping strategies have been presented by Kafatos et al. (1991), Ashburner et al. (1991), Hartl et al. (1992), and Hartl and Lozovskaya (1995). A complete physical map of the X chromosome of *D. melanogaster* has been completed (Madueño et al., 1995). Hartl et al. (1994) point out that about 80% of their P1 clones used to construct a map of *D. melanogaster* cross hybridize strongly to *obscura* group polytene chromosomes, so that once a map of *D. melanogaster* is made, maps of other species phylogenetically as distant as the *obscura* group could rapidly follow. In addition, Hartl and Lozovskaya (1995) report on progress on a genome map of *D. virilis.* This species has a much larger genome than *D. melanogaster* (table 9-1); the increase in size is due to about a 40% increase in euchromatin and the rest in heterochromatin. Reflecting back to this chapter's opening quote, as these genome projects reach fruition, we can anticipate even more delights as we prepare to wander about genomes in much greater detail than has ever been possible.

10

Molecular Evolution

By representing a composite whole as a function of its constituent parts, we are almost automatically empowered to envisage a domain of possible wholes other than that which formed the original subject of the analysis. . . . the whole world is seen reductively as one only of a whole domain of possible worlds—not necessarily the best. This way of looking at reductive analysis makes it clear that among all conceivable ways of understanding the world, reductive analysis is the one that makes it easiest to see how, if need be, the world might be changed.

P. W. and J. S. Medawar, 1983

The previous chapter presented a global view of the evolution of *Drosophila* genomes, and now we increase our magnification in examination of the genome and deal with changes in specific genes and gene systems. We will be concerned primarily with DNA variation, since protein variation, as revealed by allozyme electrophoresis, was discussed in chapter 2. However, since most studies at the DNA level have involved regions coding for proteins, we will frequently refer to issues relevant to protein evolution. In addition to describing *patterns* of variation observed in *Drosophila*, we will also examine what is known of the *mechanisms* that generate the patterns.

Measures of Variation

Before presenting the results of studies of DNA variation in *Drosophila*, we need to mention the ways of measuring variation and statistics that summarize the results. A more detailed discussion of these measures is in Tajima (1993).

The history of study of DNA variation in *Drosophila* follows the history of technological developments in DNA analysis. Initially, restriction enzymes were used to assay DNA variation; in the late 1970s J. Avise introduced this technology to evolutionary genetics (see Avise, 1994; Powell, 1994, for historical accounts). Restriction enzymes recognize specific sets of four to eight nucleotides and cleave the DNA at those sites. By using a battery of such enzymes, variation in the DNA sequences of two sources (different individuals or species) can be estimated from the number of recognition sites in common and the number that are different. The resulting data consist of sets of DNA fragments that vary in size; polymorphism for restriction enzyme sites is called RFLP, restriction fragment length polymorphism. Based on the number of nucleotides in the recognition site, the number of enzymes used, and the number of fragments in common, an estimate of the per nucleotide difference can be calculated. One measure, π, is the number of estimated nucleotide differences per site between two sequences (Nei and Li, 1979). This method does not require knowledge of the actual map positions of the recognition sites in the sequences involved.

A variation of this measure is one predicated on neutral mutation-drift theory (Engels, 1981; Ewens et al., 1981; Hudson, 1982). Under the neutral model, the expected heterozygosity in a randomly mating population at equilibrium is equal to $4N\mu$, usually symbolized θ. Tajima (1989) has suggested that a test of the neutral model is to determine if pairwise nucleotide variation actually observed (π) is significantly different from that predicted by neutral theory (θ).

A third measure of heterozygosity can be obtained if restriction site maps are available rather than simply the sizes of fragments. If a map has been established, one can record the presence or absence of specific sites. Kaplan and Langley (1979) and Nei and Tajima (1983) derived maximum likelihood estimators for this type of data. This statistic is usually denoted d.

Both π and θ are also used as measures of nucleotide diversity from sequence data. In the case of RFLP data, the diversity is estimated; in sequence data it is directly observed. It is important to note that all the measures, π, θ, and d, are meant to express the same thing, the mean per nucleotide difference between two sequences. For diploid nuclear sequences, this is equivalent to the average probability of being heterozygous at a single nucleotide position. For haploid mitochondrial DNA (mtDNA) one cannot really speak of heterozygosity, but the measure is equivalent: the per nucleotide average probability that two haploid genomes differ.

It has long been recognized that the actually observed number of nucleotide differences between sequences is a minimal estimate of the actual number of nucleotide changes that have occurred in evolution. This is because there are only four bases and there is the possibility that a position will mutate and then mutate back to the original state; also, when two sequences differ at a site, it may be due to one mutation or several. For example, a site may differ by being A in one sequence and T in another. This could have occurred by a single mutation ($A \rightarrow T$, $T \rightarrow A$) or several ($A \rightarrow G \rightarrow C \rightarrow T$, etc.).

Various measures to correct for multiple "hits" have been proposed (reviewed in Li et al., 1985; Nei, 1987). The methods differ in the amount of detail incorporated into the correction. The first and simplest case was that assuming random mutation, that is, equal probability of mutation from and to all bases; this is the Jukes and Cantor (1969) method. For many sequences, it is now known that transitions (purine to purine, or pyrimidine to pyrimidine) are more likely than transversions (purine to pyrimidine and vice versa). Kimura (1980) introduced this as the two-parameter model, in which this type of mutation bias can be corrected. Three-parameter (Kimura, 1981), four-parameter (Takahata and Kimura, 1981), and six-parameter (Gojobori et al., 1982) models have also been developed where different probabilities of other types of changes can be taken into account. Finally, two methods take into consideration the triplet genetic code and determine differently the probabilities of change in two-, three-, four-, and six-fold redundant codons (Li et al., 1985; Lewontin, 1989); obviously these latter methods are only applicable to protein-coding sequences.

In considering and comparing DNA variation studies, it is important to note whether a correction has been employed. Many of the correction methods are reasonably accurate as long as the divergences are not too great, that is, <0.5 average probability of change per nucleotide (Li et al., 1985). Once divergence becomes greater than one (i.e., expectation of at least one substitution at each site), no correction is very accurate (not surprisingly). All of the data on intraspecific polymorphisms concern sequences that differ by considerably less than 0.5 per nucleotide site, so the precise method of correction (or even making a correction) is generally not crucial. Only when comparing species from different groups of *Drosophila* do we encounter divergences greater than 1 such that serious concern for methods of correction comes into play.

Levels of Nuclear DNA Variation

D. melanogaster

D. melanogaster has been extensively studied using restriction enzyme technology. We already presented in chapter 2 one such study (figure 2-3). Table 10-1 is a summary of RFLP studies done on *D. melanogaster*. The first RFLP study is that of Langley et al. (1982), and throughout most of the 1980s RFLP was the most widely applied methodology. The studies listed in table 10-1 vary in the length of the contiguous regions studied, from a few kb to over 100 kb. Generally, the studies of larger regions employed 6-base "cutters" (enzymes with 6-base recognition sites), while those examining smaller regions employed the more frequently-cutting 4-base cutters. Altogether a little more than 700 kb of the total single-copy genome has been sur-

TABLE 10-1 Studies of levels of DNA variation in *D. melanogaster* estimated by RFLP analysis.

	Sample Size		Variation		
Chromosome/Gene	Alleles	kb	θ	π	Reference
X Chromosome					
yellow, ac-sc	64	106	0.001	0.0003	1
	49	120	0.002	0.002	2
	109	31	0.003	0.002	3
	245	23	0.001	0.0004	21
	50	6.7	0.001	0.001	22
period	66	52	0.002	0.002	4
zeste-tko	64	20	0.004	0.004	5
white	38	45	0.013	0.011	6
	64	45	0.008	0.009	7
	64	10	0.004	0.004	8
Notch	37	60	0.007	0.005	9
forked	64	25	0.004	0.002	10
vermilion	64	24	0.006	0.003	10
G6pd	126	13	0.004	0.001	11
Su (f)	64	24	0.000	0.000	10
	50	6.4	0.0008	0.0006	22
	55	6.4	0.0005	0.0002	24
Pgd	142	5.3	0.0020	0.0024	20
cut	35	40	0.0022	0.0006	23
Mean X-linked			0.0038	0.0028	

TABLE 10-1 Continued.

	Sample Size		Variation		
Chromosome/Gene	Alleles	kb	θ	π	Reference
Second chromosome					
Adh	48	13	0.007	0.006	12
	58	12	0.005	na	13
	81	3	na	0.004	14
	86	11	na	0.0049	25
	39	2.7	na	0.0034	26
	85	2.4	na	0.005	27
P6	85	2.0	na	0.0074	27
amylase	85	15	0.006	0.008	15
	86	14	na	0.0046	25
Gpdh	86	23	na	0.0078	25
Punch	86	18	na	0.0041	25
Ddc	46	65	0.004	0.005	16
Third chromosome					
87A heat shock	29	25	0.002	0.002	17
rosy (*Xdh*)	60	100	0.005	0.005	18
Est-6	42	22	0.010	na	19
Mean autosomal			0.0056	0.0055	
Total Unweighted Mean			0.0046	0.0055	
Total Weighted Mean		711	0.0043	0.0039	

References: (1) Aguade et al., 1989a; (2) Beech and Leigh Brown, 1989; Macpherson et al., 1990; (3) Eanes et al., 1989b; (4) D. Stern et al., unpublished, cited in Aquadro, 1993; (5) Aguade et al., 1989b; (6) Langley and Aquadro, 1987; (7) Miyashita and Langley, 1988 with six-cutters; (8) Miyashita and Langley, 1988 with four-cutters; (9) Schaeffer et al., 1988; (10) cited in Langley, 1990; (11) Eanes et al., 1989a; (12) Langley et al., 1982; Aquadro et al., 1986; (13) Jiang et al., 1988; (14) Kreitman and Aguade, 1986b with four-cutters; (15) Langley et al., 1988a; (16) Aquadro et al., 1992; (17) Leigh Brown, 1983; (18) Aquadro et al., 1988; (19) Game and Oakeshott, 1990; (20) Begun and Aquadro, 1994; (21) Martin-Campos et al., 1992; (22) Begun and Aquadro, 1995; (23) Begun et al., 1994; (24) Langley et al., 1993; (25) Takano et al., 1991; (26) Aguadé, 1988; (27) Bénassi et al., 1993.

Note: θ and π are two estimates of average heterozygosity per nucleotide; see text for explanation. When more than one study was done on a locus, the one with largest region is used in the calculations of the means. The weighted mean is weighted for the size of the region studied.

After Aquadro (1993) with additions.

veyed for variation, about 0.7% of the total single-copy genome (about 100,000 kb). The heterozygosity estimates, whether estimated from π or θ, comes out to be about 0.4%, with θ estimates averaging a bit higher than π estimates. This would indicate a deficiency of haplotypes in an intermediate range, although given the error associated with these estimates, one cannot take too seriously the difference between 0.0043 and 0.0039, except to note a trend.

The first study of direct sequence variation within a species was Kreitman's now-classic study of 1983. A listing of studies on *D. melanogaster* intraspecific variation determined by direct sequencing is in table 10-2. In contrast to many RFLP studies,

TABLE 10-2 Polymorphisms of nuclear genes in *D. melanogaster* based on sequence data.

Gene	Alleles[a]	Coding					Noncoding			Reference
			Total		Silent					
		Length (bps)	π	θ	π	θ	Length (bps)	π	θ	
ase	6	1068	2.06	2.46	5.07	5.57				Hilton et al., 1994
su(s)[b]	(50)						3214	1.00	0.70	Aguadé et al., 1994
su(w^a)[b]	(50)						1985	1.75	1.96	Aguadé et al., 1994
pn	8 (6)	1098	0.23	0.36	0.00	0.00	462	0.72	0.95	Simmons et al., 1994
Pgd	13	1443	1.33	0.89	4.18	2.87	2900	1.59	1.47	Begun & Aquadro, 1994
z	6	804[e]	2.09	2.74	10.11	13.27	183	0.00	0.00	Hey and Kliman, 1993
per	6	1682[e]	4.95	5.21	20.64	21.50	186	17.63	21.54	Kliman & Hey, 1993a
Yp2	6	1046[e]	4.46	3.77	15.17	12.94	68	15.92	13.07	Hey & Kliman, 1993
Zw	33	1558[e]	3.84	3.80	15.81	15.90	147	20.65	17.06	Eanes et al., 1993
Average (X-linked)			2.71	2.75	10.14	10.29		7.41	7.09	
Acp26Aa	10	792	7.35	7.14	10.71	10.34	584[c]	8.76	7.98	Aguadé et al., 1992
Acp26Ab	10	270	8.81	6.55	24.27	20.39				Aguadé et al., 1992
Adh	15	768	8.11	7.61	28.51	25.90	1210	19.10	17.02	Laurie et al., 1991
Adhr	11	816	1.34	2.51	2.08	5.89	845	4.43	5.46	Kreitman & Hudson, 1991
Lcp1Psi[d]	10	384	1.22	0.93	0.00	0.00	184	2.67	2.02	Pritchard & Schaeffer*

Pgi	11	1674	0.78	0.82	1.62	1.79	889	1.89	2.30	McDonald & Kreitman*
Amy-d	8 (5)	1482	8.82	9.63	27.29	31.86	503	26.48	27.95	Inomata et al., 1995
Amy-p	10 (6)	1482	9.75	8.11	33.51	27.19	437	24.50	20.00	Inomata et al., 1995
Sod	11	441[e]	4.37	5.42	14.99	16.41	969	17.81	18.83	Hudson et al., 1994
(CRS)[f]	25	441[e]	3.58	4.20	12.27	12.73	969	11.56	14.60	Hudson et al., 1994
Est-6	13 (12)	1632	7.21	8.89	22.05	25.06	119	17.19	19.48	Cooke & Oakeshott, 1989
tra	11	588	0.74	0.58	3.19	2.51	342	0.54	1.01	Walthour & Schaeffer, 1994
Rh3	5	1149	0.70	0.84	2.97	3.57				Ayala et al., 1993
boss	5	1566[e]	4.86	4.90	19.09	18.88				Ayala & Hartl, 1993
Mlc1	16	314[e]	0.00	0.00	0.00	0.00	656	11.79	9.00	Leicht et al., 1995
ci	10	958[e]	0.00	0.00	0.00	0.00				Berry et al., 1991
Average (autosomal)			4.43	4.41	13.40	13.29		11.72	11.53	
Average (all)[g]			4.02	4.04	13.46	13.51		10.80	10.49	

[a]If numbers of alleles differ between coding and noncoding regions, the number for noncoding region is shown in parentheses.
[b]From direct sequencing and single-strand conformation polymorphism analysis.
[c]Noncoding regions for both Acp26Aa and Acp26Ab are included.
[d]Pseudogene, excluded from calculations for Total and Average of coding regions.
[e]Partial sequence.
[f]Constructed Random Sample, which includes 22 Fast alleles and 3 Slow alleles (Hudson et al., 1994).
[g]Nucleotide diversity for X-linked genes is multiplied by 4/3 and added to that for autosomal genes.

*Names of submitters to GenBank; no publication found.

From Moriyama and Powell (1996).

sequencing studies have tended to concentrate on coding regions with relatively little noncoding adjacent DNA being assayed. In the tables for sequence data, we separate the data into three classes:

Total coding: All nucleotides coding for amino acid sequences

Silent coding: Silent or synonymous sites within coding regions

Noncoding: Introns and 5′ and 3′ DNA adjacent to coding regions

It is interesting to note in table 10-2 that there was a hiatus in sequence work of this sort until the 1990s (with the exception of Cooke and Oakeshott, 1989). This reflects evolving technological advances. The Kreitman and Cooke and Oakeshott studies relied on the laborious cloning of each allele by preparation of a library for each allele sampled. With the advent of PCR technology and its widespread dissemination by about 1990, direct DNA sequencing became much less laborious and new studies are appearing almost monthly. At the time of this compilation there were 24 genes or gene regions in *D. melanogaster* studied for intraspecific nucleotide variation using direct sequencing. *D. simulans* is the second most studied species, with 12 genes studied; the results are summarized in table 10-3. Finally, we list in table 10-4 studies of nucleotide variation in other species; we include RFLP studies of *D. simulans* and all studies of other species.

Other Species

As is evident in tables 10-1 through 10-4, there is considerable variation among genes and gene regions in level of nucleotide diversity found within species. Therefore, in comparing the level of variation among species, it is best to confine ourselves to genes studied in common. Table 10-5 shows the results for the *melanogaster* subgroup. As we will see, *Adh* in *D. melanogaster* is something of a special case, as there is strong evidence that selection in the coding region is maintaining variation. Thus, the last line of table 10-5 is average heterozygosities excluding *Adh*. *D. sechellia* is by far the least variable species studied, with an average per nucleotide heterozygosity of only 0.03%. This species is an endemic confined to the Seychelles Islands, so that population size and bottlenecks may have affected the species. However, *D. mauritiana*, another island endemic of the subgroup, is not particularly depauperate in genetic diversity, being somewhat more variable than *D. melanogaster*. *D. simulans* appears to be the most genetically diverse species in this subgroup. We will return to this issue of level of variability in this subgroup later, under the section "Out of Africa."

Outside the *melanogaster* subgroup, *D. pseudoobscura* is the best studied species. A comparison of levels of nucleotide variation at loci studied in common with *D. melanogaster* is in table 10-6. Both RFLP and sequencing studies indicate *D. pseudoobscura* is more variable than *D. melanogaster*. Only the *Adh* locus of *D. melanogaster* approaches the nucleotide diversity of its homologue in *D. pseudoobscura*, and this is the locus being maintained polymorphic by selection in *D. melanogaster*. The greater diversity of *D. pseudoobscura* is not unexpected, as this species probably has older and more stable populations than those sampled for *D. melanogaster*. *D. pseudoobscura* samples came from its native habitat, the western third of North America. *D. simulans* is nearly as variable as *D. pseudoobscura* (Moriyama

TABLE 10-3 Polymorphisms of nuclear genes in *D. simulans.*

Gene	No. Alleles[a]	Coding					Noncoding			Reference
			Total		Silent					
		Length (bps)	π	θ	π	θ	Length (bps)	π	θ	
ase	6	1068[b]	0.00	0.00	0.00	0.00				Hilton et al., 1994
pn	4	1095	6.55	6.97	16.19	16.98	484	4.22	4.74	Simmons et al., 1994
z	6	804[b]	6.09	6.04	29.51	29.38	183	16.09	16.84	Hey & Kliman, 1993
per	6	1679[b]	10.72	12.00	40.48	45.50	193	15.35	6.05	Kliman & Hey, 1993a
Yp2	6	1046[b]	0.83	0.84	3.66	3.71	66	5.24	6.95	Hey & Kliman, 1993
Zw	12	1558[b]	3.47	2.76	15.65	12.48	147	13.74	8.42	Eanes et al., 1993
Average (X-Linked)			4.61	4.77	17.58	18.01		10.93	10.60	
Adh	5	768	6.77	6.88	27.24	27.79	206	27.07	27.22	McDonald & Kreitman, 1991
Pgi	6	1674	3.95	4.71	15.53	18.44	945	5.82	6.61	McDonald & Kreitman*
Est-6[c]	4	1626	22.24	21.47	79.65	79.45	1362	24.98	24.22	Karotam et al., 1995
(5′ far)[c]	(3)						500	33.81	33.81	Karotam et al., 1995
Rh3	5	1149	12.03	12.53	51.55	53.93				Ayala et al., 1993
boss	5	1566[b]	12.45	12.26	52.13	50.95				Ayala & Hartl, 1993
ci	9	958[b]	0.23	0.38	0.00	0.00				Berry et al., 1991
Average (autosomal)			9.61	9.61	37.68	38.43		22.92	22.97	
Average (all)[d]			7.88	8.04	30.56	31.22		18.75	18.55	

[a]If numbers of alleles differ between coding and noncoding regions, the number for noncoding region is shown in parentheses.
[b]Partial sequence.
[c]Noncoding region of *Est-6* was divided into two parts; see Karotam et al. (1995).
[d]Nucleotide diversity for X-linked genes is multiplied by 4/3 and added to that for autosomal genes.

*Names of submitters to GenBank; no publication found.

From Moriyama and Powell (1996).

TABLE 10-4 Estimates of nucleotide variation in *Drosophila* species other than *D. melanogaster*.

Species/Gene	R, RFLP; S, Sequence	Sample Size		Variation		Reference
		Alleles	kb	θ	π	
D. simulans						
Adh	R	38	52	na	0.015	1
Adh (*rosey*)	R	30	100	na	0.018	2
period	R	38	52	na	0.007	1
	R	36	30	0.0052	0.007	3
y-ac	R	36	40	0.0005	0.0001	3
Pgd	R	36	16	0.0026	0.0011	3
	R	19	5.3	0.0037	0.0047	4
cut	R	34	40	0.0028	0.0030	5
su(f)	R	103	24	0.0002	0.0005	6
D. sechellia						
zeste	S	6	1.0	0	0	7
yp2	S	6	1.1	0.0003	na	7
per	S	6	1.9	0.0009	na	8
ci^D	S	4	1.1	0	0	9
asense	S	6	1.1	0	0	9
D. mauritiana						
zeste	S	6	1.0	0.0045	na	7
yp2	S	6	1.1	0.0012	na	7
period	S	6	1.9	0.0118	na	8
ci^D	S	6	1.1	0.0003	na	9
asense	S	5	1.1	0.0023	na	9
D. yakuba						
bride-of-sevenless	S	4	1.6	na	0.008	10
Rh3	S	5	1.1	na	0.0002	11
Adh (coding only)	S	12	0.8	0.006	0.006	12
jingwei	S	20	0.8	0.006	0.008	13
D. teissieri						
bride-of-sevenless	S	3	1.6	na	0.013	10
Rh3	S	5	1.1	na	0.0021	11
jingwei	S	10	0.8	0.005	0.005	13
D. ananassae						
forked	R	39	14	na	0.010	14
Om(ID)	R	60	36.5	0.0064	0.0044	15
vermillion	R	39	18	0.0017	0.0007	14,16
furrowed	R	39	21	0	0	16
D. pseudoobscura						
Adh	R	19	32	na	0.026	17
	S	99	1.3	0.033	0.018	18
Adh-dup	S	99	0.7	0.034	0.017	18
Amy	R	28	26	na	0.019	19
Xdh (*rosey*)	R	58	4.6	na	0.013	20
	S	7	5.5	na	0.012	21
Rh3	S	3	1.1	na	0.0043	11
Est-5B	S	16	1.6	na	0.0155	22
D. subobscura						
rp49	R	107	1.6	na	0.0045	23
	S	10	1.6	na	0.0108	24

TABLE 10-4 Continued.

Species/Gene	R, RFLP; S, Sequence	Sample Size		Variation		Reference
		Alleles	kb	θ	π	
D. athabasca						
period	R					
Western-northern semispecies		18	4.5	0.0009	0.0003	25
Eastern A		8	4.5	0.0019	0.0010	25
Eastern B		16	4.5	0.0029	0.0022	25
D. willistoni						
Adh	S	18	1.2	na	0.0050	26
per	S	18	1.2	0.0080	0.0120	27

References: (1) D. Stern et al., cited in Aquadro, 1993; (2) Aquadro et al., 1988; and C. F. Aquadro et al., cited in Aquadro, 1993; (3) Begun and Aquadro, 1991; (4) Begun and Aquadro, 1994; (5) Begun et al., 1994; (6) Langley et al., 1993; (7) Hey and Kliman, 1993; (8) Kliman and Hey, 1993; (9) Hilton et al., 1994; (10) Ayala and Hartl, 1993; (11) Ayala et al., 1993; (12) McDonald and Keitman, 1991; (13) Long and Langley, 1993; (14) Stephen and Langley, 1989; (15) Stephan, 1989b; (16) Stephen and Mitchell, 1992; (17) Schaeffer et al., 1987; (18) Schaeffer and Miller, 1992; (19) Aquadro et al., 1991; (20) Riley et al., 1989; (21) Riley et al., 1992; Schaeffer, 1995; (22) Veuille and King, 1995; (23) Rozas and Aquadé, 1990; (24) Rozas and Aquadé, 1993; (25) Ford et al., 1994; (26) Carew, 1993; (27) Gleason, 1996.

Sequence data for *D. simulans* are in table 10-3.

and Powell, 1996), implying that *D. simulans* maintains larger, more stable populations than does *D. melanogaster*, despite having similar biogeographic histories and ecologies.

In the context of levels of variation in nucleotide diversity within a species, the studies on intraspecific variation in *D. mercatorum* discussed in the previous chapter also indicate a high level of DNA variation (Caccone et al., 1987). While it is difficult to compare a study using DNA–DNA hybridization of total scDNA directly to studies discussed in this section, the average divergence of 2% between haploid genomes would place *D. mercatorum* in the high end of nucleotide heterozygosity, comparable to that in *D. pseudoobscura*.

Pattern of Polymorphism

Functional Differences

Rather than simply calculating overall DNA variation, it is more instructive to take a closer look at the pattern of polymorphisms displayed in the data discussed above. One very clear pattern from tables 10-2 and 10-3 is that the functionally different classes of DNA vary in their level of variation. Table 10-7 is a summary for the three best-studied species. (Here and in much of the following, we confine ourselves to sequence studies, as they contain more detailed information than RFLP studies.) As expected, coding DNA has less variation than do the surrounding noncoding regions and introns. Moreover, consistently in all three species, silent polymorphisms within coding regions are more plentiful than polymorphism in the noncoding DNA. This

TABLE 10-5 Average per-nucleotide heterozygosity based on sequence data for genes sequenced from multiple alleles in species of the *D. melanogaster* subgroup.

Gene	mel	sim	mel	sech	mel	mau	mel	yak	mel	teis
Adh (coding only)	8.1	5.4					8.1	6.0		
period	6.2	11.5	6.2	0.9	6.2	11.8				
bride-of-sevenless	5.1	19.0					5.1	8.0	5.1	13.0
Ci^D	0	0.1	0	0	0	0.3				
prune	0.2	6.6								
yolk protein2	5.2	1.1	5.2	0.3	5.2	1.2				
zeste	2.5	7.8	2.5	0	2.5	4.5				
G6pd	3.7	3.2								
asense	2.1	0	2.1	0	2.1	2.3				
Rh3	0.8	12.0					0.8	0.2	0.8	2.1
Unweighted mean	3.4	6.7	3.2	0.2	3.2	4.0	4.6	4.7	3.0	7.6
Weighted mean	3.5	7.1	3.6	0.3	3.6	4.8	4.4	5.1	3.2	8.6
Excluding *Adh*	3.1	7.2					3.3	4.8		

Note: If only one estimate (π or θ) is given it was used; when both were estimated, the mean is shown. All estimates are times 10^3. Data from tables 10-2, 10-3, and 10-4, wherein references are given. Species abbreviations: mel = *melanogaster*, sim = *simulans*, sech = *sechellia*, mau = *mauritiana*, yak = *yakuba*, teis = *teissieri*.

TABLE 10-6 Comparison of nucleotide diversity between *D. melanogaster* and *D. pseudoobscura*.

	RFLP			Sequencing		
Gene Region	Length (kb)	mel	pseudo	Length (kb)	mel	pseudo
Adh	13/52	7.0	26.0	2.0	14.1	18.8
Adhr				1.7/1.6	3.5	14.9
Xdh	100	5.0	13.0			
Amy	15/26	7.0	19.0			
Rh3				1.1	0.08	4.32
Unweighted mean		6.3	19.3		5.9	12.7

Note: Where different lengths of regions were studied in the two species, the *D. melanogaster* length is given first, with the *D. pseudoobscura* length after the slash. In the case of *Adh* and *Adhr*, the RFLP data include both loci together, whereas the sequencing data separates them. The diversity figure is the mean ($\times 10^3$) of π and θ over the total region, coding and non-coding. Data from tables 10-1 to 10-4, wherein references are given. mel = *melanogaster*, pseudo = *pseudoobscura*.

implies that there are less selective constraints on silent sites in coding regions relative to the flanking DNA. This is perhaps not surprising, since at least some of the noncoding adjacent DNA must be involved in gene regulation.

X-Linked Versus Autosomal

As can be seen in the calculation of θ ($=4N\mu$), neutral theory predicts a direct correlation between effective population size and level of heterozygosity. Assuming a 1 : 1 sex-ratio, X-linked genes have a population size 3/4 that of autosomes, and thus should have 3/4 of the heterozygosity. An examination of tables 10-2 and 10-3 show that this is the case for both *D. melanogaster* and *D. simulans*. In the case of *D. melanogaster*, the predicted 3/4 as the ratio of diversity of X-linked to autosomal genes is very close to the observed, although statistically, Moriyama and Powell (1996) could not demonstrate that this ratio is significantly different from 1. In *D.*

TABLE 10-7 Mean diversity measures; both π and θ are multiplied by 10^3.

	Coding Region					
	Total		Silent		Noncoding	
Species	π	θ	π	θ	π	θ
D. melanogaster	4.02	4.04	13.46	13.51	10.80	10.49
D. simulans	7.88	8.04	30.56	31.22	18.75	18.55
D. pseudoobscura	7.56	11.15	28.12	41.98	17.04	22.36

Note: Based on 22 gene/gene regions for *D. melanogaster*, 12 for *D. simulans*, and 5 for *D. pseudoobscura*.

From Moriyama and Powell (1996).

simulans the ratio averages just under 0.5, which is statistically significantly less than 1, but not significantly less than the predicted 3/4 (Moriyama and Powell, 1996). In summary, it is clear and consistent that X-linked genes tend to have lower nucleotide heterozygosity as predicted by neutral theory, but it is not clear whether the predicted ratio 3/4 is the case, especially in *D. simulans*; the errors associated with these data preclude any strong conclusions.

Transitions-Transversions

If the mutation process were truly random, meaning that when a base mutates it has equal probability of changing to each of the three alternatives, we would expect transversions (Tv's) to outnumber transitions (Ts's) by 2 : 1. In the polymorphism data, the opposite is observed: transitions outnumber transversions. Table 10-8 summarizes the observations for three species. In the noncoding regions, Tv's do tend to outnumber Ts's just slightly, while in coding regions the ratio of Ts : Tv is about 2 : 1, exactly the opposite predicted for random mutation. One possible reason Ts polymorphisms outnumber Tv polymorphisms in coding regions is the structure of the genetic code. Synonymous (silent) substitutions are more often Ts's than Tv's, so stronger selection against amino acid polymorphisms than for silent substitutions would lead to selection for Ts's over Tv's. One class of codons allows us to test this. Fourfold degenerate codons are those that allow any of the four bases in the third position. Thus both Ts's and Tv's are silent. In table 10-8 we indicate the Ts's and Tv's for these sites. As can be seen, the third position in four-fold degenerate codons have significantly more Tv's than for the total coding region. This frequency is not significantly different from that in noncoding DNA. One interpretation of this pattern is that the ratio of Ts : Tv at four-fold degenerate sites and in noncoding DNA represents the mutational input and the preponderance of Ts's in the coding region is due to stronger selection against replacement polymorphism relative to silent polymorphisms.

Silent Versus Replacement Polymorphism

Do silent polymorphisms outnumber replacement polymorphisms? Table 10-9 summarizes the observations. In all species studied, silent polymorphisms are more common than replacement polymorphisms, although the species are not quantitatively identical. The proportion of polymorphisms that are replacements in *D. melanogaster* is significantly greater than in *D. simulans*, whether one considers all genes ($\chi^2 = 18.1$, $p < 0.001$) or only those studied in common ($\chi^2 = 7.09$, $p < 0.01$). Assuming that most replacement polymorphisms are slightly deleterious, then this observation is consistent with the previous finding that *D. melanogaster* is less polymorphic than *D. simulans*. Both observations are expected if *D. simulans* populations are larger and more stable than *D. melanogaster* populations. Selection is more efficient at removing slightly deleterious mutations in larger more stable populations.

However, while overall, silent polymorphisms outnumber replacement polymorphisms, some individual genes do not show this pattern; table 10-10 lists some exceptions. All exceptions are for *D. melanogaster*. (Two loci with a single replacement polymorphism have been left out of this table, *pn* in *D. melanogaster* and *ci* in *D.*

TABLE 10-8 Summary of transition (Ts) and transversion (Tv) polymorphisms for different species and for different classes of DNA.

	Coding											
	Total			Fourfold			Noncoding			Chi-Squared (Fisher's Exact P)		
	Ts	Tv	Tv's	Ts	Tv	Tv's	Ts	Tv	Tv's	T/4f	T/non	4f/non
melanogaster	177	84	32.2%	55	43	43.9%	121	150	54.0%	4.26 (0.04)	25.98 (<0.001)	2.95 (0.09)
simulans	165	81	32.9%	69	55	44.4%	68	61	47.3%	4.63 (0.03)	7.42 (0.006)	0.22 (0.64)
pseudoobscura	229	142	38.7%	56	60	51.7%	296	320	51.5%	5.36 (0.02)	9.04 (0.003)	0.002 (0.97)

Note: "Fourfold" indicates fourfold degenerate, that is, any nucleotide can be in the third position. Contingency chi-squares are presented for total versus fourfold (T/4f), total versus noncoding (T/non), and fourfold versus noncoding (4f/non). Chi-squared analyses indicate that the species are homogeneous for all three classes.

From Moriyama and Powell (1996).

TABLE 10-9 Silent and replacement polymorphisms.

Species	No. Silent	No. Replacement	Percentage Replacement ±S.D.
melanogaster			
Total	192	69	26.4 ± 2.7%
In common with *simulans*	121	23	21.4 ± 3.3%
In common with *pseudoobscura*	21	6	22.2 ± 8.0%
simulans	221	29	11.6 ± 2.0%
pseudoobscura	202	41	16.0 ± 2.4%

Note: Standard deviation (S.D.) assumes a binomial distribution.

From Moriyama and Powell (1996).

simulans.) Again, this is consistent with replacement polymorphisms being slightly deleterious and with *D. melanogaster*'s population histories and size being such that selection cannot effectively remove slightly deleterious mutants. On the other hand, we might ask if there is some functional significance to the exceptions. The most extreme exception, *ref(2)P*, codes for a protein conferring resistance to infection with sigma rhabdovirus (Dru et al., 1993). It is conceivable that replacement substitutions are selectively favored if the protein needs to contend with variation in the viruses. However, none of the other exceptional loci have any known reason why replacement polymorphisms would be favored.

The *Adh* data in the lower part of table 10-10 indicate that the same gene in different species groups can vary in the relative ratio of silent and replacement polymorphisms. *D. willistoni* has a significantly greater proportion of replacement polymorphisms than do either the *melanogaster* subgroup or *D. pseudoobscura* ($G = 10.3$, $p < 0.002$). Whether this is due to population size and stability phenomena or different functional constraints is not clear; not enough genes have been studied in the *willistoni* group to know if this pattern tends to be genome-wide (as would be expected if a population level explanation is true) or if it is specific to *Adh*.

TABLE 10-10 Upper part: Examples of loci that show equal or greater replacement polymorphisms than silent polymorphisms. Lower part: Frequencies of silent and replacement polymorphisms in *Adh* from three species groups.

Gene	Species	Silent	Replacement	Reference
ase	mel	3	3	Hilton et al., 1994
Acp26Aa/b	mel	8	13	Aguadé et al., 1992
Adhr	mel	3	3	Kreitman and Hudson, 1991
ref(2)P	mel	1	7	Dru et al., 1993
Pgi	mel	2	2	McDonald and Kreitman*
Adh	pseudo	43	1	Schaeffer & Miller, 1992
Adh	mel + sim + yak	42	2	McDonald & Kreitman, 1991
Adh	*willistoni*	14	6	Carew, 1993

Note: Species abbreviations as in previous tables; * indicates GenBank deposit but no publication.

Insertions/Deletions (Indels)

So far we have been concerned with variation generated by substitution of one base for another. Another important source of variation is the insertion or deletion of segments of DNA. Because it is usually not possible to determine unequivocally whether differences in length of a stretch of DNA are due to the insertion of some bases in one sequence or a deletion of bases in another, such events are often simply referred to as *indels*.

Indels are remarkably common in some species. Referring back again to figure 2-3, we can see that the region around the *Adh* gene of *D. melanogaster* is remarkably rich in indels. The size of the indels ranges from 21 bp to 10.2 kb. In fact, 80% of all alleles examined had at least one indel relative to the consensus standard. Also, it is obvious in figure 2-3 that indels are much more common in segments away from coding regions. This produces a methodological bias. Most sequence studies of DNA polymorphisms have concentrated on coding sequences and their close vicinity, yet judging from such studies as Aquadro et al.'s (1986), illustrated in figure 2-3, this is the region least prone to have indels. Thus a more accurate view of indel frequency comes from RFLP studies covering a larger contiguous segment of the genome, including intergenic DNA. However, RFLP studies have their own problem in being able to detect only relatively large indels depending on the accuracy of the gels in size-separating fragments. The smallest indel detected by Aquadro et al. (1986), 20 bp, probably reflects this lower limit. Thus the most accurate picture of indel frequencies comes from RFLP studies, which detect only larger indels.

Considering this, Aquadro (1993) has estimated the frequency, or density, of indels for a number of genes in a few species. He confined himself to indels greater than 250 bp, the majority of which are transposable elements. The density is defined as the average number of indels/kb/chromosome studied. Table 10-11 shows the data. Unlike the case with nucleotide variation, indel variation in *D. melanogaster* is among the highest of all species studied. Only *D. ananassae* is as replete with indels as is *D. melanogaster*. In one study, *D. ananassae* is about the same, whereas the study of

TABLE 10-11 Density of indels > 250 bp in four species of *Drosophila*.

Species	Total DNA Studied	Density of Indels
D. melanogaster	229 kb	0.004
D. simulans	165 kb	0.0005
D. pseudoobscura	32 kb	0
D. ananassae		
forked:	14 kb	0.004
vermillion	18 kb	0.016

Note: Density is defined as the average number of indels/kb/chromosome examined. For *D. ananassae*, only two studies are available, which yielded very different results, probably due to an anomaly around the *vermillion* locus.

From Aquadro (1993).

vermilion gave extremely high value due largely to one indel that had very high frequency (Stephan and Langley, 1989) and, therefore, may be an anomalous case.

Leaving aside *D. ananassae*, the three well-studied species rank in perfectly opposite order with regard to nucleotide variation and indel variation. *D. melanogaster* is the least variable for nucleotide variation and *D. pseudoobscura* the most variable; for indels it is the opposite, with *D. simulans* intermediate in both cases. Whether one should place much faith in a sample of only three species is doubtful, but at least from the information available, there does seem to be a negative correlation between the two types of DNA variation. Also it should be noted that while *D. pseudoobscura* seems to have fewer large indels, the extensive sequence data of Schaeffer and Miller (1992) did detect a number of small indels. Using *D. miranda* as an outgroup, they were able to infer that insertions were smaller and more frequent than were deletions.

An Historical Aside

In chapter 2 the history of the "struggle to measure variation" was discussed at some length and the chapter ended with the hint of what DNA studies have brought. The extensive data collected in the last 12 years, and especially the last two or three years, indicate that this struggle, to some extent, has ended. We have now seen what must be the ultimate look at genetic variation—at least with today's paradigms, it is impossible to imagine how one could delve deeper into the problem than actual DNA sequence data. Those individuals who predicted that every gene in a population was polymorphic have been largely vindicated, although there are exceptions (e.g., the dot chromosome gene *ci*). Perhaps the bravest (or most rash) premolecular statement was that of Bruce Wallace (1963), who predicted that not only are populations highly polymorphic, individuals are heterozygous at nearly all their genes. An average per-nucleotide heterozygosity of 0.4% for *D. melanogaster* coding DNA (greater in most other species) means that even a small gene of about 1000 bp would have any two independent copies differing at four positions on average. Therefore, except for close relatives or highly inbred individuals, the DNA data do indicate that the majority of genes in each diploid individual are heterozygous. Because the majority of substitutions are silent, this obviously means that the high heterozygosity is for the gene, not the gene product, so one might argue this is not necessarily consistent with the balance view, although the distinction between gene and gene product was seldom broached during the debates. Most importantly, the question of the meaning of this high level of variation with respect to selection and adaptation remains open.

Perhaps a more important question is whether we have come to the end of the odyssey and whether molecular population genetics has no further to go. This subject (as well as the selectionist/neutralist controversy) will be taken up in the final chapter of the book, but suffice it to say that I believe we are far from the end. While the gathering of data of the sort discussed above has become technologically easier, and thus the data are accumulating rapidly, what remains is to understand its meaning. We are quickly approaching a situation where the amount of data is outstripping the development of theory and analytical procedures necessary to synthesize and interpret it, an unusual situation in the history of population genetics!

Recombination and Hitchhiking

In considering the variation in level of DNA polymorphism at different loci, several possible explanations can be considered. First, there may be variation in mutation rate from region to region of the genome. However, this has never been documented and, given the lack of isochores in *Drosophila*, there is no evidence of either qualitative or quantitative variation in the mutation process in the genome (Woodruff et al., 1983; Moriyama and Hartl, 1993). Second, if many of the polymorphic sites are neutral variants, then the differences in level of variation may simply reflect differences in selective constraints. That is, under the neutral theory, the level of polymorphism is expected to be $4N_e\mu$ where μ is the mutation rate *to neutral mutations.* Not all mutations are neutral, such as stop codons in the middle of a protein-coding region. If different genes and gene regions vary in what proportion of all mutations are neutral, meaning the neutral mutation rate varies from gene to gene even though the overall mutation rate is constant, then this may explain variation in level of polymorphism. This is a difficult explanation to evaluate in many cases, but we will have occasion to return to it later in this section.

A third factor, recombination, was not widely appreciated until the empirical data accumulated. Recombination can have a significant effect on polymorphism and since it is well known that different regions of the *Drosophila* genome vary in their level of recombination, this is a potential explanation for variation in level of DNA polymorphism among genes and gene regions. Interestingly, recombination effects combined with selection can both increase the level of polymorphism and decrease it.

Hitchhiking and Selective Sweeps

The importance of recombination and the strength of linkage effects on DNA polymorphism were first suggested by Aguadé et al. (1989a) in their study of the *yellow-acheate-scute* region on the tip of the X chromosome, which undergoes very little recombination (see also Begun and Aquadro, 1991). The extreme importance of this phenomenon became apparent in 1991 with the sequence data of Berry et al. (1991) on the nonrecombining *cubitus interruptus* (*ci*) gene, located on the dot fourth chromosome. This was the first *Drosophila* nuclear gene to have been found with no DNA polymorphism. The explanation proffered by these researchers was one of hitchhiking and selective sweeps: if, sometime in the recent history of a gene region, a selectively advantageous mutation arose and quickly spread through the species (or population studied) and became fixed, neutral, or nearly neutral, variants linked to the region would also become fixed. The length of chromosome affected will depend on the strength of selection (and thus the time until fixation) and the tightness of linkage (Maynard Smith and Haigh, 1974; Kaplan et al., 1989).

Since these initial suggestions of the importance of recombination rate in affecting level of nucleotide diversity, a growing data set indicates the phenomenon is not simply qualitative, but reasonably quantitative. That is, there is not simply a set of genes experiencing low recombination with an average low nucleotide diversity compared to genes with high recombination and high nucleotide diversity; there are also cases of DNA in areas of intermediate recombination with intermediate levels of polymorphism. This generally quantitative relationship has made the reality of the phenomenon much more convincing. Aquadro et al. (1994), Aguadé and Langley (1994), Stephan (1994), and Moriyama and Powell (1996) summarize the evidence, some of which will be reported here.

Quantifying the degree of recombination of a region in terms of physical length of DNA is not precise. Aquadro and colleagues have attempted to do so as follows. They selected two well-studied genes that flank the region of interest and whose recombination rate was known. They then divided this by the number of polytene bands separating the markers. The recombination rate is expressed as cM (centimorgans) per polytene band. Assuming the amount of DNA is reasonably constant from band to band, this should give an estimate of recombination rate per unit of DNA; these workers used markers separated by 20 to 50 bands, so variation in DNA per band would be somewhat controlled for by averaging over a number of bands. Figure 10-1a shows this statistic for the third chromosome of *D. melanogaster.* As expected from other studies, the centromere is a region of low recombination, as are some telomeres. Figure 10-1b gives the level of variation observed for fifteen genes on the third chromosome that had been studied for level of nucleotide variation, mostly by RFLP analysis. The correlation is clearly significantly positive.

The tips of the X chromosome are also known to be regions of low recombination. Figure 10-2, from Aguadé and Langley (1994), displays the quantitative decrease in nucleotide variation approaching the tip of this chromosome. Using the Aquadro approach to estimating degree of recombination, Begun and Aquadro (1992) and Aquadro et al. (1994) also show the phenomenon for the X chromosome.

The only gene so far studied that seems to be an exception to the pattern in *D. melanogaster* is *cut* (Begun et al., 1994). This gene is in a region of relatively high recombination, yet exhibits about a fourth of the DNA polymorphism found in genes experiencing similar amounts of recombination. It is possible that this gene, or some very closely linked region or gene, underwent a selective sweep in recent times. Such exceptions are expected as the selective sweep theory is a quantitative one, that is, regions of low recombination tend to be lower in DNA polymorphism because the hitchhiking effect is greater there than in regions of high recombination. Nevertheless, hitchhiking will still occur in regions of high recombination, just over a shorter distance. Thus it is a question of density of segments of the genome undergoing selective sweeps. One expects occasionally to find a region, even in an area of high recombination, that was subjected to a hitchhiking sweep.

The only other species of *Drosophila* for which data are available on DNA polymorphism in genes experiencing different levels of recombination is *D. ananassae* (Stephan, 1989b; Stephan and Langley, 1989; Stephan and Mitchell, 1992). In this species, reduced recombination is associated with low levels of DNA polymorphism. Genes experiencing normal amounts of polymorphism, *Om(1D)* and *forked*, were nearly ten times more variable than genes located in regions of very low recombination, *furrowed* and *vermilion* (table 10-4).

Considering that both the third chromosome and the X chromosome of *D. melanogaster* show the same phenomenon (the second chromosome has not been studied in this regard), and that the nonrecombining fourth is almost devoid of nucleotide variation, it seems safe to conclude that recombination is generally affecting levels of diversity in the *D. melanogaster* genome. Furthermore, the only other species studied,

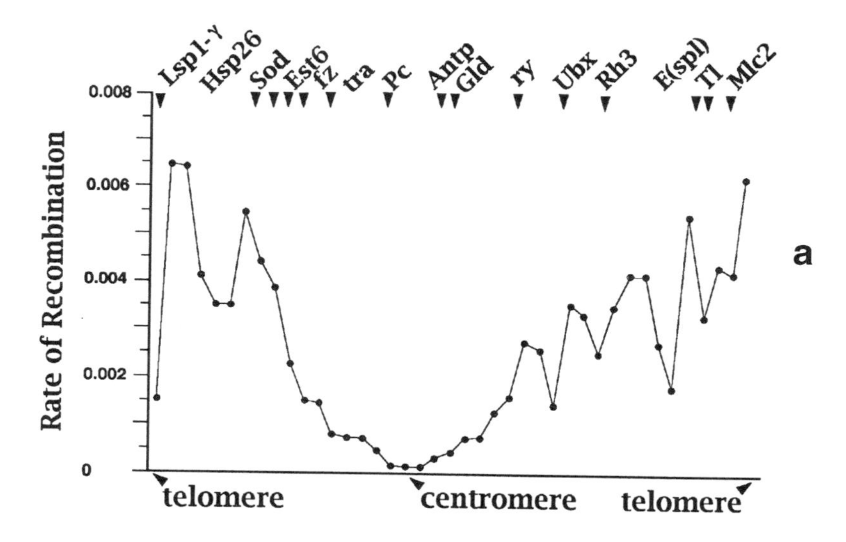

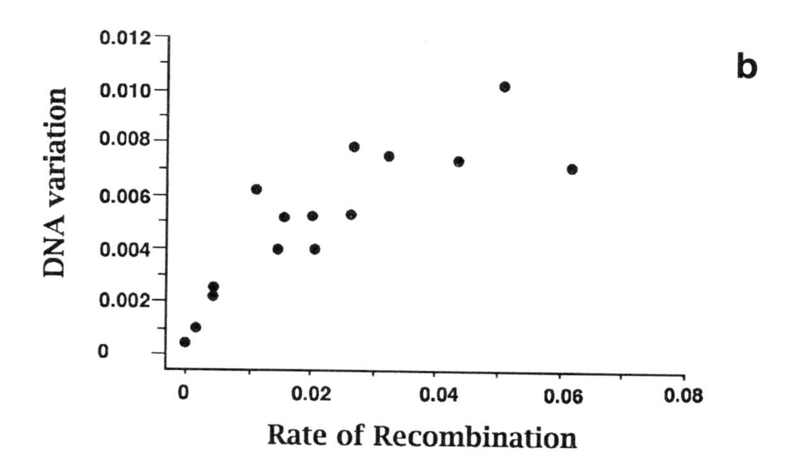

FIGURE 10-1 Illustration of relationship between rate of recombination and level of DNA variation. (a) Estimated rate of recombination along chromosome 3 of *D. melanogaster*. Recombination rate is given as cM/polytene band × 1/2 (to account for no recombination in males). (b) Estimates of DNA variation (θ) along the third chromosome. From Aquadro et al. (1994).

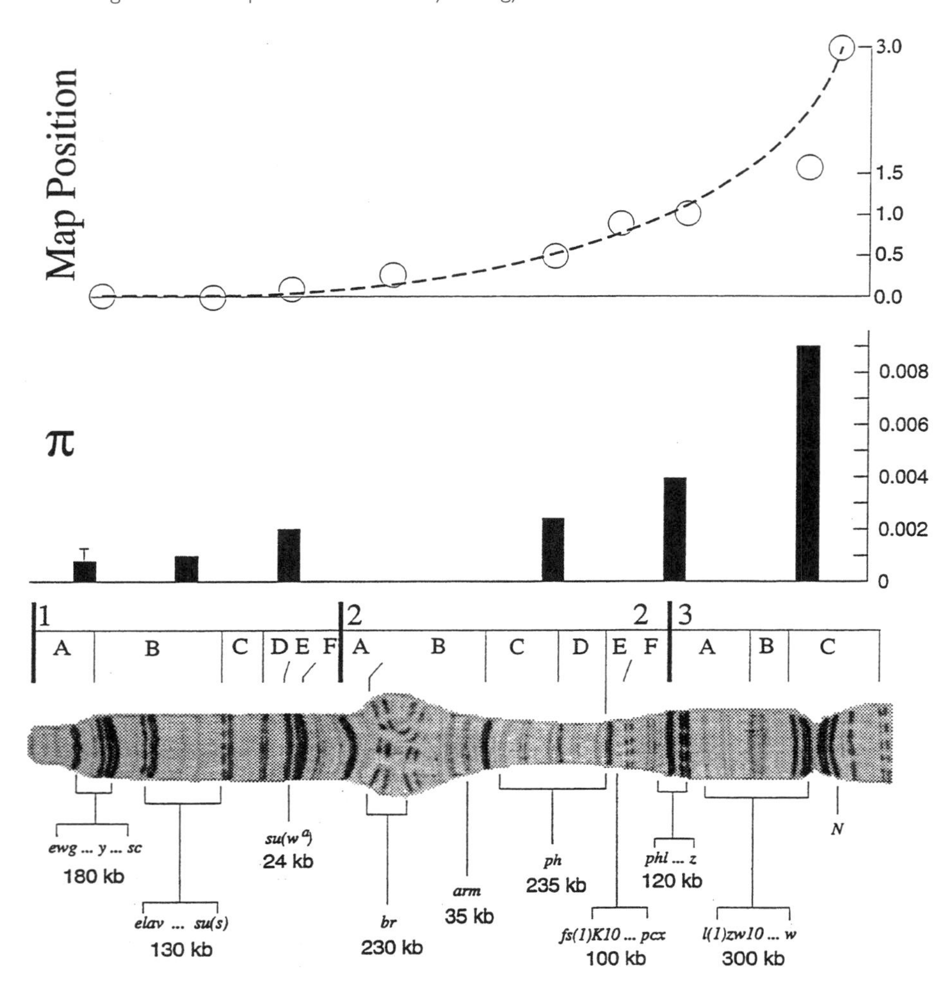

FIGURE 10-2 Recombination and level of DNA variation for the tip of the X chromosome of *D. melanogaster*. The upper graph is the map distance from *yellow* (*y*). The lower graph gives estimates of nucleotide diversity as measured by π. The cytological map is shown below with the cloned regions noted. From Aquadé and Langley (1994).

D. ananassae, also shows a similar pattern. The question that remains is whether the positive selective sweep/hitchhiking model is the correct one. An alternative would be that recombination itself is mutagenic. This possibility can be tested as the prediction would be that regions experiencing high (or low) rates of recombination should show high (or low) divergence between species. Mutation rate would affect both the level of polymorphism within a species and the degree of divergence between species. This does not seem to be the case (figure 10-3). The level of divergence between *D. melanogaster* and *D. simulans* for individual loci does not correlate to degree of recombination (Begun and Aquadro, 1992; Martin-Campos et al., 1992).

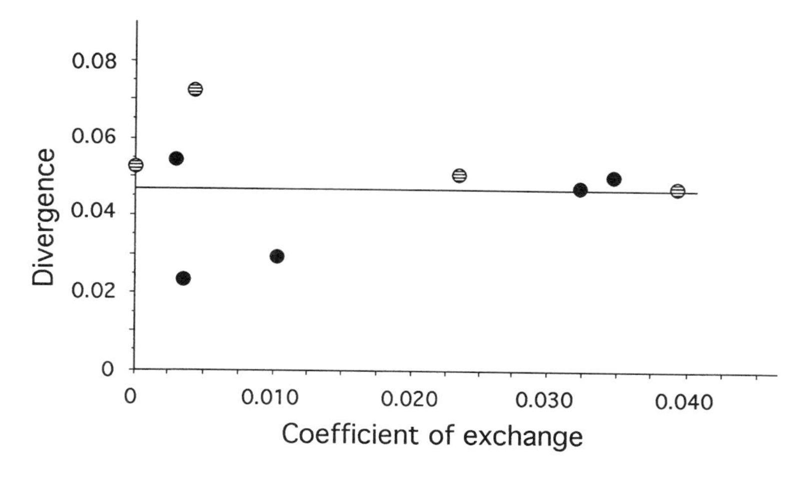

FIGURE 10-3 Measures of sequence divergence between *D. melanogaster* and *D. simulans* for genes experiencing various levels of recombination. Autosomal genes are hatched circles and X-linked genes are solid circles. From Begun and Aquadro (1992).

This interspecific comparison also addresses the issue of different selective constraints for different regions. In effect, selective constraints and decreased neutral mutation rate are two sides of the same coin, so the same argument concerning recombination increasing mutation rate pertains: if variation in selective constraints explains variation in levels of intraspecific polymorphism, it should be reflected in interspecific divergence. The lack of effect on interspecific differences is inconsistent with the selective constraints model (see also Hudson, 1994).

Another alternative to positive selection sweeps decreasing variation is negative selection, that is, selection against deleterious genes can also decrease nucleotide variation in linked regions (Charlesworth et al., 1993). Aquadro et al. (1994) and Stephan (1994) argue against this model by pointing out that X-linked genes should be less affected by negative selection because X chromosomes carry fewer deleterious alleles, due to selection against hemizygous males. The data show X-linked genes less polymorphic (see above), contrary to the negative selection model. In a similar vein, Hudson (1994) has found the negative selection model lacking, primarily because it cannot explain the extreme reduction in nucleotide variation at the tip of the X and fourth chromosomes. Chromosomes (or chromosomal segments) that are eliminated by negative selection never achieve a very high frequency and thus their removal does not sweep very much of the population, unlike the fixation of an initially rare chromosome (segment). In fact, the term fixation means a total sweep.

Another prediction of the positive selective sweep with hitchhiking model is that regions of low recombination should show more differentiation between populations than regions of high recombination. This is especially true if the selective sweeps occurred rapidly and not enough time has elapsed for alleles swept to fixation in one population to spread to others. Begun and Aquadro (1993) have observed significant differentiation of DNA segments experiencing low recombination between U.S. and

Zimbabwe populations of *D. melanogaster*. In fact, they found fixed and nearly fixed differences, much more differentiation than has been found for allozymes or DNA in regions of high recombination. Likewise, Stephan and Mitchell (1992) found in their two samples of *D. ananassae* from India and Burma that while very little or no variation was found within each population sample at two loci with very low recombination, there were fixed (or nearly fixed) differences between populations.

One aspect of the data is not consistent with our present understanding of the positive selective sweep model. After a sweep (fixation), nucleotide variation should recover slowly due to mutation. During this accumulation period (before equilibrium) there should be an excess of low frequency variants compared to the expectations of populations at equilibrium for neutral mutations. The overall agreement between π and θ in tables 10-2 and 10-3, confirmed quantitatively by Moriyama and Powell (1996), would indicate this prediction is not generally met; the genes in regions of low recombination appear no different from the rest in this regard. Braverman et al. (1995) studied this issue in greater detail and likewise reached the conclusion that the data are not compatible with a simple positive selection with hitchhiking model. Simulations indicated that the samples are sufficient that the deviations from equilibrium predicted by this model should have been detected.

While there seems no doubt that the correlation of nucleotide variation with recombination rate is real, this is an excellent example of the aforementioned phenomenon of data accumulation outstripping the theory needed to interpret it. The positive selective sweep with hitchhiking model would appear to be the best explanation (cf. Hudson, 1994), but some bothersome aspects of the data make this model not completely satisfying. Either the model is incorrect or insufficient.

Linkage and Balancing Selection (HKA)

The preceding section showed how positive selective sweeps, by fixing an initially rare variant (directional selection), can decrease linked neutral variation. The opposite can occur for balancing selection. Strobeck (1983) explored the behavior of a neutral variant linked to a selected locus in the context of inversions, that is, how unselected genes tightly linked to selected inversions (or even within inversions) will be affected by the selection. Hudson et al. (1987) formalized the theory with explicit reference to DNA data. They derived a way to test for selection by identifying regions that are significantly more heterozygous than would be expected by solely neutral behavior; the test has since become known as the HKA test. The extent of the region around which heterozygosity is elevated will depend on tightness of linkage. A region that is kept polymorphic by balancing selection will remain in a population longer than if there was no selection; more time means mutation will generate more variation. Another way of looking at this is that the effect of drift in decreasing variability will be diminished. The data required are DNA sequences for at least two genes (or gene regions) in two different species and intraspecific polymorphism data for one of the species. The interspecific comparison is necessary to derive the expected heterozygosity, based on neutral theory that predicts that intra- and interspecific variation are directly related. Usually a noncoding region is taken as the standard for expectations of neutrality against which a coding region is tested.

Hudson et al. (1987) and Kreitman and Hudson (1991) have applied the HKA test to the *Adh* gene system in *D. melanogaster.* In the first study, two parts of the same region (coding versus 5′ region of *Adh*) were used as the two loci, so they are not likely to be truly independent. In the later study, the interspecific comparison also included the *Adhr* locus (the tightly linked *Adh-related* gene). In both cases, the coding region of *Adh* in *D. melanogaster* was shown to be particularly high in nucleotide variation. Kreitman and Hudson (1991) introduced a graphic method based on a sliding window technique to illustrate the effect visually, figure 10-4. Part (a) shows the departure of observed variation from that predicted by strictly neutral behavior. This locus has two alleles in high frequency, the *F* and *S* alleles; they differ by a single amino acid substitution (threonine/lysine), which occurs at position 1490. Significantly, the peak of excess heterozygosity centers precisely on this position. Of course the selection could be within a few base pairs of this site, but it seems too much a coincidence if the amino acid changes are not involved in the selection. A considerable amount of other evidence supports the notion that selection maintains the *S/F* *Adh* polymorphism in *D. melanogaster* (see below).

Moriyama and Powell (1996) applied the HKA test to all of the polymorphism studies listed in tables 10-2 and 10-3. They used as the reference all possible noncoding regions and as the test locus all coding regions; over 1000 tests were performed. Their results showed that about a third of the loci in *D. melanogaster* failed the HKA test quite consistently when different reference regions were used. In the case of *D. simulans*, only 2 out of 12 loci showed consistent deviations from neutral expectations. These two loci in *D. simulans*, *ase* and *ci*, are in areas of particularly low recombination. However, *ase* in *D. melanogaster* conformed quite well to neutral expectations predicted by the HKA test. There is some indication that the HKA test is not particularly sensitive in detecting selection. For *D. melanogaster Adh* where we have good reason to believe selection is acting, the HKA test could detect deviations from neutrality for only 5 of 17 test loci. At the very least, this shows the importance of choosing a proper test locus.

Rates of DNA Evolution

So far we have been concerned with levels and patterns of intraspecific DNA variation. We now turn to consideration of interspecific comparisons. We will first consider estimates of rates of DNA change across *Drosophila* lineages. Stating the conclusion first, it is clear that *there is no single rate of DNA evolution in Drosophila.* This holds for comparisons among genes within a single lineage as well as a single gene in different lineages of *Drosophila.*

Absolute Estimates

In chapter 8 various estimates of the absolute times of separation of lineages of *Drosophila* were discussed. From such estimates, various conclusions about the rates of DNA divergence have been made. Table 10-12 presents some such estimates. Taken together, it would seem that an overall rate of 1% nucleotide change per million years per lineage is a reasonable average estimate. However, given the range of estimates and the inherent errors in assigning absolute dates of divergence, such an overall

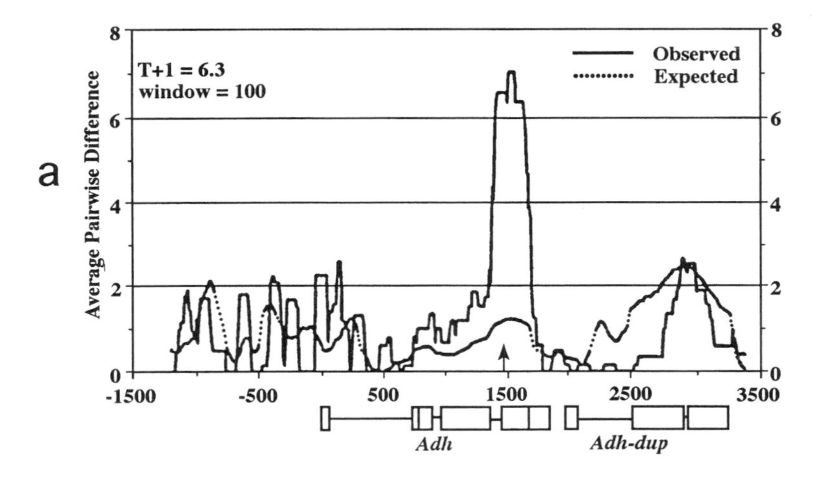

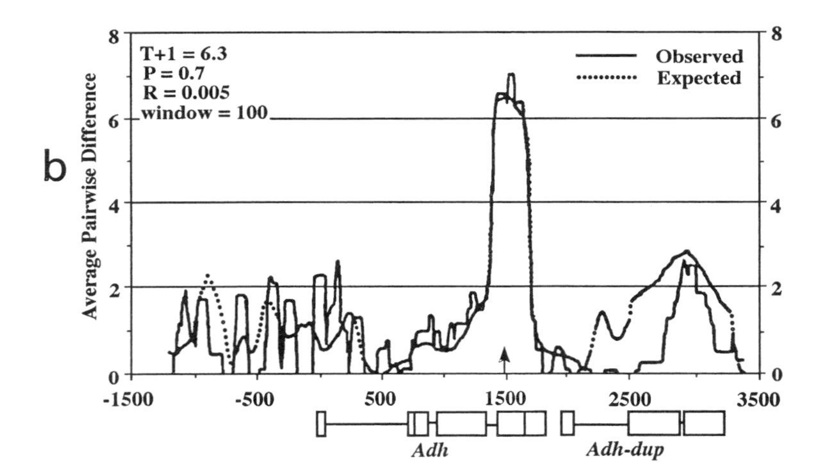

FIGURE 10-4 Sliding window graph of variation in pairwise nucleotide differences in a window of 100 silent substitutions along the *Adh* region of *D. melanogaster*. (a) The expected curve is that generated by predictions from interspecific divergences between *D. melanogaster* and *D. simulans*. $T+1$ is the estimated coalescence time (in $2N$ units) for one *D. melanogaster* and one *D. simulans* allele. Arrow points to the nucleotide position causing the single replacement polymorphism corresponding to *F* and *S* alleles. (b) Same as (a) but with the expected curve generated from a model of balancing selection at the replacement polymorphism site. From Kreitman and Hudson (1991).

TABLE 10-12 Estimates of absolute rates of DNA evolution in *Drosophila*.

Group	Percentage Base Substitutions/Million Years/Lineage	Method	Reference
melanogaster	0.8–2.0	Hybridization	Caccone et al., 1988b
Hawaiians	0.4	Hybridization	Hunt et al., 1981
mulleri	0.2–0.6	Hybridization	Schultze & Lee, 1986
obscura	0.2–1.0	Hybridization	Goddard et al., 1990
Several	0.8–1.6	Sequencing	Sharp & Li, 1989
Several	1.1–2.7	Sequencing	Moriyama & Gojobori, 1992

Note: The first four studies are based on DNA–DNA hybridization of total single-copy DNA and the last two on synonymous substitutions based on sequence data. The two measures are about the same for the same species pairs.

From Caccone and Powell (1990).

average should not be taken too seriously. Nevertheless, it is clear that virtually all these estimates indicate *Drosophila* are evolving (on the DNA level) considerably faster than are mammals (Li et al., 1985). In figure 10-5, comparison between mammal and *Drosophila* synonymous substitution rates is graphically illustrated.

Overall absolute rates, such as those just presented, mask considerable heterogeneity among sites and analysis of this heterogeneity provides important insights into mechanisms of DNA evolution. We presented data in the previous chapter indicating there is considerable difference between rate of change of coding DNA versus intergenic DNA (table 9-5). In general, coding DNA is about half as diverged as the average divergence of total single-copy nuclear DNA. Because by far the richest source of detail comes from DNA sequence data and protein coding regions are best studied in this regard, most of the remainder of these discussions will concern such genes and gene regions. However, from a global view of the genome, it should be kept in mind that these regions are relatively conserved.

Synonymous Substitution Rates

It is relatively easy to understand that different proteins, and even different parts of proteins, may vary in the rates at which amino acids change during evolution. Synonymous substitutions, which do not affect the protein phenotype, should behave much closer to neutrality and, indeed, were taken as the quintessential, obviously neutral mutations. Therefore, if there is a single mutation rate across the entire *Drosophila* genome that is the same in all lineages, then synonymous substitution rates should be uniform. This is not the case. Moriyama and Gojobori (1992) analyzed the then-available data and documented considerable heterogeneity among genes. Figure 10-5 summarizes their results. It is clear there is a considerable range of rate of synonymous substitution among the 24 genes studied. The most extreme genes are the amylase locus (*Amy*, m in figure 10-5) and an esterase (*Est-6*, s in the figure), which vary from 0.401 to 1.984 synonymous substitutions per site between the *melanogaster* and *obscura* groups. Taking this divergence time to be about 30 million years ago, the abso-

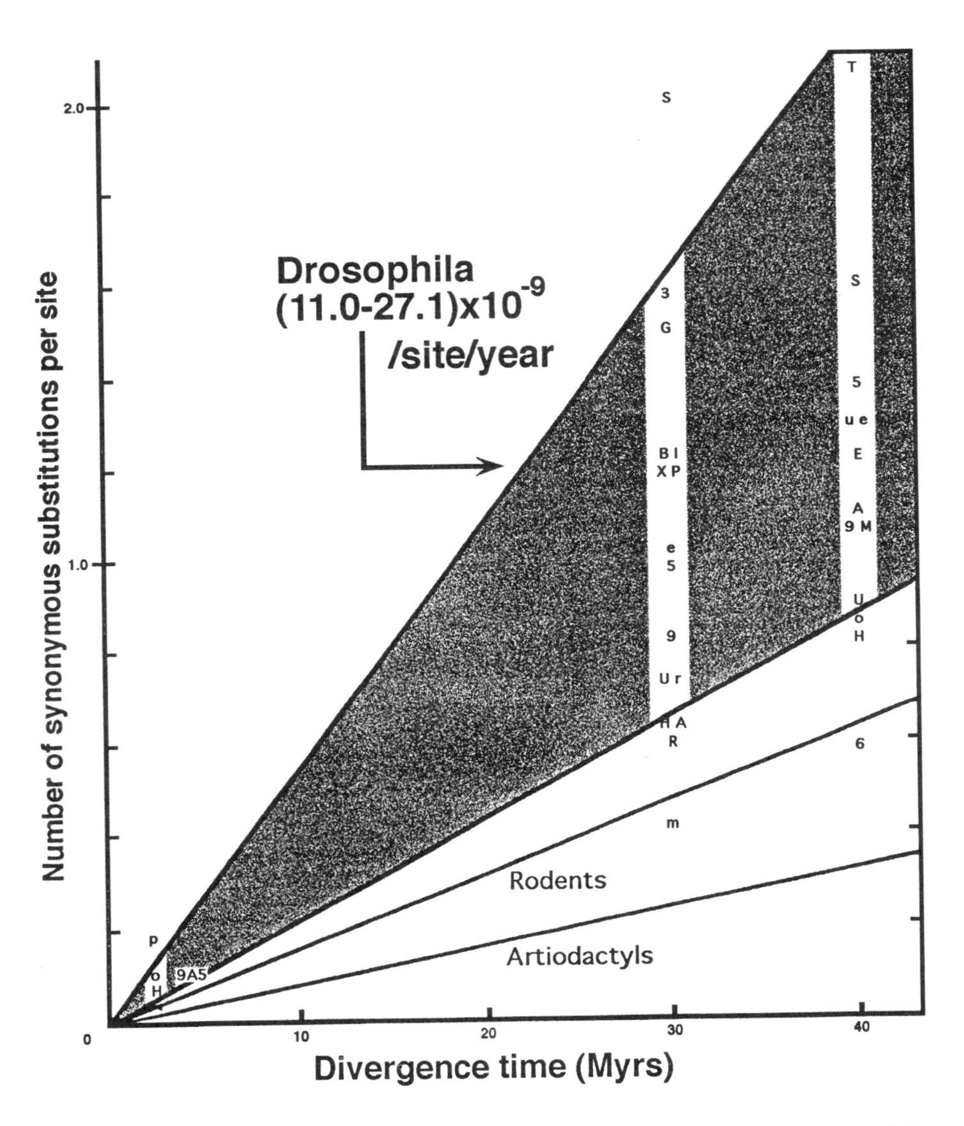

FIGURE 10-5 Estimates of divergences at silent sites for several genes and species. Each letter represents a gene. The group of letters on the left are among *melanogaster* group species, the middle set between *melanogaster* and *obscura* groups, and on the right between subgenera. Note the great range among genes. For comparison, the estimated rates of silent substitutions for two groups of mammals are shown. From Moriyama and Gojobori (1992).

lute rates are 0.6 to 3.3% per million years per lineage. Whether or not the absolute times are accurate (see figure 1-3 for indication of errors), it is clear that genes can vary five-fold in rates of synonymous substitutions.

Heterogeneity among lineages was tested by a relative rate test (Wu and Li, 1985). The main pattern seen by Moriyama and Gojobori (1992) was that, in general, the lineage leading to the *melanogaster* group was evolving slower than other lineages. They attributed this to a higher G+C content in the third positions. In fact, in comparison of individual genes, they also showed a negative correlation with rate of synonymous substitution and G+C content at third positions of codons. In the next section we will see this pattern is consistent with the effect of codon bias.

Codon Usage Bias and Effects on Rates

Evidence that synonymous substitutions can be affected by selection comes from analysis of the frequency of use of synonymous codons. If there were no selection among synonymous codons (and if mutations are random), then one would expect all codons to be used equally, that is, an amino acid with two codons would use each about half the time, one with four codons would use each about 25% of the time, and so forth. This is often not the case. Table 10-13 presents the codon usage table for nuclear genes in *D. melanogaster*. This table gives the percentages of codons used for each amino acid based on the sequence of hundreds of proteins with over 250,000 codons total. It is clear that unequal usage of codons (codon usage bias) is the rule rather than the exception. The main trend notable in this table is that codons with G and especially C in the third wobble position tend to be used more than A or T ending codons. In *D. melanogaster*, about 85% of all third positions are G or C. Recall that the total *D. melanogaster* genome is relatively A+T-rich, around 57% (Laird, 1973), which presumably reflects, at least approximately, the equilibrium mutation pressure. Thus to achieve the high G+C content in the wobble position of codons, mutation pressure toward A+T must be overcome.

Four statistics have been developed to measure overall codon usage bias for a gene: Codon Adaptation Index, CAI (Sharp and Li, 1987); a scaled chi-square, χ^2/L (Shields et al., 1988); Effective Number of Codons, ENC (Wright, 1990); and F_{op}, the frequency of use of the "optimal codon" (Sharp and Lloyd, 1993). We need not be concerned with the details of these measures as they are highly correlated with one another (Moriyama and Powell, 1996). Essentially, they measure the same thing.

There is considerable variation among genes in codon usage and thus G+C in third position. In table 10-13, in addition to the mean usage of codons, there is an indication of the range in use of codons among genes. Sharp and Lloyd (1993) have analyzed 438 genes and taken the 44 (10%) with the lowest codon usage bias and the 44 with the highest bias. The cases where these two sets are significantly different at the $p < 0.01$ level are indicated in table 10-13. (Note that the range for the "low" and "high" biased genes sometimes does not include the mean; this is because the designation "low" or "high" is based on all codons in the genes. For example, for CUC coding Leu, the highly biased genes use this codon less than average because of a great increase in use of CUG.) The 22 codons where the "low" and "high" biased genes differ significantly are considered to have "optimal" codons indicated by that used most frequently in highly biased genes.

TABLE 10-13 Frequency in percent of codon usage in nuclear genes of *D. melanogaster*.

Amino Acid	Codon	Frequency		
		Mean	Low	High
Phe	UUU	33		
	UUC	67	43*	93*
Leu	UUA	4		
	UUG	18		
	CUU	9		
	CUC	16	9*	15*
	CUA	8		
	CUG	45	22*	69*
Ile	AUU	32		
	AUC	52	26*	81*
	AUA	16		
Met	AUG	100		
Val	GUU	18		
	GUC	25	20*	37*
	GUA	10		
	GUG	47	33*	48*
Ser	UCU	8		
	UCC	25	29*	61*
	UCA	9		
	UCG	22	20*	26*
	AGU	12		
	AGC	24		
Pro	CCU	11		
	CCC	35	23*	65*
	CCA	24		
	CCG	30		
Thr	ACU	15		
	ACC	42	26*	80*
	ACA	19		
	ACG	24		
Ala	GCU	19		
	GCC	47	29*	70*
	GCA	16		
	GCG	18		
ter	UAA	53		
	UAG	28		
	UGA	18		
His	CAU	34		
	CAC	66	47*	82*
Gln	CAA	27		
	CAG	73	52*	92*
Asn	AAU	42		
	AAC	58	45*	88*
Lys	AAA	25		
	AAG	75	51*	96*
Asp	GAU	52		
	GAC	48	45*	92*
Glu	GAA	29		
	GAG	71	45*	92*
Cys	UGU	27		
	UGC	73	60*	90*
Trp	UGG	100		
Arg	CGU	18	17*	37*
	CGC	35	18*	53*
	CGA	14		
	CGG	14		
	AGA	8		
	AGG	10		
Tyr	UAU	34		
	UAC	66	49*	88*
Gly	GGU	23		
	GGC	43	27*	48*
	GGA	28		
	GGG	6		

Note: The mean is shown along with the average for the 10% of genes with the lowest and 10% highest codon bias and for which these two groups are significantly different at the $p < 0.01$ level (indicated by the asterisks) for a particular codon. Based on a total of 438 genes with 264,421 condons.

From Sharp and Lloyd (1993).

Causes of Bias

There are two possible explanations for the bias in codon usage: there could be local nonrandom mutation pressure in gene regions to increase C+G content, or the pattern could be due to selection. One prediction of the mutation explanation is that introns of genes with particularly high G+C content should also have high G+C. Shields et al. (1988) and especially Moriyama and Hartl (1993) explored this possibility by looking for correlation between base composition of exons and introns in the same gene. Figure 10-6 shows Moriyama and Hartl's results. First it can be seen that as genes become more biased in codon usage, they tend to use more C in the third position at the expense of A; to a lesser extent G also increases and T decreases. However, this is occurring independently of base composition in introns (right hand graphs in figure 10-6). This is good evidence that variation in local mutation pressure cannot account for changes in codon usage.

This is not the case, however, with warm-blooded vertebrates. In those organisms with their relatively A+T-rich and G+C-rich isochores, there is good correlation between the base composition of introns and exons, presumably due to local variation in mutation pressure (Bernardi and Bernardi, 1986).

Kliman and Hey (1994) also analyzed the correlation between base content of introns and exons. Contrary to Moriyama and Hartl (1993) they did find a weak effect, although not nearly strong enough to account for the level of codon usage bias observed. More interestingly, they found the correlation between exon and intron base content to be highest for low codon bias genes. The difference between intron base composition and the third position of codons increases with increasing codon bias. This indicates that selection for particular codon usage drives the difference between intron and exon base composition of highly biased genes.

Kliman and Hey (1993b) also examined codon usage bias and rates of recombination. The reasoning is that in regions of low recombination, selection is less effective than in regions of high recombination, the so-called Hill–Robertson effect (Hill and Robertson, 1966). Thus selection for use of particular codons should be more effective in genes in high recombination regions and should show greater codon bias than genes in regions of low recombination. They found this to be the case, as did Moriyama and Powell (1996).

Finally, in pseudogenes, G+C content tends to decrease relative to the G+C content of the active genes giving rise to them (Shields et al., 1988; Moriyama and Gojobori, 1992). Presumably selection for codons is relaxed in pseudogenes, so that mutation pressure will be more effective in controlling the base composition.

Taken together, it is hard to escape the conclusion that selection for use of particular codons is acting on some genes at least. The cause of this is far from clear. For two groups of organisms, there are reasonable explanations for codon usage bias. In warm-blooded vertebrates there are isochores and strong correlations between base content in a coding region and the surrounding DNA including introns (Bernardi and Bernardi, 1986). There are no isochores in *Drosophila* and little evidence of correlation between base content of introns and exons. In unicellular organisms in which all transcripts share a common cytoplasm and, presumably, a common tRNA pool, it is fairly well established that the relative use of synonymous codons reflects the relative abundance of isoaccepting tRNAs. Genes producing high levels of protein are particularly prone to this kind of selection, as the translation process tends to be their limiting step. There is mixed evidence that genes expressed at higher levels tend to be more biased in *Drosophila*.

Sharp and Lloyd (1993) tabulated cases where there was evidence that duplicated genes are expressed at different levels; thus this comparison minimizes complicating factors such as functional differences among completely different genes. They used

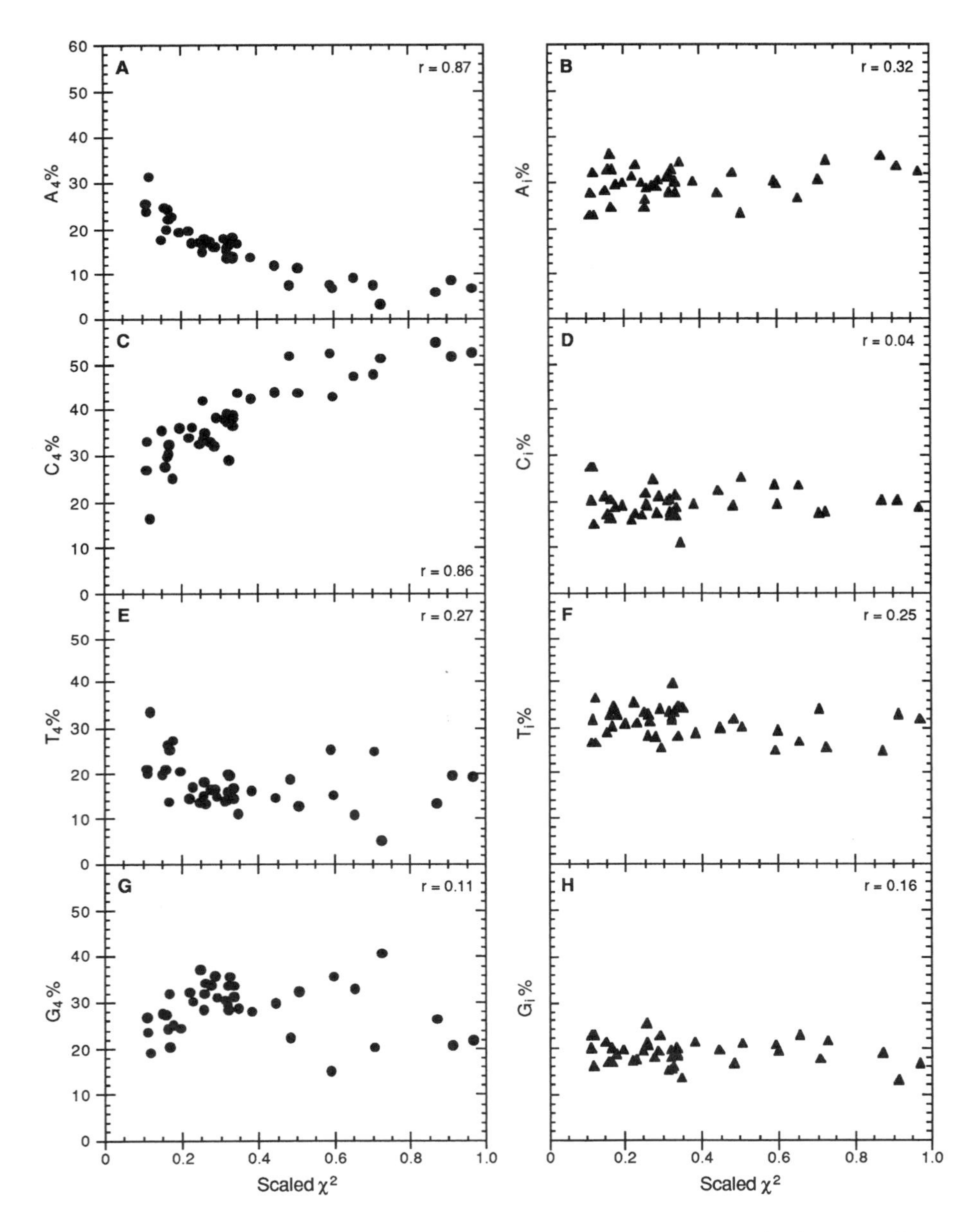

FIGURE 10-6 Plots of base composition at the four-fold degenerate sites (A, C, E, and G) and introns (B, D, F, and H) for *D. melanogaster* genes with varying degrees of codon usage bias measured by the scaled χ^2, which is positively correlated to degree of codon usage bias. As a gene becomes more biased, A and T decrease in the third position of four-fold degenerate codons, and C and G increase. Intron base composition is independent of bias. From Moriyama and Hartl (1993).

F_{op}, which is the number of times the optimal codons (defined in table 10-13) are used, divided by the total number of times an amino acid is encoded. Thus the higher the F_{op}, the more biased the gene. Table 10-14 shows the results. While the F_{op}s differ little between copies, in all four cases the direction of difference is that predicted by the level of expression theory.

However, there is also evidence that level of expression may not be related to degree of codon usage bias. In *D. simulans, Adh* has a considerably lower level of expression, yet is just as biased as in *D. melanogaster.* Another study at odds with the level of expression hypothesis is Fitch and Strausbaugh's (1993) observation that the highly expressed histones have low codon usage bias. Kliman and Hey (1993b) counter by pointing out that histones are in a region of low recombination and, according to their reasoning, this will make selection for codon usage less effective.

One other explanation for codon usage bias that may not be totally independent of those already mentioned, is that selection is acting on the accuracy of translation and thus codons enhancing accuracy are favored (Akashi, 1994). Akashi reasoned that the importance of particular amino acids for the proper function of a protein can be inferred from relative conservation in evolutionary time. Amino acids free to change during evolution presumably mark positions that would have less effect for misincorporation compared to amino acid positions conserved during evolution and thus are crucial to proper function. If use of particular codons reduces the chance of misincorporation (for which there is evidence in bacteria [Precup and Parker, 1987]), then there should be more codon usage bias at conserved positions than at variable positions. Akashi found a statistically significant relationship as predicted. Further, Akashi found that for 28 proteins with putative DNA-binding domains (zinc fingers and homeoboxes) codon bias was greater in those areas than in the rest of the gene. However, here again, the histones would seem to be a case not explained by Akashi's theory, that is, it is a protein for which almost no amino acid substitutions have occurred, yet is low in codon usage bias.

Codon Usage in Other Species

Table 10-13 applies only to *D. melanogaster* and it is of interest to examine how much codon usage varies among species. *Adh* has been sequenced in more species

TABLE 10-14 Relative codon usage bias for duplicated genes of *D. melanogaster* for which there is evidence of different levels of gene expression.

Gene	High	Low	Reference
Cytochrome c	0.77	0.57	Limbach and Wu, 1985
α-Tubulin	0.79	0.57	Kalfayan and Wensink, 1982
Elongation factor 1α	0.76	0.71	Hovemann et al., 1988
Lysozyme	0.70	0.63	Klysten et al., 1992

Note: The copy more highly expressed is on the left and the less-expressed gene on the right. The reference for difference in level of expression is given. The F_{op} statistic is shown, which is higher in more highly biased genes.

From Sharp and Lloyd (1993).

than any other gene and, at least in *D. melanogaster*, is highly biased in codon usage and thus should be a good candidate for seeing patterns of bias in other species. Starmer and Sullivan (1989), Moriyama and Gojobori (1992), and Moriyama and Hartl (1993) review the available data for *Drosophila Adh* and particularly note the variation in G+C content of third positions. This varies from a high of 80–87% for the *melanogaster* group (7 species) to 59–61% for Hawaiian *Drosophila* (6 species). The *willistoni* group has perhaps the most extreme difference, with only 53–58% G+C in the third position (Anderson et al., 1993; Carew, 1993).

A closer look at variation for particularly variable amino acids in *D. willistoni* is shown in table 10-15. Note the significant shift in predominantly C-ending codons to U-ending codons. The shift from C to U also occurs in the first position for Leu for *Adh*. Very few genes have been sequenced in *D. willistoni* and two others are shown in table 10-15. *Sod* shows the shift from C to U in the third position for some amino acids (His, Asn, Asp, and Ile), but not for others. Relative to *D. melanogaster, Per* shows only a small shift to U in most amino acids, but for Gly the shift is quite marked. Why *D. willistoni* shows this shift is not at all clear, but the fact that the same genes in different species can display very different codon usage seems very clear in comparing *D. willistoni* to *D. melanogaster*.

TABLE 10-15 Codon usage for select amino acids in *D. melanogaster* and *D. willistoni*.

Amino Acid	Codon	*D. melanogaster*			*D. willistoni*			
		Adh[1]	Sod[2]	Per[3]	Adh[4]	Sod[2]	Per[5]	*P* element[6]
Tyr	UAU	1	1	5	3	0	15	19
	UAC	3	0	25	1	0	10	7
His	CAU	1	1	4	3	8	18	9
	CAC	3	7	29	1	1	14	6
Asn	AAU	2	3	10	7	7	35	34
	AAC	14	4	27	9	3	19	14
Asp	GAU	5	5	11	10	7	31	32
	GAC	7	5	38	3	2	16	21
Ile	AUU	9	5	9	12	6	15	31
	AUC	14	4	22	10	4	7	17
	AUA	0	0	9	0	0	12	16
Leu	CUU	0	1	4	1	0	8	10
	CUC	3	1	14	2	0	9	8
	CUA	0	0	3	0	1	17	5
	CUG	20	5	33	5	4	13	9
	UUA	0	0	0	1	0	9	22
	UUG	4	0	3	17	2	16	20
Gly	GGU	6	6	17	9	8	46	8
	GGC	8	14	71	7	12	41	7
	GGA	5	4	41	2	7	22	13
	GGG	0	1	17	0	0	6	4

References: [1]Kreitman, 1983; [2]Kwiatowski et al., 1994; [3]Citri et al., 1987; [4]Anderson et al., 1993; [5]J. Gleason, unpublished; [6]Rio et al., 1986.

Given the hypothesis that *P* elements have been acquired by *D. melanogaster* from *D. willistoni* (chapter 9), we include in table 10-15 codon usage in the *P* element. *P* elements have the "signature" of *D. willistoni* genes, namely the shift from C to U in the third positions for several amino acids and for the first position for Leu. The difference in codon usage between *D. melanogaster* and *P* elements is statistically significantly different, while between *P* elements and *D. willistoni* it is not (Powell and Gleason, 1996). Assuming that the codon usage bias of transposable elements takes on the pattern of its host genome, this is evidence that *P* elements have had a longer history in *D. willistoni* and in *D. melanogaster* and thus added evidence for horizontal transfer.

While considerable differences in codon usage occur among lineages of *Drosophila*, individual amino acids in some genes can show remarkable phylogenetic persistence in codon usage bias. Table 10-16 shows the case of Ile in *Adh* for a large number of species. The codon AUA is avoided in this gene through about 100 million years of evolution. It is not the case that AUA is not used by *Drosophila* in general as other genes use this codon. *D. melanogaster* uses AUA 16% of the time for Ile (table 10-13) and the *per* locus in *D. willistoni* "prefers" AUA for Ile. It is particularly difficult to reconcile the proposed explanations for codon usage bias with this pattern of phylogenetic persistence for particular amino acids in some genes, along with the previous paragraph's observation of significant shifts between species for other amino acids. The only conclusion is that we have yet to find a satisfactory explanation for codon usage bias in *Drosophila*.

Effect on Rates

If codon usage bias is caused by selection for use of particular codons, and if this pressure varies from gene to gene, then one would expect a correlation between rate of silent substitution and bias: more biased genes are more constrained in codon usage and thus should accumulate synonymous substitutions more slowly than relatively unconstrained low bias genes. Sharp and Li (1989) were the first to note this in *Drosophila*; their results are shown in figure 10-7. Again, absolute rates are not crucial

TABLE 10-16 Codon usage for Ile in *Adh* for a large number of species.

		Isoleucine		
Species Group	No. Species/Sequences	AUU	AUC	AUA
melanogaster subgroup	8	63	121	0
obscura group	5	46	68	1
willistoni group	6	79	53	0
Subgenus *Drosophila*	13	113	188	5
Hawaiian *Idiomyia*	10	113	115	1

Note: The number of times each codon is used is shown. One sequence from each species is used except in the cases of three species in the subgenus *Drosophila*, where duplicated genes exist and both sequences from a species are included.

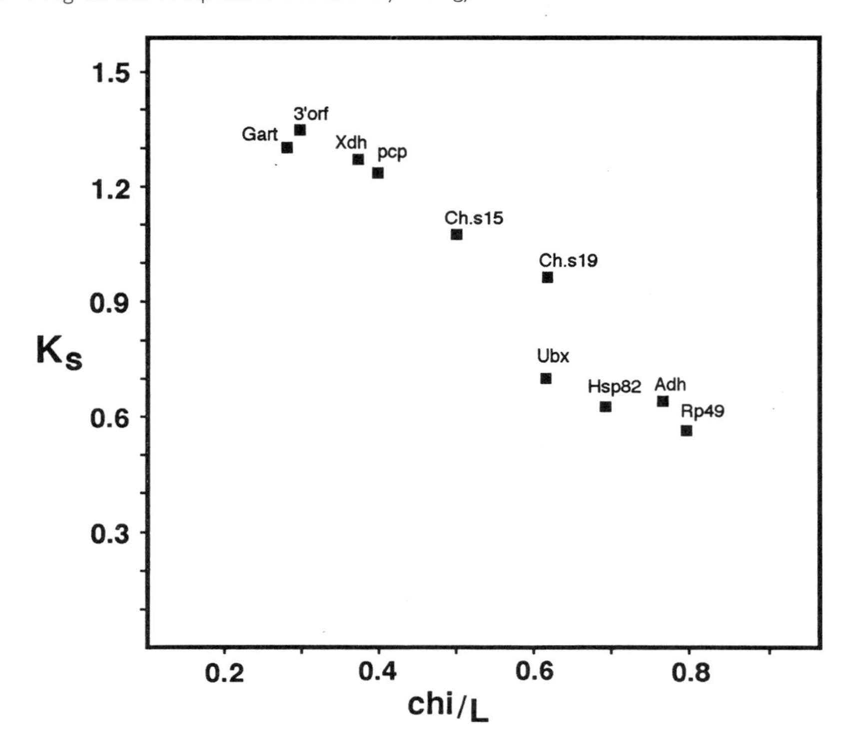

FIGURE 10-7 Silent site divergence (K_s) between the *melanogaster* and *obscura* groups for genes with varying degrees of codon usage bias measured by the χ^2/L, which is positively related to level of bias. The more biased the gene, the less the silent divergence. From Sharp and Li (1989).

here; rather, by examining genes in the same two species, *D. melanogaster* and *D. pseudoobscura*, one can see their relative divergence since the two species last had a common ancestor. This pattern has been further confirmed by Carulli et al. (1993). Moriyama and Gojobori (1992) also indirectly found the same. They found the rate of synonymous substitution to be negatively correlated with G+C in the third position, which we just saw was correlated with degree of codon usage bias (figure 10-6).

Between Versus Within

Now that we have examined both intra-and interspecific patterns of DNA variation, we turn to a comparison of the two. Several times in the preceding discussion we had cause to mention that the strictly neutral theory of molecular evolution predicts that the processes controlling intraspecific polymorphisms act likewise to cause interspecific differences. When the neutral theory was first proposed (King and Jukes, 1969), it was explicitly concerned with describing and explaining interspecific differences in protein sequences. DNA data were not then available. Kimura and Ohta (1971) made the jump to include intraspecific polymorphism in their explicitly titled paper, "Protein Polymorphism as a Phase of Molecular Evolution." Virtually all of the arguments can be extended to DNA variation (e.g., Kimura, 1983). Therefore, neutral theory predicts that within population variation and between species, differences should exhibit the same evolutionary dynamics (ironically, one of the major tenets of the neodarwinian synthesis, too!). We already discussed one such test of the neutral theory based on these predictions, the HKA test, in which the levels of variation are compared.

McDonald and Kreitman (1991) introduced a variation on this theme, namely that if the same factors control intra- and interspecific DNA variability, then the ratio of synonymous to nonsynonymous substitutions should be the same for any given locus when considering intraspecific polymorphisms and interspecific differences. Their initial application was to the *Adh* locus in the *melanogaster* subgroup. They concluded that the patterns were not the same, that relatively more amino acid substitutions occur between species than are polymorphic within the three species compared. Their data, along with examples of several other applications of this test, are in table 10-17. It is important to realize that the McDonald/Kreitman test should only be applied to situations where saturation has not been reached. Therefore, in this table only studies among the very closely related members of the *melanogaster* subgroup and *willistoni* group are shown. Comparisons between groups are likely to be at or close to saturation, in some unknown proportion of cases at least.

TABLE 10-17 Examples of application of the McDonald/Kreitman test to intra- and interspecific DNA comparisons in *D. melanogaster* subgroup species.

Gene	Species	Reference		Synonomous	Nonsynonymous	*p*
Adh	mel/sim/yak	McDonald & Kreitman, 1991	W	42	2	0.006
			B	17	7	
G6pdh	mel/sim	Eanes et al., 1993	W	36	2	$\leq$0.001
			B	26	21	
prune	mel/sim	Simmons et al., 1994	W	8	6	0.06
			B	21	4	
yp2	mel/sim	Hey & Kliman, 1993	W	11	4	0.597
	sech/maur		B	13	7	
period	mel/sim	Kliman & Hey, 1993a	W	84	10	>0.9
	sech/maur		B	35	3	
Rh3	mel/sim	Ayala et al., 1993	W	47	1	0.43
	teis/yak		B	49	3	
boss	mel/sim	Ayala & Hartl, 1993	W	106	13	0.89
	teis/yak		B	71	8	
Adh	*willistoni*	Carew, 1993	W	14	6	>0.1
	group		B	54	9	
ref(2)P	mel/erecta	Dru et al., 1993	W	1	7	0.011
			B	84	82	

Note: W = within-species polymorphisms, B = between-species fixed differences. The final study is set off as it is the only one indicating more nonsynonymous polymorphism than synonymous and may be an anomalous case.

Two tests in table 10-17 are clearly significant and four are clearly not. *Prune* is borderline significant. *Ref(2)P* is set off as this is an exceptional locus, as can be seen by the pattern of substitutions; it is the only locus where the ratio of nonsynonymous to synonymous polymorphisms (7 : 1) is greater than that between species differences (about 1 : 1). Whether this reflects some unusual consequence of its function as a locus that confers resistance to viral infections is not known.

While only 2 out of 7 McDonald-Kreitman tests on the *melanogaster* subgroup produce indications of deviations from neutral prediction, there is a tendency in most studies to find relatively more nonsynonymous substitutions between species than polymorphic within. Table 10-18 sums up all the studies, excluding the exceptional *ref(2)P* locus. Overall, there is a highly significant deviation from neutral expectation of an equal ratio. McDonald and Kreitman (1991) argue this is the expected pattern if selection were acting on the protein level to drive fixation of amino acid substitutions. The reasoning is that when an amino acid replacement is driven to fixation by selection, it happens much more rapidly than the fixation or loss of a neutral (synonymous) substitution. Therefore, the probability of catching a replacement polymorphism at one slice in time is less than that of catching a neutral polymorphism. But all selectively driven and neutral changes will be evident when comparing different species. It may also be that many of the selectively driven amino acid substitutions occur at the time of speciation, presumably when a lineage is making an adaptive peak shift in the sense of Wright (chapter 7).

One other difference in comparing intra- and interspecific nuclear DNA variation concerns the effect of selection for codon bias. We already noted that such selection slows the rate of silent substitutions when comparing different species (figure 10-7). One might expect, then, that a similar pattern should be seen on the intraspecific level, namely that more highly biased genes should show less silent polymorphism. This is not the case. In fact, for *D. melanogaster* the *opposite* was found: genes with greater codon usage bias display more silent polymorphism than do genes with less codon usage bias (Moriyama and Powell, 1996). It may be that the effect of recombination on the level of nucleotide polymorphism (higher recombination, greater polymorphism) and on codon usage bias (higher recombination, greater codon usage bias) overrides the expected effect of the constraints induced by selection for codon usage.

TABLE 10-18 McDonald/Kreitman test on combined data for *melanogaster* subgroup species listed in table 10-17, excluding *ref(2)P*, an exceptional gene.

	Synonymous	Nonsynonymous	Percentage Nonsynonymous
Within species	334	38	10.2%
Between species	232	53	18.6%

$G = 159$, $p \ll 0.001$

Mitochondrial DNA (mtDNA)

So far we have discussed evolutionary patterns of nuclear DNA only, and now we turn our attention to mtDNA. Mitochondrial DNA differs from nuclear DNA (nucDNA) in several important ways. It is a small molecule between 15 and 20 kb, is circular, and exists in the haploid state. There is little or no intergenic DNA in mtDNA. Figure 10-8 is an illustration of *D. yakuba* mtDNA, the first *Drosophila* mtDNA to be sequenced in total (Wolstenholme and Clary, 1985). The molecule codes for 13 proteins that function inside the mitochondria, as well as tRNAs and two rRNAs. Overall, the molecule is 75 to 80% A+T, not including the A+T-rich region. The A+T-rich region (called the D-loop in vertebrates) is the origin of replication and is some 90 to 95%

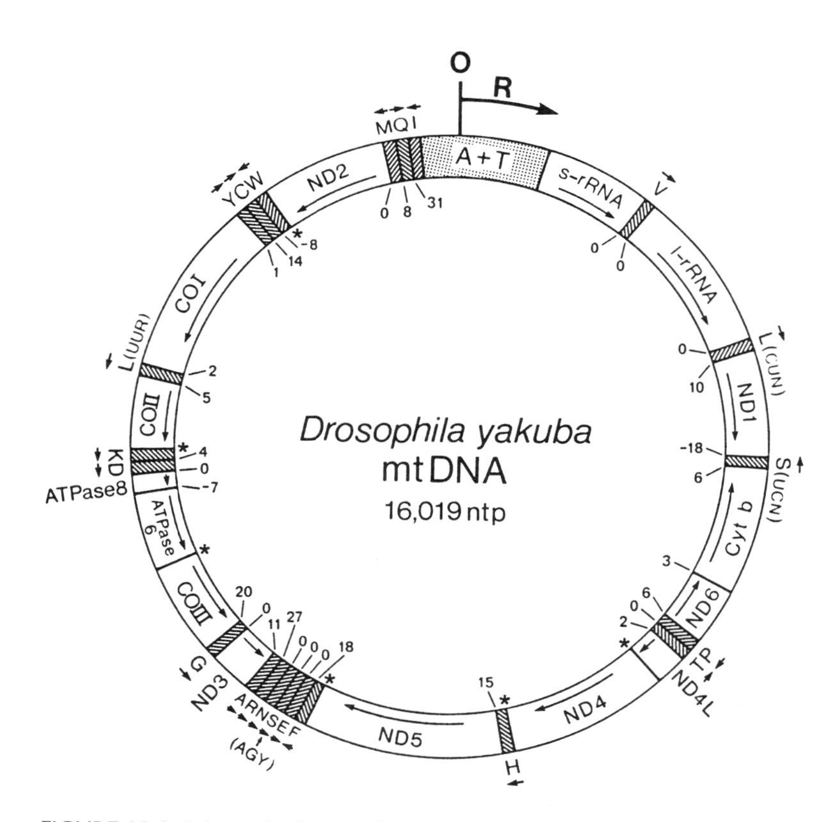

FIGURE 10-8 Schematized map of *Drosophila yakuba* mitochondrial DNA. Arrow at top is the origin of replication. Arrows indicate direction of transcription. Letters on the outside of the map are for tRNAs with the one letter amino acid code and the particular codon families for serine and leucine tRNAs in parentheses. Numbers on the inside of the map indicate numbers of nucleotides between genes, with negative numbers indicating gene overlap. Original drawing kindly provided by David Wolstenholme; reproduced with permission.

A+T. The size of this region accounts for the variation in size of the mtDNA molecule among and within species. Except for this highly variable region, the rest of the mtDNA molecule is extremely conserved in terms of structure and gene order. In fact, as far as is known, it is invariant in the genus *Drosophila.*

Of perhaps most importance from an evolutionary standpoint is that mtDNA is maternally inherited through the egg cytoplasm, and the molecule does not recombine. This latter fact makes analysis of DNA sequence data much simpler than recombining nucDNA. It is one reason this molecule has been very popular in phylogenetic studies, many of which were discussed in chapter 8. The maternal inheritance of mtDNA in *Drosophila* may not be strictly true, however. Kondo et al. (1990, 1992) have demonstrated small amounts of paternally derived mtDNA, on the order of 0.1% or less. Whether this has any evolutionary significance remains to be demonstrated, but it could certainly be a source of heteroplasmy (the presence of more than one type of mtDNA in a single individual).

Intraspecific mtDNA Variation

Two conflicting processes affect the level of nucleotide variation in mtDNA. One is that mtDNA in most organisms is thought to have a higher mutation rate due to the lack of an efficient DNA repair mechanism in mitochondria. Because of this relatively high rate of mutation, in many vertebrates at least, it has proven to be a very valuable marker for studying patterns of relatedness among populations within a species (Avise, 1994). On the other hand, we have just discussed how lack of recombination, coupled with selective sweeps, can greatly decrease nucleotide variation in a region of nucDNA, and, given the complete lack of recombination of mtDNA, one might predict mtDNA to be low in variation. Table 10-19 lists studies of intraspecific variation in mtDNA based on RFLP studies (with one exception). The diversity measure is either the estimated average pairwise per-nucleotide difference when only fragment size data are available (π). When restriction maps are available, the maximum likelihood estimator is given (d). Table 10-20 gives some mtDNA diversity measures based on DNA sequence data. The main difference between tables 10-19 and 10-20 is that the former contains studies that cover the entire molecule, but use a less accurate measure of diversity than the more limited, but detailed studies in table 10-20. While strictly speaking one cannot use the term heterozygosity for mtDNA, in fact the parameters π and d do estimate conceptually the same thing as heterozygosity. Comparing this mtDNA diversity to the heterozygosity of diploid nuclear DNA, it would seem that mtDNA is less variable. Looking at individual species, we see that *D. melanogaster* has mtDNA diversity of 0.1% or less, about one-fourth that observed for nuclear protein-coding DNA heterozygosity. *D. yakuba* mtDNA has a diversity of only 0.03%, while nuclear coding DNA heterozygosity is close to 1% (table 10-4). *D. pseudoobscura* has one of the highest mtDNA diversity estimates (0.9%), but this species was also among the highest in nuclear coding DNA heterozygosity, being around 1.1%. So in general it seems safe to conclude that mtDNA is less variable than nucDNA.

However, all of the comparisons we have just made are for species in subgenus *Sophophora.* In subgenus *Drosophila*, the level of mtDNA variation would seem to be somewhat higher (table 10-19). Excluding *D. buzzatii* in Australia, which was

TABLE 10-19 Estimates of mtDNA variation in *Drosophila*.

Species	π or d	Reference
Subgenus Sophophora		
D. melanogaster	0.00053 (overall)	Hale & Singh, 1991
	0.00106 (within Africa/Europe)	
	0.00056 (within Asia)	
	0.00033 (within New World)	
sequence data ND5	0.0016 (diverse strains)	Rand et al., 1994
	0.0002 (single California population)	
	0.0006 (Zimbabwe)	
sequence data ATPase 6	0.0045	Kaneko et al., 1993
D. simulans		
sequence data ND5	0.0018 (diverse)	Rand et al., 1994
D. teissieri	0.0007	Monnerot et al., 1990
D. yakuba	0.0003	Monnerot et al., 1990
D. subobscura	0.0081 (overall)	Alfonzo et al., 1990
	0.0025 (within populations)	
	0.0080	Gonzalez et al., 1990
D. athabasca		
Eastern A	0.0017	Yoon & Aquadro, 1994
Eastern B	0.0009	Yoon & Aquadro, 1994
Western-North	0.0034	Yoon & Aquadro, 1994
D. affinis	0.0009	Yoon & Aquadro, 1994
D. pseudoobscura	0.0090	Gonzalez et al., 1990
D. ambigua	0.0100	Gonzalez et al., 1990
10 species *obscura* group	<0.003[a]	Beckenbach et al., 1993
Subgenus Drosophila		
D. sulfurigaster		
subspecies *bilimbata*	0.0063	Tamura et al., 1991
subspecies *albostrigata*	0.0041	
D. albomicans	0.0102 (within populations)	Chang et al., 1989
	0.0097 (between populations)	
D. buzzatii Australia only	0	Halliburton & Barker, 1993
D. heteroneura	0.0170	DeSalle et al., 1986b
D. silvestris	0.0113	DeSalle et al., 1986b
	0.0270	DeSalle & Templeton, 1992

[a]DNA sequence data for one gene, COII.

Note: Most estimates are based on RFLP data; in some cases π is calculated when only size variation is available and d is calculated when restriction site map variation is available. The figure shown is the mean for all pair-wise comparisons. In all cases the estimates are of the average differences per nucleotide between two mtDNA molecules, comparable to heterozygosity for a diploid system.

found to have no variation (almost certainly due to recent introduction and bottlenecks), mtDNA diversity in subgenus *Drosophila* is around 1%. The two Hawaiian species studied, *D. heteroneura* and *D. silvestris*, seem particularly variable, being in the range of the most variable species known for nucDNA. As we will see later, there is other evidence that the evolutionary dynamics of mtDNA may differ in the two subgenera.

TABLE 10-20 Estimates of variation in mtDNA based on DNA sequence data.

Species	Gene	Sample Size	π	θ
D. melanogaster	*ND5*	9	0.0016	0.0019
	Cyt-b	16	0.0009	0.0021
D. simulans	*ND5*	6	0.0018	0.0015
	Cyt-b	18	0.0003	0.0011
D. yakuba	*Cyt-b*	13	0.0014	0.0019

Note: Data for 1515 bps of the *ND5* gene is from Rand et al. (1994); data for 1137 bps for *Cyt-b* gene is from Ballard and Kreitman (1994).

Rate of mtDNA Evolution

Presumably because of the relatively high rate of mutation in vertebrate mtDNA, mtDNA is found to be some five to ten times more diverged than nucDNA when pairs of vertebrate species are compared (Brown et al., 1979). The first indication that this may not be so in *Drosophila* came from RFLP (Shah and Langley, 1979) and DNA–DNA hybridization (Powell et al., 1986) studies on the *melanogaster* subgroup. The latter experiments (with fuller data in Caccone et al., 1988b) indicated that mtDNA and total single-copy nuclear DNA (scnDNA) have about the same level of divergence between pairs of species. Now that DNA sequence data are available for the entire mtDNA for *D. yakuba* and *D. melanogaster*, it is possible to examine this question in more detail.

Table 10-21 summarizes the divergence measurements of various sorts for *D. melanogaster* and *D. yakuba*. It can be seen that the divergence of the entire mtDNA

TABLE 10-21 Comparison of DNA divergence between *D. melanogaster* and *D. yakuba* for different types of DNA.

	Sequence/ΔT_m	Percentage Divergence	Reference
mtDNA			
Total	ΔT_m	6.3	1
Total minus A+T-rich	Sequence	6.2	2,3
Protein-coding	Sequence	7.2	2,3
tRNAs	Sequence	3.9	2,3
tRNAs including indels	Sequence	4.5	2,3
Nuclear			
Total scnDNA	ΔT_m	6.8	4
cDNA from embryo mRNA	ΔT_m	2.3	4
cCNA from adult mRNA	ΔT_m	3.8	4
Intergenic DNA	ΔT_m	15.3	4

References: (1) Caccone et al., 1988b; (2) Wolstenholme and Clary, 1985; (3) Garesse, 1988; (4) Powell et al., 1993.

Note: Some estimates are from ΔT_ms of DNA–DNA hybridization experiments and others come directly from sequence data.

TABLE 10-22 Comparison of DNA divergence for *melanogaster* subgroup species other than *melanogaster–yakuba* shown in table 10-20.

	Percentage Divergence					
Species Compared	Total Nuclear Single-Copy DNA	cDNA from Embryo mRNA	cDNA from Adult mRNA	mtDNA Total	mtDNA COI + ND2	mtDNA + tRNA
melanogaster–simulans	3.7	1.3	2.1	3.4	4.2	3.7
melanogaster–mauritiana	4.3	1.4	2.0	3.9	3.6	3.1

Note: The first four columns are from DNA–DNA hybridization experiments (Caccone et al., 1988b; Powell et al., 1993) and the two right-most columns from DNA sequences (Satta et al., 1987).

estimated from DNA–DNA hybridization studies (6.3%) corresponds extremely well with the DNA sequence data (6.2%), excluding the A+T-rich region, the divergence of which is difficult to align and estimate accurately. One can also see in this table that, within the mtDNA molecules, the coding region is evolving faster than tRNAs. When the total scnDNA of these two species is hybridized, the overall divergence is estimated to be 6.8%, and thus the conclusion was reached that mtDNA and scnDNA evolve at about the same rate. However, recall that almost all the mtDNA is coding DNA, while the minority of scnDNA is coding. From the DNA–DNA hybridization experiments using only coding DNA (cDNA) from the nucleus, we see that the divergence between the two species is only about 3% (table 10-21). For these two species, confining ourselves to protein-coding genes, it seems as if mtDNA is evolving about twice as fast as nuclear DNA.

Tables 10-22 and 10-23 extend these comparisons to other species. In comparing *D. melanogaster* to two other members of its subgroup, *D. simulans* and *D. mauritiana* (table 10-22), we see a picture very similar to the comparison to *D. yakuba.* Here there is only partial mtDNA sequence data, but again the divergence between coding mtDNA is about the same as for total scnDNA. The protein-coding fraction of the

TABLE 10-23 Estimates of DNA divergence in the *Drosophila obscura* group for total nuclear single-copy DNA (scnDNA) compared to the COII gene of the mtDNA.

	Percentage Divergence		
Species Compared	scnDNA	mtCOII	mtCOI
pseudoobscura–persimilis	1.6	0	0.2
pseudoobscura–miranda	2.6	3.1	5.2
pseudoobscura–affinis group	4.3	6.0	9.1
pseudoobscura–subobscura	6.8	8.9	12.0

Note: scnDNA divergence estimated from ΔT_ms of DNA–DNA hybridization experiments (Goddard et al., 1990). The mitochondrial COII (Beckenbach et al., 1993) and COI (Gleason et al., in press) data are from direct sequences.

nuclear genome is about half as diverged as the total scnDNA, so again we conclude that for protein-coding DNA, mtDNA is evolving about twice as fast as nucDNA. For the *obscura* group the picture is similar, although the differences would seem to be a bit greater (table 10-23). Mitochondrial DNA appears to be about 1.5 to 2.0 times as diverged as total scnDNA. While the experiments have not been done, if these *obscura* group species have a pattern similar to the *melanogaster* subgroup, namely that protein-coding DNA is about half as diverged as total scnDNA, then we would have to conclude that in this group, comparing only protein-coding DNA, mtDNA is evolving three to four times faster than scnDNA.

Satta et al. (1987) examined the *Adh* gene in the *melanogaster* subgroup compared to mtDNA sequence data and concluded that mtDNA was evolving about 1.4 times as fast as *Adh*. Sharp and Li (1989) found that mtDNA evolved about as fast as the least codon-biased genes (i.e., the fastest nuclear genes, see figure 10-7) and about twice as fast as the most biased, slowly-evolving nuclear genes.

Two studies of species in subgenus *Drosophila* indicate a somewhat different pattern. DeSalle et al. (1986b, c) found about 2% divergence of mtDNA between the Hawaiian species *D. heteroneura* and *D. silvestris*, while Hunt and Carson (1983) estimated only about a 1% divergence for total scnDNA, implying about 0.5% divergence for nuclear protein-coding DNA. This indicates a four-fold higher rate for protein-coding genes in mtDNA. Tamura (1992) studied the *sulfurigaster* group and estimated that mtDNA was evolving three times faster than nuclear DNA; however, this estimate was based on conversion of observed substitutions to absolute rate instead of direct DNA comparisons for the same species. Both these subgenus *Drosophila* studies come to similar conclusions: mtDNA evolves at least three times as fast as nuclear DNA (of the same function), but only about two times as fast in subgenus *Sophophora*. Many more such studies are needed to firmly conclude that the two subgenera have different mtDNA dynamics. Also, we need to note that if the arguments concerning the *obscura* group are correct, then there are cases within *Sophophora* where mtDNA is evolving considerably faster than protein-coding nuclear DNA.

To summarize, the overwhelming evidence is that mtDNA is evolving at the same rate or only slightly faster than total scnDNA. When only protein-coding regions are considered, mtDNA evolves two to four times faster than nucDNA. The relative rate of change of these two genomes varies from group to group, and there may be some systematic differences between the two subgenera in this regard. However, in no case has the extreme five-to-ten-fold faster rate seen in vertebrates been observed in *Drosophila.*

Pattern of Change

A second generality of vertebrate mtDNA evolution is that Ts's occur much more frequently than Tv's (e.g., Brown, 1983). In *Drosophila*, there is evidence that this is also the case, but as with rates, the differences are probably not so extreme as in vertebrates.

In the initial analyses of Wolstenholme and Clary (1985) and Garesse (1988) comparing *D. yakuba* and *D. melanogaster* mtDNA, they concluded that Ts's and Tv's were about equally frequent. Considering that with random change, Tv's are expected to be twice as frequent as Ts's, the equal observation indicates about a twofold bias in favor of Ts's.

On the other hand, DeSalle et al. (1987b) found a much higher rate of Ts's, especially for very closely related Hawaiian *Drosophila.* Figure 10-9 illustrates their results. For very closely related species, about 90% of differences are Ts's, and this number drops to about 50% and plateaus. Beckenbach et al. (1993) found a very similar picture with the *obscura* group (figure 10-10). For very closely related species within the *affinis* subgroup, Ts's outnumber Tv's about 9 : 1. As divergence increases, it drops to 50% and at the furthest distance, between the *obscura* and *melanogaster* groups, it drops to about 40%. This kind of curve, that is, the changing ratio of Ts : Tv with increasing evolutionary distance, is typical of a saturation phenomenon. Ts's occur between only two bases and after a while the amount of observed change

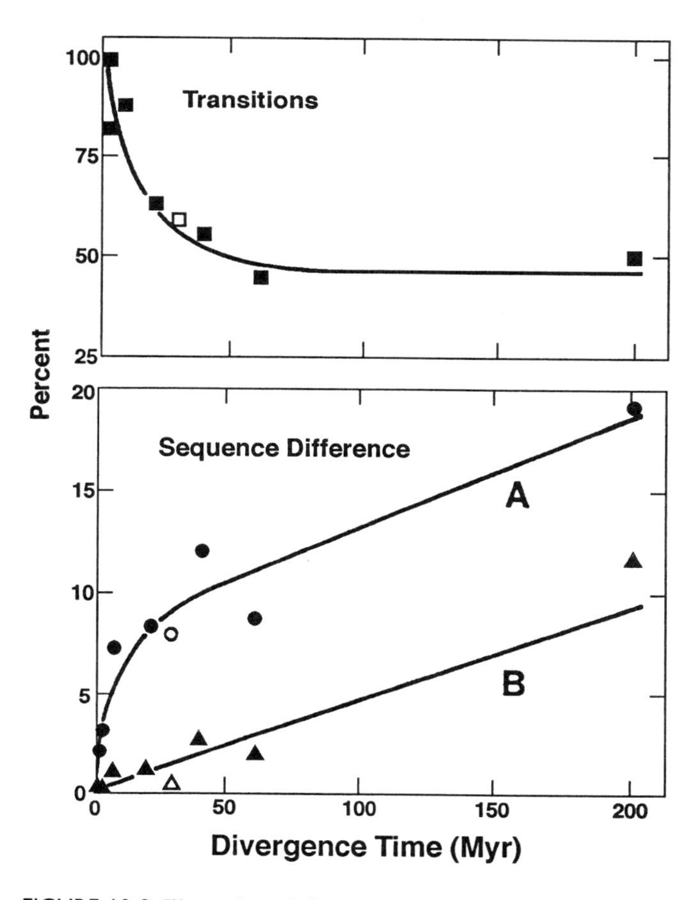

FIGURE 10-9 Illustration of dependence of divergence time on proportion of nucleotide differences that are transitions for mtDNA NADH subunit 1 among various Diptera. In the lower graph, curve A is for all substitutions (uncorrected) and curve B only for replacement substitutions. From DeSalle et al. (1987b).

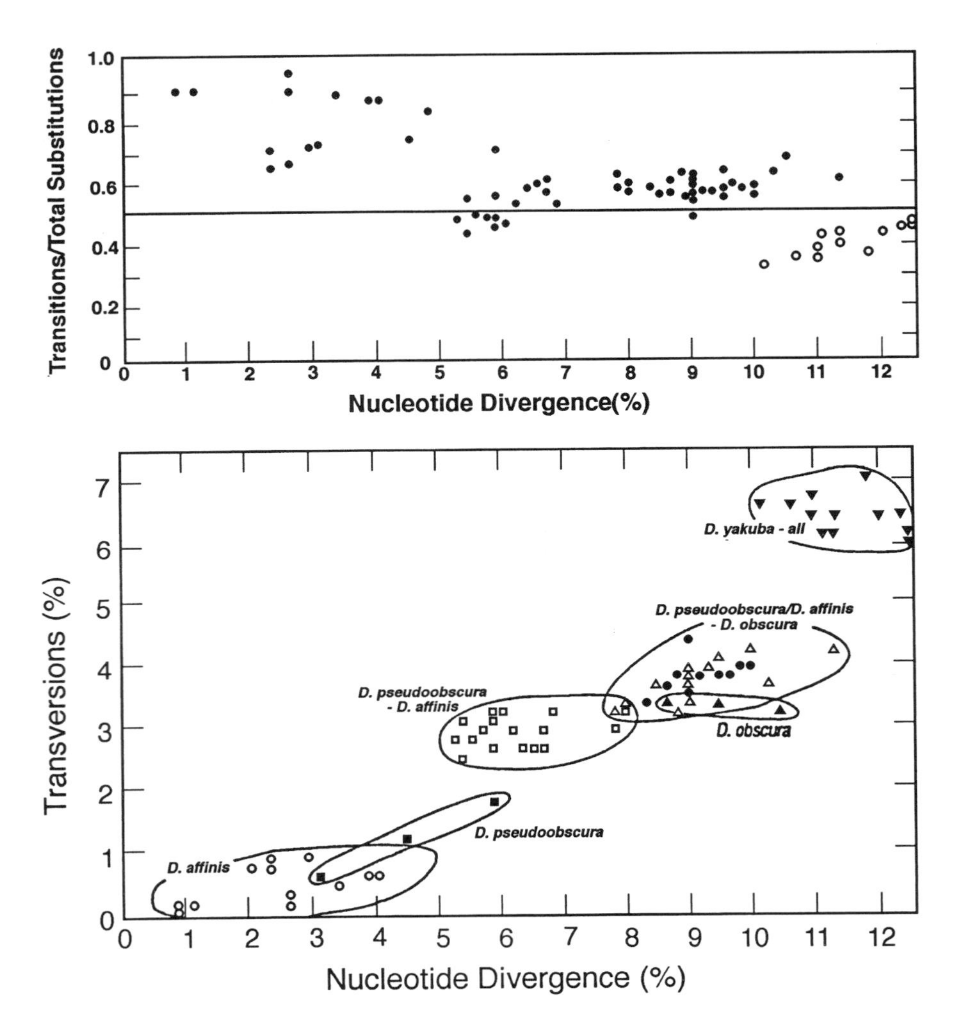

FIGURE 10-10 Relative proportions of transitions and transversions among mtDNA COII genes of the *obscura* group. Nucleotide divergence on the X axis can be thought of as time. In the upper graph the open circles on the far right and in the lower graph the inverted triangles are comparisons between the *obscura* group and *D. yakuba* of the *melanogaster* group. Note in the upper graph that early in divergence about 90% of substitutions are Ts's, whereas by about 6% divergence Tv's and Ts's are about equally frequent. From Beckenbach et al. (1993).

saturates. Since Tv's are much rarer and have two bases that may change, saturation takes much longer. Thus, the ratio of the two types of substitution changes with increasing time since divergence.

Considering these two figures (10-9 and 10-10), recall that *D. yakuba* and *D. melanogaster* have a nucleotide divergence of about 6 to 7% (table 10-21), which would place this comparison exactly at the point where the *obscura* group reaches equal numbers of Ts's and Tv's. Therefore, the conclusion of Wolstenholme and Clary (1985) and Garesse (1988) that Ts's and Tv's are about equal and DeSalle et al.'s (1987) contention that Ts's outnumber Tv's do not really conflict. It was simply that, quite unexpectedly, *Drosophila* mtDNA saturates for Ts's very rapidly, in contrast to vertebrates, where saturation occurs at about 20% divergence (Brown, 1983). Why *Drosophila* mtDNA saturates for Ts's so much sooner than does vertebrate mtDNA has not been fully explored or explained.

MtDNA in Population Cages, Cytoplasmic-Nuclear Interactions

Considering that the 13 proteins encoded in the mtDNA genome of *Drosophila* function coordinately with more than 100 nuclear-encoded proteins to form the functioning electron transport system of the inner membrane of mitochondria (Attardi and Schatz, 1988), it would not be at all surprising if the fitness of different mtDNA molecules depended on the genetic constitution of the nucleus. The most relevant experiments have concerned competition in population cages among flies having different combinations of nuclear and cytoplasmic genomes. Table 10-24 lists a number of studies on *Drosophila* mtDNA dynamics in laboratory studies.

The overall conclusion from these studies is that there can be strong nuclear-cytoplasmic interactions, the cytoplasmic component of which is presumably largely due to mtDNA variation. But as Nigro and Prout (1990) point out, other cytoplasmic factors such as the *Wolbachia* endosymbionts discussed in chapter 5 cannot be excluded in some of the experiments. Nevertheless, except for the *D. simulans* populations where it is known that *Wolbachia* is present, it does seem likely most cytoplasmic effects are due to mtDNA.

A few studies produced somewhat unexpected results and are worth special attention, as they have implications of general importance. Clark and Lyckegaard (1988) studied second chromosomes from six different populations and tested for segregation on the six cytoplasmic backgrounds. Variation in mtDNA was monitored by RFLP analysis. A significant effect of the mtDNA type on second chromosome segregation was found for interpopulation tests but not for intrapopulation tests. The implication is that there is coadaptation between mtDNA haplotypes and the second chromosome of the nucleus within a population, but that the coadaptation breaks down when mtDNA and second chromosomes from different populations interact.

Hutter and Rand (1995) studied the effect of replacing the mtDNA (cytoplasm) of one species with another, using a large number of backcrosses between *D. pseudoobscura* and *D. persimilis.* Figure 10-11 shows the results. *D. pseudoobscura* mtDNA had a clear advantage over *D. persimilis* mtDNA when placed in a population with *D. pseudoobscura* nuclear chromosomes; the same direction and intensity of selection occurred in all four replicates as well as in the perturbation populations (populations

TABLE 10-24 Laboratory studies on fitness of mtDNA genotypes.

Nature of Study	Reference
Fitness interactions between homozygous second chromosomes and mtDNA types. Interpopulational effects, no intrapopulational effects. *D. melanogaster.*	Clark & Lyckegaard, 1988
D. pseudoobscura and Bogota subspecies with different mtDNA haplotypes competed in population cages with strong evidence for selection in favor of Bogota mtDNA.	MacRae & Anderson, 1988
Above critiqued by	Singh & Hale, 1990 Nigro & Prout, 1990 MacRae & Anderson, 1990
D. simulans mtDNA haplotypes competed. Results were complicated by interaction with Wolbachia endosymbiont.	Nigro & Prout, 1990
D. subobscura mtDNA haplotypes competed on different nuclear backgrounds. Significant interactions detected but complex and seemingly dependent on the effective population size (females).	Fos et al., 1990
MtDNA from *D. pseudoobscura* and *D. persimilis* competed on native nuclear background and on other species' nuclear background. Asymmetrical results indicating strong cytoplasmic-nuclear fitness interactions.	Hutter & Rand, 1995
Unidirectional introgression of *D. simulans* mtDNA into *D. mauritiana* populations.	Aubert & Solignac, 1990
Injection experiments	
D. simulans made heteroplasmic by injection; indicated strong selection in favor of one mtDNA type.	deStordeur et al., 1989
Injection of *D. mauritiana* mtDNA into *D. melanogaster* embryos led to a complete takeover of the foreign mtDNA.	Niki et al., 1989
Temperature-dependent rates of transmission of mtDNA haplotypes observed in heteroplasmic flies.	Matsuura et al., 1991, 1993

that were manipulated to increase the mtDNA being selected against). The reciprocal did not occur. Apparently *D. pseudoobscura* and *D persimilis* mtDNAs are equally fit in populations with *D. persimilis* nuclear chromosomes. Components of fitness tests revealed that the selection on mtDNA on the *D. pseudoobscura* nuclear background was due largely to fertility differences. The intriguing point is the asymmetry in the selection effects corresponds to the asymmetry on fertility effects found by Orr (1987). When F_1 hybrids are backcrossed to *D. pseudoobscura* males, Orr found the resulting females are sterile if they have X chromosomes from different species and almost completely fertile if the X are both *D. pseudoobscura.* In the reciprocal backcross to *D. persimilis* males, this strong X effect was not observed. It is intriguing to speculate that some of the fertility deficit exhibited in interspecies hybrids and their backcrosses may be due to mtDNA and nuclear chromosome fitness interactions.

However, the generalization that interspecific fertility effects are associated with mtDNA–nucDNA interactions is called into question by the results of Niki et al. (1989). When *D. mauritiana* mtDNA was injected into *D. melanogaster* embryos, in the resulting heteroplasmic flies, the *D. mauritiana* mtDNA most often completely replaced the resident *D. melanogaster* mtDNA. Note that because these two species

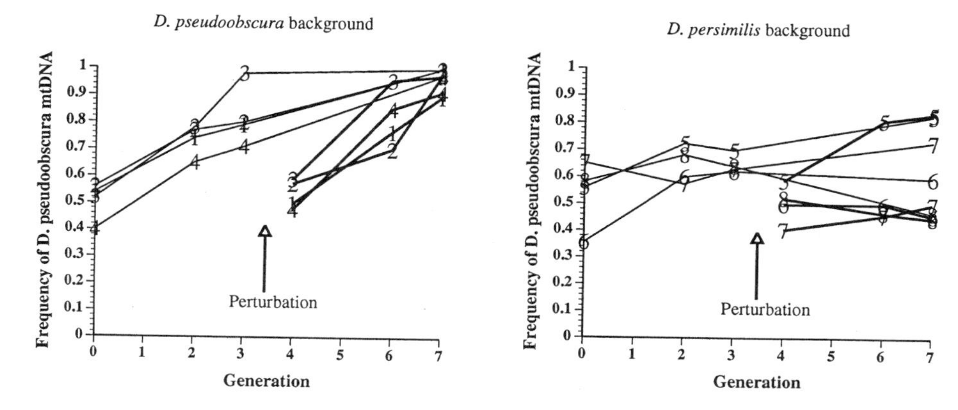

FIGURE 10-11 Population cage dynamics of mtDNA in different nuclear backgrounds. In the left graph, the results indicate that individuals with *D. pseudoobscura* mtDNA and nuclear genes outcompete those with *D. persimilis* mtDNA. The right graph indicates flies with *D. persimilis* nuclear genomes have little or no fitness difference dependent upon mtDNA type. Numbers refer to replicate populations. From Hutter and Rand (1995).

cannot produce fertile hybrids, the injection method is a way of combining mitochondrial and nuclear genomes in combinations not possible by crosses. This unexpected replacement of one species' mtDNA with another was accompanied by no detectable decrease in fitness of the flies; in fact, the selective elimination of flies with the resident mtDNA would imply the interspecific combination was superior! However, we need to add one caveat to this remarkable observation and that has to do with the level at which selection may act on mtDNA. It may be that organismal selection caused the replacement of *melanogaster* mtDNA in that individuals carrying *mauritiana* mtDNA had an advantage. Alternatively, it may be that *mauritiana* mtDNA had a cellular advantage such as simply replicating faster than *melanogaster* mtDNA in the female germline of the particular strain. Conceivably, the *mauritiana* mtDNA could even decrease the fitness of individuals with *melanogaster* nuclear genomes, yet still manage to replace the native mitochondria by intracellular selection.

Out of Africa: A Reevaluation of *D. melanogaster*

The fact that *D. melanogaster* has received more attention than any other species and that its populations have an unusual history due to association with human habitats has been repeatedly emphasized in this book. Recent molecular data have begun to clarify the history of this species and allow evaluation of how greatly the recent association with humans has affected the genetic makeup of populations. Almost certainly, *D. melanogaster* and all members of its subgroup had an origin in sub-Saharan Africa (Lachaise et al., 1988). Yet virtually every study of molecular variation in this species listed in the preceding tables in this chapter has been done on strains not from the native region. It is of some interest, then, to try to reconstruct the history of the origin of non-African *D. melanogaster* populations.

David and Capy (1988) reviewed several lines of evidence, including the then-available molecular data, and hypothesized a scenario (outlined in figure 10-12) for *D. melanogaster*'s spread. They suggest two waves of migration out of Africa. The first occurred after the last glaciation some 10 to 15 thousand years ago and established the Eurasian populations, which David and Capy refer to as "ancient" (African populations are "ancestral"). Both the New World and Australia are hypothesized to have been colonized much more recently by a second wave of migration that occurred only a couple hundred years ago, or even less. Whether the New World and/or Australia were colonized from ancestral Africa, ancient Eurasia, or both is not clear. The adaptation to temperate regions would make the Eurasian populations more adapted to the temperate New World and Australia, whereas slave trade from Africa to tropical New World regions might have transported tropical-adapted *D. melanogaster*. David and Capy (1988) review evidence that the different regions harbor *D. melanogaster* populations that are genetically differentiated with regard to pigmentation, ovariole number, and physiological traits such as desiccation resistance. However, based on allozyme studies (mostly on non-African populations, although including a few studies on strains from sub-Saharan Africa), it was thought that migration among the regions was quite common such that neutral or nearly neutral genetic variation was fairly homogeneous across the whole species.

This view has been significantly altered by studies of DNA variation. Hale and Singh (1991) reported on mtDNA variation among populations including sub-Saharan African samples from Benin and Zaire. They detected considerable differences among the regions of the world and could clearly delineate three major groups based on mtDNA variation: Afro-European, Asian, and New World. The Afro-European populations were the most diverse, the New World the least diverse, and Asia had very

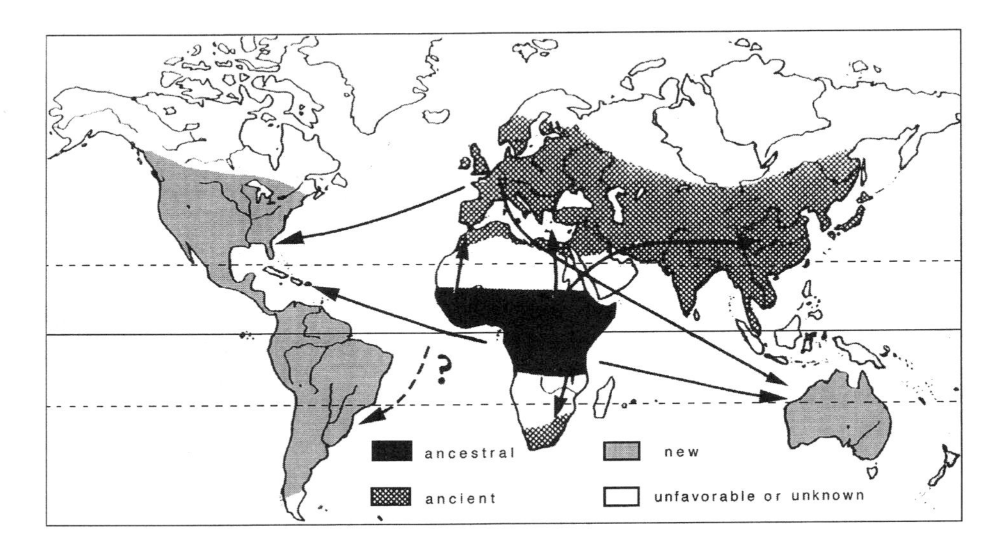

FIGURE 10-12 Map illustrating the hypothesized history of *D. melanogaster*. From David and Capy (1988).

complex patterns of variation. Table 10-25 summarizes their result and compares it to allozyme data. Clearly the geographic differentiation of mtDNA, both within the regions identified and among them, is much greater than that found for allozymes.

Begun and Aquadro (1993) added information on nuclear DNA diversity in populations from Zimbabwe. As in the pattern with mtDNA, the nuclear DNA also shows considerable differentiation among populations, especially between Zimbabwe and U.S. populations (table 10-25). In fact, Begun and Aquadro found nearly fixed differences between Africa and the United States for several loci. In addition, the overall level of diversity was significantly higher in Zimbabwe than in the United States (figure 10-13). When the same loci are compared, Zimbabwe had an average $\leq$ of 0.36% and U.S. populations 0.21%. (This included a disproportionate number of loci with very low variation due to being in regions of low recombination.) The conclusion is that ancestral populations of *D. melanogaster* are 1.5 to 2.0 times more variable at the nucleotide level than U.S. populations, and presumably than other populations outside Africa. As with geographic differentiation, this contrasts with allozymes. Allozymic variation is not only geographically relatively homogeneous in frequencies as noted by relatively low G_{st} values (table 10-25), the overall level of variation is very similar in populations from around the world (table 10-26).

Several important points need to be emphasized. The decrease in nucleotide diversity in populations outside sub-Saharan Africa argues strongly for founder effects during the colonization process. The overall relatively low DNA variation in *D. melanogaster* compared to other species may well be a misperception because most studies have excluded the most diverse ancestral populations. However, one might expect *D. simulans* to have a history similar to *D. melanogaster* with regard to founder effects and so forth, yet the latter species is considerably higher in DNA variation when comparing populations from the same regions of the world. Unfortunately, DNA variation studies using *D. simulans* from its native African region have not been reported.

Perhaps most startling is the contrast between allozymes and DNA variation. It remains difficult to explain why the two types of polymorphism produce such contrasting data indicative of very different histories. Allozyme data are consistent with

TABLE 10-25 Estimates of population differentiation in *D. melanogaster* for various genetic polymorphisms.

	Fixation Index		
Populations	Allozymes	mtDNA	nucDNA
Afro-European	0.106	0.456	
Western Hemisphere	0.079	0.441	
Far East	0.097	0.724	
Worldwide	0.126	0.660	
Zimbabwe versus United States			0.407 (mean for seven loci)

Note: Figures shown are for the fixation index G_{st} for allozymes and mtDNA and F_{st} for nucDNA, comparable measures.

From Hale and Singh (1991) and Begun and Aquadro (1993).

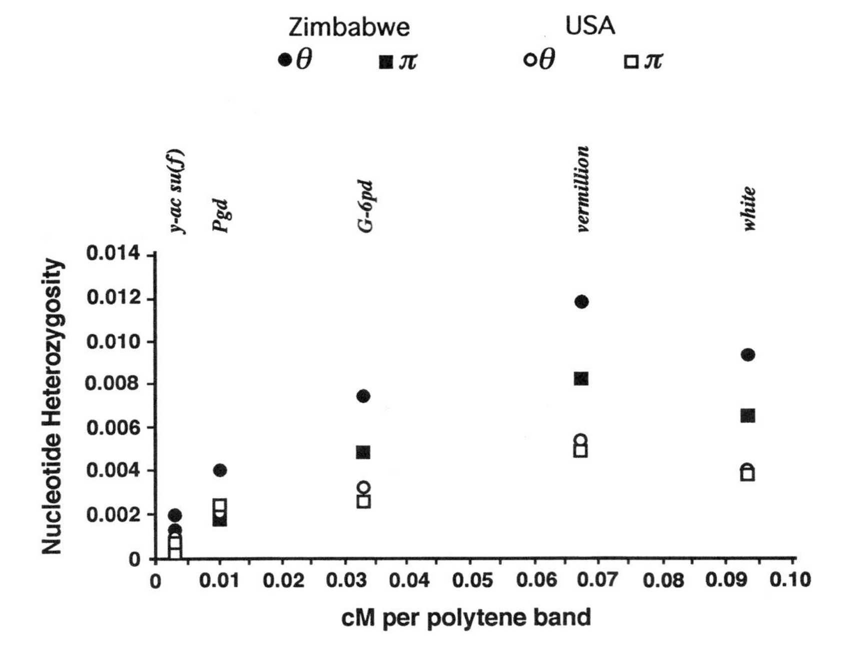

FIGURE 10-13 Scatterplot indicating the greater heterozygosity of Zimbabwe populations of *D. melanogaster* than of U.S. populations. Loci experiencing the same degrees of recombination lie over and under one another, so the greater heterozygosity of Zimbabwe populations is independent of the recombination effect. Circles and squares symbolize the two ways of calculating diversity, as indicated above the graph. From Begun and Aquadro (1993).

TABLE 10-26 Allozyme variation in U.S. and African populations of *D. melanogaster.*

	Massachusetts, U.S. ($n = 30$)	Texas, U.S. ($n = 30$)	Benin, West Africa ($n = 28$)
Proportion of loci polymorphic	0.43 (0.81)	0.36 (0.81)	0.39 (0.77)
Average number of alleles per locus	1.65 (2.48)	1.55 (2.52)	1.66 (2.64)
Heterozygosity	0.13 (0.28)	0.12 (0.31)	0.11 (0.27)

Note: The numbers not in parentheses are for a survey of 117 randomly chosen loci, while those in parentheses are for a sample of 26 loci known to be polymorphic. n is the number of strains studied.

Data summarized in Begun and Aquadro (1993), originally collected by Singh and Rhomberg (1987b) and Singh et al. (1982).

frequent migration among populations (relatively low F_{st}, table 10-25) and give no indication of founder effects (high heterozygosity, table 10-26). In contrast, DNA polymorphisms indicate very limited or no migration (high F_{st}, table 10-25), with founder effects outside native Africa (lower nucleotide diversity, figure 10-13). Recall there is behavioral reproductive isolation between Zimbabwe and U.S. populations (Wu et al., 1995, discussed in chapter 7), further evidence for significant differentiation in this species. The data on allozymes and DNA polymorphisms are especially difficult to reconcile with the notion that both types of polymorphism are neutral.

Finally, these results have considerable implications for previous and future studies of *D. melanogaster.* As Begun and Aquadro (1993) conclude:

> Our data provide convincing evidence that *D. melanogaster*, as a species, is far more variable than previously thought and that most of this variation is not segregating in the populations from which our evolutionary inferences (effective population size, frequency of lethal mutations, transposable element copy number and so on) and models are derived. It is no longer tenable to think of *D. melanogaster* as essentially panmictic for nuclear genes. Furthermore, it is no longer realistic to assume that *D. melanogaster* populations are near equilibrium or that conclusions derived from studies of USA (or similar) samples are even roughly true for the species as a whole.

DNA Variation and Inversions: Ancestry and Gene Conversions

In chapter 3, a discussion of molecular variation studies of genes associated with inversions was presented, especially allozyme studies. Some DNA-level work was also discussed, and we now turn to this latter subject in more detail.

D. pseudoobscura Subgroup

The initial study of Aquadro et al. (1991) using RFLP analysis of the *Amy* region within the breakpoints of most naturally occurring inversions of *D. pseudoobscura* was presented in figure 3-19 and again discussed and compared with other data in figure 8-18 in the context of phylogenies. Popadic and Anderson (1994) have now added DNA sequence data from this region in an attempt to understand better the history of this inversion system. They made the important observation that the molecular phylogeny derived from DNA sequences within an inversion can be used to help determine the ancestral gene arrangement. As discussed in chapter 3, almost all attention to the ancestral gene arrangement has centered on the trunk of the tree in figure 3-3, that is, the arrangements TL, SC, HY, and ST. Papadic and Anderson reason that there are six possible trees that would have each of these as the ancestor (figure 10-14a). ST and TL each can be the ancestor of only one possible tree because they are at the ends of the trunk, whereas HY and SC each have two possible trees as they are central in the trunk. Based on DNA sequences of 958 bps 3′ and 5′ to the *Amy* coding region, they produced the Neighbor-Joining tree in the lower part of figure 10-14. The topology of this lower tree is compatible with only two of the six possible scenarios, B and D1, effectively ruling out ST and HY as the ancestral arrangement and leaving only SC and TL as candidates. (This is why in figure 3-3, a double arrow was used to connect these two, whereas all other arrows indicate directionality.) Popadic and Anderson (1994) went on to consider further RFLP data from this region

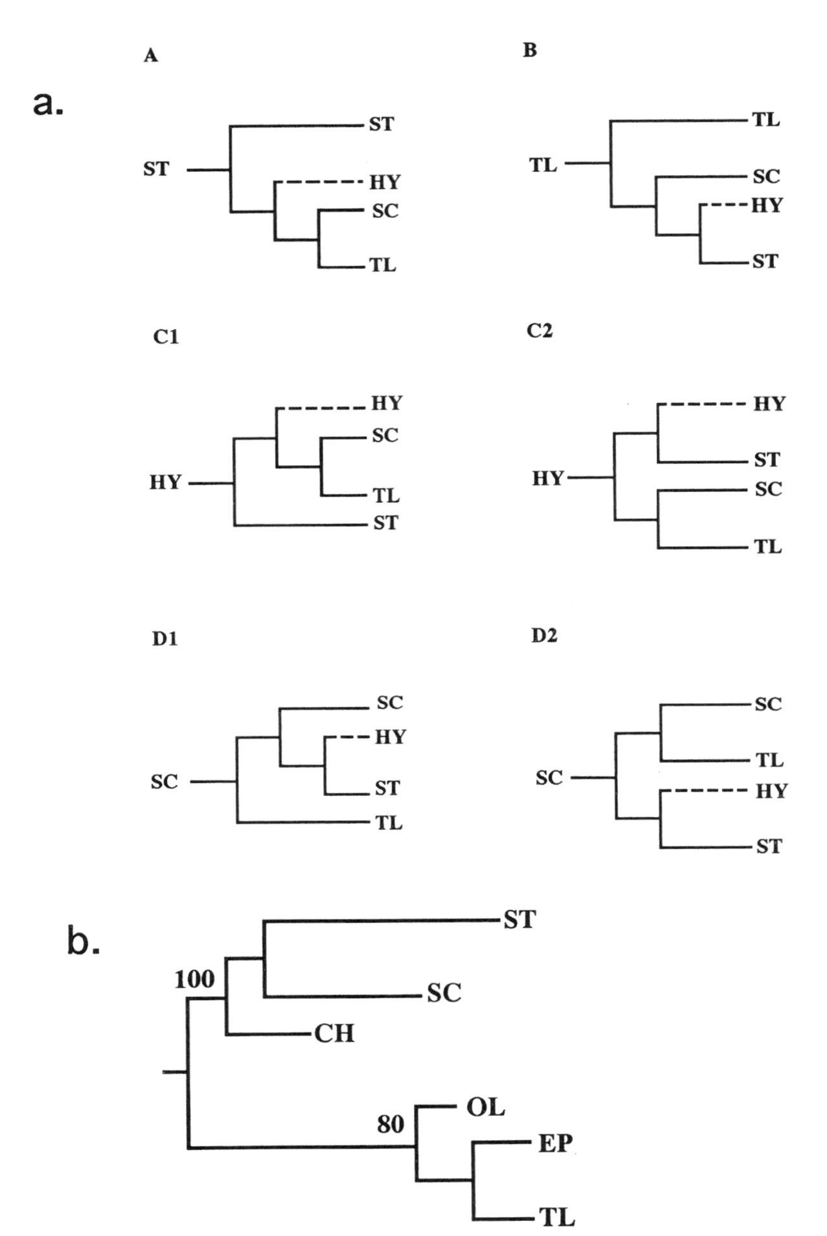

FIGURE 10-14 (a) Possible scenarios for the evolutionary relationships of the inversions of *D. pseudoobscura*. A and B are cases where gene arrangements ST or TL are ancestral; because these arrangements are at termini of the main trunk of the phylogeny, only one possible tree can be hypothesized. Since HY and SC lie at interior nodes on the trunk, two scenarios are possible with one of these as the ancestor. (b) The observed tree derived from nucleotide sequence data of the *Amy* region is shown at the bottom. This tree topology is compatible only with B and D1 above, meaning that SC and TL remain candidates for the ancestral gene arrangement. From Popadic and Anderson (1994).

and reanalyzed the results of Aquadro et al. (1991). They concluded the data moderately strongly support the ancestral status of SC over TL, as SC forms a deeper branch relative to the outgroup, *D. miranda*. On the other hand, the *Amy* gene in *D. pseudoobscura* exists in varying copies: one copy in TL, two in SC and three in ST (Brown et al., 1990). If we assume the ancestral state was a single copy, this would be added evidence that TL is ancestral.

The Sex-Ratio (SR) chromosome of *D. pseudoobscura* has also been studied for DNA variation. Babcock and Anderson (1996) sequenced an intergenic region near the *esterase* gene, which is inside the subbasal inversion of the three associated with SR in this species (chapter 3). In agreement with Keith's (1983) allozyme studies (tables 2-13, 3-14), the SR chromosomes are much less variable than the Standard X chromosome, an observation the authors attribute to the population size of SR chromosomes, which is smaller than ST. From a phylogenetic perspective, Babcock and Anderson conclude that the SR in *D. pseudoobscura* is rather ancient and arose before the split of *D. pseudoobscura* and *D. persimilis*. Both the Standard and SR chromosomes of *D. persimilis* are apparently derived from the ST of *D. pseudoobscura.*

D. subobscura

The major study of DNA variation associated with inversions of *D. subobscura* has been on the ribosomal protein gene, *rp49*, located very near the breakpoints of inversions in the O chromosome (Rozas and Aguadé, 1990, 1991a, 1991b, 1993, 1994). Initially, their RFLP studies of this gene and region revealed that different gene arrangements (mostly the standard and 3+4 inversion of the O chromosome) shared polymorphisms and thus, unlike the *D. pseudoobscura* subgroup case, the data were equivocal about the monophyly of the inversions. Later studies involved DNA sequence data and the situation was clarified. The data support monophyly in that all copies of an arrangement are more closely related to each other than to any copy of the other arrangements. As is the case with *D. pseudoobscura*, the shared sequences (polymorphisms) are small segments embedded in a larger, more diverged region. Rozas and Aguadé reach the conclusion that gene conversion is the most likely mechanism to account for the pattern. Given the degree of divergence of the gene arrangements and assuming a molecular clock, they calculate that the rate of gene conversion need be only on the order of 10^{-7} to obtain the observed patterns, a rate lower than that observed in laboratory studies of *D. melanogaster.*

Two biogeographical points of interest have also come from these studies. Rozas and Aguadé (1991a) studied the isolated *D. subobscura* populations on the Canary Islands and found them to be significantly differentiated, indicating that the island populations are long-established rather than recent derivatives of continental populations. In their study of the colonization of the Americas by *D. subobscura*, Rozas and Aguadé (1991b) found only eight haplotypes, compared to 70 in Old World populations. Furthermore, in their samples, 7 out of 8 of the New World haplotypes were found in both South America and North America, indicating the colonization of the New World was a single introduction of 8 to 12 haplotypes. This means that while there is a founder effect in the New World, it was not so drastic as a single inseminated female.

D. melanogaster

It is well-established that the *F* and *S* allozyme alleles of *Adh* in *D. melanogaster* are in linkage disequilibrium with inversions of the arm (2L) on which the locus resides. In particular, virtually all *In(2L)t* inversions carry the *S* allele, whereas the standard arrangement of this arm is polymorphic for *F* and *S Adh* alleles. Aguadé (1988) found that the *Adh* region in all *S*-bearing chromosomes (both *t* and standard) were more variable than *F* standard chromosomes. She interpreted this to indicate the *F* allele was relatively new and arose on the standard arrangement after the inversion polymorphism.

Bénassi et al. (1993) also studied the *Adh* region, as well as another related locus, *P6*, found within the same inversion of *D. melanogaster*. They sampled a presumably ancestral population from the Ivory Coast. More *Adh* haplotypes were found than were previously found in America and Spain, some of which were intermediates between previously known haplotypes, lending evidence that recombination has occurred in this region. Surprisingly, and contrary to the data discussed in the previous section, the African and American populations were found to be about equally variable. Also, they found no linkage between *P6* haplotypes and inversions, indicating enough time has passed to allow double-crossovers and/or gene conversion to destroy the presumed ancestral disequilibrium. This locus is placed near the middle of the inversion, so one might expect double-crossovers and gene conversions (due to better pairing) to be more common than for the *Adh* locus, which is located near the breakpoint where pairing is mechanically more difficult. This is consistent with Krimbas and Loukas' (1980) "middle gene hypothesis," which predicts that disequilibrium will decrease moving away from breakpoints.

Gene Duplications

Because most gene functions are crucial to the survival of the organism, the evolution of a gene with a new function is thought to depend on the generation of a duplication of a preexisting locus. One copy retains the ancestral function, and one is free to evolve a new function. Of course, taken to the extreme, this implies that all genes can be traced back to a single gene "Eve" (this also does not take into consideration recombining preexisting genes along the lines envisioned in exon shuffling). Nevertheless, gene duplication is considered an important phenomenon in evolution, and it is thus of interest to review what is known in *Drosophila*. The fate of a duplicated gene can be one of two paths: loss of function (usually transcription or translation), becoming a "dead" pseudogene; or to remain functional and diverge from the progenitor copy.

Pseudogenes

Relative to what is known in vertebrates, especially mammals, pseudogenes are rare in *Drosophila*. In fact, only four well-documented cases are known, although for some cases discussed later it is not clear if the duplicated copy is active or not. Perhaps the paucity of pseudogenes is related to the relatively small genome size of *Drosophila*: selection pressure of some sort acts to rid the genome of junk DNA such as pseudogenes—or perhaps pseudogenes are not as readily generated in *Drosophila* as in mammals.

The first documented pseudogene was for the larval cuticle protein, *Lcp* (Snyder et al., 1982). In *D. melanogaster* this gene is in five copies, four of which are active and one of which is an inactive pseudogene. The evidence that this is a pseudogene includes the fact that it has lost its TATA box, it has a stop codon 19 amino acids before termination in the other copies, and the ratio of silent and replacement substitutions between it and the other copies is the expected ratio if mutation is random with no constraints. This last observation does not hold for all *Drosophila* pseudogenes, as we will see later.

The second example of a pseudogene is *Amy* in *D. pseudoobscura.* We briefly mentioned above that this gene is found in different copies in different gene arrangements of the third chromosome (Brown et al., 1990): TL has one copy, SC two, and ST three. One copy is active in all chromosomes and the third copy is silent in ST. This third copy is a pseudogene, based on the fact that it has a stop codon in the coding region and has 4 replacement substitutions and no silent substitutions relative to the copy from which it is derived. Whether the second copy is active or a pseudogene is not clear. A second amylase protein has not been observed in this species, although microinjection of the *D. pseudoobscura* second copy of *Amy* into an amylase-null *D. melanogaster* did produce a low level of amylase activity (Hawley et al., 1990). Later in this chapter, *Amy* is further discussed as a case of concerted evolution and gene conversion.

A third pseudogene is the glycolytic gene phosphoglyceratemutase, *Pglym* (Currie and Sullivan, 1994). The two copies of this gene are on different arms of the third chromosome at positions 78 and 87, so the copies are designated *Pglym78* and *Pglym87*, respectively. The *Pglym78* copy appears to be the active gene as transcripts are detectable; it has two introns. *Pglym87* lacks both introns and no transcript can be detected; thus it seems likely that this is a pseudogene of the retroposon sort, that is, is a reverse transcribed processed mRNA. However, like other cases (to be discussed next) the proposed pseudogene retains some properties that hint that it may retain function, namely a nearly complete open reading frame and codon bias. *Pglym78* and *Pglym87* have 86% and 78% C+G in the third codon position, respectively. This partial reversion of the proposed pseudogene to random usage in the third position is consistent with a recent duplication and silencing. Obviously, analysis of related species should settle the question of when this putative pseudogene arose in the phylogeny of the group and thus shed light on the question of whether it is truly a pseudogene.

The final example of a pseudogene is *Adh* in the *repleta* group (Yum et al., 1991). In this case the duplication that generated the pseudogene is adjacent to the two active genes, and the pseudogene retains the intron/exon structure of the active genes. Figure 10-15 indicates the evolutionary history of *Adh* in this group. It is assumed that the ancestral gene had the usual *Adh* structure with two promoters and three introns, two of which are in the coding region. The distal promoter is used in adults (of most species) and the proximal one in larvae. The duplications did not include the distal promoter. Because expression of *Adh* is presumably crucial at all stages of development, in *D. mettleri* (situation C in figure 10-15) the proximal promoter is now used in both adults and larvae. The second duplication generated the three-gene situation

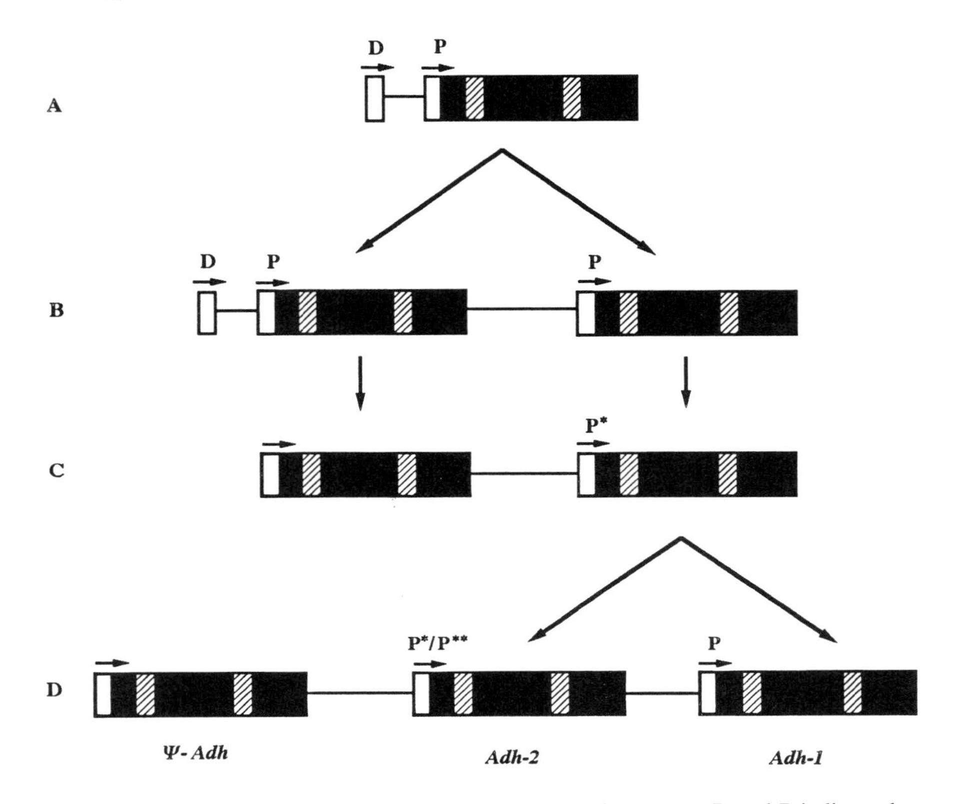

FIGURE 10-15 Model of the evolution of *Adh* in the *repleta* group. D and P indicate the distal and proximal promoters, respectively. A is the assumed ancestral state characteristic of most species. The initial duplication illustrated in B did not include the distal promoter. The 5′ gene became silent in C and the proximal promoter, now denoted P*, had evolved to be expressed in both larvae and adults; this is the existing stage in *D. mettleri*. D is the existing structure in *D. mojavensis* and *D. hydei*. *Adh-1* is expressed in larvae in both species, while *Adh-2* is expressed differently in the two species: in *D. mojavensis* it is expressed only in adults, while in *D. hydei* it is expressed in both larvae and adults. From Yum et al. (1991).

in part D of figure 10-15. In *D. mojavensis, Adh-2* is expressed only in adults and *Adh-1* only in larvae. In *D. hydei, Adh-2* retains the expression pattern of *D. mettleri* and is expressed at all stages.

The evidence is that ψ-*Adh* arose in the ancestry of the group and is a true pseudogene (Sullivan et al., 1994). All subgroups of the *repleta* group have species with ψ-*Adh*, all of which have the same mutation in codon 2 and the same deletion just 5′ to the former start signal; this is evidence for the generation of the pseudogene in the ancestral lineage. All ψ-*Adh*s have many frameshift mutations leading to stop codons, typical of true pseudogenes and indicating the gene can no longer produce a functional protein. In comparing the pseudogenes of different species, a most surprising observation was made by Sullivan et al. (1994): silent substitutions in ψ-*Adh* occur much more frequently than replacement substitutions! Table 10-27 shows the data. Recall that in coding sequences, some 75–80% of random mutations are expected to be replacement, and in some pseudogenes such as *Amy* in *D. pseudoobscura*, replacement substitutions outnumber silent substitutions. Why a nonfunctional pseudogene should have more synonymous substitutions is something of a mystery. Sullivan et al. (1994) have demonstrated that ψ-*Adh* in the *repleta* group does produce an RNA transcript. They speculate this transcript may have a function other than producing a protein. Perhaps this function requires some kind of triplet structure, so that the selective constraints are not for coding for a protein but to maintain a three-bp repeating structure which, by chance, constrains the first two positions of a codon but not the third. Sullivan et al. admit this is highly speculative and, in reality, we really do not understand the mechanism behind this pattern.

In two species of the *melanogaster* subgroup, *D. yakuba* and *D. teissieri*, another *Adh* pseudogene is thought to exist (Jeffs and Ashburner, 1991). It is very different from the *repleta* group in that the putative pseudogene has no introns and is located on a different chromosome (the third) than that on which the active gene resides (the second). This is typical of vertebrate pseudogenes, which are generated by reverse transcription of a processed mRNA, which then gets reinserted, more or less randomly, in the genome. As in the case with the *repleta* pseudogene, Jeffs and Ashburner (1991) also noted that silent substitutions outnumber replacements by about 2 : 1, not as extreme as the 10 : 1 ratio in *repleta*, but still unexpected. However, unlike the *repleta* group situation, there are not a lot of stop codons in the *melanogaster* subgroup pseudogene. There is some evidence it may not be a true pseudogene, but it may have acquired a new function (Long and Langley, 1993; discussed below in the *Adh* section). This remains a controversial contention, but if true, it has at least the potential to explain the excess of silent substitutions over that expected in a true pseudogene. On the other hand, the same unknown constraint operating in the *repleta* group may also be affecting the *Adh* pseudogene in this group.

TABLE 10-27 Ratios of silent to replacement substitutions in *Adh* between species of the *repleta* group.

Species Pair	Active Gene K_s/K_a	Pseudogene K_s/K_a
mojavensis–mulleri	16.8	9.8
mojavensis–buzzatii	17.3	11.0
mojavensis–mettleri	12.4	10.8
mojavensis–hydei	10.6	16.6
mulleri–buzzatii	22.1	10.6
mulleri–mettleri	16.0	12.5
mulleri–hydei	14.9	12.2
buzzatii–mettleri	19.2	11.7
buzzatii–hydei	18.4	14.6
mettleri–hydei	28.9	13.1

Note: K_s is the frequency of silent substitutions and K_a the frequency of replacement substitutions.

From Sullivan et al. (1994).

D. subobscura also appears to have some processed *Adh*-like pseudogenes, which Marfany and Gonzalez-Duarte (1992a) have called "retropseudogenes."

Functional Duplications

The first step in the generation of a new gene via duplication would be an intraspecific polymorphism for duplicated copies of a gene. A few such cases are known and listed in table 10-28. Of these examples, only three are definitely known to involve polymorphism for duplicated active genes: metallothionein genes in *D. melanogaster*, urate oxidase in *D. virilis*, and *Amy* in *D. melanogaster* and *D. ananassae.* These would appear to be good candidates for the early stages of evolution of new genes, assuming, of course, the trajectory for these cases is to remain in the populations and functional, rather than simply being on their way to loss.

Perhaps the most interesting and well-studied case is that of metallothionein genes, which code for proteins involved in detoxification of metals. Evidently, the ancestral state of this locus is a single copy of the gene designated $Mtn^{\cdot 3}$, as this is the allele found in all species examined, including six *melanogaster* subgroup species, *D. ananassae*, and *D. pseudoobscura* (Theodore et al., 1991; Stephan et al., 1994). Two additional derived alleles are known in *D. melanogaster*, one of which is designated Mtn^{1}, and the other is a duplication of this allele to $Dp(Mtn^{1})$. This two-step transition has been accompanied by a five-fold increase in mRNA production for this gene (Theodore et al., 1991). In presumed ancestral strains from Africa, only the $Mtn^{\cdot 3}$ allele is found, whereas populations outside Africa are polymorphic for two or three alleles. The duplicated allele exists at a frequency of 0 to 20% in some eastern U.S. populations (Lange et al., 1990) and nearly 50% in a French population (Theodore et al., 1991).

In a laboratory study, Maroni et al. (1987) showed that strains with the duplicated copies of the gene had a greater tolerance for copper sulfate (figure 10-16). These authors also demonstrated that different strains had somewhat different structures for

TABLE 10-28 Examples of intraspecific polymorphisms for gene duplications in species of *Drosophila.*

Species	Gene	Reference
D. melanogaster	*Stellate**	Lyckegaard and Clark, 1989
	Metallothionein	Maroni et al., 1987
		Lange et al., 1990
		Stephan et al., 1994
	*Gpdh**	Takano et al., 1989
		Koga et al., 1993
	Gapdh[a]	Wojtas et al., 1992
D. virilis	*Urate oxidase*	Lootens et al., 1993
D. pseudoobscura	*Amy**	Brown et al., 1990
D. ananassae group	*Amy*	Da Lage et al., 1992

[a]Both copies are active in *D. melanogaster*, but one or the other is silent in other species.

*Whether duplication is functional or a pseudogene is in doubt.

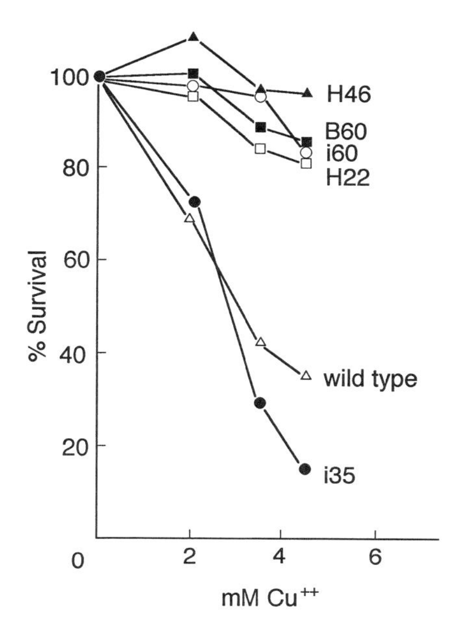

FIGURE 10-16 Survival of *D. melanogaster* larvae on medium with varying levels of copper. The wild-type strain used carried a single copy of the metallothionein gene, *mtn*. The rest of the strains carried a duplication, but in strain i35 the duplication is interrupted with an insertion. The four upper strains produce 1.7 to 2.1 times more *mtn* mRNA. From Maroni et al. (1987).

the duplications, so the evidence is that the duplication has arisen more than once. Whether the high frequency of the duplication is related to metal contamination in the environment is not clear, although Maroni et al. suggest that the practice of spraying vineyards with copper compounds to inhibit fungi and bacteria may be the selective agent. However, Lange et al. (1990) could find no differences in the frequency of the duplication between *D. melanogaster* samples collected from metal-contaminated areas and noncontaminated areas, although they admit that migration between the areas could have obscured selective differences. In any event, such a polymorphic duplication would likely suffer a selective disadvantage due to intrachromosomal recombination and/or other difficulties at meiosis in heterozygotes, so some force must be counteracting this. Clearly this case deserves more study, as it is analogous to other cases of resistance, such as drug resistance in mammalian cells and insecticide resistance in mosquitoes, in which selection has produced multiple copies of genes involved in detoxification.

Variation in the number of gene copies fixed in different species is much more commonly observed than intraspecific polymorphism for copy numbers. An interesting and well-studied case is esterase genes in *D. melanogaster* and *D. pseudoobscura.* In the former species two copies of an X-linked esterase are known, designated *Est-6* and *Est-P*, separated by only 197 bases (Collet et al., 1990), whereas in the latter species there are three copies designated *Est-5C, Est-5B*, and *Est-5A* (Brady et al., 1990). The relationships of these genes are indicated in figure 10-17; based on sequence similarity, the *pseudoobscura Est-5C* and *Est-5B* are duplicates of the ancestral *Est-6* of *melanogaster.* Brady and Richmond (1992) have analyzed this system in some detail and demonstrate its use in discerning variation in selective constraints on amino acid substitutions and also in detecting possible gene conversions. Furthermore, the duplicate copies have diverged with respect to tissue- and sex-specific expression

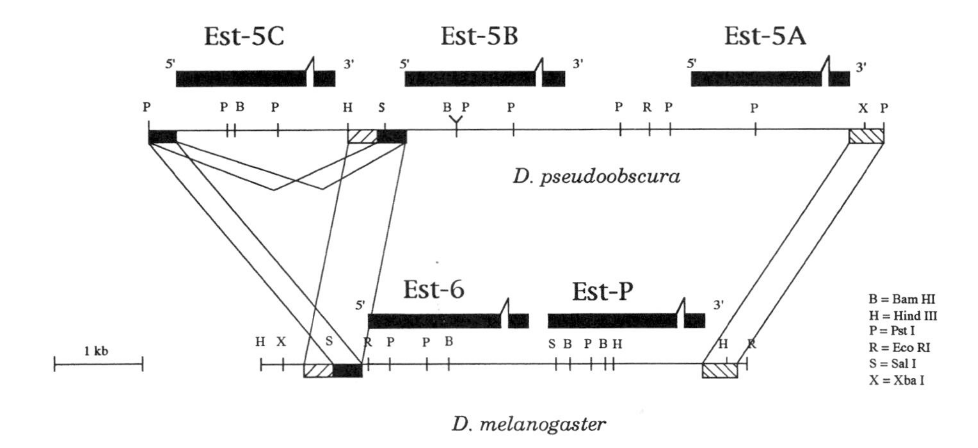

FIGURE 10-17 Model for the evolutionary relationship between *D. melanogaster* and *D. pseudoobscura* esterase genes. Shaded areas indicate regions of high similarity between species. Lines connect similar regions. The interpretation is that *Est-5C* and *Est-5B* in *D. pseudoobscura* are a duplication of *Est-6* of *D. melanogaster*, as they share a common 5′ sequence, and that *Est-5A* and *Est-P* are homologues. From Brady and Richmond (1992).

(Brady et al., 1990; Collet et al., 1990; Oakeshott et al., 1995). Given the evidence of conservation and change in flanking sequences indicated in figure 10-17, this would seem a particularly favorable model system for understanding the origin of new gene functions. (Oakeshott et al. [1993; 1995] discuss more generally the evolutionary genetics of esterases in *Drosophila*, some of which will be discussed in chapter 11.)

α-Amylase in *Drosophila* has a complex pattern with regard to duplications, some of which remain active, and some of which have become pseudogenes. *D. melanogaster* usually has two active inverted repeat copies and likely some pseudogenes (Doane et al., 1990). *D. ananassae*, another member of the *melanogaster* group, has four copies (usually active) in two clusters on different chromosomes; they are direct repeats (Da Lage and Cariou, 1993). In *D. pseudoobscura* there is yet another arrangement of the repeated copies. The evidence is this gene has duplicated independently at least three times in a very restricted sample of species within one subgenus. Whether the propensity of *Amy* to duplicate is related to adaptations (it is an enzyme that interfaces directly with the environment as it digests ingested starch) is not clear. *Amy* is one of the few well-characterized gene–enzyme systems in the genus that is known to duplicate relatively frequently and thus is a good model for evolution by duplication. (*Amy* has also been studied in the context of gene regulation evolution, discussed in chapter 11, pp. 426–427).

Gene Structure

Homologous genes sequenced from different species are often observed to differ in structure, usually in regard to presence or absence of introns. It is often not possible to determine the direction of change, that is, was there a loss or gain of introns? However, in three cases in *Drosophila*, phylogenetic analysis provides strong evidence for direction of change in gene structure, with all three indicating intron loss.

One example is *Adh* in the *willistoni* group (Anderson et al., 1993). *Adh* has three known structures in *Drosophila* (figure 10-18). Most species examined have the prototypic structure first observed in *D. melanogaster*, a single copy with three introns, the first of which interrupts the primary transcript from the distal promoter but not the coding sequence. In *D. willistoni* and six other species in the group (but not *D. prosaltans*), the third intron in the coding region is precisely deleted. In the previ-

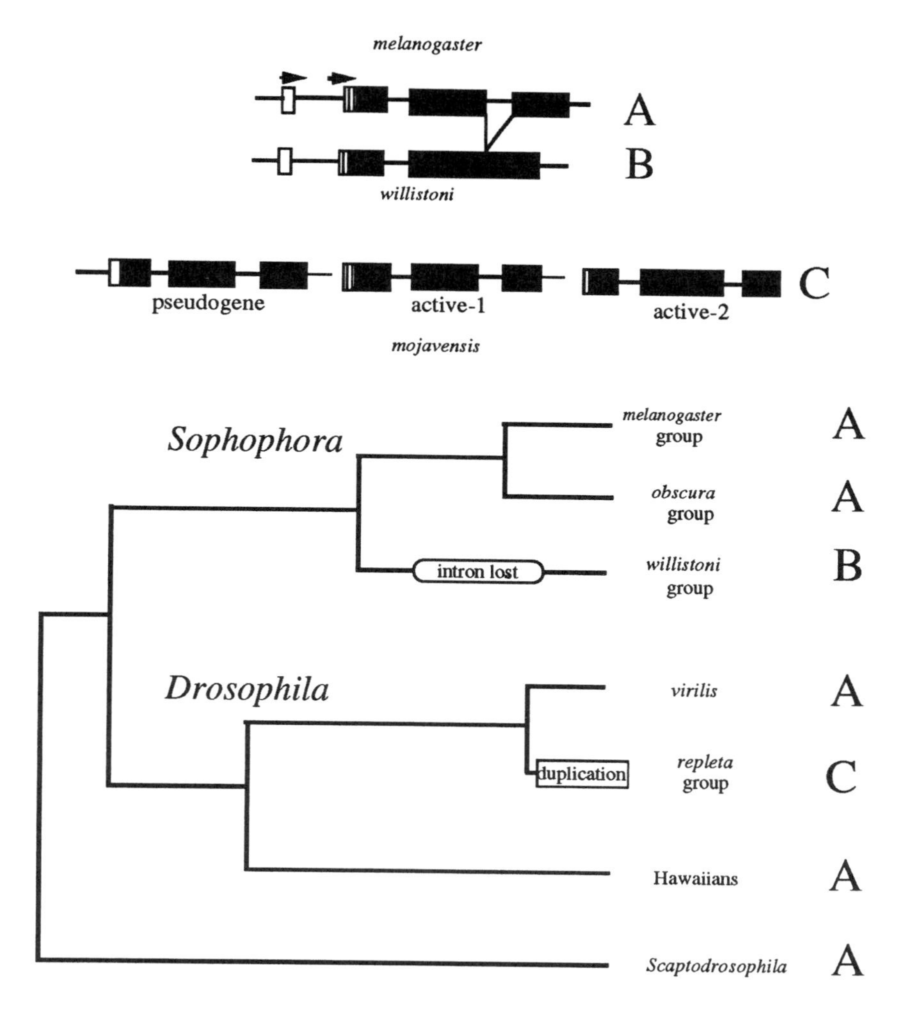

FIGURE 10-18 Evolution of *Adh* structure. Three structures have been observed and are illustrated at the top. The phylogenetic distribution of the structures is shown on the lower tree. B and C are both derived structures in the most parsimonious interpretation. From Powell and DeSalle (1995).

ous section we discussed the duplication of this gene in the *repleta* group. When these three structures are placed on a phylogeny, it becomes apparent that the missing intron in the *willistoni* group is a derived character, meaning there has been a loss of the intron. (The duplication in the *repleta* group is also a derived state.)

A second example is very similar in that the *Gapdh* gene also has three structures that differ in intron/exon structure and numbers of copies (figure 10-19; Wojtas et al., 1992). Again it appears that the intron is present in the ancestral state and the derived state has lost it. Also for this gene, the two duplicated copies differ slightly in size.

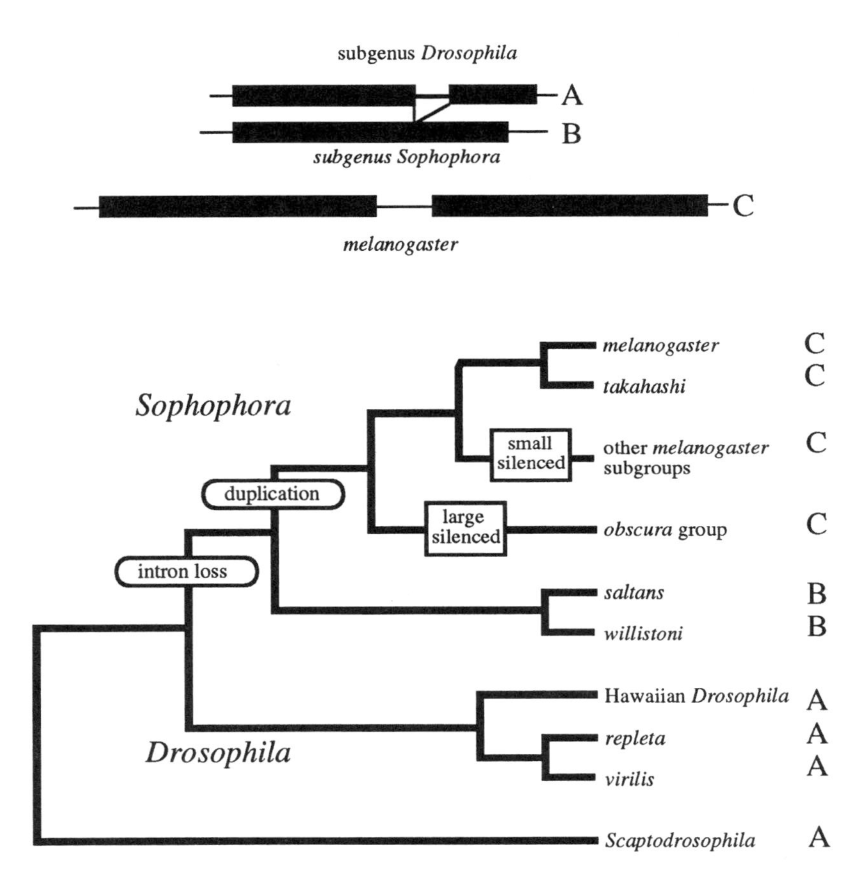

FIGURE 10-19 Phylogenetic analysis of the structure of the glyceraldehyde phosphate dehydrogenase gene, *Gapdh*. The presumed ancestral state is structure A, as it exists in the outgroup and subgenus *Drosophila*. Near the origin of the *Sophophora* the single intron was lost, as all species in this subgenus lack the intron. In the lineage leading to the *obscura* and *melanogaster* groups, a duplication occurred, which led to two copies of slightly different size. Both copies are expressed in some members of the *melanogaster* group, whereas the smaller copy is silent in other members. The larger copy is silent in the *obscura* group. Data from Wojtas et al. (1992); figure from Powell and DeSalle (1995).

One copy is silenced in the *melanogaster* and the other copy silenced in the *obscura* group.

A third example is the opsins (Carulli et al., 1994). As in the two previous cases, phylogenetic considerations led these authors to conclude intron loss rather than gain has been the direction of change. In particular, the *Rh1* gene has a specific intron in subgenus *Drosophila* that is missing in the *Sophophora*. The related duplicate of *Rh1, Rh2*, has an intron in precisely the same position as does *Rh1* in subgenus *Drosophila.* It seems more likely that the *Sophophora* have simply lost this intron than that subgenus *Drosophila* has acquired one in exactly this position.

For all three cases where strong inferences can be made, the evidence is that introns are lost during evolution with no evidence of gain. How much faith one puts in a sample of three in drawing a generality can be questioned, but the trend is clear. Perhaps of most interest is that these examples give evidence of a mechanism that can remove introns precisely without disrupting the protein reading frame. One such mechanism uses a processed mRNA, which is reverse transcribed and then reinserted into the genome. However, this cannot account for the precise deletion of only one of three introns as in the case of *D. willistoni Adh*, as the processed mRNA would have all introns removed. Whatever the mechanism, it would seem to lend support to the intron-first theory of gene evolution as there is, evidently, a mechanism for removing introns and, at least in *Drosophila*, there is no compelling evidence favoring gain of introns during evolution of this genus.

Adh: The Center of Attention

From an evolutionary perspective, no other gene has been studied as extensively in *Drosophila* as the *Alcohol dehydrogenase* locus. Throughout this book, we have had reason to cite work on this gene as prime examples of several basic principles. These examples will not be repeated here; rather, the intent of this section is to convey a sense of the literature on this gene and to highlight a few further aspects of it not already mentioned elsewhere. Reviews can be consulted for further details and citations to the extensive literature: van Delden (1982), Sofer and Martin (1987), Sullivan et al. (1990), Chambers (1988, 1991), Geer et al. (1990), and Heinstra (1993).

The Big Gene Tree

Adh was one of the first genes cloned and sequenced from *Drosophila* and has doubtless been sequenced more times than any other gene. Figure 10-20 shows an extensive gene tree for *Drosophila Adh* (Russo et al., 1995) and conveys an idea of the number of species for which the sequence is known. Some 39 species whose sequences have been deposited in GenBank are shown, and I am aware of nearly 20 more species that have been sequenced but the results not yet published. It is also worth pointing out that this single-gene phylogeny is remarkably consistent with those presented in chapter 8. In particular, the Hawaiian *Drosophila* (including the Hawaiian *Scaptomyza* and *Engyscaptomyza*) are shown to be a monophyletic sister group to subgenus *Drosophila.* The two major subgenera (*Sophophora* and *Drosophila*) are each monophyletic groups. Within *Sophophora*, the relative relationship of the three major groups is in agreement with all other evidence. The only disagreements with the conclusions

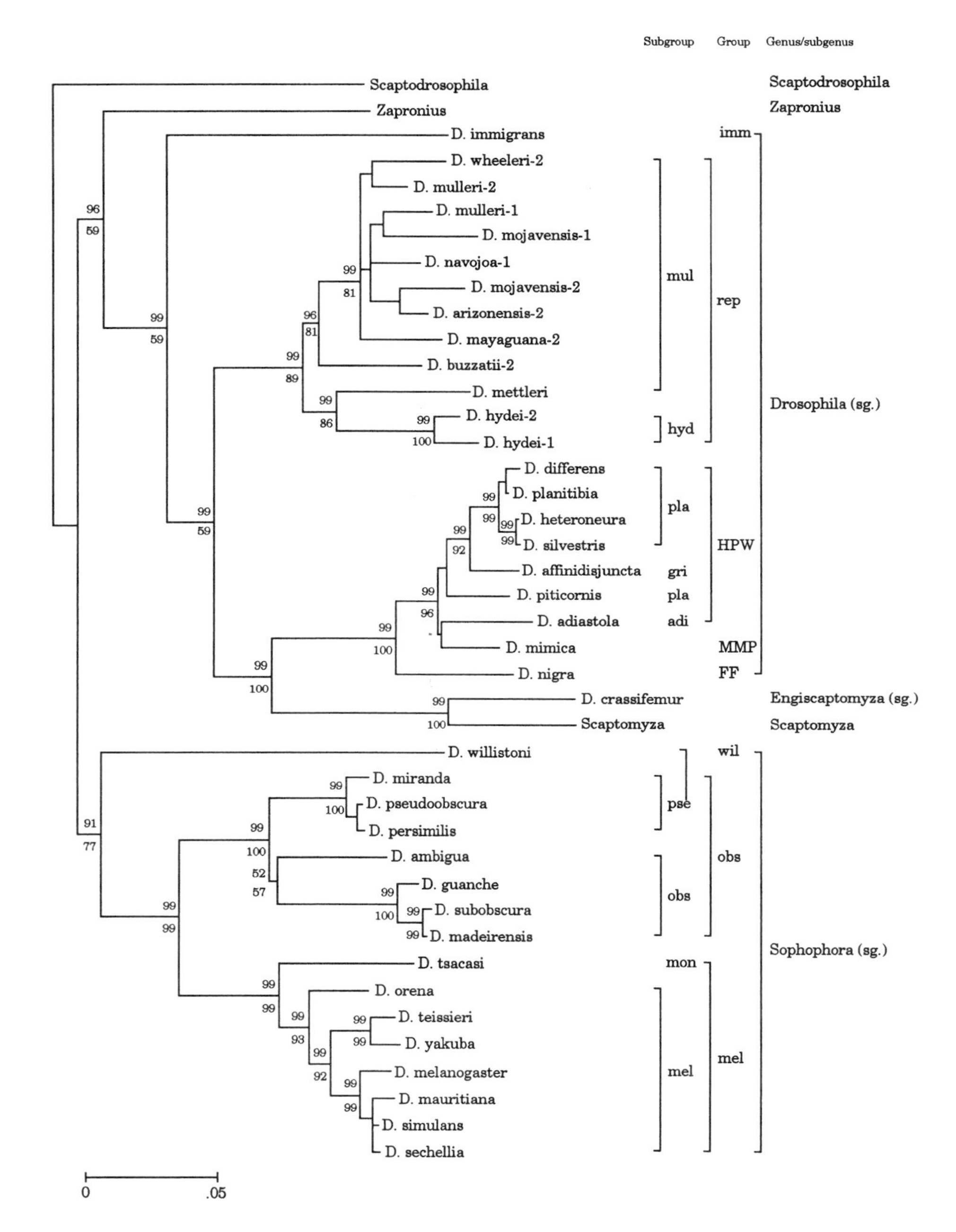

FIGURE 10-20 Neighbor-Joining tree of 42 sequences of *Adh*. Numbers above nodes are confidence probabilities and those below are the bootstrap values. From Russo et al. (1995).

reached in chapter 8 concern *Zaprionus* and *D. immigrans* and some details within subgenus *Drosophila.* Russo et al. (1995) also present alternative trees using other tree-building algorithms and other outgroups in which these problems solve themselves—*Zaprionus* joins outside the genus, *immigrans* goes inside subgenus *Drosophila*, and the details within subgenus *Drosophila* become unresolved.

Disequilibrium

The most extensive intraspecific variation study has been S. Schaeffer and colleagues' work on *Adh* in *D. pseudoobscura.* One important result of this work is the intriguing pattern of linkage disequilibrium found. Figure 10-21 shows the pattern of disequilibrium observed among 99 alleles. Because so many pair-wise tests were performed, one needs to correct for the number expected by chance to be significant. This is done by the Bonferroni method of dividing the significance level (0.05) by the total number of tests (74,278 in this case); this is 6.7×10^{-7}. At this level, only nucleotide pairs in the 331–355 region and 1454–1500 region are significantly in linkage disequilibrium. Schaeffer and Miller call these "clustered linkage disequilibrium sites." Most interestingly, the clusters are in introns and no disequilibrium is found in exons! Stephan and Kirby (1993) have suggested this may be due to compensatory changes needed to maintain pre-mRNA secondary structure. Whatever the explanation, there are obviously aspects of such data that were quite unexpected and deserve further scrutiny.

Adhr or *Adh-dup*

In the initial report of Kreitman (1983), one of the more surprising findings was that DNA sequence immediately 3′ to the coding region of *D. melanogaster Adh* was remarkably monomorphic compared to other noncoding regions. This observation was made understandable when Schaeffer and Aquadro (1987) realized in their study of *D. pseudoobscura* that there was another open reading frame within a couple hundred nucleotides of the 3′ end of *Adh.* This gene has since been characterized in a number of species and has been shown to have considerable similarity to *Adh* in both sequence and structure and probably represents an ancient duplication of the locus. Amino acid similarity between the two genes is 35–40% (depending on species), indicating the likelihood of homology, although considerably diverged. It was initially designated *Adh-dup*, but in the FlyBase it is referred to as *Adhr, Adh-related*; we use the FlyBase nomenclature.

The distribution of *Adhr* in the genus is somewhat confusing. It has been found in all species of the *melanogaster* and *obscura* groups that have been examined in sufficient detail. It has not been found in the *repleta* group in subgenus *Drosophila* (Sullivan et al., 1990) or in the *willistoni* group (Carew, 1993, and unpublished observations). However, it is found in the outgroup to the genus, *Scaptodrosophila lebanonensis* (Juan et al., 1994). Whether *Adhr* has been lost in some lineages within *Drosophila*, or whether it has moved and just not been found, remains to be determined.

Adh and *Adhr* display different patterns of molecular evolution. In both *D. melanogaster* (Kreitman and Hudson, 1991) and *D. pseudoobscura* (Schaeffer and Miller, 1993), *Adh* has a higher level of nucleotide polymorphism than does *Adhr.* Kreitman and Hudson suggest that a selective sweep may have occurred in the recent history of

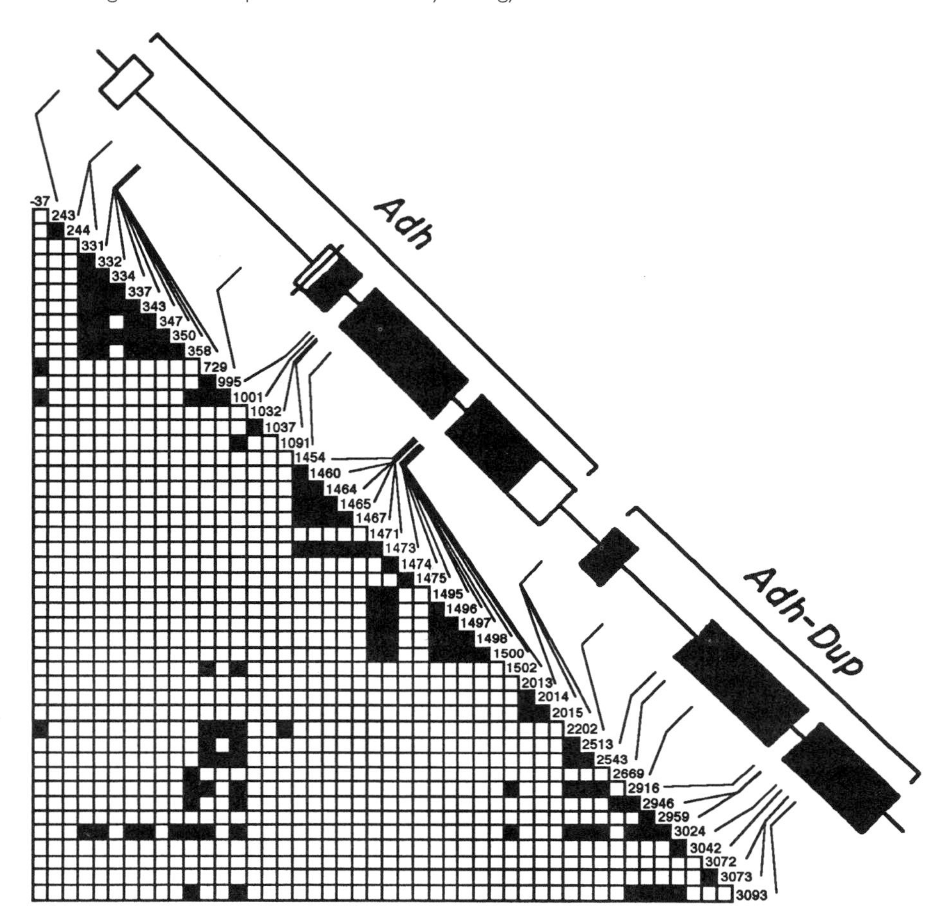

FIGURE 10-21 Analysis of nonrandom associations among pairs of 359 segregating sites in the *Adh* region of *D. pseudoobscura* based on a sample of 99 alleles. The only sites shown are those that exhibited some degree of nonrandom association by Bonferroni tests. The black boxes indicate individual pairs that show a significant association, with a Fisher exact probability of <0.05. Note that most of the sites exhibiting associations fall into introns, especially for *Adh*. From Schaeffer and Miller (1993).

Adhr in *D. melanogaster.* Schaeffer and Miller, on the other hand, show that if they remove the clustered linkage disequilibrium sites from the data set, *Adh* and *Adhr* are not significantly different in level of variation in *D. pseudoobscura.*

Juan et al. (1994) point out that in comparing *S. lebanonensis* to the *Sophophora, Adh* has more replacement substitutions (23.2%) than does *Adhr* (16.3%), but fewer silent substitutions (29.8% for *Adh* and 32.5% for *Adhr).* The latter observation is consistent with the codon usage bias of the two genes: *Adh* is more biased than *Adhr.* In a similar study, Albalat et al. (1994) show for several pairs of species that the replacement substitutions for *Adh* and *Adhr* are not significantly different, but that silent replacements are statistically fewer in *Adh* than in *Adhr.*

The Origin of *Adh*

Alcohol dehydrogenases of various sorts have been studied in a large number of organisms. In comparing *Drosophila* ADH (nonitalicized capitals refer to the protein product) to other organisms, it is clearly very unusual, exhibiting almost no similarity to any other ADH. Evidently, *Adh* arose independently in *Drosophila.* The origin of the gene may have been partly explained by the observation that it is fairly similar to another *Drosophila* gene called *P6* that is also highly expressed in the fat body (Rat et al., 1991). Both P6 and ADH have similarity to short-chained dehydrogenases known mostly from bacteria and mammals. Discounting the amino acid methionine, which seems to have been selectively enriched in P6, P6 and ADH are only slightly more different on the amino acid level than are ADH and ADHR. However, if one does not discount the methionines, ADH and P6 are only 28% similar on the amino acid level, whereas 30% is generally agreed to be the level at which true homology can be safely inferred.

Jingwei: A Reincarnation?

We mentioned that the second copies of *Adh* found by Jeffs and Ashburner (1991) were potential pseudogenes in *D. yakuba* and *D. teissieri*; these second copies have the appearance of processed pseudogenes, as they lack all introns and are on a different chromosome. Long and Langley (1993) challenged this view by proposing that the second copy has actually become part of a new functional gene they called *jingwei* after an ancient Chinese tale of reincarnation. The evidence for the continued function of this second copy came initially from polymorphism studies; Long and Langley found a ratio of 27 : 4 silent : replacement polymorphisms in this sequence in the two species, a pattern typical of active genes. They also showed this second copy is transcribed, and forms the 3′ end of a transcript; that is, it "captured" a 5′ end of another gene, and *jingwei* is a chimeric new gene with potentially an added 77 amino acids attached to the ADH protein. The chimeric transcript is very similar in the two species, so presumably the capturing occurred shortly after the transposition and before the species separated.

Two facts cloud the interpretation that *jingwei* is an active gene. First, the recent evidence of Sullivan et al. (1994) that (for unknown reasons) true pseudogenes can display higher rates of silent substitutions than replacements weakens the strength of that evidence. Second, no protein corresponding to the *jingwei* transcript has been detected. The functionality of this gene remains an open question, but it is a most intriguing situation.

Protein Polymorphism

The F/S polymorphism of *D. melanogaster Adh* has been repeatedly cited in this book as one of the most convincing cases for selection affecting an enzyme polymorphism. Recall also that in chapter 6 we discussed strong evidence that an *Adh* allozyme polymorphism is selectively maintained in *D. mojavensis.* Further discussion of the evidence for *D. melanogaster* is presented here.

Perhaps the strongest evidence indicating selection in natural populations is the fact that latitudinal clines exist in both the Northern and Southern hemispheres in the same direction, namely a high frequency of F in higher latitudes and S in lower latitudes (Oakeshott et al., 1982). This is in contrast to DNA polymorphisms (other than that causing the F/S difference) in and around *Adh*, which are geographically very uniform (Kreitman and Aguadé, 1986a; Simmons et al., 1989). Berry and Kreitman (1993) have done a detailed analysis of comparing the F/S protein polymorphism and nucleotide variation in a north-south cline along the east coast of the United States. Figure 10-22 summarizes their observations. As in previous studies, the F allele increases with increasing latitude (the upper part of figure 10-22), but the other polymorphic sites are quite uniform geographically, with the exception of Site 5 (lower part of figure 10-22). Site 5 in the first intron is a complex insertion/deletion polymorphism in strong linkage disequilibrium with the F/S polymorphism (Laurie et al., 1991) and is likely involved in controlling level of gene expression (discussed in more detail in chapter 11). In any case, the evidence is that migration is great enough to homogenize presumed neutral molecular polymorphisms, but not the selected F/S site.

On the microgeographic level, there is evidence that *D. melanogaster* populations living in and around wineries may display distribution differences for the F/S *Adh* polymorphism. In particular, the higher activity of ADH-F was found to be associated with the high level of ethanol found in wineries. Follow-up studies on several wineries have clouded this simple picture. McKechnie and Geer (1993) review these studies.

We have already discussed the evidence from laboratory studies that different F-S genotypes of *D. melanogaster Adh* have differential tolerance to ethanol (chapter 4) and also how the HKA test strongly implies balancing selection at the exact position causing the electrophoretically detected protein polymorphism. In the next chapter we will take up gene regulation variation for this gene. For now we will turn to biochemical properties of the allozymes.

Biochemistry

If selection can detect amino acid polymorphisms in enzymes, then one expects that the allozymes should have a phenotypic effect that might be detectable by biochemical analysis. This is the case with the ADH-F and ADH-S allozymes of *D. melanogaster*. The kinetic properties of the allozymes differ (Heinstra et al., 1987; Chambers, 1988), as do the thermostabilities (Vigue and Johnson, 1973). The amount of protein produced by F alleles is about double that of S alleles, and thus the ADH activity differs among genotypes (Laurie-Ahlberg, 1985).

While it is satisfying that variation can be detected in the test tube for such properties, it would be even more satisfying to be able to measure physiological effects in vivo. It has been demonstrated that different *Adh* genotypes exhibit different ethanol-to-lipid flux in living organisms (Geer et al., 1993; Heinstra, 1993; Freriksen et al., 1994a, b). The results are complex and difficult to relate to viability and fertility studies, as they depend on stage of selection and level of ethanol in the environment. Figure 10-23 is an attempt to summarize, somewhat simplistically, the results. FF genotypes have a higher ADH activity than SS genotypes. When environmental ethanol is in low concentration (around the K_m), neither survival nor fecundity is greatly

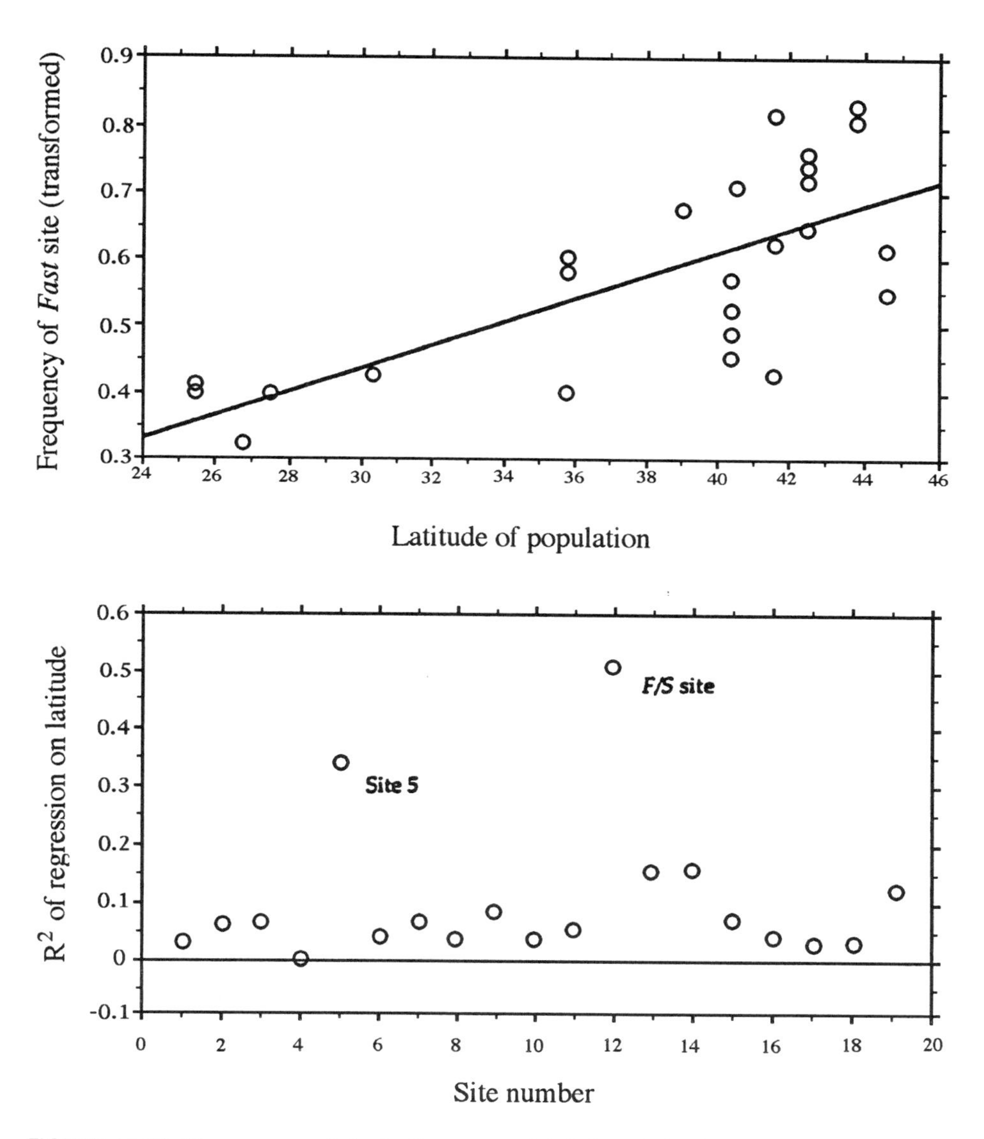

FIGURE 10-22 The upper graph is the frequency of *Adh* fast allele along a north-south transect of the east coast of the United States. The lower graph shows the correlation coefficient for individual segregating sites versus latitude. The highest correlation is for the site causing the protein polymorphism, F/S, plus one site in the 5′ intron that displays linkage disequilibrium with the F/S site. From Berry and Kreitman (1993).

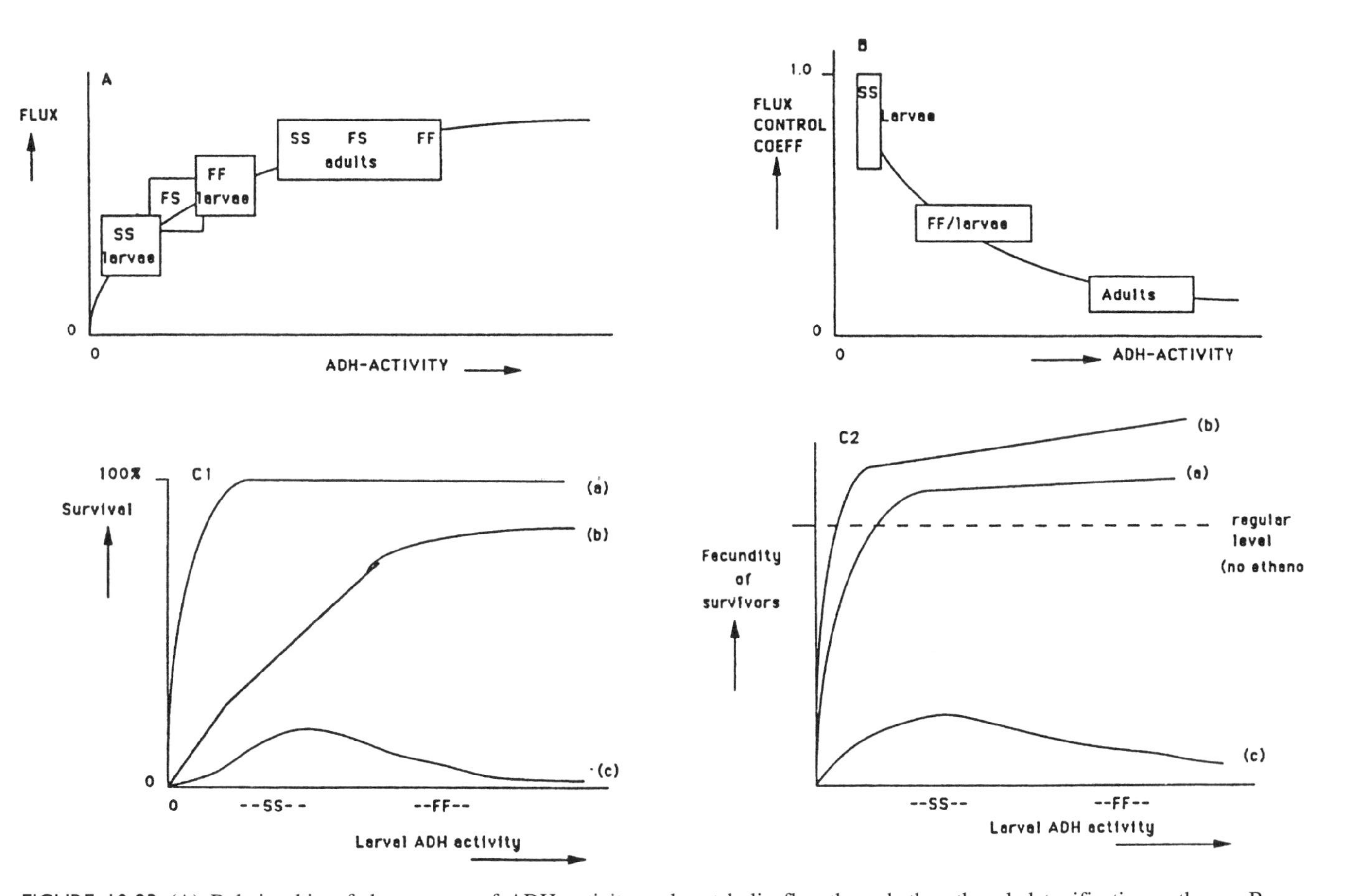

FIGURE 10-23 (A) Relationship of the amount of ADH activity and metabolic flux through the ethanol detoxification pathway. Boxes indicate ranges of wild-type alleles. (B) Relationship between flux control and ADH activity. (C1) Survival and (C2) fecundity related to ADH activity. Flies were reared in medium with three different concentrations of ethanol. Line (a) is ethanol concentration about equal to the K_m of ADH; line (b) is ethanol concentration about 20 times K_m; and line (c) is ethanol concentration much higher. From Heinstra (1993).

different between genotypes (line (a) in the figure); note that the fecundity of both genotypes is higher than in the case with no ethanol, because the conversion of ethanol to lipids allows females to develop more eggs. As ethanol increases to about $20 \times K_m$, FF have a greater survival rate due to more efficient conversion of the somewhat toxic ethanol. FF genotypes also have a slight edge in fecundity due to better utilization of ethanol as a nutrient. However, when ethanol reaches concentrations $\gg K_m$ (condition (c) in the figure), neither genotype has good survival, although SS is somewhat better as it is less efficient in converting ethanol to acetaldehyde, which is toxic if produced faster than the next enzyme. Aldehyde dehydrogenase can convert it to acetate, so at such concentrations a high activity of ADH may be detrimental. This aspect becomes a bit murky, as it is not clear if aldehyde dehydrogenase functions in larvae or only in adults.

In any case, these in vivo studies do indicate that the genotypes of *D. melanogaster Adh* produce proteins that differentially affect metabolism. Such case studies of selection, from the nucleotide polymorphism through metabolic effects to environmental factors, are rare, and despite all the effort placed in this single, relatively simple polymorphism, some questions still remain.

Other Genes

The number of *Drosophila* genes and gene systems studied from a molecular evolution perspective is too great to discuss each in detail. Rather, a description of the highlights of studies that illustrate particularly interesting patterns will be briefly mentioned and references given to allow the interested reader entrance into the literature.

Amylases

We have mentioned these genes in the context of duplicated genes, pseudogenes, and their association with inversions. One aspect not yet mentioned is the evidence for concerted evolution and gene conversion.

Popadic and Anderson (1995a) have evidence of gene conversion of *Amy* among the gene arrangements of the third chromosome of *D. pseudoobscura.* While the RFLP studies on a 26-kb region in and around *Amy* showed strong disequilibrium with inversions and allowed the construction of a tree consistent with the inversion phylogeny, more detailed DNA sequence data of the region revealed a smaller, 2.7-kb region, embedded in the larger region, that has evidently been transferred between gene arrangements. In particular they found this 2.7-kb segment to be identical in Hidalgo (HI) and Estes Park (EP), yet this sequence was about 0.6% divergent from TL. HI and EP are both derivatives of TL (figure 3-3). Thus in the absence of gene exchange, one would expect HI and EP to be at least as diverged as each is from TL. Given the relatively small region shared and the fact that it is embedded in a larger diverged region that produced the expected inversion phylogeny, gene conversion rather than double crossover would seem to be the more likely explanation.

In a similar study, Popadic and Anderson (1995b) provide strong evidence for gene conversion among the copies of *Amy* on the same chromosome. The original Brown et al. (1990) sequence data of the three copies indicated very little divergence, on the order of 0.5% or less. The data of Popadic and Anderson (1995b) argue strongly that this is due to concerted evolution caused by gene conversion. The coding sequence of the duplicate copies, *Amy2* and *Amy3*, in ST, CH, and SC are virtually identical to one another and to *Amy1*, the active gene found in all gene arrangements. However, outside the coding region, *Amy2* and *Amy3* are considerably diverged from *Amy 1*, indicating gene conversion is only for coding regions. Phylogenetic analysis indicates the data are consistent with a scenario of *Amy1* giving rise once to *Amy2*, with *Amy3* being a duplicate of *Amy2*.

Hickey et al. (1991) present evidence for gene conversion (or at least concerted evolution) between copies of *Amy* in the *melanogaster* subgroup. In particular, in comparing the two copies in *D. melanogaster* to those in *D. erecta*, there are 39 substitutions in the coding regions; the two copies in *D. erecta* share 38 out of 39 differences. Assuming that the duplication occurred in an ancestral lineage before *D. melanogaster* and *D. erecta* arose, the paralogs are clearly not evolving independently. Shibata and Yamazaki (1995) have confirmed that all species of the *melanogaster* subgroup have two copies of *Amy* confirming the duplication occurred prior to the speciation events. Furthermore, they show that the concerted evolution occurs only for the coding region. The similarity of this observation with that for *Amy* in *D. pseudoobscura* (Popadic and Anderson, 1995b) is striking, in particular the fact that the concerted evolution and gene conversion events occur only for the coding region. This suggests that gene conversion may be occurring via an mRNA intermediate.

Using DNA sequence analysis, the amino acid basis for allozymic (electrophoretic) variation for *Amy* in *D. melanogaster* was studied by Inomata et al. (1995). They show that sequential accumulation of single amino acid changes that cause a charge change can account for the electrophoretic mobility shifts.

Histones

Histones are among the most evolutionarily conserved proteins and are also among the most highly expressed genes in *Drosophila.* Fitch and Strausbaugh (1993) have studied the molecular evolution of the tandemly repeated five-histone genes in *D. melanogaster* and *D. hydei.* Two findings are of particular interest. As with other organisms, very few amino acid substitutions have occurred in the 50 million years since the two species shared a common ancestor. Four of the gene copies are the core for the histone complex and only two amino acid differences exist, yielding a K_a (average number of replacement substitutions per replacement site) of <0.001. The average number of synonymous substitutions per synonymous site, K_s, is 1.17 ± 0.12. Thus the highly conservative amino acid sequence has not inhibited considerable silent substitution; this rate of synonymous change would place histones in the moderately rapidly evolving genes as characterized by Moriyama and Gojobori (1992; see figure 10-5).

The second surprise is that codon bias in histones is quite low despite the high abundance of the protein in all cells. The codon usage bias index, CAI, is 0.39 and 0.35 for *D. melanogaster* and *D. hydei*, respectively. Shields et al. (1988) found this index to be 0.31 for the 15 lowest-biased genes and 0.44 for the 15 medium-biased genes, so histones fall between these two categories. The G+C content of silent positions in histones is very nearly 50%. This contrasts to 62% G+C in low-bias genes to 80% in high-bias genes (Shields et al., 1988). This is also a strikingly odd aspect of these highly expressed genes. Of course the efficiency of translation associated with particular codons may be compensated for by the multiple copies of histone genes, so that more mRNA can be produced.

In a study of intraspecific variation among the clusters of histone genes, Colby and Williams (1993) found evidence that the multiple copies are not identical within *D. melanogaster.* They digested 73 naturally occurring arrays of the genes with restriction enzymes that the consensus DNA sequence indicated should not cut the cluster. Almost 70% of the clusters had some variation, as indicated by minor bands. Of most interest was the observation that the variants were distributed in the arrays inversely proportional to distance along the chromosome; if a variant had spread, it had a 92% probability of being in its nearest neighbor. These observations give hints about the mechanism of spread of copies in a multigene cluster.

Chorions

Chorions are another gene cluster that stand in almost complete contrast to the histones. Chorion genes produce the proteins used to build the eggshell; in all species examined, four copies of the gene are fairly tightly linked in the same order and are selectively amplified in the ovaries (Martinez-Cruzado et al., 1988, and references therein). In comparing five species of Hawaiian *Drosophila*, Martinez-Cruzado (1990) observed about twice as many replacement substitutions as silent substitutions, a nearly unique situation. As a generality, when comparing closely related species (before saturation has set in), silent substitutions outnumber replacement substitutions, but the range of this ratio can be extremely variable as witnessed by the histones, about 1000 : 1, and chorions, 1 : 2. (The chorion genes will be discussed further in the next chapter, as they have also been studied with regard to evolution of gene regulation.)

Opsins

Vision in animals depends on light-absorbing rhodopsins, the protein portions of which are called opsins. In *Drosophila* there is a family of four opsin genes called *Rh1, Rh2, Rh3*, and *Rh4.* Each opsin confers slightly different spectral properties on the rhodopsins, each with a characteristic absorption maximum. The amino acid sequences of these genes are highly conserved, being more than 90% identical between the subgenera *Drosophila* and *Sophophora* (Neufeld et al., 1991; Carulli et al., 1994). Physiologically, the functions (i.e., absorption spectra) of the opsins have remained conserved as well (Carulli et al., 1994), implying that all species "see" about the same set of colors. The species studied were *D. melanogaster, D. simulans, D. pseudoobscura, D. virilis*, and *D. mercatorum.* Given the diverse ecologies of these species (chapter 5), such conservation of visual sensitivity is somewhat surprising.

One might also predict that genes that are so important, conservative, and closely related in terms of function, would evolve in similar manners. However, Carulli and Hartl (1992) showed that each of the four genes is evolving in its own manner, with differences in the relative rates of silent and replacement substitutions, base composition, and codon usage bias. Such results simply emphasize our ignorance of mechanisms of molecular evolution that generate the patterns observed.

Transformer—A Sex-Determining Gene

Sex determination in *Drosophila* is a well-studied subject that has uncovered several features of interest (reviewed in Baker, 1989). One gene involved in sex determination is called *transformer* or *tra*, the protein product of which is involved in RNA splicing of the transcript of another gene, *doublesex*; in turn, the transcript of *tra* is alternatively spliced by a protein product of another gene, *Sex-lethal.* The determination of sex, therefore, depends on the alternative splicing of transcripts in the different sexes.

One might predict that the amino acid sequence of a protein involved in alternatively splicing an RNA, as well as having its own transcript alternatively spliced, would be very constrained in evolution. The opposite has been observed by O'Neil and Belote (1992): *tra* genes sequenced in four species showed greater divergence at the amino acid level than any other genes for those species. Between three species of the *melanogaster* subgroup and two members of subgenus *Drosophila* (*hydei* and *virilis*) there was only about 35% amino acid conservation. For 12 other genes compiled by O'Neil and Belote, the amino acid divergence between the subgenera ranges from 2 to 45%, nowhere near the 65% divergence for *tra.* Even more unexpected is the fact that a *D. melanogaster tra*-null is rescued by transformation with the *D. virilis* gene!

Given such an unexpected pattern of interspecific evolution, Walthour and Schaeffer (1994) undertook an intraspecific polymorphism study by sequencing 10 alleles from a single U.S. population of *D. melanogaster.* Contrary to what would be expected from the interspecific study, *tra* was found to be remarkably nonpolymorphic. Only two silent mutations were detected among the 10 alleles, making *tra* among the least polymorphic genes in the species (table 10-2). Something very strange is occurring at this locus. Walthour and Schaeffer (1994) prefer the selective sweep explanation; for example, a variant of *tra*, or one closely linked, recently underwent a fixation in the population studied. Clearly, this case needs to be studied in the ancestral African populations.

Genes Within Genes: *Gart, dunce* and *sina*

The *Gart* locus encodes proteins with three enzymatic activities in the purine pathway. The largest intron in *Gart* is about 4.1 kb and has been found to contain another gene that is transcribed in the opposite direction of the *Gart* transcript (Henikoff et al., 1986). This imbedded or "nested" gene produces a transcript of about 900 bp and has a 71-bp intron, so there is also an intron within an intron. The gene is expressed at the prepupal stage and encodes a pupal cuticle protein, so it is not at all related to the *Gart* gene. The *Gart* locus is expressed throughout the life of the fly, whereas the cuticle protein gene is developmentally regulated. However, the highest level of expression of *Gart* coincides with the pupal cuticle protein expression. Henikoff and Eghtedarzadeh (1987) speculated the physical arrangement may have something to do with gene control. Partly, they based this on the fact that *D. pseudoobscura* also has the same nested arrangement of these genes, and therefore this evolutionary conservation is consistent with some functionality (a not particularly compelling argument).

Moriyama and Gojobori (1989) found sufficient similarity of sequence of the nested pupal cuticle protein gene to larval cuticle protein genes to suggest homology. Using silent rates and assuming a molecular clock, they estimated the larval and pupal genes diverged from one another about 70 million years ago. The finding of the nested pupal gene in *D. pseudoobscura* but not in *Chironomus* (Clark and Henikoff, 1992) is consistent with the nested organization being older than 25 million years and younger than 100 million years.

The second example of a gene within a gene is more complex. This involves the *dunce* or *dnc* locus (reviewed in Davis and Dauwalder, 1991; Nighorn et al., 1994). This gene was first identified as a learning-deficient mutant and later shown to encode cAMP phosphodiesterase. It is one of the largest genes known in *Drosophila*, with at least 19 exons spread over about 148 kb and nine polytene chromosome bands. Five different transcripts are produced, composed of a common 3′ end but with different 5′ exons. At least seven unrelated genes have been found in the largest 73-kb intron (Chen et al., 1987; Furia et al., 1990, 1993). Among these genes is one of the salivary gland secreted (*Sgs-4*) proteins, which acts as a glue for the pupae, and a pre-intermolt gene (*Pig-1*). Different genes are transcribed in different directions.

Given the huge size of the *dunce* gene and its intron, it is not inconceivable that the nested genes transposed into the gene. On the other hand, the speculation of Davis and Dauwalder (1991) that *dunce* may have "swallowed" the other genes when it recruited more 5′ upstream promoters would seem to be a more attractive hypothesis, or at least more graphic.

The third example of a nested gene is the *seven in absentia* (*sina*) gene, which occurs in the intron of one of the opsin genes, *Rh4* (Neufeld et al., 1991). In this case the two genes may have functional relationships: *Rh4* is expressed exclusively in the R7 photoreceptor cells and *sina* was named such because mutants in the gene cause an absence of R7 photoreceptor cells. However, this intimate association seems not to be crucial to the genes' functions, as in *D. virilis* this nested organization does not exist; the *Rh4* and *sina* genes are widely separated in this species (Neufeld et al., 1991).

Whatever the evolutionary origin of these remarkable cases, they clearly violate one of the major tenets of classical transmission genetics, stemming from Sturtevant's original demonstration that genes are linearly arrayed along a chromosome.

Period

In the context of speciation (chapter 7) we already discussed the remarkable *per* locus and its possible role in controlling species-specific male courtship songs, at least in the *melanogaster* subgroup. Recall that the gene has a section near the middle that consists of Thr-Gly repeats and that the region just downstream from this repeat region is implicated in the species-specific differences between *D. melanogaster* and *D. simulans* (Wheeler et al., 1991). The Thr-Gly repeat region is the most variable part of the gene from both an intra- and interspecific perspective, although it must be pointed out that the rest of the gene has not been well studied in the *melanogaster* subgroup.

On the intraspecific level, both *D. melanogaster* and *D. simulans* vary in the length of the Thr-Gly region. *D. melanogaster* has two major forms consisting of 17 pairs $(\text{Thr-Gly})_{17}$ and 20 $(\text{Thr-Gly})_{20}$ (Costa et al., 1991, 1992; Peixoto et al., 1992). These authors have argued that this repeat is analogous to minisatellites, and the repeating unit induces high rates of mutation. They have evidence that the length variants have arisen independently more than once, based on the observation that the same length variant can be associated with different flanking sequences. Most interesting is that the frequencies of the two major alleles vary clinally in Europe, with the (Thr-Gly)$_{17}$ allele predominating in lower latitudes and (Thr-Gly)$_{20}$ in higher latitudes (Costa et al., 1992). There is some evidence that the Thr-Gly repeat region does confer different thermal stability on the circadian phenotypes (Ewer et al., 1990).

D. simulans displays a similar polymorphism for length of the Thr-Gly repeat region, but for somewhat longer variants with between 21 and 25 Thr-Gly pairs (Rosato et al., 1994). One major difference is that in *D. melanogaster* the region downstream from the Thr-Gly repeat is monomorphic with regard to replacement substitutions, whereas in *D. simulans* three replacement polymorphisms exist in this region. They are in strong linkage disequilibria between these downstream variants and the length variants. Also unlike *D. melanogaster*, no clines or other obvious geographic patterns could be detected among the eight samples from Europe and North Africa. Other polymorphism data for this locus in *D. simulans* (and its close relatives) is in Kliman and Hey (1993a).

The part of the gene corresponding to the Thr-Gly repeat in *D. melanogaster* is also highly variable among species (Peixoto et al., 1992, 1993). Figure 10-24 illustrates this variation. This region ranges from 19 to 209 amino acids, although not all species have Thr-Gly repeats in this interval; for example, *D. pseudoobscura* has a degenerate five-amino-acid motif (Colot et al., 1988). Interestingly, the difference in length of this region correlates with the amino acid differences in flanking sequences independently of phylogenetic relatedness. Peixoto et al. argue this may be due to molecular coevolution between the flanking regions and repeat regions.

Molecular variation in the *per* locus has been studied in two other species groups. Ford et al. (1994) studied variation in the *D. athabasca* complex and could find no evidence that variation in *per* is implicated in mating song differences among the semispecies. Gleason (1996) found no intraspecific polymorphism in the Thr-Gly repeat region in the *D. willistoni* group, nor are there any differences among species within this group. Considering Dipterans outside the Drosophilidae, Nielson et al. (1994) also conclude that *per* is quite conserved. Thus the pattern of variation in *per* can vary considerably from group to group. Outside the *melanogaster* subgroup, there is as yet little support for the involvement of this gene in speciation.

species	N-terminal conserved region	variable length region	C-terminal conserved region
virilis	EGSGGSGSSGNLTTASNVRMSSVTNTSNTGTG	19	TLTEILLNK
mediostrata	...........F..G.................	-22-	....S....
immigrans	...........F..G.............A...	-22-	....S....
mohavensis	...........F..G.............A...	--24--	....S....
saltans	...........F..G..IH.........A...	---25---	....S....
willistoni	...........F..G..LH.........A...	----27----	....S....
robusta	...........F..G..I..........A...	-----45-----	....S....
melanogaster	...........F....IH........IA...	------66------	....S....
ananassae	...........F..G...H.........A...	-------84-------	....S....
serrata	..........HF..G...H.........G...	--------100--------	....S....
pseudoobscura	....A.....H..SG..IH...A.....A...	----------209----------	S...S....

FIGURE 10-24 The structure of *period* (*per*) in several species. Only the region in the vicinity of the Thr-Gly repeat unit of *D. melanogaster* is shown. Considerable homology exists on either side of the variable region, whereas the length of the sequence between the conserved region varies from 19 to 209 amino acids. In some species this variable region is composed of Thr-Gly repeats, but not in others. From Peixoto et al. (1993).

11

Development

For tho' th' immortal
species never dies,
Yet ev'ry year new maggots
make new flies

J. Dryden, 1696

The relationship between the development of organisms and their evolutionary history has long been a focus of biologists; witness Haekel's (1879) famous dictum "ontogeny is a recapitulation of phylogeny." Such concerns were part of mainstream biology well into the present century, most notably in de Beer's (1930, 1940) books. De Beer transcended Haekel's work in not only relating development to the history of life, but using development to understand mechanisms of evolution. (The notion that development could provide any meaningful insights into evolutionary history fell into disrepute by the mid-20th century [Gould, 1977].) While some attention continued to be paid to development and evolution, it is only the last 20 years that have witnessed a flourishing of experimental work attempting to connect rules of development to mechanisms and patterns of evolution. The stimulus for the reawakening of the field must be credited to the late A. C. Wilson, who, in the mid-1970s (e.g., King and Wilson, 1975; Wilson et al., 1977), documented that the degree of morphological differences among species was not always in accord with the amino acid differences in their proteins. The implication was that while the same "building blocks" (proteins) are used by related organisms, what makes them phenotypically (morphologically, physiologically, behaviorally, etc.) distinct is the "blueprint" of how the building blocks are used. A dichotomy was drawn between "regulatory genes" and "structural genes," the former being invoked as the important stuff of adaptive evolution. Since even today we do not understand in detail how gene regulation occurs in eukaryotes, much less whether there is a significant dichotomy between protein-coding structural genes and genes that regulate their expression, speculation of this sort provoked much more discussion and debate than precise empirical testing.

It is only today that we are beginning to vaguely discern the kinds of mechanistic connections that exist between development and evolution. For general treatments of several of these issues, the books of Raff and Kaufman (1983), Buss (1987), and Raff (1996) remain indispensable; for a detailed treatment of one particularly well-understood insect example, the book on development of butterfly wings by Nijhout (1991) is highly recommended. In this chapter I focus more narrowly on *Drosophila* and out of necessity, some other insect examples. With *Drosophila* serving as one of the most actively researched models of development as well as evolution, it is safe to predict this fly will be an important bridge in closing the circle of knowledge. However, it must be admitted at the outset of this chapter that the circle is far from closed. Of all the chapters in this book, this will be the least satisfying.

Broadly, we can divide studies of development and evolution into two research agendas. The first is the study of changes in gene expression: How have homologous genes changed their quantitative, temporal, and spatial expression through evolutionary time? The second is the direct study of changes in developmental programs: When we see changed morphology, how have developmental programs changed to produce this result? Neither of these questions has been definitively answered in *Drosophila* (or in any other organism), but progress is being made on both fronts.

Gene Regulation

Broadly, gene regulation can be divided into three concerns: quantitative (How much of a gene product is produced?), temporal (When is the gene actively producing a product?), and spatial (Where is the gene active?). Along with the mechanism(s) controlling gene expression, the bases of development must lie in the answers to these three questions. Phenotypic change through evolutionary time, inasmuch as it is dependent on change in development, must also rely on changes in gene regulation. Studies of variation in gene regulation, both within species and between species of *Drosophila*, have been done to a limited degree. Reviews covering these topics are Laurie-Ahlberg (1985) and Dickinson (1991).

Adh

One of the first obvious polymorphisms in level of enzyme activity was noted for the F/S alcohol dehydrogenase polymorphism in *D. melanogaster*. Generally, carriers of the F allele have three to five times the Adh enzyme activity as do carriers of the S allele; figure 11-1 illustrates the activity of 49 homozygous second chromosome lines, the chromosome on which *Adh* resides. While there is some overlap, clearly F/F homozygotes exhibit higher activity than do S/S homozygotes; not shown is the fact that F/S heterozygotes have intermediate activity. When only observing overall enzyme activity, it is not clear if differences are due to (a) catalytic differences among allozymes or (b) different amounts of the allozymes being present. If (b) is true, this could be due to (1) different levels of mRNA transcription and/or stability, (2) different efficiency of translation, or (3) different stability of the enzyme. In a long series of experiments, C. C. Laurie and colleagues have examined these questions for *Drosophila Adh* (Laurie-Ahlberg and Stam, 1987; Laurie and Stam, 1988; Laurie et al., 1990, 1991; Choudhary and Laurie, 1991). The conclusion is that the difference between ADH activity in carriers of F and S alleles are due to both (a) and (b), that is, both difference in catalytic activity of the proteins and level of gene product. The difference in catalytic activity was shown by transformation studies to be due exclusively to the F/S single amino acid Thr/Lys difference.

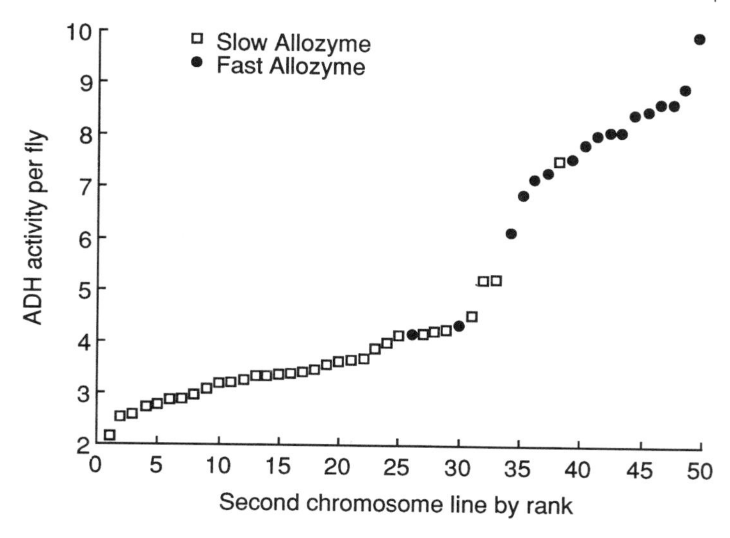

FIGURE 11-1 Alcohol dehydrogenase activity in 47 lines of *D. melanogaster*, each with a different homozygous second chromosome from a natural population, but identically homozygous for all other chromosomes. From Aquadro et al. (1986).

F/F homozygotes also produce about 1.5 times the ADH protein that S/S homozygotes do. This is not due to a difference in level of mRNA (Laurie and Stam, 1988). Part of the difference in protein level was shown, again by in vitro mutagenesis and transformations, to be associated with an insertion/deletion polymorphism in the first intron, that which separates the distal and proximal promoters (figure 11-2; Laurie and Stam, 1994). Because mRNA levels are not affected and the polymorphism is in an intron, it would seem most likely the differences are due to RNA processing differences between alleles. This polymorphism accounts for about one third of the difference in protein amounts, so it is not the sole factor involved. It is important to note

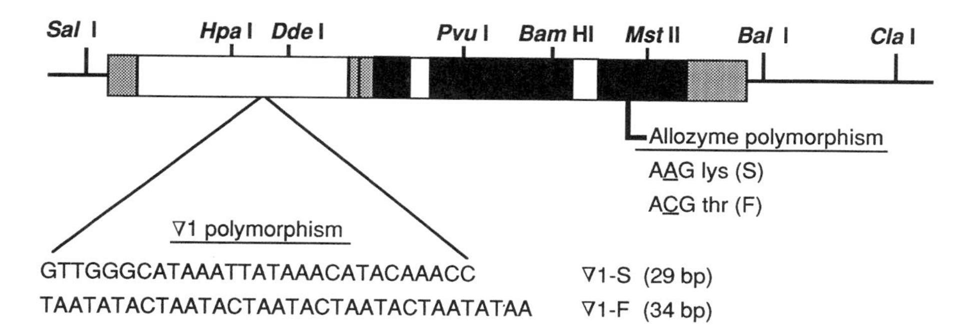

FIGURE 11-2 Schematic map of the *D. melanogaster Adh* locus with an indication of the polymorphism—∇1—that causes a change in level of enzyme. From Laurie et al. (1991).

that this indel polymorphism is in the exact position exhibiting the clines associated with latitude which parallels the F/S allozyme cline (site 5 in figure 10-22; Berry and Kreitman, 1993).

A second insertion in the first intron, that designated delta 2 by Kreitman (1983), is associated with variation in larval ADH activity in a winery population of *D. melanogaster* in Australia (Matthew et al., 1992). This polymorphism has not been associated with adult differences in ADH levels, but it is in linkage disequilibrium with F alleles worldwide.

Naturally occurring extreme variation in expression, null alleles, are also associated with indels in intronic regions of *D. melanogaster Adh* (Gibson et al., 1992). Null alleles caused by a deletion of 438 bp in the second intron persisted in a Tasmanian natural population for at least three years; this deletion is not associated with any signs ("footprint") of transposable element activity. The null activity is due to disruption of normal mRNA processing (Freeth et al., 1990). A second null allele in the Tasmanian population is associated with a 320-bp insertion in the first intron.

D. simulans is fixed for an allele differing from the *D. melanogaster* S allele by two amino acids and three amino acids from the F allele. In a comparison of several strains of each species (all *D. melanogaster* strains were homozygous for S), Thomson et al. (1991) found considerable variation among strains for level of enzyme activity throughout the life cycle, with the two species overlapping in ADH activity level at the larval and pupal stages. However, at the adult stage, *D. melanogaster* strains consistently displayed about twice the activity of *D. simulans.* This is due to about a two-fold higher level of ADH enzyme in the former species, but about equal amounts of *Adh* mRNA. Thus some factor such as increased efficiency of processing, efficiency of translation, and/or greater enzyme stability must account for the differences in the species. Further light is shed on this case by interspecific transformation experiments in which *D. melanogaster Adh* genes are placed in *D. simulans*, and the reciprocal (Laurie et al., 1990). The results are that the *D. simulans Adh* in *D. melanogaster* has a higher level of expression than it did in its parent species; likewise, *D. melanogaster Adh* in *D. simulans* shows a lower level of expression. This indicates that transacting factors are produced by the genetic background and that the control of level of *Adh* expression is not solely due to tightly linked cis-acting factors such as the intronic indel polymorphism discussed above. This is consistent with a study of McDonald and Ayala (1978) showing considerable intraspecific variation in ADH protein level in *D. melanogaster* associated with unlinked genetic factors, especially on the third chromosome (see also Mercot, 1994).

On the other hand, evidence of cis-regulatory elements causing interspecies differences comes from studies of hybrids. In *D. melanogaster/simulans* (Dickinson et al., 1984), the *virilis* group (Ranganayakulu et al., 1991), and the Hawaiian *Drosophila* (Dickinson and Carson, 1979; Dickinson, 1983; see later section for more detail), *Adh* alleles in hybrids most often take on the expression pattern of their parental species, especially in regard to tissue-specificity. This is observed when the two parental ADHs have different electrophoretic mobility (isozymes); each isozyme is expressed in the hybrid as it is in the pure parental background.

Given the evidence that many spontaneous mutations are due to transposable elements (discussed in chapters 2 and 9), it is of interest to know if the variation in expression of *Adh* is affected by transposable elements; several are known to be in the vicinity of the locus in strains from natural populations (figure 2-4). The answer to this query is yes and no. Aquadro et al. (1990) studied 47 lines of *D. melanogaster* that originally all started with a single second chromosome. After 300 generations representing some 28,200 allele generations, the lines were examined for variation in level of ADH activity; all other chromosomes were made identical so all differences among lines were due to the accumulation of mutations on the second chromosome. The amount of activity variation among strains approached that seen in surveys of natural populations. No RFLP variation could be detected in the 13-kb region around the locus, meaning that no transposon movements had caused the variation in level of expression. On the other hand, Dunn and Laurie (1995) studied a line of *D. melanogaster* from a Rhode Island population that showed unusually low ADH activity, the strain represented by the leftmost point in figure 11-1. This line was unusual in that it had a Copia insertion in the 5′ flanking region. Removal of this TE, followed by reinsertion by transformation, led to a threefold increase in ADH activity. Clearly, here was a case where a TE caused variation in expression of a gene.

In the previous chapter we discussed the situation in some species of the *repleta* group in which two active copies of *Adh* exist (figure 10-15). In the present context, the most interesting aspect of this case is the fact that the distal and proximal promoter functions have been assumed by the single proximal promoter. That is, in most species, the primary transcript with the proximal promoter is expressed in larvae and the distal promoter transcript in adults. The distal promoter has been lost in the duplication event found in some species such as *D. mojavensis*. The gene expressed in adults, *Adh-2*, now has an evolved proximal promoter (Atkinson et al., 1988). In other words, a promoter homologous to a larval promoter has assumed the analogous function of an adult promoter.

Variation among species in the embryonic stage and spatial distribution of expression of *Drosophila Adh* was studied by Ranganayakulu (1994). While some transient differences were noted among species, this study is more notable in being one of the few to explore the use of in situ hybridization to mRNA to study interspecies regulatory differences.

Other studies of interspecific variation in overall levels of expression of *Adh* have been performed with *Drosophila*: McDonald and Avise (1976), David et al. (1979), David and Herrwege (1983), and Holmes et al. (1980). The most extensive such study is that of Mercot et al. (1994), who found a 65-fold variation in ADH activity among species, *D. melanogaster* having the highest level of the 67 species studied. The relative ability to utilize ethanol as a resource was inferred from the ADH activity on ethanol compared to isopropanol. This ratio increases independently several times in *Drosophila* phylogeny and is only partially correlated with survival on high levels of ethanol. Laboratory studies on larval choice of media with various levels of alcohols was reviewed in chapter 4 (pp. 134–135).

In the foregoing, only studies directly relevant to evolutionary considerations are mentioned. *Drosophila Adh* has also served as a model system to study the molecular mechanisms of gene regulation (recent discussions with relevant references are Abel et al., 1993 and Falb et al., 1992). Clearly, connection of the two types of studies is highly desirable.

Amylase

α-Amylases were discussed in the previous chapter as an example of a gene–enzyme system particularly prone to gene duplication (pp. 399 and 416). Variation in this system has also been studied in regard to repressability and temporal and spatial patterns of expression.

Glucose is known to repress amylase expression in *Drosophila.* In *D. melanogaster* the effect may be as much as a hundred-fold and appears to be due to repression of mRNA transcription (Benkel and Hickey, 1986a, 1986b). Different strains of this species are differentially susceptible to dietary repression of *Amy* (Yamazaki and Matsuo, 1984; Hickey et al., 1989), indicating naturally occurring genetic variance for this character. Yamazaki and Matsuo (1984) found a high correlation among strains of *D. melanogaster* for inducibility/repressability of *Amy* and fitness on the inducing media (glucose and starch) but not on standard medium. On the interspecific level, all eight species of the *melanogaster* subgroup appear to have similar degrees of glucose repression for α-amylase (Payant et al., 1988). *D. ananassae* (Da Lage and Cariou, 1993) and *D. busckii* (Stamenkovic-Bojic et al., 1994) display intermediate levels of sensitivity to dietary repression. The beetle *Tribolium* does not display glucose repression of amylase activity (Hickey et al., 1989). Thus there may be a series of insects displaying the evolutionary acquisition of this regulatory change.

The primary site of expression of amylase in *Drosophila* is the midgut. Abraham and Doane (1978) demonstrated polymorphism within *D. melanogaster* with regard to the location along the midgut where *Amy* is expressed, and that this polymorphism was at least partly under simple Mendelian control (further reviewed in Doane et al., 1990). One locus, designated *mapP*, controls the activity in the posterior midgut and maps about 1 cM from the structural gene (Klarenberg et al., 1986). Thompson et al. (1992) have shown that *mapP* coordinately controls the level of enzyme activity, amount of AMY protein, and level of *Amy* mRNA. Furthermore, they confirmed that the dietary effect of glucose is to repress mRNA. This led to the suggestion that much (or even all) of the variation among strains in the reaction to dietary factors may be due to strains having different alleles at the *mapP* locus.

The possibility of correlations (due to coevolution?) of structural gene variation and regulatory elements controlling the gene's expression was investigated with the *Amy* system in *D. pseudoobscura.* First it was demonstrated that the expression of *Amy* along the midgut in this species is under strong genetic control, although specific loci could not be identified (Powell and Lichtenfels, 1979). A survey of ten natural populations revealed considerable variation in both the amylase midgut expression and *Amy* allozymes (Powell, 1979), although there was no correlation between the two polymorphic systems. This was true at the individual and population level. An interspecific survey of related *obscura* group species also did not reveal correlations between allozyme differentiation and expression patterns (Powell et al., 1980). Replicate laboratory populations maintained on media in which the only carbohydrate source was maltose or starch did diverge significantly in frequency of midgut patterns of expression in a repeatable manner, although the *Amy* allozymes sometimes did (Powell and Andjelkovic, 1983) and sometimes did not (Powell and Amato, 1984) also respond. The latter results were interpreted as evidence that the expression patterns were under selection in the different environments, but that the allozyme polymorphism was sensitive to particular combinations of strains (i.e., genetic background effects). Other studies on *Amy* allozymes with *D. pseudoobscura* (e.g., Yardley et al., 1977) reached similar conclusions. The most important conclusion from this set of experiments is that there is no evidence for coordinated evolution between *Amy* structural gene variants (allozymes) and the genetic factors controlling its expression.

D. subobscura has also been studied for both geographic variation in *Amy* allozymes and midgut patterns of expression (Andjelkovic et al., 1987; 1991). Studies of laboratory populations on starch and maltose media (Stamenkovic-Radak et al., 1987; Stamenkovic-Radak and Andjelkovic, 1992), as well as direct studies of fitness components (Marinkovic et al., 1984), have been carried out on this species. The results are very similar to those seen in *D. pseudoobscura.*

Somewhat different results have been obtained from *D. melanogaster.* Klarenberg and Scharloo (1986) did find correlations between *Amy* allozymes and midgut pattern of expression in flies from three laboratory populations originating from different localities. De Jong and Scharloo (1976) also found fitness differences among allozymes dependent on media used in laboratory populations. Clark and Doane (1984) provide evidence that, in *D. melanogaster*, apparent selection at the *Amy* locus is also dependent on the linkage phase with alleles at another locus, called *adipose.*

D. ananassae has four active copies of *Amy*: two tightly linked copies are designated 1 and 2, and two other tightly linked copies are designated 3 and 4 (Da Lage et al., 1992). One block is on the 3L (= *melanogaster* 2R, the location of *Amy* in that species) and one on the 2L (= *melanogaster* 3R). Unlike the case in *D. melanogaster*, the isozymes in this species show considerable temporal variation in expression (Da Lage and Cariou, 1993). The 1-2 block is expressed predominantly in the larval stage and the 3-4 block predominantly in the adult stage. There also appears to be some isozymic differences along the midgut, with the 1-2 block being almost entirely confined to the anterior midgut (Da Lage and Cariou, 1993). Thus, while *D. ananassae* is in the *melanogaster* group, it has evolved a very different set of regulatory patterns for amylase than *D. melanogaster.* It is difficult to imagine an ecological factor as both are, ancestrally, tropical fruit-breeders. Other closely-related species in the *ananassae* subgroup display the same two blocks of amylase genes in the same chromosomal locations, although the number of gene copies and the level of allozymic variation is higher in *D. ananassae* (Da Lage et al., 1992; Cariou and Da Lage, 1993).

Two conclusions can be drawn from these results with *Drosophila* amylases. First, there is a wealth of intra- and interspecific variation in gene regulation. This holds for quantitative, spatial, and temporal expression patterns, as well as sensitivity to dietary factors. Some of this variation has a fairly simple genetic basis. Second, despite the relative simplicity of the system, few clear evolutionary patterns or principles have yet been evident, except that there is little or no evidence that regulatory changes are correlated with amino acid changes in the protein.

Glucose Dehydrogenase (*Gld*)

D. Cavener and colleagues (reviewed in Cavener, 1992) have carried out a series of studies on the tissue- and sex-specific expression of *glucose dehydrogenase, Gld.* The function of GLD is not completely understood, but it apparently plays an important role in eclosion from the pupal case; in some unknown manner it weakens the pupal wall, allowing the imago to emerge (Cavener and MacIntyre, 1983). In the larval stage, the distribution of GLD enzyme activity is fairly general and conserved among a large number of species examined and is particularly strongly expressed in the anterior spiracular gland (Cox-Foster et al., 1990). Using a *lacZ* fusion construct, Gunaratne et al. (1994) demonstrated that the expression in this tissue is due to a palindrome conserved among species. During pupal metamorphosis, GLD is also expressed in the rectal papillae of *D. melanogaster* but not in *D. pseudoobscura* or *D. virilis.* Using transformation, this expression difference is localized to a TTAGACCA motif found in *D. melanogaster* but not in the other two species. Interestingly, the *D. melanogaster Gld* null is a lethal that is rescued by the *D. pseudoobscura* gene, even though the transgene is not expressed in the rectal papillae.

Among 50 species examined, adults display a surprising level of variation in expression pattern of *Gld*, mostly in organs associated with reproduction (Schiff et al., 1992). In males the variation is in expression in the ejaculatory duct and bulb. All females have GLD activity in spermathecae and vaginal plate, but vary in expression in the oviduct, seminal receptacle, and parovaria. The species differences have been particularly well-studied between *D. melanogaster* and *D. pseudoobscura* using transformation (table 11-1). As noted in this table, both sexes display interspecific differences in expression of GLD. In *D. melanogaster* carrying the *D. pseudoobscura* gene, the gene is often expressed as in the native species, although lines varied somewhat, indicating that chromosomal position also has an effect. Analogous experiments involving transplantation of imaginal discs that give rise to the relevant tissues produced similar results (Cavener, 1985). Taken together, these experiments indicate cell-autonomous, cis-, and trans-acting factors are causing the species differences in expression of this gene.

In an examination of the seven species of the *melanogaster* subgroup, Ross et al. (1994) found that males of all species except *D. teissieri* expressed GLD in the ejaculatory ducts. This species lacks three TTAGA regulatory elements found in *D. mela-*

TABLE 11-1 Glucose dehydrogenase expression in *D. melanogaster*, *D. pseudoobscura*, and *D. melanogaster* transformed with the *D. pseudoobscura Gld* gene.

Species/Strain	Sex	GLD Expression in Adults	Number of Transgenic Lines with Pattern
D. melanogaster	Male	ED	
	Female	SP, SA, & SRA	
D. pseudoobscura	Male	EB	
	Female	SP, SA, PO, SRA, & SR	
Transgenic flies	Male	ED	2
(*D. melanogaster* with		ED & EB	3
D. pseudoobscura gene)	Female	SP, SA, PO, & SRA	3
		SP, SA, PO, SRA, & SR	2

ED = ejaculatory duct; EB = ejaculatory bulb; SP = spermathecae; SA = spermovarial atrium; SRA = seminal receptacle atrium; SR = seminal receptable; OD = oviduct.

From Cavener (1992).

nogaster, which had been shown by transformation to be necessary for expression in the ejaculatory duct (Quine et al., 1993).

In addition to its role in eclosion, Cavener (1992) speculates that GLD may also play a role in sperm storage, as females deficient in GLD have low fertility. GLD is also transferred by the male to females during copulation (Cavener and MacIntyre, 1983). In this regard, the role of *Gld* is similar to that of *Est5/6*, which is also expressed in the male ejaculatory duct of *D. melanogaster* and transferred to females at copulation. One might think there would be some common regulatory sequence for the two genes, but none have been identified. In addition, transformation studies with *Est5/6* (Brady and Richmond, 1990) indicate this gene is regulated differently from *Gld.*

Esterases

Esterases are an example of gene duplication in the genus *Drosophila* (chapter 10). Studies have shown that duplicate copies have probably acquired different functions, as indicated by their tissue-specific expression. In particular, *Est6/C* probably was ancestrally expressed only in the hemolymph. In some species, the hemolymph retains the activity, but esterase activity is also acquired in the male ejaculatory duct. Other species have duplicate copies that no longer are expressed in the hemolymph, but only in the ejaculatory duct. Yet other species have duplicate copies expressed in the ejaculatory duct, but the gene product has lost esterase enzymatic activity. Oakeshott et al. (1995) review this literature.

Hawaiian *Drosophila*

W. J. Dickinson and colleagues have carried out the most detailed systematic survey for gene regulation variation in a single group of *Drosophila*, the Hawaiian Picture-wing *Drosophila.* The initial survey was for four enzymes encoded by five loci across 27 species (Dickinson, 1980a); later work added an additional six enzymes (Thorpe et al., 1993). The survey consisted of dissecting 13 different tissues (six larval, seven adult), running them on electrophoretic gels, and staining for activity. The relative expression of each enzyme in a tissue was estimated relative to a standard diluted by factors of two; thus only changes of at least two-fold in level of expression were recorded.

In the first study, Dickinson (1980a) conservatively estimated that about 30% of the patterns of expression differed in one or more species. Most differences were quantitative, although some were qualitative in the sense of presence or absence of activity in a particular tissue ("absence" refers to no detectable staining for activity, which was estimated to mean in excess of a 100-fold reduction). At least three cases of intraspecific polymorphism for expression were also found. In the several cases examined, differences in level of enzyme staining were shown to be due to a difference in level of enzyme protein rather than in catalytic efficiency (Dickinson, 1980b).

Three of the species studied produce viable F_1 hybrids. In a series of studies, Dickinson (1980b, 1980c; Dickinson and Carson, 1979) determined the expression pattern in hybrids. Of 17 traits differing between hybrid pairs, 11 were shown to express a pattern identical to the parent from which each allele was derived; that is, cis-acting control was found. In three cases intermediate patterns of expression were found in hybrids, indicating some cis-control with added complexity. In the final three cases, one parental expression pattern was dominant, that is, both alleles in the hybrid were expressed as in one of the parents. This indicates trans-acting control.

Fang et al. (1991), Wu and Brennan (1993), McKenzie et al. (1994), and Hu et al. (1995) have studied the expression of Hawaiian fly-derived *Adh* transformed into *D. melanogaster*, revealing further complexity of control of expression. Perhaps the most important conclusion from these studies is that a control element may not have a fixed function; rather, the effect is very much dependent on context. Furthermore, in some cases a pattern of expression has been conserved while the control element has changed. Thus there is no one-to-one relationship between the sequence of a control element and the tissue in which it is expressed.

Thorpe et al. (1993) reasoned that the Hawaiian flies may show a particularly high level of gene regulatory variation because they are fast evolving, as indicated by the considerable morphological variation and large number of species produced in a relatively short time. While we suggested in chapter 8 that the Hawaiian *Drosophila* probably originated before the extant islands appeared, most of the radiation of the Picture-wing group has likely occurred in the last five million years, producing about 100 species. They chose the *virilis* group as a contrasting lineage that has produced only 12 described species in an estimated 10–20 million years (Spicer, 1991); morphologically, *virilis* group species are very similar, to the extent that specimens have often been misidentified (Throckmorton, 1982b). The degree of regulatory difference among species of the *virilis* group was significantly less than for the Picture-wing flies. Figure 11-3 shows the distribution of divergences for the two groups. In this graph a trait is

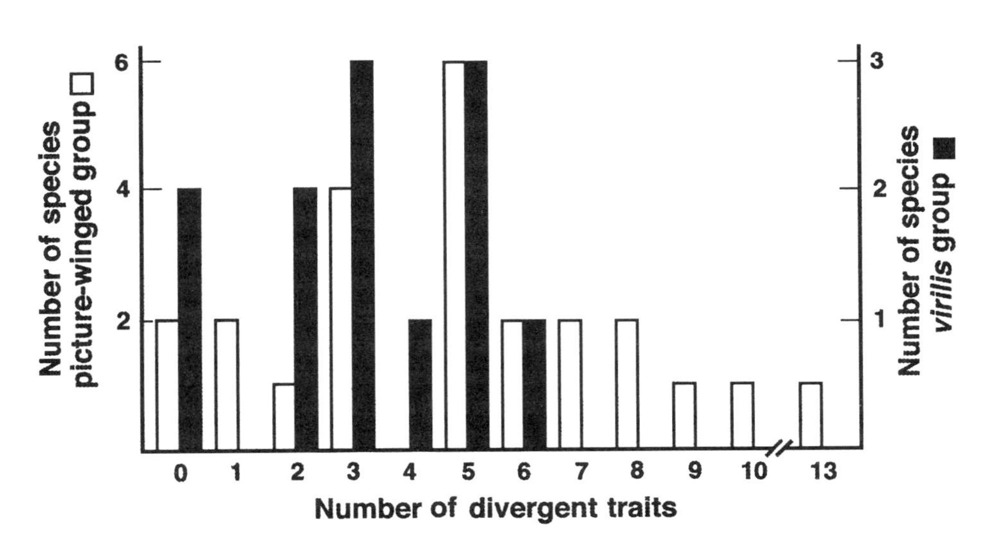

FIGURE 11-3 Distributions of the numbers of divergent traits in studies of enzyme expression in the Picture-wing group of Hawaiian *Drosophila* (open bars) and *virilis* group species (solid bars). On average, many fewer traits are divergent among the phylogenetically older, but evolutionarily more conservative, *virilis* species than in the Hawaiian flies. From Thorpe et al. (1993).

the expression of an enzyme in a particular tissue; there were 65 traits and five loci in 13 tissues. A trait is considered divergent if it varies more than ±1 step from the consensus for the group; a step is measured by the two-fold dilutions. Two loci, *Adh* and *Ao2* (aldehyde oxidase), accounted for almost all differences between groups.

The second issue addressed by Thorpe et al. (1993) was whether different classes of enzymes differed in levels of regulatory divergence. Recall that in chapter 2 (p. 40) we discussed the fact that in allozyme studies, enzymes having multiple substrates and/or substrates originating externally to the fly had more variation than enzymes having single substrates acting in intermediary metabolism. In the original study of Hawaiian flies and for the *virilis* group, all the enzymes studied were of the former type: *Adh, Odh* (octanol dehydrogenase), *Xdh* (xanthine dehydrogenase), and *Ao* (aldehyde oxidase, two loci). The additional six loci coding for enzymes in intermediary metabolism (malate dehydrogenase, malic enzyme, hexokinase, α-glycerophosphate dehydrogenase, and hydroxybuterate dehydrogenase) had levels of variation in tissue expression indistinguishable from the original set of enzymes. Thus, unlike the case for the amino acid variation in the proteins, no consistent patterns in regulatory variation could be detected between classes of enzymes categorized by the heterogeneity of substrates.

Egg Proteins

During *Drosophila* oögenesis, large amounts of protein are produced both to provision the developing embryo and to package it to protect it from environmental insults. The expression of the major proteins is both temporally and spatially specific and at very high levels. Thus they have been favored systems for understanding gene regulation; they have also received some attention from an evolutionary perspective. The two major groups of proteins are those comprising the eggshell or chorion and the vitellogenins or yolk proteins.

Chorion genes in *D. melanogaster* exist in two clusters, one on the X and one on the third chromosome; these two clusters are conserved in all species examined (see Kafatos et al., 1987; Orr-Weaver, 1991 for reviews). Two things happen to these genes during oögenesis. First they are amplified, the X cluster about 20-fold, the third chromosome cluster 60- to 80-fold. Second, they are temporally expressed, most notably in the follicular epithelial cells in ovaries. The proteins are excreted extracellularly and layered onto the developing egg. In addition to providing protection, they must also allow for gaseous exchange. (Kambysellis [1974, 1993] provides excellent photomicrographs of the chorions of a number of species, especially the Hawaiian fauna, and relates different morphological features to ecological factors of the oviposition substrate of a species.)

From an evolutionary perspective, there are two notable features of *Drosophila* chorion genes. The first was mentioned in chapter 10 (p. 417), namely that chorion genes accumulate replacement substitutions at a faster rate than silent substitutions (Martinez-Cruzado, 1990). This indicates these proteins are relatively flexible in amino acid composition with regard to performing their functions; perhaps there is even positive selection for protein changes. Perhaps more remarkable is the extreme conservation of the gene regulation mechanism for these genes. DNA sequence analysis of the third chromosome chorion gene cluster in *D. melanogaster, D. subobscura, D. virilis*, and *D. grimshawi* indicated extensive divergence in introns and coding regions, but considerable conservation in some 5′- and 3-flanking sequences (Martinez-Cruzado et al., 1988; Fenerjian et al., 1989). Subsequent transformation studies placing the *D. grimshawi s18* chorion gene into *D. melanogaster* produced flies that amplified the *grimshawi* sequence and expressed the gene in the correct temporal and spatial manner (Swimmer et al., 1990; 1992); both positive and negative control of expression was associated with only 200 bp 5′ to the coding region. Recall that *D. grimshawi* is a Hawaiian fly diverged from *D. melanogaster* for about 50 my.

F. C. Kafatos and colleagues also studied chorion genes from moths, diverged for 200 to 250 my from *Drosophila.* While still clustered, moths (specifically *Bombyx mori, Antheraea pernyi*, and *Antheraea polyphemus*) have two major families with many copies of each gene (reviewed in Kafatos et al., 1987). The structures of the genes are very different from those in *Drosophila* and they are not amplified during oögenesis. It is thought that the slower egg development (two days in moths compared to five hours in *Drosophila*) and the higher number of gene copies do not necessitate amplification in moths. Despite all these differences, when moth chorion gene promoter regions were attached to a bacterial reporter gene and transformed into *D. melanogaster*, the reporter gene was specifically expressed in the choriogenic tissue at the correct stage of oögenesis; approximately 300 bps of the 5′ region was sufficient (Mitsialis et al., 1987; 1989). One such moth promoter region was shown to be bidirectionally effective in that, when transformed into *D. melanogaster*, it drove the tissue-specific expression of reporter genes both 3′ and 5′ to it (Fenerjian and Kafatos, 1994).

Vitellogenin or yolk protein genes are expressed in the same cells as chorions at a slightly earlier time, as well as in female fat bodies (Petri et al., 1976). Three genes, designated *Yp1, Yp2*, and *Yp3*, reside in two areas of the X chromosome of *D. melanogaster* and all nine other species that have been examined (Craddock et al., 1983; Kozma and Bownes, 1986). The vitellogenin proteins produced in the fat body are excreted and taken up by the ovaries. When active, the genes can produce tremendous amounts of protein, despite the fact that they are not amplified. For example, in *D. grimshawi* fat bodies can produce up to 2 μg protein/hr and ovaries 4 μg/hr, for a total of 152 μg/day, fully 2% of the weight of the fly (Kambysellis et al., 1989). Gene expression appears to be under hormonal control (Postlethwait and Shirk, 1981). Vitellogenin synthesis seems to coincide often with female sexual receptivity (Kambysellis and Craddock, 1991), and the presence of mature males may even enhance the expression of the genes (Craddock and Boake, 1992). The genes have been isolated from *D. melanogaster* and *D. grimshawi.* They display both conservation and differences in structure (intron/exon arrangements), and differ in size, but are conserved in position. In both species *Yp1* and *Yp2* are close, with *Yp3* more distant; all remain on the X chromosome (Hatzopoulos and Kambysellis, 1987a; 1987b; 1988). Which genes are active, at what time, and in which tissues varies among species; furthermore, the three genes are coordinately regulated in *D. melanogaster* but not in *D. grimshawi* (Hung and Wensink, 1983; Kambysellis et al., 1986).

Martinez and Bownes (1992) have studied the uptake by ovaries of the vitellogenin protein secreted by fat bodies into the hemolymph. The uptake is quite specific in that only vitellogenin proteins are sequestered by ovaries, so that a specific receptor must be involved. They injected radioactively labeled vitellogenin from one species into another species and then determined whether the radioactivity was taken up by the ovaries of the new host. They studied 13 species of *Drosophila* and five other dipterans. They found all species took up the foreign proteins, indicating conservation of the recognition system throughout the cyclorrhaphan diptera.

A third set of proteins associated with egg production is vitelline membrane proteins, also synthesized in follicular epithelium of mature ovaries; they form the first layer of the eggshell, under the chorions. Four genes have been isolated for these proteins, all on the second chromosome (Lindsley and Zimm, 1992). Virtually no work from an evolutionary standpoint has been done on this system.

Although not fitting under the heading gene regulation, it is of interest to note in the context of oögenesis and evolution that several interspecific ovarian transplant studies have been performed. The most interesting result is that in mainland (i.e., non-Hawaiian) species, only very closely related species support the development of a foreign ovary. However, in Hawaiian *Drosophila*, distantly related species can serve as hosts, indicating a more uniform internal environment in these more rapidly evolving species (Kambysellis, 1970).

Morphology and Embryogenesis

The past 20 years have seen tremendous advances in our understanding of *Drosophila* development, especially embryonic development. The literature is truly voluminous and often highly technical. Attempts to connect this knowledge to evolutionary concerns has been much more limited. Before discussing such studies, a very brief treatment of *Drosophila* embryogenesis and development will be provided. For a somewhat more detailed, but highly readable, discussion the reader can consult Slack (1991); for an intermediate level of detail, Lawrence (1992) is recommended; and for those wishing the most extreme detail, consult the two volumes (>1500 pages) edited by Bate and Martinez-Arias (1993) plus an atlas of development with lovely detailed color drawings by Hartenstein (1993). To paraphrase Slack (1991), the description that follows is for readers with less than 8 megabytes of RAM; it is sufficient for understanding the evolutionary studies.

Drosophila Early Development

When an oöcyte is mature, it passes from the ovaries down the oviduct to be laid; along the way, a sperm stored in the seminal receptacle is released and enters the egg through a special hole in the chorion, called the micropyle. At normal temperature (e.g., 20–25°C), embryogenesis takes place very rapidly and, in *D. melanogaster*, a larva hatches in about 24 hours. One very important aspect of this early development is that there are several nuclear divisions in the absence of cell membrane formation. After about 2.5 hours and nine nuclear divisions, the nuclei migrate to the periphery and form the acellular syncytial blastoderm. After four more divisions, cell membranes begin to form the cellular blastoderm. One other important point about early *Drosophila* development is that the zygotic nuclei begin transcriptional activity very early, within an hour or two of fertilization, unlike many animals that rely on maternal messages for much of their early development (Davidson, 1986).

As with all development, the major problem is to understand how an initially undifferentiated egg can give rise spatially to specific structures. What are the signals that tell cells in the anterior to form anterior structures, and so forth? How do the signals get sent and received? It is in this realm that the genetic control of these early decisions is best understood in *Drosophila*. The stimulus for much of this progress was the now-classic work of Nüsslein-Volhard and Wieschaus (1980) who undertook a mammoth screen of mutations affecting early embryogenesis. The number of complementation groups found was neither very small nor very large, around 150. That is, the evidence is that there are about 150 genes (or more accurately, complementation groups) that are crucial to setting up early embryonic patterns. On the order of one hundred of these have been molecularly isolated and studied to some degree. What follows will be concerned with those that have been best-studied and/or have been examined from an evolutionary perspective; the treatment is not encyclopedic. In an undifferentiated zygote, two major axes can be defined: one distinguishing top from bottom (dorsal from ventral) and one distinguishing front from back (anterior from posterior). The genes involved in these patterns can also be divided into two groups, those involved in the initial polarization set up during oögenesis (maternally acting genes) and those transcribed from the nuclei of the embryo itself (zygotic genes). Table 11-2 lists some genes involved in these patterns.

The key to understanding *Drosophila* development is the repeating segment concept. It has long been recognized that insects consist of segments, each with its own identity. It is thought insects evolved from segmented worm-like animals, with the successive specialization of segments from initially undifferentiated to highly specialized (see below). The crucial steps in insect development are first, to set up the segment system in the embryo, and second, to tell each segment what it should become. We will confine ourselves to the anterior-posterior system. Figure 11-4 illustrates the expression patterns of genes involved in the earliest steps. In the oöcyte, two maternal genes can be singled out as crucial to establishing anterior and posterior, *bicoid* (*bic*) and *nanos* (*nos*). The mRNA and protein of *bic* exists in a gradient extending from the anterior pole to about 40% the length of the egg, or in the standard designation, region 60–100% EL (egg length). The egg is divided into percentage of length starting posteriorly (figure 11-4). At the earliest stages, *nos* product is primarily mRNA, which is concentrated at the very posterior end; later, protein is produced, also concentrated posteriorly. Several other anterior-posterior maternal-effect genes exist.

The first set of genes transcribed from the zygote are called gap genes. They begin to subdivide the developing embryo on a finer scale (figure 11-4). The maternal genes are crucial in activating and deactivating the gap genes. In this regard, the *bicoid* gene product acts like a classical morphogen with a concentration gradient that differentially activates genes along the gradient. The *nos* protein is somewhat different in only activating *knirps* (*kni*) and inhibiting the translation of *hunchback* (*hb*) in the posterior region. We need not go into much more detail, but from figure 11-4, one can see that the embryo is becoming more differentiated as the gap genes set up their region-specific patterns.

The next set of genes, called pair rule genes, actually set up the segments. They were called pair rule because mutants in them caused the deletion of every other segment in the embryo. The domain of activity of four of these genes is illustrated in figure 11-5. Two of these genes, *fushi tarazu* (*ftz*) and *even skipped* (*eve*), are ex-

TABLE 11-2 Genes that have been identified in controlling early *Drosophila* development.

Gene	Mutant Phenotype	Molecular Nature
Maternal dorsal–ventral system		
dorsal (*dl*)	Extreme dorsalization, embryo becomes tube of dorsal epidermis	Nuclear protein
Toll (*Tl*)	Similar to *dl*, gain of function alleles ventralize	Cell surface protein
Snake (*snk*)	Similar to *dl*	Serine protease
torpedo (*tor*)	Ventralizing	EGF receptor
Zygotic dorsal–ventral system		
decapentaplegic (*dpp*)	Loss of aminioserosa	Signaling molecule
zerknullt (*zen*)	Loss of aminioserosa and optic lobe	Homeobox gene
single minded (*sim*)	Loss of certain neurons	Nuclear protein
Maternal anterior–posterior axis		
bicoid (*bic*)	Deletion of head and thorax	Homeobox gene
nanos (*nos*)	Absence of abdomen	
pumilio (*pum*)	Most abdomen missing	
Bicaudal (*Bic-D*)	Symmetrical double abdomen	Maybe cytoskeletal
Zygotic anterior–posterior gap class		
huckerbein (*hkb*)	Defects in termini	
hunchback (*hb*)	Deletion labium, thorax, A7a/8p	Zinc-finger
tailless (*tll*)	Head skeleton reduced; hindgut and malphigian tubules absent.	Steroid receeptor family
giant (*gt*)	Defects or absence of labrum, labium, A5-7	Leu zipper
Krüppel (*Kr*)	Deletion thorax, A1-5, inverted repeated A6	Genetic regulator
knirps (*kni*)	Replacement of A1-7 by single A type	Genetic regulator
Zygotic pair-rule class		
fushi tarazu (*ftz*)	Deletion of even parasegments	Homeobox
even skipped (*eve*)	Deletion of odd parasegments	Homeobox
hairy (*h*)	Deletion of odd parasegments, production of denticles, loss of labral tooth	Helix-loop-helix
runt (*run*)	Deletions > one segment centered on T2, A1, 3, 5, 7, with mirror duplication	Nuclear protein
Zygotic segment polarity		
engrailed (*en*)	Ventral cuticle continuous lawn of denticles	Homeobox
wingless (*wg*)	Ventral cuticle continuous lawn of denticles indicating deletion of 3/4 of each segment	Signaling molecular
patched (*ptc*)	Deletes posterior half of each denticle belt, replaces with mirror copy of anterior half	Cell receptor
gooseberry (*gsb*)	Ventral cuticle continuous lawn of denticles formed by deletion of posterior half of each segment	Homeobox
Homeotic selector genes		
Deformed (*Dfd*)	Deletion of mandibular and maxillary segments	Homeobox
Distal-less (*Dll*)	Null mutants do not develop appendages	Homeobox
Sex combs reduced (*Scr*)	Transforms PS3 to PS4 and PS2 to PS1	Homeobox
Antennapedia (*Antp*)	Transforms PS4, 5, to PS3	Homeobox
Ultrabithorax (*Ubx*)	Transforms PS 5, 6 to PS4	Homeobox
abdominal A (*abd-A*)	Transforms PS 7-9 to PS6	Homeobox
Abdominal B (*Abd-B*)	Transforms PS 10-14 to PS9	Homeobox

Note: The table is far from complete, but contains the genes most worked on, especially from an evolutionary perspective. A followed by a number refers to an abdominal segment; PS refers to a parasegment (see figure 11-5).

Modified from Slack (1991).

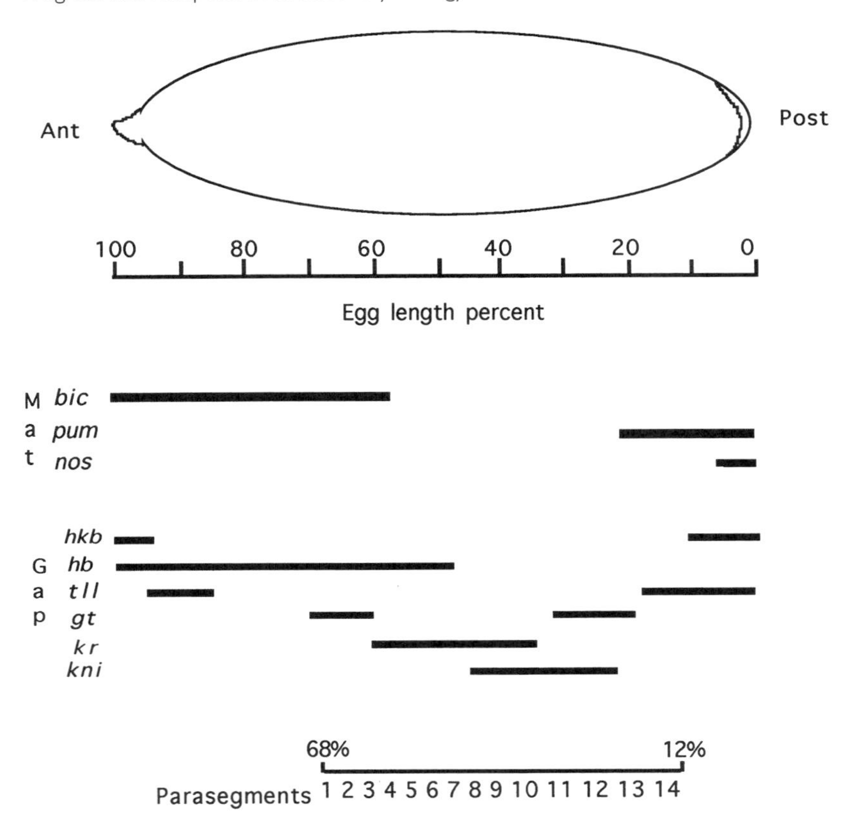

FIGURE 11-4 Schematic of the distributions of gene products for three maternal genes and six gap genes in early *D. melanogaster* development. Abbreviations are defined in table 11-2 with a brief description of their properties. Note that the standard designation of egg goes from right to left, posterior to anterior. The parasegments in figure 11-5 are derived from the part of the egg indicated in the bottom of this figure.

pressed in alternate parasegments. Mutants of *ftz* result in the deletion of even-numbered parasegments, and mutants of *eve* result in the loss of odd-numbered parasegments (not what the name suggests, due to a switch in numbering after the gene was characterized).

It is important to be clear on the concept of segments and parasegments. Segments refer to the adult structures and parasegments are defined as embryonic regions of differentiation. Fourteen parasegments can be identified in embryos and there are at least 16 adult segments. This varies depending on how many segments are considered to be in the head; evolutionarily, probably seven worm-like segments evolved into advanced insect head parts. See Schmidt-Orr et al. (1994) for recent evidence for seven head segments based on patterns of gene expression. As can be seen in figure 11-5, the embryonic parasegment and adult segments are offset by half a segment such that parasegment 1 (PS1) affects the posterior half and anterior half of the adja-

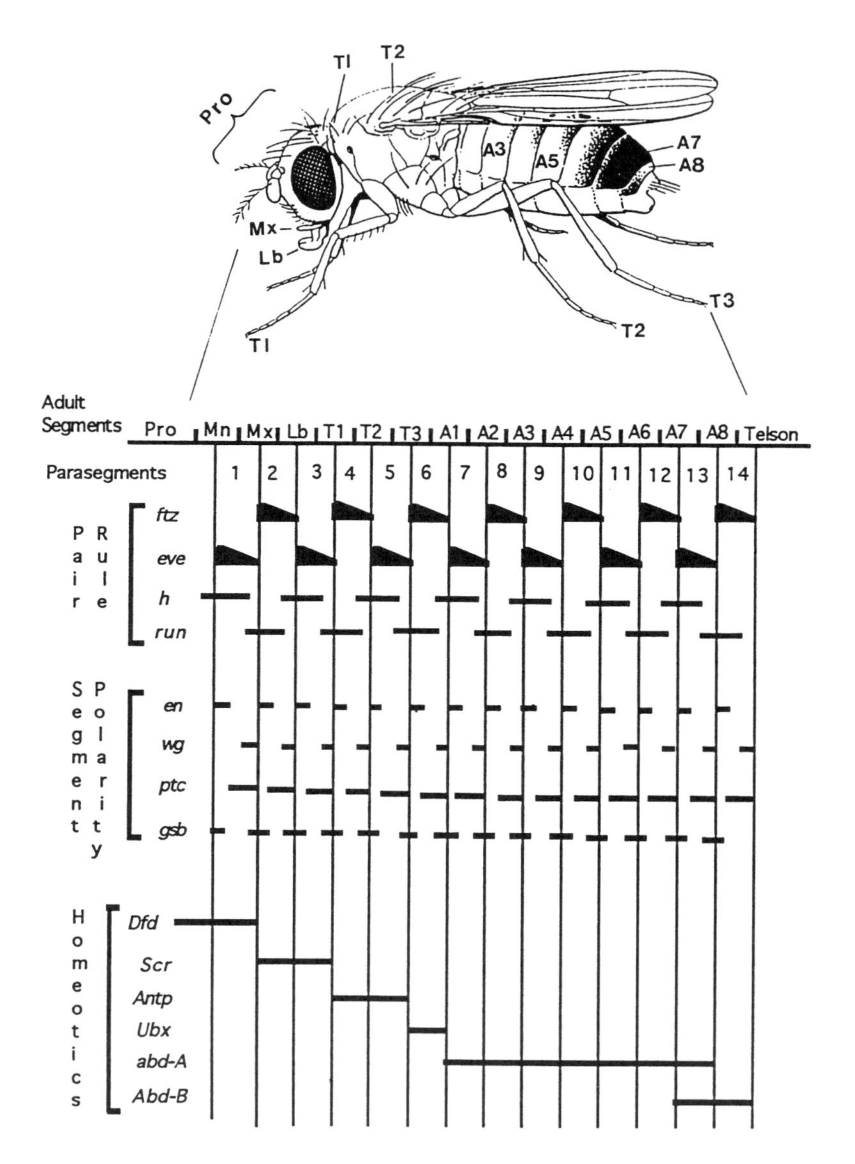

FIGURE 11-5 Schematic of expression of several pair rule, segment polarity, and homeotic selector loci in later stages of embryogenesis. In the case of *ftz* and *eve* there is a clear gradient in level of expression, as denoted by the triangles. Gene abbreviations are defined in table 11-2. Segments: Pro = procephalon, Mn = mandibular, Mx = maxillary, Lb = labial, T = thoracic, A = abdominal.

cent adult mandibular and maxillary segments, and so on. The part of the embryo giving rise to the 14 parasegments is 12–68% EL (figure 11-4).

Pair rule genes cause the embryo to switch from an aperiodic structure to a periodic structure (compare figures 11-4 and 11-5). Pair rule genes can be subdivided into primary pair rule genes that are regulated by information provided by the maternal and gap genes; these include *hairy* (*h*), *eve*, and *runt* (*run*). The secondary pair rule genes respond to the primary ones. Thus it is the primary pair rule genes that need to initially establish the periodicity based on preexisting information. These genes have very large and complex regulatory regions capable of responding to several signals and also responding quantitatively, that is, being activated at one concentration and deactivated at another concentration of a single regulatory molecule. While many mutants of pair rule genes result in deletion of a whole set of parasegments, in fact each segment may be individually determined. This has been shown for *eve* and *hairy* by transformation experiments in which individual parts of the regulatory region control the presence or absence of individual parasegments.

Once established by pair rule genes, the parasegments themselves are subdivided into anterior and posterior by the segment polarity genes. Four are illustrated in figure 11-5. The discovery of segment polarity genes was foreshadowed by earlier work on imaginal discs, in which it was clear that cell lineages did not cross a posterior-anterior boundary in many cases; this is the so-called compartment concept. Garcia-Bellido and Santamaria (1972) had found that a locus called *engrailed* (*en*) was necessary for maintaining the anterior-posterior boundaries of the wing disc. This is now considered the classic segment polarity gene in embryogenesis. Segment polarity genes are thought to be initially regulated by the pair rule genes (often in combinations), but the maintenance of proper expression depends on interaction among segment polarity genes.

At this point in the developing embryo, a repeating segmental system is established with each segment subdivided into anterior and posterior. What remains is to individualize the parasegments. This is the role of the homeotic selector genes. These genes have been known for a long time primarily based on their effect on adults: for example, *proboscipedia* (*pbp*), in which the proboscis is transformed into a leg, and *Antennapedia* (*Antp*) in which the antennae look like legs. Since legs should exist only on the three thoracic segments, in these cases a head segment has become a thoracic segment. As can be seen in table 11-2, the primary effect of mutants in these genes is to change parasegment identity. Three aspects of homeotic genes warrant special note.

First, in *Drosophila* most homeotic genes fall into two clusters on the right arm of the third chromosome (figure 11-6). One, the *Antennapedia complex* (ANT-C), is close to the centromere, and the other, the *Bithorax complex* (BX-C), is a bit more distal. Each is a complex locus with several other genes, some of which fall into other classes, for example, *ftz* and *Scr* in ANT-C. It is likely that a single cluster is the primitive state, as has been found in the beetle *Tribolium* (Beeman et al., 1989; Stuart et al., 1991). Presumed homologous clusters are also known from mammals and a large number of other creatures (Ruddle et al., 1994).

The second point is that the proximal to distal arrangement of genes in the *Drosophila* genome corresponds to the spatial anterior to posterior control of segment identity (figure 11-6). Even within a transcriptional unit such as *abd-A*, the mutations

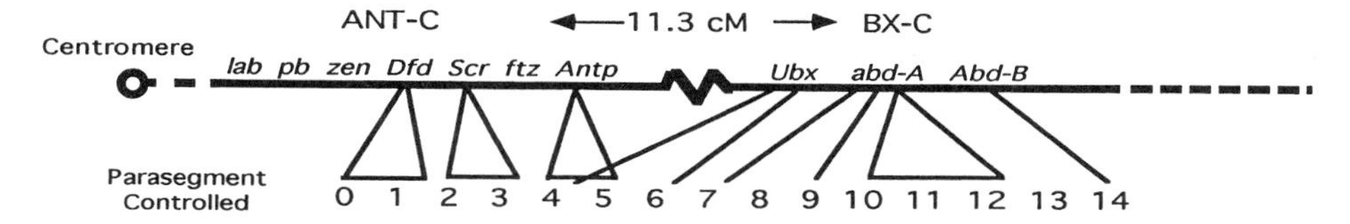

FIGURE 11-6 Maps of the Antennapedia complex (ANT-C) and Bithorax complex (BX-C) in *D. melanogaster* on chromosome 3R. The parasegments affected by the loci are noted. In the case of the very large *Ubx* and *abd-A*, different mutations within the transcriptional units affect the segments colinearly. Each complex is a few hundred kb with about 11 crossover units between them.

in the regulatory regions are linearly arrayed in the order of parasegments affected. In figure 11-5 it can be seen that the site of expression also corresponds to the order of genes, as the genes are listed in this figure from proximal to distal. This remains one of the more enigmatic aspects of homeotic gene clusters. Whether this is solely chance or has something to do with function remains unknown, but the fact that a similar pattern is found for mammals certainly gives one pause.

The third surprise in the homeotic genes was the discovery of the *homeobox*, a stretch of about 60 amino acids that occurs (with variations) in genes of both the ANT-C and BX-C, as well as elsewhere in the genome. This short sequence embedded in a larger protein forms four α-helices, the third of which fits into the major groove of DNA and thus is a DNA-binding motif, not a surprising feature of proteins controlling the expression of other genes. On the order of 100 such motifs exist in the *Drosophila* genome, most of which, when examined in detail, turn out to be embedded in genes involved in development (some are listed in table 11-2).

Surprisingly, the homeobox has been found in virtually every animal examined (there are even some claims for plants and fungi). This has given rise to a cottage industry of searching for more HOX genes (as they are now called) in ever-more exotic organisms. It is very tempting to hypothesize that all HOX genes and clusters in all animals are homologous; however, I remain skeptical. It may be that there are a limited number of ways in which a protein can efficiently, yet flexibly and reversibly, bind to a DNA helix, a fundamental property of any protein directly regulating gene expression. Selection pressure to do so may have independently led several genes to find similar solutions. While it is amusing to calculate the probability that two stretches of amino acids or nucleotides have attained a certain level of similarity by chance, nonrandom selection renders such moot. On the other hand, the similar arrangement in the chromosomes of genes with analogous function (such as the anterior–posterior expression being linear with the genetic map) is harder to explain without invoking homology. Ruddle et al. (1994) recognize some 13 paralogous groups of genes that (when present) are in very similar order in all animals, suggesting, first, that the paralogs were created by duplication to form the clusters and, second, that the clusters duplicated. Furthermore, the similarities among proposed homeobox homologs often extend beyond simply the homeobox itself (Bürglin, 1994). Finally, perhaps the most convincing argument for true evolutionary homology between vertebrate and *Drosophila* homeobox-gene clusters is the fact that transformation of vertebrate genes into flies results in expression patterns predicted for the fly homolog (Malicki et al., 1990, 1992; McGinnis et al., 1990) and vice versa (Awgulewitsch and Jacobs, 1992).

An illustration of a particular case should bring together many of the concepts of *Drosophila* development. Figure 11-7 illustrates the effect of mutations in the bithorax complex, BX-C. Normally, the second (T2) and third (T3) thoracic segments in adults have a wing and a haltere, respectively. The *bithorax* (*bx*) mutation results in the haltere becoming half a wing, the anterior half. A second mutation, *postbithorax* (*pbx*), causes the haltere to form the posterior half of a wing. Referring to figure 11-5, it is clear that *bx* must transform PS5 into PS4, which is responsible for producing the anterior half of a wing; *pbx* must transform PS6 into PS5. The combined double mutant, *pbx* plus *bx*, produces the widely illustrated four-winged fly (Lewis, 1978) which, in some sense, is the evolutionarily primitive state (the earliest flying insects had four wings). This illustrates the posterior–anterior independence of segments, the offset parasegment–segment arrangement, and that homeotic selector genes determine specificity of segments.

The above description of *Drosophila* development was primarily concerned with early embryogenesis, but inevitably the connection to adult structures quickly came into the picture. It is important to keep in mind that *Drosophila* are holometabolous insects and that the immature stages do not just become larger, modified, mature adults. Rather, there is a radical breakdown of tissue followed by a buildup of new body parts during the pupal stage, emanating from the imaginal discs. While we can illustrate the principle of parasegments and adult segments as indicated in figure

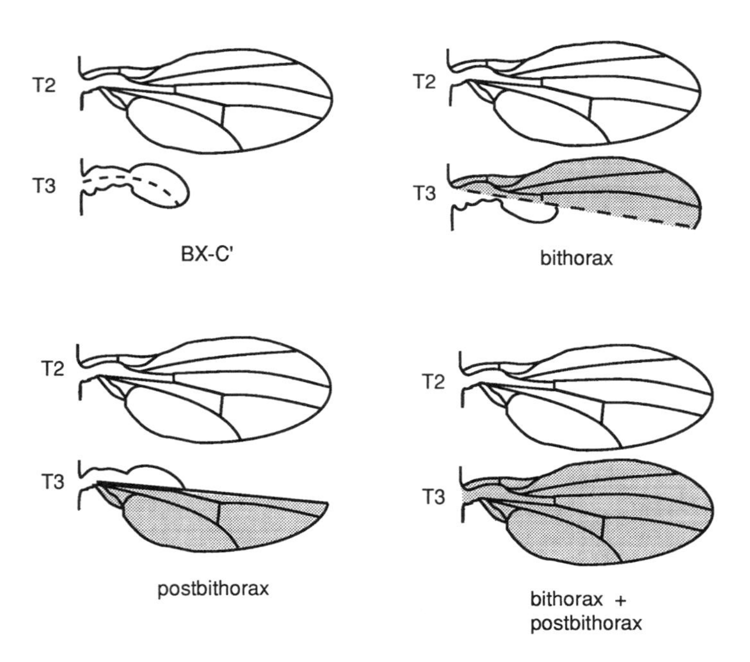

FIGURE 11-7 Effects of *Ubx* mutations on wing development. Modified from Lawrence (1992).

11-5, it is not meant to be a literal depiction of physical transformation. Rather, the same genetic factors and controlling programs that establish the embryonic parasegments are responsible for setting up the patterns in the imaginal discs that will give rise to adult structures. The earliest stages of development have received the most attention in recent years, with relatively little attention to later stages (we have virtually ignored the intermediate stage, the larva). But the above is sufficient for appreciating what follows, so let's get back to evolution.

Variation Within *Drosophila*

The most remarkable generality from comparative studies of developmental patterns and the genes controlling them is that, within the genus *Drosophila, very little variation exists in the expression patterns!* This is not to imply they are invariant structurally or in DNA sequence; they are about as variable as most other genes in that regard. But in contrast to the variation in expression pattern of genes discussed in the first part of this chapter, in every study done on more than one *Drosophila* species, developmental genes are found to be expressed in the same manner, and when examined, the stages of embryogenesis appear virtually identical. Furthermore, in a number of cases when genes listed in table 11-2 have been isolated from a species other than *D. melanogaster* and transformed into *D. melanogaster*, they almost always function like the native genes. Thus at the level of analysis so far achieved, without variation, few evolutionary conclusions can be reached. This finding is at odds with virtually every other subject discussed in this book where ample interspecific and, in most cases, intraspecific variation is the rule rather than the exception. Nevertheless, it is worthwhile discussing some of the empirical evidence behind the generality, as it serves as a guide to future research directions.

The evolutionary conservation of the genomic arrangement of the two major homeotic complexes has been best studied for the ANT-C. Figure 11-8 illustrates the known variation found within *Drosophila.* In all species studied, the linear arrangement of the genes (transcriptional units) is identical. However, two cases of inversion of the orientation (directions of transcription) of genes are known. Maier et al. (1990, 1993) found that *ftz* had been inverted in *D. hydei* and *D. virilis* and Randazo et al. (1993) found *Dfd* inverted in *D. pseudoobscura.* In the latter study, a gene closely related to *zen*, called *z2*, was found to be deleted; this gene has not been ascribed any function within *D. melanogaster*, so it is not clear what its deletion means, nor is it clear in what other species it exists, as it has not been well studied. It is also worth noting that the inverted relationship of *ftz* may follow the subgeneric distinction: all species of subgenus *Sophophora* examined (*D. pseudoobscura, D. azteca, D. willistoni*, and *D. nebulosa*) have the *melanogaster* orientation, while the only two species in subgenus *Drosophila* that were examined have the inverted orientation. Curiously, the orientations of *Dfd* in *D. pseudoobscura* and *ftz* in *D. virilis/hydei* may represent the ancestral state, as this would place all the genes in the same transcriptional orientation, which might be the expected state for tandemly repeated genes (assuming the complex arose by duplications). Furthermore, in *D. melanogaster*, part of the control region of *Scr* lies on the other side of *ftz*; if the gene's control region were adjacent in the ancestral state, then the inversion of *ftz* along with part of the *Scr* 5′-control region would be the derived state.

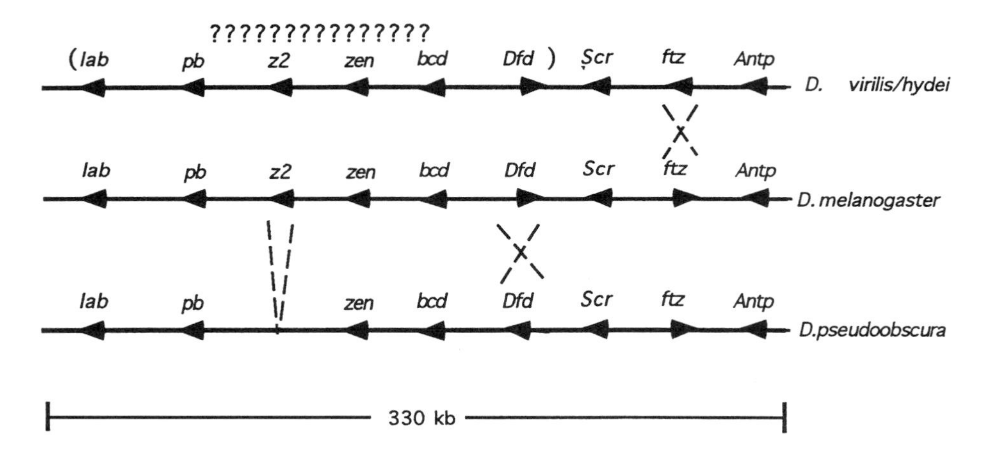

FIGURE 11-8 The known variation in the ANT-C among species. Arrowheads indicate direction of transcription. Two inversions, noted by dashed crossed lines, and a deletion are indicated. The question marks over the *D. virilis/hydei* map indicates these genes are probably located in those positions and orders, but have not been explicitly shown to be. *D. virilis/hydei* data from Maier et al. (1990, 1993) and *D. pseudoobscura* data from Randazo et al. (1993).

The two nested transcriptional units of the single gene, *Antp*, within the ANT-C have also been studied in a number of species (Hooper et al., 1992). This gene is more than 100 kb with up to 16 different transcripts due to various complexities of expression. Despite the complexity, it is very well conserved among *D. melanogaster, D. subobscura*, and *D. virilis*. Sequences both upstream and downstream are conserved, the sizes of the transcriptional units are conserved, and the intron/exon structure is conserved, as are alternate splicing signals. Similarly, the large and complex *Ubx* gene in the BX-C is highly conserved between the two subgenera (Bomze and López, 1994). *D. melanogaster, D. pseudoobscura, D. hydei*, and *D. virilis* all show the same pattern of alternative splicing of *Ubx* primary transcripts. Some exons have identical amino acid sequences and even conserved third base synonymous positions, and in all species, the temporal and spatial expression is identical.

The importance of the regulation of expression of genes controlling development is illustrated by two studies. It is known from transformation studies that there are three important, yet separable, 5′-control regions for *ftz* (Hiromi et al., 1985). Maier et al. (1993) showed that across 10 species of both subgenera, these three control regions were more conserved than the coding region. This comparative study was unusual in that it was predicted a priori from functional studies which regions should be conserved evolutionarily rather than vice versa, that is, most comparative studies find evolutionarily conserved regions and presume functionality. Furthermore, rescue of *D. melanogaster* ftz^- embryos could be achieved by the *D. hydei* sequences (Maier et al., 1990), further confirming essential conservation of proper regulation.

Li and Noll (1994) have even more direct evidence for the importance of cis-regulatory sequences in developmentally important genes. They studied three related genes, *paired* (*prd*), *gooseberry* (*gsb*), and *gooseberry neural* (*gsbn*), with different functions: *prd* is a pair rule gene, *gsb* is a segment polarity gene, and *gsbn* is activated by *gsb* and expressed only in the central nervous system, suggesting involvement in neural development. While parts of the amino acid coding region of the three genes are conserved, there is also considerable divergence. That the proteins encoded by the three genes can function for one another was shown by two types of experiments. First, when a heat shock promoter was placed 5′ to the *gsb* protein and induced to be expressed at an earlier time than normal (when pair rule genes function), it induced a pair rule phenotype. Similarly, a gene construct with the *gsb* 5′-control region and the *prd* protein could rescue normally lethal embryos completely deficient for *gsb* and *gsbn*. The essential differences among these three genes in their effects on development do not reside in the proteins produced, but in the timing of expression controlled by 5′ cis-regulatory regions. Thus, if the genes arose by duplication, their functions diverged because of differences in regulation. Of course, this does not mean the protein differences are totally meaningless, just that at the level of sensitivity of rescuing otherwise lethal embryos, the proteins function sufficiently similarly to be able to replace each other. In a similar vein, in the previous chapter we had reason to mention the extreme divergence in the protein encoded by the *transformer* gene crucial to sex determination (O'Neil and Belote, 1992). Clearly, the study of protein-coding regions of genes involved in development can reveal only a part of the functional story, perhaps even a minor part.

Another revealing example of how easily one gene may replace another during development comes from the study of the gap genes *knirps (kni)* and *knirps related (knrl).* The two differ primarily in that *kni* has a 0.7-kb intron, while *knrl* has a 19.1-kb intron. The importance of this is that when *kni* is expressed (at about the 12th division of the zygote nucleus), mitotic divisions are happening so rapidly that any transcripts larger than about 6 kb are aborted (Shermoen and O'Farrell, 1991), so that *kni* but not *knrl* can function at this stage. Two mutants, *Resurrector* (*Res*) and *Godzilla* (*God*), were found to suppress otherwise lethal mutants of *kni* (Ruden and Jäckle, 1995). These two mutants lengthen the cell cycle enough to allow complete transcription of *knrl*, which then can serve the essential function of *kni.* This shows that a gene can be fairly easily recruited into a new function by a single mutation in another gene and that the duplicated genes diverged in regard to time of expression and not protein function. Furthermore, this demonstrates that sometimes the presence and/or size of an intron is not a trivial aspect of a gene's structure.

The rapidly dividing syncytial early stages of *Drosophila* development may have other important consequences for molecular evolution. In chapter 9, studies of the divergence of cDNA prepared from mRNA in the adult stage were discussed (pp. 318–320). In parallel, cDNA prepared from embryonic mRNA was also used in DNA–DNA hybridization studies of the same species pairs (Powell et al., 1993). For the closely related species within the *melanogaster* subgroup, the divergence of embryonic cDNA was about half that for adult cDNA. Among these closely related species, about 90% or more of substitutions in protein-coding genes are silent. Because DNA–DNA hybridization studies do not distinguish replacement and silent substitutions, the results implied that silent substitutions in embryonically expressed genes may be more constrained than silent substitutions in adult expressed genes. This may be a consequence of the fact that in the embryo, at least until cellularization of the blastoderm, all genes share a common tRNA pool. Furthermore, a size constraint is imposed on mRNA by the rapid cell cycles, which could also affect rates of evolution of genes transcribed at this time. Clearly, this needs more investigation.

Finally in this section, to reemphasize the point made at the outset, the work of Dickinson et al. (1993) will be described. They chose to study several *virilis* group species for the pattern of hairs on the dorsal surface of first instar larvae. Members of these closely related species (figure 8-14) vary considerably in this trait, from 20 to 70% of the segment being covered. Six genes were chosen for careful analysis. Four were segment polarity genes (*en, gsb-d, ptc*, and *wg*), with the notion that differences in the distribution of hairs are basically a difference in restriction of expression of precursors in each segment. Furthermore, two of the genes, *gsb-d* and *wg*, are also involved in dorsal/ventral differentiation. Since the hair patterns are variable only dorsally, the independence of dorsal and ventral structures might be relevant. Two other genes, the homeotic *abd-A* and the gap gene *hb*, were also studied. A careful analysis throughout embryonic development with regard to temporal and spatial expression of the six genes revealed no consistent differences; "the conservation of detail is remarkable" (Dickinson et al., 1993). The authors concluded that the morphological differences in the cuticles are not due to differences in the patterning of the signaling molecules, but must rest in the (unknown) receptors of the information. Needless to state, this was a disappointing conclusion, given the careful planning to optimize the probability of finding differences.

Note that all the above studies have been done on members of the subgenera *Sophophora* and *Drosophila*. Species of these subgenera do vary in morphology and, between the genera, about 100 my of accumulated evolution has passed on the two lineages, yet the developmental patterns so far discerned seem to be invariant. Because a developmental basis for the diversity must exist, it may simply be that we have not yet attained the theoretical understanding and/or technical tools to observe the underlying changes in development. On the other hand, by far the greatest morphological diversity exists for the Hawaiian subgenus *Idiomyia*. In light of our inability to detect the underlying subtle changes typical of the species so far studied, perhaps it would be more fruitful to study species where the changes are not subtle. In the section on phylogeny and development below, some intriguing proposals for understanding the developmental basis of the great morphological diversity of Hawaiian flies will be discussed.

Beyond *Drosophila*

While virtually every aspect of evolution covered in this book could be discussed very well within the limits of the Drosophilidae, in the present context, it is instructive to go beyond these limits to ask whether all the knowledge about *D. melanogaster* development might shed light on evolutionary patterns in Insecta or even Arthropoda. Clearly, the morphological diversity represented in these larger groups is much greater than in even the Hawaiian *Drosophila*, and thus one might predict the variation in development must likewise be greater. Furthermore, given the evidence that some of the genes identified as important in development of *Drosophila* exist in vertebrates and, in a crude sense, are even expressed similarly, surely some of them must be important in Arthropod development and evolution. Some promising research programs have taken this approach.

Figure 11-9 is a classical representation of Snodgrass's (1935) view of the evolutionary origin of insects. The most important principle to draw from this illustration is that the most basic differences among the proposed stages of evolution are changes in function of segments, that is, segment identity. The specialized, individually unique, segments of insects initially evolved from uniform segments; the terms *homonomous trunk* and *tagmosis* are used to describe these two states. This classic view of evolution of invertebrates, at least in its most literal form, has been called into question by two observations. First, reevaluation of morphological data plus newer molecular data have tended to support mollusks as the arthropods' closest relatives, not annelids (e.g., Eernisse et al., 1992). Second, as we will discuss shortly, the genes that, at least in *Drosophila*, control the distinction of thorax from abdomen almost certainly existed in the ancestor of insects and crustaceans. This has led to an alternative proposal (Akam et al., 1994), namely that the ancestor of insects and crustaceans already displayed tagmosis; the homonomous groups within the insects and crustaceans (myriapods and remipedes, respectively) secondarily lost the individualization of the segments. This further leads to the prediction that the basis for tagmosis in insects and crustaceans is homologous and thus should be controlled by homologous genes.

In the crustacean *Artemia*, several of the genes responsible for tagmosis in *Drosophila* have homologues: *Dfd, Scr, Antp, Ubx*, and *abd-A* (Akam et al., 1994; Averof and Akam, 1993). The other homeotic selector genes in the ANT-C and BX-C are also presumed to exist in crustaceans, but have not been found. Amphipods, a distantly related crustacean, have been shown to express *engrailed* in a striped manner analogous to *Drosophila* (Scholtz et al., 1994). Thus at least all the genes needed for tagmosis likely existed before insects and crustaceans diverged. In Akam et al.'s alternative scheme, they do not doubt that tagmosis originated from an initially homonomous-trunked animal, but suggest that it may have occurred much earlier than Snodgrass's scheme would predict.

Confining our view to insects, there are three major variations on the manner in which oögenesis and embryogenesis occur (figure 11-10). First, three germ types are recognized. Long-germ embryos are typified by *Drosophila* where the complete body plan is present by the end of the blastoderm, that is, all segments are represented in the blastoderm (figure 11-5). Short-germ embryos have only the very anterior head segments determined by the blastoderm, with the addition of posterior segments during postblastoderm development. Intermediate-germ embryos are intermediate in having determination to the thorax by the end of the blastoderm. The second variable is that oögenesis can occur with (meroistic) or without (panoistic) nurse cells. One might get the impression from figure 11-10 that panoistic oöcytes are primitive and meroistic, derived. However, this is misleading in that some primitive groups (e.g., Ephemeroptera, Dermaptera, and Collembola) have meroistic modes, and some derived groups (e.g., fleas) have panoistic modes. Finally, most insects have a syncytial blastoderm without cell membranes, although some of the earliest lineages (e.g., Collembola) have cell membranes throughout embryogenesis.

With regard to the earliest stages of development, the disposition of maternal gene products before fertilization, unambiguous homologues of the maternally acting

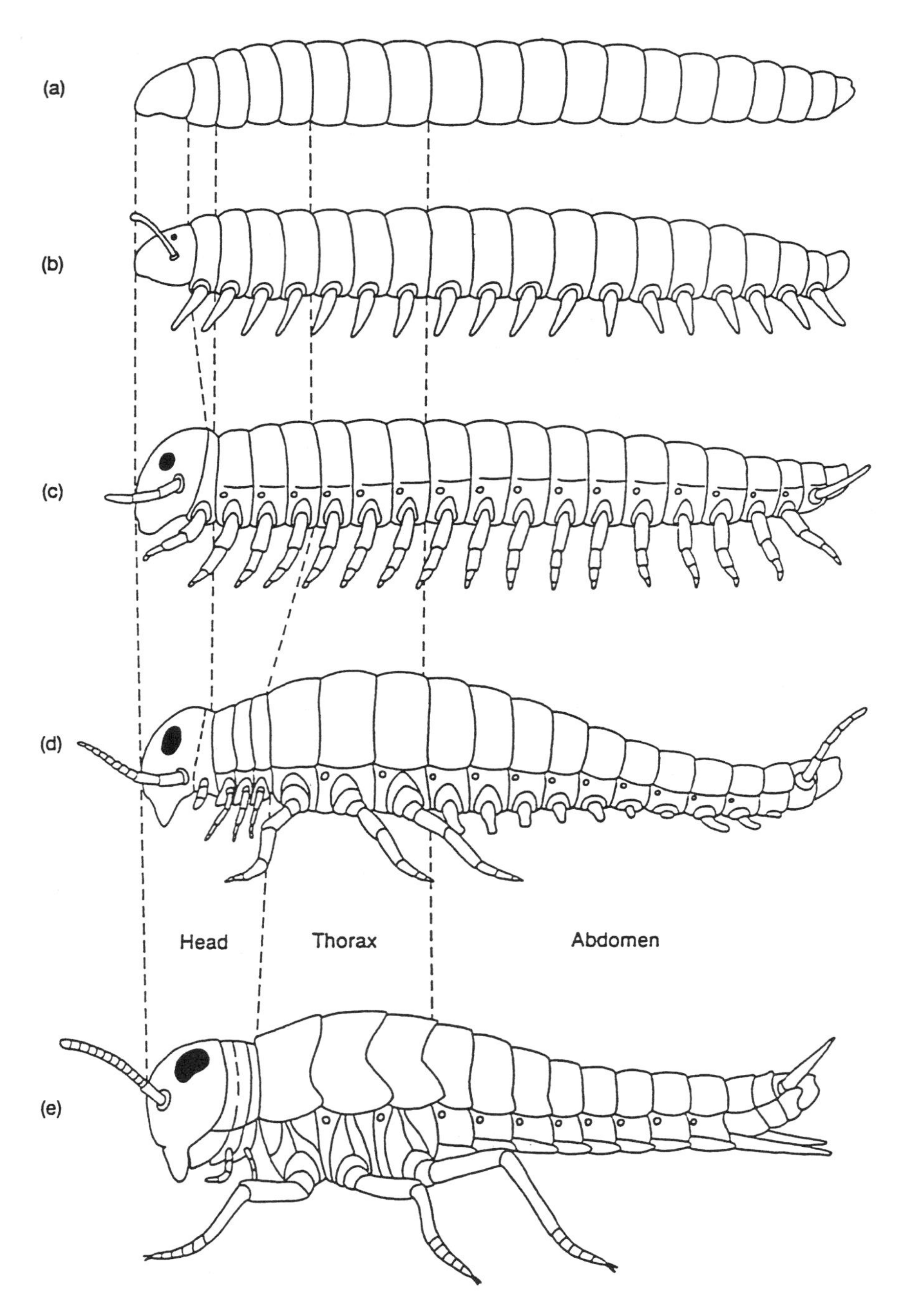

FIGURE 11-9 Hypothesized steps in the evolution of insects. (a) represents an annelid, (b) an onychophora, (c) a myriapod, (d) an apterygote insect, and (e) a pterygote insect. Modified from Snodgrass (1935) and Raff and Kaufman (1983).

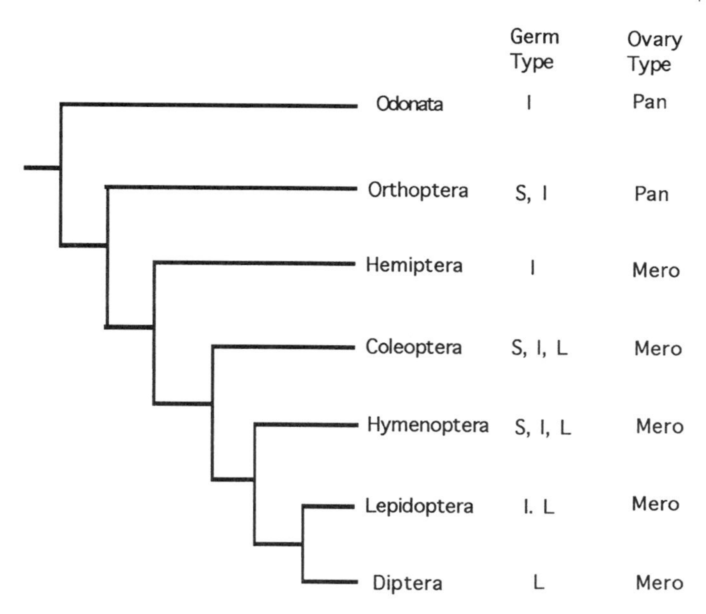

FIGURE 11-10 Distribution of diversity of oögenesis and embryos. Under germ type, the presence of short (S), intermediate (I), and long (L) germ types are noted. Under ovary type, panoistic oöcytes (Pan) without nurse cells and meroistic oöcytes (Mer) with nurse cells are indicated. After Patel et al. (1994) with additions.

genes have not been studied in any other insect other than *Drosophila*. Since it is known that *bicoid* mRNA is synthesized in the nurse cells (which lie anterior to the oöcyte) and is passed into the oöcyte, one might expect panoistic and meroistic modes of oögenesis to differ in this initial step. Indeed, Sander (1994) has speculated that lower dipterans, those primarily (though not exclusively) with panoistic oögenesis, may not use *bicoid* as the initial anterior-posterior determinant. This is partly based on centrifugation experiments: when lower dipteran eggs are centrifuged, a double abdomen or double head can be induced, but not in wild-type *Drosophila*; however, *bicoid* mutants are susceptible to doubling structures upon centrifugation (Schröder, 1992; cited in Sander, 1994). St. Johnston and Nüsslein-Volhard (1992) have proposed that lower diptera might use *hunchback* as the initial maternal determinant.

With regard to the next stages of development, the setting up of segments by gap, pair rule, and segment polarity genes, considerably more is known in other insects. Of particular interest is whether insects displaying the different germ types set up their segments differently; from the morphology of the embryo, it would seem that it must be different. One short-germ embryo insect that has been studied in some detail is the locust *Schistocerca* (Orthoptera). Two pair rule genes have been studied in *Schistocerca, even-skipped* (*eve*) (Patel et al., 1992, 1994) and *ftz* (Dawes et al., 1994); both were shown not to display a pair rule pattern of expression during embryogenesis, but are expressed in neuronal tissue in a manner analogous to *Drosophila*. Segments are formed during gastrulation in *Schistocerca*, as evidenced by the striped pattern of *engrailed* (*en*) expression, but the expression occurs in a sequential manner.

Can we conclude that all short-germ embryos do not use the pair rule pathway to set up segments? Patel et al. (1994) studied three beetles (Coleoptera), one each with a short-germ (*Tribolium*), an intermediate-germ (*Dermestes*), and a long-germ (*Callosobruchus*) embryo. They found that in all three, the expression of *eve* was as predicted for a pair rule gene, but in fewer stripes (e.g., three in *Tribolium*) in the early stages of the short-germ embryo. As segments are added at later stages of embryogenesis, more stripes are formed. The relationship of *eve* to *engrailed* (*en*) was also studied and found to be identical in all three beetles, although somewhat different than in *Drosophila*. A very similar finding for a second pair rule gene, *hairy*, was seen in *Tribolium* (Sommer and Tautz, 1993), as well as for another segment polarity gene, *wingless* (Nagy and Carroll, 1994). The conclusions are: the presence or absence of the pair rule pathway does not coincide with the germ length of the embryo, and the major difference among Coleoptera germ types is in the timing of expression of the pair rule genes. The same mechanisms of determination are used, but they occur later for the posterior segments in short-germ embryos. While long-germ embryos like *Drosophila* set up all the segments simultaneously, some short-germ embryos use the same pathway to set them up sequentially. At this level of analysis, a change in timing of expression of genes (heterochrony) would appear to be the case rather than some completely new developmental pathway. Phylogenetic affinity (figure 11-10) is a better guide to similarity in development than the morphology of the embryo.

A somewhat different finding was made by Kraft and Jäckle (1994) for the Lepidoptera, *Manduca sexta*, an insect with an intermediate-germ embryo. They studied two gap genes (*hb, Kr*), a pair rule gene (*run*), and a segment polarity gene (*wg*). They found the expression to be very similar to *Drosophila* even in regard to timing. That is, the prepatterning determined by these genes already exists for all segments in the blastoderm as in long-germ embryos; they are simply very compressed. This is in contrast to the sequential addition at later stages seen in the beetle *Tribolium*.

Two genes involved in the formation of the mesoderm in *Drosophila, snail* and *twist*, have also been studied in *Tribolium* (Sommer and Tautz, 1994). These genes are involved in dorsal–ventral patterning as well. Expression in the beetle followed the same dorsal–ventral pattern seen in *Drosophila*, the first evidence that dorsal–ventral as well as anterior–posterior pathways may be conserved between these orders.

Hox genes have been studied in a number of insects (reviewed by Akam et al., 1994). From this work, two points are particularly noteworthy. First, overall there is considerable conservation of expression, with minor differences due almost always to changes in timing of expression. For example, *abd-A* is expressed only in the abdomen of all insects studied, and its anterior boundary coincides with the parasegmental position defined by *engrailed* (Tear et al., 1990). This shows that homeotic selector genes interact with the segment-forming genes in a manner similar to that in *Drosophila*. In *Schistocerca, abd-A* is initially expressed more posteriorly than in *Drosophila*, extending all the way to the terminal (cf. figure 11-5). However, the expression is rapidly lost as *Abd-B* is expressed in the posterior, which inhibits *abd-a* expression. The difference between the locust and fly is simply in the timing of expression of *Abd-B*, which occurs sooner in the fly.

The second point of interest is that only the true homeotic selector loci of the Hox complexes seem to be well conserved in insects. Clear homologues can be found for *lab, pb, Dfd, Scr, Antp, Ubx, abd-A*, and *Abd-B* (figure 11-6). However, the "anomalous" loci *bcd, ftz*, and the two *zen* loci (none of which is a true homeotic selector locus) are very different in other insects and may not even be present. A putative homologue of *ftz* has been found in *Tribolium* (Brown et al., 1994) and *Schistocerca* (Dawes et al., 1994), but the sequences are so diverged as to call true homology into doubt. The gene is expressed in neuronal tissue as in *Drosophila*, but its expression in the embryo is very different in these two species. Akam et al. (1994) argue that these four *ftz*-related genes take on different functions in different arthropod lineages and are an exception to evolutionary conservation of the developmental program.

A particularly revealing and promising analysis has been the work of S. Carroll and colleagues on butterflies (reviewed in Carroll, 1994; Carroll et al., 1994; Williams and Carroll, 1993; Panganiban et al., 1994). Three aspects of the appendage differences between *Drosophila* and butterflies have been studied: the presence of prolegs in larvae, the presence of a second pair of wings, and the elaborate scale patterns on the wings. From figure 11-9, it is clear that the appendages of each segment largely define the differences between the animals shown, as well as among the various orders of insects. Regarding the presence of legs on abdominal segments, it is known that in *Drosophila* the homeobox gene *Distal-less* (*Dll*) is the earliest marker of limb primordia and is required for the proximal–distal axis in appendages of adults. *Dll* is not expressed in the *Drosophila* abdomen, being directly repressed by genes of the BX-C. In the Buckeye butterfly (*Precis coenia*), Panganiban et al. (1994) find *Dll* expressed in all appendages including abdominal prolegs. Based on the *Drosophila* model, they then went on to demonstrate that *Dll* was turned on in the abdomen because of the repression of *abd-A* and *Ubx* in the proleg primordia. Two points can be made. First, the presence of prolegs on Lepidoptera larvae is not due to the addition of a set of genes controlling leg development, but is simply a derepression of genes already present. From an evolutionary standpoint, it is not too surprising, as the ancestor to insects almost certainly had abdominal legs (figure 11-9). Second, the derepression of *Dll* was not caused by a change in that gene (e.g., a cis-regulatory sequence) but by a change in the field set up by homeotic selector loci.

The second character, the presence of wings on the third thoracic segment, might be predicted to be under the control of *Ubx*, as it was mutants in this gene that gave rise to four-winged *Drosophila* (figure 11-7; Lewis, 1978). However, *Ubx* is highly expressed in the hindwing primordia of *P. coenia*, just as in wild-type *Drosophila*. Carroll (1994) interprets this as meaning that *Ubx* is not sufficient to suppress wings; rather, *Drosophila* have evolved a set of genes responsible for transforming wing to haltere that have come under the control of *Ubx*. The four-winged *Drosophila* is not a reversal of evolution in the sense of reversing the many single steps that led to the network of genes involved, but is the sudden loss of the network due to a mutation in the "master switch." The presence of *Ubx* in wingless insects, crustacea, and even annelids indicates this gene existed long before winged insects evolved and is not solely a wing gene but a general controlling gene used for different purposes by different lineages.

The approach to understanding the generation of the beautiful array of wing patterns in Lepidoptera was to isolate the homologues of genes known to affect wing patterns in *Drosophila*, namely *apterous, invected, engrailed, decapentaplegic, scalloped*, and *Distal-less*. In early butterfly development these genes are expressed similarly to their expression in *Drosophila* (Carroll et al., 1994), suggesting similar function. Unexpectedly, a second round of transcription occurs in the fifth larval instar in the wing primordia, and each pattern of expression is reiterated in each wing cell, the units of the wing. The expression patterns are those typical of butterfly wings (Nijhout, 1991): stripes, chevrons, and so forth; however, none corresponded precisely to the pattern of adult *P. coenia* wings. Carroll et al. (1994) propose that the patterning seen is a dynamic one basic to all Lepidoptera, but eventually interpreted in a species-specific manner to generate species-specific patterns.

Three lessons can be learned from these *P. coenia* studies. First, sometimes the knowledge from *Drosophila* can lead to a simple explanation for differences, as in the case with the larval abdominal prolegs, although it is not the genes directly involved in formation of legs that differ but the more "upstream" controlling gene. Second, sometimes knowledge from *Drosophila* does not directly apply, as in the case with *Ubx* and the second pair of wings. This indicates different lineages may have coopted major controlling genes for different purposes. Third, the wing pattern studies indicate the conservation of use of the same genes in wing pattern development, but in a very different manner. Carroll (1994) argues all these studies indicate the major differences between Diptera and Lepidoptera are primarily a "tinkering" with the genetic machinery already in existence in their common ancestor, rather than the evolution of some whole new sets of genes.

Before ending this section, a few caveats are worth expressing. First, the majority of studies on comparative molecular embryogenesis among insects (and even more distantly related arthropods) have involved observing the expression pattern of (presumed) homologous genes. Relatively little has been done by way of proving similar (or different) function. The exceptions are some homeotic mutants of beetles (e.g., Stuart et al., 1993) and Lepidoptera (e.g., Booker and Truman, 1989) that indicate similar functionality of homologous genes. Nevertheless, similar expression of a gene in a particular tissue, by itself, does not prove homologous functionality. As Patel (1994) points out, it is strange that so many of the genes important in segmentation are also expressed in neuronal tissue or cells giving rise to neurons; it may simply be that neurons have some general transcription factors that turn on all kinds of genes that may have little or nothing to do with neuron function or development. Second, it may be that our view of the conservation of developmental pathways is due to the use of knowledge from a single species, *D. melanogaster*. All the probes used and genes studied were first defined as developmentally important genes in that species. For example, it may not be correct to conclude that *Schistocirca* does not use a pair rule system for setting up segments; it may simply be that it doesn't use genes similar to *ftz* and *eve*. If other insects (or even other species of *Drosophila*) were using different developmental programs, would we recognize it?

Development and Phylogenies

The two subjects heading this section are seldom considered simultaneously, but given the advances in both, connections are being made. In particular, as the previous section reveals, the concept of phyletic phenocopies (Stebbins and Basile, 1986) may be relevant in *Drosophila*; this concept states that the morphological or physiological changes occurring during the evolutionary diversification of a group may be understood by very careful genetic study of one member of the group (see also Hall and Clemens, 1923, for an early discussion of this principle applied to plant evolution). The question becomes: Can all the detailed genetic and developmental knowledge of *D. melanogaster* be used to understand the phylogenetic patterns in related species? In particular we can break the discussion into four related topics. One, can molecular and developmental understanding of a trait help decide questions of homology? Two, can understanding the genetic basis of a trait help in evaluating the importance of apparent convergences? Three, can knowledge of the molecular genetic basis of the control of development of structures in one species help define the kind of genetic differences responsible for between-group differences? Four, can phylogenetic knowledge help the developmental biologist by illuminating the temporal (evolutionary) order of acquisition of morphologies?

Regarding the first topic, DeSalle and Grimaldi (1992) discuss the case of the *Bobbed* or *bb* syndrome. Morphologically, this syndrome causes altered bristles, specifically an extreme bending, primarily on the thorax. Figure 11-11 shows the phylogenetic distribution of this morphology. Based solely on this character, one would have placed *D. mercatorum* and *D. hydei* as closer relatives to *D. melanogaster*, rather than close to *D. repleta* and *D. virilis*, where they almost certainly belong. Once the underlying molecular basis for the phenotype is examined, it is clear that the *bb* phenotype in the two groups is not homologous. In *D. melanogaster, bb* is caused by a major deletion of the rRNA, while in the other two species, there are insertions and differential replication (see also the discussion of the *abnormal abdomen* syndrome in chapter 6, pp. 202–204). This is also a case where knowledge of the cause of a

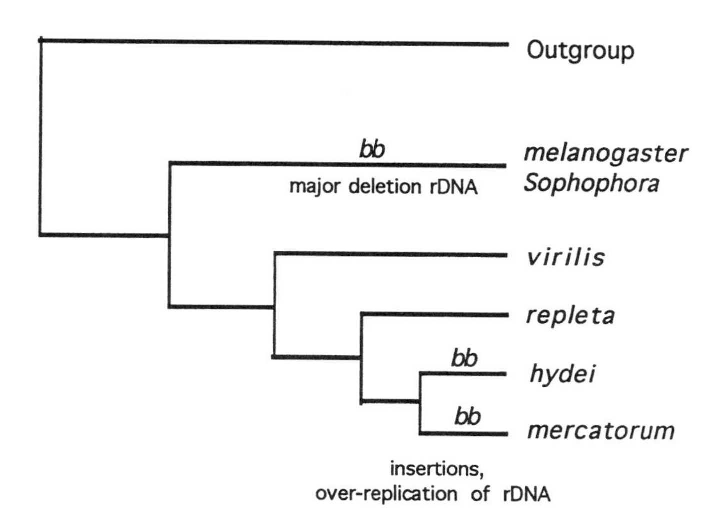

FIGURE 11-11 Phylogenetic distribution of the *Bobbed* syndrome, *bb*. Note that while producing very similar phenotypes, the cause is very different in the two clades. After DeSalle and Grimaldi (1992).

phenotype in *D. melanogaster* (in the rRNA) was a guide to where to look in other species.

The second example given by DeSalle and Grimaldi (1992, 1993) is the presence or absence of interfacetal setae, sense organs. It was largely the absence of these setae in the Hawaiian *Drosophila* that led Grimaldi (1990) to exclude them from genus *Drosophila.* However, it is now recognized that this trait can fairly easily arise, as single mutations in the *Hairless* gene can cause the absence of these setae (Bang et al., 1991), corresponding to a phyletic phenocopy. It is also known now that, in detail, the absence of setae in the Hawaiian flies is morphologically distinct from the primitive state outside the genus *Drosophila.* Knowing that what appeared to be a strong phylogenetic character has a very simple genetic basis led to a reevaluation of its use in phylogenies, and now it is widely accepted that the Hawaiian flies belong in genus *Drosophila.*

The third example concerns the Hawaiian flies. DeSalle and Carew (1992) have pointed out that many of the most spectacular morphological modifications in Hawaii resemble to some extent mutations in *D. melanogaster.* Most of the distinctive changes concern head parts and thoracic appendages, as implied by the names: modified mouthparts group, the Picture-wing flies, the modified tarsi group, and so forth. Flattened heads, such as in the Hawaiian species *D. heteroneura* (figure 7-5), have arisen six times in the family Drosophilidae (Grimaldi and Fenster, 1989; DeSalle and Grimaldi, 1993). Mutants of the *Dfd* gene and *labial* in *D. melanogaster* have broadened heads approaching that of *D. heteroneura*, although not as extreme. The *proboscipedia* mutants of *D. melanogaster* have morphologies similar to the modified-mouthparts Hawaiian flies. The *Distal-less* gene, determining the proximal–distal pattern of appendages, is a good candidate to examine in the modified tarsi group.

Moving a little farther away phylogenetically, Powell and DeSalle (1995) examined the variation in wing patterns among genera of Drosophilidae. Table 11-3 lists the wing characters recognized by Grimaldi (1990). These are mapped onto the cladogram in figure 11-12. From this analysis it would appear that characters 1, 8, and 9 have occurred only once, and thus one might predict that the underlying genetic mechanism inducing the changes is complex and not easily converged upon. No known single mutations of *D. melanogaster* produce these morphologies. On the other hand, characters 4, 6, and 7 appear to "flicker" on and off in the cladogram and thus might be predicted to have simple genetic bases. Indeed, from the listing of genes and mu-

TABLE 11-3 Wing characters mapped onto cladogram in figure 11-12.

1. R4 + 5 is convergent with M1 at the wing tip
2. Stout sharp warts on the section of costal vein between apex of R2 + 3 and R4 + 5
3. Vein R2 + 3 is turned costal
4. Thickened and darkened costal lappets—I
5. Thickened and darkened costal lappets—II
6. Short or absent vein A1 and CuA2
7. Loss of the small basal medial wing cell
8. Wing veins R4 + 5 and M1 slightly convergent
9. Crossvein dm-cu is oblique with respect to R4 + 5 and M1

From Grimaldi (1990).

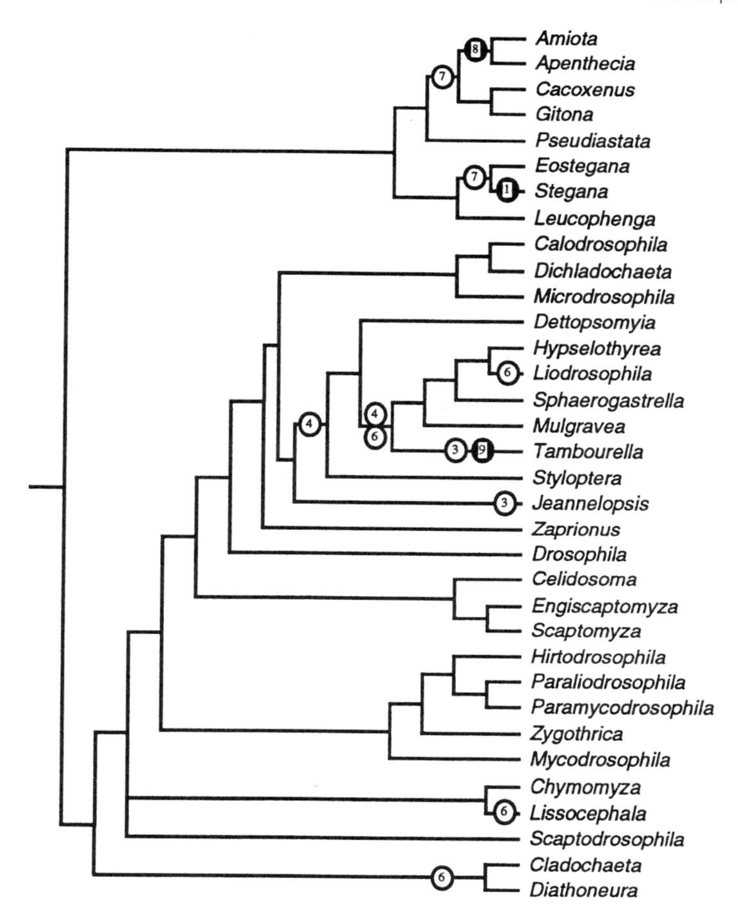

FIGURE 11-12 Phylogenetic distribution of the wing pattern traits listed in table 11-3 among the genera of Drosophilidae. Completely open circles are traits that appear more than once; partly closed circles are traits that arose only once. Cladogram is based on Grimaldi (1990) and figure is modified from Powell and DeSalle (1995).

tants controlling wing formation in *D. melanogaster* compiled by Garcia-Bellido and deCellis (1992), phyletic phenocopies can be guessed at. For example, character 4 is similar to mutations in *Notch* and *Delta*, character 6 may be related to *radius incompletus* or *abrupt* mutants, and character 7 is reminiscent of mutations in *crossveinless* and *tilt*.

On the level of families within the superfamily Ephydroidea, Powell and DeSalle (1995) analyzed changes in segmentation. The results are in figure 11-13. The most important result is that these characters are highly consistent phylogenetically with no convergences or reversals. Furthermore, sufficient variation exists within this group that analysis of the evolution of segmentation changes can be examined; one need not go further afield, and thus there is a better chance that principles and probes from *D.*

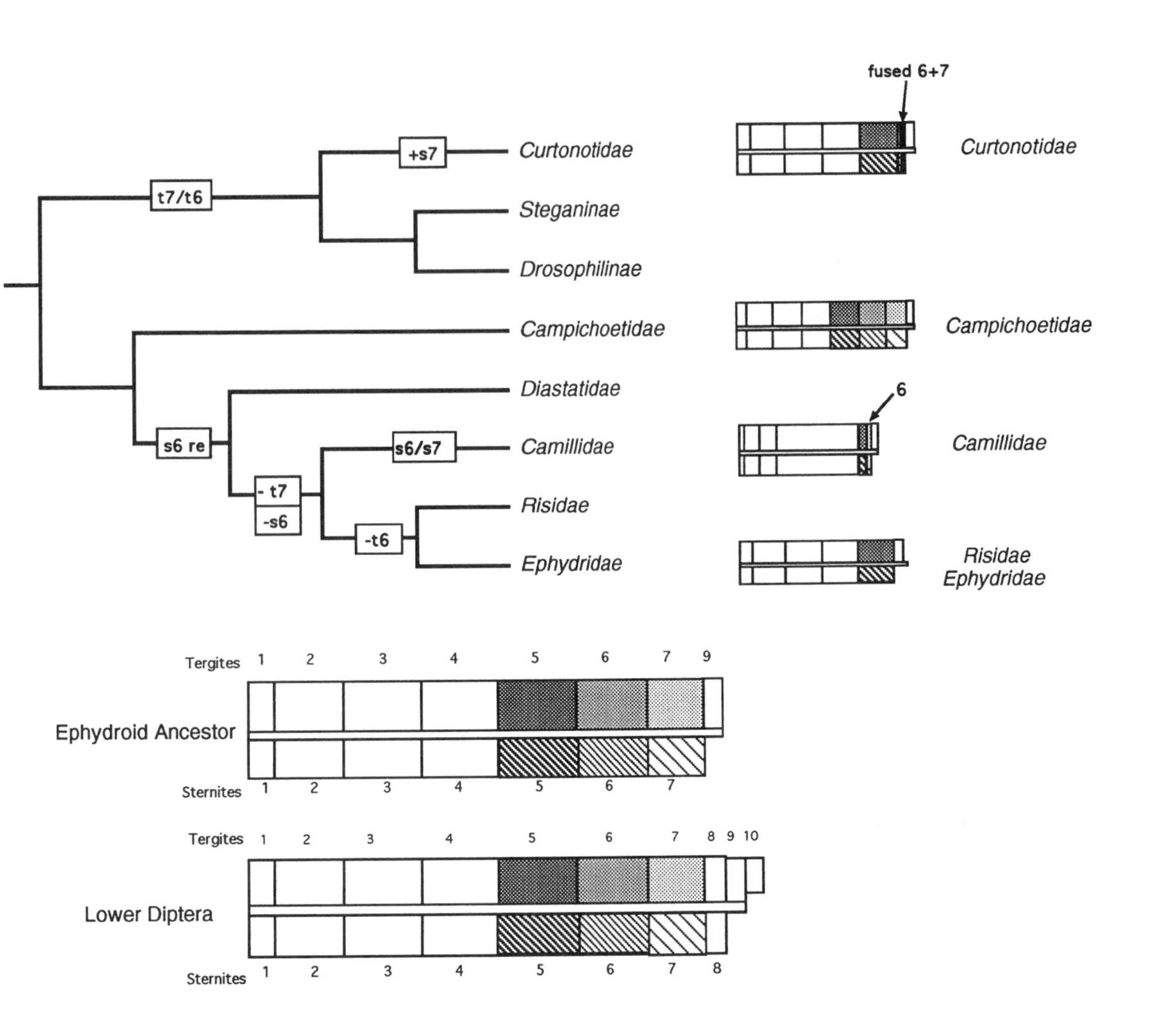

t7/t6
+s7
Curtonotidae
Steganinae
Drosophilinae
Campichoetidae
Diastatidae
Camillidae
Risidae
Ephydridae
s6 re
- t7
-s6
s6/s7
-t6
fused 6+7
Curtonotidae
Campichoetidae
6
Camillidae
Risidae
Ephydridae
Tergites
1 2 3 4 5 6 7 9
Ephydroid Ancestor
Sternites
1 2 3 4 5 6 7
Tergites
1 2 3 4 5 6 7 8 9 10
Lower Diptera
Sternites
1 2 3 4 5 6 7 8

melanogaster will be applicable. Having such phylogenetically consistent characters to work with should be an advantage.

Moving slightly out of the Diptera, the case of the Strepsiptera is highly intriguing. The females of this group are neotenic and retain larval characters throughout life. The males are more interesting in that they develop a pair of wings on the third thoracic segment and halteres on the second, precisely the complement of *Drosophila* (figure 11-7). This was considered a curious group quite distantly related to the Diptera. However, recent molecular data and a reanalysis of morphological data make a strong argument that the Strepsiptera may well be the sister taxon to Diptera (Whiting and Wheeler, 1994). By analogy to *Drosophila*, Whiting and Wheeler ask: "Are Strepsiptera *cbx Ubx/pbx bx*?" This combination of *D. melanogaster* genes should lead to an overexpression of *Ubx* in the second thoracic segment (inhibiting wing formation and promoting haltere development) and a suppression of *Ubx* product in the third segment, allowing wing development. This would indeed be a spectacular phyletic phenocopy, if true. Fortunately, it is a testable hypothesis.

Have any of these insights into evolutionary bases of development been useful as a guide to developmental biologists probing mechanisms? It is too soon to answer this definitively, but certainly there has been a healthy reawakening of interest in comparative embryology, especially molecular embryology, among developmental biologists, and it can only be hoped that insights into mechanisms will be forthcoming. If nothing else, we are certainly obtaining a much broader view of arthropod development. In chapter 12, I return to some speculations along these lines.

Canalization and Genetic Assimilation

Much of this chapter has been concerned with work done in the last 20 years. To end, we turn to a discussion of some older issues and concepts that take a rather different conceptual approach to relating development to evolution. C. H. Waddington proposed an unusual mechanism for evolutionary change based on developmental canalization and its change. Waddington (1960; see also Rendel, 1967) argued that development can be viewed as an ontogenetic landscape whereby development follows a set course that cannot be easily changed (figure 11-14); selection has acted to buffer the system from both environmental and genetic variation so that the same phenotype is achieved by many genotypes in many environments. However, some environments are so extreme as to push the embryo into a new path, forming a new phenotype; these are the classic phenocopies whereby a phenotype of a known mutation is induced by an environmental stimulus. In *Drosophila* it is well-known that environmental stimuli

---

FIGURE 11-13 (FACING PAGE) Phylogenetic analysis of segmentation changes among families of the superfamily Ephydrodea. Cladogram is based on Grimaldi's (1990) morphological analysis. The lower cartoons depict the generic patterns of segmentation thought to have existed in ancestors. Cartoons on right are present-day taxa. The changes depicted on the cladogram: +s7 is the retention of sternite 7; t7/t6 is the fusion of tergites 6 and 7; s6 re is the reduction of sternite 6; -t7 is the loss of tergite 7; -s6 is the loss of sternite g; -t6 is the loss of tergite 6; s6/s7 is the fusion of sternites 6 and 7. Modified from Powell and DeSalle (1995), in which errors occur.

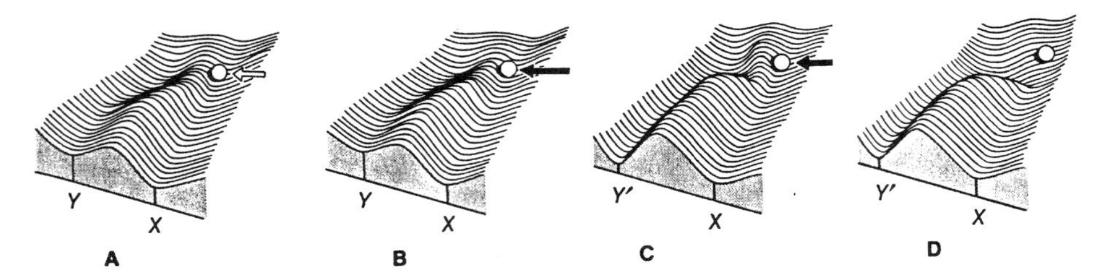

FIGURE 11-14 Epigenetic landscape envisioned by Waddington. The ball represents a developing organ. Under normal conditions, phenotype X is the outcome. If an environmental perturbation (open arrow) can sometimes push the organ into pathway Y, the ease with which this occurs (height of the barrier) may have genetic variance. If so, selection for genetic variants (solid arrows) may succeed in fixing a new canalized pathway (in D) so that Y is most often attained even in the absence of environmental perturbation. After Waddington (1956b).

like heat shock can produce phenotypes mimicking mutations. However, when such stimuli are given, not all individuals respond the same; some produce extreme phenotypes, others are hardly changed from the wild type. If this sensitivity to environmental stimuli has a genetic basis, then it should be possible to select for production of more extreme phenotypes, perhaps even to the extent that the phenocopy phenotype is produced in the absence of the stimuli. This is what Waddington called genetic assimilation.

Figure 11-15 is a cartoon that helps illustrate these concepts (inspired by Wallace, 1968). Suppose you were given the task to select *Drosophila* for short wing size, but that your major professor only allowed you a limited view of the flies, as indicated by window A in figure 11-15. You cannot distinguish any differences, and thus it is unlikely you will complete your dissertation. Now suppose your professor takes pity on you and allows you the view from window B. You impress your colleagues with a successful selection experiment, so successful in fact that now when you go back

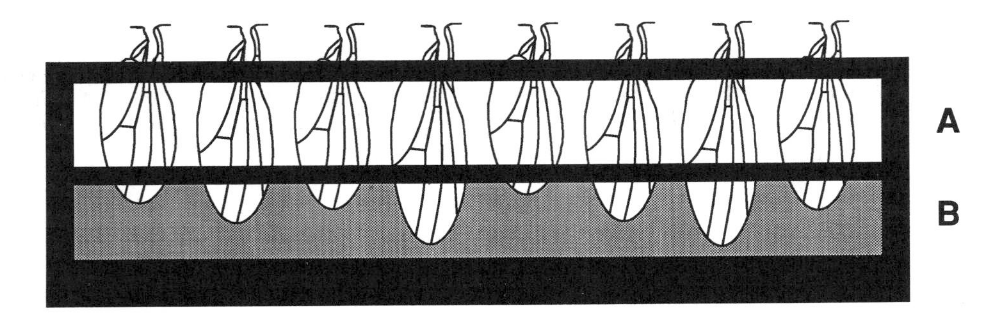

FIGURE 11-15 Cartoon illustrating the concept of canalization and genetic assimilation. Ovals represent wings and the goal is to select for short wings. If the only view were the normal one, window A, the selection could not work. By taking a new view, window B, selection can work; window B represents the view obtained by an environmental perturbation, such as heat shock. Inspired by Wallace (1968).

to window A you can see differences among individuals. Window A can be thought of as the normal environment, that is, the half-pint bottle kept at 20°C, whereas window B is the abnormal environment caused by the heat shock, for example. While at first you needed the heat shock to see the genetic variation that must have been there for your selection to eventually succeed, after some time the genetic changes brought on by selection are such as to reveal the variation in its absence. Referring to figure 11-14, the environmental change allowed a lowering of the barriers between developmental pathways, and genetic variants are selected that keep the barrier down, or even build up new barriers to push the development into a new path.

Can this happen? Yes. Several experiments have been successful. We discuss two classics from Waddington himself, as they illustrate a couple principles of note, and also employ *Drosophila.* The first experiments (Waddington, 1953) were done for the crossveinless (cv) phenotype, a phenotype that can be caused by mutations at several loci or be induced in wild-type flies by a 4-hour heat shock (40°C) 17 to 23 hours after puparium formation. After 14 generations of selection for increased expression of the cv phenotype, the strain began expressing it in the absence of heat shock. Individuals who showed the most extreme expression (without treatment) were interbred and selection carried on for seven or eight generations to produce the results in table 11-4. All four replications showed a high frequency of cv while the controls and the replicates selected for "downward" (decreased sensitivity to heat shock) showed virtually no cv. Surprisingly, the lower temperature produced more cv phenotypes, while it was the heat shock that initiated the experiment. Crosses among the lines indicated the change was polygenic in nature.

Reasoning that the loss of a crossvein is a relatively minor effect, in the next set of experiments, Waddington (1956a) set out to study a more drastic morphological change, bithorax. When 2.5- to 3.5-hr eggs are exposed to ether vapor, a high proportion produce a bithorax-like phenotype with expanded halteres and, generally, the

TABLE 11-4 Results of Waddington's (1953) genetic assimilation experiments on the crossveinless phenotype.

		Percentage Crossveinless Phenotype	
		at 25°C	at 18°C
High lines	H1	68	100
	H16	87	100
	H24	67	99
	H26	95	100
Downward lines		0	0
Wild Edinburgh		0	0

Note: "High" lines are those selected for high level of expression of cv phenotype; "downward" lines are those selected not to express cv, and "wild Edinburgh" is the original unselected stock. The results are for flies reared in the absence of the heat shock.

third thoracic segment appears more like the second. Three successful lines were eventually derived that showed a high proportion (60 to 80%) of haltere effects (He in Waddington's notation) in the absence of ether treatment. Two of these were likely single mutations in the bithorax complex. The third line was more interesting. The genetic change was multigenic, with a major effect of a recessive, which showed a strong maternal effect. Knowing what we now know about *Drosophila* development, this is not unexpected, and it would be interesting to examine the expression patterns of some of the known maternal-effect genes in these altered lines.

Waddington (1956a) went on to ask whether these genetic assimilation experiments produced lines generally more or less susceptible to environmental perturbations. The cv-assimilated stocks were subjected to the ether shock and the He-assimilated stocks were subjected to the heat shock. He stocks were more sensitive to heat shock than were unselected controls of the same original strain, indicating a greater sensitivity to phenocopy in general. On the other hand, cv-assimilated stocks were less sensitive to ether than the control, indicating perhaps a greater buffering (canalization) had occurred in these stocks. Waddington and Robertson (1966) did show that selection for increased canalization can occur, that is, starting with a trait that shows variation in expression depending on environment, selection to reduce the environmental influence can be successful. They studied *Bar* eye mutations that produce many fewer eye facets at higher temperature than when reared at a lower temperature. Waddington and Robertson reduced the difference from 50 facets to less than 15 in six generations.

Does the concept of canalization and genetic assimilation apply to the real world? Waddington (1953, p. 124) concluded that "we have to rest content with a knowledge of the categories of possible processes instead of a theory capable of detailed predictions." Stearns and Kawecki (1995) did not rest content. They reasoned that, if canalization is a reality in evolution, traits more important in fitness should be more canalized than traits less important. As part of their studies on life history characters (discussed in chapter 5), they could use the data to test the prediction just made. They estimated the relative importance of different life history parameters, for example, early fecundity is more important than late fecundity, and so forth. They then studied whether the more important traits are more canalized than the less. The results displayed a significantly positive correlation as predicted. They showed that this result was not due to differences in level of genetic variance for the traits; the relationship held whether they studied nearly homozygous lines or highly outbred lines, that is, the degree of canalization did not depend on the amount genetic variation in the line. By complicated reasoning, they could also conclude that the canalized buffering was against genetic perturbation, not environmental.

The importance of a developmental conceptual orientation toward understanding and analyzing macroevolutionary change has been well accepted for a long time. However, as Stearns and Kawecki (1995, p. 1449) conclude, "Development may play a more intimate role in microevolution than had previously been suspected."

12

Retrospective and Prospective

[I]ntellectual progress usually occurs through sheer abandonment of questions together with both of the alternatives they assume—an abandonment that results from their decreasing vitality and a change of urgent interest. We do not solve them: we get over them. Old questions are solved by disappearing, evaporating, while new questions corresponding to the changed attitude of endeavor and preference take their place.

John Dewey

It was prettily devised of Aesop, "The fly sat upon the axle-tree of the chariot-wheel and said, 'What a dust do I raise.'"

Francis Bacon

The foregoing chapters have been largely factual and reasonably objective. Here, I unabashedly change tack and will present judgments and speculations. The following is an essay rather than a well-referenced review; however, I frequently refer to tables and figures elsewhere in this book that provide evidence and references for some of the statements.

Integration of Information

For the reader who has faithfully read this volume from the start, I think it safe to state that one of the major points presented in the first chapter has been amply documented, namely, that no other organism could have served as a focal point for the presentation of research on so many different aspects of evolutionary biology. *Drosophila* have served as models for studies at every conceivable level, from the most reductionist molecular study to behavior, development, populations, ecology, and even ecosystems. To be sure, not all aspects are equally well-studied, nor are species of this genus equally suited for all these kinds of research. Nevertheless, a more diverse array of research is being done on this fly than on any other organism.

One of the major goals of such concentration on a single model system is to integrate the various levels of knowledge. It is often lamented that this has not been done in evolutionary biology, but that is certainly changing. Some examples from the foregoing chapters will illustrate this.

We started the book with a discussion of single-gene variants in natural populations, primary morphologically recognizable variants. We reviewed the literature indicating such variants are generally rare and cause a decrease in fitness, or perhaps we should state, *because* they cause a decrease in fitness, they are rare. Now that we know from molecular studies that most such mutations are caused by the loss of function of a gene, very often due to a transposable element, these results are understandable. Loss of function of a gene almost certainly must decrease fitness; why else would a gene remain functional for a long evolutionary period? When studies were first done on morphological mutants, there was no reason to believe or predict that loss of function was their overriding cause, so that it was certainly plausible that such variants could have had a much more important role in evolution. Now we understand why, by and large, they do not play much role.

Some of the laboratory studies have been well-integrated with subjects in several chapters. Examples include the selective nature of inversions, founder effects in speciation, and the role of heterogeneous environments and habitat choice in maintaining genetic variation. Perhaps the most remarkable connection of laboratory and field work is the rare male effect. This effect was totally unanticipated, but well-documented in the laboratory. Now that some evidence is accruing that it may also exist in natural populations for inversion polymorphisms (table 6-8), this takes on added meaning. It may be conjectured that its recognition in nature might never have even been tested had not the laboratory experiments given direction to the analysis.

The importance of recombination in understanding evolutionary mechanisms is another area of integration between seemingly disparate fields of study. It was long recognized that inversion polymorphisms are strongly affected by natural selection due to the suppression of recombination and a subsequent magnification of selective effect on the whole set of nonrecombining genes (so-called supergenes). Now as molecular data are accumulating, the role of recombination in understanding patterns and mechanisms at that level is becoming increasingly apparent. Even mathematical models first developed for inversions have been elaborated on to help understand nucleotide variation.

The growing knowledge of the phylogenetic relationships of *Drosophila* species and groups has been incorporated into a number of other levels of study. We discussed how phylogenies have shed light on the evolution of gene structure: introns are more often lost than gained (figures 10-18 and 10-19). Phylogenetic analysis supports the contention in karyotypic evolution that chromosomal arm fusions are more frequent than fissions (figure 9-5). Analysis of horizontal transfer of transposable elements has been aided by phylogenies (figure 9-18). On a broader level than the genus *Drosophila*, the evolution of various segmentation patterns of dipterans could be understood in a phylogenetic framework (figure 11-13).

One of the better-integrated examples is alcohol dehydrogenase polymorphism. The *D. melanogaster* polymorphism is understood from a biochemical to ecological level, at least to some degree. Perhaps even more impressive is the *Adh* polymorphism in *D. mojavensis*, where the ecological aspects are better integrated than for *D. melanogaster* (tables 6-5 and 6-6; figure 6-4). Another system that stands out is the *abnormal abdomen* (*aa*) of *D. mercatorum*, studied so well in Hawaii (table 6-7). Here a connection between a molecular change and life-history traits has been made, a rare case indeed.

Despite these successes of understanding evolutionary processes on more than one level, it must also be conceded that complete integration of understanding has not been achieved for *Drosophila*. The most conspicuous example is development. With the exception of some of the speculation on phyletic phenocopies, our increasingly detailed understanding of *Drosophila* development has not been incorporated into any level of evolutionary analysis or understanding. Except for *Adh* and *aa*, our increasingly sophisticated knowledge of molecular evolution has not been particularly well integrated with other aspects of evolution. What would appear to be a natural for integration, molecular evolution and development, has barely been attempted. However, molecular studies have provided the data for increased understanding of phylogenetic relationships. It is no coincidence that the shortest chapter of the book (except the first and last) is the one on ecological genetics. With the exceptions discussed in that chapter, little connection has been made between ecology and genetics; what connections have been made are primarily (although not exclusively) correlative rather than mechanistic.

Genetic Loads

In chapter 2, the evidence was presented that making whole chromosomes homozygous leads to a great decrease in fitness. Not only is viability decreased greatly as indicated by the classic studies (figure 2-2), but when additional components are considered, the decrease in overall fitness is about twice as great as for viability alone. We emphasized in chapter 2 that it is not known how many loci are involved in generating these loads, although the recombination studies (tables 2-7 and 2-8) would indicate at least several loci are involved. Similarly, in chapter 4 we discussed the studies indicating the density of selectively-detected polymorphic regions along a chromosome must be fairly high (figures 4-4 and 4-5). Given all the recent molecular work on *Drosophila*, it is reasonable to ask if we have found the genetic basis for these observations. Somewhere in the fairly extensive DNA polymorphism sequence data (tables 10-1 to 10-4) must be the answer to these questions.

One possibility is nonsynonymous polymorphism. In the next section, arguments are made that these polymorphisms are, in general, slightly deleterious. All one needs to hypothesize further is that the fitness effects be largely recessive, a not-unreasonable assumption. Recall that we noted a significant difference between *D. melanogaster* and *D. simulans* in regard to the proportion of nucleotide polymorphisms that are silent versus replacements (table 10-9); *D. melanogaster* has a higher proportion of replacements than *D. simulans*. But overall, *D. simulans* has greater nucleotide variation (table 10-7). It turns out the two observations cancel each other out, so the average nonsynonymous diversity of the two species is about equal (the mean nonsynonymous π for *D. melanogaster* is 0.0014 for all loci studied and 0.00076 for those studied in common with *D. simulans;* the latter species' value is 0.0010). Furthermore, recall that *D. melanogaster* has considerably more transposable elements than *D. simulans* (chapter 9), as well as an associated higher diversity of insertions/deletions (table 10-11). If any of these types of molecular variation account for genetic loads, then we can make predictions. If all nucleotide substitutions are involved, then *D. simulans* should have a greater genetic load than *D. melanogaster*. If only nonsynonymous substitutions contribute to genetic loads, then the species should have about equal loads. Finally, if indels and/or transposable elements are the cause of genetic loads, *D. melanogaster* should have the greater load. Unfortunately, we know nothing about genetic loads in *D. simulans* because the appropriate markers with inversions have been unavailable. Nevertheless, the above considerations indicate such studies would be highly valuable.

One other observation on genetic loads can be related to newer data, namely the lack of nucleotide variation in genes on the dot fourth chromosome of *D. melanogaster* (at least for the one gene studied). If this observation holds for the entire chromosome, then we would predict it has no genetic load. In fact, this is the case (Kenyon, 1967): The viability of fourth chromosome homozygotes and heterozygotes is equal, unlike any other chromosome in any other species. However, the variance in viability of fourth chromosome homozygotes is significantly different from zero, indicating that there is some genetic variation on this chromosome. This single study involved chromosomes from a laboratory cage population and has not, to my knowledge, been repeated with fourth chromosomes from nature or from the ancestral *D. melanogaster* populations in Africa.

Whither the Selectionist–Neutralist Debate?

In the population genetics chapters, we used as a theme to organize the discussion the progression of contrasting views characterized by Chetverikov–Morgan, Dobzhansky–Muller, selectionist–neutralist. The issues embodied in these dichotomies dominated evolutionary genetics for 50 years but seem no longer the foremost questions of today. Perhaps the quote opening this chapter best describes what has happened to these contentions. No one would claim that we have solved the selectionist–neutralist controversy; we have simply gotten over it. The question whether selection can detect single nucleotide changes has been solved; there is ample evidence selection can act on fitness changes caused by single nucleotide substitutions. Likewise, the question whether selection detects *all* nucleotide changes at *all* times is answered; of course selection does not detect all changes at all times. Now that the two extreme views have been discounted, the real effort and attention has turned to empirical and analytical studies of how to detect selection or true neutrality. This is not meant to imply all controversy and disagreements have been smoothed over, although certainly the acrimony of earlier days has subsided.

There is little doubt that one aspect of the traditional controversy has been solved: There is a lot of genetic variation floating in populations and species. No two individuals are genetically identical. It is intriguing to contemplate what our view of evolutionary mechanisms would be had this not been the outcome of all the studies measuring levels of genetic variation. It is conceivable that the wild-type concept, in its most naive formulation, could have been true. This would have settled things simply. The fact that so much variation has been uncovered keeps the controversy alive, although the ground keeps shifting.

What have we learned from the work on *Drosophila* in this regard? There have been some surprises. Probably the most surprising is the growing recognition that silent substitutions in coding regions are not always neutral. When the neutralist theory was first put forward, given the redundancy of the genetic code, these were clearly neutral mutations as they did not affect the gene product and thus could not possibly affect phenotype. Even the majority of selectionists would have conceded these kinds of mutations were neutral. Yet the data on codon usage bias in *Drosophila* is indicating that selection very likely is affecting silent mutations. In addition to the existence of codon usage bias, two other aspects of the data are particularly striking. First is the observation that the codon usage bias of a gene can change significantly from lineage to lineage, witness *D. melanogaster* and *D. willistoni* (table 10-15). Yet, superimposed on this overall change in usage pattern, there can be remarkable persistence of codon usage in particular cases, for example, the virtual absence of AUA for isoleucine in *Adh* throughout all *Drosophila* (table 10-16). Any theory of codon usage bias in *Drosophila* must account for both observations. I do not believe there presently exists a satisfactory theory for these patterns, although it is inconceivable to me that such a theory would not involve selection.

A second surprising empirical finding is that the ratio of nonsynonymous to silent substitutions in interspecific comparisons is higher than the ratio of nonsynonymous to silent polymorphisms within species. While this is not always provable statistically for each gene, the overall trend is clear (tables 10-17 and 10-18). This implies either something unusual is happening at speciation events to fix amino acid changes or, perhaps more likely, amino acid substitutions occur rapidly compared to neutral polymorphisms. Therefore, the probability of "catching" an amino acid substitution in the polymorphic state is much less than catching a more nearly neutral silent change. This suggests that, by and large, amino acid substitutions are being driven by selection. This is not inconsistent with another observation we made with respect to the differing ratio of nonsynonymous to silent polymorphisms in *D. melanogaster* and *D. simulans.* This ratio is higher in *D. melanogaster* and the overall level of nucleotide variation is lower. This implies that *D. melanogaster* populations are smaller than those of *D. simulans.* Selection is less effective (meaning greater fitness changes are necessary for detection by selection) in smaller populations and therefore slightly deleterious changes would go undetected. The tentative conclusion is that most mutations causing amino acid substitutions are slightly deleterious, but those that become fixed are driven by selection.

Given such observations, what is left for neutrality? Plenty. Recall, for example, the observations on *D. melanogaster Adh.* Two sites clearly seem affected by selection, the F/S amino acid polymorphism and the indel in the first intron causing expression differences. Otherwise, both the sliding window picture (figure 10-4) and the clinal data (figure 10-23) would indicate the great majority of variants are consistent with neutrality. In fact, the acceptance of neutrality has so permeated our view of molecular variation that it is used as the null hypothesis against which we test for exceptions such as those two sites in *Adh.* The beauty of the neutral theory is that it does make explicit predictions; selection does not. Selection is notorious for being able to explain any pattern. We need the neutral theory: without it, we could not detect nonneutrality.

Whither the selectionist–neutralist controversy? The rhetoric surrounding this issue seems to have changed in recent years. Those historically coming from the neutralist school have had to admit that empirical observations have shown that selection can detect individual nucleotide changes, even silent mutations. But they do not concede that selection is the driving force of molecular evolution. Rather, the term "purifying" selection has come into vogue. That is, selection is simply relegated to the role of removing deleterious changes, a view very much consistent with T. H. Morgan's: "Natural selection may then be invoked to explain the absence of a vast array of forms that appeared, but this is saying no more than that most of them have not had a survival value. The argument shows that natural selection does not play the role of a creative principle in evolution" (Morgan, 1932, p. 131).

While the distinction of purifying selection versus positive selection is a real one, it seems to me that it is a thin line, especially in regard to empirically knowing one from the other. The view of selection as solely purifying in nature is perilously close to what I have called the Michelangelo conundrum. One could argue that Michelangelo was not creative. He simply took a piece of marble and "purified" it by removing the unfit parts to reveal the statue inside. He did not actually "positively" put in place the atoms in the stature. According to a simplistic Neodarwinian view of evolution, selection takes a set of random mutations and removes the unfit. Does one say what is left was "created" by selection? Or did selection simply "purify" the gene pool? Is selection only negative in removing the unfit, or is it also positive in driving more fit variants to fixation? It seems to me that the controversy has evolved to this negative-positive distinction. The question is no longer whether selection is acting on the molecular level; rather, the question is what is the role of selection. The descendants of the Morgan–Muller neutralist school would assign an exclusive, or at least predominant, negative role to selection. Those from the Chetverikov–Dobzhansky selectionist school would assign both negative and positive roles. Superimposed on this selectively detectable variation, both schools would admit that there are also many nucleotide changes, both polymorphic within populations and fixed between species, that are effectively neutral, meaning they behave so close to neutral predictions as to be indistinguishable from absolute neutrality.

Figure 12-1 is an illustration of these views. The whole issue is simply one of the relationship of the distribution of mutational effects to the quantity $1/(2N_e)$. Neither of these is known, and the curve in figure 12-1 is one possibility that follows approximately the distribution of fitness effects of making chromosomes homozygous. The majority of newly arising mutations are deleterious, some even lethal. Most are quickly eliminated and play little or no role as either polymorphisms or interspecific differences; this is the purifying selection. The real questions are the width of the bars that indicate the situation where the absolute selective value falls into the range of being effectively neutral and the extent to which newly arising mutations fall to the positive side on the abscissa. The staunchest positivist would argue for a narrow bar as in (a). Very little of the variation is neutral, and most mutations that aren't quickly eliminated by selection (falling to the left of the bar) are positively maintained or fixed (those falling to the right of the bar). The staunchest negativist would hold the view indicated by (d), where virtually all mutations that are not subject to purifying selection are neutral. A very important point here is that the bars underlying this illustration are not temporally static. Real populations fluctuate in size and thus the width of these bars fluctuates over time. Likewise, different species will have varying bar width. This illustrates the difficulty in making blanket statements that hold for all species at all times. The contrasting views really concern a quantitative argument as to which state holds for most species most of the time.

Scales of Analysis

Much of the history of genetic, and subsequently evolutionary, analysis is concerned with scale. Gross phenotype, as revealed by classical morphological mutants, is one such level. Chromosomal analysis is another; examples from *Drosophila* include homozygous chromosome studies and inversion polymorphisms. Once the biochemical

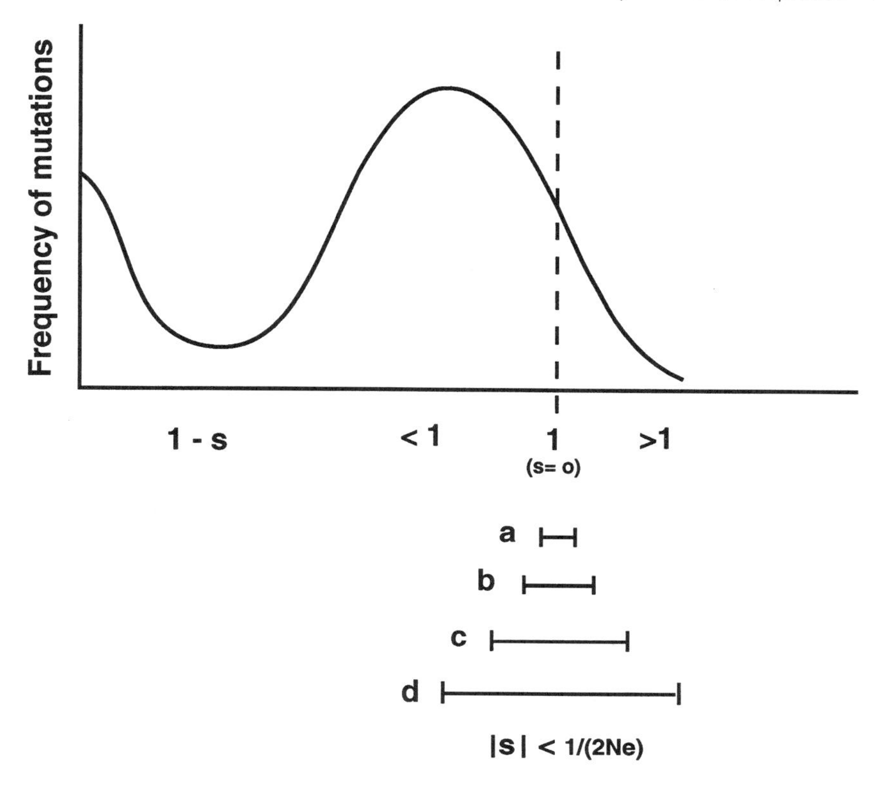

FIGURE 12-1 Hypothetical distribution of fitness effects of newly arising mutations. s is the selection coefficient; N_e is the effective population size. Bars indicate the region under which the mutations are effectively neutral.

techniques of allozyme analysis were introduced, proteins became the scale of analysis. Finally, we have hit the bottom of the trough with DNA sequencing; given today's understanding and paradigms, it is not conceivable to go below this scale. It is as though evolutionary geneticists have been using a zoom-lens microscope to delve progressively deeper and deeper into the genetic basis of evolutionary change. We now have a good, and will soon have a much better, picture of the detailed pattern of genetic variation within and between species of *Drosophila.* The challenge now is to translate pattern into processes and mechanisms. This is the time-honored Darwinian tradition that has served evolutionary biology for so long.

From a biological perspective, the most important question we can ask of the accumulating data is whether we have found the genetic basis of adaptive evolutionary change. We recounted in this book the few instances where there is clear evidence of natural selection acting on genetic variation to mold adaptations, but, I would argue, we have not attained a general theory of adaptation at the molecular genetic level. In zooming our analysis from morphology to chromosome, to proteins, to the nucleotide, I suspect we have passed right through a scale of analysis that may well be the key to understanding genetic mechanisms of adaptation. Single nucleotides and single genes act in the context of many other nucleotides and genes. Perhaps the scale of analysis we need to consider is on the order of 10^4 to 10^6 base pairs. We need to consider interactions of nucleotides, not just in the conventional sense of an amino acid change in one position being compensated by a second amino acid change in the same or different protein. We need to consider interactions at the DNA and RNA levels, interactions perhaps crucial to gene function. We know very little about the three-dimensional structure and function of nucleic acids. The observation of linkage disequilibrium among nucleotides in introns (figure 10-12) and the fact that sometimes synonymous substitutions in pseudogenes occur at a faster rate than nonsynonymous substitutions (table 10-27) should humble us: we have no real understanding of the causes of these observations, nor even testable hypotheses. The scale at which selection is acting at the molecular level is virtually *terra incognito.* To paraphrase Richard Lewontin, context and contingency is not simply detail, it is of the essence.

Can evolutionary studies transcend the Darwinian tradition of inferring process from pattern? Can it become a truly experimental science? At one level this has been achieved; witness all the manipulative studies discussed in chapter 4. However, at the more fundamental level of molecules, this has been rarely achieved (the only cases being those where the substrate of an enzyme has been artificially deleted or augmented in laboratory medium). However, it may be hoped that *Drosophila* could once again lead the way in such studies. The reason is that DNA transformation is possible. This is potentially an extremely powerful method to begin to unravel the molecular genetic basis of adaptations. However, three technical problems remain before the potential of this technique for evolutionary studies can be realized. First, to date, only *D. melanogaster* can be transformed with any efficiency. From an evolutionary perspective, it will be necessary to perform reciprocal experiments on a number of species. Second, no way of targeting the site of insertion has been developed. If context is as important as many of us think, we will only be able to realistically measure the effect of genetic manipulations if the changed gene is returned to the same site in the genome. Third, the size of the DNA molecule that can be transformed needs to be increased. Many genes are larger than the largest transformation possible today. And if the important scale is larger than the gene, clearly we will need to transform much larger molecules. It can be hoped, perhaps even expected, that these technical difficulties will be surmounted in *Drosophila.* None is theoretically out of the question. Once this has occurred, there will be the potential to make a major increase in our understanding of the evolutionary process.

Development

At the beginning of the development chapter, I stated it would be the least satisfying chapter. Why? What have been the missing links in these studies? I hinted at some in that chapter and will expand on them here.

First, our understanding of *Drosophila* development is incomplete. We only know the major mechanisms and genes involved in the first stages of development, when the overall *Bauplan* is being established. Many of the relatively subtle changes that occur in evolution very likely occur later in development, or "downstream" in the developmental cascade. We know little of what happens later in development to produce the changes that distinguish species of *Drosophila.* Possible exceptions are some of the Hawaiian flies that superficially resemble some of the homeotic mutations of *D. melanogaster.* But even if such loci are involved in the evolution of the distinctive morphologies of Hawaiian flies, many modifiers, about which we know little or nothing, surely must also be involved.

Second, our knowledge is restricted almost exclusively to *D. melanogaster.* At the level of knowledge now attained, that is, the gross laying down of the overall body plan, the mechanisms in that single species may hold throughout the genus. On the other hand, they may not. As long as we use the paradigms from *D. melanogaster* to examine other species, we may be biased to observing similarities and blind to differences. I suspect this will become a more acute problem as the later "downstream" steps in development become defined. These may well be the steps that evolution has "tinkered" with to produce the 2000-plus species of the genus.

The third point concerns developmental homology and is related to the previous point. Because transformation experiments generally find that developmentally important genes from other species can rescue null *D. melanogaster* mutants, does this prove homology? And the transformations between mouse and *Drosophila*? As my colleague Junhyong Kim points out, if you replace a human humerus with a chimpanzee femur and the arm functions fairly well, does this prove homology? We need to reevaluate what one means by homology in a functional developmental context and what criteria are needed to prove homology. We may be missing a lot of the subtle differences among species by the relatively crude methods now used. The cases of cooption of genes and gene products for roles not normally played (see p. 443) give credence to the concerns expressed here.

Speciation

Of all the major topics discussed in this book, I am most optimistic that significant breakthroughs are being made, and more are just around the corner, in regard to understanding the mechanisms of speciation in *Drosophila.* Four factors have contributed to these advances. One is the discovery of the African species closely related to *D. melanogaster* and *D. simulans.* This has begun to allow the detailed genetic database of these species to be applied to analysis of reproductive isolation. Second, the increased molecular knowledge, including detailed genetic maps, means that the level of detail will increase, as well as allowing one to identify specific loci involved in reproductive isolation. Third, the increasing knowledge of the genetic control of development, and male spermatogenesis in particular, will be very important in advancing our understanding of the earliest stages of speciation. Fourth, the area of behavior genetics is also advancing; already there is some detailed information about the genetic basis of male courtship song variation, pheromone production, and olfaction. We can anticipate more knowledge in these areas that should be directly applicable to speciation studies.

Although I am optimistic about some aspects of speciation studies, other aspects remain less likely to advance rapidly. While the genetics of reproductive isolation is likely to be understood, the populational and ecological processes that lead to speciation remain poorly understood. Perhaps there will be clues in the detailed genetic analysis concerning population level processes, but this remains to be seen. I suspect more field work is what is really required, and this is not particularly fashionable research at the present time. Thus while the veil is being lifted on aspects of this "mystery of mysteries," other aspects are likely to remain mysterious for some time to come.

Conservation

Like all living organisms, it is safe to assume that *Drosophila* species are becoming extinct at a faster rate today than in recent millennia. Considering that the origin of the genus is the tropics and the centers of highest species diversity are still in the tropics, many species are particularly in danger of extinction. Many species in remote areas have yet to be discovered and described.

Given that there are over 2000 species in the genus and only a handful have served as models in genetic and evolutionary research, it can reasonably be questioned whether it really matters if species go extinct: there are plenty of species to keep biologists happy and busy. Furthermore, *Drosophila* have virtually no economic or medical importance, nor is it well-documented that they play a very significant role in maintaining ecosystem integrity, although this has not really been explored. A few examples will serve to illustrate why we need to be concerned with preserving *Drosophila* species.

One example is the discovery over the past 20 years of the six African species very closely related to *D. melanogaster* and *D. simulans*, namely, *D. yakuba, D. erecta, D. teissieri, D. orena, D. mauritiana*, and *D. sechellia.* These species have done more to shed light on the evolutionary history of *D. melanogaster* than any other single discovery. It is only when one has very close relatives in their native habitats that one can begin to understand the origin of a human commensal. Given how inadequately the African continent has been surveyed for *Drosophila* species, it is conceivable there are more close relatives of *D. melanogaster*, if we can find them before they become extinct.

The second example is Hawaii. Throughout this book, Hawaiian *Drosophila* have been brought in as some of the most intriguing examples of a number of evolutionary principles. The diversity of this magnificent fauna outstrips all other *Drosophila* combined. We have only scraped the surface of understanding these flies, yet the ecological crises occurring on these islands are doubtless causing extinctions of *Drosophila* species (Foote and Carson, 1995). About 20 species of Picture-wing *Drosophila* will be appearing on the next endangered species lists. The endangerment of one pair of species, *D. silvestris* and *D. heteroneura*, is particularly worrisome. These species have proven extremely useful as paradigms for the early stages of speciation due to the efforts of Hampton Carson and colleagues (discussed in chapter 7). *D. heteroneura* has not been seen on the Hilo side of the island of Hawaii for six years, and both species are now much rarer than they were previously (H. L. Carson, personal communication). *D. heteroneura* is one of the species on the endangered list. Evidently, a major cause of these species' decline has been the introduction of an exotic Vespulid wasp. Considering the difficulty or impossibility of culturing most Hawaiian species, once they are extinct in nature, they are gone forever.

While threat of extinction does not yet seem imminent, the cases of *D. pseudoobscura* and *D. persimilis* are also worrisome. The introduction of *D. subobscura* from the Old World about 15 years ago has allowed this exotic to coexist with its American relatives. *D. subobscura* seems to be rapidly spreading in the New World, at first into peridomestic habitats like fruit orchards, but recently it has successfully colonized native *D. pseudoobscura* habitats. It is conceivable that *D. subobscura* could actually drive *D. pseudoobscura* into extinction. Considering the long-term monitoring of genetic changes in populations of *D. pseudoobscura*, as well as all the other information on, and lessons learned from this species, it would be a major loss to evolutionary biology if it became extinct even in parts of its natural range. Given the increasing mobility of humans and their goods, we can anticipate that this scenario will become increasingly common for other species.

Drosophila may also serve as indicator species. Some species have long records of monitoring, not only in terms of presence or absence, but for relative abundance and genetic constitution of populations. As environments deteriorate, genetic changes in populations of native species may well precede extinction. To detect such changes, one needs species for which there is sufficient long-term monitoring of genetic variation to really document changes deviating from normal fluctuations. Other than *Drosophila*, it is difficult to think of any group that could serve this function.

In the realm of experimental conservation research, *Drosophila* also have an important role. Given the genetic knowledge and ease of artificially manipulating populations, they can serve as good models of the genetic consequences of decreasing populations. The exploitation of these flies for these purposes has just begun (e.g., Briscoe et al., 1992; Borlase et al., 1993; Fowler and Whitlock, 1994).

The Model

We now come full circle from chapter 1 to consider how well *Drosophila* has served as a model organism in providing general insights into phenomena relevant to at least some other groups of organisms.

The most obvious, yet fundamental, conceptual advances first made exclusively or largely with *Drosophila* concern genetic principles: Mendelian rules, the chromosome theory of inheritance, mapping by recombination, the consequences of chromosomal aberrations, transposable elements, the ubiquity of gene polymorphisms in populations, and so on. These phenomena are virtually universal in the living world. In addition to concepts, technical developments can also be cited: in situ mapping of genes onto chromosomes, the application of molecular approaches to population genetics, eukaryotic transformation with transposable elements, and so forth. In addition, in recent years it has become increasingly clear that evolutionary changes may be more conservative than previously thought, and many other aspects of knowledge from *Drosophila* have relevance outside the genus. The clearest example is development, where homeobox genes seem to be general features of metazoans. Perhaps one of the most unexpected suggestions has been that genes involved in eye formation in flies may also be involved in eye development in mammals and mollusks (Halder et al., 1995).

As a consequence of this conservation of developmental pathways, some of the knowledge from *Drosophila* has implications for medical issues. As just one example, we can cite the evident connection between mammalian oncogenes and genes controlling development in *Drosophila.* The *ras* oncogene turns out to have homologies to the singling pathway in *Drosophila* eye development; in fact, some of the steps in this pathway were first identified in flies (Wasserman et al., 1995). Evidently, evolutionarily homologous pathways have been coopted by different groups of animals to perform different functions. The power of genetics in *Drosophila* makes it possible to define the components of such pathways more efficiently than in more complex mammals.

From a solely evolutionary perspective, there is every reason to tout *Drosophila* as a model organism. In fact, I can think of very few observations made on *Drosophila* that have not been true of many other organisms. As just three examples we can cite the ubiquity of genetic variation, the selective nature of chromosomal rearrangements, and the sterility of the heterogametic sex in interspecific hybrids. Virtually every observation made on molecular evolution has held for all organisms, for example, more synonymous polymorphisms than nonsynonymous, codon usage bias, and varying rates of evolution of different functional parts of the genome. Other observations have been more applicable to insects (at least to date); a prime example of this is the reproductive incompatibility caused by endosymbionts that are also known in beetles and mosquitoes. Clearly, from a genetic and molecular evolution standpoint, knowledge gained from *Drosophila* has been generally applicable, as we would want for a model organism.

If *Drosophila* has a weakness as a model, it has been on the ecological and ecological genetic levels. However, I believe this is due more to a lack of attention to these subjects than to any inherent inappropriateness of these insects for such studies. Indeed, as chapters 5 and 6 document, ecological knowledge and how this relates to evolutionary concerns is not insubstantial. Much of the bad reputation of *Drosophila* in this regard has been due to the myopia of thinking of *Drosophila* as being *D. melanogaster* breeding in human garbage. Throughout this book, the diversity of species of this genus has been emphasized. Many species make ideal research organisms in an ecological realm, as best documented by the Sonoran desert and Hawaiian flies. Furthermore, with the discovery of close relatives of *D. melanogaster* in Africa and the discovery of presumably truly natural populations of *D. melanogaster*, this species and its relatives should begin to assume a more prominent role in ecological genetic studies, as it already has for speciation studies.

A final thought:

> Or te riman, lettor, sovra 'l tuo banco,
> dietro pensando a ciò che si preliba,
> s'esser vuoi lieto assai prima che stanco.
> Messo t' ho innanzi: omai per te ti ciba;
> chè a sè torce tutta la mia cura
> quella materia ond' io son fatto scriba.
>
> Dante

References

Abel, T., A. M. Michelson, and T. Maniatis. 1993. A Drosophila GATA family member that binds to Adh regulatory sequences is expressed in the developing fat body. Development 119:623–633.

Abraham, I. and W. W. Doane. 1978. Genetic regulation of tissue-specific expression of amylase structural genes in Drosophila. Proc. Natl. Acad. Sci. USA 75:446–4450.

Acton, A. B. 1961. An unsuccessful attempt to reduce recombination by selection. Am. Nat. 95:119–120.

Adams, W. T. and G. T. Duncan. 1979. A maximum likelihood statistical method for analyzing frequency-dependent fitness experiments. Behav. Genet. 9:7–21.

Agnew, J. D. 1976. A case of myphily involving Drosophilidae (Diptera). J. S. African Botany 42:85–95.

Aguadé, M. 1988. Restriction map variation at the *Adh* locus of *Drosophila melanogaster* in inverted and noninverted chromosomes. Genetics 119:135–140.

Aguadé, M. and C. H. Langley. 1994. Polymorphism and divergence in regions of low recombination in Drosophila. Pp. 67–76 in *Non-neutral Evolution: Theories and Molecular Data*, ed. B. Golding. Chapman and Hall, New York.

Aguadé, M., N. Miyashita, and C. H. Langley. 1989a. Reduced variation in the *yellow-achaete-scute* region in natural populations of *Drosophila melanogaster.* Genetics 122:607–615.

Aguadé, M., N. Miyashita, and C. H. Langley. 1989b. Restriction-map variation at the *Zeste-tko* region in natural populations of *Drosophila melanogaster.* Mol. Biol. Evol. 6:123–130.

Aguadé, M., N. Miyashita, and C. H. Langley. 1992. Polymorphism and divergence in the *Mst 26A* male accessory gland region in Drosophila. Genetics 132:755–770.

Aguadé, M., W. Meyers, A. D Long, and C. H. Langley. 1994. Single-strand conformation polymorphism analysis coupled with stratified DNA sequencing reveals reduced sequence variation in the *su(s)* and $su(w^a)$ regions of the *Drosophila melanogaster* X chromosome. Proc. Natl. Acad. Sci. USA 91:4658–4662.

Ahearn, J. M. and A. R. Templeton. 1989. Interspecific hybrids of *Drosophila heteroneura* and *D. silvestris.* I. Courtship success. Evolution 43:347–361.

Ahearn, J. M., H. L. Carson, Th. Dobzhansky, and K. Y. Kaneshiro. 1974. Ethological isolation among three species of the planitibia subgroup of the Hawaiian *Drosophila.* Proc. Natl. Acad. Sci. USA 71:901–903.

Akam, M., M. Averof, J. Castelli-Gair, R. Dawes, F. Falciani, and D. Ferrier. 1994. The evolving role of Hox genes in arthropods. Development 1994 Supplement:209–215.

Akashi, H. 1994. Synonymous codon usage in *Drosophila melanogaster*: Natural selection and translational accuracy. Genetics 136:927–935.

Albalat, R., G. Marfany, and R. Gonzalez-Duarte. 1994. Analysis of nucleotide substitutions and amino acid conservation in the *Drosophila Adh* genomic region. Genetica 94:27–36.

Alberola, T. M. and R. de Frutos. 1993. Gypsy homologous sequences in *Drosophila subobscura (gypsyDS).* J. Mol. Evol. 36:127–135.

Alexander, M. L. 1949. Note on gene variability in natural populations of Drosophila. Univ. Tex. Publ. 4920:63–69.

Alexander, M. L. 1952. Gene variability in the americana-texana-novamexicana complex of the virilis group of Drosophila. Univ. Texas. Publ. 5204:73–105.

Alfonzo, J. M., A. Volz, M. Hernandez, H. Ruttkey, M. Gonzalez, J. M. Larruga, V. M. Cabrera, and D. Sperlich. 1990. Mitochondrial DNA variation and genetic structure in Old-World populations of *Drosophila subobscura.* Mol. Biol. Evol. 7:123–142.

Allen, G. E. 1975. The introduction of *Drosophila* into the study of heredity and evolution: 1900–1910. Isis 66:322–333.

Allen, G. E. 1978. *Thomas Hunt Morgan. The man and his science.* Princeton University Press, Princeton, N.J.

Alonso-Pimentel, H., L. P. Tolbert, and W. B. Heed. 1994. Ultrastructural examination of the insemination reaction in *Drosophila.* Cell Tissue Res. 275:467–479.

Alvarez, G. and A. Fontdevila. 1981. Sexual selection and random mating in *Drosophila melanogaster*. Genetica Iberica 33:1–18.

Anderson, C. 1991. Genome evolution in the *Drosophila willistoni* species group, and molecular evolution of the alcohol dehydrogenase gene in *Drosophila.* Ph.D. dissertation, Yale University, New Haven, Conn.

Anderson, C., E. A. Carew, and J. R. Powell. 1993. Evolution of the *Adh* locus in the *Drosophila willistoni* group: The loss of an intron, and shift in codon usage. Mol. Biol. Evol. 10: 605–618.

Anderson, W. W. 1966. Genetic divergence in M. Vetukhiv's experimental populations of *Drosophila pseudoobscura.* Genet. Res. 7:255–266.

Anderson, W. W. 1968. Elimination of the sex-ratio X-chromosome in experimental populations of *Drosophila pseudoobscura.* Dros. Inf. Serv. 43:110–112.

Anderson, W. W. 1969. Polymorphism arising from the mating advantage of rare male genotypes. Proc. Natl. Acad. Sci. USA 64:190–197.

Anderson, W. W. 1973. Genetic divergence in body size among experimental populations of *Drosophila pseudoobscura* kept at different temperatures. Evolution 27:278–284.

Anderson, W. W. 1974. Frequent multiple insemination in a natural population of *Drosophila pseudoobscura.* Am. Nat. 108:709–711.

Anderson, W. W. 1989. Selection in natural and experimental populations of *Drosophila pseudoobscura.* Genome 31:239–245.

Anderson, W. W. and C. J. Brown. 1984. A test for rare male mating advantage with *Drosophila pseudoobscura* karyotypes. Genetics 107:577–589.

Anderson, W. W. and L. Ehrman. 1969. Mating choice in crosses between geographic populations of *Drosophila pseudoobscura.* Am. Midl. Nat. 81:47–53.

Anderson, W. W. and P. R. McGuire. 1978. Mating pattern and mating success of *Drosophila pseudoobscura* karyotypes in large experimental populations. Evolution 32:416–423.

Anderson, W. W. and T. K. Watanabe. 1974. Selection for fertility in *Drosophila pseudoobscura.* Genetics 77:559–564.

Anderson, W. W., Th. Dobzhansky, and C. D. Kastritsis. 1967. Selection and inversion polymorphism in experimental populations of *Drosophila pseudoobscura* initiated with the chromosomal constitutions of natural populations. Evolution 21:664–671.

Anderson, W. W., C. Oshima, T. Watanabe, Th. Dobzhansky, and O. Pavlovsky. 1968. Genetics of natural populations. XXIX. A test of the possible influence of two insecticides on the chromosomal polymorphism in *Drosophila pseudoobscura.* Genetics 58:423–434.

Anderson, W. W., Th. Dobzhansky, and O. Pavlovsky. 1972. A natural population of *Drosophila* transferred to a laboratory environment. Heredity 28:101–107.

Anderson, W. W., L. Levine, O. Olvera, J. R. Powell, M. E. de la Rosa, V. M. Salceda, L. L. Gaso, and J. Guzman. 1979. Evidence for selection by male mating success in natural populations of *Drosophila pseudoobscura.* Proc. Natl. Acad. Sci. USA 76:1519–1523.

Anderson, W. W., J. Arnold, D. G. Baldwin, A. T. Beckenbach, C. J. Brown, S. H. Bryant, J. A. Coyne, L. G. Harshman, W. B. Heed, D. E. Jeffery, L. B. Klaczko, B. C. Moore, J. M. Porter, J. R. Powell, T. Prout, S. W. Schaeffer, J. C. Stephens, C. E. Taylor, M. E. Turner, G. O. Williams, and J. A. Moore. 1991. Four decades of inversion polymorphism in *Drosophila pseudoobscura.* Proc. Natl. Acad. Sci. USA 88:10367–10371.

Andjelkovic, M., M. Milanovic, and M. Stamenkovic-Radak. 1987. Adaptive significance of amylase polymorphism in *Drosophila.* I. The geographical pattern of allozyme polymorphism at the amylase locus in *Drosophila subobscura.* Genetica 74:161–171.

Andjelkovic, M., M. Stamenkovic-Radak, M. Sekulic, and M. Milanovic. 1991. Adaptive significance of amylase polymorphism in *Drosophila.* III. Geographic patterns in *Drosophila subobscura* tissue-specific expression of amylase in adult midgut. Genet. Sel. Evol. 23: 197–204.

Anxolabéhère, D. 1980. The influence of sexual and larval selection on the maintenance of polymorphism at the sepia locus in *Drosophila melanogaster.* Genetics 95:743–755.

Anxolabéhère, D. and G. Periquet. 1970. Résistance des imagos aux bases temperature chez *Drosophila melanogaster.* Bull. Soc. Zool. Fr. 95:61–70.

Anxolabéhère, D. and G. Periquet. 1987. P-homologous sequences in Diptera are not restricted to the Drosophilidae family. Genetica 39:211–222.

Aquadro, C. F. 1993. Molecular populations genetics of *Drosophila.* Pp. 222–266 in *Molecular Approaches to Fundamental and Applied Entomology*, eds. J. Oakeshott and M. J. Whitten. Springer-Verlag, New York.

Aquadro, C. F., S. F. Desse, M. M. Bland, C. H. Langley, and C. C. Laurie-Ahlberg. 1986. Molecular population genetics of the alcohol dehydrogenase gene region of *Drosophila melanogaster.* Genetics 114:1165–1190.

Aquadro, C. F., K. M. Lado, and W. A. Noon. 1988. The *rosy* region of *Drosophila melanogaster* and *D. simulans.* I. Contrasting levels of naturally occurring DNA restriction map variation and divergence. Genetics 119:875–888.

Aquadro, C. F., H. Tachida, C. H. Langley, K. Harada, and T. Mukai. 1990. Increased variation in ADH enzyme activity in Drosophila mutation-accumulation experiment is not due to transposable elements at the *Adh* structural locus. Genetics 126:915–919.

Aquadro, C. F., A. L. Weaver, S. W. Schaeffer, and W. W. Anderson. 1991. Molecular evolution of inversions in *Drosophila pseudoobscura*: the amylase gene region. Proc. Natl. Acad. Sci. USA 88:305–309.

Aquadro, C. F., R. M. Jennings, M. M. Bland, C. C. Laurie, and C. H. Langley. 1992. Patterns of naturally occurring restriction map variation, DDC activity variation and linkage disequilibrium in the dopa decarboxylase gene region of *Drosophila melanogaster.* Genetics 132:443–452.

Aquadro, C. F., D. J. Begun, and E. C. Kindahl. 1994. Selection, recombination, and DNA polymorphism in *Drosophila.* Pp. 46–56 in *Non-neutral Evolution: Theories and Molecular Data*, ed. B. Golding. Chapman and Hall, New York.

Armstrong, E. and L. Bass. 1989. *Nosema kingi*: Effects on fecundity, fertility, and longevity of *Drosophila melanogaster*. J. Exp. Zool. 250:82–86.

Arnault, C. and I. Dufournel. 1994. Genome and stresses: Reactions against aggressions, behavior of transposable elements. Genetica 93:149–160.

Arnold, J. 1982. Statistics of natural populations: Seasonal variation in inversion frequencies of Mexican *Drosophila pseudoobscura*. Ph.D. dissertation, Yale University, New Haven, Conn.

Ashburner, M. 1981. Entomophagous and other bizarre Drosophilidae. Pp. 395–429 in *The Genetics and Biology of Drosophila, Vol. 3a*, eds. M. Ashburner, H. L. Carson, and J. N. Thompson, Jr. Academic Press, New York.

Ashburner, M. 1989. *Drosophila: A Laboratory Handbook*. Cold Spring Harbor Press, Cold Spring Harbor, N.Y.

Ashburner, M. 1993. Epilogue. Pp. 1493–1506 in *The Development of Drosophila melanogaster*, eds. M. Bate and A. Martinez Arias. Cold Spring Harbor Press, Cold Spring Harbor, N.Y.

Ashburner, M. and R. Drysdale. 1994. The Drosophila genetic data base. Development 128: 2077–2079.

Ashburner, M. and F. Lemeunier. 1976. Relationships within the *melanogaster* species subgroup of the genus *Drosophila (Sophophora)*. I. Inversion polymorphism in *Drosophila melanogaster* and *Drosophila simulans*. Proc. Roy. Soc. Lond. B 193:137–157.

Ashburner, M., D. M. Glover, R. D. C. Saunders, I. Duncan, D. Hartl, J. Merriam, G. Lee, J. Johnsen, F. C. Kafatos, I. Sidén-Kiamos, C. Louis, and C. Savakis. 1991. Genome maps 1991: *Drosophila* component. Science 254:247–262.

Asmussen, M. A. and M. T. Clegg. 1982. Rates of decay of linkage disequilibrium under two-locus models of selection. J. Math. Biol. 14:37–70.

Atkinson, P. W., L. E. Mills, W. T. Starmer, and D. T. Sullivan. 1988. Structure and evolution of the *Adh* genes of *Drosophila mojavensis*. Genetics 120:713–723.

Atkinson, W. D. 1978. Ecological studies of the breeding sites and reproductive strategies of domestic species of *Drosophila*. Ph.D. thesis, University of Leeds, U.K.

Atkinson, W. D. 1979. A comparison of the reproductive strategies of domestic species of *Drosophila*. J. Anim. Ecol. 48:53–64.

Atkinson, W. D. 1982. An ecological interaction between citrus fruit, *Penicillium* moulds and *Drosophila immigrans* Sturtevant (Diptera: Drosophilidae). Ecol. Entomol. 6:339–344.

Atkinson, W. D. and J. A. Miller. 1980. Lack of habitat choice in a natural population of *Drosophila subobscura*. Heredity 44:193–199.

Atkinson, W. D. and B. Shorrocks. 1977. Breeding site specificity in domestic species of *Drosophila*. Oecologia 29:223–232.

Attardi, G. and G. Schatz. 1988. Biogenesis of mitochondria. Ann. Rev. Cell Biol. 4:289–333.

Aubert, J. and M. Solignac. 1990. Experimental evidence for mitochondrial DNA introgression between *Drosophila* species. Evolution 44:1272–1282.

Averof, M. and M. Akam. 1993. Hom/Hox genes of *Artemia*: implications for the origin of insects and crustacean body plans. Current Biology 3:73–78.

Avise, J. C. 1986. Mitochondrial DNA and the evolutionary genetics of higher animals. Philos. Trans. Roy. Soc. Lond. (Biol.) 312:325–342.

Avise, J. C. 1994. *Molecular Markers, Natural History, and Evolution*. Chapman and Hall, New York.

Awgulewitsch, A. and D. Jacobs. 1992. *Deformed* autoregulatory element from *Drosophila* functions in a conserved manner in transgenic mice. Nature 358:341–344.

Ayala, F. J. 1968a. Environmental factors limiting the productivity and size of experimental populations of *Drosophila serrata* and *D. birchii*. Ecology 49:562–565.

Ayala, F. J. 1968b. Genotype, environment, and population numbers. Science 162:1453–1459.

Ayala, F. J. 1969. Genetic polymorphism and interspecific competitive ability in *Drosophila*. Genet. Res. 14:95–102.

Ayala, F. J. 1970a. Competition, coexistence, and evolution. Pp. 121–158 in *Essays in Evolution and Genetics in Honor of Theodosius Dobzhansky*, eds. M. K. Hecht and W. C. Steere. Appleton-Century-Crofts, New York.

Ayala, F. J. 1970b. Competition between species: Frequency dependence. Science 171:820–824.

Ayala, F. J. 1972a. Frequency-dependent mating advantage in *Drosophila*. Behav. Genet. 2: 85–91.

Ayala, F. J. 1972b. Competition between species. Am. Sci. 60:348–357.

Ayala, F. J. 1975. Genetic differentiation during the speciation process. Evol. Biol. 8:1–78.

Ayala, F. J. and Th. Dobzhansky. 1974. A new subspecies of *Drosophila pseudoobscura*. Pan-Pacific Entomol. 50:211–219.

Ayala, F. J. and J. R. Powell. 1972. Enzyme variability in the *Drosophila willistoni* group. VI. Levels of polymorphism and the physiological function of enzymes. Biochem. Genet. 7: 331–345.

Ayala, F. J. and M. L. Tracey. 1973. Enzyme variability in the *Drosophila willistoni* group. VIII. Genetic differentiation and reproductive isolation between two subspecies. J. Heredity 64:120–124.

Ayala, F. J. and M. L. Tracey. 1974. Genetic differentiation within and between species of the *Drosophila willistoni* group. Proc. Natl. Acad. Sci. USA 71:999–1003.

Ayala, F. J., J. R. Powell, and Th. Dobzhansky. 1971. Polymorphisms in continental and island populations of *Drosophila willistoni*. Proc. Natl. Acad. Sci., USA 68:2480–2483.

Ayala, F. J., J. R. Powell, M. L. Tracey, C. A. Mourao, and S. Perez-Salas. 1972. Enzyme variability in the *Drosophila willistoni* group. IV. Genic variation in natural populations of *Drosophila willistoni*. Genetics 70:113–139.

Ayala, F. J., M. E. Gilpin, and J. G. Ehrenfeld. 1973. Competition between species: theoretical models and experimental tests. Theoret. Pop. Biol. 4:331–356.

Ayala, F. J., M. L. Tracey, L. G. Barr, and J. G. Ehrenfeld. 1974a. Genetic and reproductive differentiation of the subspecies, *Drosophila equinoxialis carribensis*. Evolution 28:24–41.

Ayala, F. J., M. L. Tracey, L. G. Barr, J. F. McDonald, and S. Perez-Salas. 1974b. Genetic variation in natural populations of five *Drosophila* species and the hypothesis of selective neutrality of protein polymorphisms. Genetics 77:343–384.

Ayala, F. J., M. L. Tracey, D. Hedgecock, and R. C. Richmond. 1974c. Genetic differentiation during the speciation process in *Drosophila*. Evolution 28:576–592.

Ayala, F. J., L. Serra, and A. Prevosti. 1989. A grand experiment in evolution: the *Drosophila subobscura* colonization of the Americas. Genome 31:246–255.

Ayala, F. José and D. L. Hartl. 1993. Molecular drift of the *bride-of-sevenless* (*boss*) gene in *Drosophila*. Mol. Biol. Evol. 10:1030–1040.

Ayala, F. José, B. S. W. Chang, and D. L. Hartl. 1993. Molecular evolution of the *Rh3* gene in *Drosophila*. Genome 92:23–32.

Babcock, C. S. and W. W. Anderson. 1996. Molecular evolution of the sex-ratio inversion complex in *Drosophila pseudoobscura*: analysis of the *esterase* gene region. Mol. Biol. Evol. 13:297–308.

Bächli, G. 1994. A revised list of the family Drosophilidae. Available from FlyBase as ftp: flybase/allied-data/species.txt.

Baker, B. S. 1989. Sex in flies: The splice of life. Nature 340:521–524.

Baker, R. H., R. DeSalle, W. B. Heed, and W. J. Etges. 1994. Molecular systematics of *Drosophila repleta* group species. Abst. 3rd Internat. Congr. Dipterology, Guelph, pp. 10–11.

Baker, R. H. A. 1979. Studies on the interactions between *Drosophila* parasites. Ph.D. thesis, Oxford University, U.K.

Ballard, J. W. O. and M. Kreitman. 1994. Unraveling selection in the mitochondrial genome of Drosophila. Genetics 138:757–772.

Bang, A. G., V. Hartenstein, and J. W. Posakony. 1991. Hairless is required for the development of adult sensory organ precursor cells in *Drosophila.* Development 111:89–104.

Barker, J. S. F. 1962. Studies of selective mating using the yellow mutant of *Drosophila melanogaster.* Genetics 47:623–640.

Barker, J. S. F. 1963. The estimation of relative fitness of *Drosophila* populations. III. The fitness of certain strains of *Drosophila melanogaster.* Evolution 17:138–146.

Barker, J. S. F. 1973. Natural selection for coexistence or competitive ability in laboratory populations of *Drosophila.* Egypt. J. Genet. Cytol. 2:288–315.

Barker, J. S. F. 1979. Inter-locus interactions: A review of experimental evidence. Theoret. Pop. Biol. 16:323–346.

Barker, J. S. F. 1982. Population genetics of *Opuntia* breeding *Drosophila* in Australia. Pp. 209–224 in *Ecological Genetics and Evolution: The Cactus-Yeast-Drosophila Model System,* eds. J. S. F. Barker and W. T. Starmer. Academic Press, New York.

Barker, J. S. F. 1983. Interspecific competition. Pp. 285–341 in *The Genetics and Biology of Drosophila Vol. 3c,* eds. M. Ashburner, H. L. Carson, and J. N. Thompson, Jr. Academic Press, New York.

Barker, J. S. F. 1990. Experimental analysis of habitat selection and maintenance of genetic variation. Pp. 161–175 in *Ecological and Evolutionary Genetics of Drosophila,* eds. J. S. F. Barker, W. T. Starmer, and R. J. MacIntyre. Plenum, New York.

Barker, J. S. F. 1992. Genetic variation in cactophilic *Drosophila* for oviposition on natural yeast substrates. Evolution 46:1070–1083.

Barker, J. S. F. and L. J. Cummins. 1969. The effect of selection for sternopleural bristle number on mating behaviour in *Drosophila melanogaster.* Genetics 61:713–719.

Barker, J. S. F. and L. J. E. Karlsson. 1974. Effects of population size and selection intensity on responses to disruptive selection in *Drosophila melanogaster.* Genetics 78:715–735.

Barker, J. S. F. and J. C. Mulley. 1976. Isozyme variation in natural populations of *Drosophila buzzatii.* Evolution 30:213–233.

Barker, J. S. F. and W. T. Starmer (eds.). 1982. *Ecological Genetics and Evolution: The Cactus-Yeast-Drosophila Model System.* Academic Press, New York.

Barker, J. S. F., G. J. Parker, G. L. Toll, and P. R. Widders. 1981a. Attraction of *Drosophila buzzatii* and *D. aldrichi* to species of yeasts isolated from their natural environments. I. Laboratory experiments. Aust. J. Biol. Sci. 34:593–612.

Barker, J. S. F., G. L. Toll, P. D. East, and P. R. Widders. 1981b. Attraction of *Drosophila buzzatii* and *D. aldrichi* to species of yeasts isolated from their natural environments. II. Field experiments. Aust. J. Biol. Sci. 34:613–624.

Barker, J. S. F., P. D. East, and B. S. Weir. 1986. Temporal and microgeographic variation in allozyme frequencies in natural populations of *Drosophila buzzatii.* Genetics 112:577–611.

Barker, J. S. F., W. T. Starmer, and D. C. Vacek. 1987. Analysis of spatial and temporal variation in the community structure of yeasts associated with decaying *Opuntia* cactus. Microb. Ecol. 14:267–276.

Barker, J. S. F., W. T. Starmer, and R. J. MacIntyre (eds.). 1990. *Ecological and Evolutionary Genetics of Drosophila.* Plenum, New York.

Barrio, E., A. Latorre, and A. Moya. 1994. Phylogeny of the *Drosophila obscura* species group deduced from mitochondrial DNA sequences. J. Mol. Evol. 39:478–488.

Bartelt, R. J., M. T. Arnold, A. M. Schaner, and L. L. Jackson. 1986. Comparative analysis of the cuticular hydrocarbons in the *Drosophila virilis* species group. Comp. Biochem. Physiol. 83B:731–742.

Barton, N. H. and B. Charlesworth. 1984. Genetic revolutions, founder effects, and speciation. Ann. Rev. Ecol. Syst. 15:133–164.

Bastock, M. 1956. A gene mutation which changes a behavior pattern. Evolution 10:421–439.

Bate, M. and A. Martinez Arias (eds.). 1993. *The Development of Drosophila melanogaster. Volumes I and II.* Cold Spring Harbor Press, Cold Spring Harbor, N.Y.

Bateson, W. 1909. *Mendel's Principles of Heredity.* Cambridge University Press, Cambridge, U.K.

Battaglia, B. and H. Smith. 1961. The Darwinian fitness of polymorphic and monomorphic populations of *Drosophila pseudoobscura.* Heredity 16:475–484.

Beardmore, J. A., Th. Dobzhansky, and O. Pavlovsky. 1960. An attempt to compare the fitness of polymorphic and monomorphic populations of *Drosophila pseudoobscura.* Heredity 14: 19–33.

Beckenbach, A. T. 1978. The "sex-ratio" trait in *Drosophila pseudoobscura*: fertility relations of males and meiotic drive. Am. Nat. 112:97–117.

Beckenbach, A. T. 1981. Multiple mating and sex-ratio trait in *Drosophila pseudoobscura.* Evolution 35:275–281.

Beckenbach, A. T. 1983. Fitness analysis of the sex-ratio polymorphism in experimental populations of *Drosophila pseudoobscura.* Am. Nat. 121:630–648.

Beckenbach, A. T. and A. Prevosti. 1986. Colonization of North America by the European species *Drosophila subobscura* and *D. ambigua.* Am. Midl. Nat. 115:10–18.

Beckenbach, A. T., Y. W. Wei, and H. Liu. 1993. Relationships in the *Drosophila obscura* species group, inferred from mitochondrial cytochrome oxidase II sequences. Mol. Biol. Evol. 10:619–634.

Beech, R. N. and A. J. Leigh Brown. 1989. Insertion-deletion variation at the *yellow-achaete-scute* region in two natural populations of *Drosophila melanogaster.* Genet. Res. 53:7–15.

Beeman, R. W., J. J. Stuart, M. S. Haas, and R. E. Denell. 1989. Genetic analysis of the homeotic gene complex (HOM-C) in *Tribolium castaneum.* Develop. Biol. 133:196–209.

Begon, M. 1975. The relationships of *Drosophila obscura* Fallen and *D. subobscura* Collin to naturally occurring fruits. Oecologia 20:255–277.

Begon, M. 1976a. Dispersal density and microdistribution in *Drosophila subobscura* Collin. J. Anim. Ecol. 45:441–456.

Begon, M. 1976b. Temporal variation in the reproductive condition of *Drosophila obscura* Fallen and *D. subobscura* Collin. Oecologia 23:31–47.

Begon, M. 1977. The effective size of a natural *Drosophila subobscura* population. Heredity 38:13–18.

Begon, M. 1978. Population densities in *Drosophila obscura* Fallen and *D. subobscura* Collin. Ecol. Entomol. 3:1–12.

Begon, M. 1982. Yeasts and Drosophila. Pp. 345–384 in *The Genetics and Biology of Drosophila, Vol. 3b*, eds. M. Ashburner, H. L. Carson, and J. N. Thompson, Jr. Academic Press, New York.

Begon, M. and B. Shorrocks. 1978. The feeding and breeding sites of *Drosophila obscura* Fallen and *D. subobscura* Collin. J. Nat. Hist. 12:137–151.

Begun, D. J. and C. F. Aquadro. 1991. Molecular population genetics of the distal portion of the X chromosome in *Drosophila*: evidence for genetic hitchhiking of the *yellow-achaete* region. Genetics 129:1147–1158.

Begun, D. J. and C. F. Aquadro. 1992. Levels of naturally occurring DNA polymorphism correlate with recombination rates in *D. melanogaster.* Nature 356:519–520.

Begun, D. J. and C. F. Aquadro. 1993. African and North American populations of *Drosophila melanogaster* are very different at the DNA level. Nature 365:548–550.

Begun, D. J. and C. F. Aquadro. 1994. Evolutionary inferences from DNA variation at the 6-phosphogluconate dehydrogenase locus in natural populations of Drosophila: selection and geographic differentiation. Genetics 136:155–171.

Begun, D. J. and C. F. Aquadro. 1995. Evolution at the tip and base of the X chromosome in an African population of *Drosophila melanogaster.* Mol. Biol. Evol. 12:382–390.

Begun, D. J., S. N. Boyer, and C. F. Aquadro. 1994. *cut* locus variation in natural populations of *Drosophila.* Mol. Biol. Evol. 11:806–809.

Bell, A. E., C. H. Moore, and D. C. Warren. 1955. The evaluation of new methods for the improvement of quantitative characteristics. Cold Spring Harbor Symp. Quant. Biol. 20: 197–212.

Belyaeva, E. S., E. V. Ananiev, and V. A. Gvozdev. 1984. Distribution of mobile dispersed genes (mdg-1 and mdg-3) in chromosomes of *Drosophila melanogaster.* Chromosoma 90: 16–19.

Benado, M. and D. Brncic. 1970. Adaptive superiority of inversion heterozygotes in *Drosophila pavani.* Genetica (Iberica) 22:133–147.

Bénassi, V., S. Aulard, S. Mazeau, and M. Veuille. 1993. Molecular variation of *Adh* and *P6* genes in an African population of *Drosophila melanogaster* and its relation to chromosomal inversions. Genetics 134:789–799.

Bengtsson, B. O. and W. F. Bodmer. 1976. On the increase of chromosome mutations under random mating. Theor. Pop. Biol. 9:260–281.

Benjamin, R. K. 1973. Laboulbeniomycetes. Pp. 223–246 in *The Fungi: An Advanced Treatise*, eds. G. C. Ainsworth et al. Academic Press, New York.

Benkel, B. F. and D. A. Hickey. 1986a. Glucose repression of amylase gene expression in *Drosophila melanogaster.* Genetics 114:137–144.

Benkel, B. F. and D. A. Hickey. 1986b. The interaction of genetic and environmental factors in the control of amylase gene expression in *Drosophila melanogaster.* Genetics 114: 943–954.

Bennett, J. 1960. A comparison of selective methods and a test of the pre-adaptation hypothesis. Heredity 15:65–77.

Berg, R. L. 1941. Lowering of the mutation rate as a result of intra-specific hybridization. C. R. Acad. Sci., URSS 32:213–215.

Berg, R. L. 1942. Dominance of deleterious mutations in the populations of *Drosophila melanogaster.* C. R. Acad. Sci., URSS 36:212–218.

Berger, E. M. 1971. A temporal survey of allelic variation in natural and laboratory populations of *D. melanogaster.* Genetics 67:121–136.

Bernardi, G. 1993. The vertebrate genome: Isochores and evolution. Mol. Biol. Evol. 10:186–204.

Bernardi, G. and G. Bernardi. 1986. Compositional constraints and genome evolution. J. Mol. Evol. 24:1–11.

Bernardi, G., B. Olofsson, J. Filipski, M. Zerial, J. Salinas, G. Cuny, M. Meunier-Rotival, and F. Rodier. 1985. The mosaic genome of warm-blooded vertebrates. Science 228:953–958.

Berry, A. J., and M. Kreitman. 1993. Molecular analysis of an allozyme cline: Alcohol dehydrogenase in *Drosophila melanogaster* on the East Coast of North America. Genetics 134: 869–893.

Berry, A. J., J. W. Ajioka, and M. Kreitman. 1991. Lack of polymorphism on the Drosophila fourth chromosome resulting from selection. Genetics 129:1111–1117.

Besansky, N. J. and J. R. Powell. 1992. Reassociation kinetics of *Anopheles gambiae* (Diptera: Culicidae) DNA. J. Med. Entomol. 29:125–128.

Beverley, S. M. and A. C. Wilson. 1984. Molecular evolution in *Drosophila* and the higher Diptera. II. A time scale for fly evolution. J. Mol. Evol. 21:1013.

Beverley, S. M. and A. C. Wilson. 1985. Ancient origin for Hawaiian Drosophilinae inferred from protein comparisons. Proc. Natl. Acad. Sci. USA 82:4753–4757.

Biémont, C. 1986. Polymorphism of the mdg-1 and I mobile elements in *Drosophila melanogaster.* Chromosoma 93:393–397.

Biémont, C. 1993. Population genetics of transposable DNA elements: A Drosophila point of view. Pp. 74–91 in *Transposable Elements and Evolution*, ed. J. F. McDonald. Kluwer Press, Dordrecht. *Also*: Genetica 86:67–84 (1992).

Biémont, C. and C. Gautier. 1988. Localization polymorphism of mdg-1, copia, I and P mobile elements in genomes of *Drosophila melanogaster*, from data of inbred lines. Heredity 60: 335–346.

Biémont, C., F. Lemeunier, M. P. Garcia Guerreiro, J. F. Brookfield, C. Gautier, S. Aulard, and E. G. Pasyukova. 1994. Population dynamics of the copia, mdg1, mdg3, gypsy, and P transposable elements in a natural population of *Drosophila melanogaster*. Genet. Res. 63:197–212.

Biessmann, H., K. Valgeirsdottir, A. Lofsky, C. Chin, B. Ginther, R. W. Levis, and M.-L. Pardue. 1992. HeT-A, a transposable element specifically involved in "healing" broken chromosome ends in *Drosophila melanogaster*. Mol. Cell. Biol. 12:3910–3918.

Bingham, P. M., M. G. Kidwell, and G. M. Rubin. 1982. The molecular basis of P-M hybrid dysgenesis: The role of the P element, a P strain-specific transposon family. Cell 29: 995–1003.

Birch, L. C. 1955. Selection in *Drosophila pseudoobscura* in relation to crowding. Evolution 9:389–399.

Birch, L. C. 1960. The genetic factor in population ecology. Am. Nat. 94:5–24.

Bird, A. P. and M. H. Taggart. 1980. Variable patterns of total DNA and rDNA methylation in animals. Nucleic Acids Res. 8:1485–1497.

Bock, I. 1984. Interspecific hybridization in the genus *Drosophila*. Evol. Biol. 18:41–70.

Boesiger, E. 1968. Estimation globale de l'age des femelles de *Drosophila melanogaster* capturees dans des populations naturaelles. Compte rendu Seances Soc. Biol. 162:358–361.

Booker, R. B. and J. W. Truman. 1989. *Octopod*, a homeotic mutation of the moth *Manduca sexta*, influences the fate of identifiable pattern elements within the CNS. Development 105:621–629.

Bomze, H. M. and A. Javier López. 1994. Evolutionary conservation of the structure and expression of alternatively spliced *Ultrabithorax* isoforms from Drosophila. Genetics 136: 965–977.

Bonaccorsi, S. and A. Lohe. 1991. Fine mapping of satellite DNA sequences along the Y chromosome of *Drosophila melanogaster*: Relationships between satellite sequences and fertility factors. Genetics 129:177–189.

Borisov, A. I. 1970a. Adaptive value of chromosomal polymorphism. IV. The prolonged observations on a populations of *Drosophila funebris* in Moscow town. Genetika (Moscow) 6: 115–122.

Borisov, A. I. 1970b. Disturbance of panmixis in natural populations of *Drosophila funebris* polymorphic with respect to the II-1 inversions. Genetika (Moscow) 6:61–67.

Borlase, S. C., D. A. Loebel, R. Frankham, R. K. Nurthen, D. A. Briscoe, and G. E. Daggard. 1993. Modeling problems in conservation genetics using captive Drosophila populations: Consequences of equalization of family sizes. Conserv. Biol. 7:122–131.

Bos, M., B. Burnet, R. Farrow, and R. A. Woods. 1977. Mutual facilitation between larvae of the sibling species, *Drosophila melanogaster* and *D. simulans*. Evolution 31:824–828.

Brady, J. P. and R. C. Richmond. 1990. Molecular analysis of evolutionary changes in the expression of Drosophila esterases. Proc. Natl. Acad. Sci. USA 87:8217–8221.

Brady, J. P. and R. C. Richmond. 1992. An evolutionary model for duplication and divergence of esterase genes in *Drosophila*. J. Mol. Evol. 34:506–521.

Brady, J. P., R. C. Richmond, and J. G. Oakeshott. 1990. Cloning of the esterase-5 locus from *Drosophila pseudoobscura* and comparison with its homologue in *D. melanogaster*. Mol. Biol. Evol. 7:525–546.

Braverman, J. M., R. R. Hudson, N. L. Kaplan, C. H. Langley, and W. Stephan. 1995. The hitchhiking effect on the site frequency spectrum of DNA polymorphisms. Genetics 140: 783–796.

Brezinsky, L., T. D. Humphreys, and J. A. Hunt. 1993. Evolution of the transposable element Uhu in five species of Hawaiian *Drosophila.* Pp. 275–289 in *Transposable Elements and Evolution*, ed. J. F. McDonald. Kluwer Publishers, Dordrecht.

Bridges, C. B. and K. S. Brehme. 1944. The mutations of *Drosophila melanogaster.* Carnegie Institution of Washington, Publication 552.

Bridges, C. B. and M. Demerec. 1934. Forward. Dros. Inf. Serv. 1:2.

Briscoe, D. A., J. M. Malpica, A. Robertson, G. J. Smith, R. Frankham, R. G. Banks, and J. S. F. Barker. 1992. Rapid loss of genetic variation in large captive populations of Drosophila flies: Implications for the genetic management of captive populations. Conserv. Biol. 6: 416–425.

Brncic, D. 1966. Ecological and cytogenetics of *Drosophila flavopilosa*, a neotropical species living in *Cestrum* flowers. Evolution 20:16–29.

Brncic, D. 1967. Chromosomal polymorphism in an ecologically restricted species of *Drosophila* living in Chile. Cienc. Cult. (San Paulo) 19:45–53.

Brncic, D. 1968. The effects of temperature on chromosomal polymorphism of *Drosophila flavopilosa* larvae. Genetics 59:427–432.

Brncic, D. 1970. Studies on the evolutionary biology of Chilean species of *Drosophila*. Pp. 401–436 in *Essays in Evolution and Genetics in Honor of Theodosius Dobzhansky*, eds. M. K. Hecht and W. C. Steere. Appleton-Century-Crofts, New York.

Brncic, D. 1972. Seasonal fluctuations of inversion polymorphism in *Drosophila flavopilosa* and the relationships with certain ecological factors. Univ. Texas Publ. 7213:103–116.

Brncic, D. 1973. Further studies of chromosomal polymorphism in *Drosophila pavani.* J. Heredity 64:175–180.

Brncic, D. 1976. Geographic variation of the chromosomal polymorphism in *Drosophila flavopilosa.* Can. J. Genet. Cytol. 18:111–118.

Brncic, D. 1983. Ecology of flower-breeding *Drosophila.* Pp. 333–382 in *The Genetics and Biology of Drosophila, Vol. 3d*, eds. M. Ashburner, H. L. Carson, and J. N. Thompson, Jr. Academic Press, New York.

Brower, A. V. Z. and R. DeSalle. 1994. Practical and theoretical considerations for choice of a DNA sequence region in insect molecular systematics, with a short review of published studies using nuclear gene regions. Ann. Entomol. Soc. Am. 87:702–716.

Brown, C. J., C. F. Aquadro, and W. W. Anderson. 1990. DNA sequence evolution of the amylase multigene family in *Drosophila pseudoobscura.* Genetics 126:131–138.

Brown, S. J., R. B. Hilgenfeld, and R. E. Denell. 1994. The beetle *Tribolium castaneum* has a *fushi tarazu* homolog expressed in stripes during segmentation. Proc. Natl. Acad. Sci. USA 91:12922–12926.

Brown, W. M. 1983. Evolution of animal mitochondrial DNA. Pp. 62–88 in *Evolution of Genes and Proteins*, eds. M. Nei and R. Koehn. Sinauer Associates, Sunderland, Mass.

Brown, W. M., M. George, and A. C. Wilson. 1979. Rapid evolution of animal mitochondrial DNA. Proc. Natl. Acad. Sci. USA 76:1967–1971.

Brun, G. and N. Plus. 1980. The viruses of *Drosophila*. Pp. 625–693 in *Genetics and Biology of Drosophila Vol. 2d*, eds. M. Ashburner and T. R. F. Wright. Academic Press, New York.

Brussard, P. F. 1984. Geographic patterns and environmental gradients: The central-marginal model in *Drosophila* revisited. Ann. Rev. Ecol. Syst. 15:25–64.

Bryant, E. H., A. Kence, and K. T. Kimball. 1980. A rare-male advantage in the housefly induced by wing-clipping and some general considerations for Drosophila. Genetics 96: 975–993.

Bryant, S. H. 1976. The frequency of allelism of lethal chromosomes in isolated desert populations of *Drosophila pseudoobscura.* Genetics 84:777–786.

Bryant, S. H., A. T. Beckenbach, and G. A. Cobbs. 1982. Sex-ratio, sex composition, and relative abundance in *Drosophila pseudoobscura.* Evolution 36:27–34.

Budnick, M. and D. Brncic. 1975. Effects of larval biotic residues on viability in four species of *Drosophila.* Evolution 29:777–780.

Budnick, M. and D. Brncic. 1982. Colonizacion de *Drosophila subobscura* Collin en Chile. Actas V. Congreso Latinamericano de Genetica. Universidad de Santiago, Chile, pp. 117–182.

Bull, J. 1983. *Evolution of Sex Determining Mechanisms.* Benjamin/Cummings Publishing Co., Menlo Park, Calif.

Bundgaard, J. and F. B. Christiansen. 1972. Dynamics of polymorphisms: I. Selection components in an experimental population of *Drosophila melanogaster.* Genetics 71:439–460.

Bürglin, T. R. 1994. A comprehensive classification of homeobox genes. Pp. 25–71 in *Guidebook to the Homeobox Genes*, ed. D. Duboule. Oxford University Press, Oxford.

Buri, P. 1956. Gene frequency in small populations of mutant *Drosophila.* Evolution 10:367–402.

Burla, H. and W. Goetz. 1965. Veranderlichkeit des chromosomalen Polymorphismus bei *Drosophila subobscura.* Genetica 36:83–104.

Burla, H., A. B. da Cuhna, A. G. L. Cavalcanti, Th. Dobzhansky, and C. Pavan. 1950. Population density and dispersal rates in Brazilian *Drosophila willistoni.* Ecology 31:393–404.

Burnet, B. and K. Connolly. 1974. Activity and sexual behaviour in *Drosophila melanogaster.* Pp. 201–258 in *The Genetics of Behaviour*, ed. J. H. F. van Abeelen. North-Holland, Amsterdam.

Burnett, R. G. and R. C. King. 1962. Observations on a microsporidian parasite of *Drosophila willistoni.* J. Insect. Pathol. 4:104–112.

Buss, L. W. 1987. *The Evolution of Individuality.* Princeton University Press, Princeton, N.J.

Butlin, R. K. 1989. Reinforcement of premating isolation. Pp. 158–179 in *Speciation and Its Consequences*, eds. D. Otte and J. A. Endler. Sinauer Associates, Sunderland, Mass.

Butlin, R. K. and M. G. Ritchie. 1989. Genetic coupling in mate recognition systems: What is the evidence? Biol. J. Linn. Soc. 37:237–246.

Cabot, E. L., A. W. Davis, N. A. Johnson, and C.-I. Wu. 1994. Genetics of reproductive isolation in the *Drosophila simulans* clade: Complex epistasis underlying hybrid sterility. Genetics 137:175–189.

Cabrera, V. M., A. M. Gonzalez, and A. Gullon. 1980. Enzymatic polymorphism in *Drosophila subobscura* populations from the Canary Islands. Evolution 34:875–887.

Cabrera, V. M., A. M. Gonzalez, J. M. Larruga, and A. Gullon. 1982. Electrophoretic variability in natural populations of *Drosophila melanogaster* and *Drosophila simulans.* Genetica 59: 191–201.

Cabrera, V. M., A. M. Gonzalez, J. M. Larruga, and A. Gullon. 1983. Genetic distance and evolutionary relationships in the *Drosophila obscura* group. Evolution 37:675–689.

Cabrera, V. M., A. M. Gonzalez, M. Hernandez, J. M. Larruga, and M. Martell. 1985. Microgeographic and temporal genetic differentiation in natural populations of *Drosophila subobscura.* Genetics 110:247–256.

Caccone, A. and J. R. Powell. 1987. Molecular evolutionary divergence among North American cave crickets. II. DNA–DNA hybridization. Evolution 41:1215–1238.

Caccone, A. and J. R. Powell. 1990. Extreme rates and heterogeneity in insect DNA evolution. J. Mol. Evol. 30:273–280.

Caccone, A. and J. R. Powell. 1991. A protocol for the TEACL method of DNA–DNA hybridization. Pp. 385–407 in *Molecular Techniques in Taxonomy*, eds. G. M. Hewitt, A. Johnston, and J. Young. Springer-Verlag, Heidelberg.

Caccone, A., G. D. Amato, and J. R. Powell. 1987. Intraspecific DNA divergence in *Drosophila*: Evidence from parthenogenetic *D. mercatorum.* Mol. Biol. Evol. 4:343–350.

Caccone, A., R. DeSalle, and J. R. Powell. 1988a. Calibration of the change in thermal stability of DNA duplexes and degree of base-pair mismatch. J. Mol. Evol. 27:212–216.

Caccone, A., G. D. Amato, and J. R. Powell. 1988b. Rates and patterns of scnDNA and mtDNA divergence within the *Drosophila melanogaster* subgroup. Genetics 118:671–683.

Caccone, A., J. M. Gleason, and J. R. Powell. 1992. Complementary DNA–DNA hybridization in *Drosophila.* J. Mol. Evol. 34:130–140.

Caccone, A., E. N. Moriyama, J. M. Gleason, L. Nigro, and J. R. Powell. 1996. A molecular phylogeny for the *Drosophila melanogaster* subgroup and the problem of polymorphism data. Mol. Biol. Evol. 13:1224–1232.

Camargo, R. de, and H. J. Phaff. 1957. Yeasts occurring in *Drosophila* flies and in fermenting tomato fruits in northern California. Food Res. 22:367–372.

Cannon, W. F. 1961. The impact of uniformitarianism. Proc. Am. Phil. Soc. 105:301–314.

Capy, P., J. R. David, and D. L. Hartl. 1992. Evolution of the transposable element mariner in the *Drosophila melanogaster* species group. Genetica 86:37–46. *Also* McDonald, 1993a.

Carew, E. A. 1993. Molecular evolution of the alcohol dehydrogenase gene in the *Drosophila willistoni* species group. Ph.D. dissertation, Yale University, New Haven, Conn.

Cariou, M. L. and J. L. Da Lage. 1993. Isozyme polymorphisms. Pp. 160–171 in *Drosophila ananassae: Genetical and Biological Aspects*, ed. Y. N. Tobari. Japan Scientific Society Press, Tokyo; and S. Karger AG, Basel.

Carroll, S. B. 1994. Developmental regulatory mechanisms in the evolution of insect diversity. Development 1994 Supplement:217–223.

Carroll, S. B., J. Gates, D. Keys, S. W. Paddock, G. F. Panganiban, J. Selegue, and J. A. Williams. 1994. Pattern formation and eyespot determination in butterfly wings. Science 265:109–114.

Carson, H. L. 1946. The selective elimination of inversion dicentric chromatids during meiosis in the eggs of *Sciara impatiens.* Genetics 31:95–113.

Carson, H. L. 1951. Breeding sites of *Drosophila pseudoobscura* and *Drosophila persimilis* in the transition zone of the Sierra Nevada. Evolution 5:91–96.

Carson, H. L. 1956. A female-producing strain of *D. borealis.* Dros. Inf. Serv. 30:109–110.

Carson, H. L. 1958a. The population genetics of *Drosophila robusta.* Adv. Genet. 9:1–40.

Carson, H. L. 1958b. Response to selection under different conditions of recombination in *Drosophila.* Cold Spring Harbor Symp. Quant. Biol. 23:291–305.

Carson, H. L. 1961a. Relative fitness of geographically open and closed experimental populations of *Drosophila robusta.* Genetics, 46:553–567.

Carson, H. L. 1961b. Heterosis and fitness in experimental populations of *Drosophila melanogaster*. Evolution 15:496–509.

Carson, H. L. 1967. Selection for parthenogenesis in *D. mercatorum.* Genetics 55:157–171.

Carson, H. L. 1971a. The ecology of *Drosophila* breeding sites. Harold L. Lyon Arboretum Lecture Number 2, University of Hawaii, Honolulu.

Carson, H. L. 1971b. Speciation and the founder principle. Stadler Symp. 3:51–70.

Carson, H. L. 1974. Three flies and three islands: Parallel evolution in *Drosophila.* Proc. Natl. Acad. Sci. USA 71:3517–3521.

Carson, H. L. 1975. The genetics of speciation at the diploid level. Am. Nat. 109:83–92.

Carson, H. L. 1978. Speciation and sexual selection in Hawaiian *Drosophila.* Pp. 93–107 in *Ecological Genetics: The Interface*, ed. P. F. Brussard. Springer-Verlag, New York.

Carson, H. L. 1982. Evolution of Drosophila species on the newer Hawaiian volcanoes. Heredity 48:3–25.

Carson, H. L. 1986. Sexual selection and speciation. Pp. 391–409 in *Evolutionary Processes and Theory*, eds. S. Karlin and E. Nevo. Academic Press, New York.

Carson, H. L. 1987. High fitness of heterokaryotypic individuals segregating naturally within a long-standing laboratory population of *Drosophila silvestris*. Genetics 116:415–422.

Carson, H. L. 1992. Inversions in Hawaiian *Drosophila*. Pp. 407–439 in *Drosophila Inversion Polymorphism*, eds. C. B. Krimbas and J. R. Powell. CRC Press, Boca Raton, Fla.

Carson, H. L. and P. J. Bryant. 1979. Change in a secondary sexual character as evidence of incipient speciation in *Drosophila*. Proc. Natl. Acad. Sci. USA 76:1929–1933.

Carson, H. L. and D. A. Clague. 1995. Geology and biogeography of Hawaii. Pp. 14–29 in *Hawaiian Biogeography: Evolution on a Hot Spot Archipelago*, eds. W. L. Wagner and V. A. Funk. Smithsonian Institution Press, Washington, D.C.

Carson, H. L. and R. Lande. 1984. Inheritance of a secondary sexual character in *Drosophila silvestris*. Proc. Natl. Acad. Sci. USA 81:6904–6907.

Carson, H. L. and H. D. Stalker. 1951. Natural breeding sites for some wild species of *Drosophila* in the eastern United States. Ecology 32:317–330.

Carson, H. L. and A. R. Templeton. 1984. Genetic revolutions in relation to speciation phenomena: The founding of new populations. Ann. Rev. Ecol. Syst. 15:97–131.

Carson, H. L. and M. Wasserman. 1965. A widespread chromosomal polymorphism in a widespread species, *Drosophila buzzatii*. Am. Nat. 99:111–115.

Carson, H. L., E. P. Knapp, and H. J. Phaff. 1956. Studies on the ecology of *Drosophila* in the Yosemite region of California. III. The yeast flora of the natural breeding sites of some species of *Drosophila*. Ecology 37:538–544.

Carson, H. L., W. E. Johnson, P. S. Nair, and F. M. Sene. 1975. Allozymic and chromosomal similarity in two *Drosophila* species. Proc. Natl. Sci. USA 72:4521–4525.

Carson, H. L., K. Y. Kaneshiro, and F. C. Val. 1989. Natural hybridization between the sympatric Hawaiian species *Drosophila silvestris* and *Drosophila heteroneura*. Evolution 43:190–203.

Carson, H. L., J. P. Lockwood, and E. M. Craddock. 1990. Extinction and recolonization of local populations on a growing shield volcano. Proc. Natl. Acad. Sci. USA 87:7055–7057.

Carson, H. L., F. C. Val, and A. R. Templeton. 1994. Change in male secondary sexual characters in artificial interspecific hybrids. Proc. Natl. Acad. Sci. USA 91:6315–6318.

Carton, Y. 1978. La recherche de l'habitat de l'hote par un insecte entomophage. Ann. Zool. Ecol. Anim. 10:353–355.

Carton, Y. 1983. Low temperature induced diapause still extant in a tropical parasitoid species. Pp. 13–15 in *Adaptations*, eds. N. S. Margaris, M. Arianoutsou Faraggitaka, and R. J. Feiter. Plenum, New York.

Carton, Y., M. Bouletreau, J. J. M. van Alphen, and J. C. van Lenteren. 1986. The *Drosophila* parasitic wasps. Pp. 347–394 in *The Genetics and Biology of Drosophila Vol. 3e*, eds. M. Ashburner, H. L. Carson, and J. N. Thompson, Jr. Academic Press, New York.

Carulli, J. P. and D. L. Hartl. 1992. Variable rates of evolution among Drosophila opsin genes. Genetics 132:193–204.

Carulli, J. P., D. E. Krane, D. L. Hartl, and H. Ochman. 1993. Compositional heterogeneity and patterns of molecular evolution in the Drosophila genome. Genetics 134:837–845.

Carulli, J. P., D.-M. Chen., W. S. Stark, and D. L. Hartl. 1994. Phylogeny and physiology of *Drosophila* opsins. J. Mol. Evol. 38:250–262.

Carvalho, A. B. and L. B. Klaczko. 1993. Autosomal suppressors of sex-ratio in *Drosophila mediopunctata*. Heredity 71:546–551.

Carvalho, A. B. and L. B. Klaczko. 1994. Y-linked suppressors of the sex-ratio trait in *Drosophila mediopunctata*. Heredity 73:573–579.

Cavalcanti, A. G. L., D. N. Falcao, and L. E. Castro. 1957. Sex-ratio in *D. prosaltans*, a character due to interaction between nuclear genes and a cytoplasmic factor. Am. Nat. 91: 321–325.

Cavalier-Smith, T. 1985. *The Evolution of Genome Size.* John Wiley and Sons, Chichester, U.K.

Cavener, D. 1979. Preference for ethanol in *Drosophila melanogaster* associated with the alcohol dehydrogenase polymorphism. Behav. Genet. 9:359–365.

Cavener, D. G. 1985. Coevolution of the glucose dehydrogenase gene and the ejaculatory duct of in the genus *Drosophila.* Mol. Biol. Evol. 2:141–149.

Cavener, D. G. 1992. Transgenic animal studies on the evolution of genetic regulatory circuitries. BioEssays 14:237–244.

Cavener, D. G. and R. J. MacIntyre. 1983. Biphasic expression and function of glucose dehydrogenase in *Drosophila melanogaster.* Proc. Natl. Acad. Sci. USA 80:6286–6288.

Chambers, G. K. 1988. The Drosophila alcohol dehydrogenase gene-enzyme system. Adv. Genet. 25:40–107.

Chambers, G. K. 1991. Gene expression, adaptation and evolution in higher organisms: Evidence from studies of Drosophila alcohol dehydrogenase. Comp. Biochem. Physiol. 99B:723–730.

Chang, H.-Y., D. Wang, and F. J. Ayala. 1989. Mitochondrial DNA evolution in the *Drosophila nasuta* subgroup species. J. Mol. Evol. 28:337–348.

Charlesworth, B. 1974. Inversion polymorphism in a two-locus genetic system. Genet. Res. 23: 259–280.

Charlesworth, B. 1991. The evolution of sex chromosomes. Science 251:1030–1033.

Charlesworth, B. and D. Charlesworth. 1973. Selection of a new inversion in a multilocus genetic system. Genet. Res. 21:167–183.

Charlesworth, B. and D. Charlesworth. 1985a. Genetic variation in recombination in *Drosophila.* I. Response to selection and preliminary genetic analysis. Heredity 54:71–83.

Charlesworth, B. and D. Charlesworth. 1985b. Genetic variation in recombination in *Drosophila.* II. Genetic analysis of a high recombination stock. Heredity 54:85–98.

Charlesworth, B. and C. H. Langley. 1986. The evolution of self-regulated transposition of transposable elements. Genetics 112:359–383.

Charlesworth, B. and C. H. Langley. 1989. The population genetics of *Drosophila* transposable elements. Ann. Rev. Genet. 23:251–287.

Charlesworth, B., J. A. Coyne, and N. H. Barton. 1987. The relative rates of evolution of sex chromosomes and autosomes. Am. Nat. 130:113–146.

Charlesworth, B., M. T. Morgan, and D. Charlesworth. 1993. The effect of deleterious mutations on neutral molecular variation. Genetics 134:1289–1303.

Chen, C. N., T. Malone, S. K. Beckendorf, and R. L. Davis. 1987. At least two genes reside within a large intron of the *dunce* gene of *Drosophila.* Nature 329:721–724.

Chen, C.-P. and V. K. Walker. 1993. Increase in cold-shock tolerance by selection of cold resistant lines in *Drosophila melanogaster.* Ecol. Entomol. 18:184–190.

Chetverikov, S. S. 1926. On certain aspects of the evolutionary process from the standpoint of genetics. Zhurnal Exp. Biol. 1:3–54 (Russian). English translation by Th. Dobzhansky, 1959, Proc. Amer. Phil. Soc. 105:167–195.

Chiang, H. C., D. Benoit, and J. Maki. 1962. Tolerance of adult *Drosophila melanogaster* to sub-freezing temperatures. Can. Entomol. 94:722–727.

Chinnici, J. P. 1971a. Modification of recombination frequency in Drosophila. I. Selection for increased and decreased crossing-over. Genetics 69:71–83.

Chinnici, J. P. 1971b. Modification of recombination frequency in Drosophila. II. The polygenic control of crossing-over. Genetics 69:85–96.

Choudhary, M. and C. C. Laurie. 1991. Use of *in vitro* mutagenesis to analyze the molecular basis of the difference in *Adh* expression associated with the allozyme polymorphism in *Drosophila melanogaster.* Genetics 129:481–488.

Cirera, S., J. M. Martin-Campos, C. Segarra, and M. Aguadé. 1995. Molecular characterization of the breakpoints of an inversion fixed between *Drosophila melanogaster* and *D. subobscura.* Genetics 139:321–326.

Citri, Y., H. V. Colot, A. C. Jacquier, Q. Yu, J. C. Hall, D. Baltimore, and M. Rosbash. 1987. A family of unusually spliced biologically active transcripts encoded by a Drosophila clock gene. Nature 326:42–47.

Civetta, A. and R. S. Singh. 1995. High divergence of reproductive tract proteins and their association with postzygotic reproductive isolation in *Drosophila melanogaster* and *Drosophila virilis* group species. J. Mol. Evol. 41:1085–1095.

Clark, A. G. and W. W. Doane. 1984. Interactions between the *Amylase* and *adipose* chromosomal regions of *Drosophila melanogaster.* Evolution 38:957–982.

Clark, A. G. and E. M. S. Lyckegaard. 1988. Natural selection with nuclear and cytoplasmic transmission. III. Joint analysis of segregation and mitochondrial DNA in *Drosophila melanogaster.* Genetics 118:471–481.

Clark, D. V. and S. Henikoff. 1992. Unusual organizational features of the *Drosophila gart* locus are not conserved within Diptera. J. Mol. Evol. 35:51–59.

Clark, J. B., W. P. Maddison, and M. G. Kidwell. 1994. Phylogenetic analysis supports horizontal transfer of P transposable elements. Mol. Biol. Evol. 11:40–50.

Clark, R. L., A. R. Templeton, and C. F. Sing. 1981. Studies of enzyme polymorphism in the Kamuela population of *D. mercatorum.* I. Estimation of the level of polymorphism. Genetics 98:597–611.

Clarke, J. M., J. Maynard Smith, and K. Sondhi. 1961. Asymmetrical responses to selection for rate of development in *D. subobscura.* Genet. Res. 2:70–81.

Clayton, F. E. and W. C. Guest. 1986. Overview of chromosomal evolution in the family Drosophilidae. Pp. 1–38 in *The Genetics and Biology of Drosophila, Vol. 3e,* eds. M. Ashburner, H. L. Carson, and J. N. Thompson, Jr. Academic Press, New York.

Clayton, F. E. and C. L. Ward. 1954. Chromosomal studies of several species of Drosophilidae. Univ. Texas Publ. 5422:98–105.

Clayton, F. E. and M. R. Wheeler. 1975. A catalog of Drosophila metaphase chromosome configurations. Pp. 471–512 in *Handbook of Genetics, Vol. 3, Invertebrates of Genetic Interest,* ed. R. C. King. Plenum, New York.

Clegg, M. T. 1978. Dynamics of correlated genetic systems. II. Simulation studies of chromosomal segments under selection. Theoret. Pop. Biol. 13:1–23.

Clegg, M. T. 1984. Dynamics of multilocus genetic systems. Oxford Surveys Evol. Biol. 1: 160–183.

Clegg, M. T., J. F. Kidwell, M. G. Kidwell, and N. J. Daniel. 1976. Dynamics of correlated genetic systems. I. Selection in the region of the glued locus of *Drosophila melanogaster.* Genetics 83:793–810.

Clegg, M. T., J. F. Kidwell, and M. G. Kidwell. 1978. Dynamics of correlated genetic systems. III. Behavior of chromosomal segments under lethal selection. Genetica 48:95–106.

Clegg, M. T., J. R. Kidwell, and C. R. Horch. 1980. Dynamics of correlated genetic systems. V. Rates of decay of linkage disequilibria in experimental populations of *Drosophila melanogaster.* Genetics 94:217–234.

Cobb, M. L. and J.-M. Jallon. 1990. Pheromones, mate recognition, and courtship stimulation in the *Drosophila melanogaster* species subgroup. Animal Behav. 39:1058–1067.

Cobbs, G. 1977. Multiple insemination and male sexual selection in natural populations of *Drosophila pseudoobscura.* Am. Nat. 111:641–656.

Cobbs, G. 1987. Modifier genes of the sex ratio trait in *Drosophila pseudoobscura.* Genetics 113:355–365.

Cohan, F. M. and Hoffmann, A. A. 1986. Genetic divergence under uniform selection. II.

Different responses to selection for knockdown resistance to ethanol among *Drosophila melanogaster* populations and their replicate lines. Genetics 114:145–163.

Cohen, A. J., D. L. Williamson, and K. Oishi. 1987. SpV3 virus of *Drosophila* spiroplasmas. Israeli J. Med. Sci. 23:429–433.

Cohen, E. H. and S. C. Bowman. 1979. Detection and location of three simple sequence DNA's in polytene chromosomes from *virilis* group species of *Drosophila*. Chromosoma 73:327–355.

Colby, C. and S. M. Williams. 1993. The distribution and spreading of rare variants in the histone multigene family of *Drosophila melanogaster*. Genetics 135:127–133.

Collet, C., K. M. Nielsen, R. J. Russell, M. Karl., J. G. Oakeshott, and R. C. Richmond. 1990. Molecular analysis of duplicated esterase genes in *Drosophila melanogaster*. Mol. Biol. Evol. 7:9–28.

Colot, H. V., J. C. Hall, and M. Rosbash. 1988. Interspecific comparison of the *period* gene of *Drosophila* reveals large blocks of non-conserved coding DNA. EMBO J. 7:3929–3937.

Cook, R. M., P. A. Parsons, and I. R. Bock. 1977. Australian endemic *Drosophila*. II. A new *Hibiscus*-breeding species with its description. Aust. J. Zool. 25:755–763.

Cooke, P. H. and J. G. Oakeshott. 1989. Amino acid polymorphisms for esterase-6 in *Drosophila melanogaster*. Proc. Natl. Acad. Sci. USA 86:1426–1430.

Cooper, D. M. 1960. Food preferences of larval and adult *Drosophila*. Evolution 14:41–55.

Cooper, J. W. 1964. Genetic and cytological studies of *Drosophila nigrospiracula* in the Sonoran Desert. M.S. thesis, Univ. of Arizona, Tucson, Ariz.

Cooper, K. W. 1946. The mechanism of non-random segregation of sex chromosomes in male *Drosophila miranda*. Genetics 31:181–194.

Cooper, K. W., S. Zimmering, and J. Krivshenko. 1955. Interchromosomal effects and segregation. Proc. Natl. Acad. Sci. USA 41:911–914.

Cordeiro, M., L. Wheeler, C. S. Lee, C. D. Kastritsis, and R. H. Richardson. 1975. Heterochromatic chromosomes and satellite DNAs of *Drosophila nasutoides*. Chromosoma 51:65–73.

Cory, L., P. Fjeld, and W. Serat. 1971. Environmental DDT and the genetics of natural populations. Nature 229:128–130.

Costa, R., A. A. Peixoto, J. T. Thackeray, R. Dalgleish, and C. P. Kyriacou. 1991. Length polymorphism in the threonine-glycine-encoding repeat region of the *period* gene in *Drosophila*. J. Mol. Evol. 32:238–246.

Costa, R., A. A. Peixoto, G. Barbujani, and C. P. Kyriacou. 1992. A latitudinal cline in a *Drosophila* clock gene. Proc. Roy. Soc. Lond. B 250:43–49.

Courtney, S. P. and G. K. Chen. 1988. Genetic and environmental variation in oviposition behaviour in the mycophagous *Drosophila suboccidentalis*. Func. Ecol. 2:521–528.

Cox-Foster, D. L., C. P. Schonbaum, M. T. Murtha, and D. R. Cavener. 1990. Developmental expression of the glucose dehydrogenase gene in *Drosophila melanogaster*. Genetics 124: 873–880.

Coyne, J. A. 1983. Genetic basis of differences in genital morphology among three sibling species of *Drosophila*. Evolution 37:1101–1118.

Coyne, J. A. 1984. Genetic basis of male sterility in hybrids between two closely related species of *Drosophila*. Proc. Natl. Acad. Sci. USA 81:4444–4447.

Coyne, J. A. 1985a. Genetic studies of three sibling species of *Drosophila* with relationship to theories of speciation. Genet. Res. 46:169–192.

Coyne, J. A. 1985b. The genetic basis of Haldane's rule. Nature 314:736.

Coyne, J. A. 1986. Meiotic segregation and male recombination in interspecific hybrids of *Drosophila*. Genetics 114:485–494.

Coyne, J. A. 1989. Genetics of sexual isolation between two sibling species *Drosophila simulans* and *Drosophila mauritiana*. Proc. Natl. Acad. Sci. USA 86:5464–5468.

Coyne, J. A. 1992a. Genetics of speciation. Nature 355:511–515.

Coyne, J. A. 1992b. Genetics of sexual isolation in females of the *Drosophila simulans* species complex. Genet. Res. 60:25–31.

Coyne, J. A. 1993. The genetics of an isolating mechanism between two sibling species of *Drosophila.* Evolution 47:778–788.

Coyne, J. A. and B. Charlesworth. 1986. Location of an X-linked factor causing sterility in male hybrids of *Drosophila simulans* and *D. mauritiana.* Heredity 57:243–246.

Coyne, J. A. and B. Charlesworth. 1989. Genetic analysis of X-linked sterility in hybrids between three sibling species of *Drosophila.* Heredity 62:97–106.

Coyne, J. A. and B. Grant. 1972. Disruptive selection on I-maze activity in *Drosophila melanogaster.* Genetics 71:185–188.

Coyne, J. A. and M. Kreitman. 1986. Evolutionary genetics of two sibling species of *Drosophila, D. simulans* and *D. mauritiana.* Evolution 40:673–691.

Coyne, J. A. and B. Milstead. 1987. Long-distance migration of *Drosophila.* 3. Dispersal of *D. melanogaster* alleles from a Maryland orchard. Am. Nat. 130:70–82.

Coyne, J. A. and H. A. Orr. 1989a. Patterns of speciation in Drosophila. Evolution 43:362–381.

Coyne, J. A. and H. A. Orr. 1989b. Two rules of speciation. Pp. 180–207 in *Speciation and Its Consequences*, eds. D. Otte and J. A. Endler. Sinauer Associates, Sunderland, Mass.

Coyne, J. A., A. A. Felton, and R. C. Lewontin. 1978. Extent of genetic variation at a highly polymorphic esterase locus in *Drosophila pseudoobscura.* Proc. Natl. Acad. Sci. USA 75: 5090–5093.

Coyne, J. A., I. A. Boussy, T. Prout, S. H. Bryant, J. S. Jones, and J. A. Moore. 1982. Long-distance migration of *Drosophila.* Am. Nat. 119:589–595.

Coyne, J. A., I. A. Boussy, and S. Bryant. 1984. Is *Drosophila pseudoobscura* a garbage species? Pan Pacific Entomol. 60:16–20.

Coyne, J. A., S. H. Bryant, and M. Turelli. 1987. Long-distance migration of *Drosophila.* II. Presence in desolate sites and disperal near a desert oasis. Am. Nat. 129:847–861.

Coyne, J. A., B. Charlesworth, and H. A. Orr. 1991a. Haldane's rule revisited. Evolution 45: 1710–1714.

Coyne, J. A., S. Aulard, and A. Berry. 1991b. Lack of underdominance in a naturally occurring pericentric inversion in *Drosophila melanogaster* and its implications for chromosome evolution. Genetics 129:791–802.

Coyne, J. A., B. C. Moore, J. A. Moore, J. R. Powell, and C. E. Taylor. 1992. Temporal stability of third chromosome inversion frequencies in *Drosophila persimilis.* Evolution 46:1558–1563.

Coyne, J. A., W. Meyers, A. P. Crittenden, and P. Sniegowski. 1993. The fertility effects of pericentric inversions in *Drosophila melanogaster.* Genetics 134:487–496.

Coyne, J. A., A. P. Crittenden, and K. Mah. 1994. Genetics of a pheromonal difference contributing to reproductive isolation in *Drosophila.* Science 265:1461–1464.

Craddock, E. M. and C. R. B. Boake. 1992. Onset of vitellogenesis in female *Drosophila silvestris* is accelerated in the presence of sexually mature males. J. Insect Physiol. 38: 643–650.

Craddock, E. M. and H. L. Carson. 1989. Chromosomal inversion patterning and population differentiation in a young insular species: *Drosophila silvestris.* Proc. Natl. Acad. Sci. USA 86:4798–4802.

Craddock, E. M., M. P. Kambysellis, and P. Hatzopoulos. 1983. Synthesis and characterization of the three vitellogenin genes of *Drosophila grimshawi.* Genetics 104:s18.

Crossley, S. A. 1974. Changes in mating behavior produced by selection for ethological isolation between ebony and vestigial mutants of *Drosophila melanogaster.* Evolution 28:631–647.

Crossley, S. A. 1986. Courtship sounds and behaviour in the four species of the *Drosophila bipectinata* complex. Anim. Behav. 34:1146–1159.

Crow, J. F. 1942. Cross fertility and isolating mechanisms in the *Drosophila mulleri* group. Univ. Texas Publ. 4228:53–67.

Crow, J. F. 1979. Genes that violate Mendel's rules. Sci. Am. 240:134–146.

Crow, J. F. 1988. The ultraselfish gene. Genetics 118:389–391.

Crow, J. F. 1991. Why is mendelian segregation so exact? BioEssays 13:1–8.

Crow, J. F. and N. E. Morton. 1955. Measurement of gene frequency drift in small populations. Evolution 9:202–214.

Crow, J. F. and M. L. Simmons. 1983. The mutation load in *Drosophila.* Pp. 1–36 in *Genetics and Biology of Drosophila, Vol. 3c*, eds. M. Ashburner, H. L. Carson, and J. N. Thompson, Jr. Academic Press, New York.

Crow, J. F. and R. G. Temin. 1964. Evidence for the partial dominance of recessive lethal genes in natural populations of *Drosophila*. Am. Nat. 98:21–23.

Crumpacker, D. W. 1974. The use of micronized fluorescent dusts to mark adult *Drosophila pseudoobscura.* Amer. Midl. Nat. 91:118–129.

Crumpacker, D. W. and D. Marinkovic. 1967. Preliminary evidence of cold temperature resistance in *Drosophila pseudoobscura.* Am. Nat. 101:505–514.

Crumpacker, D. W. and J. S. Williams. 1973. Density, dispersion and population structure in *Drosophila pseudoobscura.* Ecol. Monographs 43:499–538.

Crumpacker, D. W., J. Pyati, and L. Ehrman. 1977. Ecological genetics and chromosomal polymorphism in Colorado populations of *Drosophila pseudoobscura*. Evol. Biol. 10:437–469.

Cuesta, L. and M. A. Comendor. 1982. Comparación de la resistencia al frío de cuatro especies del género *Drosophila.* Medio Ambiente 6:33–42.

Currie, P. D. and D. T. Sullivan. 1994. Structure, expression and duplication of genes which encode phosphoglyceromutase of *Drosophila melanogaster.* Genetics 138:353–363.

Curtsinger, J. W. and M. W. Feldman. 1980. Experimental and theoretical analysis of the "sex-ratio" polymorphism in *Drosophila pseudoobscura.* Genetics 94:445–466.

Curtsinger, J. W., H. H. Fudui, D. R. Townsend, and J. W. Vaupel. 1992. Demography of genotypes: Failure of the limited life-span paradigm in *Drosophila melanogaster.* Science 258:461–464.

da Cunha, A. B. and Th. Dobzhansky. 1954. A further study of chromosomal polymorphism in *Drosophila willistoni* in relation to environment. Evolution 8:119–134.

da Cunha, A. B., Th. Dobzhansky, and A. Sokoloff. 1951. On food preferences of sympatric species of *Drosophila*. Evolution 5:97–101.

da Cunha, A. B., A. M. E. Shehata, and W. de Oliveira. 1957. A study of the diets and nutritional preferences of tropical species of Drosophila. Ecology 38:98–106.

da Cunha, A. B., Th. Dobzhansky, O. Pavlovsky, and B. Spassky. 1959. Genetics of natural populations. XXVIII. Supplemental data on the chromosomal polymorphism in *Drosophila willistoni* in relation to the environment. Evolution 13:389–404.

Da Lage, J.-L. and M.-L. Cariou. 1993. Organization and structure of the amylase gene family. Pp. 171–181 in *Drosophila ananassae: Genetical and Biological Aspects*, ed. Y. N. Tobari. Japan Scientific Society Press, Tokyo; and S. Karger AG, Basel.

Da Lage, J.-L., F. Lemeunier, M.-L. Cariou, and J. R. David. 1992. Multiple amylase genes in *Drosophila ananassae* and related species. Genet. Res. 59:85–92.

Dal Molin, C. 1979. An external scent as the basis for a rare-male mating advantage in *Drosophila melanogaster.* Am. Nat. 113:951–954.

Daniels, S. B. and L. D. Strausbaugh. 1986. The distribution of P-element sequences in *Drosophila*: The *willistoni* and *saltans* species groups. J. Mol. Evol. 23:138–148.

Daniels, S. B., L. D. Strausbaugh, L. Ehrman, and R. Armstrong. 1984. Sequences homologous to P elements occur in *Drosophila paulistorum.* Proc. Natl. Acad. Sci. USA 81:6794–6797.

Daniels, S. B., K. R. Peterson, L. D. Strausbaugh, M. G. Kidwell, and A. Chovnick. 1990a.

Evidence for horizontal transmission of the P transposable element between *Drosophila* species. Genetics 124:339–355.

Daniels, S. B., A. Chovnick, and I. A. Boussy. 1990b. Distribution of hobo transposable elements in the genus *Drosophila.* Mol. Biol. Evol. 7:589–606.

Darlington, C. D. and Th. Dobzhansky. 1942. Temperature and "sex-ratio" in *Drosophila pseudoobscura*. Proc. Natl. Acad. Sci. USA 28:45–48.

David, J. R. and P. Capy. 1988. Genetic variation of *Drosophila melanogaster* natural populations. Trends in Genetics 4:106–111.

David, J. R. and J. van Herrwege. 1983. Adaptation to alcohol fermentation in *Drosophila* species: Relationship between alcohol tolerance and larval habitat. Comp. Biochem. Physiol. 74A:283–288.

David, J. R., J. van Herrwege, M. Monclus, and A. Prevosti. 1979. High ethanol tolerance in two distantly related *Drosophila* species: A probable case of recent convergent adaptation. Comp. Biochem. Physiol. 636:53–56.

David, J. R., R. Allemand, J. van Herrwege, and Y. Cohet. 1983. Ecophysiology: Abiotic factors. Pp. 105–170 in *Genetics and Biology of Drosophila Vol. 3d*, eds. M. Ashburner, H. L. Carson, and J. N. Thompson, Jr. Academic Press, New York.

Davidson, E. H. 1986. *Gene Activity in Early Development.* Academic Press, New York.

Davis, A. and I. Hardy. 1994. Hares and tortoises in *Drosophila* community ecology. Trends Ecol. Evol. 9:119–120.

Davis, A. W., E. Noonberg, and C.-I. Wu. 1994. Complex genic interactions between conspecific chromosomes underlying hybrid female sterility in the *Drosophila simulans* clade. Genetics 137:191–199.

Davis, R. L., and B. Dauwalder. 1991. The *Drosophila dunce* locus: Learning and memory genes in the fly. Trends Genet. 7:224–229.

Dawes, R., I. Dawson, F. Falciani, G. Tear, and M. Akam. 1994. *Dax*, a locust Hox gene related to *fushi-tarazu* but showing no pair-rule expression. Development 120:1561–1572.

Deakin, M. A. B., and R. B. Teague. 1974. A generalized model of inversion polymorphism. J. Theor. Biol. 48:105–117.

de Beer, G. R. 1930. *Embryology and Evolution.* Clarendon Press, Oxford, U.K.

de Beer, G. R. 1940. *Embryos and Ancestors.* Clarendon Press, Oxford, U.K.

De Jong, G. and W. Scharloo. 1976. Environmental determination of selective significance or neutrality of amylase variants in *Drosophila melanogaster.* Genetics 84:77–94.

del Solar, E. 1966. Sexual isolation caused by selection for positive and negative phototaxis and geotaxis. Proc. Natl. Acad. Sci. USA 56:484–487.

de Magalhaes, L. E. and M. A. Q. Rodregues-Pereira. 1976. Frequency-dependent mating success with ebony of *Drosophila melanogaster.* Experientia 32:309–310.

Demerec, M. (ed.). 1950. *Biology of Drosophila.* J. Wiley and Sons, New York.

de Oliveira, A. K. and A. R. Cordeiro. 1980. Adaptation of *Drosophila willistoni* experimental populations to extreme pH medium. Heredity 44:123–130.

DeSalle, R. 1992a. The phylogenetic relationships of flies in the family Drosophilidae deduced from mtDNA sequences. Molec. Phylogenetics Evol. 1:31–40.

DeSalle, R. 1992b. Origin and possible time of divergence of the Hawaiian Drosophilidae. Mol. Biol. Evol. 9:905–916.

DeSalle, R. 1995. Molecular approaches to biogeographic analysis of Hawaiian Drosophilidae. Pp. 72–89 in *Hawaiian Biogeography: Evolution on a Hot Spot Archipelago*, eds. W. L. Wagner and V. A. Funk. Smithsonian Institution Press, Washington, D.C.

DeSalle, R. and E. Carew. 1992. Phyletic phenocopy and the role of developmental genes in morphological evolution in the Drosophilidae. J. Evol. Biol. 5:363–374.

DeSalle, R. and D. A. Grimaldi. 1991. Morphological and molecular systematics of the Drosophilidae. Ann. Rev. Ecol. Syst. 22:447–475.

DeSalle, R. and D. A. Grimaldi. 1992. Characters and systematics of the Drosophilidae. J. Heredity 83:182–188.

DeSalle, R. and D. A. Grimaldi. 1993. Phylogenetic pattern and developmental process in *Drosophila*. Syst. Biol. 42:458–475.

DeSalle, R. and A. R. Templeton. 1986. The molecular through ecological genetics of abnormal abdomen in *Drosophila mercatorum*. III. Tissue-specific differential replication of ribosomal genes modulates the abnormal abdomen phenotype in *Drosophila mercatorum*. Genetics 112:877–886.

DeSalle, R. and A. R. Templeton. 1987. Comments on "The significance of asymmetrical sexual isolation." Evol. Biol. 21:21–27.

DeSalle, R. and A. R. Templeton. 1992. The mtDNA genealogy of closely related *Drosophila silvestris*. J. Heredity 83:211–216.

DeSalle, R., J. Slighton, and E. Zimmer. 1986a. The molecular through ecological genetics of abnormal abdomen in *Drosophila mercatorum*. II. Ribosomal DNA polymorphism is associated with the abnormal abdomen syndrome in *Drosophila mercatorum*. Genetics 112:861–875.

DeSalle, R., L. V. Giddings, and A. R. Templeton. 1986b. Mitochondrial DNA variability in natural populations of Hawaiian *Drosophila*. I. Methods and levels of variability in *D. silvestris* and *D. heteroneura* populations. Heredity 56:75–85.

DeSalle, R., L. V. Giddings, and A. R. Templeton. 1986c. Mitochondrial DNA variability in natural populations of Hawaiian Drosophila. II. Genetic and phylogenetic relationships of natural populations of *D. silvestris* and *D. heteroneura*. Heredity 56:87–96.

DeSalle, R., A. R. Templeton, I. Mori, S. Pletscher, and J. S. Johnston. 1987a. Temporal and spatial heterogeneity of mtDNA polymorphisms in natural populations of *Drosophila mercatorum*. Genetics 116:215–223.

DeSalle, R., T. Freedman, E. M. Prager, and A. C. Wilson. 1987b. Tempo and mode of sequence evolution in mitochondrial DNA of Hawaiian *Drosophila*. J. Mol. Evol. 26:157–164.

de Souza, H. L., A. B. da Cunha, and E. P. dos Santos. 1968. Adaptive polymorphism of behavior developed in laboratory populations of *Drosophila willistoni*. Am. Nat. 102:583–586.

de Souza, H. L., A. B. da Cunha, and E. P. dos Santos. 1970. Adaptive polymorphism of behavior evolved in laboratory populations of *Drosophila willistoni*. Am. Nat. 124:175–189.

de Stordeur, E., M. Solignac, M. Monnerot, and J.-C. Mounolou. 1989. The generation of transplasmic *Drosophila simulans* by cytoplasmic injection: Effects of segregation and selection on the perpetuation of mitochondrial DNA heteroplasmy. Mol. Gen. Genet. 220: 127–132.

Detlefsen, J. A. and E. Roberts. 1921. Studies on crossing over. I. The effect of selection on crossover values. J. Expt. Zool. 32:333–354.

Dicke, M., J. C. van Lenteren, G. J. F. Boskamp, and E. van Dongen-van Leeuwen. 1984. Chemical stimuli in host habitat location by *Leptopilina heterotoma* (Thompson) (Hymenoptera: Eucoilidae), a parasite of *Drosophila*. J. Chem. Ecol. 10:695–715.

Dickinson, W. J. 1980a. Evolution of patterns of gene expression in Hawaiian picture-winged *Drosophila*. J. Mol. Evol. 16:73–94.

Dickinson, W. J. 1980b. Complex cis-acting regulatory genes demonstrated in Drosophila hybrids. Develop. Genet. 1:229–240.

Dickinson, W. J. 1980c. Tissue specificity of enzyme expression regulated by diffusable factors: Evidence in *Drosophila* hybrids. Science 207:995–997.

Dickinson, W. J. 1983. Tissue-specific allelic isozyme patterns and *cis*-acting developmental regulators. Isozymes 9:107–122.

Dickinson, W. J. 1991. The evolution of regulatory genes and patterns in *Drosophila.* Evol. Biol. 25:127–173.

Dickinson, W. J. and H. L. Carson. 1979. Regulation of the tissue specificity of an enzyme by a *cis*-acting genetic element: Evidence from Drosophila hybrids. Proc. Natl. Acad. Sci. USA 76:4559–4562.

Dickinson, W. J., R. G. Rowan, and M. D. Brennan. 1984. Regulatory gene evolution: adaptive differences in expression of alcohol dehydrogenase in *Drosophila melanogaster* and *Drosophila simulans.* Heredity 52:215–223.

Dickinson, W. J., Y. Yang, K. Schuske, and M. Akam. 1993. Conservation of molecular prepatterns during the evolution of cuticle morphology in *Drosophila* larvae. Evolution 47:1396–1406.

Diederich, G. W. 1941. Nonrandom mating between yellow-white and wild type in *Drosophila melanogaster*. Genetics 26:148–156.

Doane, W. W., D. B. Thompson, R. A. Norman, and S. A. Hawley. 1990. Molecular genetics of a three-gene cluster in the *Amy* region of *Drosophila.* Pp. 19–48 in *Isozymes: Structure, Function, and Use in Biology and Medicine*, eds. Z.-I. Ogita and C. L. Markert. Allan Liss, New York.

Dobzhansky, Th. 1935. *Drosophila miranda*, a new species. Genetics 20:377–391.

Dobzhansky, Th. 1936. Studies on hybrid sterility. II. Localization of sterility factors in *D. pseudoobscura* hybrids. Genetics 21:113–135.

Dobzhansky, Th. 1937. *Genetics and the Origin of Species*, 1st edition, Columbia University Press, New York. Second edition, 1941; third edition, 1951; first edition republished, 1982.

Dobzhansky, Th. 1943. Genetics of natural populations. IX. Temporal changes in the composition of populations of *Drosophila pseudoobscura.* Genetics 28:162–186.

Dobzhansky, Th. 1946. Genetics of natural populations. XIII. Recombination and variability in populations of *Drosophila pseudoobscura.* Genetics 31:269–290.

Dobzhansky, Th. 1948a. Genetics of natural populations. XVI. Altitudinal and seasonal changes produced by natural selection in certain populations of *Drosophila pseudoobscura.* Genetics 33:158–176.

Dobzhansky, Th. 1948b. Genetics of natural populations. XVIII. Experiments on chromosomes of *Drosophila pseudoobscura* from different geographic regions. Genetics 33:588–602.

Dobzhansky, Th. 1950. Genetics of natural populations. XIX. Origin of heterosis through natural selection. Genetics 35:288–302.

Dobzhansky, Th. 1951. Experiments on sexual isolation in Drosophila. X. Reproductive isolation between *Drosophila pseudoobscura* and *Drosophila persimilis* under natural and under laboratory conditions. Proc. Natl. Acad. Sci. USA 37:792–796.

Dobzhansky, Th. 1955. A review of some fundamental concepts and problems of population genetics. Cold Spring Harbor Symp. Quant. Biol. 20:1–15.

Dobzhansky, Th. 1957. Genetics of natural populations. XXVI. Chromosomal variability in island and continental populations of *Drosophila willistoni* from Central America and the West Indies. Evolution 11:280–293.

Dobzhansky, Th. 1962. Rigid versus flexible chromosomal polymorphism in *Drosophila.* Am. Nat. 96:321–328.

Dobzhansky, Th. 1965. "Wild" and "domestic" species of *Drosophila*. Pp. 533–552 in *Genetics of Colonizing Species*, eds. H. G. Baker and G. L. Stebbins. Academic Press, New York.

Dobzhansky, Th. 1968. On some fundamental concepts of Darwinian biology. Evol. Biol. 2: 1–34.

Dobzhansky, Th. 1970. *Genetics of the Evolutionary Process*. Columbia University Press, New York.

Dobzhansky, Th. 1971. Evolutionary oscillations in *Drosophila pseudoobscura.* Pp. 109–133 in *Ecological Genetics and Evolution*, ed. R. Creed. Blackwell, Oxford.

Dobzhansky, Th. 1973a. *Genetic Diversity and Human Equality.* Basic Books, New York.

Dobzhansky, Th. 1973b. Active dispersal and passive transport in *Drosophila.* Evolution 27: 565–575.

Dobzhansky, Th. 1974. Genetic analysis of hybrid sterility within the species *Drosophila pseudoobscura.* Hereditas 77:81–88.

Dobzhansky, Th. and F. J. Ayala. 1973. Temporal frequency changes of enzyme and chromosomal polymorphisms in natural populations of *Drosophila.* Proc. Natl. Acad. Sci. USA 70:680–683.

Dobzhansky, Th. and A. B. da Cunha. 1955. Differentiation of nutritional preferences in Brazilian species of Drosophila. Ecology 36:34–39.

Dobzhansky, Th. and A. Dreyfus. 1943. Chromosomal aberrations in Brazilian *Drosophila ananassae.* Proc. Natl. Acad. Sci. USA 29:368–375.

Dobzhansky, Th. and C. Epling. 1944. Contributions to the genetics, taxonomy, and ecology of *Drosophila pseudoobscura* and its relatives. Carnegie Institution of Wash. Publ. 554.

Dobzhansky, Th. and C. Pavan. 1943. Chromosomal complements of some South-Brazilian species of *Drosophila.* Proc. Natl. Acad. Sci. USA 29:301–305.

Dobzhansky, Th. and O. Pavlovsky. 1953. Indeterminate outcome of certain experiments of *Drosophila* populations. Evolution 7:198–210.

Dobzhansky, Th. and O. Pavlovsky. 1955. An extreme case of heterosis in a Central American population of *Drosophila tropicalis.* Proc. Natl. Acad. Sci. USA 41:289–295.

Dobzhansky, Th. and O. Pavlovsky. 1957. An experimental study of interaction between genetic drift and natural selection. Evolution 11:311–319.

Dobzhansky, Th. and O. Pavlovsky. 1967. Experiments on the incipient species of the *Drosophila paulistorum* complex. Genetics 55:141–156.

Dobzhansky, Th. and O. Pavlovsky. 1971. Experimentally created incipient species of *Drosophila.* Nature 230:289–292.

Dobzhansky, Th. and O. Pavlovsky. 1975. Unstable intermediates between Orinocan and Interior semispecies of *D. paulistorum.* Evolution 29:242–248.

Dobzhansky, Th. and J. R. Powell. 1974. Rates of dispersal of *Drosophila pseudoobscura* and its relatives. Proc. Roy. Soc. Lond. B 187:281–298.

Dobzhansky, Th. and J. R. Powell. 1975. *Drosophila pseudoobscura* and its American relatives. Pp. 537–587 in *Handbook of Genetics Vol. 3,* ed. R. C. King. Plenum, New York.

Dobzhansky, Th. and M. L. Queal. 1938. Genetics of natural populations. II. Genic variation in populations of *Drosophila pseudoobscura* inhabiting isolated mountain ranges. Genetics 23:463–484.

Dobzhansky, Th. and B. Spassky. 1953. Genetics of natural populations. XXI. Concealed variability in two sympatric species of Drosophila. Genetics 38:471–484.

Dobzhansky, Th. and B. Spassky. 1954. Genetics of natural populations. XXII. A comparison of concealed variability in *Drosophila prosaltans* with that in other species. Genetics 39: 472–487.

Dobzhansky, Th. and N. P. Spassky. 1962. Genetic drift and natural selection in experimental populations of *Drosophila pseudoobscura.* Proc. Natl. Sci. USA 48:148–156.

Dobzhansky, Th. and B. Spassky. 1967. Geotactic and phototactic behavior in Drosophila. I. Proc. Roy. Soc. Lond., Ser. B 168:27–47.

Dobzhansky, Th. and B. Spassky. 1969. Artificial and natural selection for two behavioral traits in *Drosophila pseudoobscura.* Proc. Natl. Acad. Sci. USA 62:75–80.

Dobzhansky, Th. and C. C. Tan. 1936. Studies on hybrid sterility. III. A comparison of the gene arrangement in two species, *Drosophila pseudoobscura* and *Drosophila miranda.* Z. Indukt. Abstamm.-Vererbungsl. 72:88–114.

Dobzhansky, Th. and S. Wright. 1941. Genetics of natural populations. V. Relations between mutation rate and accumulation of lethals in populations of *Drosophila pseudoobscura.* Genetics 26:23–51.

Dobzhansky, Th. and S. Wright. 1943. Genetics of natural populations X. Dispersion rates in *Drosophila pseudoobscura.* Genetics 28:304–340.

Dobzhansky, Th. and S. Wright. 1947. Genetics of natural populations. XV. Rate of diffusion of a mutant gene through a population of *Drosophila pseudoobscura.* Genetics 32:303–324.

Dobzhansky, Th., H. Levene, B. Spassky, and N. Spassky. 1959. Release of genetic variability through recombination III. *D. prosaltans.* Genetics 44:75–92.

Dobzhansky, Th., A. S. Hunter, O. Pavlovsky, B. Spassky, and B. Wallace. 1963. Genetics of natural populations. XXXI. Genetics of an isolated marginal population of *D. pseudoobscura.* Genetics 48:91–103.

Dobzhansky, Th., R. C. Lewontin, and O. Pavlovsky. 1964. The capacity for increase in chromosomally polymorphic and monomorphic populations of *Drosophila pseudoobscura.* Heredity 19:597–614.

Dobzhansky, Th., W. W. Anderson, and O. Pavlovsky. 1966. Genetics of natural populations. XXVIII. Continuity and change in populations of *Drosophila pseudoobscura* in Western United States. Evolution 20:418–427.

Dobzhansky, Th., B. Spassky, and J. Sved. 1969. Effects of selection and migration on geotactic and phototactic behavior of *Drosophila.* II. Proc. Roy. Soc. (London) Series B 173:191–207.

Dobzhansky, Th., O. Pavlovsky, and J. R. Powell. 1976. Partially successful attempts to enhance reproductive isolation between semispecies of *Drosophila paulistorum.* Evolution 30:201–292.

Dobzhansky, Th., J. R. Powell, C. E. Taylor, and M. Andregg. 1979. Ecological variables affecting the dispersal behavior of *Drosophila pseudoobscura* and its relatives. Am. Nat. 114:32–334.

Dodd, D. M. B. 1984. Behavioral correlates of the adaptive divergence of *Drosophila* populations. Ph.D. dissertation, Yale University, New Haven, Conn.

Dodd, D. M. B. 1989. Reproductive isolation as a consequence of adaptive divergence in *Drosophila pseudoobscura.* Evolution 43:1308–1311.

Dodd, D. M. B. and J. R. Powell. 1985. Founder-flush speciation: An update of experimental results with *Drosophila.* Evolution 39:1388–1392.

Dolan, R. and A. Robertson. 1975. The effect of conditioning the medium in *Drosophila,* in relation to frequency-dependent selection. Heredity 35:311–316.

Dorit, R. L., L. Shoenbach, and W. Gilbert. 1990. How big is the exon universe? Science 250: 1377–1382.

Dowsett, A. P. and M. W. Young. 1982. Differing levels of dispersed repetitive DNA among closely related species of *Drosophila.* Proc. Natl. Acad. Sci. USA 79:4570–4574.

Drake, J. 1969. Comparative rates of spontaneous mutations. Nature 221:1132.

Drosophila Information Service. 1994. The Bibliography of *Drosophila*, Vol. 74. Drosophila Information Service.

Dru, P., F. Bras, S. Dezelee, P. Gay, A.-M. Petitjean, Anne Pierre-Deneubourg, D. Teninges, and D. Contamine. 1993. Unusual variability of the *Drosophila melanogaster ref(2)P* protein which controls the multiplication of Sigma rhabdovirus. Genetics 133:943–854.

Druger, M. 1962. Selection and body size in *Drosophila pseudoobscura.* Genetics 47:209–222.

Dubinin, N. P. 1931. Genetico-automatical processes and their bearing on the mechanism of organic evolution. J. Exp. Biol. (Russian) 7:463–479.

Dubinin, N. P. and G. G. Tiniakov. 1946. Inversion gradients and natural selection in ecological races of *Drosophila funebris.* Genetics 31:537–545.

Dubinin, N. P. and G. G. Tiniakov. 1947. Inversion gradients and selection in ecological races of *Drosophila funebris.* Am. Nat. 81:148–153.

Dubinin, N. P., and 14 collaborators. 1934. Experimental study of the ecogenotypes of *Drosophila melanogaster.* Bio. Zh. Mosk. 3:166–216.

Dubinin, N. P., M. A. Heptner, Z. A. Demidova, and L. I. Djachkova. 1936. The genetical structure of the population and its dynamics in wild *Drosophila melanogaster.* Bio. Zh. Mosk. 5:939–976.

Dubinin, N. P., D. D. Romashov, M. A. Haptner, and Z. A. Demidova. 1937. Aberrant polymorphism in *Drosophila fasciata* (Syn-*melanogaster* Meig.). Bio. Zh. Mosk. 6:311–354.

Dumouchel, W. H. and W. W. Anderson. 1968. The analysis of selection in experimental populations. Genetics 58:435–449.

Dunn, L. C. 1965. *A Short History of Genetics.* Chapter 20, The rise of population genetics. McGraw-Hill, New York.

Dunn, R. C. and C. C. Laurie. 1995. Effects of a transposable element insertion on alcohol dehydrogenase expression in *Drosophila melanogaster.* Genetics 140:667–677.

Eanes, W. F., C. Wesley, J. Hey, and D. Houle. 1988. Fitness consequences of P element insertions in *Drosophila melanogaster.* Genet. Res. 52:17–26.

Eanes, W. F., J. W. Ajioka, J. Hey, and C. Wesley. 1989a. Restriction-map variation associated with the G6PD polymorphism in natural populations of *Drosophila melanogaster.* Mol. Biol. Evol. 6:384–397.

Eanes, W. F., J. Labate, and J. W. Ajioka. 1989b. Restriction-map variation within the *yellow-achaete-scute* region in five populations of *Drosophila melanogaster.* Mol. Biol. Evol. 6: 492–502.

Eanes, W. F., C. Wesley, and B. Charlesworth. 1992. Accumulation of P elements in minority inversions in natural populations of *Drosophila melanogaster.* Genet. Res. 59:109.

Eanes, W. F., M. Kirchner, and J. Yoon. 1993. Evidence for adaptive evolution of the *G6pd* gene in the *Drosophila melanogaster* and *Drosophila simulans* lineages. Proc. Natl. Acad. Sci. USA 90:7475–7479.

Ebbert, M. A. 1988. The *Drosophila*-spiroplasma symbiosis: Inter- and intraspecific variation in host fitness and bacterial transmission rates. Ph.D. dissertation, Yale University, New Haven, Conn.

Ebbert, M. A. 1991. The interaction phenotype in the *Drosophila willistoni*-spiroplasma symbiosis. Evolution 45:971–988.

Ebbert, M. A. 1993. Endosymbiotic sex ratio distorters in insects and mites. Pp. 150–191 in *Evolution and Diversity of Sex Ratio in Haplodiploid Insects and Mites*, eds. D. L. Wrensch and M. A. Ebbert. Chapman and Hall, New York.

Ebbert, M. A. 1995. Variable effects of crowding on *Drosophila* hosts of male-lethal and non-male-lethal spiroplasmas in laboratory populations. Heredity 74:227–240.

Edwards, A. W. F. 1961. The population genetics of "sex-ratio" in *Drosophila pseudoobscura.* Heredity 16:291–304.

Eernisse, D. J., J. S. Albert, F. E. Anderson. 1992. Annelida and Arthropoda are not sister taxa: A phylogenetic analysis of spiralian metazoan morphology. Syst. Biol. 41:305–330.

Ehrman, L. 1960. The genetics of hybrid sterility in *Drosophila paulistorum.* Evolution 14: 212–223.

Ehrman, L. 1961. The genetics of sexual isolation in *Drosophila paulistorum.* Genetics 46: 1025–1038.

Ehrman, L. 1964. Genetic divergence of M. Vetukhiv's experimental populations of *Drosophila pseudoobscura.* Genet. Res. 5:150–157.

Ehrman, L. 1965. Direct observation of sexual isolation between allopatric and between sympatric strains of the different *Drosophila paulistorum* races. Evolution 19:459–464.

Ehrman, L. 1966. Mating success and genotype frequency in *Drosophila*. Anim. Behav. 14: 332–339.

Ehrman, L. 1967. Further studies on genotype frequency and mating success in *Drosophila*. Am. Nat. 101:415–424.

Ehrman, L. 1968. Frequency dependence of mating success in *Drosophila pseudoobscura*. Genet. Res. 11:135–140.

Ehrman, L. 1969a. The sensory basis of mate selection in *Drosophila*. Evolution 23:59–64.

Ehrman, L. 1969b. Genetic divergence of M. Vetukhiv's experimental populations of *Drosophila pseudoobscura*. 5. A further study of rudiments of sexual isolation. Am. Midl. Nat. 82: 272–276.

Ehrman, L. 1970a. Sexual isolation versus mating advantage of rare *Drosophila* males. Behav. Genet. 1:111–118.

Ehrman, L. 1970b. A release experiment testing the mating advantage of rare *Drosophila* males. Behavioral Science 15:363–365.

Ehrman, L. 1970c. Simulation of the mating advantage of rare *Drosophila* males. Science 167: 905–906.

Ehrman, L. 1971. Natural selection for the origin of reproductive isolation. Am. Nat. 105: 479–483.

Ehrman, L. 1972a. A factor influencing the rare male mating advantage in *Drosophila*. Behav. Genet. 2:69–78.

Ehrman, L. 1972b. Rare male advantages and sexual isolation in *Drosophila immigrans*. Behav. Genet. 2:79–84.

Ehrman, L. 1973. More on natural selection for the origin of reproductive isolation. Am. Nat. 107:318–319.

Ehrman, L. 1979. Still more on natural selection for the origin of reproductive isolation. Am. Nat. 113:148–150.

Ehrman, L. and P. R. Kernaghan. 1972. Infectious heredity in *Drosophila paulistorum*. Pp. 227–250 in *Pathogenic Mycoplasmas*, Ciba Foundation Symposium, Associated Science Publishers, Amsterdam.

Ehrman, L. and C. Petit. 1968. Genotype frequency and mating success in the willistoni species group of *Drosophila*. Evolution 22:649–658.

Ehrman, L. and J. R. Powell. 1982. Evolution and speciation in the *Drosophila willistoni* group. Pp. 193–225 in *The Genetics and Biology of Drosophila, Vol. 3b*, eds. M. Ashburner, H. L. Carson, and J. N. Thompson, Jr. Academic Press, New York.

Ehrman, L. and E. B. Spiess. 1969. Rare-type mating advantage in *Drosophila*. Am. Nat. 103: 675–680.

Ehrman, L. and M. Wasserman. 1987. The significance of asymmetrical sexual isolation. Evol. Biol. 21:1–20.

Ehrman, L., B. Spassky, O. Pavlovsky, and Th. Dobzhansky. 1965. Sexual selection, geotaxis, and chromosomal polymorphism in experimental populations of *Drosophila pseudoobscura*. Evolution 19:227–246.

Ehrman, L., S. Koref-Santibanez, and C. T. Falk. 1972. Frequency-dependent mating in the two species of mesophragmatica species group of *Drosophila*. Dros. Inf. Serv. 48:36–37.

Ehrman, L., W. W. Anderson, and L. Blatte. 1977. A test for rare male mating advantage at an "enyzme locus" in *Drosophila*. Behav. Genet. 7:427–432.

Ehrman, L., N. L. Somerson, and F. J. Gottlieb. 1986. Reproductive isolation in a neotropical insect: behavior and microbiology. Pp. 97–108 in *Evolutionary Genetics of Invertebrate Behavior*, ed. M. D. Huettel. Plenum, New York.

Ehrman, L., J. R. Factor, N. Somerson, and P. Manzo. 1989. The *Drosophila paulistorum* endosymbiont in an alternative species. Am. Nat. 134:890–896.

Ehrman, L., M. M. White, and B. Wallace. 1991. A long-term study involving *Drosophila melanogaster* and toxic media. Evol. Biol. 25:175–209.

El-Helw, M. R. and A. M. M. Ali. 1970. Competition between *Drosophila melanogaster* and *D. simulans* on media supplemented with *Saccharomyces* and *Schizosaccharomyces*. Evolution 24:531–537.

El-Helw, M. R., A. M. M. Ali, and H. Moawad. 1972. Fitness of *Drosophila melanogaster* and *D. simulans* in relation to natural genera of yeasts. Egypt. J. Genet. Cytol. 1:196–202.

Ellison, J. R. 1968. A study of some species of the genus *Samoaia* (Diptera: Drosophilidae). Univ. Texas. Publ. 6818:423–430.

Endler, J. 1973. Gene flow and population differentiation. Science 179:243–250.

Endler, J. 1977. *Geographic Variation, Speciation and Clines.* Monographs in Population Biology No. 10. Princeton University Press, Princeton, N.J.

Endler, J. 1979. Gene flow and life history patterns. Genetics 93:263–284.

Engels, W. R. 1981. Estimating genetic divergence and genetic variability with restriction endonucleases. Proc. Natl. Acad. Sci. USA 78:6329–6333.

Engels, W. R. and C. R. Preston. 1984. Formation of chromosome rearrangements by P factors in Drosophila. Genetics 107:657–678.

Etges, W. J. 1984. Genetic structure and change in natural populations of *Drosophila robusta*: systematic inversion and inversion association frequency shifts in the Great Smoky Mountains. Evolution 38:675–688.

Etges, W. J. 1990. Direction of life history evolution in *Drosophila mojavensis*. Pp. 37–56 in *Ecological and Evolutionary Genetics of Drosophila*, eds. J. S. F. Barker, W. T. Starmer, and R. J. MacIntyre. Plenum, New York.

Etges, W. J., W. E. Johnson, G. A. Duncan, G. Huckins, and W. B. Heed. (in press). Ecological genetics of cactophilic *Drosophila*: Inversion polymorphism in *Drosophila mojavensis* and *Drosophila pachea*. In *Ecology and Conservation of the Sonoran Desert*, ed. R. Robichaux. Univ. Arizona Press, Tucson, Ariz.

Evgen'ev, M. B., D. H. Lankenau, and V. G. Croces. 1993. Cloning and characterization of a novel class of retrotransposon involved in hybrid dysgenesis and the mobilization of other elements in *D. virilis*. Abstracts 34th Ann. Dros. Res. Conf., San Diego.

Ewens, W. J., R. S. Spielman, and H. Harris. 1981. Estimation of genetic variation at the DNA level from restriction endonuclease data. Proc. Natl. Acad. Sci. USA 78:3748–3750.

Ewers, J., M. Hamblen-Coyle, M. Rosbash, and J. C. Hall. 1990. Requirement for *period* gene expression in the adult and not during development for locomotor activity rhythms of imaginal *Drosophila melanogaster*. J. Neurogen. 7:31–73.

Ewing, A. W. 1969. The genetic basis of sound production in *Drosophila pseudoobscura* and *D. persimilis*. Animal Behav. 17:555–560.

Ewing, A. W. and J. A. Miyan. 1986. Sexual selection, sexual isolation and the evolution of song in the *Drosophila repleta* group of species. Animal Behav. 34:421–429.

Falb, D., J. Fischer, and T. Maniatis. 1992. Rearrangement of upstream regulatory elements leads to ectopic expression of the *Drosophila mulleri Adh-2* gene. Genetics 132:1071–1079.

Falconer, D. S. 1981. *Introduction to Quantitative Genetics*, 2nd ed. Longman, London and New York.

Fallen, C. F. 1823. *Diptera Sueciae Geomyzides*. Vol. 2, p. 4. Berlingianis Literis. Lund, Sweden.

Fang, X.-M., C.-Y. Wu, and M. D. Brennan. 1991. Complexity in evolved regulatory variation for alcohol dehydrogenase genes in Hawaiian *Drosophila*. J. Mol. Evol. 32:220–226.

Feldman, M. W. 1972. Selection for linkage modification. I. Random mating populations. Theor. Pop. Biol. 3:324–346.

Feldman, M. W., F. B. Christiansen, and L. D. Brooks. 1980. Evolution of recombination in a constant environment. Proc. Natl. Acad. Sci. USA 77:4838–4841.

Felger, I. and W. Pinsker. 1987. Histone gene transposition in the phylogeny of the *Drosophila obscura* group. Z. Zool. Syst. Evolutionsforsch. 25:127–140.

Felix, R., J. Guzman, and A. de Garay-Arellano. 1971. CO_2 sensitivity of *Drosophila* flies from a location in the outskirts of Mexico City. Dros. Inf. Serv. 47:110–112.

Felsenstein, J. 1981. Skepticism towards Santa Rosalia, or why are there so few kinds of animals? Evolution 35:124–138.

Felsenstein, J. 1985. Confidence limits on phylogenies: an approach using bootstrap. Evolution 39:783–791.

Felsenstein, J. 1988. Phylogenies from molecular sequences: Inference and reliability. Ann. Rev. Gent. 22:521–565.

Fenerjian, M. G. and F. C. Kafatos. 1994. Developmental specificity of a bi-directional moth chorion promoter in transgenic *Drosophila.* Develop. Biol. 161:37–47.

Fenerjian, M. G., J. C. Martinez-Cruzado, C. Swimmer, D. King, and F. C. Kafatos. 1989. Evolution of the autosomal chorion cluster in *Drosophila.* II. Chorion gene expression and sequence comparisons of the *s16* and *s19* genes in evolutionarily distant species. J. Mol. Evol. 29:108–125.

Ferrar, P. 1987. *A Guide to the Breeding Habits and Immature Stages of Diptera Cyclorrhapha.* E. J. Brill/Scandinavian Science Press, Leiden-Copenhagen.

Ferrari, J. A. and C. E. Taylor. 1981. Hierarchical patterns of chromosome variation in *Drosophila subobscura.* Evolution 35:391–394.

Ferveur, J.-F. 1991. Genetic control of pheromones in *Drosophila simulans.* I. *Ngbo*, a locus on the second chromosome. Genetics 128:293–301.

Ferveur, J.-F. and J.-M. Jallon. 1993. Genetic control of pheromones in *Drosophila simulans.* II. *kete*, a locus on the X chromosome. Genetics 133:561–567.

Filatova, L. P. 1973. The effect of temperature on chromosomal polymorphism in *Drosophila funebris.* Genetika (USSR) 9:62–74.

Finnegan, D. J. 1989. Eukaryotic transposable elements and genome evolution. Trends Genet. 5:103–107.

Finnegan, D. J. and D. H. Fawcett. 1986. Transposable elements in *Drosophila melanogaster.* Oxford Surveys on Eukaryotic Genetics 3:1–62.

Fitch, D. H. A. and L. D. Strausbaugh. 1993. Low codon bias and high rates of synonymous substitutions in *Drosophila hydei* and *D. melanogaster* histone genes. Mol. Biol. Evol. 10: 397–413.

Fleuriet, A. 1988. Maintenance of a hereditary virus: The sigma virus in populations of its host, *Drosophila melanogaster.* Evol. Biol. 23:1–30.

Flyg, C. and H. G. Boman. 1988. *Drosophila* genes *cut* and *miniature* are associated with the susceptibility to infection by *Serratia marcescens.* Genet. Res. 52:51–56.

Flyg, C., G. Dalhammer, B. Rasmuson, and H. G. Boman. 1987. Insect immunity: Inducible antibacterial activity in *Drosophila.* Insect Biochem. 17:153–160.

Fogelman, J. C. 1982. The role of volatiles in the ecology of cactophilic *Drosophila.* Pp. 191–206 in *Ecological Genetics and Evolution: The Cactus-Yeast-Drosophila Model System*, eds. J. S. F. Barker and W. T. Starmer. Academic Press, Sydney.

Fogelman, J. C., W. T. Starmer, and W. B. Heed. 1981. Larval selectivity for yeast species by *Drosophila mojavensis* in natural substrates. Proc. Natl. Acad. Sci. USA 78:4435–4439.

Fontdevila, A. and H. L. Carson. 1978. Spatial distribution and dispersal in a population of *Drosophila.* Am. Nat. 112:365–394.

Fontdevila, A. and J. Mendez. 1979. Frequency-dependent mating in a modified allozyme locus of *Drosophila pseudoobscura.* Evolution 33:634–640.

Fontdevila, A., W. T. Starmer, W. B. Heed, and J. S. Russel. 1977. Migrant selection in a natural population of *Drosophila.* Experientia 33:1447–1448.

Fontdevila, A., A. Ruiz, G. Alonso, and J. Ocana. 1981. Evolutionary history of *Drosophila buzzatii*. I. Natural chromosomal polymorphism in colonized populations of the Old World. Evolution 35:148–157.

Fontdevila, A., A. Ruiz, J. Ocana, and G. Alonso. 1982. Evolutionary history of *Drosophila buzzatii*. II. How much has chromosomal polymorphism changed in colonization? Evolution 36:843–851.

Fontdevila, A., C. Zapata, G. Alverez, L. Sanchez, J. Mendez, and I. Enriquez. 1983. Genetic coadaptation in the chromosomal polymorphism of *Drosophila subobscura*. I. Seasonal changes of gametic disequilibrium in a natural population. Genetics 105:935–955.

Foote, D. and H. L. Carson. 1995. Drosophila as monitors of change in Hawaiian ecosystems. Pp. 368–372 in *Our Living Resources*, eds. E. T. LaRoe et al. U.S. Department of the Interior, National Biological Service, Washington, D.C.

Ford, M. J., C. K. Yoon, and C. F. Aquadro. 1994. Molecular evolution of the *period* gene in *Drosophila athabasca*. Mol. Biol. Evol. 11:169–182.

Fos, M., M. A. Dominguez, A. Latorre, and A. Moya. 1990. Mitochondrial DNA evolution in experimental populations of *Drosophila subobscura*. Proc. Natl. Acad. Sci. USA 87: 4198–4201.

Fowler, K. and M. C. Whitlock. 1994. Fluctuating asymmetry does not increase with moderate inbreeding in *Drosophila melanogaster*. Heredity 73:373–376.

Frank, S. A. 1991. Divergence of meiotic drive-suppression systems as an explanation for sex-biased hybrid sterility and inviability. Evolution 45:262–267.

Franz, G. and W. Kunz. 1981. Intervening sequences in ribosomal RNA genes and bobbed phenotype in *Drosophila hydei*. Nature 292:638–640.

Fraser, A., D. Burnell, and D. Miller. 1966. Simulation of genetic systems. X. Inversion polymorphism. J. Theor. Biol. 13:1–14.

Freeth, A. L., J. B. Gibson, A. V. Wilks. 1990. Aberrant splicing of a naturally occurring alcohol dehydrogenase null activity allele in *Drosophila melanogaster*. Genome 33:873–877.

Freire-Maia, N. 1961. Peculiar gene arrangements in Brazilian natural populations of *Drosophila ananassae*. Evolution 15:486–495.

Freriksen, A., B. L. A. deRuiter, H.-J. Groenenberg, W. Scharloo, and P. W. H. Heinstra. 1994a. Multilevel approach to the significance of genetic variation in alcohol dehydrogenase of Drosophila. Evolution 48:781–790.

Freriksen, A., D. Seykens, and P. W. H. Heinstra. 1994b. Differences between larval and adult Drosophila in metabolic degradation of ethanol. Evolution 48:504–508.

Frydenberg, O. 1962. The modification of polymorphism in some artificial populations of *Drosophila melanogaster*. Hereditas 48:423–441.

Frydenberg, O. 1964. Long term instability of an ebony polymorphism in artificial populations of *Drosophila melanogaster*. Hereditas 51:198–206.

Furia, M., F. A. Digilio, D. Artiaco, E. Giordano, and L. C. Polito. 1990. A new gene nested within the dunce genetic unit of *Drosophila melanogaster*. Nucl. Acids. Res. 18:5837–5841.

Furia, M., P. P. D'Avino, S. Crispi, D. Artiaco, and L. Polito. 1993. A dense cluster of genes is located at the ecdysone-regulated 3C puff of *Drosophila melanogaster*. J. Mol. Biol. 231:531–538.

Futch, D. G. 1966. A study of speciation in South Pacific populations of *Drosophila ananassae*. Univ. Texas Publ. 6615:79–120.

Futuyma, D. J. 1970. Variation in genetic response to interspecific competition in laboratory populations of *Drosophila*. Am. Nat. 104:239–252.

Galiana, A., A. Moya, and F. J. Ayala. 1993. Founder-flush speciation in *Drosophila pseudoobscura*: A large-scale experiment. Evolution 47:432–444.

Gall, J. G., E. H. Cohen, and M. L. Polan. 1971. Repetitive DNA sequences in *Drosophila*. Chromosoma 33:319–344.

Game, A. Y. and J. G. Oakeshott. 1990. The association between restriction site polymorphism and enzyme activity variation for Esterase 6 in *Drosophila melanogaster.* Genetics 126: 1021–1031.

Ganguly, R., K. D. Swanson, K. Ray, and R. Krishnan. 1992. A BamHI element is predominantly associate with the degenerating neo-Y chromosome of *Drosophila miranda* but absent in the *Drosophila melanogaster* genome. Proc. Natl. Acad. Sci. USA 89:1340–1344.

Garcia-Bellido, A. and J. F. deCellis. 1992. Developmental genetics of the veination pattern of Drosophila. Ann. Rev. Genet. 26:277–304.

Garcia-Bellido, A. and P. Santamaria. 1972. Developmental analysis of the wing disc in the mutant *engrailed* of *Drosophila melanogaster.* Genetics 72:87–104.

Garesse, R. 1988. *Drosophila melanogaster* mitochondrial DNA: Gene organization and evolutionary consequences. Genetics 118:649–663.

Gartner, L. P. 1986. *Aging in Drosophila: A selectively annotated bibliography.* Jen House, Baltimore, Md.

Gatti, M. and S. Pimpinelli. 1992. Functional elements in *Drosophila melanogaster* heterochromatin. Ann. Rev. Genet. 26:239–275.

Gay, P. 1978. Les gènes de la *Drosophila* qui interviennent dans la multiplication du virus sigma. Mol. Gen. Genet. 159:269–283.

Gebhardt, M. D. and S. C. Stearns. 1993a. Phenotypic plasticity for life history traits in *Drosophila melanogaster.* I. Effect on phenotypic and environmental correlations. J. Evol. Biol. 6:1–16.

Gebhardt, M. D. and S. C. Stearns. 1993b. Phenotypic plasticity for life history traits in *Drosophila melanogaster.* II. Epigenetic mechanisms and the scaling of variances. J. Evol. Biol. 6:17–29.

Geer, B. W., P. W. H. Heinstra, A. M. Kapoun, and A. van der Zel. 1990. Alcohol dehydrogenase and alcohol tolerance in *Drosophila melanogaster.* Pp. 231–252 in *Ecological and Evolutionary Genetics*, eds. J. S. F. Barker, W. T. Starmer, and R. J. MacIntyre. Plenum, New York.

Geer, B. W., P. W. H. Heinstra, and S. W. McKechnie. 1993. The biological basis of ethanol tolerance in Drosophila. Comp. Biochem. Physiol. 105B:203–229.

Gershenson, S. M., Y. N. Alexandrov, and S. S. Muliuta. 1975. *Mutagenic Action of DNA and Viruses in Drosophila.* Naukova Dumka, Kiev.

Gibson, J. B., A. V. Wilks, and A. Agrotis. 1992. Molecular relationships between alcohol dehydrogenase null-activity alleles from natural populations of *Drosophila melanogaster.* Mol. Biol. Evol. 9:250–260.

Gillespie, J. H. 1991. *The Causes of Molecular Evolution.* Oxford University Press, New York, Oxford.

Gillespie, J. H. and K.-I. Kojima. 1968. The degree of polymorphism in enzymes involved in energy production compared to that in nonspecific enzymes in two *Drosophila ananassae* populations. Proc. Natl. Acad. Sci. USA 61:582–601.

Gilpin, M. E. and F. J. Ayala. 1973. Global models of growth and competition. Proc. Nat. Acad. Sci. USA 70:3590–3593.

Glass, B. (ed.). 1980. *The Roving Naturalist: Travel Letters of Theodosius Dobzhansky.* The American Philosophical Society, Philadelphia.

Gleason, J. M. 1996. Molecular evolution of the period locus and evolution of courtship song in the *Drosophila willistoni* sibling species. Ph.D. dissertation, Yale University, New Haven, Conn.

Gleason, J. M., A. Caccone, E. N. Moriyama, K. P. White, and J. R. Powell. (1997). Mitochondrial DNA phylogenies for the *Drosophila obscura* group. Evolution (April issue).

Goddard, K., A. Caccone, and J. R. Powell. 1990. Evolutionary implications of DNA divergence in the *Drosophila obscura* group. Evolution 44:1656–1670.

Gojobori, T., K. Ishii, and M. Nei. 1982. Estimation of average number of nucleotide substitution when the rate of substitution varies with nucleotide. J. Mol. Evol. 18:414–423.

Gomariz-Zilber, E. and M. Thomas-Orillard. 1993. *Drosophila* C virus and *Drosophila* hosts: a good association in various environment. J. Evol. Biol. 6:677–689.

Gonzalez, A. M., V. M. Cabrera, J. M. Larruga, and A. Gullon. 1982. Genetic distance in the sibling species *Drosophila melanogaster, Drosophila simulans*, and *Drosophila mauritiana.* Evolution 36:517–522.

Gonzalez, A. M., M. Hernandez, A. Volz, J. Pestano, J. M. Larruga, D. Sperlich, and V. M. Cabrera. 1990. Mitochondrial DNA evolution in the *obscura* subgroup of *Drosophila.* J. Mol. Evol. 31:122–131.

Goodman, D. 1979. Competitive hierarchies in laboratory *Drosophila*. Evolution 33: 207–219.

Gordon, C. 1936. The frequency of heterozygosis in free-living populations of *Drosophila subobscura.* J. Genet. 33:25–60.

Gordon, C., H. Spurway, and P. A. R. Street. 1939. An analysis of three wild populations of *Drosophila subobscura.* J. Genetics 38:37–90.

Gould, S. J. 1977. *Ontogeny and Phylogeny.* Belknap/Harvard Press, Cambridge.

Gould, S. J. 1982. Introduction. Pp. xvii–xli in *Genetics and the Origin of Species* by Th. Dobzhansky. Columbia University Press, New York.

Gould, S. J. and E. S. Vrba. 1982. Exaptation—a missing term in the science of form. Paleobiology 8:4–15.

Graber, H. 1957. Afrikanishe Drosophiliden als Blutenbesucher. Zool. Jahrb. 85, Abt. Syst. 305–316.

Grant, B. and L. E. Mettler. 1969. Disruptive and stabilizing selection on the "escape" behavior of *Drosophila melanogaster.* Genetics 62:625–637.

Green, M. M. 1959. The discrimination of wild-type iso-alleles at the *white* locus of *Drosophila melanogaster.* Proc. Natl. Acad. Sci. USA 45:549–553.

Green, M. M. 1963. Unequal crossing over and the genetical organization of the *white* locus of *Drosophila melanogaster.* Z. Vererbungslehere 94:200–214.

Green, M. M. 1969a. Mapping a *Drosophila melanogaster* "controlling element" by interallelic crossing over. Genetics 61:423–428.

Green, M. M. 1969b. Controlling element mediated transpositions of the *white* gene in *Drosophila melanogaster.* Genetics 61:429–441.

Green, M. M. 1976. Mutable and mutator loci. Pp. 929–946 in *The Genetics and Biology of Drosophila Vol. 1b*, eds. M. Ashburner and E. Novitski. Academic Press, New York.

Green, M. M. 1987. Mobile DNA elements and spontaneous gene mutation. Pp. 41–50 in *Eukaryotic Transposable Elements as Mutagens*, eds. M. E. Lambert, J. F. McDonald, and I. B. Weinstein. Cold Spring Harbor Press, Cold Spring Harbor, N.Y.

Greenacre, M. L., M. G. Ritchie, B. C. Byrne, and C. P. Kyriacou. 1993. Female song preference and the *period* gene in *Drosophila*. Behav. Genet. 23:85–90.

Gregg, T. G., A. McCrate, G. Reveal, S. Hall, and A. L. Rypstra. 1990. Insectivory and social digestion in *Drosophila.* Biochem. Genet. 28:197–207.

Grimaldi, D. A. 1987. Amber fossil Drosophilidae (Diptera), with particular reference to the Hispanolan taxa. Amer. Mus. Novitates 2880:1–23.

Grimaldi, D. A. 1988. Relicts in the Drosophilidae (Diptera). Liebherr 1988:183–213.

Grimaldi, D. A. 1990. A phylogenetic, revised classification of genera in the Drosophilidae. Bull Am. Mus. Nat. Hist. 197:1–139.

Grimaldi, D. A. and G. Fenster. 1989. Evolution of extreme sexual dimorphisms: Structural and behavioral convergence among broad-headed male Drosophilidae (Diptera). Am. Mus. Novit. 2939:1–25.

Grimaldi, D. A. and J. Jaenike. 1984. Competition in natural populations of mycophagous *Drosophila.* Ecology 65:113–1120.

Gunaratne, P., J. L. Ross, Q. Zhang, E. L. Organ, and D. R. Cavener. 1994. An evolutionarily conserved palindrome in the *Drosophila Gld* promoter directs tissue-specific expression. Proc. Natl. Acad. Sci. USA 91:2738–2742.

Gutknecht, J., D. Sperlich, and L. Bachmann. 1995. A species specific satellite DNA family of *Drosophila subsilvestris* appearing predominantly in B chromosomes. Chromosoma 103: 539–544.

Guzman, J., L. Levine, O. Olvera, J. R. Powell, M. E. de la Rosa, V. M. Salceda, Th. Dobzhansky, and R. Felix. 1975. Population genetics of Mexican *Drosophila*. III. Preliminary report on the chromosomal variation of *Drosophila pseudoobscura* in geographic zones of Central Mexico (in Spanish). Ann. Inst. Biol. Univ. Natl. Auton. Mexico 46:75–88.

Hackstein, J. H. P. 1987. Spermatogenesis in *Drosophila*. Pp. 63–116 in *Spermatogenesis: Genetic Aspects*, ed. W. Hennig. Springer Verlag, Berlin.

Haekel, E. 1879. *The Evolution of Man: A Popular Exposition of the Principal Points of Human Ontogeny and Phylogeny*. D. Appleton and Co., New York.

Hagemann, S., W. J. Miller, and W. Pinsker. 1990. P-related sequences in *Drosophila bisfasciata*: A molecular clue to understanding of P-element evolution in the genus *Drosophila*. J. Mol. Evol. 31:478–484.

Haldane, J. B. S. 1922. Sex ratio and unisexual sterility in hybrid animals. J. Genetics 12: 101–109.

Halder, G., P. Callaerts, and W. J. Gehring. 1995. New perspectives on eye evolution. Cur. Opinion Genet. Develop. 5:602–609.

Hale, L. R. and R. S. Singh. 1991. A comprehensive study of genic variation in natural populations of *Drosophila melanogaster*. IV. Mitochondrial DNA variation and the role of history vs. selection in the genetic structure of geographic populations. Genetics 129:103–117.

Haley, C. S. and A. J. Birley. 1983. The genetical response to natural selection by varied environments. II. Observations on replicate populations in spatially varied laboratory environments. Heredity 51:581–606.

Hall, H. M. and F. E. Clemens. 1923. The phylogenetic method in taxonomy. Carnegie Institution of Washington Publication 326. Washington, D.C.

Halliburton, R. and J. S. F. Barker. 1993. Lack of mitochondrial DNA variation in Australian *Drosophila buzzatii*. Mol. Biol. Evol. 10:484–487.

Harden, N. and M. Ashburner. 1990. Characterization of FB-NOF transposable element of *Drosophila melanogaster*. Genetics 126:387–400.

Hardy, D. E. and K. Y. Kaneshiro. 1981. Drosophilidae of the Pacific Oceania. Pp. 309–348 in *Genetics and Biology of Drosophila, Vol. 3a*, eds. M. Ashburner, H. L. Carson, and J. N. Thompson, Jr. Academic Press, New York.

Harris, H. 1966. Enzyme polymorphism in man. Proc. Roy. Soc. Ser. B 164:298–310.

Hartenstein, V. 1993. *Atlas of Drosophila Development*. Cold Spring Harbor Press, Cold Spring Harbor, N.Y.

Hartl, D. L. and Y. Hiraizumi. 1976. Segregation distortion. Pp. 615–666 in *The Genetics and Biology of Drosophila Vol. 1B*, eds. M. Ashburner and E. Novitski. Academic Press, New York.

Hartl, D. L. and E. R. Lozovskaya. 1995. *The Drosophila Genome Map: A Practical Guide*. R. G. Landes Publishing Co., Austin, Tex.

Hartl, D. L., J. Ajioka, H. Cai, A. R. Lohe, E. R. Lozovskaya, D. A. Smoller, and I. W. Duncan. 1992. Towards a *Drosophila* genome map. Trends Genet. 8:70–75.

Hartl, D. L., D. I. Nurminsky, R. W. Jones, and E. R. Lozovskaya. 1994. Genome structure and evolution in *Drosophila*: Application of the framework P1 map. Proc. Natl. Acad. Sci. USA 91:6824–6829.

Hatsumi, M. 1987. Karyotype polymorphism in *Drosophila albomicana*. Genome 29:395–400.

Hatzopoulos, P. and M. P. Kambysellis. 1987a. Isolation and structural analysis of *Drosophila grimshawi* vitellogenin genes. Mol. Gen. Genet. 206:475–484.

Hatzopoulos, P. and M. P. Kambysellis. 1987b. Differential and temporal expression of the vitellogenin genes in *Drosophila grimshawi.* Mol. Gen. Genet. 210:564–571.

Hatzopoulos, P. and M. P. Kambysellis. 1988. Comparative biochemical and immunological analysis of the three vitellogenins from *Drosophila grimshawi.* Comp. Biochem. Physiol. 89B:557–564.

Hauschteck-Jungen, E. 1990. Postmating reproductive isolation and modification of the "sex ratio" trait in *Drosophila subobscura.* Genetica 83:31–44.

Hauschteck-Jungen, E. and G. Maurer. 1976. Sperm dysfunction in sex ratio males of *Drosophila subobscura*. Genetica 46:459–477.

Hawley, R. S. 1980. Chromosomal sites necessary for normal levels of meiotic recombination in *Drosophila melanogaster.* I. Evidence for and mapping of the sites. Genetics 94:625–646.

Hawley, S. A., R. A. Norman, C. J. Brown, W. W. Doane, W. W. Anderson, and D. A. Hickey. 1990. Amylase gene expression in intraspecific and interspecific somatic transformants of *Drosophila.* Genome 33:501–508.

Hawley, R. S., H. Irick, A. E. Zitron, D. A. Haddox, A. Lohe, C. New, M. D. Whitley, T. Arbel, J. Jang, K. McKim, and G. Childs. 1993. There are two mechanisms of achiasmate segregation in *Drosophila*, one of which requires heterochromatic homology. Dev. Genet. 13:440–467.

Hedrick, P. W. 1981. The establishment of chromosomal variants. Evolution 35:322–332.

Hedrick, P. W. 1990. Theoretical analysis of habitat selection and the maintenance of genetic variation. Pp. 209–227 in *Evolutionary Genetics of Drosophila*, eds. J. S. F. Barker, W. T. Starmer, and R. J. MacIntyre. Plenum, New York.

Hedrick, P. W. and E. Murray. 1983. Selection and measures of fitness. Pp. 61–104 in *The Genetics and Biology of Drosophila, Vol. 3d*, eds. M. Ashburner, H. L. Carson, and J. N. Thompson, Jr. Academic Press, New York.

Hedrick, P. W., S. Jain, and L. Holden. 1978. Multilocus systems in evolution. Evol. Biol. 11: 101–184.

Heed, W. B. 1957. Ecological and distributional notes on the Drosophilidae (Diptera) of El Salvador. Univ. Texas Publ. 5721:62–78.

Heed, W. B. 1968. Ecology of the Hawaiian *Drosophila*. Univ. Texas Publ. 6818 Studies in Genetics 4:387–419.

Heed, W. B. 1971. Host plant specificity and speciation in Hawaiian *Drosophila*. Taxon 20: 115–121.

Heed, W. B. 1977. A new cactus-feeding but soil breeding species of *Drosophila* (Diptera: Drosophilidae). Proc. Entomol. Soc. Wash. 79:649–654.

Heed, W. B. 1978. Ecology and genetics of Sonoran Desert *Drosophila.* Pp. 110–126 in *Ecological Genetics: The Interface*, ed. P. F. Brussard. Springer-Verlag, New York.

Heed, W. B. 1981. Central and marginal populations revisited. Dros. Inf. Serv. 56:60–61.

Heed, W. B. and H. W. Kircher. 1965. Unique sterol in the ecology and nutrition of *Drosophila pachea.* Science 149:758–761.

Heed, W. B. and R. L. Mangan. 1986. Community ecology of the Sonoran desert *Drosophila.* Pp. 311–345 in *The Genetics and Biology of Drosophila, Vol. 3e*, eds. M. Ashburner, H. L. Carson, and J. N. Thompson, Jr. Academic Press, New York.

Heed, W. B., D. W. Crumpacker, and L. Ehrman. 1969. *Drosophila lowei*, a new American member of the *obscura* species group. Ann. Entomol. Soc. Am. 62:388–393.

Heed, W. B., W. T. Starmer, M. Miranda, M. W. Miller, and H. J. Phaff. 1976. An analysis of the yeast flora associated with cactiphilic *Drosophila* and their host plants in the Sonoran Desert and its relation to temperate and tropical associations. Ecology 57:151–160.

Heinstra, P. W. H. 1993. Evolutionary genetics of the *Drosophila* alcohol dehydrogenase gene-enzyme system. Genetica 92:1–22.

Heinstra, P. W., W. Scharloo, and G. E. W. Thorig. 1987. Physiological significance of the alcohol dehydrogenase polymorphism in larvae of Drosophila. Genetics 117:75–84.

Heitz, E. 1933. Dies somatische Heteropyknose bei *Drosophila melanogaster* und ihre genetische Bedeutung. Z. Zellforsch. Mikrosk. Anat. 20:237–287.

Heitz, E. 1934. Uber α und β-Heterochromatin sowie konstanz und bauder Chromomeren bei *Drosophila.* Biol. Zentralbl. 45:588–609.

Helman, B. 1949. Etude de la vitalite relative du genotype sauvage Oregon et de genotype comprotant le gene Stubble chez *Drosophila melanogaster.* Compt. Rend. Acad. Sci. 228: 2057–2058.

Henikoff, S. and M. K. Eghtedarzadeh. 1987. Conserved arrangement of nested genes at the Drosophila *Gart* locus. Genetics 117:711–725.

Henikoff, S., M. A. Keene, K. Fechtel, and J. W. Fristrom. 1986. Gene with a gene: Nested Drosophila genes encode unrelated proteins on opposite DNA strands. Cell 44:33–42.

Hennig, W. 1965. Die Acalyptratae Baltischen Bernsteins. Stuttg. Beitr. Naturkd. 145:1–213.

Hennig, W. 1973. Part 31. Diptera (Aweiflugler). *Kukenthal's Handbuch der Zoologie* 4:1–200.

Hennig, W. 1977. Gene interactions in germ cell differentiation of *Drosophila.* Pp. 363–371 in *Advances in Enzyme Regulation, Vol. 15,* ed. G. Weber. Pergamon Press, Oxford, U.K.

Hess, O. 1976. Genetics of *Drosophila hydei* Sturtevant. Pp. 1343–1363 in *The Genetics and Biology of Drosophila, Vol. 1c,* eds. M. Ashburner and E. Novitski. Academic Press, New York.

Hey, J. 1989. The transposable portion of the genome of *Drosophila algonquin* is very different from that in *D. melanogaster.* Mol. Biol. Evol. 6:66–79.

Hey, J. 1994. Bridging phylogenetics and population genetics with gene tree models. Pp. 435–449 in *Molecular Approaches to Ecology and Evolution,* eds. B. Schierwater, B. Streit, G. Wagner, and R. DeSalle. Birkhauser, Basel.

Hey, J. and D. Houle. 1987. Habitat choice in the *Drosophila affinis* group. Heredity 58:463–471.

Hey, J. and R. M. Kliman. 1993. Population genetics and phylogenetics of DNA sequence variation at multiple loci within the *Drosophila melanogaster* species complex. Mol. Biol. Evol. 10:804–822.

Hickey, D. A., L. Bally-Cuif, S. Abukashawa, V. Payant, and B. F. Benkel. 1991. Concerted evolution of duplicated protein-coding genes in *Drosophila.* Proc. Natl. Acad. Sci. USA 88:1611–1615.

Hickey, D. A., B. F. Benkel, and C. Magoulas. 1989. Molecular biology of enzyme adaptations in higher eukaryotes. Genome 31:272–283.

Hill, W. G. and A. Robertson. 1966. The effect of linkage on limits to artificial selection. Genet. Res. 8:269–294.

Hillis, D. M., M. W. Allard, and M. M. Miyamoto. 1993. Analysis of DNA sequence data: phylogenetic inferences. Methods Enzymol. 224:456–487.

Hilton, H., R. M. Kliman, and J. Hey. 1994. Using hitchhiking genes to study adaptation and divergence during speciation within the *Drosophila melanogaster* species complex. Evolution 48:1900–1913.

Hinton, C. W. 1967. Genic modifiers of recombination in *Drosophila melanogaster.* Can. J. Genet. Cytol. 9:711–716.

Hiraizumi, Y. 1971. Spontaneous recombination in *Drosophila melanogaster* males. Proc. Natl. Acad. Sci. USA 68:268–270.

Hiraizumi, Y., L. Sandler, and J. R. Crow. 1960. Meiotic drive in natural populations of *Drosophila melanogaster.* III. Population implications of the segregation-distorter locus. Evolution 24:415–423.

Hiromi, Y., A. Kuroiwa, and W. J. Gehring. 1985. Control elements of the *Drosophila* segmentation gene *fushi tarazu.* Cell 43:603–613.

Hoenigsberg, H. F., I. J. Palomino, C. Chiappe, G. G. Rojas, and B. M. Canas. 1977. Population genetics in the American tropics. XI. Seasonal and temporal variations in the relative frequencies of species belonging to the *willistoni* group in Colombia. Ecological 27:295–304.

Hoffmann, A. A. 1985. Effects of experience on oviposition and attraction in *Drosophila*: comparing apples and oranges. Am. Nat. 126:41–51.

Hoffmann, A. A. and P. A. Parsons. 1991. *Evolutionary Genetics of Environmental Stress.* Oxford University Press, Oxford, U.K.

Hoffmann, A. A. and M. Turelli. 1985. Distribution of *Drosophila melanogaster* on alternative resources: effects of experience and starvation. Am. Nat. 126:662–679.

Hoffmann, A. A. and M. Turelli. 1988. Unidirectional incompatibility in *Drosophila simulans*: Inheritance, geographic variation and fitness effects. Genetics 119:435–444.

Hoffmann, A. A. and M. Watson. 1993. Geographical variation in the acclimation responses of *Drosophila* to temperature extremes. Am. Nat. 142:S93–S113.

Hoffmann, A. A., M. Turelli, and G. M. Simmons. 1986. Unidirectional incompatibility between populations of *Drosophila simulans.* Evolution 40:692–701.

Hoffmann, A. A., D. J. Clancy, and E. Merton. 1994. Cytoplasmic incompatibility in Australia populations of *Drosophila melanogaster.* Genetics 136:993–999.

Holden, P. R., P. Jones, and J. F. Y. Brookfield. 1993. Evidence for a *Wolbachia* symbiont in *Drosophila melanogaster.* Genet. Res. 62:23–29.

Hollocher, H. and C.-I. Wu. 1996. The genetics of reproductive isolation in the *Drosophila simulans* clade: X *vs.* autosomal effects and male *vs.* female effects. Genetics 143:1243–1255.

Holmes, R. S., L. N. Moxon, and P. A. Parsons. 1980. Genetic variability of alcohol dehydrogenase among Australian *Drosophila* species: Correlation of ADH biochemical phenotype with ethanol resource utilization. J. Exp. Zool. 214:199–204.

Holmquist, G. 1975. Organization and evolution of *Drosophila virilis* heterochromatin. Nature 257:503–505.

Hooper, J. E., M. Pérez-Alonso, J. R. Bermingham, M. Prout, B. A. Rocklein, M. Wagenbach, J.-E. Edstrom, R. de Frutos, and M. P. Scott. 1992. Comparative studies of Drosophila *Antennapedia* genes. Genetics 132:453–469.

Houck, M. A., J. B. Clark, K. R. Peterson, and M. G. Kidwell. 1991. Possible horizontal transfer of *Drosophila* genes by the mite *Proctolaelaps regalis.* Science 253:1125–1129.

Hovemann, B., S. Richter, U. Walldorf, and C. Czielpluch. 1988. Two genes encode related cytoplasmic elongation factors 1-α (EF-1) in *Drosophila melanogaster* with continuous and stage-specific expression. Nucl. Acids Res. 16:3175–3194.

Hsiang, W. 1949. The distribution of heterochromatin in *Drosophila tumiditarus*. Cytologia 15: 149–153.

Hu, J., H. Qazzaz, and M. D. Brennen. 1995. A transcriptional role for conserved footprinting sequences within the larval promoter of a *Drosophila* alcohol dehydrogenase. J. Mol. Biol. 249:259–269.

Huang, S. L., M. Singh, and K.-I. Kojima. 1971. A study of frequency-dependent selection observed in the esterase-6 locus of *Drosophila melanogaster.* Genetics 68:97–104.

Hubby, J. L. and R. C. Lewontin. 1966. A molecular approach to the study of genic heterozygosity in natural populations. I. The number of alleles at different loci in *Drosophila pseudoobscura.* Genetics 54:577–594.

Hubby, J. L. and L. H. Throckmorton. 1965. Protein differences in Drosophila. II. Comparative species genetics and evolutionary problems. Genetics 52:203–215.

Hudson, R. R. 1982. Estimating genetic variability with restriction endonucleases. Genetics 100:711–719.

Hudson, R. R. 1994. How can the low levels of DNA sequence variation in regions of the

Drosophila genome with low recombination rates be explained? Proc. Natl. Acad. Sci. USA 91:6815–6818.

Hudson, R. R., M. Kreitman, and M. Aguadé. 1987. A test of neutral molecular evolution based on nucleotide data. Genetics 116:153–159.

Hudson, R. R., K. Bailey, D. Skarecky, J. Kwiatowski, and F. J. Ayala. 1994. Evidence for positive selection in the superoxide dismutase (*Sod*) region of *Drosophila melanogaster.* Genetics 136:1329–1340.

Huey, R. B. and J. G. Kingsolver. 1993. Evolution of resistance to high temperature in ectotherms. Am. Nat. 142:S21–S46.

Huey, R. B., L. Partridge, and K. Fowler. 1991. Thermal sensitivity of *Drosophila melanogaster* responds rapidly to laboratory natural selection. Evolution 45:751–756.

Hung, M. C. and P. C. Wensink. 1983. Sequence and structure conservation in yolk proteins and their genes. J. Mol. Biol. 164:481–492.

Hunt, J. A. and H. L. Carson. 1983. Evolutionary relationship of four species of Hawaiian *Drosophila* measured by DNA reassociation. Genetics 104:353–364.

Hunt, J. A., T. J. Hall, and R. J. Britten. 1981. Evolutionary distance in Hawaiian *Drosophila* measured by DNA reassociation. J. Mol. Evol. 17:361–367.

Hurst, G. D. and M. E. N. Majerus. 1993. Why do maternally inherited microorganisms kill males? Heredity 71:81–95.

Hurst, L. D. 1992. Is *Stellate* a relict meiotic driver? Genetics 130:229–230.

Hurst, L. D. and A. Pomiankowski. 1991. Causes of sex ratio bias may account for unisexual sterility in hybrids: A new explanation of Haldane's rule and related phenomena. Genetics 128:841–858.

Hutter, C. M. and D. M. Rand. 1995. Competition between mitochondrial haplotypes in distinct nuclear genetic environments: *Drosophila pseudoobscura* versus *D. persimilis.* Genetics 140:537–548.

Hutter, P. and M. Ashburner. 1987. Genetic rescue of inviable hybrids between *Drosophila melanogaster* and its sibling species. Nature 327:331–333.

Hutter, P., J. Roote, and M. Ashburner. 1990. A genetic basis for the inviability of hybrids between sibling species of *Drosophila.* Genetics 124:909–920.

Ikeda, H. 1970. The cytoplasmically-inherited "sex-ratio" condition in natural and experimental populations of *D. bifasciata.* Genetics 65:311–333.

Inomata, N., H. Shibata, E. Okuyama, and T. Yamazaki. 1995. Evolutionary relationships and sequence variation of α-amylase variants encoded by duplicated genes in the *Amy* locus of *Drosophila melanogaster.* Genetics 141:237–244.

Inoue, Y., T. Watanabe, and T. K. Watanabe. 1984. Evolutionary change of the chromosomal polymorphism in *D. melanogaster* populations. Evolution 38:753–765.

Irick, H. 1994. A new function for heterochromatin. Chromosoma 103:1–3.

Ives, P. T. 1970. The genetic structure of American populations of *Drosophila melanogaster.* Genetics 24:507–518.

Izquierdo, J. I. 1991. How does *Drosophila melanogaster* overwinter? Entomol. Exp. Appl. 59: 51–58.

Jablonka, E. and M. J. Lamb. 1991. Sex chromosomes and speciation. Proc. Roy. Soc. Lond. B 243:203–208.

Jaenike, J. 1978. Ecological genetics in *Drosophila athabasca*: Its effects on local abundance. Am. Nat. 112:287–299.

Jaenike, J. 1982. Environmental modification of oviposition behavior in *Drosophila.* Am. Nat. 119:784–802.

Jaenike, J. 1985. Parasite pressure on the evolution of amanitin tolerance in *Drosophila.* Evolution 39:1295–1301.

Jaenike, J. 1986. Genetic complexity of host-selection behavior in *Drosophila*. Proc. Natl. Acad. Sci. USA 83:2148–2151.

Jaenike, J. 1987. Genetics of oviposition-site preference in *Drosophila tripunctata*. Heredity 59:363–369.

Jaenike, J. 1989. Genetic population structure of *Drosophila tripunctata*: Patterns of variation and covariation affecting resource use. Evolution 43:1467–1482.

Jaenike, J. 1990. Factors maintaining genetic variation for host preference in Drosophila. Pp. 195–207 in *Ecological and Evolutionary Genetics of Drosophila*, eds. J. S. F. Barker, W. T. Starmer, and R. J. MacIntyre. Plenum, New York.

Jaenike, J. and D. Grimaldi. 1983. Genetic variation for host preference within and among populations of *Drosophila tripunctata*. Evolution 37:1023–1033.

Jaenike, J., D. Grimaldi, A. Sluder, and A. L. Greenleaf. 1983. α-Amanitin tolerance in mycophagous *Drosophila*. Science 221:165–167.

Jallón, J. M. and J. David. 1987. Variations in cuticular hydrocarbons among the eight species of the *Drosophila melanogaster* group. Evolution 41:294–302.

Jeffery, D. E., J. L. Farmer, and M. D. Piley. 1988. Identification of Mullerian chromosomal elements in Hawaiian *Drosophila* by *in situ* hybridization. Pacific Sci. 42:48–50.

Jeffs, P. S. and M. Ashburner. 1991. Processed pseudogenes in Drosophila. Proc. R. Soc. Lond. Ser. B 244:151–159.

Jeffs, P. S., E. C. Holmes, and M. Ashburner. 1994. The molecular evolution of the alcohol dehydrogenase and alcohol dehydrogenase-related genes in the *Drosophila melanogaster* species subgroup. Mol. Biol. Evol. 11:287–304.

Jiang, C., J. B. Gibson, A. V. Wilks, and A. L. Freeth. 1988. Restriction endonuclease variation in the region of the alcohol dehydrogenase gene: a comparison of null and normal alleles from natural populations of *Drosophila melanogaster*. Heredity 60:101–107.

John, B. and G. Miklos. 1988. *The Eukaryote Genome in Development and Evolution*. Allen and Unwin, London.

Johnson, F. M. 1971. Isozyme polymorphism in *Drosophila ananassae*. Genetic diversity among island populations in the South Pacific. Genetics 68:77–95.

Johnson, G. B. 1974. Enzyme polymorphism and metabolism. Science 184:28–37.

Johnson, N. A. and C.-I. Wu. 1993. Evolution of postmating reproductive isolation: Measuring the fitness effects of chromosomal regions containing hybrid sterility factors. Am. Nat. 142:213–223.

Johnson, N. A., D. E. Perez, E. L. Cabot, H. Hollocher, and C.-I. Wu. 1992. A test of the reciprocal X-Y interaction as a cause of hybrid sterility in *Drosophila*. Nature 358:751–753.

Johnson, N. A., H. Hollocher, E. Noonburg, and C.-I. Wu. 1993. The effects of interspecific Y chromosome replacements on hybrid sterility within the *Drosophila simulans* clade. Genetics 135:443–453.

Johnson, W. R. 1973. Chromosome variation in natural populations of *Drosophila mojavensis*. M.S. thesis, Univ. of Arizona, Tucson.

Johnson, W. R. 1980. Chromosomal polymorphism in natural populations of the desert adapted species, *D. mojavensis*. Ph.D. dissertation, Univ. of Arizona.

Johnston, J. S. and J. Ellison. 1982. Exact age determination in laboratory and field-caught *Drosophila*. J. Insect Physiol. 28:773–780.

Johnston, J. S. and W. B. Heed. 1975. Dispersal of *Drosophila*: The effect of baiting on the behavior and distribution of natural populations. Am. Nat. 109:207–216.

Johnston, J. S. and W. B. Heed. 1976. Dispersal of desert-adapted *Drosophila*: The Saguaro-breeding *D. nigrospiracula*. Am. Nat. 110:629–651.

Johnston, J. S. and A. R. Templeton. 1982. Dispersal and clines in *Opuntia* breeding *Drosophila mercatorum* and *D. hydei* at Kamuela, Hawaii. Pp. 241–256 in *Ecological Genetics and*

Evolution: The Cactus-Yeast-Drosophila Model System, eds. J. S. F. Barker and W. T. Starmer. Academic Press, Sydney.

Joly, D., M. L. Cariou, D. Lachaise, and J. David. 1989. Variation of sperm length and heteromorphism in Drosophilid species. Genet. Sel. Evol. 21:283–293.

Joly, D., M. L. Cariou, and D. Lachaise. 1991a. Can sperm competition explain sperm polymorphism in *Drosophila teissieri?* J. Evol. Biol. 5:25–44.

Joly, D., C. Bressac, J. Deveux, and D. Lachaise. 1991b. Sperm length diversity in Drosophilidae. Dros. Inf. Serv. 70:104–108.

Jones, J. S. and R. F. Probert. 1980. Habitat selection maintains a deleterious allele in a heterogeneous environment. Nature 287:632–633.

Jones, J. S. and T. Yamazaki. 1974. Genetic background and the fitness of allozymes. Genetics 78:1185–1189.

Jones, J. S., S. H. Bryant, R. C. Lewontin, J. A. Moore, and T. Prout. 1981. Gene flow and the geographical distribution of a molecular polymorphism in *Drosophila pseudoobscura.* Genetics 98:157–178.

Jost, E. and M. Mameli. 1972. DNA content in nine species of Nematocera with special reference to the sibling species of the *Anopheles maculipennis* group and the *Culex pipiens* group. Chromosoma 37:201–208.

Jousset, F. X. 1972. Le virus Iota de *Drosophila immigrans* etudie chez *D. melanogaster*: symptome de la sensibilite au CO_2, descriptions des anomalies provoquees che l'hote. Ann. Inst. Pasteur 121:275–288.

Juan, E., M. Papaceit, and A. Quintana. 1994. Nucleotide sequence of the genomic region encompassing Adh and Adh-Dup genes of *D. lebanonensis* (Scaptodrosophila): Gene expression and evolutionary relationships. J. Mol. Evol. 38:455–467.

Jukes, T. H. and C. R. Cantor. 1969. Evolution of protein molecules. Pp. 21–120 in *Mammalian Protein Metabolism*, ed. H. W. Munro. Academic Press, New York.

Kafatos, F. C., N. Spoerel, S. A. Mitsialis, H. T. Nguyen, C. Romano, J. R. Lingappa, B. D. Mariani, G. C. Rodakis, R. Lecanidou, and S. G. Tsitilou. 1987. Developmental control and evolution in the chorion gene families of insects. Adv. Genet. 24:223–242.

Kafatos, F. C., C. Louis, C. Savakis, D. M. Glover, M. Ashburner, A. J. Link, I. Sidén-Kiamos, and R. D. C. Saunders. 1991. Integrated maps of the *Drosophila* genome: progress and prospects. Trends Genet. 7:155–161.

Kalfayan, L. and P. C. Wensink. 1982. Developmental regulation of *Drosophila* α-tubulin genes. Cell 29:91–98.

Kambysellis, M. P. 1970. Compatibility in insect tissue transplantations. I. Ovarian transplantations and hybrid formation between *Drosophila* species endemic to Hawaii. J. Exp. Zool. 175:169–180.

Kambysellis, M. P. 1974. Ultrastructure of the chorion in very closely related *Drosophila* species endemic to Hawaii. Syst. Zool. 23:507–512.

Kambysellis, M. P. 1993. Ultrastructural diversity in the egg chorion of Hawaiian *Drosophila* and *Scaptomyza*: Ecological and phylogenetic considerations. Int. J. Insect Morphol. Embryol. 22:417–446.

Kambysellis, M. P. and E. M. Craddock. 1991. Insemination patterns in Hawaiian *Drosophila* species (Diptera: Drosophilidae) correlate with ovarian development. J. Insect Behav. 4: 83–100.

Kambysellis, M. P. and W. B. Heed. 1971. Studies of oögenesis in natural populations of Drosophilidae: I. Relation of ovarian development and ecological habitats of the Hawaiian species. Am. Nat. 105:31–49.

Kambysellis, M. P., P. Hatzopoulos, E. W. Seo, and E. M. Craddock. 1986. Noncoordinate synthesis of the vitellogenin proteins in tissues of *Drosophila grimshawi.* Develop. Genet. 7:81–97.

Kambysellis, M. P., P. Hatzopoulos, and E. M. Craddock. 1989. The temporal pattern of vitellogenin synthesis in *Drosophila grimshawi*. J. Exp. Zool. 251:339–348.

Kaneko, M., Y. Satta, E. T. Matusuura, and S. I. Chigusa. 1993. Evolution of mitochondrial ATPase 6 gene in *Drosophila*: unusually high level of polymorphism in *D. melanogaster*. Genet. Res. 61:195–204.

Kaneshiro, K. Y. 1976. Ethological isolation and phylogeny in the *planitibia* subgroup of Hawaiian *Drosophila*. Evolution 30:259–274.

Kaneshiro, K. Y. 1989. The dynamics of sexual selection and founder effects in species formation. Pp. 279–296 in *Genetics, Speciation, and the Founder Principle*, eds. L. V. Giddings, K. Y. Kaneshiro, and W. W. Anderson. Oxford University Press, Oxford, U.K.

Kaneshiro, K. Y. and L. V. Giddings. 1987. The significance of sexual isolation and the formation of new species. Evol. Biol. 21:29–43.

Kaneshiro, K. Y. and J. S. Kurihara. 1982. Sequential differentiation of sexual behavior in populations of *Drosophila silvestris*. Pacific Science 35:177–183.

Kaneshiro, K. Y., H. L. Carson, F. E. Clayton, and W. B. Heed. 1973. Niche separation in a pair of homosequential *Drosophila* species from the island of Hawaii. Am. Nat. 107: 766–774.

Kaneshiro, K. Y., R. G. Gillespie, and H. L. Carson. 1995. Chromosomes and male genetalia of Hawaiian *Drosophila*. Pp. 57–71 in *Hawaiian Biogeography: Evolution on a Hot Spot Archipelago*, eds. W. L. Wagner and V. A. Funk. Smithsonian Institution Press, Washington, D.C.

Kaplan, N. and C. H. Langley. 1979. A new estimate of sequence divergence of mitochondrial DNA using restriction endonuclease mapping. J. Mol. Evol. 13:295–304.

Kaplan, N., R. R. Hudson, and C. H. Langley. 1989. The "hitchhiking effect" revisited. Genetics 123:887–899.

Karotam, J., T. M. Boyce, and J. G. Oakeshott. 1995. Nucleotide variation at the hypervariable esterase 6 isozyme locus of *Drosophila simulans*. Mol. Biol. Evol. 12:113–122.

Kaufmann, B. P. 1937. Morphology of the chromosomes of *Drosophila ananassae*. Cytologia, Fuji Jubilee Vol. pp. 1043–1055.

Kawanishi, M. and T. K. Watanabe. 1977. Ecological factors controlling the coexistence of the sibling species *Drosophila simulans* and *D. melanogaster*. Jap. J. Ecol. 27:279–283.

Keith, T. P. 1983. Frequency distribution of esterase-5 alleles in two populations of *Drosophila pseudoobscura*. Genetics 105:135–155.

Keith, T. P., L. D. Brooks, R. C. Lewontin, J. C. Martinez-Cruzado, and D. L. Rigby. 1985. Nearly identical allelic distributions of xanthine dehydrogenase in two populations of *Drosophila pseudoobscura*. Mol. Biol. Evol. 2:206–216.

Kekic, V., C. E. Taylor, and M. Andjelkovic. 1980. Habitat choice and resource utilization by *Drosophila subobscura*. Genetika (Yugoslavia) 12:219–225.

Kenyon, A. 1967. Comparison of frequency distribution of viability of second and fourth chromosomes from caged *Drosophila melanogaster*. Genetics 55:123–130.

Kernaghan, R. P. and L. Ehrman. 1970. Antimycoplasmal antibiotics and hybrid sterility in *Drosophila paulistorum*. Science 169:63–66.

Kerr, W. E. and S. Wright. 1954a. Experimental studies of the distribution of gene frequencies in very small populations of *Drosophila melanogaster*. I. Forked. Evolution 8:172–177.

Kerr, W. E. and S. Wright. 1954b. Experimental studies of the distribution of gene frequencies in very small populations of *Drosophila melanogaster*. III. Aristapedia and spineless. Evolution 8:293–301.

Kessler, S. 1966. Selection for and against ethological isolation between *Drosophila pseudoobscura* and *Drosophila persimilis*. Evolution 20:634–645.

Kidwell, M. G. 1972a. Genetic change of recombination value in *Drosophila melanogaster*. I.

Artificial selection for high and low recombination and some properties of recombination-modifying genes. Genetics 70:419–432.

Kidwell, M. G. 1972b. Genetic change of recombination value in *Drosophila melanogaster.* II. Simulated natural selection. Genetics 70:433–443.

Kidwell, M. G. 1983. Evolution of hybrid dysgenesis determinants in *Drosophila melanogaster.* Proc. Natl. Acad. Sci. USA 80:1655–1659.

Kidwell, M. G. 1987. Regulatory aspects of the expression of P-M hybrid dysgenesis in *Drosophila.* Pp. 183–194 in *Eukaryotic Transposable Elements as Mutagens*, eds. M. E. Lambert, J. F. McDonald, and I. B. Weinstein. Cold Spring Harbor Press, Cold Spring Harbor, N.Y.

Kidwell, M. G. 1993. Lateral transfer in natural populations of eukaryotes. Ann. Rev. Genet. 27:235–256.

Kidwell, M. G., J. F. Kidwell, and J. A. Sved. 1977. Hybrid dysgenesis in *Drosophila melanogaster*: A syndrome of aberrant traits including mutation, sterility and male recombination. Genetics 86:813–833.

Kidwell, M. G., J. B. Novy, and S. M. Feeley. 1981. Rapid unidirectional change of hybrid dysgenesis potential in *Drosophila.* J. Heredity 72:32–38.

Kikkawa, H. H. 1938. Studies on the genetics and cytology of *Drosophila ananassae.* Genetica 20:458–516.

Kilias, G. and S. N. Alahiotis. 1982. Genetic studies on sexual isolation and hybrid sterility in long-term cage populations of *Drosophila melanogaster.* Evolution 36:121–131.

Kilias, G. and S. N. Alahiotis. 1985. Indirect thermal selection in *Drosophila melanogaster* and adaptive consequences. Theoret. Appl. Genet. 69:645–650.

Kilias, G., S. N. Alahiotis, and M. Pelecanos. 1980. A multifactorial genetic investigation of speciation theory using *Drosophila melanogaster.* Evolution 34:730–738.

Kimura, M. 1980. A simple method for estimating evolutionary rates of base substitutions through comparative studies of nucleotide sequences. J. Mol. Evol. 16:111–120.

Kimura, M. 1981. Estimation of evolutionary distances between homologous nucleotide sequences. Proc. Natl. Acad. Sci. USA 78:454–458.

Kimura, M. 1983. *The Neutral Theory of Molecular Evolution.* Cambridge Univ. Press, Cambridge, U.K.

Kimura, M. and T. Ohta. 1971. Protein polymorphism as a phase of molecular evolution. Nature 229:467–469.

Kimura, M. T. and K. Beppu. 1993. Climatic adaptations in the *Drosophila immigrans* species group: seasonal migration and thermal tolerance. Ecol. Entomol. 18:141–149.

King, J. L. and T. H. Jukes. 1969. Non-Darwinian evolution: Random fixation of selectively neutral mutations. Science 164:788–798.

King, L. M. 1994. Associations among nucleotide polymorphisms as evidence of balancing selection at the *Esterase-5B* locus of *Drosophila pseudoobscura.* Unpublished manuscript.

King, M.-C. and A. C. Wilson. 1975. Evolution at two levels in humans and chimpanzees. Science 188:107–116.

Kircher, H. W. 1982. Chemical composition of cacti and its relationship to Sonoran Desert *Drosophila.* Pp. 143–158 in *Ecological Genetics and Evolution: The Cactus-Yeast-Drosophila Model System*, eds. J. S. F. Barker and W. T. Starmer. Academic Press, New York.

Klaczko, L. B., J. R. Powell, and C. E. Taylor. 1983. *Drosophila* baits and yeasts: Species attracted. Oecologia 59:411–413.

Klaczko, L. B., C. E. Taylor, and J. R. Powell. 1986. Genetic variation for dispersal by *Drosophila pseudoobscura* and *Drosophila persimilis.* Genetics 112:229–235.

Klarenberg, A. J. and W. Scharloo. 1986. Nonrandom association between structural *Amy* and regulatory *map* variants in *Drosophila melanogaster.* Genetics 114:875–884.

Klarenberg, A. J., A. J. S. Visser, M. F. M. Willemse, and W. Scharloo. 1986. Genetic localization and action of regulatory genes and elements for tissue-specific expression of α-amylase in *Drosophila melanogaster.* Genetics 114:1131–1145.

Kliman, R. M. and J. Hey. 1993a. DNA sequence variation at the *period* locus within and among species of the *Drosophila melanogaster* complex. Genetics 133:375–387.

Kliman, R. M. and J. Hey. 1993b. Reduced natural selection associated with low recombination in *Drosophila melanogaster.* Mol. Biol. Evol. 10:1239–1258.

Kliman, R. M. and J. Hey. 1994. The effects of mutation and natural selection on codon bias in genes of Drosophila. Genetics 137:1049–1056.

Klysten, P., D. A. Kimbrell, S. Daffre, C. Samakovlis, and D. Hultmark. 1992. The lysozyme locus in *Drosophila melanogaster*: different genes are expressed in midgut and salivary glands. Mol. Gen. Genet. 232:335–343.

Knibb, W. R. 1982. Chromosome inversion polymorphisms in *D. melanogaster*. II. Geographic clines and climatic associations in Australasia, North America and Asia. Genetica 58: 213–222.

Knibb, W. R. 1986. Temporal variation of *D. melanogaster* Adh allele frequencies, inversion frequencies and population sizes. Genetica 71:175–184.

Knight, G. R., A. Robertson, and C. H. Waddington. 1956. Selection for sexual isolation within a species. Evolution 10:14–22.

Knoppien, P. 1984. No evidence for rare male mating advantage in *Drosophila melanogaster* for strains raised at different temperatures. Dros. Inf. Serv. 60:131–132.

Knoppien, P. 1985. Rare male mating advantage: a review. Biol. Rev. Cambridge Phils. Soc. 60:81–117.

Koepfer, H. R. 1987a. Selection for sexual isolation between geographic forms of *Drosophila mojavensis.* I. Interactions between the selected forms. Evolution 41:37–48.

Koepfer, H. R. 1987b. Selection for sexual isolation between geographic forms of *Drosophila mojavensis.* II. Effects of selection on mating preference and propensity. Evolution 41: 1409–1412.

Koga, A., H. Baba, S. Kusakabe, M. Hattori, and T. Mukai. 1993. Distribution of polymorphic gene duplication at the *Gpdh* locus in natural populations of *Drosophila melanogaster.* J. Mol. Evol. 36:532–535.

Kohler, R. E. 1994. *Lords of the Fly: Drosophila Genetics and the Experimental Life.* Univ. Chicago Press, Chicago.

Kojima, K.-I. 1971. Is there a constant fitness for a given genotype? No! Evolution 25:281–285.

Kojima, K.-I. and Y. N. Tobari. 1969. Selective modes associated with karyotypes in *Drosophila ananassae*. II. Heterosis and frequency-dependent selection. Genetics 63:639–651.

Kojima, K.-I., J. Gillespie, and Y. N. Tobari. 1970. A profile of *Drosophila* species enzymes assayed by electrophoresis. I. Number of alleles, heterozygosities, and linkage disequilibrium in glucose-metabolizing systems and some other enzymes. Biochem. Genet. 4:627–637.

Koller, P. 1936. Structural hybridity in *Drosophila pseudoobscura.* J. Genet. 32:79–102.

Kondo, R., Y. Satta, E. T. Matsuura, H. Ishiwa, N. Takahata, and S. I. Chigusa. 1990. Incomplete maternal transmission of mitochondrial DNA in *Drosophila.* Genetics 126:657–664.

Kondo, R., E. T. Matsuura, and S. I. Chigusa. 1992. Further evidence of paternal transmission of *Drosophila* mitochondrial DNA by PCR selective amplification method. Genet. Res. 59:81–84.

Konopka, R. J. and S. Benzer. 1971. Clock mutants of *Drosophila melanogaster.* Proc. Natl. Acad. Sci. USA 68:2112–2116.

Koopman, K. F. 1950. Natural selection for reproductive isolation between *Drosophila pseudoobscura* and *Drosophila persimilis.* Evolution 4:135–148.

Koref-Santibañez, S. 1972a. Courtship behavior of the semispecies of the superspecies *Drosophila paulistorum.* Evolution 26:108–115.

Koref-Santibañez, S. 1972b. Courtship interaction in the semispecies of *Drosophila paulistorum.* Evolution 26:326–333.

Koref-Santibañez, S. and C. H. Waddington. 1958. The origin of sexual isolation between different lines within a species. Evolution 12:485–493.

Korol, A. and K. G. Iliadi. 1994. Increased recombination frequencies resulting from directional selection for geotaxis in *Drosophila.* Heredity 72:64–68.

Kozma, R. and M. Bownes. 1986. Yolk protein induction in males of several *Drosophila* species. Insect Biochem. 16:263–271.

Kraaijeveld, A. R. and J. J. M. Van Alphen. 1994a. Geographical variation in resistance of the parasitoid *Asobara* against encapsulation by *Drosophila melanogaster* larvae: the mechanism explored. Physiol. Entomol. 19:9–14.

Kraaijeveld, A. R. and J. J. M. Van Alphen. 1994b. Geographic variation in reproductive success of the parasitoid *Asobara tabida* in larvae of several *Drosophila species.* Ecol. Entomol. 19:221–229.

Kraaijeveld, A. R. and J. J. M. Van Alphen. 1995. Geographical variation in encapsulation ability of *Drosophila melanogaster* larvae and evidence for parasitoid-specific components. Evol. Ecology 9:10–17.

Kraft, R. and H. Jäckle. 1994. *Drosophila* mode of metamerization in the embryogenesis of the lepidopteran *Manduca sexta.* Proc. Natl. Acad. Sci. USA 91:6634–6638.

Krebs, R. A. 1990. Genetics of sexual isolation in *Drosophila mojavensis.* Behav. Genet. 20: 535–543.

Krebs, R. A. and J. S. F. Barker. 1993. Coexistence of ecologically similar colonising species. II. Population differentiation in *Drosophila aldrichi* and *D. buzzatii* for competitive effects and responses at different temperatures and allozyme variation in *D. aldrichi.* J. Evol. Biol. 6:281–298.

Krebs, R. A. and T. A. Markow. 1989. Courtship behavior and control of reproductive isolation in *Drosophila mojavensis.* Evolution 43:908–912.

Kreitman, M. 1983. Nucleotide polymorphism at the alcohol dehydrogenase locus of *Drosophila melanogaster.* Nature 304:412–417.

Kreitman, M. and M. Aguadé. 1986a. Genetic uniformity in two populations of *Drosophila melanogaster* as revealed by filter hybridization of 4-nucleotide-recognizing restriction enzyme digests. Proc. Natl. Acad. Sci. USA 83:3562–3566.

Kreitman, M. and M. Aguadé. 1986b. Excess polymorphism at the *Adh* locus in *Drosophila melanogaster.* Genetics 114:93–110.

Kreitman, M. and R. R. Hudson. 1991. Inferring the evolutionary histories of the *Adh* and *Adh-dup* loci in *Drosophila melanogaster* from patterns of polymorphism and divergence. Genetics 127:565–582.

Kress, H. 1993. The salivary gland chromosomes of *Drosophila virilis*: a cytological map, pattern of transcription and aspects of chromosomal evolution. Chromosoma 102:734–742.

Krimbas, C. B. 1961. Release of genetic variability through recombination. IV. *D. willistoni.* Genetics 46:1323–1334.

Krimbas, C. B. 1992. The inversion polymorphism of *Drosophila subobscura.* Pp. 127–220 in *Drosophila Inversion Polymorphism*, eds. C. B. Krimbas and J. R. Powell. CRC Press, Boca Raton, Fla.

Krimbas, C. B. 1993. *Drosophila subobscura: Biology, Genetics and Inversion Polymorphism.* Verlag Dr. Kovac, Hamburg.

Krimbas, C. B. 1995. Resistance and acceptance: Tracing Dobzhansky's influence. Pp. 23–38 in *Genetics of Natural Populations: The continuing influence of Theodosius Dobzhansky*, ed. L. Levine. Columbia University Press, New York.

Krimbas, C. B. and V. Alevizos. 1973. The genetics of *Drosophila subobscura.* IV. Further data on inversion polymorphism in Greece—evidence of microdifferentiation. Egypt. J. Genet. Cytol. 2:121–132.

Krimbas, C. B. and M. Loukas. 1980. The inversion polymorphism of *Drosophila subobscura.* Evol. Biol. 12:163–234.

Krimbas, C. B. and M. Loukas. 1984. Evolution of the *obscura* group *Drosophila* species. I. Salivary chromosomes and quantitative characters in *Drosophila subobscura* and two closely related species. Heredity 53:469–482.

Krimbas, C. B. and J. R. Powell (eds.). 1992a. *Drosophila Inversion Polymorphism.* CRC Press, Boca Raton, Fla.

Krimbas, C. B. and J. R. Powell. 1992b. Introduction. Pp. 1–52 in *Drosophila Inversion Polymorphism*, eds. C. B. Krimbas and J. R. Powell. CRC Press, Boca Raton, Fla.

Krivshenko, J. D. 1959. New evidence for the homology of the short euchromatic elements of the X and Y chromosomes of *Drosophila busckii* with the microchromosome of *Drosophila melanogaster.* Genetics 44:1027–1040.

Kunze-Muehl, E., E. Mueller, and D. Sperlich. 1958. Qualitative, quantitative und jahreszeitliche Untersuchungen uber den chromosomalen Polymorphismus naturlicher Populationen von *Drosophila subobscura* in der Umgebung von Wein. Z. Indukt Abstamm.-Vererbungsl. 89:636–646.

Kwiatowski, J., D. Skarecky, K. Bailey, and F. J. Ayala. 1994. Phylogeny of *Drosophila* and related genera inferred from the nucleotide sequence of the Cu, Zn, *Sod* gene. J. Mol. Evol. 38:443–454.

Kyriacou, C. P. and J. C. Hall. 1980. Circadian rhythm mutations in *Drosophila* affect short-term fluctuations in the male's courtship song. Proc. Natl. Sci. USA 77:6929–6933.

Kyriacou, C. P. and J. C. Hall. 1982. The function of courtship song rhythms in *Drosophila.* Anim. Behav. 30:794–801.

Kyriacou, C. P. and J. C. Hall. 1986. Inter-specific genetic control of courtship song production and reception in *Drosophila.* Science 232:494–497.

Kyriacou, C. P. and J. C. Hall. 1994. Genetic and molecular analysis of *Drosophila* behavior. Adv. Genet. 31:139–186.

Labeyrie, V. 1978. The significance of the environment in the control of insect fecundity. Ann. Rev. Entomol. 23:69–89.

Lachaise, D. and L. Tsacas. 1983. Breeding sites in tropical African Drosophilids. Pp. 221–332 in *The Genetics and Biology of Drosophila, Vol. 3d*, eds. M. Ashburner, H. L. Carson, and J. N. Thompson, Jr. Academic Press, New York.

Lachaise, D., M. C. Pignal, and J. Roualt. 1979. Yeast flora partitioning by Drosophilid species coexisting in a tropical African savanna. Ann. Soc. Ent. France 4:659–680.

Lachaise, D., L. Tsacas, and G. Coututier. 1982. The Drosophilidae associated with tropical African figs. Evolution 36:141–151.

Lachaise, D. L., M.-L. Cariou, J. R. David, F. Lemeunier, and M. Ashburner. 1988. Historical biogeography of the *D. melanogaster* species subgroup. Evol. Biol. 22:159–226.

Lai, C. and T. F. C. Mackay. 1993. Mapping and characterization of P-element-induced mutations at quantitative trait loci in *Drosophila melanogaster.* Genet. Res. 61:177–193.

Laird, C. D. 1973. DNA of *Drosophila* chromosomes. Ann. Rev. Genet. 7:177–204.

Laird, C. D. and B. J. McCarthy. 1969. Molecular characterization of the *Drosophila* genome. Genetics 63:865–882.

Lakovaara, S. and A. Saura. 1971a. Genic variation in marginal populations of *Drosophila subobscura.* Hereditas 69:77–82.

Lakovaara, S. and A. Saura. 1971b. Genetic variation in natural populations of *Drosophila obscura.* Genetics 69:377–384.

Lakovaara, S. and A. Saura. 1972. Location of enzyme loci in chromosomes of *Drosophila willistoni*. Experientia 28:355–356.

Lakovaara, S. and A. Saura. 1982. Evolution and speciation in the *Drosophila obscura* group. Pp. 2–59 in *The Genetics and Biology of Drosophila, Vol. 3b*, eds. M. Ashburner, H. L. Carson, and J. N. Thompson, Jr. Academic Press, New York.

Lakovaara, S., A. Saura, and C. T. Falk. 1972. Genetic distance and evolutionary relationships in the *Drosophila obscura* group. Evolution 26:177–184.

Lakovaara, S., A. Saura, P. Lankinen, L. Pohjola, and J. Lokki. 1976. The use of isoenzymes in tracing evolution and in classifying Drosophilidae. Zoologica Scripta 5:173–179.

Land, B. 1973. *Evolution of a Scientist: The Two Worlds of Theodosius Dobzhansky*. Thomas Y. Crowell Company, New York.

Lande, R. 1984. The expected fixation rate of chromosomal inversions. Evolution 38:743–752.

Lande, R. 1985. The fixation of chromosomal rearrangements in a subdivided population with local extinction of groups and colonization. Heredity 54:323–332.

Lange, B. W., C. H. Langley, and W. Stephen. 1990. Molecular evolution of Drosophila metallothionein genes. Genetics 126:921–932.

Langley, C. H. 1990. The molecular population genetics of Drosophila. Pp. 75–91 in *Population Biology of Genes and Molecules*, eds. N. Takahata and J. F. Crow. Baifukan Co. Ltd., Tokyo.

Langley, C. H. and C. F. Aquadro. 1987. Restriction map variation in natural populations of *Drosophila melanogaster*: white locus region. Mol. Biol. Evol. 4:651–663.

Langley, C. H., Y. N. Tobari, and K.-I. Kojima. 1974. Linkage disequilibrium in natural populations of *Drosophila melanogaster*. Genetics 8:921–936.

Langley, C. H., R. A. Voelker, A. J. Leigh Brown, S. Ohnishi, B. Dickson, and E. Montgomery. 1981. Null allele frequencies at allozyme loci in natural populations of *Drosophila melanogaster*. Genetics 99:151–156.

Langley, C. H., E. Montgomery, and W. F. Quattlebaum. 1982. Restriction map variation in the *Adh* region of *Drosophila*. Proc. Natl. Acad. Sci. USA 79:5631–5635.

Langley, C. H., A. E. Shrimpton, T. Yamazaki, N. Miyashita, Y. Matsuo, and C. F. Aquadro. 1988a. Naturally occurring variation in the restriction map of the *Amy* region of *Drosophila melanogaster*. Genetics 119:619–629.

Langley, C. H., E. A. Montgomery, R. Hudson, N. Kaplan, and B. Charlesworth. 1988b. On the role of unequal exchange in the containment of transposable element copy number. Genet. Res. 52:223–236.

Langley, C. H., J. MacDonald, N. Miyashita, and M. Aguadé. 1993. Lack of correlation between interspecific divergence and intraspecific polymorphism at the suppressor of forked region in *Drosophila melanogaster* and *Drosophila simulans*. Proc. Natl. Acad. Sci. USA 90:1800–1803.

Larruga, J. M., J. Rozas, M. Hernandez, A. M. Gonzalez, and V. M. Cabrera. 1993. Latitudinal differences in sex chromosome inversions, sex linked allozymes, and mitochondrial DNA variation in *Drosophila subobscura*. Genetica 92:67–74.

Latorre, A., E. Barrio, A. Moya, and F. J. Ayala. 1988. Mitochondrial DNA evolution in the *Drosophila obscura* group. Mol. Biol. Evol. 5:717–728.

Latter, B. H. D. and J. A. Sved. 1994. A reevaluation of data from competitive tests shows high levels of heterosis in *Drosophila melanogaster*. Genetics 137:509–511.

Laurie, C. C., and L. F. Stam. 1988. Quantitative analysis of RNA produced by slow and fast alleles of *Adh* in *Drosophila melanogaster*. Proc. Natl. Acad. Sci. USA 85:5161–5165.

Laurie, C. C. and L. F. Stam. 1994. The effect of an intronic polymorphism on alcohol dehydrogenase expression in *Drosophila melanogaster*. Genetics 138:379–385.

Laurie, C. C., E. M. Heath, J. W. Jacobson, and M. S. Thomson. 1990. Genetic basis of the difference in alcohol dehydrogenase expression between *Drosophila melanogaster* and *Drosophila simulans*. Proc. Natl. Acad. Sci. USA 87:9674–9678.

Laurie, C. C., J. T. Bridgham, and M. Choudhary. 1991. Association between DNA sequence variation and variation in expression of the *Adh* gene in natural populations of *Drosophila melanogaster*. Genetics 129:489–499.

Laurie-Ahlberg, C. C. 1985. Genetic variation affecting the expression of enzyme-coding genes in Drosophila: An evolutionary perspective. Isozymes Curr. Top. Biol. Med. Res. 12: 33–88.

Laurie-Ahlberg, C. C. and L. F. Stam. 1987. Use of P-element-mediated transformation to identify the molecular basis of naturally occurring variants affecting *Adh* expression in *Drosophila melanogaster*. Genetics 115:129–140.

Lawrence, P. A. 1992. *The Making of a Fly: The genetics of animal design*. Blackwell Scientific Publications, Oxford, U.K.

Legal, L., B. Chappe, and J. M. Jallon. 1994. Molecular basis of *Morinda citrifolia* (L.): Toxicity on Drosophila. J. Chem. Ecol. 28:1931–1943.

Leicht, B. G., S. V. Muse, M. Hanczyc, and A. G. Clark. 1995. Constraints on intron evolution in the gene encoding the myosin alkali light chain in *Drosophila*. 139:299–309.

Leigh Brown, A. J. 1983. Variation at the 87A heat shock locus in *Drosophila melanogaster*. Proc. Natl. Acad. Sci. USA 80:5350–5354.

Lemeunier, F. and M. Ashburner. 1976. Relationships in the *melanogaster* species subgroup of the genus *Drosophila* (Sophophora). II. Phylogenetic relationships between six species based upon polytene banding sequences. Proc. R. Soc. Lond. B 193:257–294.

Lemeunier, F. and M. Ashburner. 1984. Relationships in the *melanogaster* species subgroup of the genus *Drosophila* (Sophophora). IV. The chromosomes of two new species. Chromosoma 89:343–351.

Lemeunier, F. and S. Aulard. 1992. Inversion polymorphism in *Drosophila melanogaster*. Pp. 339–406 in *Drosophila Inversion Polymorphism*, eds. C. B. Krimbas and J. R. Powell. CRC Press, Boca Raton, Fla.

Lemeunier, F., B. Dutrillaux, and M. Ashburner. 1978. Relationships within the *melanogaster* species subgroup of the genus *Drosophila*. III. The mitotic chromosomes and quinicrine fluorescent patterns of the polytene chromosomes. Chromosoma 69:349–361.

Lemeunier, F., J. R. David, and L. Tsacas. 1986. The *melanogaster* species group. Pp. 148–256 in *Genetics and Biology of Drosophila Vol. 3e*, eds. M. Ashburner, H. L. Carson, and J. N. Thompson. Academic Press, New York.

Leonard, J. E. and L. Ehrman. 1983. Does the rare male advantage result from faulty experimental design. Genetics 104:713–716.

Leonard, J. E., L. Ehrman, and M. Schorsch. 1974. Bioassay of a *Drosophila* pheromone in sexual selection. Nature 104:261–262.

Levene, H. 1953. Genetic equilibrium when more than one ecological nice is available. Am. Nat. 87:331–333.

Levene, H., O. Pavlovsky, and Th. Dobzhansky. 1954. Interaction of the adaptive values in polymorphic experimental populations of *Drosophila pseudoobscura*. Evolution 8:335–349.

Levene, H., Th. Dobzhansky, and O. Pavlovsky. 1958. Dependence of the adaptive values of certain genotypes of *Drosophila* on the composition of the gene pool. Evolution 12:10–23.

Levine, L. (ed.). 1995. *Genetics of Natural Populations: The continuing importance of Theodosius Dobzhansky*. Columbia University Press, New York.

Levine, L., M. Asmussen, O. Olvera, J. R. Powell, M. E. de la Rosa, V. M. Salceda, M. I. Gasa, J. Guzman, and W. W. Anderson. 1980. Population genetics of Mexican *Drosophila*. V. A high rate of multiple insemination in a natural population of *Drosophila pseudoobscura*. Am. Nat. 116:493–503.

Levine, L., O. Olvera, J. R. Powell, R. F. Rockwell, M. E. de la Rosa, V. M. Salceda, W. W. Anderson, and J. Guzmán. 1995. Studies on Mexican populations of *Drosophila pseudoobscura.* Pp. 120–139 in *Genetics of Natural Populations: The continuing importance of Theodosius Dobzhansky*, ed. L. Levine. Columbia University Press, New York.

Levins, R. 1969. Thermal acclimation and heat resistance in Drosophila species. Am. Nat. 103: 483–499.

Levitan, M. 1958. Non-random associations of inversions. Cold Spring Harbor Symp. Quant. Biol. 23:251–268.

Levitan, M. 1992. Chromosome variation in *Drosophila robusta.* Pp. 221–338 in *Drosophila Inversion Polymorphism*, eds. C. B. Krimbas and J. R. Powell. CRC Press, Boca Raton, Fla.

Lewis, E. B. 1978. A gene complex controlling segmentation in *Drosophila.* Nature 276:565–570.

Lewontin, R. C. 1955. The effects of population density and composition on viability in *Drosophila melanogaster.* Evolution 9:27–41.

Lewontin, R. C. 1957. The adaptation of populations to varying environments. Cold Spring Harbor Symp. Quant. Biol. 22:395–408.

Lewontin, R. C. 1965. Selection for colonizing ability. Pp. 77–224 in *The Genetics of Colonizing Species*, eds. H. G. Baker and G. L. Stebbins. Academic Press, New York.

Lewontin, R. C. 1974. *The Genetic Basis of Evolutionary Change.* Columbia University Press, New York.

Lewontin, R. C. 1981. The scientific work of Th. Dobzhansky. Pp. 93–115 in *Dobzhansky's Genetics of Natural Populations*, eds. R. C. Lewontin, J. A. Moore, W. B. Provine, and B. Wallace. Columbia University Press, New York.

Lewontin, R. C. 1989. Inferring the number of evolutionary events from DNA coding sequence differences. Mol. Evol. Biol. 6:15–32.

Lewontin, R. C. and J. L. Hubby. 1966. A molecular approach to the study of genic heterozygosity in natural populations. II. Amount of variation and degree of heterozygosity in natural populations of *Drosophila pseudoobscura.* Genetics 54:595–609.

Lewontin, R. C., L. R. Ginzberg, and S. D. Tuljapurkar. 1978. Heterosis as an explanation for large amounts of genic polymorphism. Genetics 88:149–169.

Lewontin, R. C., J. A. Moore, W. B. Provine, and B. Wallace (eds.). 1981. *Dobzhansky's Genetics of Natural Populations.* Columbia University Press, New York.

L'Heritier, P. 1970. *Drosophila* viruses and their roles as evolutionary factors. Evol. Biol. 4: 185–208.

L'Heritier, P. 1975. The *Drosophila* viruses. Pp. 813–818 in *Handbook of Genetics, Vol. 3*, ed. R. C. King. Plenum, New York.

L'Heritier, P. and F. Hugon de Scoeux. 1947. Transmission par greffe et ejection de la sensibilite hereditaire au gaz cabonique chez la Drosophile. Bull. Biol. Fr. Belg. 81:70–91.

L'Heritier, P. L. and G. Teissier. 1933. Etude d'une population de *Drosophila* en equilibre. Compt. Rend. Acad. Sci. 197:1765.

L'Heritier, P. L. and G. Teissier. 1934. Une experience de selection naturelle: Courbe d'elimination du gene "Bar" das une population de *Drosophila* en equilibre. Compt. Rend. Acad. Sci. 117:1049–1051.

L'Heritier, P. L. and G. Teissier. 1935. Recherches sur la concurrence vitale. Etude de populations mixtes de *Drosophila melanogaster* et *Drosophila funebris.* Compt. Rend. Seanc. Soc. Biol. 118:1396–1398.

L'Heritier, P. L. and G. Teissier. 1937a. Elimination des formes mutantes dans les populations de *Drosophila.* Compt. Rend. Acad. Sci. 124:880–884.

L'Heritier, P. L. and G. Teissier. 1937b. Une anomalie physiologique hereditaire chez la Drosophile. C. R. Acad. Sci. Paris 205:1099–1101.

L'Heritier, P. L., Y. Neefs, and G. Teissier. 1937. Apterisme de insectes et selection naturelle. Compt. Rend. Acad. Sci. 204:907–909.

Li, W.-H., C.-C. Luo, and C.-I. Wu. 1985. Evolution of DNA sequences. Pp. 1–94 in *Molecular Evolutionary Genetics*, ed. R. J. MacIntyre. Plenum, New York.

Li, X. and M. Noll. 1994. Evolution of distinct developmental functions of three *Drosophila* genes by acquisition of different cis-regulatory regions. Nature 367:83–87.

Limbach, K. J. and R. Wu. 1985. Characterization of two *Drosophila melanogaster* cytochrome c genes and their transcripts. Nucl. Acids Res. 13:631–644.

Limbach, K. J. and R. Wu. 1985. Characterization of two *Drosophila melanogaster* cytochrome c genes and their transcripts. Nucl. Acids Res. 13:631–644.

Lindsay, S. L. 1958. Food preferences of *Drosophila* larvae. Am. Nat. 92:279–285.

Lindsley, D. L. and E. H. Grell. 1967. Genetic variations of *Drosophila melanogaster.* Carnegie Inst. Washington, Publ. 627. Washington.

Lindsley, D. L. and E. Lifschytz. 1972. The genetic control of spermatogenesis in Drosophila. Pp. 203–222 in *The Genetics of Spermatozoon*, eds. R. A. Beatty and S. Gluecksohn-Waelsch. University of Edinburgh, Edinburgh.

Lindsley, D. L. and K. T. Tokuyasu. 1980. Spermatogenesis. Pp. 226–294 in *The Genetics and Biology of Drosophila, Vol. 2d*, eds. M. Ashburner and T. R. F. Wright. Academic Press, New York.

Lindsley, D. L. and G. Zimm. 1992. *The Genome of Drosophila melanogaster.* Academic Press, New York.

Lints, F. A. and M. Hani Soliman (eds.). 1988. *Drosophila as a Model Organism for Aging Studies.* Blackie, Glasgow.

Liu, J., J. M. Mercer, L. F. Stam, G. C. Gibson, Z.-B. Zeng, and C. C. Laurie. 1996. Genetic analysis of a morphological shape difference in the male genitalia of *Drosophila simulans* and *D. mauritiana.* Genetics 142:1129–1145.

Lofdahl, K. L. 1986. A genetic analysis of habitat selection in the cactophilic species, *Drosophila mojavensis.* Pp. 153–162 in *Evolutionary Genetics of Invertebrate Behavior*, ed. M. D. Huettel. Plenum, New York.

Lohe, A. and P. Roberts. 1988. Evolution of satellite DNA sequences in *Drosophila.* Pp. 148–186 in *Heterochromatin*, ed. R. Verma. Cambridge University Press, Cambridge, U.K.

Lohe, A. R., E. N. Moriyama, D.-A. Lidholm, and D. L. Hartl. 1995. Horizontal transmission, vertical inactivation, and stochastic loss of mariner-like transposable elements. Mol. Biol. Evol. 12:62–72.

Long, M. and C. H. Langley. 1993. Natural selection and the origin of *jingwei*, a chimeric processed functional gene in *Drosophila.* Science 260:91–95.

Lootens, S., J. Burnett, and T. B. Friedman. 1993. An intraspecific gene duplication polymorphism of the urate oxidase gene of *Drosophila virilis*: A genetic and molecular analysis. Mol. Biol. Evol. 10:635–646.

Lopez-Saurez, C., M. A. Toro, and C. Garcia. 1993. Genetic heterogeneity increases viability in competing groups of *Drosophila hydei.* Evolution 47:977–981.

Louis, C. and L. Nigro. 1989. Ultrastructural evidence of *Wolbachia* Rickettsiales in *Drosophila simulans* and their relationship with unidirectional cross-compatibility. J. Invert. Pathol. 54:39–44.

Loukas, M. and F. C. Kafatos. 1986. The actin loci in the genus *Drosophila*: Establishment of chromosomal homologies among distantly related species by in situ hybridization. Chromosoma 94:287–308.

Loukas, M. and F. C. Kafatos. 1988. Chromosomal locations of actin genes are conserved between the *melanogaster* and *obscura* groups of *Drosophila.* Genetica 76:33–41.

Loukas, M., C. B. Krimbas, and J. Sourdis. 1980. The genetics of *Drosophila subobscura* populations. XIII. A study of lethal allelism. Genetica 197–207.

Loukas, M., C. B. Krimbas, and Y. Vergini. 1984. Evolution of the *obscura* group *Drosophila* species. II. Phylogeny of ten species based on electrophoretic data. Heredity 53:483–493.

Lowenhaupt, K. Y., A. Rich, and M. L. Pardue. 1989. Nonrandom distribution of long mono- and dinucleotide repeats in *Drosophila* chromosomes: Correlations with dosage compensation, heterochromatin, and recombination. Mol. Cell. Biol. 9:1173–1182.

Lozovskaya, E. R., V. S. Scheinker, and M. B. Evgen'ev. 1990. A hybrid dysgenesis syndrome in *Drosophila virilis*. Genetics 126:619–623.

Lozovskaya, E. R., D. A. Petrov, and D. L. Hartl. 1993. A combined molecular and cytogenetic approach to genome evolution in *Drosophila* using large-fragment DNA cloning. Chromosoma 102:253–366.

Lucchesi, J. C. 1977. Dosage compensation: Transcription-level regulation of X-linked genes in *Drosophila*. Am. Zool. 17:685–693.

Lucchesi, J. C. 1978. Gene dosage compensation and the evolution of sex chromosomes. Science 202:711–716.

Luckinbill, L., R. Arking, M. G. Clare, W. C. Cirocco, and S. A. Buck. 1984. Selection for delayed senescence in *Drosophila melanogaster*. Evolution 38:996–1003.

Lumme, J. 1981. Localization of the genetic unit controlling the photoperiodic adult diapause in *Drosophila littoralis*. Hereditas 94:241–244.

Lumme, J. and L. Keränen. 1978. Photoperiodic diapause in *Drosophila lummei* Hackman is controlled by an X-chromosomal factor. Hereditas 89:261–262.

Lumme, J. and S. Lakovaara. 1983. Seasonality and diapause in Drosophilids. Pp. 171–220 in *The Genetics and Biology of Drosophila, Vol. 3d*, eds. M. Ashburner, H. L. Carson, and J. N. Thompson, Jr. Academic Press, New York.

Lund, D. 1959. Further observations on the incidence of CO_2 sensitivity in North American species of *Drosophila*. Dros. Inf. Serv. 33:145.

Lutz, F. E. 1911. Experiments with *Drosophila ampelophila* concerning evolution. Carnegie Inst. Washington Publ. 143:1–40.

Lyckegaard, E. M. S. and A. G. Clark. 1989. Ribosomal DNA and *stellate* gene copy number variation on the Y chromosome of *Drosophila melanogaster*. Proc. Natl. Acad. Sci. USA 86:1944–1848.

Lynch, M. 1988. The rate of polygenic mutation. Genet. Res. 51:137–148.

Lyttle, T. W. 1977. Experimental population genetics of meiotic drive systems. I. Pseudo-Y chromosomal drive as a means of eliminating cage populations of *Drosophila melanogaster*. Genetics 86:413–445.

Lyttle, T. W. 1979. Experimental population genetics of meiotic drive systems. II. Accumulation of genetic modifiers of segregation distorter (SD) in laboratory populations. Genetics 91:339–357.

Lyttle, T. W. 1981. Experimental population genetics of meiotic drive systems. III. Neutralization of sex-ratio distortion in Drosophila through sex chromosome aneuploidy. Genetics 98:317–334.

Lyttle, T. W. 1991. Segregation distorters. Ann. Rev. Genet. 25:511–557.

Lyttle, T. W. and D. S. Haymer. 1992. The role of the transposable element hobo in the origin of endemic inversions in wild populations of *Drosophila melanogaster*. Genetica 86:113–126.

Maca, J. 1982. Parasitizing and transported macroorganisms dependent on Drosophilidae (Diptera) in Czechoslovakia. Folia Fac. Sci. Nat. Univ. Purkynianae Brunenesis (Biologia 74) 23:69–74.

MacAlpine, J. F. and J. E. H. Martin. 1966. Systematics of Sciodoceridae and relatives with a description of two new genera and species from Canadian amber: An erection of family Ironomyiidae (Diptera: Phoroidae). Canad. Entomol. 98:527–544.

MacArthur, R. H. and E. O. Wilson. 1967. *The Theory of Island Biogeography*. Princeton University Press, Princeton, N.J.

MacIntyre, R. J. 1985. Preface. Pp. vii–xiii in *Molecular Evolutionary Genetics*, ed. R. J. MacIntyre. Plenum, New York.

MacIntyre, R. J. and T. R. F. Wright. 1966. Responses of esterase 6 alleles of *Drosophila melanogaster* and *D. simulans* to selection in experimental populations. Genetics 53:371–387.

Mackay, T. F. C. 1986. Transposable element-induced fitness mutations in *Drosophila melanogaster*. Genet. Res. 48:77–87.

Mackay, T. F. C. 1988. Transposable element-induced quantitative genetic variation in *Drosophila*. Pp. 219–235 in *Proceedings of the Second International Conference on Quantitative Genetics*, eds. B. S. Weir, E. J. Eisen, M. M. Goodman, and G. Namkoong. Sinauer Associates, Sunderland, Mass.

Mackay, T. F. C. and C. H. Langley. 1990. Molecular and phenotypic variation in the *achaete-scute* region of *Drosophila melanogaster*. Nature 348:64–66.

Mackay, T. F. C., R. F. Lyman, and M. S. Jackson. 1992. Effects of P element insertions on quantitative traits in *Drosophila melanogaster*. Genetics 130:315–332.

MacKnight, R. H. 1939. The sex-determining mechanism of *Drosophila miranda*. Genetics 24: 180–201.

Macpherson, J. N., B. S. Weir, and A. J. Leigh Brown. 1990. Extensive linkage disequilibrium in the *achaete-scute* complex of *Drosophila melanogaster*. Genetics 126:121–129.

MacRae, A. F. and W. W. Anderson. 1988. Evidence for non-neutrality of mitochondrial DNA haplotypes in *Drosophila pseudoobscura*. Genetics 120:485–494.

MacRae, A. F. and W. W. Anderson. 1990. Can mating preferences explain changes in mtDNA haplotype frequencies? Genetics 124:999–1001.

Maddison, W. P. and D. R. Maddison. 1992. *MacClade: Analysis of Phylogeny and Character Evolution*. Sinauer Associates, Inc., Sunderland, Mass.

Madueño, E., G. Papagiannakis, G. Rimmington, R. D. C. Saunders, C. Savakis, I. Sidén-Kiamos, G. Skavdis, L. Spanos, J. Trennear, P. Adam, M. Ashburner, P. Benos, V. N. Bolshakov, D. Coulson, D. M. Glover, S. Herrmann, F. C. Kafatos, C. Louis, T. Majerus, and J. Modolell. 1995. A physical map of the X chromosome of *Drosophila melanogaster*: cosmid contigs and sequence tagged sites. Genetics 139:1631–1647.

Maier, D., A. Preiss, and J. R. Powell. 1990. Regulation of the segmentation gene *fushi tarazu* has been functionally conserved in *Drosophila*. EMBO Journal 9:3957–3966.

Maier, D., D. Sperlich, and J. R. Powell. 1993. Conservation and change of the developmentally crucial *fushi tarazu* gene in *Drosophila*. J. Mol. Evol. 36:315–326.

Malicki, J., K. Schughart, and W. McGinnis. 1990. Mouse *Hox-2.2* specifies thoracic segmental identity in Drosophila embryos and larvae. Cell 63:961–967.

Malicki, J., L. C. Cianetti, C. Peschle, and W. McGinnis. 1992. A human HOX4B regulatory element provides head specific expression in *Drosophila* embryos. Nature 358:345–347.

Malogolowkin-Cohen, C. and M. A. Q. Rodrigues-Pereira. 1975. Sexual drive of normal and SR flies of *Drosophila nebulosa*. Evolution 29:579–580.

Malogolowkin-Cohen, C., H. Levene, N. P. Dobzhansky, and A. S. Simmons. 1964. Inbreeding and the mutational and balanced loads in natural populations of *Drosophila willistoni*. Genetics 50:1299–1311.

Mangan, R. L. 1978. Competitive interactions among host plant specific *Drosophila* species. Ph.D. dissertation, University of Arizona, Tucson.

Marfany, G. and R. Gonzàlez-Duarte. 1992a. Evidence for retrotranscription of protein-coding genes in the *Drosophila subobscura* genome. J. Mol. Evol. 35:492–501.

Marfany, G. and R. Gonzàlez-Duarte. 1992b. The *Drosophila subobscura Adh* genomic region contains valuable evolutionary markers. Mol. Biol. Evol. 9:261–277.

Marinkovic, D. 1967. Genetic loads affecting fecundity in natural populations of *D. pseudoobscura.* Genetics 56:61–71.
Marinkovic, D., F. J. Ayala, and M. Andjelkovic. 1978. Genetic polymorphism and phylogeny of *Drosophila subobscura.* Evolution 32:164–173.
Marinkovic, D., M. Milosevic, and M. Andjelkovic. 1984. Regulatory polymorphism in midgut α-amylase activity and developmental rate of *Drosophila subobscura.* Genetica 64:115–122.
Markert, C. L. and F. Moller. 1959. Multiple forms of enzymes: Tissue, ontogenetic and species specific patterns. Proc. Natl. Acad. Sci. USA 45:753–763.
Markow, T. A. 1975. A genetic analysis of phototactic behavior in *Drosophila melanogaster.* I. Selection in the presence of inversions. Genetics 79:527–534.
Markow, T. A. 1978. A test for the rare male mating advantage in coisogenic strains of *Drosophila.* Gent. Res. 32:123–127.
Markow, T. A. 1980. Rare male advantages among *Drosophila* of the same laboratory strain. Behav. Genet. 10:553–556.
Markow, T. A. 1981a. Courtship behavior and control of reproductive isolation between *Drosophila mojavensis* and *Drosophila arizonensis.* Evolution 35:1022–1026.
Markow, T. A. 1981b. Mating preferences are not predictive of the direction of evolution in experimental populations of *Drosophila.* Science 213:1405–1407.
Markow, T. A. 1991. Sexual isolation among populations of *Drosophila mojavensis.* Evolution 45:1525–1529.
Markow, T. A. and E. C. Toolson. 1990. Temperature effects on epicuticular hydrocarbons and sexual isolation in *Drosophila mojavensis.* Pp. 315–331 in *Ecological and Evolutionary Genetics of Drosophila*, eds. J. S. F. Barker, W. T. Starmer, and R. J. MacIntyre. Plenum, New York.
Markow, T. A., R. C. Richmond, L. Mueller, I. Sheer, S. Roman, C. Laetz, and L. Lorenz. 1980. Testing for rare male mating advantages among various *Drosophila melanogaster* genotypes. Genet. Res. 35:59–64.
Markow, T. A., J. C. Fogleman, and W. B. Heed. 1983. Reproductive isolation in Sonoran desert *Drosophila.* Evolution 37:649–652.
Maroni, G., J. Wise, J. E. Young, and E. Otto. 1987. Metellothionein gene duplications and metal tolerance in natural populations of *Drosophila melanogaster.* Genetics 117:739–744.
Martin, C. H. and E. M. Meyerowitz. 1986. Characterization of the boundaries between adjacent rapidly and slowly evolving regions in *Drosophila.* Proc. Natl. Acad. Sci. USA 83:8654–8658.
Martin-Campos, J. M., J. M. Comeron, N. Miyashita, and M. Aguadé. 1992. Intraspecific and interspecific variation at the *y-ac-sc* region of *Drosophila simulans* and *Drosophila melanogaster.* Genetics 130:805–816.
Martinez, A. and M. Bownes. 1992. The specificity of yolk protein uptake in cyclorrhphan diptera is conserved through evolution. J. Mol. Evol. 35:444–453.
Martinez-Cruzado, J. C. 1990. Evolution of the autosomal chorion cluster in *Drosophila.* IV. The Hawaiian *Drosophila*: Rapid protein evolution and constancy in the rate of DNA divergence. J. Mol. Evol. 31:402–423.
Martinez-Cruzado, J. C., C. Swimmer, M. G. Fenerjian, and F. C. Kafatos. 1988. Evolution of the autosomal chorion locus in Drosophila. I. General organization of the locus and sequence comparisons of genes s15 and s19 in evolutionarily distant species. Genetics 119: 663–677.
Maruyama, K. and D. L. Hartl. 1991. Evidence for interspecific transfer of the transposable element mariner between *Drosophila* and Zaprionus. J. Mol. Evol. 33:514–524.
Masry, A. M. 1981. The evolutionary changes of the populations structure. I. Seasonal changes in the frequencies of chromosomal inversions in natural populations of *D. melanogaster.* Egypt. J. Genet. Cytol. 10:261–272.

Mather, K. 1983. Response to selection. Pp. 155–223 in *The Genetics and Biology of Drosophila, Vol. 3c*, eds. M. Ashburner, H. L. Carson, and J. N. Thompson, Jr. Academic Press, New York.

Mather, K. and B. S. Harrison. 1949. The manifold effect of selection. Heredity 3:1–52.

Mather, W. B. 1962. Patterns of chromosomal evolution in the *immigrans* group of *Drosophila.* Evolution 16:20–26.

Mather, W. B. 1964. Temporal variation in *Drosophila rubida* inversion polymorphism. Heredity 19:231–334.

Matsuda, M., H. Sato, and Y. N. Tobari. 1993. Crossing over in males. Pp. 53–72 in *Drosophila ananassae: Genetical and Biological Aspects*, ed. Y. N. Tobari. Japan Scientific Societies Press, Tokyo and Karger AG, Basel.

Matsuura, E. T., Y. Niki, and S. I. Chigusa. 1991. Selective transmission of mitochondrial DNA in heteroplasmic lines for intraspecific and interspecific combinations in *Drosophila melanogaster.* Jap. J. Genetics 66:197–208.

Matsuura, E. T., Y. Niki, and S. I. Chigusa. 1993. Temperature-dependent selection in the transmission of mitochondrial DNA in *Drosophila.* Jap. J. Genetics 68:127–135.

Matthew, P., A. Agrotis, A. C. Taylor, and S. W. McKenzie. 1992. An association between ADH protein levels and polymorphic nucleotide variation in the Adh gene of *Drosophila melanogaster.* Mol. Biol. Evol. 9:526–536.

Matthews, T. C. and L. E. Munstermann. 1994. Chromosomal repatterning and linkage group conservation in mosquito karyotypic evolution. Evolution 48:146–154.

Maynard Smith, J. 1958. The effects of temperature and egg-laying on the longevity of *Drosophila subobscura.* J. Exper. Biol. 35:832–842.

Maynard Smith, J. and J. Haigh. 1974. The hitchhiking effect of a favorable gene. Genet. Res. 23:23–35.

Maynard Smith, J. and K. C. Sondhi. 1960. The genetics of pattern. Genetics 45:1039–1050.

Mayr, E. 1954. Change of genetic environment and evolution. Pp. 157–180 in *Evolution as a Process*, eds. J. Huxley, A. C. Hardy, and E. B. Ford. Allen and Unwin, London.

Mayr, E. 1963. *Animal Species and Evolution.* Harvard, Belknap Press, Cambridge, Mass.

Mayr, E. 1976. *Evolution and the Diversity of Life.* Harvard University Press, Cambridge, Mass.

Mayr, E. 1982. *The Growth of Biological Thought.* Harvard University Press, Cambridge, Mass.

Mayr, E. and W. B. Provine (eds.). 1980. *The Evolutionary Synthesis.* Harvard University Press, Cambridge, Mass.

McClintock, B. 1978. Mechanisms that rapidly reorganize the genome. Stadler Genetics Symp. 10:25–37.

McClintock, B. 1984. The significance of responses of the genome to challenge. Science 226: 792–801.

McDonald, J. F. 1986. Physiological tolerance and behavioral avoidance of alcohol in *Drosophila*: coadaptation or pleiotrophy? Pp. 247–254 in *Evolutionary Genetics of Invertebrate Behavior*, ed. M. D. Huettel. Plenum, New York.

McDonald, J. F. 1990. Macroevolution and retroviral elements. Bioscience 40:183–191.

McDonald, J. F. (ed.). 1993a. *Transposable Elements in Evolution*, Kluwer Academic Publishers, Dordrecht. Most of the articles in this book were first published in Genetica, 86 (1992).

McDonald, J. F. 1993b. Evolution and consequences of transposable elements. Curr. Opinion Genet. Develop. 3:855–864.

McDonald, J. F. and J. C. Avise. 1976. Evidence for the adaptive significance of enzyme activity levels: Interspecific variation in αGPDH and ADH in Drosophila. Biochem. Genet. 14:347–355.

McDonald, J. F. and F. J. Ayala. 1974. Genetic response to environmental heterogeneity. Nature 250:572–574.

McDonald, J. F. and F. J. Ayala. 1978. Genetic and biochemical basis of enzyme activity variation in natural populations. I. Alcohol dehydrogenase in *Drosophila melanogaster.* Genetics 89:371–388.

McDonald, J. F., G. K. Chambers, J. David, and F. J. Ayala. 1977. Adaptive response due to changes in gene regulation: a study with Drosophila. Proc. Natl. Acad. Sci. USA 74: 4562–4566.

McDonald, J. H. and M. Kreitman. 1991. Adaptive protein evolution at the *Adh* locus in *Drosophila.* Nature 351:652–654.

McGinnis, N., M. A. Kuziora, and W. McGinnis. 1990. Human *Hox-4.2* and Drosophila *Deformed* encode similar regulatory specificities in Drosophila embryos and larvae. Cell 63: 969–976.

McKechnie, S. W. and B. W. Geer. 1993. Micro-evolution in a wine cellar population: an historical perspective. Genetica 90:201–215.

McKee, B. D., L. Habera, and J. L. Verna. 1992. Evidence that the intergenic spacer of *Drosophila melanogaster* rRNA genes function as X-Y pairing sites in male meiosis, and a general model for achiasmate pairing. Genetics 132:529–544.

McKenzie, J. A. and Parsons, P. A. 1974. Microdifferentiation in a natural population of *Drosophila melanogaster* to alcohol in the environment. Genetics 77:385–394.

McKenzie, R. W., J. Hu, and M. D. Brennan. 1994. Redundant cis-acting elements control expression of the *Drosophila affinidisjuncta Adh* gene in the larval fat body. Nucl. Acids Res. 22:1257–1264.

Medawar, P. B. and J. S. Medawar. 1983. *Aristotle to Zoos: A philosophical dictionary of biology.* Harvard University Press, Cambridge, Mass.

Medvedev, Z. 1969. *The Rise and Fall of T. D. Lysenko.* Columbia University Press, New York.

Menozzi, P. and C. B. Krimbas. 1992. The inversion polymorphism of *Drosophila subobscura* revisited: synthetic maps of gene arrangement frequencies and their interpretation. J. Evol. Biol. 5:625–641.

Mercot, H. 1994. Phenotypic expression of ADH regulatory genes in *Drosophila melanogaster*: A comparative study between a paleoarctic and a tropical population. Genetica 94:37–41.

Mercot, H., D. Defaye, P. Capy, E. Pla, and J. R. David. 1994. Alcohol tolerance, Adh activity, and ecological niche of *Drosophila* species. Evolution 48:746–757.

Mercot, H., A. Atlan, M. Jacques, and C. Montchamp-Moreau. 1995. Sex-ratio distortion in *Drosophila simulans*: co-occurrence of a meiotic drive and a suppressor of drive. J. Evol. Biol. 8:283–300.

Merrell, D. J. 1949. Selective mating in *Drosophila melanogaster.* Genetics 34:370–389.

Merrell, D. J. 1951. Interspecific competition between *Drosophila funebris* and *Drosophila melanogaster.* Am. Nat. 85:159–169.

Merrell, D. J. 1965. Competition involving dominant mutants in experimental populations of *Drosophila melanogaster.* Genetics 52:165–189.

Merrell, D. J. and J. C. Underhill. 1956. Competition between mutants in experimental populations of *Drosophila melanogaster.* Genetics 41:469–485.

Miklos, G. L. G. 1985. Localized highly repetitive DNA sequences in vertebrates and invertebrate genomes. Pp. 241–321 in *Molecular Evolutionary Genetics*, ed. R. J. MacIntyre. Plenum, New York.

Miklos, G. L. G. and A. C. Gill. 1981. The DNA sequences of cloned complex satellite DNAs from Hawaiian *Drosophila* and their bearing on satellite DNA sequence conservation. Chromosoma 82:409–427.

Milkman, R. D. 1964. The genetic basis of natural variation. V. Selection for crossveinless polygenes in new wild strains of *Drosophila melanogaster.* Genetics 50:625–632.

Milkman, R. D. and R. R. Zietler. 1974. Concurrent multiple paternity in natural and laboratory populations of *Drosophila melanogaster.* Genetics 78:1191–1193.

Miller, D. D. 1939. Structure and variation of the chromosomes of *Drosophila algonquin.* Genetics 24:694–708.

Miller, D. D. and R. Roy. 1964. Further data on Y chromosome types in *Drosophila athabasca.* Can. J. Genet. Cytol. 6:334–348.

Miller, D. D. and L. E. Stone. 1962. A reinvestigation of karyotype in *Drosophila affinis* Sturtevant and related species. J. Heredity 53:12–24.

Minawa, A. and A. J. Birley. 1978. The genetical response to natural selection by varied environments. I. Short-term observations. Heredity 40:39–50.

Misra, R. K. and E. C. R. Reeve. 1964. Clines in body dimensions in populations of *Drosophila subobscura.* Genet. Res. 5:240–256.

Mitchell, P. and W. Arthur. 1990. Resource turnover time and consumer generation time as factors affecting the stability of coexistence: An experiment with *Drosophila.* J. Animal Ecol. 59:121–133.

Mitrofanov, V. G. and N. V. Sidorova. 1981. Genetics of the sex ratio anomaly in *Drosophila* hybrids of the *virilis* group. Theoret. Appl. Genet. 59:17–22.

Mitsialis, S. A., N. Spoerel, M. Leviten, and F. C. Kafatos. 1987. A short 5′-flanking DNA region is sufficient for developmentally correct expression of moth chorion genes in *Drosophila.* Proc. Natl. Acad. Sci. USA 84:7987–7991.

Mitsialis, S. A., S. Veletza, and F. C. Kafatos. 1989. Transgenic regulation of moth chorion gene promoters in *Drosophila*: Tissue, temporal and quantitative control of four bidirectional promoters. J. Mol. Evol. 29:486–495.

Miyamoto, M. M. and J. Cracraft (eds.). 1991. *Phylogenetic Analysis of DNA Sequences.* Oxford University Press, New York.

Miyashita, N. and C. H. Langley. 1988. Molecular and phenotypic variation of the *white* locus region in *D. melanogaster.* Genetics 120:199–212.

Mizrokhi, L. J. and A. M. Mazo. 1990. Evidence for horizontal transmission of the mobile element jockey between distant *Drosophila* species. Proc. Natl. Acad. Sci. USA 87:9216–9220.

Mohn, N. and E. B. Spiess. 1963. Cold resistance of karyotypes in *Drosophila persimilis* from Timberline of California. Evolution 17:548–563.

Monnerot, M., M. Solignac, and D. R. Wolstenholme. 1990. Discrepancy in divergence of the mitochondrial and nuclear genomes of *Drosophila teissieri* and *Drosophila yakuba.* J. Mol. Evol. 30:500–508.

Montague, J. R. 1984. The ecology of Hawaiian flower-breeding Drosophilids. I. Selection in the larval habitat. Am. Nat. 124:712–722.

Montague, J. R. and J. Jaenike. 1985. Nematode parasitism in natural populations of mycophagous Drosophilids. Ecology 66:524–626.

Montague, J. R., R. L. Mangan, and W. T. Starmer. 1981. Reproductive allocation in the Hawaiian Drosophilidae: egg size vs. number. Am. Nat. 118:865–871.

Montchamp-Moreau, C., J.-F. Ferveur, and M. Jacques. 1991. Geographic distribution and inheritance of three cytoplasmic incompatibility types in *Drosophila simulans.* Genetics 129: 399–407.

Montgomery, E. A., B. Charlesworth, and C. H. Langley. 1987. A test for the role of natural selection in the stabilization of transposable element copy number in a population of *Drosophila melanogaster.* Genet. Res. 49:31–41.

Montgomery, S. L. 1975. Comparative breeding site ecology and the adaptive radiation of picture-winged *Drosophila.* Proc. Hawaiian Entomol. Soc. 22:65–102.

Moore, J. A. 1952. Competition between *Drosophila melanogaster* and *Drosophila simulans.* II. The improvement of competitive ability through selection. Proc. Natl. Acad. Sci. USA 38:813–817.

Moore, J. A. and B. C. Moore. 1984. The *Drosophila* of Southern California. II. Isolation of populations in the Death Valley region. Am Nat. 124:738–744.
Moore, J. A., C. E. Taylor, and B. C. Moore. 1979. The *Drosophila* of Southern California. I. Colonization after a fire. Evolution 33:156–171.
Moran, C. and A. Torkamanzehi. 1990. P-elements and quantitative variation in *Drosophila.* Pp. 99–117 in *Ecological and Evolutionary Genetics of Drosophila*, eds. J. S. F. Barker, W. T. Starmer, and R. J. MacIntyre. Plenum, New York.
Morgan, C. L. 1927. *Emergent Evolution.* Williams and Norgate, London.
Morgan, T. H. 1903. *Evolution and Adaptation.* Macmillan, New York.
Morgan, T. H. 1925. *Evolution and Genetics.* Princeton University Press, Princeton, N.J.
Morgan, T. H. 1932. *The Scientific Basis of Evolution.* W. W. Norton, New York.
Moriwaki, D. and Y. N. Tobari. 1975. *Drosophila ananassae.* Pp. 513–535 in *Handbook of Genetics, Vol. 3*, ed. R. C. King. Plenum, New York.
Moriwaki, D., M. Ohnishi, and Y. Nakajima. 1956. Analysis of heterosis in populations of *Drosophila ananassae.* Proc. Int. Genet. Symp. Cytologia supplement:370–379.
Moriyama, E. N. 1987. Higher rates of nucleotide substitutions in Drosophila than in mammals. Jap. J. Genet. 62:139–147.
Moriyama, E. N. and T. Gojobori. 1989. Evolution of nested genes with special reference to cuticle proteins in *Drosophila melanogaster.* J. Mol. Evol. 28:391–397.
Moriyama, E. N. and T. Gojobori. 1992. Rates of synonymous substitution and base composition of nuclear genes in Drosophila. Genetics 130:855–864.
Moriyama, E. N. and D. L. Hartl. 1993. Codon usage bias and base composition of nuclear genes in Drosophila. Genetics 134:847–858.
Moriyama, E. N. and J. R. Powell. 1996. Intraspecific nuclear DNA variation in Drosophila. Mol. Biol. Evol. 13:261–277.
Moth, J. J. and J. S. F. Barker. 1976. Interspecific competition between *Drosophila melanogaster* and *Drosophila simulans.* Reduction in fecundity and destruction of eggs when the medium is inhabited with larvae. Oecologia 23:151–164.
Mount, S. M. and G. M. Rubin. 1985. Complete nucleotide sequence of the *Drosophila* transposable element copia: homology between copia and retroviral proteins. Mol. Cell. Biol. 5:1630–1638.
Moyer, S. E. 1964. Selection for modification of recombination frequency of linked genes. Ph.D. thesis. University of Minnesota. Dis. Abst. 25:1508–1509.
Mueller, L. D. 1985. The evolutionary ecology of *Drosophila.* Evol. Biol. 19:37–98.
Mukai, T. and C. C. Cockerham. 1977. Spontaneous mutation rates at enzyme loci in *Drosophila melanogaster.* Proc. Natl. Acad. Sci. USA 74:2514–2517.
Mukai, T., S. I. Chigusa, L. E. Mettler, and J. F. Crow. 1972. Mutation rate and dominance of genes affecting viability in *Drosophila melanogaster*. Genetics 72:335–355.
Muller, H. J. 1940. Bearings of the *Drosophila* work on systematics. Pp. 185–268 in *The New Systematics*, ed. J. S. Huxley. Oxford University Press (Clarendon), London, New York.
Muller, H. J. 1950. Our load of mutations. Am. J. Hum. Genet. 2:111–176.
Murray, N. D. 1982. Ecology and evolution of the *Opuntia-Cactoblastis* ecosystem in Australia. Pp. 17–30 in *Ecological Genetics and Evolution: The Cactus-Yeast-Drosophila Model System*, eds. J. S. F. Barker and W. T. Starmer. Academic Press, Sydney.
Nagy, L. M. and S. Carroll. 1994. Conservation of wingless patterning functions in the short-germ embryos of *Tribolium castaneum.* Nature 367:460–463.
Nair, P. S. and D. Brncic. 1971. Allelic variations within identical chromosomal inversions. Amer. Nat. 105:291–294.
Nair, P. S., D. Brncic, and K.-I. Kojima. 1971. Isozyme variation and evolutionary relationships in the mesophragmatica species group of *Drosophila.* Univ. Texas Publ. 7103, pp. 17–28.

Napp, M. and D. Brncic. 1978. Electrophoretic variability in two closely related Brazilian species of the *flavopilosa* species group of *Drosophila*. Brazilian J. Genet. 1:1–10.

Nassar, R. 1979. Frequency-dependent selection at the LAP locus in *Drosophila melanogaster*. Genetics 91:327–338.

Naveira, H. 1992. Location of X-linked polygenic effects causing sterility in male hybrids of *Drosophila simulans* and *D. mauritiana*. Heredity 68:211–217.

Naveira, H. and A. Fontdevila. 1986. The evolutionary history of *Drosophila buzzatii*. XII. The genetic basis of sterility in hybrids between *D. buzzatii* and *D. serido* from Argentina. Genetics 114:841–857.

Naveira, H. and A. Fontdevila. 1991. The evolutionary history of *Drosophila buzzatii*. XXII. Chromosomal and genic sterility in male hybrids of *D. buzzatii* and *D. koepferae*. Heredity 66:233–240.

Nei, M. 1967. Modification of linkage intensity by natural selection. Genetics 57:625–641.

Nei, M. 1972. Genetic distance between populations. Am. Nat. 106:282–292.

Nei, M. 1975. *Molecular Population Genetics and Evolution*. North Holland Publications, Amsterdam.

Nei, M. 1987. *Molecular Evolutionary Genetics*. Columbia University Press, New York.

Nei, M. and W.-H. Li. 1979. Mathematical model for studying genetic variation in terms of restriction endonucleases. Proc. Natl. Acad. Sci. USA 76:5269–5273.

Nei, M. and F. Tajima. 1983. Maximum likelihood estimation of the number of nucleotide substitutions from restriction sites data. Genetics 105:207–217.

Nei, M., K.-I. Kojima, and H. E. Schaffer. 1967. Frequency changes of new inversions in populations under mutation-selection equilibria. Genetics 57:741–750.

Neufeld, T. P., R. W. Carthew, and G. M. Rubin. 1991. Evolution of gene position: Chromosomal arrangement and sequence comparison of the *Drosophila melanogaster* and *Drosophila virilis sina* and *Rh4* genes. Proc. Natl. Acad. Sci. USA 88:10203–10207.

Nevo, E., A. Beiles, and R. Ben-Shlomo. 1984. The evolutionary significance of genetic diversity: ecological, demographic and life history correlates. Lecture Notes in Biomathematics (G. S. Mani, ed.) 53:13–213.

Nielsen, J., A. A. Peixoto, A. Piccin, R. Costa, C. P. Karyiacou, and D. Chalmers. 1994. Big flies, small repeats: The "Thr-Gly" region of the *period* gene in Diptera. Mol. Biol. Evol. 11:839–853.

Nighorn, A., Y. Qiu, and R. L. Davis. 1994. Progress in understanding the *Drosophila dnc* locus. Comp. Biochem. Physiol. 108B:1–9.

Nigro, L. and T. Prout. 1990. Is there selection on RFLP differences in mitochondrial DNA? Genetics 125:551–555.

Nijhout, H. F. 1991. *The Development and Evolution of Butterfly Wings*. Smithsonian Press, Washington, D.C.

Niki, Y., S. I. Chigusa, and E. T. Matusuura. 1989. Complete replacement of mitochondrial DNA in *Drosophila*. Nature 341:551–552.

Noor, M. 1995. Speciation driven by natural selection in *Drosophila*. Nature 375:674–675.

Novitski, E. 1947. Genetic analysis of an anomalous sex-ratio condition in *Drosophila affinis*. Genetics 32:526–534.

Nozawa, K. 1958. Competition between the brown gene and its wild type allele in *Drosophila melanogaster*. I. Effect of population density. Japan. J. Genet. 58:262–271.

Numinsky, D. I., E. N. Moriyama, E. R. Lozovskaya, and D. L. Hartl. 1996. Molecular phylogeny and genome evolution in the *Drosophila virilis* species subgroup: Duplications of the *Alcohol dehydrogenase* gene. Mol. Biol. Evol. 13:132–149.

Nüsslein-Volhard, C. and E. Wieschaus. 1980. Mutations affecting segment number and polarity in *Drosophila*. Nature 287:795–801.

Oakeshott, J. G. 1979. Selection affecting enzyme polymorphisms in laboratory populations of *Drosophila melanogaster*. Oecologia 43:341–354.

Oakeshott, J. G., J. B. Gibson, P. R. Anderson, W. R. Knibb, D. G. Anderson, and G. K. Chambers. 1982. Alcohol dehydrogenase and glycerol-3-phosphate dehydrogenase clines in *Drosophila melanogaster* in different continents. Evolution 39:86–96.

Oakeshott, J. G., E. A. van Papenrecht, T. M. Boyce, M. J. Healy, and R. J. Russell. 1993. Evolutionary genetics of *Drosophila* esterases. Genetica 90:239–268.

Oakeshott, J. G., T. M. Boyce, R. J. Russell, and M. J. Healy. 1995. Molecular insights into the evolution of an enzyme: esterase6 in *Drosophila.* Trends Ecol. Evol. 10:103–110.

O'Donald, P. 1978. Rare male mating advantage. Nature 272:188–189.

O'Donald, P. 1980. *Genetic Models of Sexual Selection.* Cambridge University Press, Cambridge, U.K.

O'Donnell, A. E. and R. C. Axtel. 1965. Predation by *Fuscuropoda vegetans* (Acarina: Uropodidae) on the house fly (*Musca domestica*). Ann. Entomol. Soc. Am. 58:403–404.

Oguma, Y. 1993. Sexual behavior. Pp. 199–207 in *Drosophila ananassae: Genetical and Biological Aspects*, ed. Y. N. Tobari. Japan Scientific Society Press, Tokyo; and S. Karger AG, Basel.

Ohba, S. 1967. Chromosomal polymorphism and capacity for increase under near optimal conditions. Heredity 22:169–189.

Ohba, S. 1970. Isozyme polymorphisms in *Drosophila virilis* populations. Jap. J. Genet. 45: 490–503.

Ohta, T. and K.-I. Kojima. 1968. Survival probabilities of new inversions in large populations. Biometrics 24:501–516.

Oishi, K., D. F. Poulson, and D. L. Williamson. 1984. Virus-mediated change in clumping propensities of *Drosophila* SR spiroplasmas. Current Microbiol. 10:153–158.

Okada, T. 1963. Caenogenetic differentiation of mouthhooks in Drosophilid larvae. Evolution 17:84–98.

Oliver, C. P. 1976. Historical introduction. Pp. 1–30 in *Genetics and Biology of Drosophila Vol. 1a*, eds. M. Ashburner and E. Novitski. Academic Press, New York.

Olvera, O., J. R. Powell, M. E. de la Rosa, V. M. Salceda, J. I. Gaso, J. Guzman, W. W. Anderson, and L. Levine. 1979. Population genetics of Mexican Drosophila VI. Cytogenetic aspects of the inversion polymorphism in *Drosophila pseudoobscura.* Evolution 33: 381–395.

O'Neil, M. T. and J. M. Belote. 1992. Interspecific comparison of the *transformer* gene of Drosophila reveals an unusually high degree of evolutionary divergence. Genetics 131: 113–128.

O'Neill, S. L. and T. L. Karr. 1990. Bidirectional incompatibility between conspecific populations of *Drosophila simulans.* Nature 348:178–180.

O'Neill, S. L., R. Giordano, A. M. E. Colbert, T. L. Karr, and H. M. Robertson. 1992. 16S rRNA phylogenetic analysis of the bacterial endosymbionts associated with cytoplasmic incompatibility in insects. Proc. Natl. Acad. USA 89:2699–2702.

Ornduff, R. 1974. *California Plant Life.* University of California Press, Berkeley, Calif.

Orr, H. A. 1987. Genetics of male and female sterility in hybrids of *Drosophila pseudoobscura* and *D. persimilis.* Genetics 116:555–563.

Orr, H. A. 1989. Genetics of sterility in hybrids between two subspecies of *Drosophila.* Evolution 43:180–189.

Orr, H. A. 1992. Mapping and characterization of a "speciation gene" in *Drosophila.* Genet. Res. 59:73–80.

Orr, H. A. and J. A. Coyne. 1989. The genetics of postzygotic isolation in the *Drosophila virilis* group. Genetics 121:527–537.

Orr-Weaver, T. 1991. *Drosophila* chorion genes: Cracking the eggshell's secrets. BioEssays 13:97–105.

Otte, D. and J. A. Endler (eds.). 1989. *Speciation and Its Consequences*. Sinauer Associates, Sunderland, Mass.

Paik, Y. K. and K. C. Sung. 1969. Behavior of lethals in *Drosophila melanogaster* populations. Jap. J. Genet. 44:180–192.

Paika, I. J. and D. D. Miller. 1974. Distribution of heterochromatin in type III Y and other chromosomes of *Drosophila affinis*. J. Heredity 65:238–240.

Painter, T. S. 1933. A new method for the study of chromosome rearrangements and the plotting of chromosome maps. Science 78:585–586.

Painter, T. S. 1935. The morphology of the third chromosome in the salivary gland of *Drosophila melanogaster* and a new cytological map of the element. Genetics 20:301–320.

Palopoli, M. F. and C.-I. Wu. 1994. Genetics of hybrid male sterility between Drosophila sibling species: A complex web of epistasis is revealed in interspecific studies. Genetics 138:329–341.

Pamilo, P. and M. Nei. 1988. Relationship between gene trees and species trees. Mol. Biol. Evol. 5:568–583.

Panganiban, G., L. Nagy, and S. B. Carroll. 1994. The development and evolution of insect limb types. Current Biology 4:671–675.

Pantazidis, A. C. and E. Zouros. 1988. Location of an autosomal factor causing sterility in *D. mojavensis* males carrying the *D. arizonensis* Y chromosome. Heredity 60:299–304.

Pantazidis, A. C., V. K. Galanopoulos, and E. Zouros. 1993. An autosomal factor from *Drosophila arizonae* restores normal spermatogenesis in *Drosophila mojavensis* males carrying the *D. arizonae* Y chromosome. Genetics 134:309–318.

Papaceit, M. and E. Juan. 1993. Chromosomal homologies between *Drosophila lebanonensis* and *D. melanogaster* determined by in situ hybridization. Chromosoma 102:361–368.

Paquin, L. and R. C. Baumiller. 1978. Sigma virus induced mutation in inversion entire X-chromosome and in the second chromosome of *Drosophila melanogaster*. XIVth Int. Cong. Genet., Moscow. Abst. 2:265.

Pardue, M. L., K. Lowenhaupt, A. Rich, and A. Nordheim. 1987. $(dC\text{-}dA)_n$ and $(dG\text{-}dT)_n$ sequences have evolutionarily conserved chromosomal locations in Drosophila with implications for roles in chromosome structure and function. EMBO J. 6:1781–1789.

Parsons, P. A. 1958. Selection for increased recombination in *Drosophila melanogaster*. Am. Nat. 92:255–256.

Parsons, P. A. 1977. Cosmopolitan, exotic and endemic *Drosophila*: Their comparative evolutionary biology especially in southern Australia. Proc. Ecol. Soc. Aust. 10:62–75.

Parsons, P. A. 1982. Evolutionary ecology of Australian *Drosophila*: a species analysis. Evol. Biol. 14:297–328.

Partridge, L. and M. Farquhar. 1981. Sexual activity reduces longevity of male fruitflies. Nature 294:580–582.

Partridge, L. and K. Fowler. 1992. Direct and correlated responses to selection for age at reproduction in *Drosophila melanogaster*. Evolution 46:76–91.

Partridge, L. and A. Gardner. 1983. Failure to replicate the results of an experiment on the rare male effect in *Drosophila melanogaster*. Am. Nat. 122:422–427.

Partridge, L. and W. G. Hill. 1984. Mechanisms for frequency-dependent mating success. Biol. J. Linn. Soc. 23:113–132.

Partridge, L., A. Green, and K. Fowler. 1987. Effect of egg-production and of exposure to males on female survival in *Drosophila melanogaster*.

Pascual, L. and G. Periquet. 1991. Distribution of *hobo* transposable elements in natural populations of *Drosophila melanogaster*. Mol. Biol. Evol. 8:282–296.

Patau, K. 1935. Chromosomenmorphologie bei *Drosophila melanogaster* und *Drosophila simulans* und ihre genetische Bedeutung. Naturwissenschaften 23:537–543.

Patel, N. H. 1994. Developmental evolution: Insights from studies of insect segmentation. Science 266:581–589.

Patel, N. H., E. E. Ball, and C. S. Goodman. 1992. Changing role of *even skipped* during the evolution of insect pattern formation. Nature 357:339–342.

Patel, N. H., B. G. Condron, and K. Zinn. 1994. Pair rule expression patterns of *even-skipped* are found in both short- and long-germ beetles. Nature 367:429–434.

Paterson, H. E. H. 1978. More evidence against speciation by reinforcement. S. African J. Sci. 74:369–371.

Patterson, J. T. and A. B. Griffen. 1944. A genetic mechanism underlying species isolation. Univ. Texas Publ. 4415:212–223.

Patterson, J. T. and W. S. Stone. 1949. The relationship of *novamexicana* to the other members of the *virilis* group. Univ. Texas Publ. 4920:7–17.

Patterson, J. T. and W. S. Stone. 1952. *Evolution in the Genus Drosophila.* Macmillan, New York.

Patterson, J. T., W. S. Stone, and A. B. Griffen. 1942. Genetic and cytological analysis of the *virilis* species group. Univ. Texas Publ. 4228:160–200.

Pavan, C. and E. N. Knapp. 1954. The genetic structure of Brazilian *Drosophila willistoni.* Evolution 8:303–313.

Pavlovsky, O. and Th. Dobzhansky. 1966. Genetics of natural populations. XXXVIII. The coadapted system of chromosomal variants in a populations of *Drosophila pseudoobscura.* Genetics 53:843–854.

Payant, V., S. Abukashawa, M. Sasseville, B. F. Benkel, D. A. Hickey, and J. David. 1988. Evolutionary conservation of the chromosomal configuration and regulation of amylase genes among eight species of the *Drosophila melanogaster* species subgroup. Mol. Biol. Evol. 5:560–567.

Payne, F. 1910. Forty-nine generations in the dark. Biol. Bull. 18:188–190.

Peixoto, A. A., R. Costa, D. A. Wheeler, J. C. Hall, and C. P. Kyriacou. 1992. Evolution of the Threonine-Glycine region of the *period* gene in the *melanogaster* species subgroup of *Drosophila.* J. Mol. Evol. 35:411–419.

Peixoto, A. A., S. Campersan, R. Costa, and C. B. Kyriacou. 1993. Molecular evolution of a repetitive region within the *per* gene of *Drosophila.* Mol. Biol. Evol. 10:127–139.

Pelandakis, M. and M. Solignac. 1993. Molecular phylogeny of *Drosophila* based on ribosomal RNA sequences. J. Mol. Evol. 37:525–543.

Pentzos-Duponte, A. 1964. Qualitative und quantitative Untersuchungen uber den chromosomalen Polymorphismus naturlicher Populationen von *Drosophila subobscura* in der Umgebung von Thessaloniki/Griechenland. Z. Indukt. Abstamm.-Vererbungsl. 95:129–144.

Perez, D. E., C.-I. Wu, N. A. Johnson, and M.-L. Wu. 1993. Genetics of reproductive isolation in the *Drosophila simulans* clade: DNA-marker assisted mapping and characterization of a hybrid male sterility genes, *Odysseus* (*Ods*). Genetics 134:261–275.

Petit, C. 1951. La role de l'isolement sexuel dans l'evolution des populations de *Drosophila melanogaster.* Bull. Biol. France Belg. 85:352–418.

Petit, C. 1958. Le determinisme genetique et psycho-physiologique de la competition sexuelle chez *Drosophila melanogaster.* Bull. Biologique 92:248–329.

Petit, C. and D. Nouaud. 1975. Ecological competition and the advantage of the rare type in *Drosophila melanogaster*. Evolution 29:763–776.

Petri, W. H., A. R. Wyman, and F. C. Kafatos. 1976. Specific protein synthesis in cellular differentiation. III. The eggshell proteins of *Drosophila melanogaster* and their program of synthesis. Develop. Biol. 49:185–199.

Phaff, H. J., M. W. Miller, J. A. Recca, M. Shifrine, and E. M. Mrak. 1956. Studies on the ecology of *Drosophila* in the Yosemite region of California. II. Yeasts found in the alimentary canal of *Drosophila.* Ecology 37:533–538.

Pignal, M. C. and D. Lachaise. 1979. Les levures des Drosophilides de savane d'Afrique intertropicale (savane de Lamto-Cote d'Ivorie). Mycopathologia 68:155–165.

Pimpinelli, S. and P. Dimitri. 1989. Cytogenetic analysis of the responder (*Rsp*) locus in *Drosophila melanogaster.* Genetics 121:765–772.

Pinsker, W. and D. Sperlich. 1984. Cytogenetic mapping of enzyme loci on chromosome J and U of *Drosophila subobscura.* Genetics 108:913–926.

Pinsker, W., P. Lankinen, and D. Sperlich. 1978. Allozyme and inversion polymorphism in a central European population of *Drosophila subobscura.* Genetica 48:207–214.

Pipkin, S. B. 1965. The influence of adult and larval food habits on population size of neotropical ground-feeding *Drosophila.* Amer. Midl. Nat. 74:1–27.

Plus, N., G. Croizier, F. X. Jousset, and J. David. 1975. Picornaviruses of laboratory and wild *Drosophila melanogaster*: geographical distribution and serotypic composition. Ann. Microbiol (Paris) 126A:107–117.

Poiner, G. O. 1984. First fossil record of parasitism by insect parasite Tylenchida (Allantomatidae: Nematoda). J. Parasitol. 70:306–308.

Policansky, D. 1974. "Sex-ratio," meiotic drive, and group selection in *Drosophila pseudoobscura.* Am. Nat. 108:75–90.

Policansky, D. 1979. Fertility differences as a factor in the maintenance of the "sex-ratio" polymorphism in *Drosophila pseudoobscura.* Am. Nat. 114:672–680.

Policansky, D. and J. Ellison. 1970. "Sex ratio" in *Drosophila pseudoobscura*: Spermiogenic failure. Science 169:888–890.

Pontecorvo, G. 1943. Viability interactions between chromosomes of *Drosophila melanogaster* and *D. simulans.* J. Genet. 45:51–66.

Popadic, A. and W. W. Anderson. 1994. The history of a genetic system. Proc. Natl. Acad. Sci. USA 91:6819–6823.

Popadic, A., D. Popadic, and W. W. Anderson. 1995a. Interchromosomal transfer of genetic information between gene arrangements on the third chromosome of *Drosophila pseudoobscura.* Mol. Biol. Evol. 12:938–943.

Popadic, A. and W. W. Anderson. 1995b. Evidence for gene conversion in the amylase multigene family of *Drosophila pseudoobscura.* Mol. Biol. Evol. 12:564–572.

Postlethwait, J. H. and P. D. Shirk. 1981. Genetic and endocrine regulation of vitellogenesis in Drosophila. Am. Zool. 21:687–700.

Pot, W., W. van Delden, and J. P. Kruijt. 1980. Genotypic differences in mating success and the maintenance of the alcohol dehydrogenase polymorphism in *Drosophila melanogaster*: no evidence for overdominance or rare genotype mating advantage. Behav. Genet. 10: 43–58.

Poulson, D. F. 1968. Nature, stability, and expression of hereditary SR infections in *Drosophila.* XII Int. Congr. Genet. Proc. 2:91–92.

Powell, J. R. 1971. Genetic polymorphisms in varied environments. Science 174:1035–1036.

Powell, J. R. 1973. Apparent selection of enzyme alleles in laboratory populations of *Drosophila.* Genetics 75:557–570.

Powell, J. R. 1974. Temperature related genetic divergence in *Drosophila* body size. J. Heredity 65:257–258.

Powell, J. R. 1975. Protein variation in natural populations of animals. Evol. Biol. 8:79–119.

Powell, J. R. 1978. The founder-flush speciation theory: An experimental approach. Evolution 32:465–474.

Powell, J. R. 1979. Population genetics of Drosophila amylase. II. Geographic patterns in *D. pseudoobscura.* Genetics 92:613–622.

Powell, J. R. 1982. Genetic and non-genetic mechanisms of speciation. Pp. 67–74 in *Mechanisms of Speciation*, ed. C. Barigozzi. Liss Inc., New York.

Powell, J. R. 1983. Interspecific cytoplasmic gene flow in the absence of nuclear gene flow: Evidence from *Drosophila*. Proc. Natl. Acad. Sci. USA 80:492–495.

Powell, J. R. 1987. "In the Air"—Theodosius Dobzhansky's *Genetics and the Origin of Species*. Genetics 117:363–366.

Powell, J. R. 1989. The effects of founder-flush cycles on ethological isolation in laboratory populations of *Drosophila*. Pp. 239–251 in *Genetics, Speciation, and the Founder Principle*, eds. L. V. Giddings, K. Y. Kaneshiro, and W. W. Anderson. Oxford University Press, Oxford, U.K.

Powell, J. R. 1991. Monophyly/paraphyly/polyphyly and gene/species trees: An example from *Drosophila*. Mol. Biol. Evol. 8:892–896.

Powell, J. R. 1992. Inversion polymorphism in *Drosophila pseudoobscura* and *Drosophila persimilis*. Pp. 73–126 in *Drosophila Inversion Polymorphism*, eds. C. B. Krimbas and J. R. Powell. CRC Press, Boca Raton, Fla.

Powell, J. R. 1994. Molecular techniques in population genetics: A brief history. Pp. 131–156 in *Molecular Ecology and Evolution: Approaches and Applications*, eds. B. Shierwater, B. Streit, G. Wagner, and R. DeSalle. Birkhauser Verlag, Basel.

Powell, J. R. and G. Amato. 1984. Population genetics of Drosophila amylase. V. Genetic background and selection on different carbohydrates. Genetics 106:625–629.

Powell, J. R. and M. Andjelkovic. 1983. Population genetics of Drosophila amylase. IV. Selection in laboratory populations maintained on different carbohydrates. Genetics 103:675–689.

Powell, J. R. and A. Caccone. 1989. Intra- and interspecific genetic variation in *Drosophila*. Genome 31:233–238.

Powell, J. R. and A. Caccone. 1991. DNA–DNA hybridization: Principles and results. Pp. 117–132 in *Molecular Techniques in Taxonomy*, eds. G. M. Hewitt, A. W. B. Johnston, and J. P. W. Young. Springer-Verlag, Berlin.

Powell, J. R. and R. DeSalle. 1995. *Drosophila* molecular phylogenies and their uses. Evol. Biol. 28:87–138.

Powell, J. R. and J. M. Gleason. 1996. Codon usage and the origin of P elements. Mol. Biol. Evol. 13:278–279.

Powell, J. R. and J. M. Lichtenfels. 1979. Population genetics of Drosophila amylase. I. Genetic control of tissue-specific expression in *D. pseudoobscura*. Genetics 92:603–612.

Powell, J. R. and L. Morton. 1979. Inbreeding and mating patterns in *Drosophila pseudoobscura*. Behav. Genet. 9:425–429.

Powell, J. R. and R. C. Richmond. 1974. Founder effects and linkage disequilibrium in experimental populations of *Drosophila*. Proc. Natl. Acad. Sci. USA 71:1663–1665.

Powell, J. R. and C. E. Taylor. 1979. Genetic variation in ecologically diverse environments. Am. Sci. 67:590–596.

Powell, J. R. and H. Wistrand. 1978. The effect of heterogeneous environments and a competitor on genetic variation in *Drosophila*. Am. Nat. 112:935–947.

Powell, J. R., H. Levene, and Th. Dobzhansky. 1972. Chromosomal polymorphism in *Drosophila pseudoobscura* used for diagnosis of geographical origin. Evolution 26:553–559.

Powell, J. R., Th. Dobzhansky, J. E. Hook, and H. E. Wistrand. 1976. Genetics of natural populations. XLIII. Further studies on rates of dispersal of *Drosophila pseudoobscura* and its relatives. Genetics 82:493–506.

Powell, J. R., M. Rico, and M. Andjelkovic. 1980. Population genetics of Drosophila amylase. III. Interspecific variation. Evolution 34:209–213.

Powell, J. R., L. B. Klazcko, D. B. Dodd, and C. E. Taylor. 1984. Ecological genetics of *Drosophila pseudoobscura*: Habitat choice in heterogeneous environments. Pp. 263–271 in *Genetics: New Frontiers*, Vol. 4. Mohan Primlani, Oxford and IBH Publishing, New Delhi.

Powell, J. R., A. Caccone, G. D. Amato, and C. K. Yoon. 1986. Rates of nucleotide substitution in *Drosophila* mitochondrial DNA and nuclear DNA are similar. Proc. Natl. Acad. Sci. USA 83:9090–9093.

Powell, J. R., A. Caccone, J. M. Gleason, and L. Nigro. 1993. Rates of DNA evolution in Drosophila depend on function and developmental stage of expression. Genetics 133:291–298.

Prakash, S. 1969. Genic variation in a natural population of *Drosophila persimilis.* Proc. Natl. Acad. Sci. USA 62:778–784.

Prakash, S. 1972. Origin of reproductive isolation in the absence of apparent genic differentiation in a geographic isolate of *Drosophila pseudoobscura*. Genetics 72:143–155.

Prakash, S. 1973. Patterns of gene variation in central and marginal populations of *Drosophila robusta.* Genetics 75:347–369.

Prakash, S. 1977a. Gene polymorphism in natural populations of *Drosophila persimilis.* Genetics 85:513–520.

Prakash, S. 1977b. Genetic divergence in closely related sibling species *Drosophila pseudoobscura, Drosophila persimilis* and *Drosophila miranda.* Evolution 31:14–23.

Prakash, S. 1977c. Further studies of gene polymorphism in the main body and geographically isolated populations of *Drosophila pseudoobscura.* Genetics 85:713–719.

Prakash, S. and R. C. Lewontin. 1968. A molecular approach to the study of genic heterozygosity. III. Direct evidence of co-adaptation in gene arrangements of *Drosophila*. Proc. Natl. Acad. Sci. USA 59:398–405.

Prakash, S. and R. B. Merritt. 1972. Direct evidence of genic differentiation between sex ratio and standard gene arrangements of X-chromosome in *Drosophila pseudoobscura.* Genetics 72:169–175.

Prakash, S., R. C. Lewontin, and J. L. Hubby. 1969. A molecular approach to the study of genic heterozygosity in natural populations. IV. Patterns of genic variation in central, marginal, and isolated populations of *Drosophila pseudoobscura.* Genetics 61:841–858.

Precup, J. and J. Parker. 1987. Missense misreading of asparagine codons as a function of codon identity and context. J. Biol. Chem. 262:11351–11356.

Prevost, G. 1985. Etude experimentale des interactions entre parasitisme et competition larvaire chez *Drosophila melanogaster.* Ph.D. thesis, University of Lyon, France.

Prevosti, A. 1955. Geographical variability in quantitative traits in populations of *Drosophila subobscura.* Cold Spring Harbor Symp. Quant. Biol. 20:294–299.

Prevosti, A. 1964. Chromosomal polymorphism in *Drosophila subobscura* populations from Barcelona. Genet. Res. 5:27–38.

Prevosti, A., L. Serra, G. Ribo, M. Aguadé, E. Sagarra, M. Monclus, and M. P. Garcia. 1985. The colonization of *Drosophila subobscura* in Chile. II. Clines in the chromosomal arrangements. Evolution 39:838–844.

Prevosti, A., G. Ribo, L. Serra, M. Aguadé, J. Balana, M. Monclus, and F. Mestres. 1988. Colonization of America by *Drosophila subobscura*: experiment in natural populations that supports the adaptive role of chromosomal-inversion polymorphism. Proc. Natl. Acad. Sci. USA 85:5597–5600.

Propping, P. 1972. Comparison of point mutation rates in different species with human mutation rates. Humangenetik 16:43–48.

Prout, T. 1962. The effect of stabilizing selection on time of development in *Drosophila melanogaster.* Genet. Res. 3:364–382.

Prout, T. 1965. The estimation of fitness from genotype frequencies. Evolution 19:546–551.

Prout, T. 1971a. The relation between fitness components and population prediction in Drosophila. I. The estimation of fitness components. Genetics 68:127–149.

Prout, T. 1971b. The relation between fitness components and population prediction in Drosophila. II. Population predictions. Genetics 68:151–167.

Prout, T. and J. S. F. Barker. 1993. F statistics in *Drosophila buzzatii*: Selection, population size, and inbreeding. Genetics 134:369–375.

Prout, T. and J. Bundgaard. 1977. The population genetics of sperm displacement. Genetics 85: 95–124.

Provine, W. B. 1981. Origins of the Genetics of Natural Population Series. Pp. 5–76 in *Dobzhansky's Genetics of Natural Populations*, eds. R. C. Lewontin, J. A. Moore, W. B. Provine, and B. Wallace. Columbia University Press, New York.

Quine, J. A., P. Gunaratne, K. Asito, E. L. Organ, B. A. Cavener, and D. R. Cavener. 1993. Tissue-specific regulatory elements of the *Drosophila Gld* gene. Mech. Dev. 42:3–13.

Raff, R. A. 1996. *The Shape of Life*. Univ. Chicago Press, Chicago.

Raff, R. A. and T. C. Kaufman. 1983. *Embryos, Genes, and Evolution*. Macmillan, New York.

Ramachandra, N. B. and H. A. Ranganath. 1987. Characterization of heterochromatin in the B chromosomes of *Drosophila nasuta albomicans*. Chromosoma 95:223–226.

Rand, D. M., M. Dorfsman, and L. M. Kann. 1994. Neutral and non-neutral evolution of *Drosophila* mitochondrial DNA. Genetics 138:741–756.

Randazo, F. M., M. A. Seeger, C. A. Huss, M. A. Sweeney, J. K. Cecil, and T. C. Kaufman. 1993. Structural changes in the Antennapedia complex of *Drosophila pseudoobscura*. Genetics 134:319–330.

Ranganath, H. A. and N. B. Krishnamurthy. 1978. Chromosomal morphism in Drosophila *nasuta* II. Coexistence of heteroselection and flexibility in polymorphic system of South India populations. Genetica 48:215–221.

Ranganayakulu, G. 1994. The in situ localization of Adh transcripts in *Drosophila* species reveals evolved regulatory differences in spatially restricted expression. Genome 37:984–991.

Ranganayakulu, G., R. B. Kirkpatrick, P. F. Martin, and A. R. Reddy. 1991. Species-specific differences in tissue-specific expression of alcohol dehydrogenase are under the control of complex cis-acting loci: evidence from *Drosophila* hybrids. Biochem. Genet. 29:577–592.

Rasch, E., H. J. Barr, and W. Rasch. 1971. The DNA content of sperm of *Drosophila melanogaster*. Chromosoma 33:1–18.

Rasmuson, M. and A. Ljung. 1973. Fitness and fitness components for a two-allele system in *Drosophila melanogaster*. Hereditas 73:71–84.

Rasmussen, J. E. 1958. Persistence of mutants in laboratory populations of *Drosophila melanogaster*. Proc. X Int. Congr. Genet. 2:227–228.

Rat, L., M. Veulle, and J. A. Lepesant. 1991. Drosophila fat body protein P6 and alcohol dehydrogenase are derived from a common ancestral protein. J. Mol. Evol. 33:194–203.

Reddy, G. S. and N. B. Krishnamurthy. 1974. Altitudinal gradients in frequencies of three common inversions in *Drosophila ananassae*. Dros. Infor. Serv. 51:136–137.

Reed, S. C. and E. W. Reed. 1950. Natural selection in laboratory populations of *Drosophila*: II. Competition between a white eye gene and its wild type allele. Evolution 4:34–42.

Rendel, J. M. 1959. Canalization of the scute phenotype of Drosophila. Evolution 13:425–439.

Rendel, J. M. 1967. *Canalization and Gene Control*. Academic, Logos Press, London.

Rice, W. R. 1985. Disruptive selection on habitat preference and the evolution of reproductive isolation: an exploratory experiment. Evolution 39:645–656.

Rice, W. R. 1987. The accumulation of sexually antagonistic genes as a selective agent promoting the evolution of reduced recombination between primitive sex chromosomes. Evolution 41:911–914.

Rice, W. R. 1992. Sexually antagonistic genes: experimental evidence. Science 256:1436–1439.

Rice, W. R. and E. E. Hostert. 1993. Laboratory experiments on speciation: What have we learned in 40 years? Evolution 47:1637–1653.

Rice, W. R. and G. W. Salt. 1988. Speciation via disruptive selection on habitat preference: experimental evidence. Am. Nat. 131:911–917.
Rice, W. R. and G. W. Salt. 1990. The evolution of reproductive isolation as a correlated character under sympatric conditions: experimental evidence. Evolution 44:1140–1152.
Richardson, R. H. 1968. Migration and isozyme polymorphisms in natural populations of *Drosophila.* Proc. XII Int. Congr. Genet. 2:155–156.
Richardson, R. H. 1969. Migration and enzyme polymorphisms in natural populations of *Drosophila*. Jap. J. Genet. 44:172–179.
Richardson, R. H. and J. S. Johnston. 1975. Behavioral components of dispersal in *Drosophila mimica.* Oecologia 20:287–299.
Richardson, R. H. and P. E. Smouse. 1975. Ecological specialization of Hawaiian *Drosophila.* II. The community matrix, ecological complementation, and phyletic species packing. Oecologia 22:1–13.
Richardson, R. H., R. J. Wallace, S. J. Gage, G. P. Bouchey, and M. Denell. 1969. Neutron activation techniques for labeling *Drosophila* in natural populations. Univ. Texas Publ. 6918:171–186.
Richmond, R. C. 1972. Enzyme variability in the *Drosophila willistoni* group. III. Amounts of variability in the superspecies, *D. paulistorum.* Genetics 70:87–112.
Richmond, R. C. 1978. Microspatial genetic differentiation in natural populations of *Drosophila*. Pp. 127–142 in *Ecological Genetics: The Interface*, ed. P. F. Brussard. Springer-Verlag, New York.
Richmond, R. C. and J. R. Powell. 1970. Evidence of heterosis associated with an enzyme locus in a natural population of *Drosophila*. Proc. Natl. Acad. Sci. USA 67:1264–1267.
Richmond, R. C., M. D. Sabath, and J. M. Jones. 1977. Patterns of allozyme polymorphism in three species of the *Drosophila affinis* subgroup. Personal communication.
Riles, L. 1965. Inversion polymorphism and embryonic lethality in *Drosophila robusta.* Genetics 52:1335–1343.
Riley, M. A., M. E. Hallas, and R. C. Lewontin. 1989. Distinguishing the forces controlling genetic variation at the *Xdh* locus of *Drosophila pseudoobscura.* Genetics 123:359–369.
Riley, M. A., S. R. Kaplan, and M. Veuille. 1992. Nucleotide polymorphism at the xanthine dehydrogenase locus in *Drosophila pseudoobscura.* Mol. Biol. Evol. 9:56–69.
Ringo, J. M. 1986. The effect of successive founder events on mating propensity of *Drosophila.* Pp. 79–88 in *Evolutionary Genetics of Invertebrate Behavior*, ed. M. D. Huetel. Plenum, New York.
Rio, D. C., F. A. Laski, and G. M. Rubin. 1986. Identification and immunochemical analysis of biologically active *Drosophila P* element transposase. Cell 44:21–32.
Ritchie, M. G. and J. M. Gleason. 1995. Rapid evolution of courtship song pattern in *Drosophila willistoni* siblings. J. Evol. Biol. 8:463–479.
Rizki, T. M. 1957. Alterations in the haemocyte population of *Drosophila melanogaster.* J. Morphol. 100:437–458.
Rizki, T. M. 1968. Hemocyte encapsulation of streptococci in *Drosophila.* J. Invert. Pathol. 12: 339–343.
Roberts, D. B. (ed.). 1986. *Drosophila: A Practical Approach.* IRL Press, Oxford, U.K.
Roberts, P. A. 1967. A positive correlation between crossing over within heterozygous pericentric inversions and reduced egg hatch of *Drosophila* females. Genetics 56:179–187.
Robertson, A. 1962. Selection for heterozygotes in small populations. Genetics 47:1291–1300.
Robertson, F. W. 1959. Studies in quantitative inheritance. XII. Cell size and number in relation to genetic and environmental variation of body size in Drosophila. Genetics 44:869–896.
Robertson, F. W. and E. Reeve. 1952. Studies in quantitative inheritance. I. The effects of selection of wing and thorax length in *Drosophila melanogaster.* J. Genet. 50:414–448.
Robertson, F. W., M. Shook, G. Takei, and H. Gaines. 1968. Observations on the biology and

nutrition of *Drosophila disticha* Hardy, an indigenous Hawaiian species. Univ. Texas Publ., Studies in Genet. 4:279–299.

Robertson, H. M. 1993. The mariner transposable element is widespread in insects. Nature 362: 241–245.

Robertson, M. and J. H. Postlethwait. 1986. The humoral antibacterial response of *Drosophila* adults. Dev. Comp. Immunol. 10:167–179.

Rockwood-Sluss, E. S., J. S. Johnston, and W. B. Heed. 1973. Allozyme genotype-environment relationships. Variation in natural populations of *Drosophila pachea.* Genetics 73:135–146.

Rodriguez, L., M. B. Sokolowski, and J. S. Shore. 1992. Habitat selection by *Drosophila melanogaster* larvae. J. Evol. Biol. 5:61–70.

Roff, D. A. and T. A. Mousseau. 1987. Quantitative genetics and fitness: lessons from *Drosophila.* Heredity 58:103–118.

Ronsseray, S. and D. Anxolabéhère. 1986. Chromosomal distribution of P and I transposable elements in a natural population of *Drosophila melanogaster.* Chromosoma 94:433–440.

Ronsseray, S., M. Lehmann, and D. Anxolabéhère. 1989. Distribution of P and I mobile elements in *Drosophila melanogaster* populations. Chromosoma 98:207–214.

Rosato, E., A. A. Peixoto, G. Barbujani, R. Costa, and C. P. Kyriacou. 1994. Molecular polymorphism in the *period* gene in *Drosophila simulans.* Genetics 138:693–707.

Rose, M. R. 1984. Laboratory evolution of postponed senescence in *Drosophila melanogaster.* Evolution 38:1004–1010.

Rose, M. R. 1991. *Evolutionary Biology of Aging.* Oxford University Press, Oxford, U.K.

Ross, J. L., P. P. Fong, and D. R. Cavener. 1994. Correlated evolution of the cis-acting regulatory elements and developmental expression of the *Drosophila Gld* gene in seven species from the subgroup *melanogaster.* Develop. Genet. 15:38–50.

Rotondo, G. M., V. G. Springer, G. A. J. Scott, and S. O. Schlanger. 1981. Plate movement and island integration—A possible mechanism in the formation of endemic biotas, with special reference to the Hawaiian Islands. Syst. Zool. 30:12–21.

Rouault, J. 1979. Role des parasites entomophages dans la competition entre especes jumelles de Drosophiles: Approche experimentale. C. R. Seances Acad. Sci. 289:643–546.

Rowton, E. D., W. B. Lushbaugh, and R. B. McGhee. 1981. Ultrastructure of the flagella apparatus and attachment of *Herpetomonas ampelophilae* in the gut and Malpighian tubules of *Drosophila melanogaster.* J. Protozool. 28:297–301.

Rozas, J. and M. Aguadé. 1990. Evidence of extensive genetic exchange in the *rp49* region among polymorphic chromosome inversions in *Drosophila subobscura.* Genetics 126:417–426.

Rozas, J. and M. Aguadé. 1991a. Study of an isolated population at the nucleotide level: *rp49* region of a Canarian population of *Drosophila subobscura.* Mol. Biol. Evol. 8:202–211.

Rozas, J. and M. Aguadé. 1991b. Using restriction-map analysis to characterize the colonization process of *Drosophila subobscura* on the American continent. I. *rp49* region. Mol. Biol. Evol. 8:447–457.

Rozas, J. and M. Aguadé. 1993. Transfer of genetic information in the *rp49* region of *Drosophila subobscura* between different chromosomal gene arrangements. Proc. Natl. Acad. Sci. USA 90:8083–8087.

Rozas, J. and M. Aguadé. 1994. Gene conversion is involved in the transfer of genetic information between naturally occurring inversions of *Drosophila.* Proc. Natl. Acad. Sci. 91: 11517–11521.

Rubin, G. M. and A. C. Spradling. 1982. Genetic transformation of *Drosophila* with transposable element vectors. Science 218:348–353.

Ruddle, F. H., J. L. Bartels, K. L. Bentley, C. Kappen, M. T. Murtha, and J. W. Pendleton. 1994. Evolution of *HOX* genes. Ann. Rev. Genet. 28:423–442.

Ruden, D. M. and H. Jäckle. 1995. Mitotic delay dependent survival identifies components of cell cycle control in the *Drosophila* blastoderm. Development 121:63–73.

Ruiz, A. and W. B. Heed. 1988. Host-plant specificity in the cactophilic *Drosophila mulleri* species group. J. Anim. Ecol. 57:237–249.

Russell, S. R. H. and K. Kaiser. 1993. *Drosophila melanogaster* male germ line-specific transcripts with autosomal and Y-linked genes. Genetics 134:293–308.

Russo, C. A. M., N. Takezaki, and M. Nei. 1995. Molecular phylogeny and divergence times of Drosophila species. Mol. Biol. Evol. 12:391–404.

Ruttkey, H., M. Solignac, and D. Sperlich. 1992. Nuclear and mitochondrial ribosomal RNA variability in the *obscura* group of *Drosophila.* Genetica 85:131–138.

Salceda, V. M. and W. W. Anderson. 1988. Rare male mating advantage in a natural population of *Drosophila pseudoobscura.* Proc. Natl. Acad. Sci. USA 85:9870–9874.

Salt, G. 1970. The cellular defence reactions of insects. Cambridge Mongr. Exp. Biol. 16: 11–118.

Samenkovic-Radak, M. and M. Andjelkovic. 1992. Adaptive significance of amylase polymorphism in *Drosophila.* VII. Phenotypic expression of amylase in *Drosophila subobscura.* Arch. Biol. Sci. 44:155–160.

Samenkovic-Radak, M., M. Andjelkovic, and M. Milosevic. 1987. Adaptive significance of amylase polymorphism in *Drosophila.* II. The characterization of polymorphism of alpha-amylase tissue-specific expression in *Drosophila subobscura.* Arch. Biol. Sci. 39:51–62.

Samols, D. and H. Swift. 1979. Genome organisation in the fleshfly *Sarcophaga bullata.* Chromosoma 75:129–143.

Sander, K. 1994. The evolution of insect patterning mechanisms: a survey of progress and problems in comparative molecular embryology. Development 1994 Supplement:187–191.

Sandler, L. and K. Golic. 1985. Segregation distortion in *Drosophila.* Trends Genet. 1:181–185.

Sandler, L. and E. Novitski. 1957. Meiotic drive as an evolutionary force. Am. Nat. 41:105–110.

Sandler, L., Y. Hiraizumi, and I. Sandler. 1959. Meiotic drive in natural populations of *Drosophila melanogaster.* I. The cytogenetic basis of segregation distortion. Genetics 44:233–250.

Sang, J. 1978. The nutritional requirements of *Drosophila.* Pp. 159–192 in *The Genetics and Biology of Drosophila, Vol. 2a*, eds. M. Ashburner and T. R. F. Wright. Academic Press, New York.

Sankaranarayahan, R. Further data on the genetic loads in irradiated experimental populations of *Drosophila melanogaster.* Genetics 52:153–164.

Santos, M., A. Ruiz, J. E. Quezada-Diaz, A. Barbadilla, and A. Fontdevila. 1992. The evolutionary history of *Drosophila buzzatii.* XX. Positive phenotypic covariance between field adult fitness components and body size. J. Evol. Biol. 5:403–422.

Satta, Y., I. Hiromi, and S. I. Chigusa. 1987. Analysis of nucleotide substitutions of mitochondrial DNAs in *Drosophila melanogaster* and its sibling species. Mol. Biol. Evol. 4:638–650.

Saura, A. 1974. Genic variation in Scandinavian populations of *Drosophila bifasciata.* Hereditas 76:161–172.

Saura, A., S. Lakovaara, J. Lokki, and P. Lankinen. 1973. Genic variation in central and marginal populations of *Drosophila subobscura.* Hereditas 75:33–46.

Sawamura, K., T. Taira, and T. K. Watanabe. 1993a. Hybrid lethal systems in the *Drosophila melanogaster* complex. I. The maternal hybrid rescue (mhr) gene of *Drosophila simulans.* Genetics 133:299–305.

Sawamura, K., M. Yamamoto, and T. K. Watanabe. 1993b. Hybrid lethal systems in the *Drosophila melanogaster* complex. II. The zygotic hybrid rescue (Zhr) gene of *Drosophila melanogaster.* Genetics 133:307–313.

Sawyer, S. and D. Hartl. 1981. On the evolution of behavioral reproductive isolation: The Wallace effect. Theoret. Pop. Biol. 19:261–273.

Schaeffer, S. W. 1995. Population genetics in *Drosophila pseudoobscura*: A synthesis based on nucleotide sequence data for the *Adh* gene. Pp. 329–352 in *Genetics of Natural Populations: The continuing Importance of Theodosius Dobzhansky*, ed. L. Levine. Columbia University Press, New York.

Schaeffer, S. W. and C. F. Aquadro. 1987. Nucleotide sequence of the *Adh* gene region of *Drosophila pseudoobscura*: evolutionary change and evidence of an ancient duplication. Genetics 117:61–73.

Schaeffer, S. W. and E. L. Miller. 1992. Estimates of gene flow in *Drosophila pseudoobscura* determined from nucleotide sequence analysis of the alcohol dehydrogenase region. Genetics 132:471–480.

Schaeffer, S. W. and E. L. Miller. 1993. Estimates of linkage disequilibrium and the recombination parameter determined from segregating nucleotide sites in the alcohol dehydrogenase region of *Drosophila pseudoobscura.* Genetics 135:541–552.

Schaeffer, S. W., C. F. Aquadro, and W. W. Anderson. 1987. Restriction-map variation in the alcohol dehydrogenase region of *Drosophila pseudoobscura.* Mol. Biol. Evol. 4:254–265.

Schaeffer, S. W., C. F. Aquadro, and C. H. Langley. 1988. Restriction-map variation in the *Notch* region of *Drosophila melanogaster.* Mol. Biol. Evol. 5:30–40.

Scharloo, W. 1964. The effect of disruptive and stabilizing selection on the expression of cubitus interruptus in Drosophila. Genetics 50:553–562.

Scharloo, W. 1971. Reproductive isolation by disruptive selection: Did it happen? Am. Nat. 105:83–86.

Scheinker, S. V., E. R. Lozovskaya, J. G. Bishop, V. G. Croce, and M. B. Evgen'ev. 1990. A long terminal repeat-containing retrotransposon is mobilized during hybrid dysgenesis in *Drosophila virilis.* Proc. Natl. Acad. Sci. USA 87:9615–9619.

Schiff, N. M., Y. Feng, J. A. Quine, P. A. Krasney, and D. R. Cavener. 1992. Evolution of expression of the *Gld* gene in the reproductive tract of *Drosophila.* Mol. Biol. Evol. 9: 1029–1049.

Schilcher, F. von and A. Manning. 1975. Courtship song and mating speed in hybrids between *Drosophila melanogaster* and *Drosophila simulans.* Behav. Genet. 5:395–404.

Schmidt-Orr, U., M. González-Gaitán, H. Jäckle, and G. M. Technau. 1994. Number, identity, and sequence of the *Drosophila* head segments as revealed by neural elements and their deletion patterns in mutants. Proc. Natl. Acad. Sci. USA 91:8363–8367.

Scholtz, G., N. H. Patel, and W. Dohle. 1994. Serially homologous engrailed stripes are generated via different cell lineages in the germ band of the amphiod crustaceans (Malacostraca, Peracarida). Int. J. Develop. Biol. 38:471–478.

Schulze, D. H. and C. S. Lee. 1986. DNA sequence comparison among closely related *Drosophila* species of the *mulleri* complex. Genetics 113:287–303.

Segarra, C. and M. Aguadé. 1992. Molecular organization of the X chromosome in different species of the *obscura* group of *Drosophila.* Genetics 130:513–521.

Sene, F. M. and H. L. Carson. 1977. Genetic variation in Hawaiian *Drosophila.* IV. Allozymic similarity between *D. silvestris* and *D. heteroneura* from the island of Hawaii. Genetics 86:187–198.

Sevenster, J. G. and J. J. M. van Alphen. 1993a. Coexistence in stochastic environments through a life history trade off in *Drosophila.* Lecture Notes in Biomathematics 98:155–172.

Sevenster, J. G. and J. J. M. van Alphen. 1993b. A life history trade-off in *Drosophila* species and community structure in variable environments. J. Animal Ecol. 62:720–736.

Shah, D. M. and C. H. Langley. 1979. Inter- and intraspecific variation in restriction maps of *Drosophila* mitochondrial DNAs. Nature 281:696–699.

Sharp, P. M. and W.-H. Li. 1987. The codon adaptation index—a measure of directional synonymous codon usage bias and its potential applications. Nucl. Acids Res. 15:1281–1295.

Sharp, P. M. and W.-H. Li. 1989. On the rate of DNA sequence evolution in *Drosophila*. J. Mol. Evol. 28:398–402.

Sharp, P. M. and A. T. Lloyd. 1993. Codon usage. Pp. 378–397 in *An Atlas of Drosophila Genes: Sequences and Molecular Features*, ed. G. Maroni. Oxford University Press, New York.

Shaw, M. W. and G. M. Hewitt. 1990. B chromosomes, selfish DNA and theoretical models: where next? Oxford Surv. Evol. Biol 7:197–223.

Sheen, F. M., J. K. Lim, and M. J. Simmons. 1993. Genetic instability in *Drosophila melanogaster* mediated by *hobo* transposable elements. Genetics 133:315–334.

Shehata, A. M. El.-T. and E. M. Mrak. 1952. Intestinal yeast floras of successive populations of *Drosophila*. Evolution 6:325–335.

Shermoen, A. V. and P. H. O'Farrell. 1991. Progression of the cell cycle through mitosis leads to abortion of nascent transcripts. Nature 67:303–310.

Shibata, H. and T. Yamazaki. 1995. Molecular evolution of the duplicated *Amy* locus in the *Drosophila melanogaster* species subgroup: Concerted evolution only in the coding region and an excess of nonsynonymous substitutions in speciation. Genetics 141:223–236.

Shields, D. C. and P. M. Sharp. 1989. Evidence that mutation patterns vary among *Drosophila* transposable elements. J. Mol. Biol. 207:843–846.

Shields, D. C., P. M. Sharp, D. G. Higgins, and F. Wright. 1988. "Silent" sites in *Drosophila* genes are not neutral: Evidence of selection among synonymous codons. Mol. Biol. Evol. 5:704–716.

Shorrocks, B. 1982. The breeding sites of temperate woodland *Drosophila*. Pp. 385–428 in *The Genetics and Biology of Drosophila, Vol. 3b*, eds. M. Ashburner, H. L. Carson, and J. N. Thompson, Jr. Academic Press, New York.

Shorrocks, B. and P. Charlesworth. 1980. The distribution and abundance of the British fungal breeding *Drosophila*. Ecol. Entomol. 5:61–78.

Shorrocks, B. and L. Nigro. 1981. Microdistributions and habitat selection in *Drosophila subobscura* Collin. Biol. J. Linn. Soc. 16:293–301.

Shrimpton, A. E., T. F. C. Mackay, and A. J. Leigh Brown. 1990. Transposable element-induced response to artificial selection in *Drosophila melanogaster*: molecular analysis of selection lines. Genetics 125:803–311.

Shroyer, D. A. and L. Rosen. 1983. Extrachromosomal inheritance of carbon dioxide sensitivity in the mosquito *Culex quiquefasciatus*. Genetics 104:649–659.

Simmons, G. M. 1992. Horizontal transfer of hobo transposable elements within the *Drosophila melanogaster* species complex: evidence from DNA sequencing. Mol. Biol. Evol. 9:1050–1060.

Simmons, G. M., M. E. Kreitman, W. F. Quattlebaum, and N. Miyashita. 1989. Molecular analysis of the alleles of alcohol dehydrogenase along a cline in *Drosophila melanogaster*: I. Maine, North Carolina, and Florida USA. Evolution 43:393–409.

Simmons, G. M., W. Kwok, P. Matulonis, and T. Venkatesh. 1994. Polymorphism and divergence at the *prune* locus in *Drosophila melanogaster* and *D. simulans*. Mol. Biol. Evol. 11:666–671.

Simmons, M. J. and J. F. Crow. 1977. Mutations affecting fitness in Drosophila populations. Ann. Rev. Genet. 11:49–78.

Simon, C., F. Frati, A. Beckenbach, B. Crespi, H. Liu, and P. Flook. 1994. Evolution, weighting, and phylogenetic utility of mitochondrial gene sequences and a compilation of conserved polymerase chain reaction primers. Ann. Entomol. Soc. Am. 87:651–701.

Singh, R. S. and M. B. Coulthart. 1982. Genic variation in abundant soluble proteins of *Drosophila melanogaster* and *Drosophila pseudoobscura*. Genetics 102:437–453.

Singh, R. S. and L. R. Hale. 1990. Are mitochondrial DNA variants selectively non-neutral? Genetics 124:995–997.

Singh, R. S. and L. R. Rhomberg. 1987a. A comprehensive study of genic variation in natural populations of *Drosophila melanogaster.* I. Estimates of gene flow from rare alleles. Genetics 115:313–322.

Singh, R. S. and L. R. Rhomberg. 1987b. A comprehensive study of genic variation in natural populations of *Drosophila melanogaster.* II. Estimates of heterozygosity and patterns of geographic differentiation. Genetics 117:255–271.

Singh, R. S., D. A. Hickey, and J. David. 1982. Genetic differentiation between geographically distant populations of *Drosophila melanogaster.* Genetics 101:235–256.

Slack, J. M. W. 1991. *From Egg to Embryo: Regional specification in early development.* Cambridge University Press, Cambridge, U.K.

Sluss, E. S. 1975. Enzyme variability in natural populations of two species of cactophilic *Drosophila.* Ph.D. dissertation, Univ. of Arizona, Tucson, Ariz.

Smathers, K. M. 1961. The contribution of heterozygosity of certain gene loci to fitness of laboratory populations of *Drosophila melanogaster.* Am. Nat. 95:27–38.

Sniegowski, P. D. and B. Charlesworth. 1994. Transposable element numbers in cosmopolitan inversions from a natural population of *Drosophila melanogaster*. Genetics 137:815–827.

Snodgrass, R. E. 1935. *Principles of Insect Morphology.* McGraw-Hill, London and New York.

Snook, R. R., T. A. Markow, and T. L. Karr. 1994. Functional non-equivalence of sperm in *Drosophila pseudoobscura.* Proc. Natl. Acad. Sci. USA 91:11222–11226.

Snyder, M., M. Hunkapiller, D. Yuen, D. Silvert, J. Fristrom, and N. Davidson. 1982. Cuticle protein genes in *Drosophila*: Structure, organization and evolution of four clustered genes. Cell 29:1027–1040.

Sofer, W. and P. F. Martin. 1987. Analysis of alcohol dehydrogenase gene expression in Drosophila. Ann. Rev. Genet. 21:203–235.

Sokal, R. R. 1986. Phenetic taxonomy: Theory and methods. Ann. Rev. Ecol. Syst. 17:423–442.

Sokal, R. R., N. L. Oden, and J. S. F. Barker. 1987. Spatial structure in *Drosophila buzzatii* populations: Simple and directional spatial autocorrelation. Am. Nat. 129:122–142.

Sokoloff, A. 1965. Geographic variation of quantitative characters in populations of *Drosophila pseudoobscura.* Evolution 19:300–310.

Sokoloff, A. 1966. Morphological variation in natural and experimental populations of *Drosophila pseudoobscura* and *Drosophila persimilis.* Evolution 20:49–71.

Sokolowski, M. B., S. J. Bauer, V. Wai-Ping, L. Rodriguez, J. L. Wong, and C. Kent. 1986. Ecological genetics and behavior of *Drosophila melanogaster* larvae in nature. Anim. Behav. 32:403–408.

Solima-Simmons, A. 1966. Experiments on random genetic drift and natural selection. Evolution 20:100–103.

Sommer, R. J. and D. Tautz. 1993. Involvement of an orthologue of the *Drosophila* pair rule gene *hairy* in segment formation of the short germ band embryo of *Tribolium.* Nature 361: 448–450.

Sommer, R. J. and D. Tautz. 1994. Expression patterns of *twist* and *snail* in *Tribolium* (Coleoptera) suggest a homologous formation of mesoderm in long and short germ band insects. Dev. Genet. 15:32–37.

Soulé, M. 1973. The epistasis cycle: A theory of marginal populations. Ann. Rev. Ecol. Syst. 4:165–187.

Spassky, B., N. Spassky, H. Levene, and Th. Dobzhansky. 1958. Release of genetic variability through recombination. I. *Drosophila pseudoobscura.* Genetics 43:844–867.

Spassky, B., R. C. Richmond, S. Pérez-Salas, O. Pavlovsky, C. A. Mournão, A. S. Hunter, H. Hoenigsberg, Th. Dobzhansky, and F. J. Ayala. 1971. Geography of the sibling species related to *Drosophila willistoni*, and of the semispecies of the *Drosophila paulistorum* complex. Evolution 25:129–143.

Spencer, H. G., B. H. McArdle, and D. M. Lambert. 1986. A theoretical investigation of speciation by reinforcement. Am. Nat. 128:241–262.

Spencer, W. P. 1932. The vermilion mutant of *Drosophila hydei* breeding in nature. Am. Nat. 66:474–479.

Spencer, W. P. 1944. Iso-alleles at the bobbed locus in *Drosophila hydei* populations. Genetics 29:520–536.

Spencer, W. P. 1946. High mutant gene frequencies in a population of *Drosophila immigrans.* Ohio Jour. Sci. 46:143–151.

Spencer, W. P. 1947. Mutations in wild populations in Drosophila. Adv. Genetics 1:359–402.

Spencer, W. P. 1949. Gene homologies and the mutants of *Drosophila hydei.* Pp. 23–44 in *Genetics, Paleontology, and Evolution*, eds. G. G. Jepsen, E. Mayr, and G. G. Simpson. Princeton University Press, Princeton, N.J.

Spencer, W. P. 1957. Genetic studies on *Drosophila mulleri.* I. The genetic analysis of a population. Univ. Texas Publ. 5721:186–205.

Sperlich, D. 1966. Equilibria for inversions induced by X-ray in isogenic strains of *Drosophila pseudoobscura.* Genetics 53:835–842.

Sperlich, D. and P. Pfriem. 1986. Chromosomal polymorphism in natural and experimental populations. Pp. 257–309 in *The Genetics and Biology of Drosophila, Vol. 3e*, eds. M. Ashburner, H. L. Carson, and J. N. Thompson, Jr. Academic Press, New York.

Spicer, G. S. 1991. Molecular evolution and phylogeny of *Drosophila virilis* species group as inferred by two-dimensional electrophoresis. J. Mol. Evol. 33:379–394.

Spicer, G. S. 1992. Re-evaluation of the phylogeny of the *Drosophila virilis* species group (Diptera: Drosophilidae). Ann. Ent. Soc. Amer. 85:11–25.

Spiess, E. B. 1957. Relation between frequencies and adaptive values of chromosomal arrangements in *Drosophila persimilis*. Evolution 11:84–93.

Spiess, E. B. 1959. Release of genetic variability through recombination. II. *D. persimilis.* Genetics 44:43–58.

Spiess, E. B. 1966. Chromosomal fitness changes in experimental populations of *Drosophila persimilis* from the timberline of the Sierra Nevada. Evolution 20:82–91.

Spiess, E. B. 1968. Low frequency advantage in mating of *Drosophila pseudoobscura* karyotypes. Am. Nat. 102:363–379.

Spiess, E. B. 1970. Mating propensity and its genetic basis in *Drosophila.* Pp. 315–379 in *Essays in Evolution and Genetics in Honor of Theodosius Dobzhansky*, eds. M. K. Hecht and W. C. Steere. Appleton-Century-Crofts, New York.

Spiess, E. B. 1982. Do female flies choose their mates? Am. Nat. 119:675–693.

Spiess, E. B. 1987. Discrimination among prospective mates in Drosophila. Pp. 75–119 in *Kin Recognition in Animals*, eds. D. J. C. Fletcher and C. D. Michener. John Wiley, Chichester, U.K.

Spiess, E. B. and H. L. Carson. 1981. Sexual selection in *Drosophila silvestris.* Proc. Natl. Acad. Sci. USA 78:3088–3092.

Spiess, E. B. and C. C. Dapples. 1981. A model of fly mating intensity, not behaviour. Am. Nat. 118:307–315.

Spiess, E. B. and J. F. Kruckeberg. 1980. Minority advantage of certain eye color mutants of *Drosophila melanogaster.* II. A behavioral basis. Am. Nat. 115:307–327.

Spiess, E. B. and C. M. Wilke. 1984. Still another attempt to achieve assortative mating by disruptive selection in *Drosophila.* Evolution 38:505–515.

Spiess, L. D. and E. B. Spiess. 1969. Minority advantage in interpopulational matings of *Drosophila persimilis.* Am. Nat. 103:155–172.

Spieth, H. T. 1966. Courtship behavior of endemic Hawaiian *Drosophila*. Univ. Texas Publ. 6615:133–145.

Spieth, H. T. 1974. Courtship behavior in *Drosophila*. Ann. Rev. Entomol. 19:395–405.

Spieth, H. T. 1978. Courtship patterns and evolution of the *Drosophila adiastola* and *planitibia* species subgroups. Evolution 32:435–451.

Spieth, H. T. 1979. The *virilis* group of *Drosophila* and the beaver *Castor.* Am. Nat. 114: 312–316.

Spieth, H. T. 1981. *Drosophila heteroneura* and *Drosophila silvestris*: Head shapes, behavior and evolution. Evolution 35:921–930.

Spieth, H. T. 1984. Courtship behavior of the Hawaiian Picture-winged *Drosophila.* Univ. Calif. Pubs. Zool. 103:1–92.

Spieth, H. T. 1988. Special note. Dros. Inf. Serv. 67:1–2.

Spieth, H. T. and J. M. Ringo. 1983. Mating behavior and sexual isolation in *Drosophila.* Pp. 223–284 in *The Genetics and Biology of Drosophila, Vol. 3c*, eds. M. Ashburner, H. L. Carson, and J. N. Thompson, Jr. Academic Press, New York.

Springer, M., E. H. Davidson, and R. J. Britten. 1992. Calculation of sequence divergence from the thermal stability of DNA heteroduplexes. J. Mol. Evol. 34:379–382.

Stacey, S. N., R. A. Lansman, H. W. Brock, and T. A. Grigliatti. 1986. Distribution and conservation of mobile elements in the genus *Drosophila.* Mol. Biol. Evol. 3:522–534.

Stalker, H. D. 1942. Triploid intersexuality in *Drosophila americana.* Genetics 27:504–518.

Stalker, H. D. 1976. Chromosome studies in wild populations of *Drosophila melanogaster.* Genetics 82:323–347.

Stalker, H. D. and H. L. Carson. 1947. Morphological variation in natural populations of *Drosophila robusta* Sturtevant. Evolution 1:237–248.

Stalker, H. D. and H. L. Carson. 1948. An altitudinal transect of *Drosophila robusta* Sturtevant. Evolution 2:295–305.

Stalker, H. D. and H. L. Carson. 1963. A very serious parasite of laboratory *Drosophila.* Dros. Inf. Serv. 38:96.

Stamenkovic-Bojic, G., M. Milanogic, and M. Andjelkovic. 1994. Adaptive significance of amylase polymorphism in *Drosophila.* VIII. Effect of carbohydrate dietary components on α-amylase activity and amy-electromorph frequency in *Drosophila busckii.* Genetica 92: 101–106.

Stamenkovic-Radak, M. Andjelkovic, and M. Milosevic. 1987. Adaptive significance of amylase polymorphism in *Drosophila.* II. The characteristics of polymorphism of alpha-amylase tissue-specific expression in *Drosophila subobscura.* Arch. Biol. Sci. (Belgrade) 39: 51–62.

Starmer, W. T. 1982. Associations and interactions among yeasts, *Drosophila*, and their habitats. Pp. 159–183 in *Ecological Genetics and Evolution: The Cactus-Yeast-Drosophila Model System*, eds. J. S. F. Barker and W. T. Starmer. Academic Press, Sydney.

Starmer, W. T. and V. Aberdeen. 1990. The nutritional importance of pure and mixed cultures of yeasts in the development of *Drosophila mulleri* larvae in *Opuntia* tissues and its relationship to host plant shifts. Pp. 145–160 in *Ecological and Evolutionary Genetics of Drosophila*, eds. J. S. F. Barker, W. T. Starmer, and R. J. MacIntyre. Plenum, New York.

Starmer, W. T. and D. T. Sullivan. 1989. A shift in the third-codon-position nucleotide frequency in alcohol dehydrogenase genes in the genus *Drosophila.* Mol. Biol. Evol. 6:546–552.

Starmer, W. T., W. B. Heed, and E. S. Rockwood-Sluss. 1977. Extension of longevity in *Drosophila mojavensis* by environmental ethanol: Differences between subraces. Proc. Natl. Acad. Sci. USA 74:387–391.

Starmer, W. T., H. J. Phaff, M. Miranda, M. W. Miller, and W. B. Heed. 1982. The yeast flora associated with the decaying stems of columnar cacti and *Drosophila* in North America. Evol. Biol. 14:269–295.

Stearns, S. C. 1976. Life-history tactics: a review of the ideas. Quart. Rev. Biol. 51:3–47.

Stearns, S. C. 1989. Trade-offs in life-history evolution. Functional Ecology 3:259–68.

Stearns, S. C. and T. J. Kawecki. 1995. Fitness sensitivity and the canalization of life-history traits. Evolution 48:1438–1450.

Stearns, S. C., M. Kaiser, and E. Hillesheim. 1993. Effects on fitness components of enhanced expression of elongation factor EF-1a in *Drosophila melanogaster.* I. The contrasting approaches of molecular and population biologists. Am. Nat. 142:961–993.

Stebbins, G. L. and D. V. Basile. 1986. Phyletic phenocopies: A useful technique for probing the genetic developmental basis of evolutionary change. Evolution 40:422–425.

Steinemann, M. 1982a. Multiple sex chromosomes in *Drosophila miranda*: A system to study the degeneration of a chromosome. Chromosoma 86:59–76.

Steinemann, M. 1982b. Analysis of chromosomal homologies between two species of the subgenus Sophophora: *D. miranda* and *D. melanogaster* using cloned DNA segments. Chromosoma 87:77–88.

Steinemann, M. and S. Steinemann. 1990. Evolutionary changes in the organization of the major LCP gene cluster during sex chromosomal differentiation in the sibling species *Drosophila persimilis, D. pseudoobscura*, and *D. miranda.* Chromosoma 99:424–434.

Steinemann, M. and S. Steinemann. 1991. Preferential Y chromosomal location of TRIM, a novel transposable element of *Drosophila miranda, obscura* group. Chromosoma 101: 169–179.

Steinemann, M. and S. Steinemann. 1992. Degenerating Y chromosome of *Drosophila miranda*: A trap for retroposons. Proc. Natl. Acad. Sci. USA 89:7591–7595.

Steinemann, M. and S. Steinemann. 1993. A duplication including the Y allele of Lcp2 and the TRIM retrotransposon at the Lcp locus on the degenerating neo-Y chromosome of *Drosophila miranda*: Molecular structure and mechanisms by which it may have arisen. Genetics 134:497–505.

Steinemann, M., W. Pinsker, and D. Sperlich. 1984. Chromosome homologies within the *Drosophila obscura* group. Chromosoma 91:46–53.

Steinemann, M., S. Steinemann, and F. Lottspeich. 1993. How Y chromosomes become genetically inert. Proc. Natl. Acad. Sci. USA 90:5737–5741.

Steiner, W. W. M. 1975. Enzyme variability in exotic and endemic *Drosophila* in Hawaii. Genetics 79:s78.

Steiner, W. W. M., K. C. Sung, and Y. K. Paik. 1976. Electrophoretic variability in island populations of *Drosophila simulans* and *Drosophila immigrans.* Biochem. Genet. 14:495–506.

Stephan, W. 1989a. Tandem-repetitive noncoding DNA: Forms and forces. Mol. Biol. Evol. 6: 198–212.

Stephan, W. 1989b. Molecular genetic variation in the centromeric region of the X chromosome in three *Drosophila ananassae* populations. II. The *Om(1D)* locus. Mol. Biol. Evol. 6: 624–635.

Stephan, W. 1994. Effects of genetic recombination and population subdivision on nucleotide sequence variation in *Drosophila ananassae.* Pp. 57–66 in *Non-neutral Evolution*, ed. B. Golding. Chapman Hall, New York.

Stephan, W. and D. A. Kirby. 1993. RNA folding in Drosophila shows a distance effect for compensatory fitness interactions. Genetics 135:97–103.

Stephan, W. and C. F. Langley. 1989. Molecular genetic variation in the centromeric region of the X chromosome in three *Drosophila ananassae* populations. I. Contrasts between the *vermilion* and *forked* loci. Genetics 121:89–99.

Stephan, W. and S. J. Mitchell. 1992. Reduced levels of DNA polymorphism and fixed between-population differences in the centromeric region of *Drosophila ananassae.* Genetics 132:1039–1045.

Stephan, W., V. S. Rodriguez, B. Zhou, and J. Parsch. 1994. Molecular evolution of the metallothionein gene *Mtn* in the *melanogaster* species group: Results from *Drosophila ananassae.* Genetics 138:135–143.

Stevens, L. and M. J. Wade. 1990. Cytoplasmically inherited reproductive incompatibility in *Tribolium* flour beetles: the rate of spread and effect on population size. Genetics 124: 367–372.

St. Johnson, D. and C. Nüsslein-Volhard. 1992. The origin of pattern and polarity in the *Drosophila* embryo. Cell 68:201–219.

Stone, W. S., W. C. Guest, and F. D. Wilson. 1960. The evolutionary implications of the cytological polymorphism and phylogeny of the *virilis* group of *Drosophila.* Proc. Natl. Acad. Sci. USA 46:350–361.

Strickberger, M. W. 1963. Evolution of fitness in experimental populations of *Drosophila pseudoobscura.* Evolution 17:40–55.

Strickberger, M. W. 1965. Experimental control over the evolution of fitness in laboratory populations of *Drosophila pseudoobscura.* Genetics 51:795–800.

Strickberger, M. W. and C. J. Wills. 1966. Monthly frequency changes of *Drosophila pseudoobscura* third chromosome gene arrangements in a California locality. Evolution 20:592–602.

Strobeck, C. 1983. Expected linkage disequilibrium for a neutral locus linked to a chromosomal arrangement. Genetics 103:545–555.

Strobel, E., P. Dunsmuir, and G. M. Rubin. 1979. Polymorphism in the chromosomal locations of elements of the 412, copia and 297 dispersed repeated gene families in *Drosophila.* Cell 17:429–439.

Strobel, E., C. Pelling, and N. Arnheim. 1978. Incomplete dosage compensation in an evolving *Drosophila* sex chromosome. Proc. Natl. Acad. Sci. USA 75:931–935.

Stuart, J. J., S. J. Brown, R. W. Beeman, and R. E. Denell. 1991. A deficiency of the homeotic complex of the beetle *Tribolium.* Nature 350:72–74.

Stuart, J. J., S. J. Brown, R. W. Beeman, and R. E. Dennell. 1993. The *Tribolium* homeotic gene *Abdominal* is homologous to *abdominal-A* of *Drosophila* bithorax complex. Development 117:233–243.

Sturtevant, A. H. 1915. A sex-linked character in *Drosophila repleta.* Am. Nat. 49:189–192.

Sturtevant, A. H. 1917. Genetic factors affecting the strength of linkage in *Drosophila.* Proc. Natl. Acad. Sci. USA 3:555–558.

Sturtevant, A. H. 1921. The North American species of *Drosophila.* Carnegie Inst. Washington Pub. No. 301.

Sturtevant, A. H. 1926. A crossover reducer in *Drosophila melanogaster* due to inversion of a section of the third chromosome. Biol. Zentralbl. 46:697–702.

Sturtevant, A. H. 1940. Genetic data on *Drosophila affinis*, with a discussion of the relationships in the subgenus *Sophophora.* Genetics 25:337–353.

Sturtevant, A. H. 1965. Chapter 17 in *A History of Genetics.* Harper and Row, New York.

Sturtevant, A. H. and G. W. Beadle. 1936. The relations of inversions in the X chromosome of *Drosophila melanogaster* to crossing-over and disjunction. Genetics 21:554–605.

Sturtevant, A. H. and Th. Dobzhansky. 1936. Inversions in the third chromosome of wild races of *Drosophila pseudoobscura* and their use in the study of the history of the species. Proc. Natl. Acad. Sci. USA 22:448–450.

Sturtevant, A. H. and E. Novitski. 1941. The homologies of the chromosome elements in the genus *Drosophila.* Genetics 26:517–541.

Sullivan, D. T., P. W. Atkinson, and W. T. Starmer. 1990. Molecular evolution of the alcohol dehydrogenase genes in the genus *Drosophila.* Evol. Biol. 24:107–147.

Sullivan, D. T., W. T. Starmer, S. W. Curtiss, M. Menotti-Raymond, and J. Yum. 1994. Unusual molecular evolution of an *Adh* pseudogene in *Drosophila.* Mol. Biol. Evol. 11:443–458.

Susman, M. and H. L. Carson. 1958. Development of balanced polymorphism in laboratory populations of *Drosophila melanogaster.* Am. Nat. 93:359–364.

Sved, J. A. 1971. An estimate of heterosis in *Drosophila melanogaster.* Genet. Res. 18:97–105.

Sved, J. A. 1981a. A two-sex polygenic model for the evolution of premating isolation. I. Deterministic theory for natural populations. Genetics 97:197–215.

Sved, J. A. 1981b. A two-sex polygenic model for the evolution of premating isolation. II. Computer simulation of experimental selection procedures. Genetics 97:217–235.

Sved, J. A. and F. J. Ayala. 1970. A population cage test for heterosis in *Drosophila pseudoobscura.* Genetics 66:97–113.

Swimmer, C., M. G. Fenerjian, J. C. Martinez-Cruzado, and F. C. Kafatos. 1990. Evolution of the autosomal chorion cluster in *Drosophila.* III. Comparison of the *s18* gene in evolutionarily distant species and heterospecific control of chorion gene amplification. J. Mol. Biol. 215:225–235.

Swimmer, C., H. Kashevsky, G. Mao, and F. C. Kafatos. 1992. Positive and negative DNA elements of the *Drosophila grimshawi s18* chorion gene assayed in *Drosophila melanogaster.* Develop. Biol. 152:103–112.

Swofford, D. L. and G. J. Olsen. 1990. Phylogeny reconstruction. Pp. 411–501 in *Molecular Systematics*, eds. D. M. Hillis and C. Moritz. Sinauer Associates, Sunderland, Mass.

Tabachnick, W. J. and J. R. Powell. 1977. Adaptive flexibility of "marginal" versus "central" populations of *Drosophila willistoni.* Evolution 31:692–694.

Tajima, F. 1989. Statistical method for testing the neutral mutation hypothesis by DNA polymorphism. Genetics 123:585–595.

Tajima, F. 1993. Measurements of DNA polymorphism. Pp. 37–60 in *Mechanisms of Molecular Evolution*, eds. N. Takahata and A. G. Clark. Sinauer Press, Sunderland, Mass.

Takahata, N. and M. Kimura. 1981. A model of evolutionary base substitutions and its application with special reference to rapid change of pseudogenes. Genetics 98:641–657.

Takano, T. S., S. Kasakabe, A. Koga, and T. Mukai. 1989. Polymorphism for the number of tandemly multiplicated glycerol-3-phosphate dehydrogenase genes in *Drosophila melanogaster.* Proc. Natl. Acad. Sci. USA 86:5000–5004.

Takano, T. S., S. Kusakabe, and T. Mukai. 1991. The genetic structure of natural populations of *Drosophila melanogaster.* Genetics 129:753–761.

Takano, T. S., S. Kusakabe, and T. Mukai. 1993. DNA polymorphism and the origin of protein polymorphism at the *Gpdh* locus of *Drosophila melanogaster.* Pp. 179–190 in *Mechanisms of Molecular Evolution*, eds. N. Takahata and A. G. Clark. Sinauer Press, Sunderland, Mass.

Tamura, K. 1992. The rate and pattern of nucleotide substitution in *Drosophila* mitochondrial DNA. Mol. Biol. Evol. 9:814–825.

Tamura, K., T. Aotsuda, and O. Kitagawa. 1991. Mitochondrial DNA polymorphisms in the two subspecies of *Drosophila sulfurigaster*: Relationship between geographic structure of population and nucleotide diversity. Mol. Biol. Evol. 8:104–114.

Tan, C. C. 1935. Salivary gland chromosomes in the two races of *Drosophila pseudoobscura.* Genetics 20:392–402.

Tantawy, A. O. and G. S. Mullah. 1961. Studies on natural populations of *Drosophila* I. Heat resistance and geographical variation in *Drosophila melanogaster* and *D. simulans.* Evolution 15:1–14.

Tardif, G. N. and M. R. Murnik. 1975. Frequency-dependent sexual selection among wild-type strains of *Drosophila melanogaster.* Behav. Genet. 5:373–379.

Taylor, C. E. 1986. Habitat choice by *Drosophila pseudoobscura*: The roles of genotype and of experience. Behav. Genet. 16:271–279.

Taylor, C. E. 1987. Habitat selection within species of *Drosophila*: a review of experimental findings. Evol. Ecol. 1:389–400.

Taylor, C. E. and C. Condra. 1980. r- and K-selection in *Drosophila pseudoobscura.* Evolution 34:1183–1193.

Taylor, C. E. and C. Condra. 1983. Resource partitioning among genotypes of *Drosophila pseudoobscura*. Evolution 37:135–149.

Taylor, C. E. and J. R. Powell. 1977. Microgeographic differentiation of chromosomal and enzyme polymorphisms in *Drosophila persimilis.* Genetics 85:681–695.

Taylor, C. E. and J. R. Powell. 1978. Habitat choice in natural populations of *Drosophila.* Oecologia 37:69–75.

Taylor, C. E. and J. R. Powell. 1983. Population structure of *Drosophila*: genetics and ecology. Pp. 29–59 in *The Genetics and Biology of Drosophila, Vol. 3d*, eds. M. Ashburner, H. L. Carson, and J. N. Thompson, Jr. Academic Press, New York.

Tear, G., M. Akam, and A. Martinez-Arias. 1990. Isolation of an *abdominal-A* gene from the locust *Schistocerca gregaria* and its expression during early embryogenesis. Development 110:915–926.

Teissier, G. 1942. Persistence d'un gene lethal dans une population de *Drosophila.* Comp. Rend. Acad. Sci. 214:327–330.

Temin, R. G., B. Ganetzky, P. Powers, T. W. Lyttle, and S. Pimpinelli, et al. 1991. Segregation distorter (SD) in *Drosophila melanogaster*: Genetic and molecular analyses. Am. Nat. 137: 287–331.

Templeton, A. R. 1977. Analysis of head shape differences between two interfertile species of Hawaiian *Drosophila*. Evolution 31:630–641.

Templeton, A. R. 1979. The unit of selection in *Drosophila mercatorum.* II. Genetic revolution and the origin of coadapted genomes in parthenogenetic strains. Genetics 92:1265–1282.

Templeton, A. R. 1980. The theory of speciation *via* the founder principle. Genetics 94:1011–1038.

Templeton, A. R. 1981. Mechanisms of speciation—A population genetic approach. Ann Rev. Ecol. Syst. 12:23–48.

Templeton, A. R. 1983. Natural and experimental parthenogenesis. Pp. 343–298 in *The Genetics and Biology of Drosophila, Vol. 3c*, eds. M. Ashburner, H. L. Carson, and J. N. Thompson, Jr. Academic Press, New York.

Templeton, A. R. and J. S. Johnston. 1982. Life history evolution under pleiotrophy and k-selection in a natural population of *Drosophila mercatorum.* Pp. 225–240 in *Ecological Genetics and Evolution: The Cactus-Yeast-Drosophila Model System*, eds. J. S. F. Barker and W. T. Starmer. Academic Press, New York.

Templeton, A. R. and J. S. Johnston. 1988. The measured genotype approach to ecological genetics. Pp. 138–146 in *Population Genetics and Evolution*, ed. G. de Jong. Springer-Verlag, Berlin.

Templeton, A. R. and M. A. Rankin. 1978. Genetic revolutions and control of insect populations. Pp. 83–112 in *The Screwworm Problem*, ed. R. H. Richardson. University Texas Press, Austin, Texas.

Templeton, A. R., T. J. Crease, and F. Shah. 1985. The molecular through ecological genetics of abnormal abdomen in *Drosophila mercatorum*. I. Basic genetics. Genetics 111:805–818.

Templeton, A. R., J. S. Johnston, and C. F. Sing. 1987. The proximate and ultimate control of aging in *Drosophila* and humans. Pp. 123–133 in *Evolution of Longevity in Animals*, eds. A. D. Woodhead and K. H. Thompson. Plenum, New York.

Templeton, A. R., H. Hollocher, S. Lawler, and J. S. Johnston. 1990a. Natural selection and ribosomal DNA in Drosophila. Genome 31:296–303.

Templeton, A. R., H. Hollocher, S. Lawler, and J. S. Johnston. 1990b. The ecological genetics of abnormal abdomen in *Drosophila mercatorum.* Pp. 17–35 in *Ecological and Evolutionary Genetics of Drosophila*, eds. J. S. F. Barker, W. T. Starmer, and R. J. MacIntyre. Plenum, New York.

Templeton, A. R., H. Hollocher, and J. S. Johnston. 1993. The molecular through ecological genetics of *abnormal abdomen* in *Drosophila mercatorum.* V. Female phenotypic expression on natural genetic backgrounds and in natural populations. Genetics 134:475–485.

Terzaghi, E. and D. Knapp. 1960. Patterns of chromosomal variability in *Drosophila pseudoobscura.* Evolution 14:347–350.

Theodore, L., A.-S. Ho, and G. Maroni. 1991. Recent evolutionary history of the metallothionein gene *Mtn* in *Drosophila.* Genet. Res. 58:203–210.

Thoday, J. M. and T. B. Boam. 1961. Regular responses to selection. I. Description of responses. Genet. Res. 2:161–176.

Thoday, J. M. and J. B. Gibson. 1962. Isolation by disruptive selection. Nature 193:1164–1166.

Thoday, J. M. and J. B. Gibson. 1970. The probability of isolation by disruptive selection. Am. Nat. 104:219–230.

Thomas, R. H. and J. A. Hunt. 1991. The molecular evolution of the alcohol dehydrogenase locus and the phylogeny of the Hawaiian *Drosophila.* Mol. Biol. Evol. 8:687–702.

Thomas, R. H. and J. A. Hunt. 1993. Phylogenetic relationships in *Drosophila*: A conflict between molecular and morphological data. Mol. Biol. Evol. 10:362–374.

Thomas, S. and R. S. Singh. 1992. A comprehensive study of genic variation in natural populations of *Drosophila melanogaster.* VII. Varying rates of genic divergence as revealed by 2-dimensional electrophoresis. Mol. Biol. Evol. 9:507–525.

Thomas-Orillard, M. 1984. Modification of mean ovariole number, fresh weight of adult females and developmental time in *Drosophila melanogaster* induced by Drosophila C virus. Genetics 107:635–644.

Thompson, D. B., L. G. Treat-Clemons, and W. W. Doane. 1992. Tissue-specific and dietary control of alpha-amylase gene expression in the adult midgut of *Drosophila melanogaster.* J. Exp. Zool. 262:122–134.

Thompson, G. and M. W. Feldman. 1975. Population genetics of modifiers of meiotic drive. IV. On the evolution of sex-ratio distortion. Theor. Pop. Biol. 8:202–211.

Thompson-Stewart, D., G. H. Karpen, and A. C. Spradling. 1994. A transposable element can drive the concerted evolution of tandemly repetitious DNA. Proc. Natl. Acad. Sci. USA 91:9042–9046.

Thomson, J. A. 1961. Interallelic selection in experimental populations of *Drosophila melanogaster*: White and satsuma. Genetics 40:1435–1442.

Thomson, M. S., J. W. Jacobson, and C. C. Laurie. 1991. Comparison of alcohol dehydrogenase expression in *Drosophila melanogaster* and *D. simulans.* Mol. Biol. Evol. 8:31–48.

Thorpe, P. A., J. Loye, C. A. Rote, and W. J. Dickinson. 1993. Evolution of regulatory genes and patterns: Relationships to evolutionary rates and to metabolic functions. J. Mol. Evol. 37:590–599.

Thorpe, W. H. 1930. The biology of the petroleum fly (*Psilopa petrolei* Coq.). Trans. R. Entomol. Soc. Lond. 78:331–343.

Throckmorton, L. H. 1966. The relationships of some of the endemic Hawaiian Drosophilidae. Univ. Texas Publ. 6615:335–396.

Throckmorton, L. H. 1975. The phylogeny, ecology, and geography of *Drosophila.* Pp. 421–469 in *Handbook of Genetics, Vol. 3*, ed. R. C. King. Plenum, New York.

Throckmorton, L. H. 1982a. Pathways of evolution in the genus *Drosophila* and the founding of the *repleta* group. Pp. 33–47 in *Ecological Genetics and Evolution: The Cactus-Yeast-Drosophila Model System*, eds. J. S. F. Barker and W. T. Starmer. Academic Press, Sydney.

Throckmorton, L. H. 1982b. The *virilis* species group. Pp. 227–296 in *The Genetics and Biology of Drosophila, Vol. 3b*, eds. M. Ashburner, H. L. Carson, and J. N. Thompson, Jr. Academic Press, New York.

Timofeeff-Ressovsky, N. W. 1932. Mutations of the gene in different directions. Proc. 6th. Int. Congr. Genetics 1:308–330.

Timofeeff-Ressovsky, H. and N. W. Timofeeff-Ressovsky. 1927. Genetische Analyse einer freilebenden *Drosophila melanogaster* Population. Arch. Entwicklungsmech. Organ. 109: 70–109.

Timofeef-Ressovsky, H. and N. W. Timofeef-Ressovsky. 1940a. Population-genetische Versuche an Drosophila II. Aktionsbereiche von *Drosophila funebris* und *Drosophila melanogaster*. Z. indukt. Abstanm. Verebungslehre 79:35–49.

Timofeef-Ressovsky, H. and N. W. Timofeef-Ressovsky. 1940b. Populations-genetische Versuche an Drosophila III. Quantitative Untersuchengen an einegen *Drosophila* Populations. Z. indukt. Abstanm. Verebungslehre 79:44–49.

Tobari, Y. N. (ed.). 1993a. *Drosophila ananassae: Genetical and Biological Aspects*. Japan Scientific Society Press, Tokyo; and S. Karger AG, Basel.

Tobari, Y. N. 1993b. Heterosis and frequency dependent selection. Pp. 151–160 in *Drosophila ananassae: Genetical and Biological Aspects*, ed. Y. N. Tobari. Japan Scientific Society Press, Tokyo; and S. Karger AG, Basel.

Tobari, Y. N. and K.-I. Kojima. 1967. Selective modes associated with inversion karyotypes in *Drosophila ananassae*. I. Frequency-dependent selection. Genetics 57:179–188.

Tobari, Y. N. and K. Kojima. 1972. A study of spontaneous mutation rates at ten loci detectable by starch gel electrophoresis in *Drosophila melanogaster*. Genetics 70:397–403.

Tobari, Y. N. and D. Moriwaki. 1993. Life cycle. Pp. 1–6 in *Drosophila ananassae: Genetical and Biological Aspects*, ed. Y. N. Tobari. Japan Scientific Society Press, Tokyo; and S. Karger AG, Basel.

Tomimura, Y., M. Matsuda, and Y. N. Tobari. 1993. Population genetics, part I. Pp. 139–151 in *Drosophila ananassae: Genetical and Biological Aspects*, ed. Y. N. Tobari. Japan Scientific Society Press, Tokyo; and S. Karger AG, Basel.

Tompkins, L., S. P. McRobert, and K. Y. Kaneshiro. 1993. Chemical communication in Hawaiian *Drosophila*. Evolution 47:1407–1419.

Tonzetich, J. H. and C. L. Ward. 1973. Adaptive chromosomal polymorphism in *Drosophila melanica*. Evolution 27:486–494.

Tonzetich, J., S. Hayashi, and T. A. Grigliatte. 1990. Conservation of sites of tRNA loci among the linkage groups of several *Drosophila* species. J. Mol. Evol. 30:182–188.

Toolson, E. C. and R. Kuper-Simbron. 1989. Laboratory evolution of epicuticular hydrocarbon composition and cuticular permeability in *Drosophila pseudoobscura*: Effects on sexual dimorphism and thermal-acclimation. Evolution 43:486–473.

Torkamanzehi, A., C. Moran, and F. W. Nicholas. 1992. P element transposition contributes substantial new variation for a quantitative trait in *Drosophila melanogaster*. Genetics 131: 73–78.

True, J. R., B. S. Weir, and C. C. Laurie. 1996. A genome-wide survey of hybrid incompatibility factors by the introgression of marked segments of *Drosophila mauritiana* chromosomes into *Drosophila simulans*. Genetics 142:819–837.

Tsacas, L., D. Lachaise, and J. R. David. 1981. Composition and biogeography of the Afrotropical Drosophilid fauna. Pp. 197–260 in *The Genetics and Biology of Drosophila, Vol. 3a*, eds. M. Ashburner, H. L. Carson, and J. N. Thompson, Jr. Academic Press, New York.

Tucic, N. 1979. Genetic capacity for adaptation to cold resistance at different developmental stages of *Drosophila melanogaster*. Evolution 33:350–358.

Tucic, N. and M. Krunic. 1975. Genotype dependent ability of *Drosophila melanogaster* to cold hardiness at different developmental stages. Genetika (Yugoslavia) 7:123–132.

Turelli, M. and A. A. Hoffmann. 1988. Effects of starvation and experience on the response of *Drosophila* to alternative resources. Oecologia 77:497–505.

Turelli, M. and A. A. Hoffmann. 1991. Rapid spread of an inherited incompatibility factor in California Drosophila. Nature 353:440–442.

Turelli, M. and H. A. Orr. 1995. The dominance theory of Haldane's rule. Genetics 140:389–402.

Turelli, M., J. A. Coyne, and T. Prout. 1984. Resource choice in orchard populations of *Drosophila*. Biol. J. Linn. Soc. 22:95–106.

Turelli, M., A. A. Hoffmann, and S. W. McKechnie. 1992. Dynamics of cytoplasmic incompatibility and mtDNA variation in natural *Drosophila simulans* populations. Genetics 132: 713–723.

Urieli-Shoval, S., Y. Gruenbaum, J. Sedat, and A. Razin. 1982. The absence of detectable methylated bases in *Drosophila melanogaster* DNA. FEBS Lett. 146:148–152.

Uyenoyama, M. K. and M. W. Feldman. 1978. The genetics of sex ratio distortion by cytoplasmic infection under maternal and contagious transmission: An epidemiological study. Theor. Pop Biol. 14:471–479.

Vacek, D. C. 1982. Interactions between microorganisms and cactophilic *Drosophila* in Australia. Pp. 175–206 in *Ecological Genetics and Evolution: The Cactus-Yeast-Drosophila Model System*, eds. J. S. F. Barker and W. T. Starmer. Academic Press, Sydney.

Vacek, D. C., W. T. Starmer, and W. B. Heed. 1979. The relevance of the ecology of *Citrus* yeasts to the diet of *Drosophila*. Microb. Ecol. 5:43–49.

Vacek, D. C., P. D. East, J. S. F. Barker, and M. H. Soliman. 1985. Feeding and oviposition preferences of *Drosophila buzzatii* for microbial species isolated from its natural environment. Biol. J. Linn. Soc. 24:175–187.

Val, F. C. 1977. Genetic analysis of the morphological differences between two interfertile species of Hawaiian *Drosophila*. Evolution 31:611–629.

Val, F. C., C. R. Vilela, and M. D. Marques. 1981. Drosophilidae of the Neotropical region. Pp. 123–168 in *The Genetics and Biology of Drosophila, Vol. 3a*, eds. M. Ashburner, H. L. Carson, and J. N. Thompson, Jr. Academic Press, New York.

Valente, V. L. S. and A. M. Araújo. 1985. Observations on the chromosomal polymorphism of natural populations of *Drosophila willistoni* and its association with the choice of feeding and breeding sites. Rev. Brasil. Genet. 8:271–284.

van Delden, W. 1970. Selection for competitive ability. Dros. Inf. Serv. 45:169.

van Delden, W. 1982. The alcohol dehydrogenase polymorphism in *Drosophila melanogaster*: Selection at an enzyme locus. Evol. Biol. 15:187–222.

van Delden, W. and A. Kamping. 1989. The association between the polymorphisms at the *Adh* and *aGpdh* loci and the *In(2L)t* inversion in *Drosophila melanogaster* in relation to temperature. Evolution 43:77–793.

van Dijken, F. R. and W. Scharloo. 1979. Divergent selection on locomotor activity in *Drosophila melanogaster*. II. Test for reproductive isolation between selected lines. Behav. Genet. 9:555–561.

Vazeille-Falcoz, M., H. Ohayon, P. Gounon, and L. Rosen. 1992. Unusual morphology of a virus which produces carbon dioxide sensitivity in mosquitoes. Virus Res. 24:235–247.

Vet, L. E. M. 1983. Host-habitat location through olfactory cues by *Leptopilina clavipes* (Hartig) (Hym: Eucoilidae) a parasitoid of fungivorous *Drosophila*: influence of conditioning. Neth. J. Zool. 33:225–248.

Vet, L. E. M., C. Janse, C. van Achtenberg, and J. J. M. van Alphen. 1983a. Microhabitat location and niche segregation in two sibling species of drosophilid parasitoids: *Asobara tabida* (Nels) and *A. rufescens* (Foerster) (Braconidae: Alysiinae). Oecologia 61: 182–188.

Vet, L. E. M., J. C. van Lenteren, M. Heymans, and E. Meelis. 1983b. An airflow olfactometer for measuring olfactory responses of hymenopterous parasitoids and other small insects. Physiol. Entomol. 8:97–106.

Veuille, M. and L. M. King. 1995. Molecular basis of polymorphism at the Esterase-5B locus in *Drosophila pseudoobscura.* Genetics 141:255–262.

Vigue, C. L. and F. M. Johnson. 1973. Isozyme variability in species of the genus Drosophila. VI. Frequency-property-environment relationships of allelic alcohol dehydrogenases in *D. melanogaster.* Biochem. Genet. 9:213–227.

Voelker, R. A. 1972. Preliminary characterization of "sex ratio" and rediscovery and reinterpretation of "male sex ratio" in *Drosophila affinis.* Genetics 71:597–606.

Voelker, R. A., C. C. Cockerham, F. M. Johnson, H. E. Schaffer, T. Mukai, and L. E. Mettler. 1978. Inversions fail to account for allozyme clines. Genetics 88:515–527.

Voelker, R. A., C. H. Langley, A. J. Leigh Brown, S. Ohnishi, B. Kickson, E. Montgomery, and S. C. Smith. 1980. Enzyme null alleles in natural populations of *Drosophila melanogaster.* Proc. Natl. Acad. Sci. USA 77:1091–1095.

Waddington, C. H. 1953. Genetic assimilation of an acquired character. Evolution 7:118–126.

Waddington, C. H. 1956a. Genetic assimilation of the *Bithorax* phenotype. Evolution 10:1–13.

Waddington, C. H. 1956b. *Principles of Embryology.* Allen and Unwin, London.

Waddington, C. H. 1960. Experiments on canalizing selection. Genet. Res. 1:140–150.

Waddington, C. H. and E. Robertson. 1966. Selection for developmental canalization. Genet. Res. 7:303–312.

Wagner, R. P. 1944. The nutrition of *Drosophila mulleri* and *D. aldrichi.* Growth of the larvae on cactus extract and the microorganisms found in cactus. Univ. Texas Publ. 4445:104–128.

Wagner, R. P. 1949. Nutritional differences in the mulleri group. Univ. Texas Publ. 4920: 39–41.

Walker, I. 1959. Die Abwehrreaktion des Wirtes *Drosophila melanogaster* gegen die zoophage Cynipidae *Pseudeucoila bochei* Weld. Rev. Suisse Zool. 66:569–632.

Wallace, B. 1948. Studies on "sex-ratio" in *Drosophila pseudoobscura.* I. Selection and "sex-ratio." Evolution 2:189–217.

Wallace, B. 1953. Genetic divergence of isolated populations of *Drosophila melanogaster.* Proc. 9th Int. Congress Genet. 9:761–764.

Wallace, B. 1959. The role of heterozygosity in Drosophila populations. Proc. 10th Int. Congr. Genetics 1:408–419.

Wallace, B. 1963. Genetic diversity, genetic uniformity, and heterosis. Can. J. Genet. Cytol. 5: 239–253.

Wallace, B. 1966a. On the dispersal of *Drosophila.* Am. Nat. 100:551–563.

Wallace, B. 1966b. Distance and the allelism of lethals in a tropical population of *Drosophila melanogaster.* Am. Nat. 100:565–578.

Wallace, B. 1968. *Topics in Population Genetics.* Norton, New York.

Wallace, B. 1970a. *Genetic Load: Its Biological and Conceptual Aspects.* Prentice Hall, Englewood Cliffs, N.J.

Wallace, B. 1970b. Observations on the microdispersion of *Drosophila melanogaster.* Pp. 381–399 in *Essays in Evolution and Genetics in Honor of Theodosius Dobzhansky*, eds. M. K. Hecht, W. S. Steere. Appleton-Century-Crofts, New York.

Wallace, B. 1974. Studies on intra- and inter-specific competition in *Drosophila.* Ecology 55: 227–244.

Wallace, B. 1975. The biogeography of laboratory islands. Evolution 29:622–635.

Wallace, B. 1978. The adaptation of *Drosophila virilis* to life on an artificial crab. Am. Nat. 112:971–973.

Wallace, B. 1984. A possible explanation for observed differences in the geographical distributions of chromosomal arrangements of plants and *Drosophila.* Egypt. J. Genet. Cytol. 13: 121–136.

Wallace, B. 1991. *Fifty years of Genetic Load: An Odyssey.* Cornell University Press, Ithaca, N.Y.

Wallace, B., J. C. King, C. V. Madden, B. Kaufman, and E. C. McGunnigle. 1953. An analysis of variability arising through recombination. Genetics 38:272–307.

Walthour, C. S. and S. W. Schaeffer. 1994. Molecular population genetics of sex determination genes: The *transformer* gene of *Drosophila melanogaster*. Genetics 136:1367–1372.

Ward, B. L., W. T. Starmer, J. S. Russell, and W. B. Heed. 1975. The correlation of climate and host plant morphology with a geographic gradient of an inversion polymorphism in *Drosophila pachea*. Evolution 28:565–575.

Ward, C. L. 1952. Chromosome variation in *Drosophila melanica*. Univ. Texas Publ. 5204: 137–157.

Wasserman, D. A., M. Therrien, and G. M. Rubin. 1995. The Ras singling pathway in Drosophila. Cur. Opinion Genet. Devlop. 5:44–50.

Wasserman, M. 1982. Evolution of the *repleta* group. Pp. 61–140 in *Genetics and Biology of Drosophila, Vol. 3b*, eds. M. Ashburner, H. L. Carson, and J. N. Thompson, Jr. Academic Press, New York.

Wasserman, M. 1992. Cytological evolution of the *Drosophila repleta* group. Pp. 455–552 in *Drosophila Inversion Polymorphism*, eds. C. B. Krimbas and J. R. Powell. CRC Press, Boca Raton, Fla.

Wasserman, M. and R. H. Koepfer. 1977. Character displacement for sexual isolation between *Drosophila mojavensis* and *Drosophila arizonensis*. Evolution 31:812–823.

Watanabe, T. K. 1979. A gene that rescues the lethal hybrids between *Drosophila melanogaster* and *D. simulans*. Jap. J. Genet. 54:325–331.

Watanabe, T. K. and M. Kawanishi. 1979. Mating preference and the direction of evolution in *Drosophila*. Science 205:906–907.

Watanabe, T. K., W. W. Anderson, Th. Dobzhansky, and O. Pavlovsky. 1970. Selection in experimental populations of *Drosophila pseudoobscura* with different frequencies of chromosomal variants. Genet. Res. 15:123–133.

Watson, G. 1960. The cytoplasmic "sex-ratio" condition in *Drosophila*. Evolution 14:256–265.

Weinberg, R. 1954. The chromosomes of *Drosophila macrospina* and comparisons of the chromosome elements with other species. Univ. Texas Publ. 5722:153–162.

Weisbrot, D. R. 1966. Genotype interactions among competing strains and species of *Drosophila*. Genetics 53:427–435.

Welbergen, P., F. R. van Duken, W. Scharloo, and W. Kohler. 1992. The genetic basis of sexual isolation between *Drosophila melanogaster* and *D. simulans*. Evolution 46:1385–1398.

Welch, H. E. 1959. Taxonomy, life cycle, development, and habits of two new species of Allantonematidae (Nematoda) parasitic in drosophilid flies. Parasitology 49:83–103.

Werman, S. D., M. S. Springer, and R. J. Britten. 1990a. DNA–DNA hybridization. Pp. 204–249 in *Molecular Systematics*, eds. D. M. Hillis and C. Moritz. Sinauer Press, Sunderland, Mass.

Werman, S. D., E. H. Davidson, and R. J. Britten. 1990b. Rapid evolution in a fraction of the *Drosophila* nuclear genome. J. Mol. Evol. 30:281–289.

Werren, J. H. 1987. The coevolution of autosomal and cytoplasmic sex ratio factors. J. Theoret. Biol. 124:317–334.

Wesley, C. S. and W. F. Eanes. 1994. Isolation and analysis of the breakpoint sequences of chromosome inversion *In(3L)Payne* in *Drosophila melanogaster*. Proc. Natl. Acad. Sci. USA 91:3132–3136.

Wharton, L. T. 1943. Analysis of the metaphase and salivary chromosome morphology within the genus *Drosophila*. Univ. Texas. Publ. 4313:282–319.

Wheeler, D. A., C. P. Kyriacou, M. L. Greenacre, Q. Yu, J. E. Rutila, M. Rosbash, and J. C. Hall. 1991. Molecular transfer of a species-specific behavior from *Drosophila simulans* to *Drosophila melanogaster*. Science 251:1082–1085.

Wheeler, L. L. and L. S. Altenburg. 1977. Hoechst 33258 banding of *Drosophila nasutoides* metaphase chromosomes. Chromosoma 62:351–360.

Wheeler, L. L., F. Arrighi, M. Cordeiro-Stone, and C. S. Lee. 1978. Localisation of *Drosophila nasutoides* satellite DNAs in metaphase chromosomes. Chromosoma 70:41–50.

Wheeler, M. R. 1963. A note on some fossil Drosophilidae (Diptera) from the amber of Chiapas, Mexico. J. Paleontol. 37:123–124.

Wheeler, M. R. 1981. The Drosophilidae: A taxonomic overview. Pp. 1–97 in *The Genetics and Biology of Drosophila, Vol. 3a*, eds. M. Ashburner, H. L. Carson, and J. N. Thompson, Jr. Academic Press, New York.

Wheeler, M. R. 1986. Additions to the catalog of the world's Drosophilidae. Pp. 395–409 in *The Genetics and Biology of Drosophila, Vol. 3e*, eds. M. Ashburner, H. L. Carson, and J. N. Thompson, Jr. Academic Press, New York.

White, M. J. D. 1978. *Modes of Speciation.* Freeman Press, San Francisco.

Whiting, J. H., M. D. Pliley, J. L. Farmer, and D. E. Jeffery. 1989. *In situ* hybridization analysis of chromosomal homologies in *Drosophila melanogaster* and *Drosophila virilis.* Genetics 122:99–109.

Whiting, M. F. and W. C. Wheeler. 1994. Insect homeotic transformation. Nature 368:696.

Wiedemann, G. 1936. Modellversuch zum Selektionswirking von Faktormutationen bei *Drosophila melanogaster.* Genetica 18:277–290.

Williams, J. A. and S. B. Carroll. 1993. The origin, patterning and evolution of insect appendages. BioEssays 15:567–578.

Williamson, D. L. 1961. Carbon dioxide sensitivity in *Drosophila affinis* and *Drosophila athabasca.* Genetics 46:1053–1060.

Williamson, D. L. and D. F. Poulson. 1979. Sex ratio organisms (Spiroplasmas) of *Drosophila.* Pp. 175–208 in *The Mycoplasmas Vol. III*, eds. R. F. Whitcomb and J. G. Tully. Academic Press, New York.

Williamson, D. L., L. Ehrman, and R. P. Kernaghan. 1971. Induction of sterility in *Drosophila paulistorum*: Effect of cytoplasmic factors. Proc. Natl. Acad. Sci. USA 68:2158–2161.

Wilson, A. C., S. S. Carlson, and T. J. White. 1977. Biochemical evolution. Ann. Rev. Biochem. 46:573–639.

Wojtas, K. M., L. von Kaim, J. R. Weaver, and D. T. Sullivan. 1992. The evolution of duplicate glycerate-3-dehydrogenase genes in Drosophila. Genetics 132:789–797.

Wolstenholme, D. R. and D. O. Clary. 1985. Sequence evolution of *Drosophila* mitochondrial DNA. Genetics 109:725–744.

Woodruff, R. 1993. Transposable DNA elements and life history traits I. Transposition of P DNA elements in somatic cells reduces the life span of *Drosophila melanogaster.* Pp. 218–229 in *Transposable Elements and Evolution*, ed. J. F. McDonald. Kluwer Publishers, Dordrecht.

Woodruff, R. C., B. E. Slatko, and J. N. Thompson, Jr. 1983. Factors affecting mutation rates in natural populations. Pp. 37–124 in *The Genetics and Biology of Drosophila Vol. 3c*, eds. M. Ashburner, H. L. Carson, and J. N. Thompson, Jr. Academic Press, New York.

Wright, F. 1990. The effective number of codons used in a gene. Gene 87:23–39.

Wright, S. 1977. *Evolution and the Genetics of Populations, Volume 3: Experimental Results and Evolutionary Deductions.* Univ. Chicago Press, Chicago.

Wright, S. 1978. *Evolution and the Genetics of Populations, Volume 4: Variability within and among Natural Populations.* Univ. of Chicago Press, Chicago.

Wright, S. 1980. Genic and organismic evolution. Evolution 34:825–843.

Wright, S. 1982. The shifting balance theory and macroevolution. Ann. Rev. Genet. 16:1–19.

Wright, S. and Th. Dobzhansky. 1946. Genetics of natural populations. XII. Experimental reproduction of some of the changes caused by natural selection in certain populations of *Drosophila pseudoobscura.* Genetics 31:125–156.

Wright, S. and W. E. Kerr. 1954. Experimental studies of the distribution of gene frequencies in very small populations of *Drosophila melanogaster*. II. Bar. Evolution 8:225–240.

Wu, C.-I. 1983a. Virility deficiency and the sex ratio trait in *Drosophila pseudoobscura*. I. Sperm displacement and sexual selection. Genetics 105:651–662.

Wu, C.-I. 1983b. Virility deficiency and the sex-ratio trait in *Drosophila pseudoobscura*. II. Multiple mating and overall virility selection. Genetics 105:663–679.

Wu, C.-I. and A. T. Beckenbach. 1983. Evidence for extensive genetic differentiation between the sex-ratio and the standard arrangement of *Drosophila pseudoobscura* and *D. persimilis* and identification of hybrid sterility factors. Genetics 105:71–86.

Wu, C.-I. and A. W. Davis. 1993. Evolution of postmating reproductive isolation: The composite nature of Haldane's rule and its genetic bases. Am. Nat. 142:187–212.

Wu, C.-I. and M. F. Hammer. 1991. Molecular evolution of ultraselfish genes of meiotic drive systems. Pp. 177–203 in *Evolution at the Molecular Level*, eds. R. K Selander, T. Whittam, and A. Clark. Sinauer Press, Sunderland, Mass.

Wu, C.-I. and W.-H. Li. 1985. Evidence for higher rates of nucleotide substitution in rodents than in man. Proc. Natl. Acad. Sci. USA 82:1741–1745.

Wu, C.-I., J. R. True, and N. Johnson. 1989. Fitness reduction associated with the deletion of a satellite DNA array. Nature 341:248–251.

Wu, C.-I., D. E. Perez, A. W. Davis, N. A. Johnson, E. L. Cabot, M. F. Palopoli, and M.-L. Wu. 1993. Molecular genetic studies of postmating reproductive isolation in *Drosophila*. Pp. 199–212 in *Mechanisms of Molecular Evolution*, eds. N. Takahata and A. G. Clark. Sinauer Associates, Sunderland, Mass.

Wu, C.-I., F. Liu, D. J. Begun, C. F. Aquadro, Y. Xu, D. E. Perez, and H. Hollocher. 1995. Sexual isolation in *Drosophila melanogaster*: A possible case of incipient speciation. Proc. Natl. Acad. Sci. USA 92:2519–2523.

Wu, C.-I., N. A. Johnson, and M. F. Palapoli. 1996. Haldane's rule and its legacy: Why are there so many sterile males? Trends Ecol. Evol. 11:281–284.

Wu, C.-Y. and M. D. Brennan. 1993. Similar tissue-specific expression of the *Adh* genes from different *Drosophila* species is mediated by distinct arrangements of cis-acting sequences. Mol. Gen. Genet. 240:58–64.

Xiong, Y. and T. H. Eickbush. 1990. Origin and evolution of retroelements based upon their reverse transcriptase sequences. EMBO J. 9:3353–3362.

Yamaguchi, O. and T. Mukai. 1974. Variation of spontaneous occurrence rates of chromosomal aberrations in the second chromosome of *Drosophila melanogaster*. Genetics 78:1209–1221.

Yamaguchi, O., R. A. Cardellino, and T. Mukai. 1976. High rates of occurrence of spontaneous chromosome aberrations in *Drosophila melanogaster*. Genetics 83:409–422.

Yamaguchi, O., T. Yamazaki, K. Saigo, T. Mukai, and A. Robertson. 1987. Distribution of three transposable elements, P, 297, and copia in natural populations of *Drosophila melanogaster*. Japan. J. Genet. 62:205–216.

Yamazaki, T. 1971. Measurement of fitness at the esterase-5 locus in *Drosophila pseudoobscura*. Genetics 67:579–603.

Yamazaki, T. S. and Y. Matsuo. 1984. Genetic analysis of natural populations of *Drosophila melanogaster* in Japan. III. Genetic variability of inducing factors of amylase and fitness. Genetics 108:223–235.

Yamazaki, T., S. Kusakabe, H. Tachida, M. Ichinose, H. Yoshimaru, Y. Matssuo, and T. Mukai. 1983. Reexamination of diversifying selection of polymorphic allozyme genes by using population cages in *Drosophila melanogaster*. Proc. Natl. Acad. Sci. USA 80:5789–5792.

Yang, S. Y., L. L. Wheeler, and I. R. Bock. 1972. Isozyme variation and phylogenetic relationships in the *Drosophila bipectinata* species complex. Univ. Texas Publ. 7213:213–227.

Yardley, D. G., W. W. Anderson, and H. E. Schaffer. 1977. Gene frequency changes at the α-amylase locus in experimental populations of *Drosophila pseudoobscura.* Genetics 87: 357–369.

Yen, J. H. and A. R. Barr. 1974. Incompatibility in *Culex pipiens*. Pp. 97–118 in *The Use of Genetics in Insect Control*, eds. R. Pal and M. J. Whitten. Elsevier, Amsterdam.

Yoon, C. K. and C. F. Aquadro. 1994. Mitochondrial DNA variation among the *Drosophila athabasca* semi-species and *Drosophila affinis.* J. Heredity 85:421–426.

Yu, Q., H. V. Colot, C. P. Kyriacou, J. C. Hall, and M. Rosbash. 1987. Behavior modification by *in vitro* mutagenesis of a variable region within the *period* gene of *Drosophila*. Nature 326:765–769.

Yum, J., W. T. Starmer, and D. T. Sullivan. 1991. The structure of the *Adh* locus of *Drosophila mettleri*: An intermediate in the evolution of the *Adh* locus in the repleta group of *Drosophila.* Mol. Biol. Evol. 8:857–867.

Zacharias, H. 1979. Underreplication of a polytene chromosome arm in the chironomid *Prodiamesa olivacea.* Chromosoma 72:23–51.

Zacharias, H. 1986. Tissue-specific schedule of selective replication in *Drosophila nasutoides.* Wilhelm Roux's Arch. Dev. Biol. 195:378–388.

Zacharias, H., W. Hennig, and O. Leoncini. 1982. Microspectrophotometric comparison of the genome sizes of *Drosophila hydei* and some related species. Genetica 58:153–157.

Zacharopoulou, A. and M. Pelecanos. 1980. Seasonal and year-to-year inversion polymorphism in a southern Greek *Drosophila melanogaster* wild population. Genetica 54:105–111.

Zeng, L.-W. and R. S. Singh. 1993. A combined classical genetic and high resolution two-dimensional electrophoresis approach to the assessment of the number of genes affecting hybrid sterility in *Drosophila simulans* and *Drosophila sechellia.* Genetics 135:135–147.

Zhuohua, D. and D. Fushan. 1986. Chromosomes study in *Drosophila auraria* complex species found in China. I. A study of karyotype between *D. auraria* and *D. triauraria* found in China. Acta Genet. Sinensis 13:285–294. (In Chinese with English summary.)

Zouros, E. 1973. Genic differentiation associated with the early stages of speciation in the *mulleri* subgroup of Drosophila. Evolution 27:601–621.

Zouros, E. 1981. The chromosomal basis of sexual isolation in two sibling species of Drosophila: *D. arizonensis* and *D. mojavensis.* Genetics 97:703–718.

Zouros, E. and C. J. d'Entremont. 1980. Sexual isolation among populations of *Drosophila mojavensis*: response to pressure from a related species. Evolution 34:421–430.

Zouros, E., C. B. Krimbas, S. Tsakas, and M. Loukas. 1974. Genic versus chromosomal variation in natural populations of *Drosophila subobscura.* Genetics 78:1223–1244.

Zwaan, J., R. Bijlsma, and R. F. Hoekstra. 1991. On the developmental theory of aging. I. Starvation resistance and longevity in *Drosophila melanogaster* in relation to preadult breeding conditions. Heredity 66:29–39.

Subject Index

Taxonomic Index

Note: All entries beginning with a lowercase letter are species within the genus *Drosophila.*